FARADAY DISCUSSIONS
NO. 120   2001

Nonlinear Chemical Kinetics: Complex Dynamics and Spatiotemporal Patterns

The Faraday Division
The Royal Society of Chemistry
London

Organising Committee
Professor S. K. Scott (*Chairman*)
Professor P. Coveney
Professor R. Hillman
Professor D. King
Professor P. K. Maini
Professor K. C. Showalter
Professor K. Waugh

ISBN: 0-85404-986-X
ISSN: 1359-6640

Typeset by Santype International Ltd., Netherhampton Road, Salisbury, Wiltshire and printed and bound in Great Britain by Black Bear Press, Cambridge, UK.

A General Discussion

on

Nonlinear Chemical Kinetics: Complex Dynamics and Spatiotemporal Patterns

10th, 11th and 12th September, 2001

A General Discussion on Nonlinear Chemical Kinetics was held at UMIST, Manchester, UK on 10th, 11th and 12th September, 2001.

Contents

1 Introductory Lecture: Nonlinear kinetics: At the crossroads of chemistry, physics and life sciences
   Gregoire Nicolis

11 A new chemical system for studying pattern formation: Bromate–hypophosphite–acetone–dual catalyst
   Miklós Orbán, Krisztina Kurin-Csörgei, Anatol M. Zhabotinsky and **Irving R. Epstein**

21 HPLC analysis of complete BZ systems. Evolution of the chemical composition in cerium and ferroin catalysed batch oscillators: experiments and model calculations
   László Hegedűs, Mária Wittmann, Zoltán Noszticzius, Shuhua Yan, Atchara Sirimungkala, Horst-Dieter Försterling and **Richard J. Field**

39 Effects of non-ionic micelles on transient chaos in an unstirred Belousov–Zhabotinsky reaction
   M. Rustici, R. Lombardo, M. Mangone, C. Sbriziolo, V. Zambrano and **M. L. Turco Liveri**

53 Nonlinear behaviour of simple ionic systems in hydrogel in an electric field
   Dalimil Šnita, Martin Pačes, Jiří Lindner, Juraj Kosek and **Miloš Marek**

67 Self-oscillating polymer chain in a laser field
   Hiroyuki Mayama and **Kenichi Yoshikawa**

85 General Discussion

105 Investigation of nonlinear dynamical properties by the observed complex behaviour as a basis for construction of dynamical models of atmospheric photochemical systems
   Alexander M. Feigin, Yaroslav I. Molkov, Dmitrii N. Mukhin and **Eugenii M. Loskutov**

125 Low-dimensional manifolds in tropospheric chemical systems
   Alison S. Tomlin, Louise Whitehouse, Richard Lowe and **Michael J. Pilling**

147 A numerical study of spatial structure during oscillatory combustion in closed vessels in microgravity
   Roger Fairlie and **John F. Griffiths**

165 Spatial bifurcations of fixed points and limit cycles during the electrochemical oxidation of H_2 on Pt ring-electrodes
   Peter Grauel, Hamilton Varela and **Katharina Krischer**

179 Identification of the intermittency-I route to chaos in oscillating CO oxidation on zeolite-supported Pd
   Marina M. Slin'ko, Anatolii A. Ukharskii, Nikolai V. Peskov and **Nils I. Jaeger**

197 General Discussion

215	Oscillatory dynamics protect enzymes and possibly cells against toxic substances **Marcus J. B. Hauser, Ursula Kummer, Ann Z. Larsen** and **Lars F. Olsen**
229	pH oscillations in the hemin–hydrogen peroxide–sulfite reaction **Marcus J. B. Hauser, Anika Strich, Rebeka Bakos, Zsuzsanna Nagy-Ungvarai** and **Stefan C. Müller**
237	Control of the excitability of neuronal tissue by weak external forces **Wolfgang Hanke, Meike Wiedemann** and **Vera M. Fernandes de Lima**
249	Spatio-temporal dynamics in glycolysis **Thomas Mair, Christian Warnke** and **Stefan C. Müller**
261	Synchronization of glycolytic oscillations in a yeast cell population **Sune Danø, Finn Hynne, Silvia De Monte, Francesco d'Ovidio, Preben Graae Sørensen** and **Hans Westerhoff**
277	Complex morphogenesis of surfaces: theory and experiment on coupling of reaction–diffusion patterning to growth **Lionel G. Harrison, Stephan Wehner** and **David M. Holloway**
295	Chemical waves in open flows of active media: Their relevance to axial segmentation in biology **Mads Kærn, Michael Menzinger, Razvan Satnoianu** and **Axel Hunding**
313	Excitability in chemical and biochemical pH-autocatalytic systems **J. Zagora, M. Voslař, L. Schreiberová** and **I. Schreiber**
325	General Discussion
353	Spatial bistability and waves in a reaction with acid autocatalysis **J. Boissonade, E. Dulos, F. Gauffre, M. N. Kuperman** and **P. De Kepper**
363	Pattern formation and spatial self-entrainment in bistable chemical systems **G. Dewel, M. Bachir, S. Métens** and **P. Borckmans**
371	Turbulent fronts in resonantly forced oscillatory systems **Christopher Hemming** and **Raymond Kapral**
383	Experimental and theoretical studies of feedback stabilization of propagating wave segments **Eugene Mihaliuk, Tatsunari Sakurai, Florin Chirila** and **Kenneth Showalter**
395	Control and coupling of spiral waves in excitable media **Michael Seipel, Friedemann W. Schneider** and **Arno F. Münster**
407	General Discussion
421	Concluding Remarks **I. R. Epstein**
427	List of Posters
429	List of Participants
431	Index of Contributors

■ Electronic supplementary information is available on http://www.rsc.org/esi See article for further information.

Introductory Lecture

Nonlinear kinetics: At the crossroads of chemistry, physics and life sciences

Gregoire Nicolis

Center for Nonlinear Phenomena and Complex Systems, Université Libre de Bruxelles, C.P. 231, Bd. du Triomphe, 1050 Brussels, Belgium

Received 19th October 2001
First published as an Advance Article on the web 20th December 2001

1. Introduction

It is a great honor to deliver this Introductory Lecture to the 120th Faraday Discussion devoted to Nonlinear Chemical Kinetics, especially after having had the privilege to contribute a paper to its famous predecessor that took place in London 27 years ago. At that time nonlinear kinetics was in its beginnings. Experimental evidence of complex behavior was limited to a few isolated examples giving essentially rise to periodic oscillations or to propagating waves in batch reactors, the mathematical analysis of the equations of nonlinear chemical kinetics was still at a rather primitive stage, and the field was represented by a handful of people. Today nonlinear kinetics is a recognized branch of chemistry. It has made its way to textbooks and university curricula and is represented by a sizable community. We dispose of a host of experimental data on whole classes of systems and on a vastly larger variety of behaviors including deterministic chaos, as well as of sophisticated analytic and experimental techniques complemented by powerful simulation and data analysis algorithms.

But the contribution of nonlinear kinetics to Science during the past three decades goes far beyond its own evolution from infancy to maturity within the discipline of chemistry. Nonlinear kinetics has indeed served over the years as an invaluable testing ground for new insights, new ideas and new techniques in nonlinear science in general, in probability and stochastic processes, in thermodynamics and statistical mechanics. This historical role could be accomplished thanks to some unique features of nonlinear kinetics as compared to other fields where nonlinearity is also prominent, on which we comment more amply later. Finally, and owing again to these unique features, nonlinear kinetics has acquired a marked interdisciplinary dimension. It has gradually become a paradigm for understanding the origins of complex behavior in different scientific disciplines outside chemistry not the least of which is biology and for exploring potential applications in technology in connection, for instance, with chemical synthesis and materials science.

The objective of this paper is to outline some key developments underlying these multiple facets of nonlinear kinetics and to illustrate them on representative case studies. After a brief historical survey (Section 2) we turn to nonlinear kinetics viewed as part of dynamical systems theory, first (Section 3) in the idealized setting of homogeneous media and next (Section 4) in a more realistic setting accounting for complexity inherent in the medium. The role of nonlinear kinetics in life sciences is discussed in Section 5. Section 6 is devoted to the foundations of nonlinear kinetics, with emphasis on the connections between mean field and microscopic descriptions. As a case study, some recent results on the mesoscopic modeling of nonlinear kinetic processes on low-dimensional supports are reported in Section 7. The main conclusions are summarized in Section 8.

DOI: 10.1039/b109569m

2. A brief history

The birthday of nonlinear kinetics can safely be situated somewhere in the 1960's. The thermodynamic evolution and stability criteria of Glansdorff and Prigogine, self-organization and dissipative structures, the experiments on the Belousov–Zhabotinski reaction and the Field–Körös–Noyes mechanism of this reaction, Turing's seminal paper, the Lotka–Volterra and the Brusselator models, are some of the cornerstones of the early developments that contributed to the awareness of the scientific community that a new branch of science in its own sight was pointing in the horizon.[1,2]

Motivated by this body of knowledge a large amount of theoretic work was carried in the 1970's, culminating in the construction of bifurcation diagrams covering an impressive variety of situations and providing a unifying frame for understanding the onset of complex behavior.[3] The Faraday symposium of 1974 provided a unique opportunity for the researchers involved to acquire a global view of the field as it was shaping up. At that time it was clear that theory was ahead of experiment, as most of the subtle and sophisticated phenomena anticipated (including deterministic chaos) had not been observed in the laboratory.

Experimentalists did take their revenge starting in the mid 1970's, through three major developments:[4–6]

(i) The design of new oscillators and the experimental construction of quantitative phase diagrams thanks to the systematic use of CSTR, the continuous stirred tank reactor (mid 1970's to mid 1980's).

(ii) The experimental observation of chemical chaos in all its known forms (early 1980's to early 1990's).

(iii) The taming of spatial patterns, thanks to the design of open unstirred reactors (starting in the late 1980's).

Each of these developments was, in turn, at the origin of a revival of the intense theoretical activity of the 1970's. Previously established results were further specified in the light of the characteristics of the experimental set ups and new results became available in connection, for instance, with patterns in two and three dimensions.

Since the late 1980's a series of novel, exciting developments has been initiated following the encounter of nonlinear kinetics with surface science as it is manifested, for instance, in heterogeneous catalysis. In particular, the development of sophisticated techniques such as the field ion microscopy opens the tantalizing perspective of monitoring chemical dynamics at the mesoscopic (nanoscale) level.[7]

Throughout its development, nonlinear kinetics has searched to strike the right balance between the quest for generic results unifying whole classes of phenomena, and the need to account in a number of key situations for the specificity of the problem at hand. This is especially crucial in biological applications as we see further below.

3. Dynamical systems and nonlinear kinetics

The essence of nonlinear kinetics is captured by the reaction–diffusion equations[2,8]

$$\frac{\partial X_i}{\partial t} = v_i(\{X_j\}, k_{i\alpha}, \Delta H_\alpha, \ldots) + D_i \nabla^2 X_i \tag{1}$$

where X_i ($i = 1, \ldots, n$) denote the concentrations or the temperature, $k_{i\alpha}$ the rate constants, ΔH_α the heats of reaction and D_i the mass or heat diffusivity coefficients. For ideal systems the rate function v_i is given by the law of mass action and accounts for the nonlinearities, whereas the contribution of transport processes is linear.

In writing eqn. (1) one anticipates a decoupling between the (closed form) evolution laws of the macroscopic observables and the dynamics at the molecular level, which merely provides the values of the phenomenological coefficients $k_{i\alpha}$, D_i etc. We refer to this as the *mean field* description, a description in which the intrinsically generated fluctuations of X_i arising from microscopic level processes are discarded.

One is now in the position to identify some unique features of nonlinear kinetics as anticipated in the Introduction.

(i) Eqn. (1) exhibits nonlinearity in its simplest—and purest—expression, namely, as a property arising from intrinsic and local cooperative events such as reactive collisions. Because of this complex behavior may arise in the absence of spatial degrees of freedom contrary to hydrodynamics and related fields, and persist even when few variables are present. This has allowed for substantial progress in dynamical systems theory by eliminating a host of effects that would obscure the mechanisms at the origin of chaos and other forms of complexity. One can express these facts in a thermodynamic language, by realizing that reactions are purely *dissipative* processes, whereas in hydrodynamics and classical mechanics inertia plays a very important role in the onset of complex behavior. Understanding how purely dissipative systems can come to terms with the restrictions imposed by the laws of thermodynamics and statistical mechanics has stimulated several fundamental developments. It has also led to the design of canonical models that are being used widely with success to test ideas and to assess the limits of validity of approximations.

(ii) The intrinsic parameters k and D in eqn. (1) have dimensions of, respectively, $[\text{time}]^{-1}$ and $[(\text{length})^2/\text{time}]$. It follows that a reaction–diffusion system possesses intrinsic time (k^{-1}) and space ($(D/k)^{1/2}$) scales contrary, again, to hydrodynamics where one has a whole spectrum of time and space scales for the selection of which the boundary conditions and the size—two extrinsic factors—play the key role. This places nonlinear kinetics at the forefront for understanding the origin of endogenous rhythmic and patterning phenomena as observed, in particular, in biology and in materials science. Notice that in thermodynamic equilibrium these intrinsic time and length scales remain dormant, owing to detailed balance. Nonequilibrium allows, on the contrary, for the excitation and eventual stabilization of finite amplitude disturbances bearing these characteristic scales.

Let us briefly survey some of the highlights of the analysis of reaction–diffusion equations.

In the absence of spatial degrees of freedom eqn. (1) reduces to a set of coupled nonlinear ordinary differential equations. In this form they have provided some of the earliest and most widely used models of chaotic attractors. They have also helped to reveal connections between chaos and multimodal oscillations which are frequently encountered in experimental situations, especially in electrochemical and biochemical systems. Of special relevance in this respect is chaos generated by the mechanism of homoclinicity, also shown to be relevant in Belousov–Zhabotinski type of reactions.[9]

In the presence of spatial degrees of freedom one deals with coupled nonlinear partial differential equations. The most characteristic single recent development in connection with these equations is the discovery of an impressive variety of intrinsically generated spatial and spatio-temporal patterns, including spatio-temporal chaos, when two or more mechanisms of instability are interfering (Turing–Hopf, Turing–zero mode, localized structures, *etc.*) as illustrated by a number of papers in this issue. The pronounced multiplicity of these solutions in systems of great spatial extent raises the problem of selection. A variety of defects are also generated under these conditions and follow an interesting dynamics of their own.[10,11]

The classical tools that have prevailed throughout these analyses are the reduction to normal form (amplitude) equations using perturbation techniques and/or symmetry arguments, complemented by interactive numerical simulations. Normal forms along with recent dynamical reconstruction techniques from time series data, giving direct access to some of the indicators of dynamical complexity (Lyapunov exponents *etc.*) also highlight the possibility of new approaches to modeling illustrated by some of the papers of this issue: instead of relying entirely on an explicit kinetic scheme some of the steps and parameters of which might be poorly known one may anticipate the validity of canonical, symmetry-inspired equations generating the type of behavior of interest and displaying only few parameters, and fit the values of these parameters from suitably designed experiments. We here find again the interplay between genericity and specificity evoked at the end of Section 2.

4. Nonlinear kinetics in complex media

After focusing for a long time on idealized systems in which the reaction medium was homogeneous, isotropic and in mechanical equilibrium and the reactive mixture could be treated as ideal, nonlinear kinetics is now addressing more realistic situations. Common to all of them is the introduction of "complexity" of some sort arising from the host medium and interfering in a nontrivial way with the kinetics itself.

A first extension relates to the inclusion of background flow. In most of the cases this flow can be considered as remaining practically unaffected by the kinetics. It therefore suffices to augment eqn. (1) by a term of the form $u \cdot \nabla X_i$, u being the (given) bulk velocity. In this setting the variables $\{X_i\}$ behave as "passive scalars" entrained by the flow, the most marked new qualitative effect being then the occurrence of two types of instabilities: absolute instabilities of a nature similar to those in the absence of flow, and convective instabilities in which a local disturbance in entrained by the flow and evolves in time in a non-trivial way. An interesting limiting case is stirring.[12] Perfect stirring should lead to complete mixing, and thus to the reduction of the space-dependent evolution laws to a set of homogeneous (0-dimensional) rate equations. The influence of incomplete mixing on the dynamical behaviors displayed in nonlinear kinetics has been intensively studied in the 1980's but several questions remain open.

A second series of attempts is based on the principle, similar to the one underlying atomic and molecular spectroscopy, that to probe the richness of behaviors allowed by nonlinear kinetics one needs to solicit the underlying system by external fields.[12] Light perturbed oscillating reactions (sometimes with deliberately adding noise to the electromagnetic field) have been intensely studied from this perspective. In a similar vein microgravity experiments allow one to eliminate density effects and thus bulk flow, and to focus on kinetic effects only as seen in one of the papers in this issue. In all cases, the most characteristic effect of the external field is to select a particular subset of solutions. This is also at the origin of its use for the purposes of control.

Parameter variability is yet another effect incorporated in recent nonlinear kinetics studies. This type of variability is ubiquitous when the kinetics takes place in solid support in the form of a porous material, or on a catalyst surface where different crystallographic planes may have different affinities toward reaction and/or mobility.

Originally contrasted with equilibrium-mediated phenomena, complex behaviors induced by nonlinear kinetics out of equilibrium are now realized to interfere in a most interesting way with phase transitions. To account for such interferences one needs to amend the classical reaction–diffusion equations (eqn. (1)) by *nonlinear* diffusion terms. As it turns out this new nonlinearity depends in a very sensitive way on the intermolecular interactions. We therefore have, already at the level of the mean-field description afforded by eqn. (1), a first instance of coupling between macroscopic-level behavior and microscopic properties. Among the effects predicted by this enlarged description is the chemical freezing of the phase transition allowing for the stabilization of spatial patterns with characteristic lengths in the range of the nanometer, as observed in a variety of phenomena of self-assembly. Based on similar mechanisms and principles, mechano-chemical couplings are expected to be at the origin of a new class of materials ("intelligent" gels, *etc.*) possessing potentially interesting physico-chemical properties.

Many of the above elements of complexity of the embedding medium are found when the reaction proceeds in a natural environment. One instance is atmospheric chemistry, the subject of two papers in this issue, and geochemistry.[13] A second instance is biology, to which we now turn.

5. Nonlinear kinetics in the life sciences

As mentioned already, early developments in nonlinear kinetics contributed significantly to the resolution of the long-standing riddle, how complex evolutionary and organizational processes that are so ubiquitous in biology can find their origin in the basic laws of chemistry and physics when nonlinearity and nonequilibrium constraints are present.

Beyond this conceptual advance, research in nonlinear kinetics has led to a semi-quantitative interpretation of a wide spectrum of dynamical behaviors in biochemistry.[14] This has been possible thanks to the development of elegant models in which the involvement of cooperative enzymes in some key steps provides the principal source of nonlinearity and feedback. Glycolytic oscillations, calcium oscillations and waves, the cell division cycle, cAMP induced aggregation in amoebae, synchronization in cell populations, are among the main achievements of this effort. In this context one has also witnessed an interesting coevolution of the models and the experiments, which helped to identify the principal mechanisms behind the observed behavior.

Nonlinear kinetics has also been a source of inspiration for approaching dynamical phenomena of crucial importance in biology, in which a modeling involving few variables and/or well-established molecular mechanisms is still not available. Circadian rhythms, immune response, the electrical activity of the brain, embryonic development, cooperative processes such as food recruitment or building activity in social insects,[15] large scale vegetation patterns and, last but not least, chemical and biochemical evolution itself[16] have been visited in one way of the other in the light of the concepts and techniques of nonlinear kinetics. Turing's seminal work on the origin of morphogenetic patters and its follow-ups are parts of this effort. After a first period of successful developments they have been superseded by the impressive experimental discoveries on gene regulation during embryonic development that have marked molecular biology since the late 1980's. A comeback is likely, but it will most probably have to integrate new information such as the complex structure of the cellular medium.

The question, what is the "constructive" role of the complex behaviors observed in biology and associated with nonlinear kinetics has been repeatedly raised; the alternative being that these phenomena are, rather, to be viewed as inevitable consequences of biological activity. Answers have been proposed in particular instances, one of which is the subject of a contribution to this issue, but the question is far from being fully settled. For instance, the advent of chaos theory and the availability of a wealth of data showing the occurrence of this phenomenon in biology forces us to reassess the idea that normal "healthy" behavior is associated with regular patterns like strictly periodic oscillations, and that irregularity in *e.g.* the form of chaos is therefore an indicator of "pathological" behavior.

6. Mean-field *versus* microscopic views: foundations of nonlinear kinetics

So far we have adopted a mean-field view of chemical kinetics (*cf.* eqn. (1)). Traditionally one tends to identify this view with a macroscopic level description—the type of description that is the most closely associated with our everyday experience. In this section we are concerned with the passage from microscopic-level dynamics to such a macroscopic description. Ordinarily, this passage is studied by appealing to the Boltzmann equation. Early work along these lines has shown that the equations of kinetics can be derived from this equation as long as the *local equilibrium hypothesis* (the state functions are, locally, related to the observables by the same relations as in equilibrium) can be justified.[17] This approach can now be reassessed and/or complemented thanks to two types of new developments.

6.1. The mean-field description can be justified and/or completed starting directly from the Liouvillian dynamics

There are two principal ingredients behind this development: the prevalence of dynamical chaos at the molecular level (in the classical limit) as a result of the defocusing character of the collisions; and the connection between, on the one side, the principal indicators of this chaotic dynamics (Lyapunov exponents *etc.*) and, on the other side, transport coefficients and reaction rates.[18]

The crux of the argument is that transport or reaction are conditioned, up to a multiplicative constant depending on geometry only, by the rate γ at which the particle trajectories escape from a certain part of phase space. We illustrate this on the Lorentz gas model of Fig. 1, where a beam of light particles is injected against a lattice of heavy immobile scatterers. When hitting the bright disks the particles undergo elastic collisions generating on a larger scale diffusion and homogenization, but a collision with the dark disks gives rise to a change of chemical identity. Clearly, in this conservative, time-reversible dynamics there exist exceptional trajectories that cannot participate to either of these events because they remain trapped in a portion of phase space. The set of these unstable trajectories forms a (multi)fractal repellor, F which is specific to the dynamical quantity that is associated to the process (and is, therefore, different for diffusion and for reaction or for different types of reaction). Conversely, diffusion or reaction imply escaping from this repellor, say at a rate γ.

By solving the initial value problem of the rate equation of a kinetic process or of the diffusion equation with boundary conditions expressing the influx of particles on some surface say $x = 0$ and the outflux on $x = L$, one obtains relations of the type

$$\gamma \approx k$$
$$\gamma \approx D/L^2 \qquad (2)$$

Fig. 1 Schematic representation of the 2-dimensional periodic Lorentz gas. Upon collision with the bright or the dark scatterers an incoming light particle undergoes, respectively, an elastic collision or a reactive transformation.

k and D being a rate constant or a diffusion coefficient. On the other hand, γ can be related to microscopic dynamics using the methods of modern nonequilibrium statistical mechanics complemented by those of dynamical systems and chaos theories. The result is a relation of the form

$$\gamma \approx \sum_{\sigma_i > 0} \sigma_i(F) - h_{\text{ks}}(F) \tag{3}$$

where σ_i and h_{ks} are the Lyapunov exponents and the Kolmogorov–Sinai entropy of the repellor. Eqns. (2) and (3) provide the basis of a fully microscopic justification of reaction–diffusion equations for the model at hand, free from phenomenological assumptions. An equivalent form of eqn. (3) displaying the Hausdorff dimension of the repellor can also be derived.[19–21]

Starting from the Liouvillian dynamics one can finally derive dispersion relations for the chemo-hydrodynamic modes in model systems,

$$\omega_q = -2k - Dq^2 + \mathrm{O}(q^4) \tag{4}$$

where q is the wave number. The modes themselves turn out to be fractal functions, whose characteristics are related to q and D.[22]

6.2. The mean-field description can be compromised

Overlooked for a long time as a theoretician's curiosity, this possibility begins now to be seriously studied by molecular dynamics and Monte Carlo simulations, Langevin and master equations, and the experimental techniques of surface science. We hereafter give a partial list of instances where such a breakdown, reflected by the generation of anomalous fluctuations, is expected to occur.

(i) *Homogeneous-phase systems containing small numbers of particles.* This situation often arises in biology where, for instance, the number of copies of a regulatory unit in a cell is typically small, entailing that the standard deviation of the fluctuations may be comparable to the mean.

Related to this is the interesting issue of persistence of oscillations at the cellular level, which in small-size systems tend to be decorrelated by phase diffusion.

(ii) Homogeneous-phase systems of large spatial extent possessing a high multiplicity of states and/or chaotic dynamics. The point is here that the distance between the available states can be so small that internal fluctuation-induced (or external noise-induced) transitions may blur coherent behavior.[23] Even when mean-field is secured, one realizes that in such systems the macroscopic description typically applies to the *most probable* value rather than to the mean, as the latter can correspond to an unrepresentative (*e.g.* unstable) state.

(iii) Systems evolving according to widely different time scales. The principal mechanism here is that minor fluctuations building up during a slow stage may be amplified by the fast stage to a point that they reach a macroscopic level. This will be reflected by the dispersion of certain dynamical quantities such as the ignition time in a combustion process or the switching time in an optical or an electronic device.[24] When the fast stage is accompanied by a substantial energy release (as, *e.g.*, the heat of reaction in combustion) this also tends to compromise the validity of the local equilibrium hypothesis, another pillar ensuring the validity of reaction–diffusion equations in their traditional form.

(iv) Under the combined action of fluctuations and asymmetric external potentials. This class of systems, referred as *ratchets*, attracts increasing attention as it exhibits counter-intuitive behavior such as the occurrence of a flux opposing the external force. It may be relevant for the understanding of certain processes of energy transduction in biology.[25]

(v) In systems of restricted geometry and/or low dimensionality such as catalytic surfaces, micelles, or Langmuir–Blodgett films. Limited possibilities of mobility or of chemical bonding may favor here segregation through the development of strong inhomogeneous fluctuations. This will compromise the effective mixing implicit in the mean-field view.

In all the above cases the description that imposes itself as a substitute of mean field is not the purely microscopic description which turns out to be impracticable but, rather, a mesoscopic-level one whose merit is to involve variables that are directly related to the macroscopic observables. This description is illustrated in the next section in connection with item (v) above.

7. A case study: mesoscopic modeling of nonlinear kinetic processes on low-dimensional supports

Striking evidence of non mean-field behavior for nonlinear kinetic processes in low dimensions is the formation of islands of homologous particles on a catalytic surface in contact with an open fluid phase. This is what happens, for instance, in the reduction of NO on a Pt emitter tip where observations using field ion microscopy reveal the existence of OH islands involving as few as $10–10^2$ particles.[26] Reaction now proceeds at the interface rather than in the bulk, contrary to the very basis of the classical mean-field picture.

To model such a *fluctuation-driven dynamics* a mesoscopic-level description will be adopted. The state variables are the occupation numbers of a given site of the surface by a reactant. On the one side, such variables are related directly to the traditional concentration variables which are merely the averages of these numbers over a volume element. And on the other side, they incorporate quite naturally the effects of microscopic dynamics since their instantaneous values fluctuate owing to the fact that the different dynamical processes involved occur with a certain probability. In what follows we shall adopt the limit of immobile reactants in order to sort out more clearly the effects due to the kinetics. We shall account for the characteristics of intermolecular interactions by precluding the simultaneous occupation of a lattice site by more than one particle (hard core potential) and shall allow for first neighbor interactions only. Under these conditions chemical transformations can be viewed as "color" changes or "spin" flips and one can write a master equation for the probability density of an instantaneous configuration $\{\sigma\}$, $P(\{\sigma\}, t)$ of the form

$$\frac{dP(\{\sigma\}, t)}{dt} = -\sum_{j=1}^{N} \omega_j(\{\sigma\} \rightarrow \{\sigma'\}) P(\{\sigma\}, t)$$

$$+ \sum_{j=1}^{N} \omega_j(\{\sigma'\} \rightarrow \{\sigma\}) P(\{\sigma'\}, t) \quad (5)$$

Here $\{\sigma\} = \{\sigma_1, \ldots, \sigma_N\}$ with $\sigma = \pm 1$ when only two species (*e.g.* A, X) are present, $\sigma = 0, \pm 1$ when three species (*e.g.* A, X and empty sites S) are present, *etc.* j runs over the lattice sites and ω_j is the probability per unit time that the occupation number of this site undergoes a transition.

We now summarize some results obtained using the above formalism.

7.1. Closed systems: the reactions $A+2X \underset{k}{\overset{k}{\rightleftharpoons}} 3X$ and $A+X \underset{k}{\overset{k}{\rightleftharpoons}} 2X$ on a fully occupied lattice

From a mean-field point of view both processes should tend to an equilibrium state where the mean occupation numbers $\langle A \rangle$ and $\langle X \rangle$ are in the ratio

$$r = \frac{\langle A \rangle}{\langle X \rangle} = 1 \quad \text{(mean field value, } r_{MF}) \tag{6}$$

Starting from a lattice fully occupied by X and counting all allowed subsequent states containing n particles of species A in a lattice of total size N one finds, on the contrary, that in a one-dimensional space the trimolecular system actually stabilizes in a locally frozen invariant state in which[27]

$$r = \frac{5 - \sqrt{5}}{5 + \sqrt{5}} \approx 0.38 \quad (r \neq r_{MF}) \tag{7}$$

The same trimolecular system reaches a mean-field ratio $r = 1$ in a two-dimensional square lattice. As for the bimolecular reaction it agrees with mean-field prediction in both one and two space dimensions. The analysis can be extended to a variety of other sets such as fractal sets spanning a wide range of (noninteger) dimensionalities up to 2, as well as regular two-dimensional lattices other than the square lattice.[28] It brings out the determining influence of the mean coordination number (rather than the dimensionality) in the deviations of r from r_{MF}. For instance, in the hexagonal 2-dimensional lattice the coordination number is 3 and one finds $r = 0.92$, definitely different from the $r = 1$ of the square lattice whose coordination number is 4.

7.2. Open systems: the cooperative desorption process $A+A \overset{k}{\rightarrow} S+S$ in a 1-dimensional lattice

The mean-field equations predict an asymptotic inverse power low decay (here t^{-1}) to a state corresponding to an empty lattice. A more microscopic view incorporating the prescriptions specified earlier shows that all configurations where isolated A particles are separated by an arbitrary number of "dimers" SS are invariant states. We therefore attain a dynamics-induced disordered state in which ergodicity breaks down. By extremizing the number of the above invariant configurations with respect to A one finds a coverage of the lattice equal to

$$\theta_{A,\infty} = \lim_{N \to \infty} \frac{\langle A \rangle}{N} = 0.177 \quad (\theta_{A,\infty} \neq \theta_{MF}) \tag{8}$$

But since ergodicity does not hold, there is *a priori* no reason for eqn. (8) to be correct. To find the correct coverage one has, rather, to solve the master equation (eqn. (5)) with the appropriate form of ω_j,

$$\omega_j(\{\sigma\} \to \{\sigma'\}) = \frac{k}{4}(\sigma_j + 1)(\sigma_{j+1} + 1) \tag{9}$$

One can derive in this way[29] a hierarchy of equations for clusters containing l contiguous sites filled by A, which can be solved exactly and leads, for $l = 1$, to the doubly exponential law

$$\theta_A = \theta_A(0) \exp[-2\theta_A(0)(1 - e^{-t})] \tag{10}$$

which for $t \to \infty$ is different from both θ_{MF} and $\theta_{A,\infty}$ (eqn. (8)). We see, in particular, that the result is initial condition-dependent as should be expected retrospectively, owing to the lack of ergodicity. As a by-product one sees that there is no closed form description for the mean value θ_A, nor for any finite-order moment of $\{\sigma\}$, for that matter: the master equation description constitutes the minimal, "irreducible" description. These conclusions are fully confirmed by Monte Carlo simulations.

7.3. More complex nonlinear kinetic processes

The classical Lotka–Volterra model[30] and a minimal model of NO reduction on a Pt surface[31] have been implemented on one- and two-dimensional regular lattices. In both cases, failure of mean-field description is concomitant to the nucleation and propagation in space of wave fronts of chemical activity. These fronts exist despite the fact that the reactants are immobile. They are, therefore, qualitatively different from the conventional wave fronts associated with mass transport, for which particle mobility (diffusion) is necessary.

In summary, the microscopic dynamics and the macroscopic behavior of nonlinear systems operating in low dimensions become intimately intertwined. Such systems are therefore ideally suited for a systematic experimental study of fluctuations. Theoretical analyses of fluctuations in nonlinear kinetics have a long history.[32–34] Until recently they had focused on homogeneous phase systems, and contact with experiment has proved difficult. A shift of emphasis toward low-dimensional systems as found in heterogeneous catalysis should lead to an interesting and long-awaited cross-fertilization.

8. Conclusions and perspectives

We have seen how, in the three decades of its existence as a branch of chemistry, nonlinear kinetics has accomplished a number of realizations of historic significance. It has established beyond doubt the generality and the relevance of complex dynamical behaviors in chemistry; it has contributed to the development of nonlinear science as a whole by providing some of the most relevant known examples of oscillations, pattern formation and deterministic chaos; and it has served as a paradigm for the modeling of complex systems encountered in areas of great concern outside the field of chemistry. Its flexible, pluridisciplinary tools and its balanced approach in which the search for unified generic results is carried out in parallel with a modeling respecting the specificity of the problem at hand confer also to it a considerable training value.

Long term projections on the future of an active field of research are perilous, as their likelihood of failure is a strongly increasing function of the vitality of the field. Still, it seems reasonable to suggest that the future of nonlinear kinetics lies, to a great extent, in the exploration and enhancement of privileged interfaces with certain other fields. We have already stressed the interest of phenomena in which chemical dynamics takes place in a complex medium characterized, for instance, by the presence of a bulk flow or by the presence of a porous material. Beyond these instances two cases deserve special mention.

First, the interference between nonequilibrium complex phenomena and equilibrium properties, especially phase transitions, is likely to lead to further insights in the emerging field of nano-sciences. In a similar vein, elucidating the mechanisms of mechano-chemical couplings should lead not only to the elaboration of interesting new materials but also to the understanding of a number of biological processes of great concern.

The interface with statistical mechanics would also need to be further enhanced. Statistical mechanics has experienced recently a renaissance, following cross-fertilization with the ideas and tools of dynamical systems theory and microscopic simulation techniques. This provides a new basis for the microscopic foundations of nonlinear kinetics, a question whose relevance is becoming increasingly obvious in view of present day possibilities of experimental monitoring of mesoscopic scale chemical processes. In this respect chemical dynamics on low-dimensional supports, through the spectacular advances of surface science associated with such fine tuned techniques as field ion microscopy, constitutes now a natural testing ground between theory and experiment. Theoretical ideas concerning fluctuations and other manifestations of mesoscopic-level dynamics, whose relevance could so far not be checked under traditional homogeneous phase conditions, can be taken up again. This should lead to new synergies in a field in which macroscopic level phenomena and microscopic level processes were so far being largely dissociated.

Acknowledgements

This work was supported by the Interuniversity Attraction Poles program of the Belgian Federal Government.

References

1 For a history of the subject see A. Pacault and J.-J. Perraud, *Rythmes et formes en chimie*, Presses Universitaires de France, Paris, 1997.
2 The field of nonlinear kinetics was first reviewed in (*a*) G. Nicolis and J. Portnow, *Chem. Rev.*, 1973, **73**, 365; (*b*) R. Noyes and R. Field, *Annu. Rev. Phys. Chem.*, 1974, **25**, 95; (*c*) G. Nicolis and I. Prigogine, *Self-organization in nonequilibrium systems*, Wiley, New York, 1977.
3 *Bifurcation theory and applications in scientific disciplines*, ed. O. Gurel and O. Rössler, *Ann. New York Acad. Sci.*, 1977, vol. 316.
4 *Nonlinear phenomena in chemical dynamics*, ed. C. Vidal and A. Pacault, Springer, Berlin, 1981. See especially articles by P. De Kepper, I. Epstein, H. Swinney, J.-C. Roux and coworkers.
5 S. Müller, Th. Plesser and B. Hess, *Physica D*, 1987, **27**, 71.
6 (*a*) V. Castets, E. Dulos, J. Boissonade and P. De Kepper, *Phys. Rev. Lett.*, 1990, **64**, 2953; (*b*) Q. Ouyang and H. Swinney, *Nature*, 1991, **352**, 610.
7 (*a*) G. Ertl and H.-J. Freund, *Phys. Today*, 1999, **52**, 32; (*b*) C. Sachs, M. Hildebrand, S. Völkening, J. Wintterlin and G. Ertl, *Science*, 2001, **293**, 1635.
8 P. Gray and S. Scott, *Chemical oscillations and instabilities*, Clarendon Press, Oxford, 1990.
9 (*a*) *Homoclinic Chaos*, Proceedings of the NATO Advanced Research Workshop, Belgium, 1991, ed. P. Gaspard, A. Arnéodo, R. Kapral and C. Sparrow, *Physica D*, 1993, **62**(1–4), 1–372.
10 (*a*) *Chemical waves and patterns*, ed. R. Kapral and K. Showalter, Kluwer Academic, Dordrecht, 1995; (*b*) Y. Kuramoto, *Chemical oscillations, waves, and turbulence*, Springer, Berlin, 1984.
11 A. De Wit, *Adv. Chem. Phys.*, 1999, **109**, 435.
12 I. Epstein and J. Pojman, *An introduction to nonlinear chemical dynamics*, Oxford University Press, Oxford, 1998.
13 P. Ortoleva, *Geochemical self-organization*, Oxford University Press, Oxford, 1994.
14 A. Goldbeter, *Biochemical oscillations and cellular rhythms*, Cambridge University Press, Cambridge, 1996.
15 S. Camazine, J.-L. Deneubourg, N. Franks, J. Sneyd, G. Theraulaz and E. Bonabeau, *Self-organization in biological systems*, Princeton University Press, Princeton, 2001.
16 M. Eigen and P. Schuster, *The hypercycle*, Springer, Berlin, 1979.
17 (*a*) I. Prigogine and F. Xhrouet, *Physica*, 1949, **15**, 913; (*b*) R. Present, *J. Chem. Phys.*, 1959, **31**, 747.
18 P. Gaspard, *Chaos, scattering and statistical mechanics*, Cambridge University Press, Cambridge, 1998.
19 P. Gaspard and G. Nicolis, *Phys. Rev. Lett.*, 1990, **65**, 1693.
20 P. Gaspard and F. Baras, *Phys. Rev. E*, 1995, **51**, 5332.
21 I. Claus and P. Gaspard, *J. Stat. Phys.*, 2000, **101**, 161.
22 P. Gaspard, I. Claus, T. Gilbert and J. Dorfman, *Phys. Rev. Lett.*, 2001, **86**, 1506.
23 F. Baras, *Phys. Rev. Lett.*, 1996, **77**, 1398.
24 F. Baras, G. Nicolis, M. Malek Mansour and J. W. Turner, *J. Stat. Phys.*, 1983, **32**, 1.
25 F. Julicher, A. Ajdari and J. Prost, *Rev. Mod. Phys.*, 1997, **69**, 1269.
26 N. Kruse, C. Voss, V. Medvedev, C. Bodenstein, D. Hanon and J.-P. Boon, *J. Stat. Phys.*, 2000, **101**, 621.
27 A. Provata, J. W. Turner and G. Nicolis, *J. Stat. Phys.*, 1993, **70**, 1195.
28 A. Tretyakov, A. Provata and G. Nicolis, *J. Phys. Chem.*, 1995, **99**, 2770.
29 F. Baras, F. Vikas and G. Nicolis, *Phys. Rev. E*, 1999, **60**, 3797.
30 A. Provata, G. Nicolis and F. Baras, *J. Chem. Phys.*, 1999, **110**, 8361.
31 Y. De Decker, F. Baras, G. Nicolis and N. Kruse, *J. Chem. Phys.*, submitted.
32 (*a*) G. Nicolis and A. Babloyantz, *J. Chem. Phys.*, 1969, **51**, 2632; (*b*) G. Nicolis and I. Prigogine, *Proc. Natl. Acad. Sci. USA*, 1971, **68**, 2102; (*c*) H. Lemarchand and G. Nicolis, *Physica A*, 1976, **82**, 251.
33 (*a*) Y. Kuramoto, *Prog. Theor. Phys.*, 1974, **52**, 711; (*b*) C. Gardiner, K. Mc Neil, D. Walls and I. Matheson, *J. Stat. Phys.*, 1976, **14**, 309.
34 (*a*) J. Portnow, *Phys. Lett. A*, 1975, **51**, 370; (*b*) P. Ortoleva and S. Yip, *J. Chem. Phys.*, 1976, **65**, 2045; (*c*) J. Boissonade, *Phys. Lett. A*, 1979, **74**, 285.

A new chemical system for studying pattern formation: Bromate–hypophosphite–acetone–dual catalyst

Miklós Orbán,[*a] **Krisztina Kurin-Csörgei,**[a] **Anatol M. Zhabotinsky**[b] **and Irving R. Epstein**[b]

[a] *Department of Inorganic and Analytical Chemistry, L. Eötvös University, H-1518 Budapest 112, P.O.Box 32, Hungary*
[b] *Department of Chemistry, MS 015 Brandeis University, Waltham, Massachusetts 02454-9110, USA*

Received 29th March 2001
First published as an Advance Article on the web 7th November 2001

A modified version of the short-lived BrO_3^-–$H_2PO_2^-$–Mn(II)–N_2 oscillator, the BrO_3^-–$H_2PO_2^-$–acetone–dual catalyst system, where the catalyst pair can be Mn(II)–Ru(bpy)$_3$SO$_4$, or Mn(II)–ferroin, or Mn(II)–diphenylamine, shows long-lasting batch oscillations in the potential of a Pt electrode and in colour, accompanying periodic transitions between the oxidised and reduced forms of the catalysts. Experimental conditions for the oscillations are established. The origin of the batch oscillations and the role of the catalyst pair in the oscillatory behaviour are discussed. The new system is ideally suited to the study of waves and patterns in reaction–diffusion systems, since in addition to the longevity of its spatial behaviour in batch, it produces no gaseous or solid products and exhibits significant photosensitivity.

Introduction

Interest in nonlinear chemical dynamics has recently focused on seeking new kinds of reaction–diffusion phenomena. Approaches to this goal include manipulation by external forces (*e.g.* light, electric or magnetic fields) of target patterns, spirals and Turing structures observed in the Belousov–Zhabotinsky (BZ) or chlorite–iodide–malonic acid (CIMA) systems, and development of new reactions capable of forming two-dimensional patterns.

While the BZ and CIMA reactions and their variants provide powerful tools for studying chemical waves and patterns, neither is ideal. The standard BZ reaction (BrO_3^-–MA–H_2SO_4–ferroin) generates bubbles of carbon dioxide, which hamper observation in experiments of long duration. Some uncatalysed bromate oscillators, variants of BZ, are bubble-free, but short-lived, and tend to form precipitates. The CIMA system requires an unstirred flow reactor, a cumbersome technology, because the patterns last for only a relatively short time without the inflow of fresh reactants. Also, the traveling waves and Turing patterns in this system must be studied at temperatures (typically at 4–6 °C) significantly below ambient.

Until now, little progress has been achieved toward eliminating bubble or precipitate formation, the two major factors that militate against long-lasting pattern evolution. Ouyang *et al.*[1] suggested that it might be possible to generate a bubble-free bromate oscillator by employing *two* substrates, one being oxidised to a product other than CO_2, the other binding Br_2, an intermediate of bromate reduction. One possibility they considered was the Mn(II)-catalysed bromate–

DOI: 10.1039/b102885p

hypophosphite–acetone system. In this reaction, $H_2PO_2^-$ is oxidised to H_3PO_3, and the acetone serves to bind bromine. They had only limited success, however, in finding suitable compositions for oscillations and pattern formation.

Aside from the recently published results of Kurin-Csörgei et al.,[2] who observed long lasting traveling waves in the bromate–cyclohexanedione(CHD)–ferroin system without formation of bubbles or precipitate, we are unaware of work directed at developing novel systems with the desired properties.

Our plan is to find new chemical systems for studying two-dimensional patterns in a thin layer of solution. Ideally, a new system would have the following properties:

Oscillate in batch at room temperature for an extended period of time, several hours at a minimum; produce no gaseous or solid products; give rise to easily visible spatial waves and patterns with a convenient wavelength and velocity; be susceptible to perturbation or control, e.g., photochemically; and have a relatively simple mechanism to facilitate modelling.

Here we report a new chemical system, a highly improved version of the bromate–hypophosphite–manganese(II)–N_2 batch oscillator, for the investigation of pattern formation in a thin unstirred solution layer. The bromate–hypophosphite–acetone–dual catalyst reaction meets almost all of the requirements listed above. It is well suited to the generation of high quality traveling waves that are close to ideal for studying sustained pattern evolution.

Experimental

Chemicals

$NaBrO_3$ (Aldrich), $NaH_2PO_2 \cdot H_2O$ (Sigma), $MnSO_4 \cdot H_2O$ (Aldrich), $Ru(bpy)_3Cl_2 \cdot 6H_2O$ (Aldrich), $Ce(SO_4)_2$ (Aldrich), 1,10-phenanthroline (Aldrich), diphenylamine sulfate (Aldrich), and acetone (Fisher) were of the highest grade commercially available and were used without purification.

The other reagents were prepared as follows:

$Ru(bpy)_3SO_4$ was obtained from $Ru(bpy)_3Cl_2 \cdot 6H_2O$ by converting the chloride salt to the sulfato form using the recipe suggested by Gao and Försterling.[3] The removal of chloride is necessary because of its known inhibitory effect on bromate oscillators. The concentration of the aqueous stock solution of $Ru(bpy)_3SO_4$ ($c = 3.75 \times 10^{-2}$ M) was determined spectrophotometrically ($\lambda = 452$ nm, $\varepsilon = 14\,800$).

$Fe(phen)_3SO_4$ (ferroin) stock solution ($c = 2.5 \times 10^{-2}$ M) was prepared by mixing 1,10-phenanthroline and $FeSO_4 \cdot 7H_2O$ in stoichiometric ratio. The stock solutions of the Ru(II) and Fe(II) complexes were found to be stable for days when stored in amber glass.

$Ru(bpy)_3^{3+}$ sulfate and $Fe(phen)_3^{3+}$ sulfate were obtained by oxidation of the Ru(II) and Fe(II) complexes with solid PbO_2 in 1 M H_2SO_4. The oxidation is complete within a few minutes: the orange Ru(II) compound turns to green Ru(III); the red Fe(II) complex is transformed into the dark blue Fe(III) complex. The excess PbO_2 is removed using a glass filter. The Ru(III) and Fe(III) complexes were stored in darkness and used within 1 h.

Diphenylamine (DPA) stock solution ($c = 2 \times 10^{-2}$ M) was made by dissolving diphenylamine sulfate in hot 10 M H_2SO_4.

$Mn_2(SO_4)_3$ stock solution ($c = 10^{-2}$ M) was prepared by reacting a known amount of $KMnO_4$ with a 50-fold excess of $MnSO_4$ in 2 M H_2SO_4.

Methods

Standard methods for monitoring the oscillations and observing pattern formation were employed. First, conditions for long-lasting batch oscillations and wave formation were established. Then, kinetic runs were carried out to estimate the relative rates of several composite reactions in order to reveal, at least qualitatively, the chemistry underlying the dynamical behaviour of the title system.

Oscillations. Chemical oscillations in the system were followed visually by the change in colour due to the periodic transition between the oxidised and reduced forms of the catalyst. The oscillations were recorded by measuring the potential of a Pt vs. Hg | Hg_2SO_4 | K_2SO_4 electrode pair immersed in the reaction mixture. The reaction was carried out in a thermostatted beaker ($V = 30$

cm^3) covered with a Teflon cap. There was an air gap between the cap and the surface of the reaction mixture. The amplitude and frequency of the oscillations were dependent on the stirring rate. Therefore, a constant rate of 300 rpm was maintained in all batch experiments using a 2 cm long, 8 mm id Teflon coated stirring bar.

Patterns. When the oscillatory reaction mixture was spread in a covered Petri dish (id 9 cm) to form a 1–2 mm solution layer, target patterns or spirals began to develop. The waves were observable by eye, but for better contrast and for better recording appropriate optical filters were used. Pictures were taken with a CCD video camera connected to a camera controller and a PC. The pictures were processed with image analysis software.

Kinetic runs. The relative rates of some reactions were measured by recording the absorption spectra of the reaction mixture at different times. The ZITA program package[4] was used for data evaluation.

Results

Oscillations

The reaction between bromate and hypophosphite ions in the presence of a metal ion catalyst (Mn(II), Ce(III), ferroin) in sulfuric acid was reported by Adamčikova and Sevčik[5] to show a few (4–8) oscillations in batch, if a flow of N_2 was used to purge the intermediate bromine. As a system for pattern formation, this reaction suffers from two drawbacks: the oscillations persist only for a short time, and the N_2 flow would be as disruptive to pattern formation as the production of bubbles.

An alternative approach is to remove the bromine by adding acetone, with which the Br_2 reacts to form an inert product. This oscillator was tested for pattern formation by Ouyang et al.[1] In the BrO_3^-–$H_2PO_2^-$–acetone reaction they could use only Mn(II) as the oscillatory catalyst; all other BZ catalysts, Ce(IV), ferroin and Ru(bpy)$_3$SO$_4$, were ineffective. This finding is in contrast to the report of Adamčikova and Sevčik, but it is in agreement with our results. Using Mn(II), however, did not allow visual observation of patterns—if they formed at all—owing to the pale colour changes accompanying the Mn(II)↔Mn(III) transitions. When Ouyang et al. used a mixture of Mn(II) and ferroin, where Mn(II) served as the catalyst and ferroin as an indicator for visualisation of the periodic behaviour, they noted the appearance of two-dimensional patterns, but gave no further details.

Following up on the idea of employing two BZ catalysts, we were able to identify compositions at which BrO_3^-–$H_2PO_2^-$–acetone–Mn(II)–second catalyst systems produce long-lasting batch oscillations and patterns. We have found that with a dual catalyst sustained oscillations in the BrO_3^-–$H_2PO_2^-$–acetone batch system can be obtained in the potential of a Pt electrode and in the colour. The presence of Mn(II) as a member of the catalyst pair is essential. As the second catalyst, Ru(bpy)$_3$SO$_4$ is by far the most effective, but ferroin or diphenylamine can also be used, Ce(IV), however, is ineffective in any combination.

Only a small amount of the second catalyst (10^{-5}–10^{-4} M Ru(bpy)$_3$SO$_4$, ferroin or diphenylamine compared to [Mn(II)] > 10^{-3} M) is required to generate long-lasting oscillations. This amount can be added either together with the Mn(II) or when the short-lived Mn(II)-catalyzed oscillations cease.

Typical sequences of oscillations are presented in Fig. 1. Fig. 1A shows the potential of the Pt electrode vs. time recorded when the two catalysts, MnSO$_4$ and Ru(bpy)$_3$SO$_4$, are added separately to the BrO_3^-–$H_2PO_2^-$–acetone–H_2SO_4 system. Oscillations start immediately after adding BrO_3^- to the rest of the mixture, resulting in 5 cycles at this composition. The amplitude and period increase in time from 50 to 60 mV and from 60 to 120 s, respectively. This is identical to the time evolution of the system when only Mn(II) is added. When these oscillations terminate (after about 10 min), introducing 2 drops of 1.56×10^{-2} M Ru(bpy)$_3$SO$_4$ solution (the final concentration is 4×10^{-5} M) generates new oscillations which last for more than 4 h. During that time, the period gradually becomes longer (7–8 min) and the amplitude increases and reaches a value of 370 mV after 4 h. The colour changes between green and yellow during the cycles result

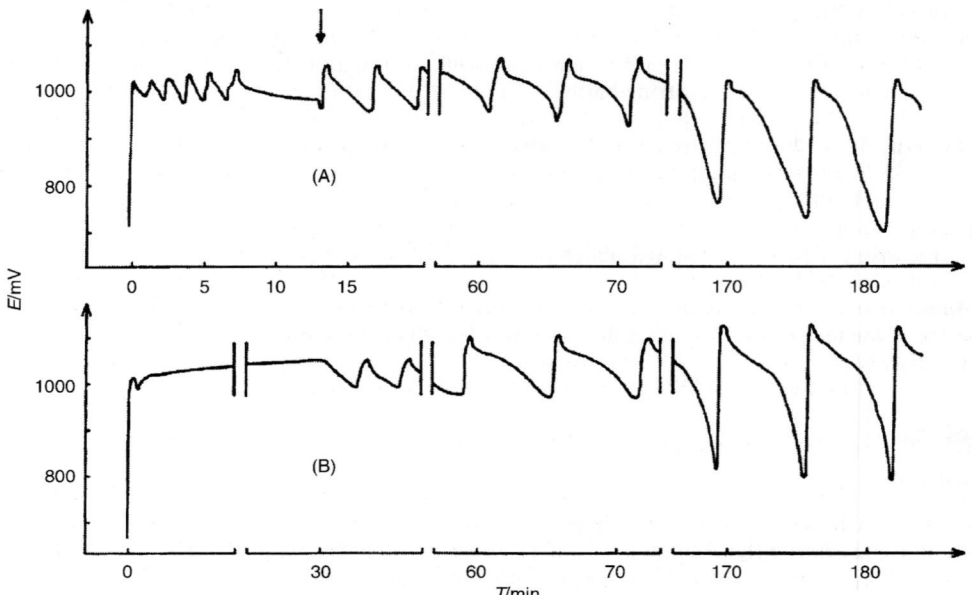

Fig. 1 Oscillations in the BrO_3^-–$H_2PO_2^-$–acetone–H_2SO_4–dual catalyst system. A: The catalysts are added separately. $MnSO_4$ is introduced at $t = 0$, $Ru(bpy)_3SO_4$ at the time (in min) shown by the arrow. B: The catalysts are applied together at $t = 0$. Compositions: $[BrO_3^-] = 2.1 \times 10^{-2}$ M; $[H_2PO_2^-] = 1.13 \times 10^{-1}$ M; $[Mn(II)] = 3.1 \times 10^{-3}$ M; [acetone] $= 1.2 \times 10^{-1}$ M; $[H_2SO_4] = 1.33$ M; $[Ru(bpy)_3SO_4] = 4 \times 10^{-5}$ M. $T = 25\,°C$. Stirring rate: 300 rpm.

from the periodic oxidation and reduction of the second catalyst, the Ru-complex. Fig. 1B shows the time course of the system when the two catalysts are added together to the reaction mixture. In that case, an induction period of 30 min is followed by sustained oscillations, which strongly resemble in amplitude and frequency those seen after the arrow in Fig. 1A.

In most cases, the long-lasting oscillations begin after an induction period whose length depends upon the identity and the quantity of second catalyst and on the order in which the members of the dual catalyst pair are introduced into the reaction mixture. For example, when $[Rubpy)_3SO_4] = 4 \times 10^{-5}$ M was applied separately from Mn(II) (Fig. 1A), the second sequence of oscillations started immediately, but an induction period of 30 min appeared if the $Ru(bpy)_3SO_4$ and Mn(II) were added together (Fig. 1B). The induction time further increased to 50 and 100 min when larger quantities of $[Ru(bpy)_3SO_4]$, 5×10^{-5} and 1×10^{-4} M, respectively, were premixed with Mn(II). In general, an induction time always occurs if the two catalysts are introduced simultaneously. Separate addition of the second catalyst may eliminate the induction time if $Ru(bpy)_3SO_4$ or DPA, but not ferroin, is employed at low concentration ($<10^{-4}$ M). When [ferroin] $= 10^{-4}$ and 2×10^{-4} M were applied instead of $Ru(bpy)_2SO_4$ under the conditions of Fig. 1A, induction times of 70 and 110 min, respectively, were measured. A higher concentration of the second catalyst results in a higher amplitude, longer period and longer sequence of oscillations. For example, with [ferroin] $= 10^{-4}$ and 2×10^{-4} M (other concentrations as in Fig. 1), the following data were recorded: period $=$ 15 and 36 min; initial amplitude of oscillations $=$ 170 and 190 mV, respectively; duration of oscillations $>$6 h in both experiments. The data above are informative about the time expected for the appearance and the persistence of two-dimensional patterns when a mixture of oscillatory composition is spread in a Petri dish.

Effect of composition on the oscillatory behaviour

Long-lasting batch oscillations in the BrO_3^-–$H_2PO_2^-$–acetone–H_2SO_4 system with dual catalyst pairs Mn(II)–$Ru(bpy)_3SO_4$, Mn(II)–ferroin or Mn(II)–DPA were observed in the following ranges of concentration: $[BrO_3^-] = 8 \times 10^{-3}$–$3.2 \times 10^{-2}$ M, $[H_2PO_2^-] = 9 \times 10^{-2}$–$1.7 \times 10^{-1}$ M, [acetone] $= 9 \times 10^{-2}$–2.5×10^{-1} M, $[MnSO_4] = 2 \times 10^{-3}$–$3.5 \times 10^{-3}$ M, $[H_2SO_4] = 1.0$–2.5 M, $[Ru(bpy)_3SO_4] = 10^{-5}$–$3 \times 10^{-4}$ M, [ferroin] $= 5 \times 10^{-5}$–3.3×10^{-4} M, [DPA] $= 5 \times 10^{-5}$–10^{-4} M. The best compositions for producing long-lasting oscillations are similar to those shown in Fig. 1. Significant changes in the characteristics of the oscillations (induction time, amplitude, period, duration of oscillations) are observed near the ends of the concentration ranges.

When DPA served as the second catalyst, the duration of the second sequence of oscillations was much shorter than with $Ru(bpy)_3SO_4$ or ferroin. For example, with $[BrO_3^-] = 8.5 \times 10^{-3}$ M, $[H_2PO_2^-] = 9.2 \times 10^{-2}$ M, $[H_2SO_4] = 2.0$ M, [acetone] $= 8.9 \times 10^{-2}$ M, [Mn(II)] $= 2 \times 10^{-3}$ M, [DPA] $= 5 \times 10^{-5}$ M and Mn(II) and DPA added simultaneously as in Fig. 1A, the DPA-induced oscillations started as soon as DPA was introduced but died away after about 2 h. During this time the amplitude of Pt oscillations first increased from 50 to 190 mV (period is 4–6 min), then gradually decreased to zero. The solution oscillated between dark blue and colourless. The termination of oscillations is a consequence of the slow decomposition of DPA by BrO_3^- oxidation; a new addition of DPA engenders a new (much shorter-lived) oscillatory state.

Effect of light and stirring rate on the oscillatory behavior

Chemical oscillators that contain photosensitive species usually show sensitivity to illumination. Because Ru(II) complexes are easily excited by visible light, the BrO_3^-–$H_2PO_2^-$–acetone–Mn(II)–Ru(bpy)$_3$SO$_4$ system is expected to be sensitive to light. When the oscillator shown in Fig. 1 was illuminated with the 300 W halogen lamp of a Fiber Optic Illuminator (Dolan-Jenner Ind. Inc., Model 190), the period and amplitude immediately decreased by 25% and 50%, respectively. When the light source was turned off, the system soon recovered to its original amplitude and frequency. The effect of light on the oscillatory state is illustrated in Fig. 2.

We found a strong dependence on the stirring rate of the induction period, amplitude and frequency of the oscillations, a dependence that is not characteristic of other bromate oscillators. Sensitivity to the stirring rate is probably associated with the escape of volatile reagents (acetone) and/or products (bromoacetone, bromine) from the reaction mixture. In order to assure reproducibility, a constant stirring rate (300 rpm) was used and the air gap between the surface of the reaction mixture and the cap was always kept the same.

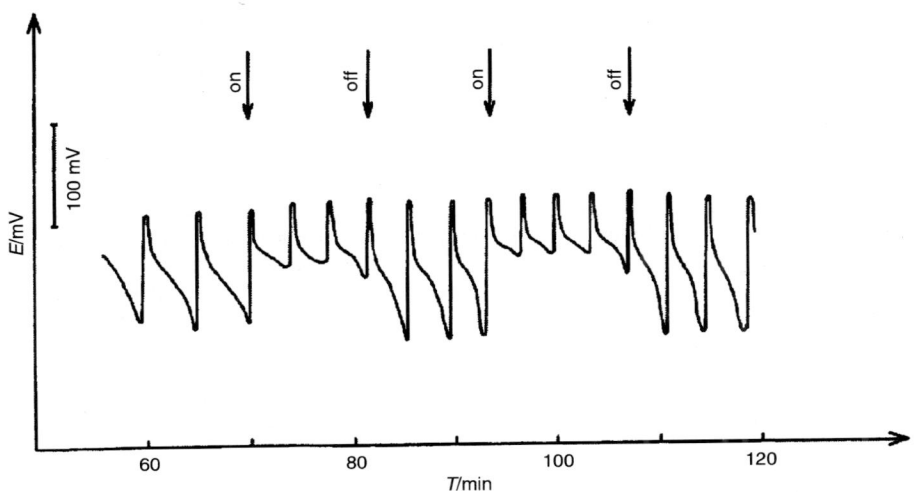

Fig. 2 Effect of light on oscillations in the potential of a Pt electrode in the BrO_3^-–$H_2PO_2^-$–acetone–H_2SO_4–Mn(II)–Ru(bpy)$_3$SO$_4$ system. Light source is switched on and off at the times shown by the arrows. Composition as in Fig. 1.

Fig. 3 Traveling chemical waves and spirals in a thin (~1 mm) solution layer. Composition: $[BrO_3^-] = 3.2 \times 10^{-2}$ M; $[H_2PO_2^-] = 1.13 \times 10^{-1}$ M; $[Mn(II)] = 3.1 \times 10^{-3}$ M; $[Ru(bpy)_3SO_4] = 5 \times 10^{-5}$ M; [acetone] $= 1.2 \times 10^{-1}$ M; $[H_2SO_4] = 1.33$ M.

Wave and spiral formation

In the oscillatory BrO_3^-–$H_2PO_2^-$–acetone–H_2SO_4–Mn(II)–second catalyst system, the changes in colour due to the Ru(II) ↔ Ru(III), ferroin ↔ ferriin or diphenylamine ↔ diphenylbenzidine transitions are visible only if the amplitude of oscillations exceeds 100 mV. This requirement can easily be fulfilled by using a 10^{-4} M or higher concentration of the second catalyst.

When a solution which is in an oscillatory state is spread in a Petri dish, target patterns develop and move with a speed of about 1 mm min^{-1} and a wavelength of 1–3 cm. Mechanical disturbance of a target pattern at an early stage results in the formation of a spiral wave. For generation of patterns the Mn(II)–Ru(bpy)$_3$SO$_4$ and the Mn(II)–ferroin catalyst pairs are equally good but poor results were obtained with Mn(II)–DPA. The evolution of the waves continues for as long as the batch oscillations persist, often many hours. We observed formation of new traveling waves 7 h after the first wave appeared in the Petri dish. The wave fronts are quite sharp, as illustrated in Fig. 3. The patterns are free of bubbles and precipitate during their entire lifetime, and they readily reform after being destroyed by swirling the solution layer.

Discussion

The overall reaction between the two main components of the BrO_3^-–$H_2PO_2^-$ oscillators is described by eqn. (1):

$$BrO_3^- + 3H_2PO_2^- + 3H^+ = Br^- + 3H_3PO_3 \qquad (1)$$

Reaction (1) is reported to be quite slow[6,7] and is autocatalytic in bromide ions, with an induction period inversely proportional to the concentrations of the reactants.[7] Reaction (1) was found to be catalysed and oscillatory only with Mn(II). No catalysis or oscillations were observed using other BZ-catalysts: Ce(III), ferroin or Ru(II)-complex.

In order to understand better the short-lived batch oscillations in the BrO_3^-–$H_2PO_2^-$–Mn(II)–N_2 or BrO_3^-–$H_2PO_2^-$–acetone–Mn(II) systems and the sustained batch oscillations in the BrO_3^-–$H_2PO_2^-$–acetone–Mn(II)–second catalyst systems, we consider the following questions:

(i) What is the origin of the batch oscillations?

(ii) Why is Mn(II) the only effective catalyst in the BrO_3^-–$H_2PO_2^-$ reaction?

(iii) How is a small amount of $Ru(bpy)_3SO_4$ or ferroin added to an exhausted BrO_3^-–$H_2PO_2^-$–Mn(II)–acetone system able to reinitiate and to maintain the oscillations for many hours, while these catalysts are completely ineffective in the absence of Mn(II)?

Questions (i)–(iii) may be, at least qualitatively, answered as follows:

(i) Studies have shown that when $H_2PO_2^-$ is oxidised to H_3PO_3 by a variety of oxidising agents (Br_2, IO_3^-, Cl_2, $HgCl_2$) the rate law and rate constant for the reaction are about the same and the velocity is independent of the concentration of the oxidant.[8] This observation was explained by assuming H_3PO_2 to exist in two forms, which differ strongly in reactivity: the "normal" form, four-coordinated $H_2PO(OH)$, is unreactive, and the "active" form, three-coordinated $HP(OH)_2$, reacts rapidly with oxidants, according to the following mechanism:

$$[H_3PO_2] \text{ normal} \xrightarrow{k_1} [H_3PO_2] \text{ active} \quad \text{(slow)} \quad (2)$$

$$[H_3PO_2] \text{ active} \xrightarrow{k_2} [H_3PO_2] \text{ normal} \quad \text{(rapid)} \quad (3)$$

$$[H_3PO_2] \text{ active} + \text{Ox} \longrightarrow H_3PO_3 + \text{Red} \quad \text{(very rapid)} \quad (4)$$

The "active" H_3PO_2 is formed in the slow acid-catalysed tautomerisation reaction in process (2) from "normal" H_3PO_2 and serves as a constant source of reactive reductant required for all variants of batch oscillators based on the reaction between BrO_3^- and $H_2PO_2^-$.

The appearance of the sustained batch oscillations shown in Fig. 1 suggests a very low conversion of H_3PO_2 to H_3PO_3 during each oscillatory period. Therefore, even if the "active" H_3PO_2 is completely consumed in one cycle, the initial concentration of reagent $H_2PO_2^-$ can support many oscillations via step (2) without the need for a continuous supply by input mass transport in a CSTR.

(ii) The effectiveness of the Mn(II) catalyst in the reaction between BrO_3^- and $H_2PO_2^-$ results from the ability of its oxidised forms to exchange two electrons during bromate oxidation of H_3PO_2 to H_3PO_3, according to the following steps:

$$BrO_3^- + 4Mn(II) + 5H^+ \rightarrow HOBr + 4Mn(III) + 2H_2O \quad (5)$$

$$2Mn(III) \rightarrow Mn(II) + Mn(IV) \quad (6)$$

$$Mn(IV) + H_3PO_2 + H_2O \rightarrow Mn(II) + H_3PO_3 + 2H^+ \quad (7)$$

Ferroin, Ce(III) and $Ru(bpy)_3SO_4$ are also readily oxidised by BrO_3^-, but their oxidised forms can exchange only a single electron with the substrate, which does not favour oxidation of H_3PO_2. We have measured the rate of oxidation of H_3PO_2 by Mn(III), Ce(IV), ferriin and Ru(III) complex and found the following sequence: Mn(III) > Ce(IV) > Ru(III) > ferriin; with pseudo-first order rate constants $k' = 0.013 > 0.0012 > 9.9 \times 10^{-5} > 8.3 \times 10^{-6}$ s^{-1}, respectively.

We also found that soluble Mn(IV) (stabilised with polyphosphate) oxidises $H_2PO_2^-$ several orders of magnitude faster than Mn(III). The reaction is autocatalytic, shows a dependence on [H$^+$], and needs further study to establish k'. These rates show the superiority of manganese species over the other ions as catalysts for the oxidation of $H_2PO_2^-$.

(iii) The role of the second catalyst in inducing a new sequence of oscillations in the exhausted BrO_3^-–$H_2PO_2^-$–acetone–Mn(II)–$Ru(bpy)_3SO_4$ system still awaits explanation. Even the mechanism of the short-lived oscillator, the BrO_3^-–$H_2PO_2^-$–Mn(II)–acetone reaction, is not yet clear.

In Table 1 we suggest a skeleton mechanism that accounts for the "exotic" behaviour observed

Table 1 Skeleton mechanism for the BrO_3^-–$H_2PO_2^-$–acetone–Mn(II)–Ru(bpy)$_3$SO$_4$ oscillatory system (the reactions are unbalanced)

$BrO_3^- + Mn(II) \rightarrow Br(I) + Mn(III)$	(M1)
$2Mn(III) \rightarrow Mn(II) + Mn(IV)$	(M2)
$Mn(IV) + H_2PO_2^- \rightarrow Mn(II) + H_3PO_3$	(M3)
$Br(I) + Br^- \rightarrow Br_2$	(M4)
$Br_2 + Acetone \rightarrow Br\text{-}Ac + Br^-$	(M5)
$Mn(III) + Ru(II) \rightarrow Mn(II) + Ru(III)$	(M6)
$Ru(III) + Br^- \rightarrow Ru(II) + Br_2$	(M7)

in the BrO_3^-–$H_2PO_2^-$–acetone–Mn(II)–Ru(bpy)$_3$SO$_4$ system and that may explain the function of the second catalyst Ru(bpy)$_3$SO$_4$ in maintaining the oscillatory state.

In the skeleton mechanism, reactions (M1)–(M5) explain the short-lived Mn(II)-catalyzed batch oscillations, which turn into long-lasting ones when steps (M6) and (M7) are included. The reagents in step (M1), together with Br^- inflow, constitute the simplest bromate oscillator (or minimal bromate oscillator) when the reaction is run in a continuous-flow stirred tank reactor (CSTR).[9] This system can oscillate between two kinetic states associated with low and high concentrations of Br^-. Bromide ions act as the control intermediate in the oscillatory mechanism.

Step (M3) makes the oscillations possible in batch. The "active" $H_2PO_2^-$ supplied continuously by process (2) reacts with Mn(IV) in a slow reaction that restores Mn(II). The Br_2 produced in (M4) is chemically bound in (M5). The oscillations continue as long as the concentration of the control intermediate Br^- lies within a critical range, but cease when (M5) causes $[Br^-]$ to increase above this range. The existence of short-lived oscillations in the BrO_3^-–$H_2PO_2^-$–Mn(II)–acetone system suggests that the proper balance can be maintained only briefly and the oscillations stop as (M5) becomes dominant.

Steps (M6) and (M7) account for the role of the second catalyst Ru(bpy)$_3$SO$_4$ in the oscillatory cycle. Ru(II) is readily oxidised in (M6) [$k_{M6} = 267$ M^{-1} s^{-1}], and Ru(III) consumes Br^- in (M7), bringing its level back into the critical range where oscillations can occur.

We performed kinetic runs to establish the relative rates of oxidation of Br^- by the oxidised form of the BZ-catalysts, and found the following order: Mn(III) > Ru(III) ~ ferriin ≫ Ce(IV). This order, plus the sequence of k' values for oxidation of $H_2PO_2^-$, suggest that Ru(III) and ferriin are not consumed in the reaction with $H_2PO_2^-$, but contribute significantly to the removal of Br^-, adjusting $[Br^-]$ to an optimum range where long-lasting oscillations occur.

Simulations are in progress to check the validity of the skeleton mechanism and to work out the detailed description of the reactions in order to explain the "exotic" phenomena observed in the BrO_3^-–$H_2PO_2^-$–acetone–dual catalyst system.

Conclusion

The bromate–hypophosphite–acetone–dual catalyst oscillatory batch system may well turn out to be superior to either of the current systems for producing and studying two-dimensional pattern formation. The only disadvantage of the patterns produced by the present system is the relatively long (1–3 cm) wavelength, which prevents the development of large numbers of waves and spirals in the Petri dish. This drawback, however, is more than compensated by the advantages of long-lasting, gas- and precipitate-free pattern evolution—much longer than in the other systems—and the fact that the wave fronts appearing in the clear layer of reaction mixture can easily be observed, recorded and/or manipulated by external forces.

Acknowledgements

This work was supported by grants from the Hungarian Academy of Sciences (HAS) (OTKA No. T029791), the Bolyai Research Fund of HAS (BO/00096/98), the US National Science Foundation (NSF) and a US–Hungarian Co-operative Research Grant from NSF and HAS.

References

1. Q. Ouyang, W. J. Tam, P. De Kepper, W. D. McCormick, Z. Noszticzius and H. L. Swinney, *J. Phys. Chem.*, 1987, **91**, 2181.
2. K. Kurin-Csörgei, M. Orbán, A. Zhabotinsky and I. R. Epstein, *J. Phys. Chem.*, 1997, **101**, 6827.
3. Y. Gao and H.-D. Försterling, *J. Phys. Chem.*, 1995, **99**, 8638.
4. G. Peintler, *ZITA Version 5.0.*, A Comprehensive Program Package for Fitting Parameters of Chemical Reaction Mechanism, JATE, Szeged, Hungary, 1989–99.
5. L. Adamčikova and P. Sevčik, *Int. J. Chem. Kinet.*, 1982, **14**, 735.
6. P. Sevčik and L. Adamčikova, *Collect. Czech. Chem. Comm.*, 1987, **52**, 2125.
7. *Gmelin's Handbook: Phosphorus*, Verlag Chemie, Weinheim, 1965, vol. III, p. 99 and related references.
8. W. A. Jenkins and D. M. Yost, *J. Inorg. Nucl. Chem.*, 1959, **11**, 297.
9. M. Orbán, P. De Kepper and I. R. Epstein, *J. Am. Chem. Soc.*, 1982, **104**, 2657.

HPLC analysis of complete BZ systems. Evolution of the chemical composition in cerium and ferroin catalysed batch oscillators: experiments and model calculations

László Hegedűs,[a] Mária Wittmann,[a] Zoltán Noszticzius,*[a] Shuhua Yan,[b] Atchara Sirimungkala,[b] Horst-Dieter Försterling[b] and Richard J. Field[c]

[a] *Center for Complex and Nonlinear Systems and the Department of Chemical Physics, Budapest University of Technology and Economics, H-1521 Budapest, Hungary*
[b] *Fachbereich Chemie, Philipps-Universität Marburg, D-35032 Marburg/Lahn, Germany*
[c] *Department of Chemistry, The University of Montana, Missoula, Montana 59812, USA*

Received 17th April 2001
First published as an Advance Article on the web 7th November 2001

In the last few years many new reaction routes and intermediates have been discovered in the mechanism of the Belousov–Zhabotinsky (BZ) reaction with the aid of high performance liquid chromatography (HPLC). These previous HPLC studies, however, were limited to the Ce^{4+}–organic substrate (malonic or bromomalonic acid) systems only. Very recently some measurements were made on a cerium catalysed full BZ system but only in its induction period. The present work follows the evolution of the main chemical components in a cerium and in a ferroin catalysed full BZ system from the start until the end of the oscillatory regime in a batch reactor. While recording the potential oscillations of a bromide selective electrode we measured from time to time the concentration of the following components: malonic and bromomalonic acids and bromate as main components; malonyl malonate, ethanetetracarboxylic and bromoethenetricarboxylic acids which are recombination products of organic free radicals; oxidized intermediates: tartronic, oxalic (OA) and mesoxalic (MOA) acids, and brominated products: dibromoacetic and tribromoacetic acids. Recombination products are generated in the intervals when the autocatalytic reaction is "switched off". In the course of the autocatalytic periods, however, the organic radicals react with the inorganic bromine dioxide radical mainly which leads to the formation of MOA and OA. Due to a very fast Ce^{4+}–MOA reaction, MOA can be detected in the ferroin catalysed BZ system only. Our model calculations deal exclusively with the cerium catalysed system. The suggested new Marburg–Budapest–Missoula (MBM) model includes both negative feedback loops (bromous acid–bromide ion Oregonator type and bromine dioxide–organic free radicals Radicalator type feedback) and the recently discovered radical–radical recombination reactions. Comparison of the experimental data with the model calculations shows a good qualitative agreement but some open problems still remain. To overcome these problems oxygen atom transfer and other redox reactions are proposed.

Introduction

The classical BZ reaction,[1,2] the cerium ion catalysed oxidation and bromination of malonic acid by acidic bromate, is the most studied chemical oscillator and a prime example for both temporal

DOI: 10.1039/b103432b

and spatial nonlinear phenomena in chemistry.[3–6] In spite of that, due to the complexity of the so-called organic reaction subset, there are still important processes in the negative feedback loops which are not known or at least not well understood. To explain the nature of these problems first a brief historical account is presented.

The FKN mechanism and bromide controlled oscillations

The basic mechanism of the BZ reaction was elucidated in 1972 by Field, Kőrös and Noyes[7] (FKN). According to the FKN theory, oscillations are due to an interplay of a positive and a delayed negative feedback loop. The positive feedback is the autocatalytic bromous acid and bromine dioxide production in the course of the oxidation of Ce^{3+} to Ce^{4+} by acidic bromate. The first step in the negative feedback is the bromide ion generation by Ce^{4+} in a reaction with bromomalonic acid. Bromide ion then reacts rapidly with bromous acid which is an autocatalytic intermediate. In this way bromide ion controls switches between an oxidized and a reduced state. The FKN mechanism is thus referred to as bromide controlled. The mechanism was generally accepted and its simplified version, the Oregonator[8] was applied successfully to model oscillations and other nonlinear phenomena in the BZ reaction. Actually, the system of the inorganic reactions (the "inorganic subset")— after a revision of its rate constants[9–11]—is still used basically in the same form as it was suggested by FKN. The main problem of the FKN mechanism was the organic subset, especially the so-called "Process C", the bromide production by Ce^{4+} in a reaction with a mixture of bromomalonic and malonic acids. Jwo and Noyes[12] and later Försterling and Stuk[13] studied the problem experimentally but the mechanism and the stoichiometry of this process remained unclear. An interesting feature of the process is that the oxidation of the two substrates does not proceed independently: more bromide is produced than would be expected from an independent oxidation of malonic (MA) and bromomalonic (BrMA) acids.[12]

The Radicalator and non-bromide controlled oscillations

Bromide ion plays the role of a control intermediate because it reacts rapidly with the autocatalytic intermediate $HBrO_2$. Beside bromous acid, bromine dioxide radical is also a part of the autocatalytic cycle, consequently any intermediate reacting with BrO_2 can also act as a control intermediate. In 1989 Försterling and Noszticzius[14] reported that malonyl radicals react with bromine dioxide at a diffusion controlled rate. Thus an additional negative feedback loop was discovered in the BZ reaction. Later Försterling et al.[15] found non-bromide controlled oscillations and suggested a new mechanistic model, the Radicalator, in which malonyl radical plays the role of the control intermediate. The same model was applied successfully to explain certain stirring effects of the BZ reaction.[16] An open problem in the chemical mechanism of the Radicalator was the further fate of the malonyl bromite ($MABrO_2$), a recombination product formed in the control reaction between malonyl and bromine dioxide radicals.

The Györgyi–Turányi–Field (GTF) mechanism

The next important development in the theory of the BZ reaction was the mechanism proposed by Györgyi, Turányi and Field[17,18] in 1990 and 1993. The GTF mechanism aimed to incorporate all the experimental information available at that time. Free radicals also play an important role in this mechanism but GTF assumed that organic free radicals, when they react with each other, disproportionate rather than recombine. For example malonyl radicals disproportionate according to reaction (41) in their scheme:

$$2\,MA^{\cdot} + H_2O \rightarrow MA + TA \qquad (GTF\ 41)$$

where TA is tartronic acid.
Radical transfer reactions like

$$MA^{\cdot} + BrMA \rightarrow MA + BrMA^{\cdot} \qquad (GTF\ 61)$$

were also regarded to be important, strengthening the negative feedback *via* bromide. (GTF 61) is a vital step of the mechanism as no oscillations can be observed without it. Försterling and Stuk,[13] however, found by EPR measurements that this reaction is probably unimportant. In their second paper[18] GTF added the hydrolysis reaction of BrMA· to their scheme:

$$BrMA^{\cdot} + H_2O \rightarrow TA + Br^- \quad\quad (GTF\ 84)$$

to explain these EPR results of Försterling and Stuk.[13]

An additional source of bromide ions in the GTF mechanism is the reduction of bromomalonic acid by carboxyl radicals:

$$BrMA + COOH^{\cdot} \rightarrow Br^- + MA^{\cdot} + CO_2 \quad\quad (GTF\ 71)$$

Finally we mention one more problem of the GTF mechanism: while it reproduces Ce^{4+} oscillations with high fidelity if the initial substrate is malonic acid the model fails to oscillate with bromomalonic acid as initial substrate in a parameter range where oscillations are observed experimentally.

HPLC measurements

Recently, as HPLC became a standard analytical technique, a systematic program was initiated in Marburg to identify various organic products and intermediates of the BZ reaction.[19–23]

HPLC of the organic subsystems. The research began with HPLC studies on the products of the Ce^{4+}–malonic acid and the Ce^{4+}–bromomalonic acid reactions. These organic subsystem studies have shown that the primary organic radicals do not disproportionate but recombine. In the case of malonic acid radicals the recombination products are ethanetetracarboxylic acid (ETA)[19] and malonyl malonate (MAMA)[20] because the malonic acid radical is capable of mesomerism.[24] When bromomalonyl radicals react with each other the final product is bromoethenetricarboxylic acid (BrEETRA).[21]

It is also important to mention here that no tartronic acid was found in the Ce^{4+}–bromomalonic acid reaction thus reaction (GTF 84), the proposed hydrolysis of bromomalonyl radicals, cannot play a role in the mechanism.

HPLC in the induction period. The next step was to study the reaction products of a complete BZ system but first only in its induction period.[23] The presence of oxalic acid was proven combining HPLC with various tests. Beside oxalic acid another new intermediate, ethenetetracarboxylic acid (EETA) was also identified.

Aims of the present work

In the first part of this work we report HPLC analysis of cerium and ferroin catalysed BZ systems in their oscillatory regime as it was measured in a batch reactor. Batch reactor experiments were chosen because accumulation of certain intermediates can be monitored this way. As oxygen has a dramatic effect on the BZ reaction[25] due to its fast reaction with the organic free radicals[26,27] all experiments were carried out in a nitrogen atmosphere. Higher than normal catalyst concentrations (5×10^{-3} M) were applied to produce elevated intermediate concentrations and accelerate consumption of the organic substrate. In this way the oscillatory regime and consequently the length of the experiment was shortened. Some additional experiments were also performed to check the validity of certain reactions of the GTF mechanism.

In the second part of this paper, after collecting all the available experimental data for the Ce^{4+} catalysed system, we performed calculations with a revised and updated GTF model which will be referred to as the MBM model. Finally some open problems of the new model together with the possible solutions are also discussed.

Experimental

Chemicals

Malonic acid (Fluka, puriss.), $Ce(SO_4)_2 \cdot H_2O$ (Merck, pro analysi), 0.025 M ferroin solution (Fluka, pro analysi), $NaBrO_3$ (Fluka, puriss.) and H_2SO_4 97% (J. T. Baker) were used as received. All solutions were prepared with doubly distilled water.

The BZ reaction with Ce^{4+} catalyst in N$_2$ atmosphere

The BZ reaction was examined under anaerobic conditions at room temperature (22 $\pm$ 2 °C) in a batch reactor. 1 mL 1 M malonic acid solution (in 1 M sulfuric acid), 1 mL 1 M NaBrO$_3$ solution (in water), 1.7 mL 5 M sulfuric acid and 5.8 mL (doubly distilled) water were mixed in a 30 mL reaction vessel and bubbled with nitrogen for 2 min. (At lower catalyst concentrations this time is usually 15 min or more. Here, due to the fast oxygen consumption this precaution was unnecessary and we also wanted to avoid any small effects caused by the slow uncatalysed malonic acid–bromate reaction.) The reaction was started by injecting the catalyst, 0.5 mL 0.1 M Ce(SO$_4$)$_2$ solution (in 1 M sulfuric acid). Thus the initial concentrations were: [MA]$_0$ = 0.1 M, [BrO$_3^-$]$_0$ = 0.1 M, [Ce^{4+}]$_0$ = 5 $\times$ 10^{-3} M and [H$_2$SO$_4$] = 1 M. A constant nitrogen stream (about 20 mL min^{-1}) created anaerobic conditions and also mixed the solution. To follow the composition of the reaction mixture 0.25 mL samples were taken from time to time and quenched abruptly by a hundredfold dilution with distilled water and were analysed by HPLC. The dilution stopped the reaction and also established the same sulfuric acid concentration in the sample as in the eluent.

The BZ reaction with ferroin catalyst in N$_2$ atmosphere

The BZ reaction with ferroin was performed in a similar way to that with Ce^{4+}. The only difference was the composition of the reaction mixture. 1.5 mL 1 M malonic acid solution (in 1 M sulfuric acid), 1.5 mL 1 M NaBrO$_3$ solution (in water), 0.7 mL 5 M sulfuric acid and 4.3 mL (doubly distilled) water were mixed in the reaction vessel and bubbled with nitrogen. The reaction was started by injecting the catalyst, 2 mL 0.025 M ferroin solution. Thus the initial concentrations were: [MA]$_0$ = 0.15 M, [BrO$_3^-$]$_0$ = 0.15 M, [ferroin]$_0$ = 5 $\times$ 10^{-3} M and [H$_2$SO$_4$] = 0.5 M. As in the Ce-system, a constant nitrogen stream was applied. The sampling procedure was similar except for an ion exchange before HPLC. To remove ferroin from the sample 0.5 g Varion KS III cation exchange resin was added to the whole diluted solution (25 mL), stirred and filtered *via* a G4 glass filter after 10 min. To avoid contamination the applied ion exchange resin was always freshly washed with the eluent.

Analytical methods

HPLC. Experiments were performed with Shimadzu equipment using an ion exchange column at 45 °C and a UV detector working at 220 nm. The eluent was 0.01 M H$_2$SO$_4$, flow rate 0.40 mL min^{-1}. The very same HPLC conditions had already been applied successfully in a series of previous investigations to separate various organic acid intermediates of the BZ reaction. For further details see ref. 20–23. The eluent was filtered with a membrane filter (Poraphil, pore size 0.2 μm) to remove the solid particles. All the measured solutions were also filtered (Sartorius Minisart NML, pore size 0.2 μm).

Potentiometric measurements. The potential signal was measured between a bromide selective electrode (Radiometer type) and a Ag/AgCl electrode in saturated KCl as a reference (Metrohm type) *via* a salt bridge containing 0.5 M sulfuric acid. The measurement was performed with a Digitalmeter Digi 610 type potentiometer, registered with a computer.

Spectrophotometric measurements. The absorbance of Ce^{4+} was followed in a 2 cm optical cell (applying a stream of nitrogen through the solution) using a dual wavelength spectrophotometer (measuring wavelength 400 nm, reference 650 nm).

Experimental results

Experiments with the cerium catalysed BZ oscillator

Potentiometric measurements. First the oscillatory behaviour of the system was recorded with a bromide selective electrode as described in the Experimental (Fig. 1(a)). To quantify the signal of a bromide selective electrode is somewhat difficult in a BZ system as beside bromide the electrode also responds to the corrosive hypobromous acid.[28] On the other hand, when comparing cerium

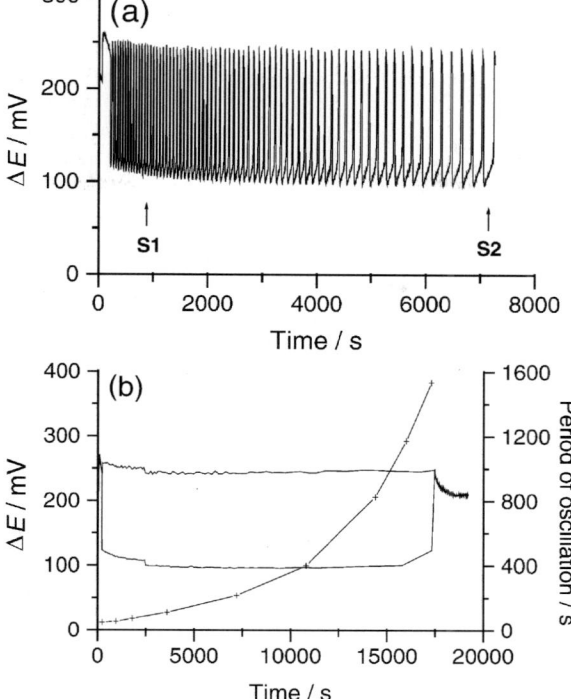

Fig. 1 Evolution of Ce-catalysed BZ oscillations in batch. $[MA]_0 = 0.1$ M, $[NaBrO_3]_0 = 0.1$ M, $[Ce^{4+}]_0 = 5 \times 10^{-3}$ M, $[H_2SO_4] = 1$ M. (a) Potential oscillations of a bromide selective electrode. Arrows indicate times when HPLC samples (S1, S2) were taken. The corresponding chromatograms are shown in Fig. 2 (S1) and Fig. 3 (S2). (b) Evolution of bromide electrode potential minima and maxima and the time period of the oscillations.

and ferroin catalysed BZ systems it is an advantage that bromide electrode oscillations can be observed in both systems. In the case of the cerium experiments the Br⁻ potential region is the one below 200 mV and the minimum potentials of 100 mV indicate about 5×10^{-5} M bromide concentration maximums. The HOBr potential region is above 200 mV and the maximums near to 250 mV correspond to roughly 7×10^{-6} M HOBr concentration. In Fig. 1(b) the complete oscillatory regime is depicted indicating the potential minima and maxima and the time period in the oscillatory regime. As the diagram shows, oscillations appear and disappear with finite amplitude in this batch experiment.

Evaluation of the HPLC data. Fig. 2 and 3 are HPLC chromatograms displayed as examples to explain the evaluation methods. Fig. 2(a) and (b) shows the same chromatogram recorded with a lower and a higher sensitivity simultaneously. Two channel recording was necessary to measure different components of the sample which are present in a wide concentration and absorbance range. The first peak of Fig. 2(a) at 480 s is the injection peak. In our present experiments this can be assigned to bromate. The contribution of other inorganic components like bromide to the injection peak is negligible. The strong tailing of the bromate peak hides any small peaks of dibromomalonic (Br_2MA) or ethenetetracarboxylic acids, both of them having a retention time of 500 s. These components, however, should be only minor ones. This is because by the time of the HPLC analysis most of the dibromomalonic acid decomposes to dibromoacetic acid (Br_2AcA). The other component, EETA was found, in independent experiments, to be also a minor one.[23] The first peak appearing after the bromate clearly (but still sitting on the bromate tailing) is due to BrEETRA at 510 s.

The next three peaks in Fig. 2(a) are due to three major components without any interference: bromomalonic (BrMA, $t_r = 723$ s), malonic (MA, $t_r = 1049$ s) and dibromoacetic (Br_2AcA, $t_r =$

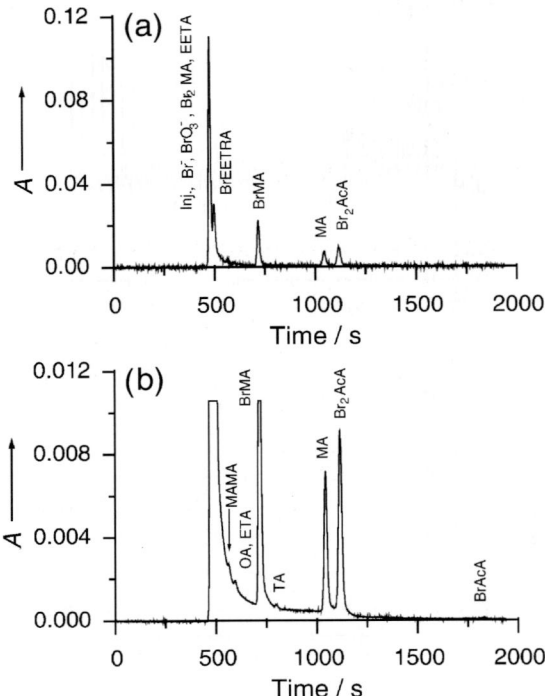

Fig. 2 HPLC chromatogram of sample S1. See Table 1 for abbreviations of the various components assigned for the peaks. (a) Low and (b) high sensitivity record.

1123 s) acids. At higher sensitivity as shown in Fig. 2(b) further components can be discovered: malonyl malonate (MAMA, $t_r = 570$ s), oxalic (OA, $t_r = 600$ s), ethanetetracarboxylic (ETA, $t_r = 600$ s), tartronic (TA, $t_r = 800$ s) and bromoacetic (BrAcA, $t_r = 1810$ s) acids. At even higher sensitivity (not shown here) these minor peaks are well separated and sharp and can be easily utilized for concentration determinations. There is an interference, however, between OA and ETA. Fortunately the contribution of ETA to the peak at 600 s can be calculated from the MAMA peak as the ETA/MAMA peak ratio is a stable value of 1.6 due to a mesomeric equilibrium between the two different forms of the malonyl radical.[24] OA concentrations can be calculated from the remaining absorbance. This method of calculation was checked with a heat treatment procedure[23] during which ETA decomposes but OA does not. OA concentrations calculated with correction due to ETA and measured directly after decomposition of ETA agreed well.

Fig. 3 shows similar chromatograms of a second sample taken at a later stage of the reaction. Time evolution of the various components is clear when Fig. 2 and 3 are compared. For example bromate, BrMA and MA peaks decreased while BrEETRA and Br$_2$AcA peaks have grown during this period. A further complication is caused by the appearance of tribromoacetic acid (Br$_3$AcA, $t_r = 605$ s) in the middle stage of the reaction because its retention time is rather close to that of the oxalic acid. Fortunately in the early stage the amount of Br$_3$AcA is negligible and in the late stage the same holds for OA and ETA thus uncertainties are limited to a short time interval where the concentrations of these components are comparable.

Development of chemical concentrations measured by HPLC. Fig. 4 displays the concentrations of the various components as a function of the reaction time. Concentrations were calculated from HPLC chromatograms similar to those shown in Fig. 2 and 3.

In Fig. 4(a) we can see the concentration changes of the main components, namely bromate, malonic and bromomalonic acids. Bromomalonic acid concentration has a maximum of about

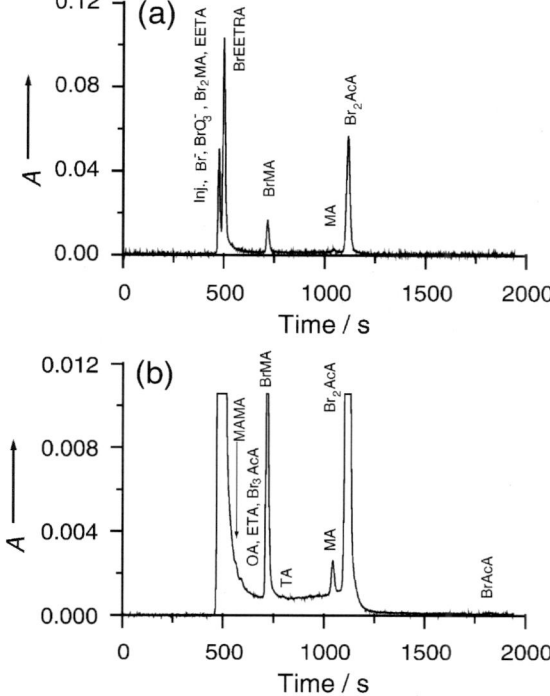

Fig. 3 HPLC chromatogram of sample S2. (a) Low and (b) high sensitivity record.

0.02 M at around 2000 s and decreases afterward slowly. Malonic acid concentration decreases more rapidly and after 8000 s reaction time the malonic and bromomalonic acid concentrations, while decreasing further, remain close to each other. As bromate is applied in excess, its concentration, after a sharp initial fall, approaches to a finite value gradually.

In Fig. 4(b) concentration changes of the radical–radical recombination products ETA, MAMA and BrEETRA are shown in the course of the reaction. Malonyl and bromomalonyl radicals can recombine mainly in the intervals between two autocatalytic periods in the reduced state of the system. When the autocatalytic reaction is "switched off" the concentration of bromine dioxide is small and recombination of organic free radicals can proceed unperturbedly. It is known[21] that BrEETRA is an inert endproduct of the BZ reaction thus it accumulates monotonically. Its concentration reaches about 0.013 M after 20 000 s of reaction time. The other two recombination products ETA and MAMA can be oxidized further thus after a sharp increase at the start their concentration is falling indicating that consumption surmounts production. Then a quasi-steady state is reached where the concentration is stabilized at a lower level. Finally, after a long stagnation, the concentrations continue to fall to zero as the oscillatory regime nears to its end.

In Fig. 4(c) concentration of certain oxidized intermediates (OA and TA) are shown in the course of the reaction. TA is interesting because it can be a decomposition product of malonyl bromite. The same is true for OA which can be, however, a decomposition product of both malonyl and bromomalonyl bromite. The maximum concentration of these oxidized intermediates (about 3×10^{-4} M for OA and 6×10^{-4} M for TA) are reached at the end of the induction period. Shortly after this OA falls to a low level the exact value of which is uncertain because of the interference with Br_3AcA. TA decreases more gradually and maintains a well measurable level throughout the oscillatory regime. It is important to remark that the concentration of another possible oxidized intermediate namely mesoxalic acid (MOA) remains under our HPLC detection limit which is 2×10^{-5} M for MOA.

The amount of brominated products, Br_2AcA and Br_3AcA shown in Fig. 4(d) is growing monotonically, the first reaching a rather high concentration (close to 0.02 M) while the latter reaches a

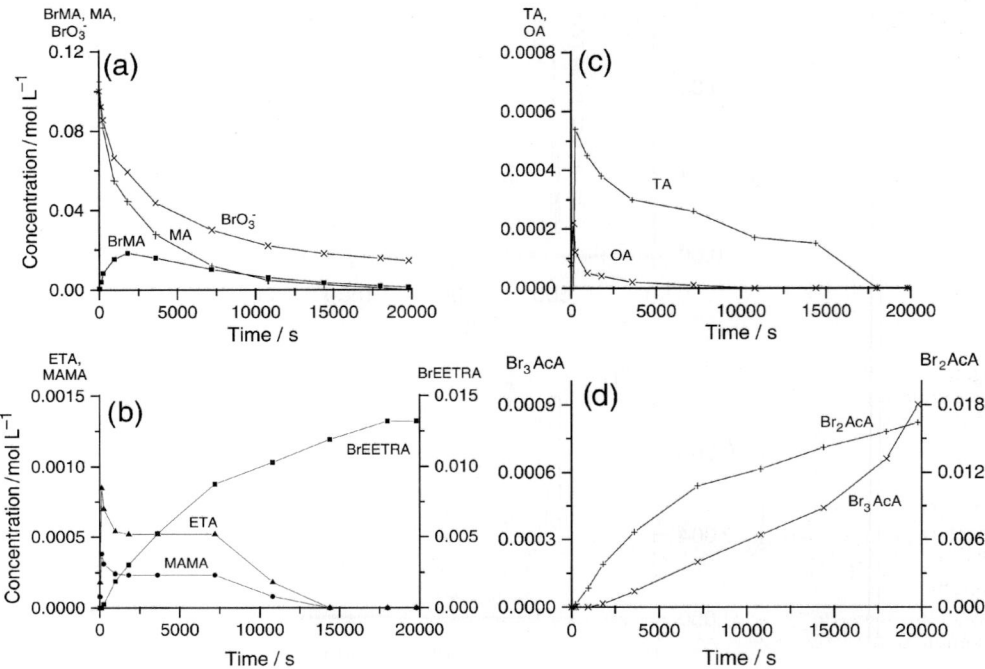

Fig. 4 Concentration evolution of bromate and various organic components in the course of the Ce-catalysed batch reaction. (a) Main components: acidic bromate, malonic and bromomalonic acids. (b) Radical recombination products: bromoethenetricarboxylic acid (right scale), malonyl malonate and ethanetetracarboxylic acid (left scale). (c) Oxidized intermediates: oxalic and tartronic acids. (d) Brominated products: Br_2AcA (right scale) and Br_3AcA (left scale).

relatively minor (about 8×10^{-4} M) concentration after 20 000 s of reaction time. BrAcA was detected in traces only (less than 10^{-5} M).

Experiments with the ferroin catalysed BZ oscillator

Potentiometric measurements. In the present ferroin catalysed BZ oscillator the catalyst concentration was the same (5×10^{-3} M), but the bromate and the malonic acid concentration was higher by 50% (both were 0.15 M) than in the cerium catalysed system. The sulfuric acid concentration was 0.5 M. For the potentiometric measurements the same bromide selective electrode was applied as with Ce^{4+} and the potential of the electrode was shifted down by about 50 mV. As Fig. 5 shows, the potential oscillations start without any induction period, with small amplitude that grows rapidly to a certain level. Then the oscillations continue but the amplitude is changing irregularly. After 2 h oscillations disappear with a finite amplitude. Compared to the cerium system both the time period of the oscillations and the length of the oscillatory regime is shorter. Also the growth of the time period during the oscillatory regime is less pronounced.

Development of chemical concentrations as measured by HPLC. In Fig. 6(a) concentration of the main components (MA, BrMA and BrO_3^-) is displayed in the course of the ferroin catalysed oscillatory reaction. While the pattern in some respects is similar to that of the cerium catalysed system shown in Fig. 4(a), there are important differences as well. In both cases the malonic and bromomalonic acid concentrations meet close to the middle of the oscillatory regime but now the bromomalonic acid concentration maximum (5×10^{-2} M) is 2.5 times larger than in Fig. 4(a). This is partly due to the fact that the initial malonic acid and bromate concentration is also higher by a factor of 1.5, which accelerates the bromomalonic acid production. The bromate curve is also similar but the organic substrate reducing the bromate is clearly different above 10 000 s. With

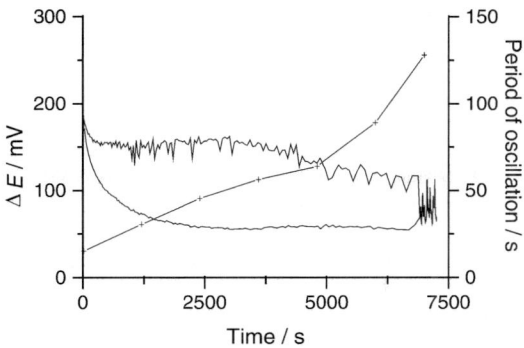

Fig. 5 Evolution of bromide electrode potential minima and maxima and the time period of the oscillations in a ferroin-catalysed BZ reaction in batch. $[MA]_0 = 0.15$ M, $[NaBrO_3]_0 = 0.15$ M, $[ferroin] = 5 \times 10^{-3}$ M, $[H_2SO_4] = 0.5$ M.

Ce^{4+} catalyst bromate is reduced by malonic and bromomalonic acids whose concentrations decrease continuously between 10 000 and 20 000 s indicating this process. With ferroin catalyst after 10 000 s of reaction time, the malonic acid concentration is practically zero and the bromomalonic acid concentration is constant. Thus these components cannot be responsible for the bromate consumption above 10 000 s. As Fig. 6(c) reveals it is OA and MOA which are the bromate consuming organic substrates in this case.

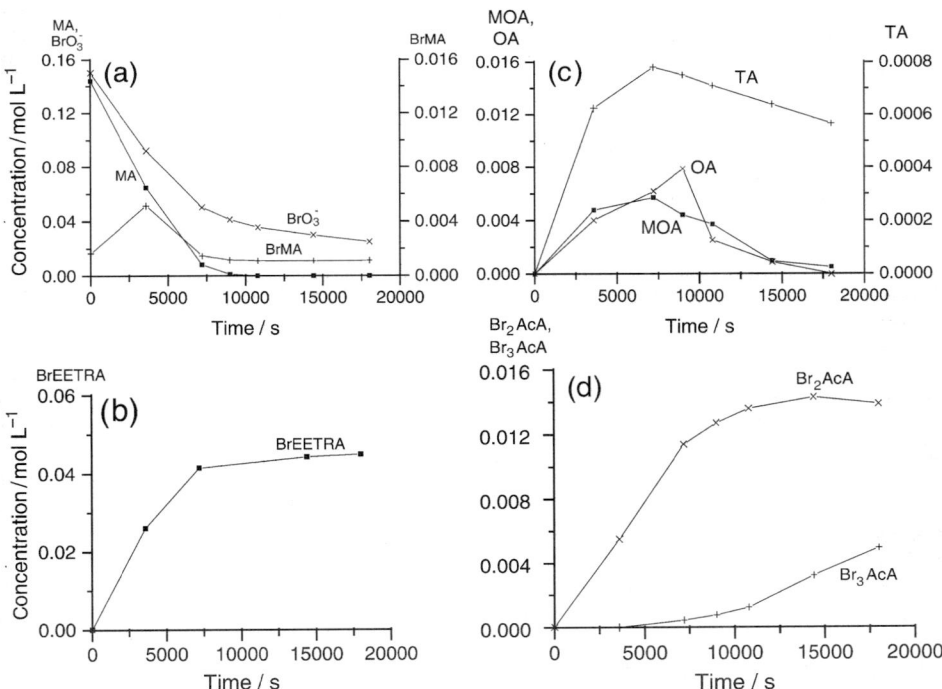

Fig. 6 Concentration evolution of bromate and various organic components in the course of the ferroin-catalysed batch reaction. (a) Main components: acidic bromate, malonic acid (left scale) and bromomalonic acid (right scale). (b) The only radical recombination product found in this system, bromoethenetricarboxylic acid. (c) Oxidized intermediates: oxalic, mesoxalic (left scale) and tartronic acids (right scale). (d) Brominated products: Br_2AcA and Br_3AcA.

In Fig. 6(b) where the evolution of the organic recombination products is shown there are two features which deserve comments. First, the measured BrEETRA concentration is more than 3 times higher here than in the cerium catalysed system (see Fig. 4(b)). This is, again, partly due to the 1.5 times higher initial malonic acid concentration but the effect is more than proportional. An even more important difference is that recombination products of malonyl radicals, that is MAMA and ETA, are absent. This indicates that a direct attack of the oxidized form of the catalyst on malonic acid, which produces malonyl radicals in the case of Ce^{4+}, is missing in the case of ferriin or any small amount of malonyl radicals react further with ferriin. Independent experiments[29] show, however, that while the ferriin–malonic acid reaction is slow or negligible the ferriin–bromomalonic acid reaction is fast and its main product is BrEETRA.

In Fig. 6(c) the concentration of the measured oxidized intermediates (OA, MOA and TA) is depicted as a function of time in the ferroin catalysed system. The most important difference compared to the cerium catalysed system is the appearance of a relatively high concentration of mesoxalic acid (the maximum concentration is about 6×10^{-3} M). The maximal oxalic acid concentration is about 8×10^{-3} M which is more than one order of magnitude higher than in the cerium catalysed system. Both OA and MOA maintain this high concentration in the course of the oscillations then reach the maximum concentration at the end of the oscillatory regime and decrease only afterward. This difference between the cerium and ferroin systems is probably due to the fact that while Ce^{4+} reacts rapidly with OA and MOA, ferriin is not able to oxidize these substrates. Thus in the ferroin catalysed systems OA and MOA are oxidized by acidic bromate directly. The average tartronic acid concentration is also higher here by a factor of two or three and its maximum occurs not at the beginning but close to the end of the oscillatory regime.

Fig. 6(d) shows the brominated products of the ferroin system. The Br_2AcA concentration is about the same as in the Ce^{4+} system but the concentration of Br_3AcA is higher by a factor of four.

All of these results indicate that there are many differences in the mechanism of the cerium and the ferroin catalysed systems. As it is also evident that there are more unknown details in the ferroin system in the present work our model calculations will concentrate on the "classical" cerium catalysed system which was also the subject of most previous modeling efforts.

Model calculations for the cerium catalysed BZ reaction

The MBM model of the BZ reaction

We discuss this model (shown in Table 1) using a three-step procedure. First, we focus on the inorganic subset. Second, we include all reactions necessary to model a BZ system starting with pure bromomalonic acid as a substrate; in this case oscillations start immediately with a finite amplitude after the addition of the catalyst, without showing an induction period. Third, we add all reactions occurring in a BZ system starting with pure malonic acid or with a mixture of malonic acid and bromomalonic acid; in this case an induction period appears indicating that oscillations can only start if additional organic intermediates appear. Finally, we will compare model calculations with experimental data for selected initial conditions.

Inorganic reactions. The set of inorganic reactions is mainly that reported by Field and Försterling.[11] Modifications were made in reactions (R4), (R5), the mechanism of the disproportionation of bromous acid.[33] Reaction (R10) is a recently discovered new decomposition path of BrO_2.[31] The products of reaction (R9), the decomposition of bromate in acidic medium, could be identified as $HBrO_2$ and oxygen[30] rather than as $HBrO_2$ and perbromate.

The most recent values of the rate constants for reactions (R1)–(R9) are used,[30,32,33] a small correction on k_2 was applied.[31]

A decisive change had to be made on rate constant k_3, the reaction of bromide with bromate in excess. So far it was believed from direct measurements that k_3 is 1.2 M^{-1} s^{-1} at 20 °C.[34] However, these experiments were carried out with initial bromide concentrations much higher than the typical range in a BZ system. Yan[31] reinvestigated this reaction using initial concentrations down to 2×10^{-6} M. Surprisingly it was found that k_3 depends on the bromide initial concentration c_0, slowing from $k_3 = 1.4$ M^{-1} s^{-1} at $c_0 = 2 \times 10^{-4}$ M to $k_3 = 0.5$ M^{-1} s^{-1} at

Table 1 Reactions and rate constants of the MBM mechanism of the cerium catalysed BZ reaction[a]

No.	Reaction		Rate constants					
			Forward		Ref.	Reverse		Ref.
	Inorganic reactions							
R1	$Br^- + HOBr + H^+$	$\uparrow\uparrow$	$Br_2 + H_2O$	$M^{-2} s^{-1}$	11, 30	110	s^{-1}	11, 30
R2	$Br^- + HBrO_2 + H^+$	$\uparrow\uparrow$	2 HOBr	$M^{-2} s^{-1}$	11, 30	2.0×10^{-5}	$M^{-1} s^{-1}$	11, 30
R3	$Br^- + BrO_3^- + 2 H^+$	$\uparrow\uparrow$	$HOBr + HBrO_2$	$M^{-3} s^{-1}$	11, 31	3.2	$M^{-1} s^{-1}$	11, 31
R4	$HBrO_2 + H^+$	$\uparrow\uparrow$	$H_2BrO_2^+$	$M^{-1} s^{-1}$	11, 30	$1.0 \times 10^{+8}$	s^{-1}	11, 30
R5	$HBrO_2 + H_2BrO_2^+$	$\uparrow\uparrow$	$HOBr + BrO_3^- + 2 H^+$	$M^{-1} s^{-1}$	11, 30	0		
R6	$HBrO_2 + BrO_3^- + H^+$	$\uparrow\uparrow$	$Br_2O_4 + H_2O$	$M^{-2} s^{-1}$	11, 30	$3.2 \times 10^{+3}$	s^{-1}	11, 30
R7	Br_2O_4	$\uparrow\uparrow$	$2 BrO_2^\cdot$	s^{-1}	11, 30	$1.4 \times 10^{+9}$	$M^{-1} s^{-1}$	11, 30
R8	$Ce^{3+} + {}^\cdot BrO_2 + H^+$	$\uparrow\uparrow$	$Ce^{4+} + HBrO_2$	$M^{-2} s^{-1}$	11, 32, 33	$1.3 \times 10^{+4c}$	$M^{-1} s^{-1}$	33
R9	$2 BrO_3^- + 2 H^+$	$\uparrow\uparrow$	$2 HBrO_2 + O_2$	$M^{-3} s^{-1}$	30	0		
R10	$\cdot BrO_2$	$\uparrow\uparrow$	$\frac{1}{2} Br_2 + O_2$	s^{-1}	31	0		
	Reactions in the BrMA subsystem							
R11	$BrMA + Ce^{4+}$	$\uparrow\uparrow$	$BrMA^\cdot + Ce^{3+} + H^+$	$M^{-1} s^{-1}$	21, 31	400	$M^{-2} s^{-1}$	31
R12	$2 BrMA^\cdot$	$\uparrow\uparrow$	$BrEETRA + Br^- + CO_2 + H^+$	$M^{-1} s^{-1}$	21, 34	0		
R13	$BrMA$	$\uparrow\uparrow$	$BrMA(enol)$	s^{-1}	35	800	s^{-1}	35
R14	$BrMA(enol) + Br_2$	$\uparrow\uparrow$	$Br_2MA + Br^- + H^+$	$M^{-1} s^{-1}$	35	0		
R15	$BrMA(enol) + HOBr$	$\uparrow\uparrow$	$Br_2MA + H_2O$	$M^{-1} s^{-1}$	35	0		
R16	$BrMA^\cdot + {}^\cdot BrO_2$	$\uparrow\uparrow$	$BrMABrO_2$	$M^{-1} s^{-1}$	31, 34	0		
R17	$BrMABrO_2$	$\uparrow\uparrow$	$OA + HOBr + Br^- + CO_2 + H^+$	s^{-1}	31, 34	0		
R18	$BrMABrO_2$	$\uparrow\uparrow$	$BrTA + HBrO_2$	s^{-1}	31, 34	0		
R19	$BrTA$	$\uparrow\uparrow$	$Br^- + MOA + H^+$	s^{-1}	34	0		
R20	$MOA + Ce^{4+} + H_2O$	$\uparrow\uparrow$	$OA + Ce^{3+} + COOH^\cdot + H^+$	$M^{-1} s^{-1}$	31	0		
R21	$OA + Ce^{4+}$	$\uparrow\uparrow$	$COOH^\cdot + Ce^{3+} + CO_2 + H^+$	$M^{-1} s^{-1}$	31	0		
R22	$2 COOH^\cdot$	$\uparrow\uparrow$	OA	$M^{-1} s^{-1}$	17, 31	0		
R23	$COOH^\cdot + Ce^{4+}$	$\uparrow\uparrow$	$Ce^{3+} + CO_2 + H^+$	$M^{-1} s^{-1}$	17	0		
R24	$COOH^\cdot + BrMA$	$\uparrow\uparrow$	$MA^\cdot + Br^- + CO_2 + H^+$	$M^{-1} s^{-1}$	17, 31	0		
R25	$COOH^\cdot + BrMA^\cdot$	$\uparrow\uparrow$	$BrMA + CO_2$	$M^{-1} s^{-1}$	17, 31	0		
R26	$COOH^\cdot + {}^\cdot BrO_2$	$\uparrow\uparrow$	$HBrO_2 + CO_2$	$M^{-1} s^{-1}$	17	0		

Forward rate constants: R1: $8 \times 10^{+9}$; R2: $2.9 \times 10^{+6}$; R3: $f(conc.)^b$; R4: $2 \times 10^{+6}$; R5: $1.7 \times 10^{+5}$; R6: 48; R7: $7.5 \times 10^{+4}$; R8: $6.0 \times 10^{+4}$; R9: 6.0×10^{-10}; R10: 0.06; R11: 0.1; R12: $1 \times 10^{+9}$; R13: 1.2×10^{-2}; R14: $3.5 \times 10^{+6}$; R15: $1.1 \times 10^{+6}$; R16: $2 \times 10^{+9}$; R17: 0.62; R18: 0.46; R19: 1.5; R20: $7.0 \times 10^{+3}$; R21: 28; R22: $5 \times 10^{+9}$; R23: $1 \times 10^{+7}$; R24: $1 \times 10^{+7}$; R25: $3 \times 10^{+9}$; R26: $5 \times 10^{+9}$.

Table 1 Continued

No.	Reaction		Rate constants				
			Forward		Ref.	Reverse	Ref.
	Reactions in the MA subsystem						
R27	MA$^\cdot$ + Ce^{4+}	$\rightarrow$	MA$^\cdot$ + Ce^{3+} + H$^+$	0.23	M^{-1} s^{-1}	36	2.2 × 10^{+4}	M^{-2} s^{-1}	36
R28	2 MA$^\cdot$	$\rightarrow$	ETA	3.2 × 10^{+9}	M^{-1} s^{-1}	19, 20	0		
R29	MA	$\rightarrow$	MA(enol)	2.6 × 10^{-3}	s^{-1}	35	180	s^{-1}	35
R30	MA(enol) + Br$_2$	$\rightarrow$	BrMA + Br$^-$ + H$^+$	2.0 × 10^{+6}	M^{-1} s^{-1}	35	0		
R31	MA(enol) + HOBr	$\rightarrow$	BrMA + H$_2$O	6.7 × 10^{+5}	M^{-1} s^{-1}	35	0		
R32	MA$^\cdot$ + $^\cdot$BrO$_2$	$\rightarrow$	MABrO$_2$	5 × 10^{+9}	M^{-1} s^{-1}	14	0		
R33	MABrO$_2$	$\rightarrow$	MOA + HOBr	0.55	s^{-1}	37	0		
R34	MABrO$_2$	$\rightarrow$	TA + HBrO$_2$	1.0	s^{-1}	37	0		
R35	MA$^\cdot$ + BrMA$^\cdot$	$\rightarrow$	Br$^-$ + EETA + H$^+$	2 × 10^{+9}	M^{-1} s^{-1}	17, 31	0		
R36	MA$^\cdot$ + COOH$^\cdot$	$\rightarrow$	MA + CO$_2$	4 × 10^{+9}	M^{-1} s^{-1}	17, 31	0		
R37	TA + Ce^{4+}	$\rightarrow$	TA$^\cdot$ + Ce^{3+} + H$^+$	0.66	M^{-1} s^{-1}	38	1.7 × 10^{+4}	M^{-2} s^{-1}	38
R38	2 TA$^\cdot$	$\rightarrow$	CO$_2$ + EEHTRA + H$_2$O	1 × 10^{+9}	M^{-1} s^{-1}	31	0		
R39	TA	$\rightarrow$	TA(enol)	2.3 × 10^{-5}	s^{-1}	35	1.5	s^{-1}	35
R40	TA(enol) + Br$_2$	$\rightarrow$	BrTA + Br$^-$ + H$^+$	3 × 10^{+5}	M^{-1} s^{-1}	35	0		
R41	TA(enol) + HOBr	$\rightarrow$	BrTA + H$_2$O	2 × 10^{+5}	M^{-1} s^{-1}	35	0		
R42	TA$^\cdot$ + MA$^\cdot$	$\rightarrow$	EETA + H$_2$O	1 × 10^{+9}	M^{-1} s^{-1}	17	0		
R43	TA$^\cdot$ + BrMA$^\cdot$	$\rightarrow$	CO$_2$ + BrEETRA + H$_2$O	1 × 10^{+9}	M^{-1} s^{-1}	17	0		
R44	TA$^\cdot$ + COOH$^\cdot$	$\rightarrow$	TA + CO$_2$	3 × 10^{+9}	M^{-1} s^{-1}	17	0		
R45	TA$^\cdot$ + $^\cdot$BrO$_2$	$\rightarrow$	TABrO$_2$	2 × 10^{+9}	M^{-1} s^{-1}	17	0		
R46	TABrO$_2$	$\rightarrow$	MOA + HBrO$_2$	0.1	s^{-1}	31	0		
R47	TA + BrO$_3^-$	$\rightarrow$	HBrO$_2$ + MOA	5 × 10^{-5}	M^{-1} s^{-1}	31	0		
R48	MA$^\cdot$ + BrO$_3^-$ + H$^+$	$\rightarrow$	$^\cdot$BrO$_2$ + TA	160	M^{-1} s^{-1}	14	0		

[a] Abbreviations: BrMA = bromomalonic acid, BrEETRA = bromoethenetricarboxylic acid, BrTA = bromotartronic acid, BrMA$^\cdot$ = bromomalonyl radical, COOH$^\cdot$ = carboxyl radical, EEHTRA = ethenehydroxytricarboxylic acid, EETA = ethenetetracarboxylic acid, ETA = ethanetetracarboxylic acid, MA = malonic acid, MA$^\cdot$ = malonyl radical, MOA = mesoxalic acid, OA = oxalic acid, TA = tartronic acid, TA$^\cdot$ = tartronyl radical. [b] k_3 can be described by a power series: $k_3 = c_0 + c_1 \times [\text{Br}^-/\text{M}] + c_2 \times [\text{Br}^-/\text{M}]^2 + c_3 \times [\text{Br}^-/\text{M}]^3$ for $[\text{Br}^-] > 2 \times 10^{-6}$ M; $k_3 = 0.3$ for $[\text{Br}^-] < 2 \times 10^{-6}$ M ($c_0 = 0.48789$ M^{-1} s^{-1}, $c_1 = 0.143911 \times 10^{+5}$ M^{-1} s^{-1}, $c_2 = -0.7076958 \times 10^{+8}$ M^{-1} s^{-1}, $c_3 = 0.116310 \times 10^{+12}$ M^{-1} s^{-1}). In the range of bromide concentrations typical for the BZ system (2×10^{-6} to 2×10^{-5} M) k_3 can be approximated by the mean value $k_3 = 0.6$ M^{-1} s^{-1}. [c] k_{-8} depends on the ionic strength μ of the solution: $k_{-8} = c_0 + c_1 \times (\mu/\text{M}) + c_2 \times (\mu/\text{M})^2 + c_3 \times (\mu/\text{M})^3$ ($c_0 = -1.74743 \times 10^{+4}$ M^{-1} s^{-1}, $c_1 = 6.75949 \times 10^{+4}$ M^{-1} s^{-1}, $c_2 = -4.30066 \times 10^{+4}$ M^{-1} s^{-1}, $c_3 = 0.797993 \times 10^{+4}$ M^{-1} s^{-1}). For a typical BZ system with an initial concentration of 0.1 M bromate and 1 M sulfuric acid the ionic strength is $\mu = 1.6$ M^{-1} s^{-1} and $k_{-8} = 1.3 \times 10^{+4}$ M^{-1} s^{-1}.

$c_0 = 2 \times 10^{-6}$ M. Apparently, this means that reaction (R3) is not an elementary reaction, but follows a complex kinetics. So far it was not possible to elucidate the reaction mechanism. For modelling the BZ reaction we treat k_3 as a parameter, depending on the actual bromide concentration in the system. Moreover, k_3 can be approximated by the mean value 0.6 M^{-1} s^{-1} for the bromide concentrations typical in a BZ system. Using this lower rate constant two severe discrepancies found in earlier kinetic research could be resolved. First, the low k_3 can quantitatively account for the induction time in the bromide inhibited autocatalytic Ce^{3+}—bromate reaction (the high k_3 gives a value too low by a factor of two). Second, the low k_3 can quantitatively account for the rate of disappearance of bromine in the presence of acidic bromate (the high k_3 leads to a decay of bromine too fast by a factor of two).

Reactions involving bromomalonic acid. Reactions (R11)–(R19) have been discussed in the past.[34,35] The rate constants for reactions (R20)–(R21) have been redetermined in this work. Reactions (R22)–(R26) describe further reactions of the carboxyl radical. A key reaction in the model is reaction (R24), the generation of malonyl radicals and bromide by the reaction of carboxyl radicals with bromomalonic acid. We could verify this reaction in a batch experiment by oxidizing a mixture of oxalic acid and bromomalonic acid by Ce^{4+}; all the expected reaction products could be detected by HPLC. Details will be published elsewhere.[31]

Reactions in the malonic acid subsystem. Reactions (R27)–(R34) have been discussed in the past.[19,20] Reaction (R35) could be verified in a similar experiment as that designed to follow reaction (R24).[31] The importance of reaction (R35) is that it can give a realistic explanation for the "extra bromide" observed first by Jwo and Noyes[12] when Ce^{4+} reacts with a mixture of malonic and bromomalonic acid. In this way reaction (R35) can substitute reactions (GTF 61) and/or (GTF 84). Reaction (R36), a reduction of malonyl radicals by carboxyl radicals to malonic acid, is in analogy to reaction (R25). An experimental proof for the (R24)–(R36) reaction sequence is that MA was found[31] as a product when a mixture of oxalic and bromomalonic acids was oxidized by Ce^{4+}. Reactions (R37)–(R41) have been discussed in the past.[35,36] Reactions (R42)–(R46) are chosen in analogy to the corresponding reactions with malonic acid or bromomalonic acid. Reaction (R47) could be verified by following BrO$_2$ as an intermediate.[31] Reaction (R48) was detected by EPR spectroscopy;[14] in that work a rate constant of 800 M^{-1} s^{-1} was obtained as an upper limit; we use a smaller value (160 M^{-1} s^{-1}). In the case of high cerium BZ systems reactions (R47) and (R48) help to start the autocatalytic reaction.

Tests for the model

We test the model in two typical cases: (i) for a BZ system starting with pure bromomalonic acid (Fig. 7) and (ii) for a BZ system starting with pure malonic acid (Fig. 8). Regarding the figures for bromomalonic acid substrate the agreement is excellent while for malonic acid substrate there is one disagreement: the time period of oscillations in the model calculations is somewhat longer than in the experiments. The amplitudes of the Ce^{4+} oscillations agree well in both cases.

Discussion

Comparison of experimental and calculated concentration—time diagrams

After testing the model simulations were performed with the initial concentrations used for the HPLC experiments. As a detailed comparison for all components would require too many diagrams as figures these diagrams are not shown here but put on the web.[39] Here we discuss the agreements or disagreements between the theoretical and experimental curves only qualitatively.

Ce^{4+} oscillations. As can be seen in Fig. 7 and 8 at low Ce^{4+} concentration (3.56 × 10^{-4} M) the agreement between theory and experiment is rather good. For HPLC measurements, however, we applied a high catalyst concentration (5 × 10^{-3} M) to have a higher yield of products and intermediates. Now, unknown reactions of these intermediates are also enhanced and due to them a less good agreement can be expected in this case. Really, the calculated induction time (380 s) is about 1.5 times higher and especially the calculated time period (300 s) is about 6 times longer

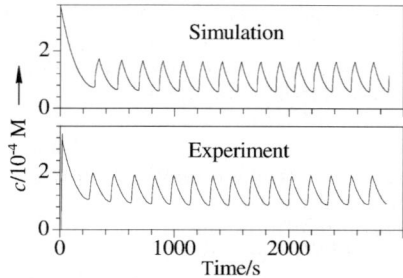

Fig. 7 Experimental and calculated Ce^{4+} oscillations in a BZ system with pure bromomalonic acid as initial substrate. Initial concentrations: $[BrMA]_0 = 0.05$ M, $[BrO_3^-]_0 = 0.15$ M, $[Ce^{4+}]_0 = 3.56 \times 10^{-4}$ M, 1 M sulfuric acid, temperature 20 °C.

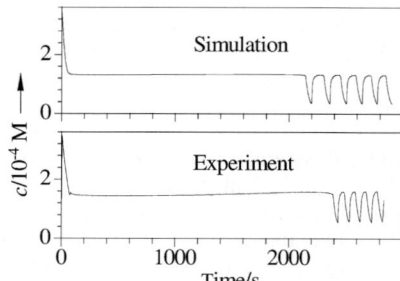

Fig. 8 Experimental and calculated Ce^{4+} oscillations in a BZ system with pure malonic acid as initial substrate. Initial concentrations: $[MA]_0 = 0.05$ M, $[BrO_3^-]_0 = 0.15$ M, $[Ce^{4+}]_0 = 3.56 \times 10^{-4}$ M, 1 M sulfuric acid, temperature 20 °C.

than the experimental value. The calculated minimum Ce^{4+} concentrations are too low and the model spends a too long a time both in the oxidized and reduced states. Especially the longer oxidized state deforms the shape of the oscillations. Incorporating certain oxygen atom transfer reactions (see later) can lessen these problems.

The length of the oscillatory regime in the calculations is close to the observed one: the last oscillation occurs at 18 000 s in the model while this is around 17 500 s in the experiment.

Bromide and hypobromous acid oscillations. The calculated amplitude of the bromide oscillations is somewhat higher ($0.8-1.2 \times 10^{-4}$ M) than the observed one. At least a part of the difference is due to the relatively slow response of the bromide selective electrode as the calculated bromide peaks are rather sharp. The same is true for the hypobromous acid whose peaks are extremely sharp and where the calculated peaks can exceed the experimental ones by nearly one order of magnitude.

Main components: BrO_3^-, MA and BrMA. Both bromate and malonic acid are consumed more slowly (about two times slower) in the simulations than in the real experiments. Also production of the bromomalonic acid is slower and the calculated maximum of the bromomalonic acid is only half the experimental value. The time of the maximum is roughly the same in both cases but the experimental error is rather large here as the HPLC measurements are not continuous.

Recombination products: ETA + MAMA and BrEETRA. The ETA + MAMA concentration is growing monotonically in the calculations. (Model calculations give only the sum of the two recombination products. Individual concentrations could be calculated from this value but such a calculation was not reasonable, see later.) In the experiments, however, the ETA and the MAMA concentrations exhibit a maximum right after the induction period and decrease afterward. This disagreement is due to the fact that in BZ experiments ETA and MAMA are not inert end-products but they are rapidly oxidized further. These oxidation processes are not known, however, thus they are not included in the theoretical scheme. In spite of that, the amount of ETA + MAMA right after the induction period is in the same order of magnitude (about 10^{-3} M) both in experiments and model calculations. On the other hand, in the case of BrEETRA we have a serious disagreement: its experimental concentration exceeds the calculated value by nearly two orders of magnitude. Obviously, there should be other reactions which produce BrEETRA or its precursor, the bromomalonyl radical. For example, the following reaction between dibromomalonic acid and a carboxyl radical

$$Br_2MA + COOH^\cdot \rightarrow Br^- + H^+ + BrMA^\cdot + CO_2 \tag{NR1}$$

could convert some Br_2MA first to bromomalonyl radical then to BrEETRA. This process would decrease the final Br_2AcA concentration which component would be the only decomposition product of Br_2MA otherwise. It is important to mention that an analogous reaction ((R24) in Table 1) has already been observed experimentally.[31]

Oxidized intermediates: OA, MOA and TA. The amount of oxalic acid is the highest right after the induction period according to both model calculations and experiments. Nevertheless calculations overestimate the average OA concentration by a factor of two. (OA oscillations appearing in the calculated diagram cannot be detected by HPLC because of the limited time resolution of the latter.)

The maximum mesoxalic acid concentration predicted by the model calculations is 10^{-6} M which is 20 times lower than its HPLC detection limit. In agreement with this prediction no MOA was found with HPLC in the cerium catalysed BZ system. In sharp contrast a significant amount of MOA (more than 4×10^{-3} M, that is 200 times above the detection limit) was found in the ferroin catalysed system. This result is in agreement with the assumption that MOA is formed in both systems but in the case of cerium catalyst its concentration is kept below the detection limit by the fast Ce^{4+}-MOA reaction (reaction (R20) in Table 1.)

Tartronic acid plays an important role in the latest models of the BZ reaction. For example in the GTF mechanism 21 reactions out of the total 80 describe reactions where tartronic acid or tartronyl radicals are products or reactants. The same ratio is kept in the present MBM model where 12 out of the total 48 are a similar type of reaction. In spite of that it seems that in real experiments TA is less important. HPLC analysis shows that the TA concentration is nearly two orders of magnitude smaller than that predicted by the MBM model. In the GTF model there were two main sources of TA: (i) disproportionation of malonyl radicals, and (ii) hydrolysis of malonyl bromite (R34). As it turned out malonyl radicals do not disproportionate but rather recombine. Thus in the MBM model we kept only the second source but, as was already mentioned, this single source still produces too much TA when compared with experiments. (Actually, even a part of the small TA found in the experiments is not coming from reaction (R34) but from the hydrolysis of MAMA,[20] a reaction not incorporated in our present model.) Thus reaction (R34) is probably less important than was assumed. The solution of this dilemma is not easy. If we drop reaction (R34) from the model then most (if not all) of the induction period disappears, which contradicts the experiments. Thus another reaction route is needed which allows the autocatalytic $HBrO_2$ production like reaction (R34) but, unlike reaction (R34), is free of TA generation. If the only product of the malonyl radical–BrO_2 radical reaction is $MABrO_2$ as shown in reaction (R32) then the possibilities are limited to a few decomposition routes. These routes, e.g. reaction (R33), however, would interrupt the autocatalytic cycle thus they cannot provide the required induction period. A possible solution for this problem is if, beside recombination, we assume a route for disproportionation in the reaction between malonyl and BrO_2 radicals:

$$MA^\cdot + BrO_2^\cdot + H^+ \rightarrow HBrO_2 + {}^+CH(COOH)_2 \tag{NR2}$$

where the organic product is a very reactive intermediate[40] that can react rapidly e.g. with acidic bromate. In this case the next step could be the following reaction (which also includes an oxygen atom transfer):

$$^+CH(COOH)_2 + BrO_3^- \rightarrow HBrO_2 + MOA \tag{NR3}$$

If only a part of the radical–radical reaction follows this route that would be enough to insure the induction period without tartronic acid production.

Bromination products: Br_2AcA and Br_3AcA. Dibromoacetic acid is mainly a decomposition product of dibromomalonic acid:

$$Br_2MA \rightarrow Br_2AcA + CO_2 \qquad (NR4)$$

If we compare the calculated amount of Br_2MA with the measured amount of Br_2AcA (as decarboxylation of Br_2MA is not included in the model) then the measured concentration is about half the calculated one. Most probably beside decarboxylation, Br_2MA can take part in other reactions as well, *e.g.* in reaction (NR1). This would decrease Br_2AcA and increase BrEETRA simultaneously.

It was somewhat unexpected for us to find tribromoacetic acid in the reaction mixture. Neither the present nor the previous models deal with this product. Presently it would be too early to speculate about the possible sources of the observed Br_3AcA formation.

Further reactions that can improve the MBM model. There are many possible new reactions beside (NR1)–(NR4) which could improve the model. Among these, reactions playing a role in the negative feedback loops can have the greatest impact on the oscillatory dynamics. In this respect it would be important to know more about the decomposition routes of malonyl bromite and bromomalonyl bromite. Here we discuss two possible routes for the decomposition of $MABrO_2$. We propose that oxygen atom transfer reactions[41] might play a role here. (In inorganic redox reactions oxygen transfer is rather common, *e.g.* (R2) and (R3) are such reactions.) First we discuss reaction (NR5):

$$MABrO_2 \rightarrow OA + CO_2 + Br^- + H^+ \qquad (NR5)$$

where the suggested mechanism of the intramolecular oxygen atom transfer is the following:

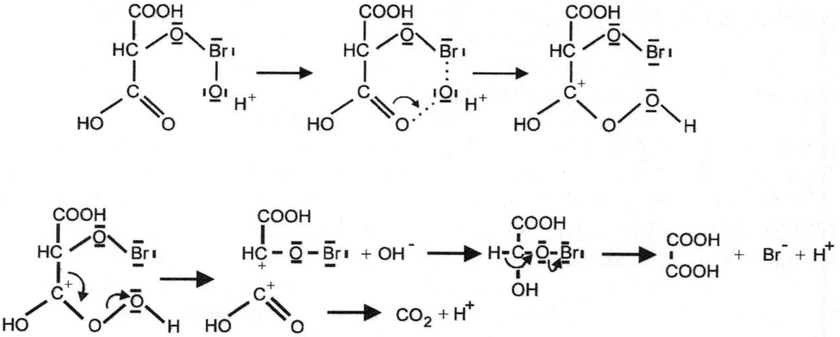

Inclusion of reaction (NR5) into the MBM mechanism with a rate constant of $1 \ s^{-1}$ results in a shorter time period and a better agreement with the experiments regarding Ce^{4+} and bromide oscillations.[39]

Another possible oxygen atom transfer reaction can occur between $HBrO_2$ and MOA (also a decomposition product of $MABrO_2$):

$$HBrO_2 + MOA \rightarrow HOBr + OA + CO_2 \qquad (NR6)$$

Inclusion of reaction (NR6) into the MBM mechanism with a rate constant of 10^5 or $10^6 \ s^{-1}$ also results in a better agreement with the experiments.[39]

Conclusions

1. HPLC analysis

The results of the HPLC analysis show that while all the stable components of the BZ oscillators can be measured with this technique there are still unknown processes in the mechanism of these complex chemical systems. This is because various radicals play an important role in the mechanism and only their relatively stable recombination products can be measured by HPLC.

(i) The reactions between organic free radicals are mostly clarified. These reactions are mainly recombinations in the case of malonyl and bromomomalonyl radicals but disproportionation also occurs in reactions involving carboxyl radical as this species acts as a strong reducing agent.

(ii) The reactions between malonyl or bromomalonyl and BrO_2 radicals are less discovered. It is usually assumed that in the first step malonyl bromite or bromomalonyl bromite intermediates are formed which decompose rapidly. Unfortunately only the possible decomposition routes are known, their relative weight is not.

(iii) HPLC analysis revealed important differences between the cerium and ferroin catalysed systems. No recombination products of malonyl radicals were found in the ferroin systems showing that a direct reaction between ferriin and malonic acid does not occur or it is negligible. In spite of that, ferroin catalysed oscillations start immediately without any induction period even if the substrate is pure malonic acid. This is not the case in cerium systems. Also the ferroin catalysed system contains much more oxalic and mesoxalic acid than the cerium catalysed one.

2. Model calculations

Due to the differences between the cerium and ferroin systems the MBM model suggested here is focussed on the cerium catalysed BZ oscillators exclusively:

(i) At low catalyst concentrations there is a good agreement between the model and experiments regarding the Ce^{4+} oscillations.

(ii) Calculated concentration–time diagrams were compared with the experimental ones for various components measured with HPLC in a BZ system with higher Ce^{4+} concentration. While a qualitative agreement was found for many components the model seriously overestimates tartronic acid and underestimates BrEETRA.

(iii) Six new hypothetical reactions are suggested for future improvements of the model. Nevertheless, it is hoped that the MBM model in its present form can already provide a realistic description of many cerium catalysed BZ systems.

Acknowledgement

This work was partially supported by OTKA (T-030110), FKFP-0090/2001 grants, by the Deutsche Forschungsgemeinschaft, the Fonds der Chemischen Industrie and the ESF Program: "Reactor". M. Wittmann acknowledges the support of the Bolyai Research Stipendium. A part of the simulations was made with the KINAL program package.[42]

References

1 B. P. Belousov, in *Sbornik Referatov po Radiatsionnoi Meditsine*, Medgiz, Moscow, 1958, p. 145.
2 A. M. Zhabotinsky, *Biofizika*, 1964, **9**, 306.
3 *Oscillations and Traveling Waves in Chemical Systems*, ed. R. J. Field and M. Burger, Wiley, New York, 1985.
4 P. Gray and S. Scott, *Chemical Oscillations and Instabilities. Nonlinear Chemical Kinetics*, Clarendon, Oxford, 1994.
5 *Chemical Waves and Patterns*, ed. R. Kapral and K. Showalter, Kluwer, Dordrecht, 1995.
6 I. R. Epstein and J. A. Pojman, *An Introduction to Nonlinear Chemical Dynamics*, Oxford University Press, New York, 1998.
7 R. J. Field, E. Kőrös and R. M. Noyes, *J. Am. Chem. Soc.*, 1972, **94**, 8649.
8 R. J. Field and R. M. Noyes, *J. Chem. Phys.*, 1974, **60**, 1877.
9 Z. Noszticzius, E. Noszticzius and Z. A. Schelly, *J. Phys. Chem.*, 1983, **87**, 510.
10 F. Ariese and Zs. Ungvárai-Nagy, *J. Phys. Chem.*, 1986, **90**, 1.
11 R. J. Field and H. D. Försterling, *J. Phys. Chem.*, 1986, **90**, 5400.
12 J. J. Jwo and R. M. Noyes, *J. Am. Chem. Soc.*, 1975, **97**, 5422.
13 H. D. Försterling and L. Stuk, *J. Phys. Chem.*, 1991, **95**, 7320.
14 H. D. Försterling and Z. Noszticzius, *J. Phys. Chem.*, 1989, **93**, 2740.
15 H. D. Försterling, Sz. Murányi and Z. Noszticzius, *J. Phys. Chem.*, 1990, **94**, 2915.
16 Z. Noszticzius, Zs. Bodnár, L. Garamszegi and M. Wittmann, *J. Phys. Chem.*, 1991, **95**, 6575.
17 L. Györgyi, T. Turányi and R. J. Field, *J. Phys. Chem.*, 1990, **94**, 7162.
18 L. Györgyi, T. Turányi and R. J. Field, *J. Phys. Chem.*, 1993, **97**, 1931.
19 Y. Gao, H. D. Försterling, Z. Noszticzius and B. Meyer, *J. Phys. Chem.*, 1994, **98**, 8377.

20 A. Sirimungkala, H. D. Försterling and Z. Noszticzius, *J. Phys. Chem.*, 1996, **100**, 3051.
21 J. Oslonovitch, H. D. Försterling, M. Wittmann and Z. Noszticzius, *J. Phys. Chem.*, 1998, **102**, 922.
22 I. Szalai, H. D. Försterling and Z. Noszticzius, *J. Phys. Chem.*, 1998, **102**, 3118.
23 L. Hegedűs, H. D. Försterling, E. Kókai, K. Pelle, G. Taba, M. Wittmann and Z. Noszticzius, *Phys. Chem. Chem. Phys.*, 2000, **2**, 4023.
24 W. F. Wang, M. N. Schuchmann, H. P. Schuchmann and C. von Sonntag, *Chem.-Eur. J.*, 2001, 7, 791.
25 P. Ruoff and R. M. Noyes, *J. Phys. Chem.*, 1989, **93**, 7394.
26 B. Neumann and S. C. Müller, *J. Am. Chem. Soc.*, 1995, **117**, 6372.
27 L. Hegedűs, H. D. Försterling, M. Wittmann and Z. Noszticzius, *J. Phys. Chem. A*, 2000, **104**, 9914.
28 Z. Noszticzius, E. Noszticzius and Z. A. Schelly, *J. Am. Chem. Soc.*, 1982, **104**, 6194.
29 L. Hegedűs, H. D. Försterling and Z. Noszticzius, unpublished results.
30 I. Szalai, J. Oslonovitch and H. D. Försterling, *J. Phys. Chem.*, 2000, **104**, 1495.
31 S. Yan, *PhD Thesis*, Philipps-Universität Marburg, 2001, in preparation.
32 H. D. Försterling, Sz. Murányi and H. Schreiber, *Z. Naturforsch. A*, 1989, **44**, 555.
33 H. D. Försterling and M. Varga, *J. Chem. Phys.*, 1993, **97**, 7932.
34 Y. Gao and H. D. Försterling, *J. Phys. Chem.*, 1995, **99**, 8638.
35 A. Sirimungkala and H. D. Försterling, *J. Phys. Chem.*, 1999, **103**, 1038.
36 H. D. Försterling, R. Pachl and H. Schreiber, *Z. Naturforsch. A*, 1987, **42**, 963.
37 H. D. Försterling and Sz. Murányi, *Z. Naturforsch. A*, 1990, **45**, 1259.
38 H. D. Försterling, H. Idstein, R. Pachl and H. Schreiber, *Z. Naturforsch. A*, 1984, **39**, 993.
39 http://www.phy.bme.hu/deps/chem_ph/Research/BZ_Simulation/index.html.
40 Sz. Nagygyőry, M. Wittmann, Sz. Pintér, A. Visegrády, A. Dancsó, Nguyen Bich Thuy, Z. Noszticzius, L. Hegedűs and H. D. Försterling, *J. Phys. Chem. A*, 1999, **103**, 4885.
41 J. Halpern and H. Taube, *J. Am. Chem. Soc.*, 1952, **74**, 380.
42 T. Turányi, *Comput. Chem.*, 1990, **14**, 253.

Effects of non-ionic micelles on transient chaos in an unstirred Belousov–Zhabotinsky reaction

M. Rustici,[a] R. Lombardo,[b] M. Mangone,[a] C. Sbriziolo,[b] V. Zambrano[a] and M. L. Turco Liveri*[b]

[a] *Dipartimento di Chimica, Via Vienna 2, 07100 Sassari, Italy*
[b] *Dipartimento di Chimica Fisica, Viale delle Scienze—Parco d'Orleans II, 90128 Palermo, Italy. E-mail: tliveri@unipa.it*

Received 19th April 2001
First published as an Advance Article on the web 12th November 2001

The behaviour of the Ce(IV)-catalyzed Belousov–Zhabotinsky (BZ) system has been monitored at 20.0 °C in unstirred batch conditions in the absence and presence of different amounts of the non-ionic micelle-forming surfactants hexaethylene glycol monodecyl ether ($C_{10}E_6$) and hexaethylene glycol monotetradecyl ether ($C_{14}E_6$). The influence of the non-ionic surfactants on both the kinetics of the oxidation of malonic acid (MA) by Ce(IV) species and the behaviour of the BZ reaction in stirred batch conditions has also been studied over a wide surfactant concentration range. The experimental results have shown that, in unstirred batch conditions, at surfactant concentrations below the critical micelle concentration (c.m.c.) no significant change in the dynamics of the Belousov–Zhabotinsky system occurs. Beyond this critical concentration the presence of micelles forces the BZ system to undergo a chaos → quasi-periodicity → period-1 transition. Thus, the surfactant concentration has been considered as a bifurcation parameter for a Ruelle–Takens–Newhouse (RTN) scenario. Addition of increasing amounts of non-ionic surfactants has no significant effect on the kinetics of the reaction between MA and Ce(IV), but it influences the oscillatory parameters of the stirred BZ system. At surfactant concentrations below the c.m.c. all the oscillatory parameters are practically unaffected by the presence of surfactant, while beyond this critical value the induction period is the same as in aqueous solution but both the oscillation period and the duration of the rising portion of the oscillatory cycle decrease. In all cases, the experimental trends have been ascribed to the enhancement in the medium viscosity due to the presence of micelles.

Introduction

The numerous theoretical and experimental studies on oscillating reactions indicate a continuing interest in these systems. Multistability, periodicity, multiperiodicity or chaotic oscillations are observed[1–4] in perfectly open stirred systems. Thermodynamic considerations dictate that closed chemical systems cannot undergo sustained oscillations because as the reactants are used up the systems settle into thermodynamic equilibrium.[5] However, in a closed reactor, the initial reactants are often present in large excess making the intermediates behave like an open subsystem for a substantial period of time. This phenomenon can be sustained for significant periods of time before the system reaches chemical equilibrium. It is well established that closed chemical systems

DOI: 10.1039/b103532k

can exhibit various dynamic regimes. Nevertheless, these are strictly transient because closed systems naturally move through the parametric space.

Transient behavior may be sustained for a significant period of time and the chemical mixture can be considered to evolve in consecutive different pseudo-steady states by spontaneous transitions. Not only simple oscillations but also chaotic behavior can occur in closed Belousov–Zhabotinsky (BZ) systems.[6–10] Transient scenarios in the BZ system have also been the object of investigation.[11–13] For example Wang et al.[14] showed experimental evidence of successive transient chaos in a closed well mixed BZ system. Our attention has been turned to the closed unstirred BZ system. It is well known that an autocatalytic reaction performed in an unstirred batch reactor can support a constant velocity wavefront resulting from the coupling between diffusion and kinetics.[15] As a front propagates both concentration and thermal gradients are formed that alter the density of the solution, often causing convection.[16,17] The chemical wave may have features that differ significantly from the familiar reaction–diffusion wave when convective mass transport plays a relevant role. The effects of convection on chemical wave propagation have been observed experimentally in several oscillating systems.[16,18–20] In these systems convection enhances the chemical wave speed and affects the curvature of the front. Natural convection disturbs a travelling wave, and the disturbed wave affects the convection pattern.[21] This hydrodynamic effect, which sometimes perturbs the initial spatial patterns caausing them to bifurcate into more complex structures, can be either practically eliminated by stirring the solution or diminished by increasing the viscosity of the medium.[22] The addition of micelle-forming surfactants, polymers and other macromolecules, to an aqueous solution allows one to achieve increased viscosity and the extent of the increase depends[23] on the nature of the additive used. In fact, the contribution to the solution viscosity is greater for non-ionic additives than for anionic ones. Among the non-ionic surfactants the alkyl polyoxyethylene glycol monoethers are the most extensively used and studied. There is evidence[24] that, for these systems, the effect of the monomers on the medium viscosity is negligible due to their very low c.m.c. while beyond this critical value a sharp increase in the viscosity can be detected with increasing surfactant concentration. Moreover, the viscosity of non-ionic surfactants that have different alkyl chain lengths[24] is almost the same, whereas for systems having different head groups, i.e., different polyoxyethylene chain length, the viscosity is enhanced when the number of ethylene oxides is increased.

The present work followed from a previous[25] study on dynamic behaviour of the BZ reaction where we found that the system, before reaching equilibrium in an unstirred batch reactor, spontaneously[7,26] gives the following sequence: period-1 → quasiperiodicity → chaos → quasiperiodicity → period-1. Two transition scenarios i.e. at the onset of chaos and at its end, are observed. One appears as the mirror image of the other. This transition is known[27] as the Ruelle–Takens–Newhouse (RTN) scenario. The onset of chaos starts spontaneously as soon as convection motion couples to diffusion and local kinetics. We have also shown[25] that when different amounts of the non-ionic polymer polyethyleneglycol with high molecular weight is added to the BZ system the chaotic dynamics interval disappears by a RTN scenario. The polyethyleneglycol concentration, and as a consequence the medium viscosity, has been set as a bifurcation parameter for the chaos–periodicity transition in an unstirred batch reactor. Nevertheless, it has not been excluded that the polymer also acts on the kinetics of the reactions.

In this paper we report the results of a kinetic study of the behaviour of the Ce(IV)-catalyzed Belousov–Zhabotinsky system at 20.0 °C in unstirred batch conditions in the absence and the presence of different amounts of the non-ionic micelle-forming surfactants hexaethylene glycol monodecyl ether ($C_{10}E_6$) and hexaethylene glycol monotetradecyl ether ($C_{14}E_6$). These two surfactants allow us to vary[24] the medium viscosity and the extent of the increase is not dependent on the alkyl chain length. In addition, they do not react with the components of the BZ systems. This has been verified by monitoring the UV–VIS spectra of the Ce(IV) species in sulfuric acid solution as a function of both the surfactant concentration and the time. It has been found that, in all cases examined, the spectra do not show any significant change with respect to that obtained in aqueous acidic solution.

In order to obtain further useful information we have studied the kinetics of oxidation of malonic acid by cerium(IV) species in the absence and presence of the two non-ionic surfactants over a wide concentration range. This latter process, as is well known, is an important component of the oscillatory BZ reaction. Moreover, the effects of the addition of increasing amount of surfactant on the behaviour of the stirred BZ system have also been monitored.

Experimental methods

Potassium bromate, $Ce(SO_4)_2 \cdot 4H_2O$, malonic acid, hexaethylene glycol monodecyl ether, hexaethylene glycol monotetradecyl ether and sulfuric acid were of commercial analytical quality (Fluka) and used without further purification. Deionized water from reverse osmosis (Elga, model Option 3), having a resistivity higher than 1 MΩ cm, was used to prepare all solutions. Stock solutions of sulfuric acid were standardized by acid–base titration.

All the oscillating mixtures for the kinetic runs and the Ce(IV)–MA subsystems were obtained from freshly prepared stock solutions in 1.00 mol dm^{-3} sulfuric acid. The unstirred experiments were performed in a batch reactor and the dynamics of the system were monitored by following the Ce(IV) absorbance changes at 320 nm using quartz UV grade spectrophotometer cuvettes (1 × 1 × 4 cm^3). A double beam spectrophotometer (Varian, series 634) was used. The cross-sectional area of the spectrophotometer light beam was 30 mm^2. The volume spanned by the beam was 300 mm^3 (7.5% of the total volume) and was located 2 cm away from the liquid/air interface, 1 cm away from the bottom of the cuvette and about 0.4 cm away from the sides. The spectrophotometer was connected to an IBM compatible PC for data acquisition by an analog to digital board converter with a 12 bit resolution. The absorbance was recorded with a sampling time (τ_s) of 1 s. Time series points were recorded and stored in the computer for data analysis.

The spectra analysis was performed by fast Fourier transform (FFT) on sequential 1024 point portions of a time series. In order to increase the spectral resolution, we applied the zero filling technique. The short time fragments (1024 points) were transformed into longer time series sequences (2048 points) by adding a constant value equal to the minimum of the signal amplitude at the end of the fragments. In this way the spectral resolution $\Delta f = 1/N\,\tau_s$, where N is the number of points of the considered time fragment and τ_s is the sampling time, was doubled. Discontinuities at the window edges have been reduced by multiplying the considered data by a Hanning window function. This operation suppressed side lobes, which would otherwise be produced in the power spectrum of the signal.

The following concentrations of reactants stock solutions were used: Ce(IV) 0.004 mol dm^{-3}, malonic acid 0.30 mol dm^{-3}, KBrO$_3$ 0.09 mol dm^{-3}.

The surfactant concentration was varied in the range 1×10^{-4}–4.3×10^{-4} mol dm^{-3}.

The oscillator was started by mixing equal volumes of reactant solutions in a flask. This solution was stirred for 10 min without surfactant with a Teflon-coated magnetic stirrer of 1 cm length, at a constant high stirring rate; then the surfactant solution at the desired concentration was added and the final solution was stirred for about 2 min. The solution was then poured into the cuvette until the sample reached the top, and measurement of the signal began.

The kinetics of the oxidation of the malonic acid by cerium(IV) were studied in the absence and presence of the two non-ionic surfactants. In all cases studied, a large excess of organic substrate over the Ce(IV) species was used. The observed pseudo-first-order rate constants, k_{obs}, were obtained from the linear least-squares fit of the plot of $\ln(A_t - A_\infty)$ vs. time (A = absorbance) and were reproducible to within ±3%. The k_{obs} values were found to be independent of the initial Ce(IV) concentration and, consequently, the oxidation rates are first order with respect to the oxidizing species. The kinetic runs in stirred batch conditions were carried out spectrophotometrically by recording the changes in the Ce(IV) absorbance at 320 nm with a computer-controlled Beckman model DU-640 spectrophotometer, equipped with thermostatted compartments for 1.00 cm cuvettes and an appropriate magnetic stirring apparatus. The following concentrations of the stock solutions were used: Ce(IV) 1.60×10^{-3} mol dm^{-3}, malonic acid 0.400 mol dm^{-3}, KBrO$_3$ 0.12 mol dm^{-3}. The surfactant concentration was varied in the range 1×10^{-4}–4×10^{-2} mol dm^{-3}.

For these kinetic measurements a wider surfactant concentration range has been used in order to better establish the role played by the surfactant on the behaviour of the BZ system.

The temperature of all the experiments was regulated to 20.0 ± 0.1 °C with a thermostat Heto model DT HetoTerm.

Critical micelle concentrations of the two non-ionic surfactants used were determined, in the presence of the components of the BZ mixture without the catalyst, by surface tension measurements with a KSV–Sigma 70 automatic tensiometer.

Results and discussions

1. Critical micelle concentrations

The critical micelle concentrations of $C_{10}E_6$ and $C_{14}E_6$ were obtained from the typical plots of the surface tension *vs.* the logarithm of the surfactant concentration. The c.m.c. data obtained are 1.6×10^{-4} and 1.3×10^{-6} mol dm^{-3} for $C_{10}E_6$ and $C_{14}E_6$, respectively. Under the experimental conditions of this work the c.m.c. values are slightly lower than those obtained[23c,28] in aqueous solution. The lower c.m.c. values can be attributed both to the presence of organic compounds and the very high value of the ionic strength.

2. Ce(IV)-catalyzed BZ system behaviour in unstirred batch conditions

Fig. 1(a) shows a typical spectrophotometric time series recording of the BZ system in the absence of both stirring and surfactant. Perusal of this figure suggests that two transitions occurs. At the first transition the system passes from a periodic to an aperiodic behaviour and, then, at the second one the periodic behaviour appears once again. This behaviour can be referred to as a periodicity → chaos → periodicity transition, since it has been previously[6] shown that the aperiodic behaviour can be considered as an example of chaotic transient. This is confirmed by a FFT, shown in Fig. 1(c), of the spectrophotometric recording evidenced in Fig. 1(b). A broadband spectrum typical of chaotic systems is obtained. The broadband spectrum does not discriminate between chaos and random motion. Nevertheless, this system[6] shows the dependence on the initial conditions which is the major distinctive signature of chaos. Actually the chaotic regime observed in Fig. 1(b) is bounded by two quasiperiodic zones. The existence of a torus can be detected at the onset of chaos and at its end. Moreover, a Ruelle–Takens–Newhouse scenario[7] occurs. The chaotic behaviour has been attributed[7] to the interplay between chemical kinetics and diffusion-convective processes.

As for the oscillatory behaviour of the BZ system in the presence of $C_{10}E_6$, it has been found that the spectrum is not distinguishable from that obtained in the absence of surfactant until the surfactant concentration reaches a given concentration value. It is interesting to note that this value coincides with the critical micelle concentration obtained by surface tension measurement. It has been found, also in this case, that the Fourier transform of the signal shows a typical broadband spectrum like that depicted in Fig. 1(c). Since either chaos or random motion give broadband spectra it is not possible to distinguish between them from the Fourier transform alone; even if there is no reason to believe that the presence of surfactant induces a transition from chaos to random motion. Thus, it can be deduced that the behaviour of the BZ system in the presence of $C_{10}E_6$ at a concentration below the c.m.c. is a manifestation of chaos. Beyond the critical micelle concentration a new feature in the behaviour of the BZ system appears. In fact, as can be seen in Fig. 2(a)–(c) in the presence of 3.2×10^{-4} mol dm^{-3} of $C_{10}E_6$ the BZ system shows periodic behaviour instead of the aperiodic pattern. Fig. 2(c) is a typical pattern of a periodic signal and it is also possible to observe the fundamental frequencies and a few harmonics. In order to investigate the transition scenario as a function of surfactant concentration, series of measurements have been performed over the concentration range $1.8 \times 10^{-4} < [C_{10}E_6]/\text{mol dm}^{-3} < 2.8 \times 10^{-4}$. The temporal series monitored in this concentration range is shown in Fig. 3(a). The Fourier transform depicted in Fig. 3(c) clearly shows the existence of two frequencies, not rationally related, and their combinations. This pattern is characteristic of quasi-periodic motion.

The effects on the behaviour of the Belousov–Zhabotinsky system due to the addition of different amounts of surfactant can be explained taking into account that the solution viscosity[24] increases with increasing surfactant concentration. At very low surfactant concentration the contribution of the monomers to the viscosity is negligible (see Introduction) and the BZ system shows the same behaviour as in aqueous solution. When the solution viscosity increases the BZ system shows different patterns.

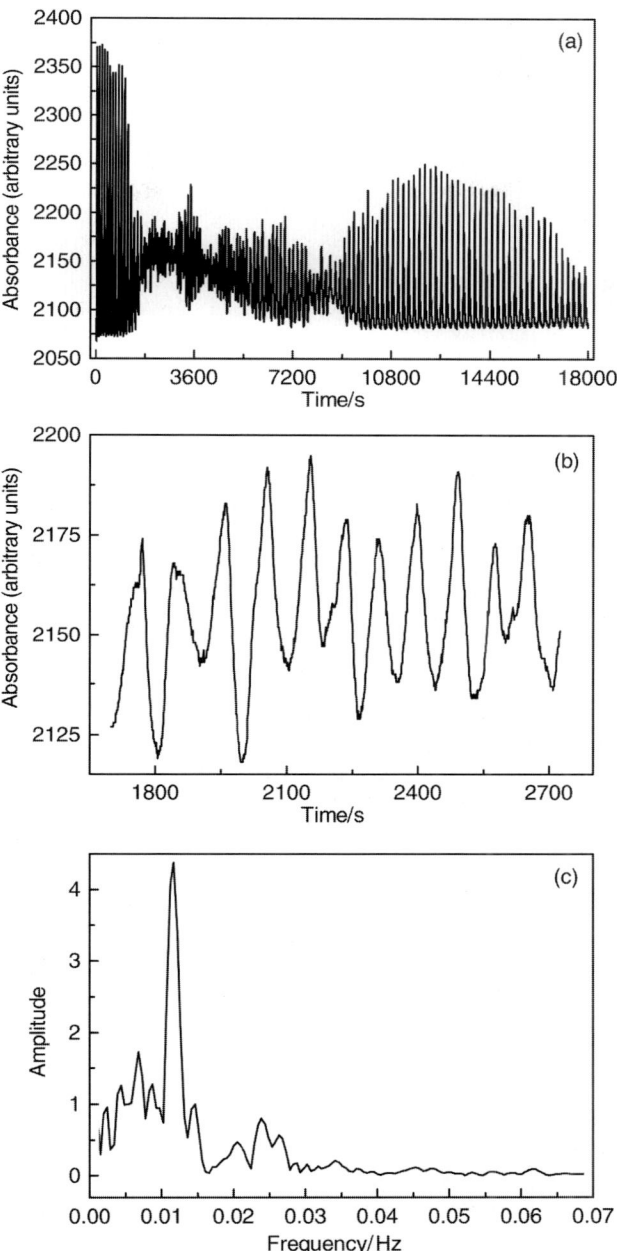

Fig. 1 (a) Typical spectrophotometric recording of the Ce(IV) absorbance in the absence of stirring and without $C_{10}E_6$. (b) Detail of (a) in the range 1800–2700 s. (c) Fourier transform of (b) evidencing a broadband spectrum typical of chaotic signal. $\lambda = 320$ nm, $\tau_s = 1$ s, $T = 20\,°C$.

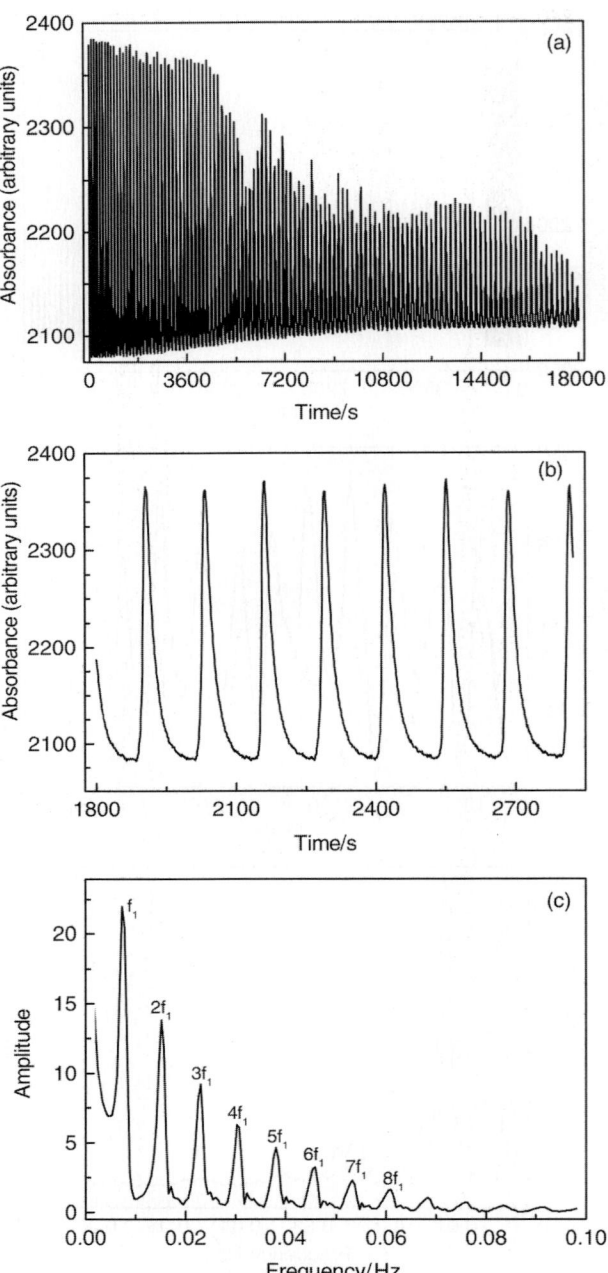

Fig. 2 (a) Typical spectrophotometric recording of the Ce(IV) absorbance in the absence of stirring with 3.2×10^{-4} mol dm^{-3} of $C_{10}E_6$. (b) Detail of (a) in the range 1800–2700 s. (c) Fourier transform of (b) showing the fundamental frequencies and a few harmonics. This is a typical pattern of a periodic signal. Experimental conditions are the same as in Fig. 1.

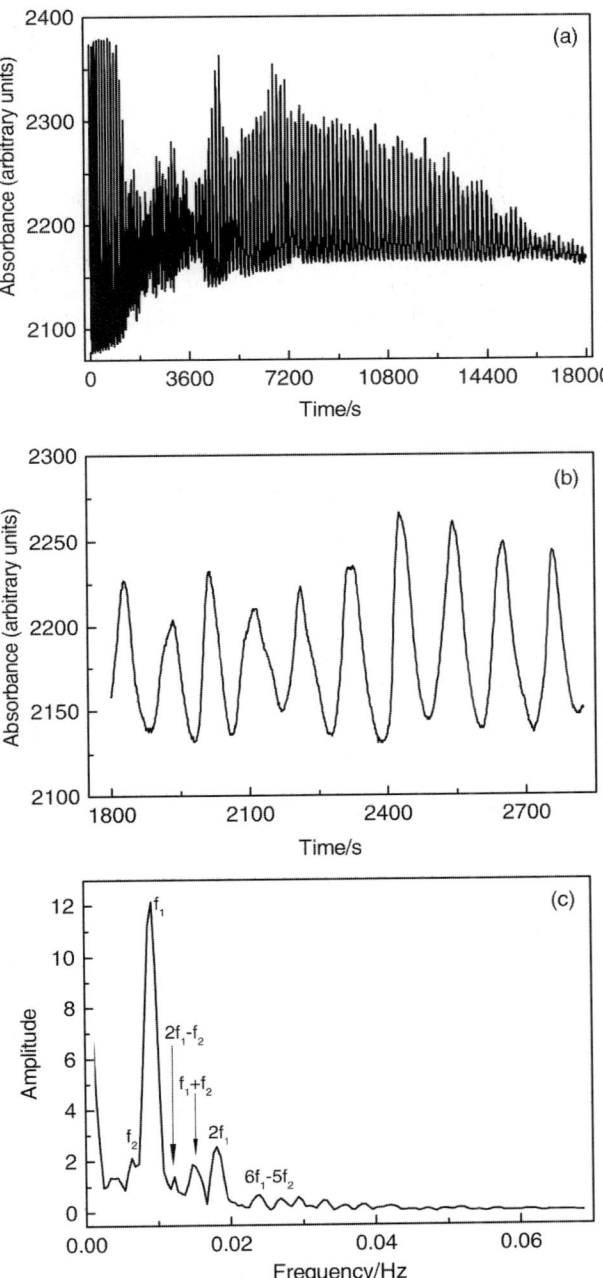

Fig. 3 (a) Typical spectrophotometric recording of the Ce(IV) absorbance in the absence of stirring with 3.1×10^{-4} mol dm^{-3} of $C_{10}E_6$. (b) Detail of (a) in the range 1800–2700 s. (c) Fourier transform of (b) showing two frequencies not rationally related and their combinations; typical of a quasi-periodic dynamics. Experimental conditions are the same as in Fig. 1.

We can conclude that the increase in medium viscosity induces the disappearance of the chaotic behaviour and this sets the viscosity as an important control parameter for the chaos of the system.

The experimental results can be summarized in the following scheme

chaotic oscillation

$$1 \times 10^{-4} \leqslant [C_{10}E_6]/\text{mol dm}^{-3} \leqslant 2.9 \times 10^{-4}$$

$$\downarrow$$

quasi-periodic oscillation

$$3.0 \times 10^{-4} \leqslant [C_{10}E_6]/\text{mol dm}^{-3} \leqslant 3.1 \times 10^{-4}$$

$$\downarrow$$

periodic oscillation

$$3.2 \times 10^{-4} \leqslant [C_{10}E_6]/\text{mol dm}^{-3} \leqslant 4.3 \times 10^{-4}$$

The observed sequence sets the surfactant concentration as a bifurcation parameter of the unstirred BZ system and, as the parameter is changed, a chaotic attractor appears or disappears depending on whether the surfactant concentration decreases or increases. This scenario in the phase plane corresponds to the strange attractor → torus → cycle transition. This type of behaviour could be interpreted using a RTN scenario where the system undergoes three Hopf bifurcations.

Analogous effects on the behaviour of the unstirred BZ system have been monitored in the presence of different amounts of the non-ionic surfactant $C_{14}E_6$ (the figures are not shown for the sake of simplicity).

3. Kinetics of the Ce(IV)–malonic acid subsystem

In order to better ascertain the role played by the surfactant in inducing the behaviour discussed in the above section we have studied the kinetics of the cerium(IV) oxidation of malonic acid in the presence of the surfactants $C_{10}E_6$ and $C_{14}E_6$ over a wide surfactant concentration range. As is well known[29–32] organized surfactant assemblies can affect the rates and the mechanism of a chemical reaction by selectively sequestering the reagent substrates by means of electrostatic and/or hydrophobic interactions. The study of the kinetics of reactions offers the twofold advantage of affording insight into the reactivity and the reaction mechanism in these reaction media[30–32] and of obtaining useful information on the solubilization[30–32] of the reagents and, in some cases, the location of the solubilizate in the micellar structure.

First, we have studied the kinetics of the redox process in the absence of surfactant at various substrate concentrations. Table 1 collects the observed pseudo-first-order rate constants obtained for this reaction. It can be noted that the k_{obs} values depend on the MA concentration. The plot of

Table 1 Observed pseudo-first-order rate constants $k_{obs}/10^{-2}$ s^{-1} for the Ce(IV) oxidation of MA at varying substrate concentrations. $t = 20.0\,°C$

[MA]/10^{-2} mol dm^{-3}	$k_{obs}{}^a/10^{-2}$ s^{-1}
0.5	0.461
1.00	0.807
2.00	1.48
5.00	3.3
10.0	6.7

^{a} $k_2 = 0.101 \pm 0.001$ s^{-1}; $K_m = 0.107 \pm 0.003$ mol dm^{-3}.

k_{obs} vs. [MA] is curved, whereas a linear trend with positive intercept is obtained when $1/k_{obs}$ is plotted as a function of $1/[MA]$. This indicates that the reaction order with respect to substrate lies between zero and unity and that the oxidation process can be interpreted by the sequential Michaelis–Menten-like mechanism given in eqn. (1). This mechanism involves[33–38] initial complex formation between the organic substrate and the reactive cerium(IV) species. Subsequently, the intermediate complex decomposes in the rate-determining electron-transfer step yielding a free carbon radical and the cerium(III) species. The final reaction products are obtained by the subsequent fast oxidation of the organic radical.

This type of mechanism has been proposed previously[33–38] for the oxidation of a variety of organic substrates by cerium(IV) in sulfuric acidic media (i.e., malonic acid and its alkyl derivatives, tartronic, tartaric, citric, glycolic, benzilic, lactic, substituted mandelic acids, etc.).

$$Ce(IV) + MA \underset{k_{-1}}{\overset{k_1}{\rightleftharpoons}} complex \overset{k_2}{\longrightarrow} Ce(III) + other\ products \quad (1)$$

According to the reaction mechanism (1) the rate constant k_{obs} is equal to $k_2[MA]/(K_m + [MA])$, where $K_m = (k_{-1} + k_2)/k_1$, and, consequently, the $1/k_{obs}$ vs. $1/[MA]$ plot yields $1/k_2 (=\text{intercept})$ and $K_m/k_2 (=\text{slope})$. The k_2 and K_m values obtained from this plot are also reported in Table 1. It should be noted that, if the rate constant k_2 is much smaller than k_{-1}, the parameter K_m reduces to k_{-1}/k_1 and represents the reciprocal of the equilibrium constant for the intermediate complex formation.

For the Ce(IV)–MA reaction previously[34a] studied at 30 °C in 1.5 mol dm^{-3} sulfuric acid medium, Kasperek et al. obtained k_2 and K_m values of 0.53 s^{-1} and 0.53 mol dm^{-3}, respectively. The k_2 and K_m values obtained in the present work are lower than those obtained by Kasperek et al. The difference can be attributed to the different experimental conditions used, i.e., lower temperature and sulfuric acid concentration. As to the effects of added surfactants on the redox reaction under examination, it has been found that addition of increasing amounts up to a concentration value of 0.04 mol dm^{-3} of either $C_{10}E_6$ or $C_{14}E_6$ has no significant influence on the rate of the redox process (the data are not shown for sake of simplicity).

The observed surfactant effects on the redox reactions can be qualitatively interpreted by the pseudophase model[29,30] bearing in mind that the reactive Ce(IV) species, as has been previously suggested,[37,39] in sulfuric (or sulfate) acidic media with $[HSO_4^-]$ (or $[SO_4^{2-}]$) $\gg$ [Ce(IV)]), may exist in a number of sulfate and protonated sulfate complex forms, which may be neutral species [e.g., $Ce(SO_4)_2$] or bear either positive [e.g., $Ce(HSO_4)_3^+$] or negative [e.g., $Ce(SO_4)_3^{2-}$, $H_2Ce(SO_4)_4^{2-}$] charges. From a kinetic point of view, the existence of various cerium(IV) complexes implies serious difficulties[37] for the identification of the actual reacting complex involved in the oxidation process. However, all these complexes do not seem to take part significantly in the overall redox rate[35,37] and only limited complex species, usually the anionic ones, have been assumed by different investigators as the reactive cerium(IV) species. This hypothesis has been corroborated[40] by Cavasino et al. They have substantiated the view that negatively charged cerium(IV) species are involved in the redox process by examining the effects of either cationic or anionic micelle-forming surfactant micelles on the reaction rate of Ce(IV) with benzyl malonic acid. These two types of differently charged micelles influence the rate of the redox process in an opposite way. Therefore, on the basis of the pK_a values[41] for the first and second dissociations of the malonic acid and the experimental conditions used it is reasonable to consider the undissociated acid to be the predominant species present in solution. Since it has been previously[42] suggested that the Ce(IV) species do not interact with non-ionic micelles only the partition of the neutral substrate between the aqueous pseudophase and the micellar pseudophase has to be considered. This implies that two reaction pathways involving the hydrophilic Ce(IV) species in the aqueous phase and the MA solubilized in both the aqueous region and the micellar pseudo-phase contribute to the progress of the redox reaction. According to this reaction scheme it has to be expected that the distribution of MA will cause a small reduction in the substrate concentration in water and, as a consequence, determine a small inhibitory surfactant effect on the reaction rate. Moreover, this reduction would increase as the surfactant concentration increases. Since this effect has not been detected we have to draw the conclusion that the redox reaction takes place only in the aqueous pseudophase and no partition of the malonic acid occurs.

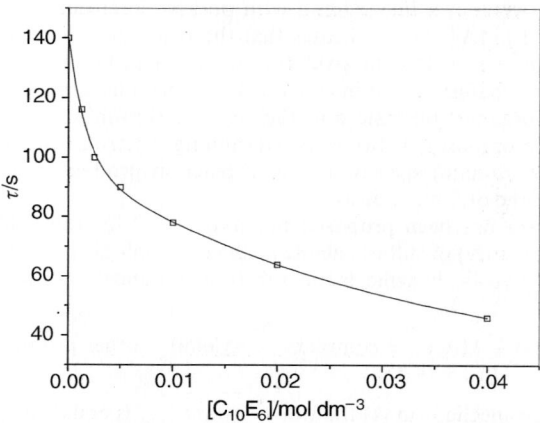

Fig. 4 Plot of τ against $[C_{10}E_6]$ for the stirred BZ system. $\lambda = 320$ nm, $T = 20\,°C$.

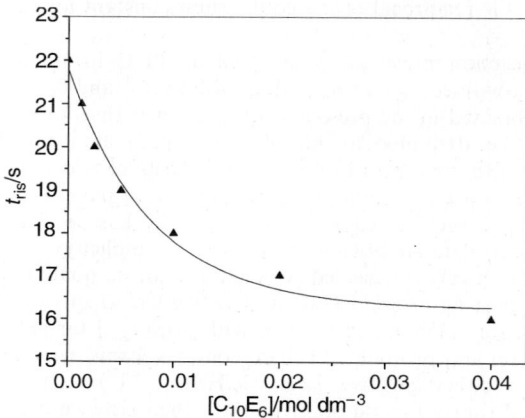

Fig. 5 Plot of t_{ris} against $[C_{10}E_6]$ for the stirred BZ system. Experimental conditions are the same as in Fig. 4.

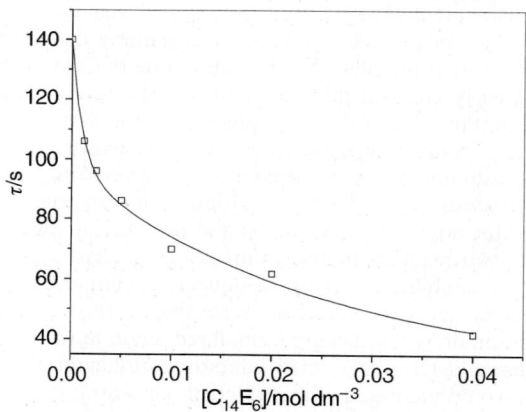

Fig. 6 Plot of τ against $[C_{14}E_6]$ for the stirred BZ system. Experimental conditions are the same as in Fig. 4.

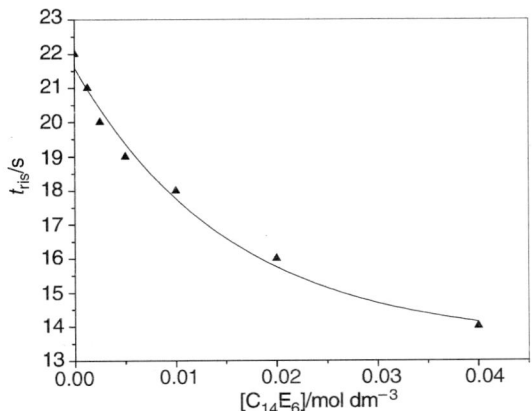

Fig. 7 Plot of t_{ris} against $[C_{14}E_6]$ for the stirred BZ system. Experimental conditions are the same as in Fig. 4.

This result is reasonable and in accordance with previous[40] suggestions. In fact, it has been observed that binding of monosubstituted malonic acid to different types of micelles depends significantly on the hydrophobicity of the group present in the malonic acid. Thus, MA, having the more hydrophilic character, is the less solubilized substrate in the micellar pseudophase as compared to the other monosubstituted malonic acid.

4. Ce(IV)-catalyzed BZ system behaviour in stirred batch conditions

In order to provide additional information on the effect of non-ionic micelles, hence the medium viscosity, we have monitored the behaviour of the BZ system in the absence and presence of different amounts of non-ionic surfactants in stirred batch conditions. In the absence of surfactants, oscillations in the BZ system have been found to begin after an induction period (I. P.) of 16 min with an oscillation period (τ) and the duration of the rising portion (t_{ris}) of an oscillatory cycle of 140 and 22 s, respectively. We wish to point out that the oscillatory parameters values reported in this work are the mean of five different experiments and both the τ and t_{ris} values represent[43] the means of the first five oscillatory cycles.

The presence of surfactant does not influence the oscillatory behaviour of the stirred BZ system until the surfactant concentration reaches a given concentration value. In fact, the oscillatory parameters keep the same values as those obtained in the absence of surfactants. As for the BZ system in unstirred batch conditions, the concentration at which the change in the behaviour occurs coincides with the critical micelle concentration. Once the c.m.c. is reached the oscillatory behaviour of the BZ system changes. The presence of micelles in the reaction medium has no dramatic effects on the I. P. while either $C_{10}E_6$ or $C_{14}E_6$ markedly influence, to the same extent, τ and t_{ris}.

Fig. 4–7 show that both τ and t_{ris} decrease and tend to reach a plateau value at high surfactant concentrations.

The findings can be explained by considering that many substances react simultaneously during the induction period but, due to their hydrophilic nature, the reactions take place, as we have found for the MA oxidation by Ce(IV), only in the aqueous pseudophase.

As to the decrease in both τ and t_{ris} as a function of surfactant concentration, it has to be take into account that, as shown by Noszticzius et al.,[44] the reaction of the malonyl radicals, which are formed during the reaction of Ce(IV) with malonic acid, with the $BrO_2 \cdot$ radicals (which are formed during the autocatalytic reaction step) occurs at a nearly diffusion-controlled rate. As a consequence, an increase in the medium viscosity hampers this process and, at the same time, promotes the oxidation of Ce(III) by $BrO_2 \cdot$ radicals.

This aforementioned result together with those found above leads to the concluding suggestion that significant changes in the oscillatory behaviour of the Belousov–Zhabotinsky system occur when the medium viscosity is increased.

Acknowledgement

This work was supported by CNR (Roma) and by MURST.

References

1. A. M. Zhabotinsky, *Chaos*, 1991, **1**, 379.
2. S. K. Scott, *Chemical Chaos*, Oxford University Press, Oxford, 1991.
3. I. R. Epstein, *Nature*, 1995, **374**, 321.
4. I. R. Epstein and J. A. Pojman, *An Introduction to Nonlinear Chemical Dynamics*, Oxford University Press, Oxford, 1998.
5. G. Nicolis and J. Portonow, *Chem. Rev.*, 1973, **73**, 365.
6. M. Rustici, M. Branca, C. Caravati and N. Marchettini, *Chem Phys. Lett.*, 1996, **263**, 429.
7. M. Rustici, C Caravati, E. Petretto, M. Branca and N. Marchettini, *J. Phys. Chem. A*, 1999, **103**, 6564.
8. S. K. Scott, B. Peng, A. S. Tomlin and K. Showalter, *J. Chem. Phys.*, 1991, **94**, 1134.
9. B. R. Johnson, S. K. Scott and B. W. Thompson, *Chaos*, 1997, **7**, 350.
10. P. V. Coveney and A. N. Chaudry, *J. Chem. Phys.*, 1992, **97**, 7448.
11. P. Ruoff and R. M. Noyes, *J. Phys. Chem.*, 1985, **89**, 1339.
12. J. Kawahito and Fujieda, *Thermochim. Acta*, 1992, **210**, 1.
13. P. E. Strzhak and A. L. Kawczynski, *J. Phys. Chem.*, 1995, **99**, 10830.
14. J. Wang, P. G. Sorensen and F. Hynne, *J. Phys. Chem.*, 1994, **98**, 725.
15. J. A. Pojman and I. Epstein, *J. Phys. Chem.*, 1990, **94**, 1966.
16. J. A. Pojman, I. R. Epstein and I. P. Nagy, *J. Phys. Chem.*, 1991, **95**, 1306.
17. J. A. Pojman, A. Komlosi and I. P. Nagy, *J. Phys. Chem.*, 1996, **100**, 16209.
18. G. Bazsa and I. R. Epstein, *J. Phys. Chem.*, 1985, **89**, 3050.
19. I. Nagypal, G. Bazsa and I. R. Epstein, *J. Am. Chem. Soc.*, 1986, **108**, 3635.
20. H. Miike, S. C. Muller and B. Hess, *Chem. Phys. Lett.*, 1988, **144**, 515.
21. B. Legawiec and A. L. Kawczynski, *J. Phys. Chem. A*, 1997, **101**, 8063.
22. T. Yamaguchi, L. Kuhnert, Zs. Nagy-Ungvarai, S. C. Muller and B. Hess, *J. Phys. Chem.*, 1991, **95**, 5831.
23. (a) P. Mukerjee, *J. Colloid Sci.*, 1964, **19**, 722; (b) D. C. Robins and L. Thomas, *J. Colloid Interface Sci.*, 1968, **26**, 415; (c) P. Becher, in *Nonionic Surfactants*, ed. M. J. Schick, Marcel Dekker, New York, 1966.
24. M. Abe and K. Ogino, *Mixed Surfactant Systems*, Marcel Dekker, New York, 1993, p. 1.
25. N. Marchettini and M. Rustici, *Chem. Phys. Lett.*, 2000, **317**, 647.
26. M. Rustici, M. Branca, A. Brunetti, C. Caravati and N. Marchettini, *Chem. Phys. Lett.*, 1988, **293**, 145.
27. J. P. Eckmann and D. Ruelle, *Rev. Mod. Phys.*, 1985, **57**, 617.
28. V. Degiorgio, in *Physics of Amphiphiles: Micelles Vesicles and Microemulsioni*, ed. V. Degiorgio and M. Conti, North-Holland, Amsterdam, 1985.
29. (a) I. V. Berezin, K. Martinek and A. K. Yatsimirsckii, *Russ. Chem. Rev.*, 1973, **42**, 787; (b) J. H. Fendler and E. J. Fendler, *Catalysis in Micellar and Macromolecular Systems*, Academic Press, New York, 1975; (c) L. S. Romsted, in *Micellization, Solubilization and Microemulsions*, ed. K. L. Mittal, Plenum Press, New York, 1977, vol. 2, p. 509; (d) C. A. Bunton, in *Solution Chemistry of Surfactants*, ed. K. L. Mittal, Plenum Press, New York, 1979, vol. 2, p. 519; (e) J. H. Fendler, *Membrane Mimetic Chemistry. Characterization and Application of Micelles, Microemulsions, Monolayers, Bilayers, Vesicles, Host–Guest Systems and Polyions*, Wiley, New York, 1982; (f) L. S. Romsted, in *Surfactants in Solution*, ed. K. L. Mittal and B. Lindman, Plenum Press, New York, 1983, vol. 2, p. 1015; (g) C. A. Bunton and G. Savelli, *Adv. Phys. Org. Chem.*, 1986, **22**, 213; (h) J. Burgess and E. Pelizzetti, *Gazz. Chim. Ital.*, 1988, **12**, 803.
30. (a) G. Calvaruso, F. P. Cavasino, C. Sbriziolo and M. L. Turco Liveri, *J. Chem. Soc., Faraday Trans.*, 1995, **91**, 1075 and references therein; (b) G. Calvaruso, F. P. Cavasino and C. Sbriziolo, *J. Chem. Soc., Faraday Trans.*, 1996, **92**, 2263 and references therein.
31. F. P. Cavasino, S. Di Stefano and C. Sbriziolo, *J. Chem. Soc., Faraday Trans.*, 1997, **93**, 1585.
32. F. P. Cavasino, C. Sbriziolo and M. L. Turco Liveri, *J. Phys. Chem. B*, 1998, **102**, 5050 and references therein.
33. Y. F. Chen, H. P. Lin, S. S. Sun and J. J. Jwo, *Int. J. Chem. Kinet.*, 1996, **28**, 345.
34. (a) G. J. Kasperek and and T. C. Bruice, *Inorg. Chem.*, 1971, **10**, 382; (b) J. J. Jwo and R. M. Noyes, *J. Am. Chem. Soc.*, 1975, **97**, 5422; (c) S. Barkin, M. Bixon, R. M. Noyes and K. Bar-Eli, *Int. J. Chem. Kinet.*, 1978, **10**, 619.
35. P. Ruoff and G. Nevdal, *J. Phys. Chem.*, 1989, **93**, 7802.
36. P. O. Kvernberg, E. W. Hansen, B. Pedersen, A. Rasmussen and P. Ruoff, *J. Phys. Chem. A*, 1997, **101**, 2327 and references therein.
37. (a) G. Arcoleo, G. Calvaruso, F. P. Cavasino and C. Sbriziolo, *Inorg. Chim. Acta*, 1977, **23**, 227 and references therein; (b) G. Calvaruso, F. P. Cavasino, C. Sbriziolo and R. Triolo, *Int. J. Chem. Kinet.*, 1983, **15**, 417 and references therein; (c) G. Calvaruso, F. P. Cavasino and C. Sbriziolo, *Int. J. Chem. Kinet.*, 1984, **16**, 1201.
38. S. B. Hanna and S. A. Sarac, *J. Org. Chem.*, 1977, **42**, 2063 and references therein.

39 (a) T. J. Hardwick and Robertson, *Can. J. Chem.*, 1951, **29**, 828; (b) L. T. Bugaenko and H. Kuan-Lin, *Russ. J. Inorg. Chem.*, 1963, **8**, 1299.
40 F. P. Cavasino, R. Cervellati, R. Lombardo and M. L. Turco Liveri, *J. Phys. Chem. B*, 1999, **103**, 4285.
41 (a) G. Arcoleo, F. P. Cavasino and E. Di Dio, *Gazz. Chim. Ital.*, 1978, **108**, 597; (b) G. Calvaruso, A. I. Carbone and F. P. Cavasino, *J. Chem. Soc., Dalton Trans.*, 1985, 1683 and references therein.
42 G. Calvaruso, F. P. Cavasino and C. Sbriziolo, *J. Chem. Soc., Faraday Trans.*, 1991, **87**, 3033.
43 A. M. Zhabotinsky, F. Buchholtz, A. B. Kiyatkin and I. R. Epstein, *J. Phys. Chem.*, 1993, **97**, 7578.
44 H. D. Försterling and Z. Noszticzius, *J. Phys. Chem.*, 1989, **93**, 2740.

Nonlinear behaviour of simple ionic systems in hydrogel in an electric field

Dalimil Šnita, Martin Pačes, Jiří Lindner, Juraj Kosek and Miloš Marek

Department of Chemical Engineering and Center for Nonlinear Dynamics, Prague Institute of Chemical Technology, Technická 5, 166 28 Prague 6, Czech Republic.
E-mail: marek@vscht.cz

Received 19th April 2001
First published as an Advance Article on the web 13th November 2001

The stationary behavior of ionic reaction–transport systems contained in hydrogel located between two reservoirs of electrolytes is investigated. The effects of the applied voltage, the composition of the electrolytes in the reservoirs and the distribution of fixed charge in the hydrogel on the spatial patterns of charge, electric potential, temperature and concentrations of the individual components are studied systematically by mathematical modelling. Such systems, containing only several ions (for example, Cl^-, K^+, OH^- and H^+) can function as electrolyte diodes and transistors and can exhibit electric oscillations and hysteresis. The mathematical description of the studied systems is based on balances of species, enthalpy and charge, on Poisson's equation and on the finite rate description of the water dissociation/recombination reaction, without assuming local electroneutrality. The modeling results are compared with experiments, including the system with hydrogel connecting reservoirs of strong base and acid studied by Noszticzius and co-workers. [L. H. Hegedus, N. Kirschner, M. Wittmann and Z. Noszticzius, *J. Phys. Chem. A*, 1998, **102**, 6491 (ref. 1); L. H. Hegedus, N. Kirschner, M. Wittmann, P. Simon and Z. Noszticzius, *Chaos*, 1999, **9**, 283 (ref. 2).]

1. Introduction

Here we shall consider electrolyte systems, where we use the term "electrolyte" to mean electrochemical systems placed far from electrodes in which diffusion, migration, near to equilibrium chemical reactions but no surface electrochemical reactions take place. These systems can exhibit complex behavior, *e.g.*, nonlinear coupling between the applied potential difference and electric current. The observed phenomena follow from the non-uniform spatial distributions of charge density, mobile species and fixed ions concentrations and from chemical reactions taking place in the system. The formation of temporal and spatial patterns on electrodes has been reviewed by Krischer.[3] The complex behaviour of the system electrode–electrolyte–electrode could arise from the interaction of the non-linear phenomena at electrodes and the nonlinear behaviour of electrolytes. The Joule heat accompanying the conduction of electric current through the electrolyte can be effectively dissipated in micro-systems. Lovreček *et al.*[4] demonstrated experimentally in 1958 that a bipolar membrane consisting of cation exchange and anion exchange layers has rectifying properties and can act as a diode. Experimental observations of oscillations in ion transfer through perm-selective membranes were reported in 1961.[5] Noszticzius *et al.*[1] recently also demonstrated experimentally that a simple chemical system consisting of nearly inert hydrogel layer with one boundary exposed to an acidic (HCl solution) environment and the other to an

DOI: 10.1039/b103530b

alkaline (KOH solution) environment can behave like an electrolyte diode as the conductivity depends on the orientation of the applied voltage. Two oppositely oriented electrolyte diodes then behave like an electrolyte transistor.

In the past most modelling studies of experimentally observed nonlinear phenomena (multiple steady states, hysteresis, oscillations) in ionic reaction–transport systems were based on a number of simplifying assumptions (*e.g.*, electroneutrality and assumptions of equilibria) enabling one to use approximate semi-analytical methods for solution of the relevant equations. Here we avoid some simplifications and present examples of results of numerical solution of balance equations for several configurations of a thin layer of hydrogel subjected to an applied electric field. The obtained spatial profiles of concentration and electric variables are then used to interpret the function of the "electrolyte diode" and other observed nonlinear phenomena.

Approximate analytical solutions of the acid–base electrolyte diode,[2] confirmed by our numerical simulations[6,7] show that the stationary solution depends qualitatively on the ratio of the diffusion coefficients. The concentration distribution in the system determines the intensity of the electric current (when assuming constant applied potential difference), which determines the amount of Joule heat released (the heat is proportional to the electric current intensity). Besides other possible sources of instabilities, it is interesting to know when the heat release and transfer could lead to multiple steady states or oscillations (in analogy to a non-isothermal reactor with an exothermic chemical reaction).

Another important property of the studied systems is the "geometrical singularity". At the interface between the stagnant layers of the solution and the hydrogel membrane and/or between particular layers of composite membranes, a very sharp, nearly discontinuous change in properties such as diffusion coefficients and concentration of fixed charged groups may occur. It is often *a priori* not known, whether these discontinuities can be described on the basis of a local equilibrium (*e.g.*, Donnan equilibrium potential) or have to be described as locally non-equilibrium phenomena resulting from intensive fluxes through membrane interfaces.

The in-depth analysis of electric phenomena in systems with ion-exchange membranes has been developed for years, *cf.* references in Rubinstein *et al.*,[8–10] where the effects of electric fields in membranes are discussed as well as the validity of commonly used simplifications in the physical and mathematical descriptions of the problem. Some commonly used concepts, for example, the equilibrium description of interfaces, the presence of a stagnant diffusion layer at the membrane surface *etc.*, are not applicable in all situations. Moreover, the electric field in the presence of fixed charges causes electro-osmotic flows that are not exactly perpendicular to the membrane surface.[9] Therefore the classical 1D description of membrane systems might not be appropriate for detailed studies. The interaction of the autocatalytic reaction fronts with the electric field was systematically studied by Šnita *et al.*[11]

In this paper we present a mathematical model describing the behavior of a spatially one-dimensional system of several ionic components in a hydrogel with suppressed convection.[12,13] Some of the ionic components could be immobilized in the gel. We consider simple dissociation chemical reactions, an externally applied electric field and non-isothermal effects. The nature of the exhibited nonlinearities is discussed.

2. Mathematical model and system specification

Mass balances of mobile components in the one-dimensional system without convection can be written in the form

$$\frac{\partial c_i}{\partial t} = -\frac{\partial J_i}{\partial x} + \sum_{j=1}^{M} v_{i,j} r_j, \qquad i = 1, \ldots, N_m, \qquad (1)$$

and for fixed (immobile) components as

$$\frac{\partial c_i}{\partial t} = \sum_{j=1}^{M} v_{i,j} r_j, \qquad i = N_m + 1, \ldots, N, \qquad (2)$$

where $N = N_m + N_f$ is the number of all components, N_m and N_f are the numbers of mobile and fixed components, respectively, M is the number of reactions, $c_i(x, t)$ is the molar concentration of the ith component, t is the time, x is the spatial coordinate, $J_i(x, t)$ is the intensity of the molar flux of the ith component described as the sum of the diffusion $J_i^{\text{diff}}(x, t)$ and migration $J_i^{\text{migr}}(x, t)$ terms by the Nernst–Planck equation

$$J_i = J_i^{\text{diff}} + J_i^{\text{migr}} = -D_i \frac{\partial c_i}{\partial x} - D_i c_i \frac{\partial \phi}{\partial x} \frac{F z_i}{RT}, \tag{3}$$

where x is the spatial coordinate, $D_i = D_i(T)$ is the temperature-dependent diffusion coefficient of the ith component, z_i is the charge number of the ith component, $v_{i,j}$ is the stoichiometric coefficient of the ith component in the jth reaction, $r_j = r_j(c_1, \ldots, c_N, T, |E|)$ is the reaction rate of the jth reaction (a function of the composition, the temperature and the electric field intensity E), $\phi(x, t)$ is the electric potential, F is the Faraday constant, R is the gas constant and T is the absolute temperature. At the steady state the right-hand side of eqn. (2) equals zero, which implies the establishment of chemical equilibria for fixed groups.

A linear combination of eqn. (1) and (2) is the charge balance equation

$$\frac{\partial q}{\partial t} = -\frac{\partial I}{\partial x}, \tag{4}$$

where

$$q = F \sum_{i=1}^{N} z_i c_i \quad \text{and} \quad I = F \sum_{i=1}^{N} z_i J_i \tag{5}$$

are the electric charge density q and the electric current intensity I, respectively. Heat balance is considered in the form

$$\rho c_p \frac{\partial T}{\partial t} = \lambda \frac{\partial^2 T}{\partial x^2} + IE - \sum_{j=1}^{M} \Delta h_{r_j} r_j \tag{6}$$

where ρ is the density, c_p is the heat capacity, λ is the heat conductivity, (IE) is the density of the Joule heat generation due to the electric current,

$$E = -\frac{\partial \phi}{\partial x} \tag{7}$$

is the electric field intensity and $\Delta h_{r_j}(T)$ is the reaction heat of the jth reaction (which depends on temperature). Eqn. (6) is suitable for the description of membranes with a large surface to thickness ratio. If the studied system is in the shape of a thin cylinder, for example of diameter 0.1 mm and length 1 mm, the heat transport in the radial direction becomes more important than heat transport in the longitudinal direction. The complete analysis of the temperature distribution then requires spatial 2D modelling. In the systems studied here the inequality $(\lambda/(\rho c_p)) \gg D_i$ holds, so that isothermal behavior can be considered, but the temperature depends on the amount of Joule heat generated

$$\alpha(T - T_0) = |\phi(0) - \phi(L)| - \int_0^L \sum_{j=1}^{M} \Delta h_{r_j} r_j, \tag{8}$$

where α is the formal heat transfer coefficient. The rate of heat generation by the chemical reaction can often be neglected when compared to the Joule heat.

Gauss's law of electrostatics (Poisson's equation) is considered in the form

$$\frac{\partial \varepsilon E}{\partial x} = q, \tag{9}$$

where q is the charge density, $\varepsilon(c_1, \ldots, c_N, T, |E|)$ is the permittivity (generally a function of composition, temperature and electric field intensity, but a constant value $\varepsilon = \varepsilon_0 \varepsilon_r$, where $\varepsilon_r = 78.5$ and ε_0 is the permittivity of vacuum is considered in this paper).

Systems of several components, namely H_2O (in excess), mobile components ($N_m = 4$) H^+, OH^-, K^+ (potassium cation), Cl^- (chloride anion) and immobile components ($N_f = 2$) FC^+ (fixed cation), FA^- (fixed anion) in the one-dimensional configurations shown schematically in Fig. 1

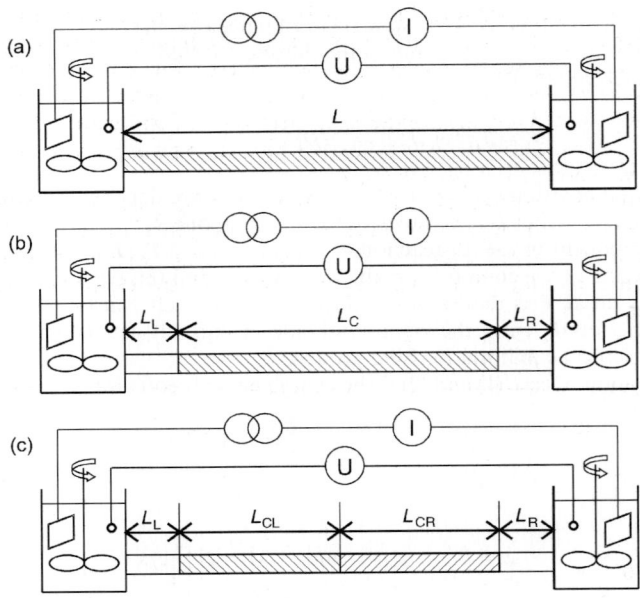

Fig. 1 Schematic configurations of the studied membrane systems: (a) Hydrogel of length L in direct contact with two well-stirred compartments, (b) hydrogel of length L_C separated by hydrodynamic boundary layers of thickness L_L and L_R from the left and right compartment, respectively, (c) the membrane consists of two layers of length L_{CL} and L_{CR} separated from the stirred compartments by boundary layers. The potential difference $\Delta\phi = \phi_{\text{left}} - \phi_{\text{right}}$ is applied to the studied membrane systems.

were studied. One reversible chemical reaction takes place in the system

$$H_2O \leftrightarrow H^+ + OH^-, \qquad r_1 = k_1(K_W - c_{H^+}c_{OH^-}). \tag{10}$$

The water dissociation reaction (10) is endothermic, relatively fast and its rate depends on temperature[14] and electric field intensity (according to Wien's second law[15])

$$\Delta h_{r_1} = \Delta h_{r_1}(T), \qquad \Delta h_{r_1}(298.15) \cong 57 \text{ kJ mol}^{-1} \tag{11}$$

$$k_1 \cong 1.3 \times 10^{11} \text{ kmol}^{-1} \text{ m}^3 \text{ s}^{-1} \tag{12}$$

$$K_W = K_W(T, |E|), \qquad K_W(298.15, 0) \cong 10^{-14} \text{ kmol}^2 \text{ m}^{-6} \tag{13}$$

$$K_W(T, E)/K_W(T, 0) = a \exp(bp|E|/kT) \tag{14}$$

where p is the electric dipole of water ($p \approx 10^{-29}$ C m) and k is the Boltzmann constant.

The mathematical model consists of five ($N_m + 1$) parabolic partial differential equations (PDEs) (1) and (6) describing evolution of concentrations c_{K^+}, c_{Cl^-}, c_{H^+}, c_{OH^-} and temperature T, one elliptic PDE (9) for electric potential ϕ and two (N_f) ordinary differential equations (ODEs) (2) for concentrations of fixed components c_{FC^+} and c_{FA^-}. At steady state, all PDEs turn out to be elliptic and the ODEs (2) for fixed components will reduce to simple algebraic equations. The water and fixed group dissociation equilibria, local electroneutrality in the bulk and Donnan's equilibria at the boundaries will be established in the isolated system due to very fast processes like protonisation–deprotonisation reactions and the attraction of oppositely charged mobile ions. Significant deviations from these equilibria can be observed in open systems. Spatially very focused structures arise in macroscopic systems. Numerical difficulties arising in the mathematical modelling can be solved by using an adaptive non-equidistant spatial grid, which is reconstructed during the integration in time (dynamic simulation) or during the use of continuation (parameter mapping of steady states).

Three different configurations of the studied systems are shown in Fig. 1: (a) the hydrogel containing no fixed charge is in direct contact with two stirred compartments, (b) the cation- or anion-exchange membrane is separated from well-stirred compartments by hydrodynamic boundary layers of thickness L_L and L_R, respectively, and (c) the bipolar membrane consists of cation- and anion-exchange layers separated by boundary layers from mixed reservoirs with constant concentrations of electrolytes. Experimental observation of current–voltage characteristics is a straightforward way of studying membrane electrolyte systems and mathematical modelling could give an explanation for the unusual characteristics with large parametric sensitivities or hysteresis phenomena that were observed experimentally.[2]

Let us first discuss two qualitatively different situations, where simple electrolyte systems exhibit nonlinear characteristics: (i) the acid–base diode and (ii) the bipolar membrane diode. The nonlinear characteristic of anion or cation exchange membranes immersed in a solution of salt are described by Rubinstein et al.[9] The nonlinear phenomena in the electrolyte systems studied here are discussed later.

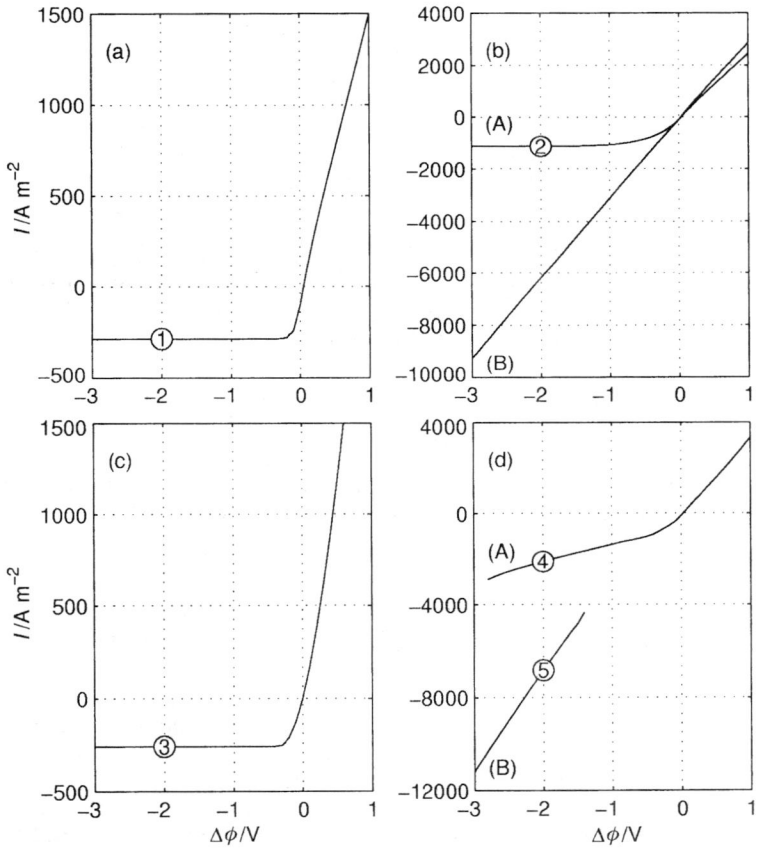

Fig. 2 Current–voltage characteristics of studied membrane systems: (a) Acid–base diode consisting of a hydrogel of length $L = 1$ mm containing no fixed charge with alkaline solution (0.1 M KOH) on the left and acidic solution (0.1 M HCl) on the right, cf. configuration in Fig. 1(a). (b) The same system as in (a), but the right compartment is contaminated by 0.15 M KCl (curve (A)) or 0.225 M KCl (curve(B)). (c) Bipolar membrane diode, cf. configuration in Fig. 1(c), left and right reservoirs contain 0.1 M KCl, left and right parts of the membrane have lengths $L_{CL} = L_{CR} = 0.49$ mm, left part has fixed anionic charge ($c_{FA^-} = 0.033$ M, $c_{FC^+} = 0$) and right part has fixed cationic charge ($c_{FC^+} = 0.033$ M, $c_{FA^-} = 0$), boundary layers $L_L = L_R = 0.01$ mm. (d) Hydrogel of length $L_C = 0.9$ mm with anionic fixed charge $c_{FA^-} = 0.033$ M, separated from the left reservoir (0.1 M KOH) and the right one (0.1 M KCl and 0.225 M KCl) by boundary layers $L_L = L_R = 0.05$ mm, cf. Fig. 1(b). Branches (A) and (B) represent multiplicity of stationary solutions. Spatial profiles of concentrations, reaction rate and electric variables describing the internal structure of studied systems corresponding to points 1, 2, 3, 4 and 5 are plotted in Fig. 3, 4, 5, 6 and 8, respectively.

3. Acid–base electrolyte diode

The "pure" hydrogel containing no fixed charge is in contact with a strong alkaline solution (0.1 M KOH) and a strong acid (0.1 M HCl) on the left- and right-hand sides, respectively, cf. scheme in Fig. 1(a). When subjected to the external potential difference $\Delta\phi = \phi_{\text{left}} - \phi_{\text{right}}$ this system exhibits a diode-like current–voltage characteristic, cf. Fig. 2(a). The electric current intensity I passing through is proportional to $\Delta\phi > 0$ in the forward direction (the open mode) but is negative, small and almost constant in the backward direction (the closed mode, $\Delta\phi < 0$). Let us discuss the cause of the diode-like action of this simple system, where only diffusion and migration of ionic species and the water dissociation/recombination reaction take place.

The profiles of relevant physical quantities in the "pure" diode realized in the hydrogel with no fixed charge corresponding to the point 1 from Fig. 2(a) are plotted in Fig. 3. The K^+ and Cl^- ions diffuse from their reservoirs to the opposite compartments. When the negative potential

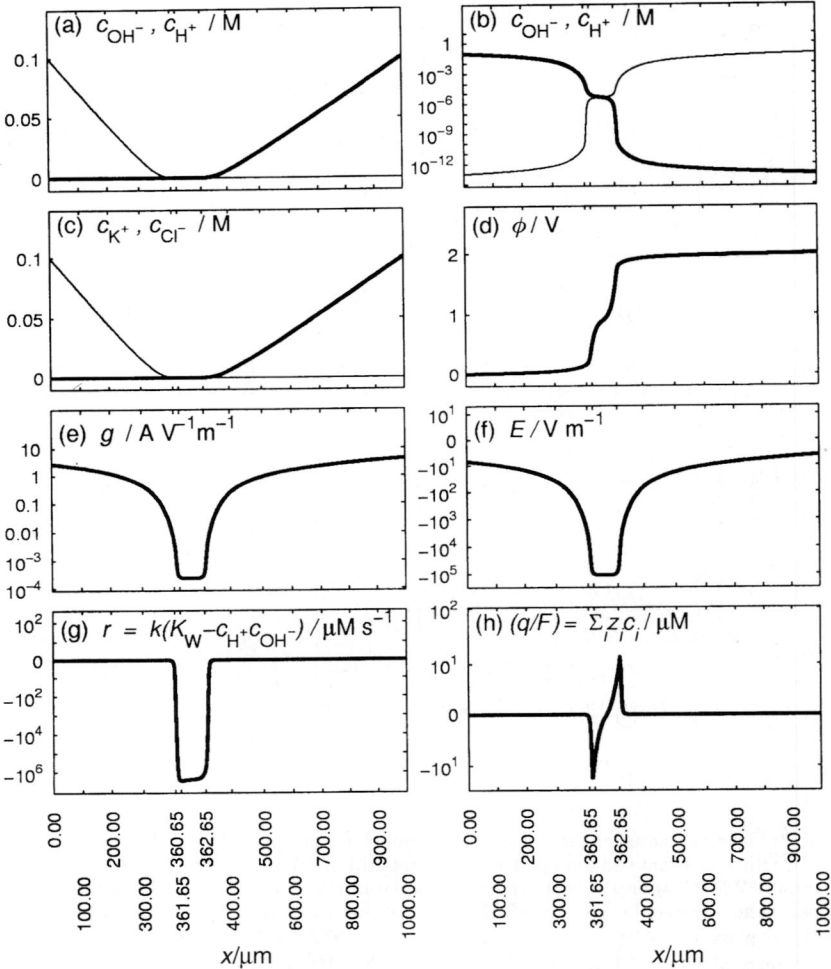

Fig. 3 The structure of the "pure" acid–base electrolyte diode for $\Delta\phi = \phi_{\text{left}} - \phi_{\text{right}} = -2$ V, cf. point 1 in Fig. 2(a). Left compartment (0.1 M KOH), right compartment (0.1 M HCl), hydrogel length $L = 1$ mm, cf. configuration in Fig. 1(a). Spatial profiles of: (a) c_{H^+} (thick) and c_{OH^-} (thin line), (b) c_{H^+} (thin) and c_{OH^-} (thick line) in logarithmic scale, (c) c_{K^+} (thin) and c_{Cl^-} (thick line), (d) electric potential ϕ, (e) ionic conductivity $g = (F^2/RT)\Sigma z_i^2 D_i c_i$, (f) electric field intensity $E = -\partial\phi/\partial x$, (g) reaction rate of water dissociation reaction r, (h) scaled charge density q/F. The spatial axis x has a non-equidistant scale.

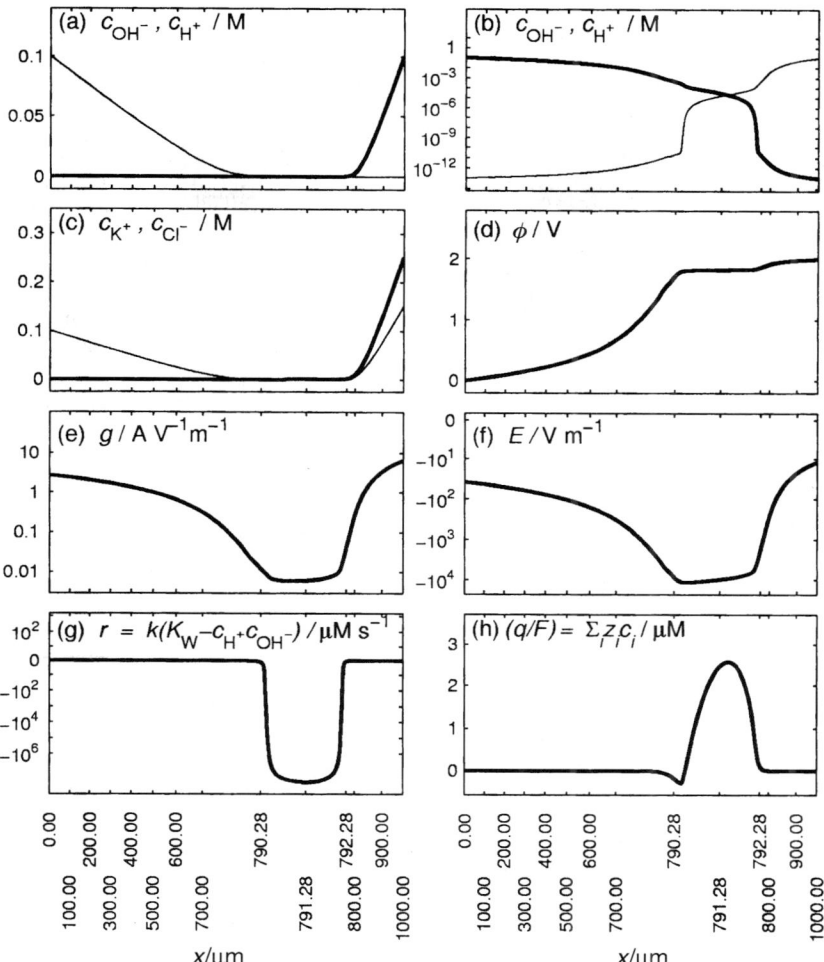

Fig. 4 The structure of the "contaminated" acid–base electrolyte diode for $\Delta\phi = \phi_{\text{left}} - \phi_{\text{right}} = -2$ V, cf. point 2 in Fig. 2(b). Left compartment (0.1 M KOH), right compartment (0.1 M HCl contaminated with 0.15 M KCl), hydrogel length $L = 1$ mm, cf. configuration in Fig. 1(a). Spatial profiles of: (a) c_{H^+} (thick) and c_{OH^-} (thin line), (b) c_{H^+} (thin) and c_{OH^-} (thick line) on a logarithmic scale, (c) c_{K^+} (thin) and c_{Cl^-} (thick line), (d) electric potential ϕ, (e) ionic conductivity $g = (F^2/RT)\Sigma z_i^2 D_i c_i$, (f) electric field intensity $E = -\partial\phi/\partial x$, (g) reaction rate of water dissociation reaction r, (h) scaled charge density q/F. The spatial axis x has a non-equidistant scale.

difference $\Delta\phi = \phi_{\text{left}} - \phi_{\text{right}} = -2$ V is applied, the electric field forces K$^+$ and Cl$^-$ ions to migrate to their original compartments (K$^+$ to the left and Cl$^-$ to the right), cf. Fig. 3(c). Thus a thin low-conductivity zone with a strong electric field $E = (-\partial\phi/\partial x)$ is formed in the hydrogel, cf. Fig. 3(d)–(f). The H$^+$ and OH$^-$ ions diffuse and migrate, for $\Delta\phi < 0$, to the opposite compartments and a water recombination reaction H$^+$ + OH$^- \to$ H$_2$O takes place at the acid–base boundary, compare Fig. 3(b) and (g). Let us note that the product $(c_{\text{H}^+} c_{\text{OH}^-})$ at the acid–base boundary is significantly larger than the water ionic product ($K_w = 10^{-14}$ M^2) due to the finite rate of the water recombination reaction, cf. Fig. 3(b). The characteristic width of the electric double layer formed at the acid–base interface of the diode is approximately 2 µm, cf. Fig. 3(h).

Now the acid–base diode consisting of the pure hydrogel containing no fixed charge, corresponding to the configuration of Fig. 1(a), is considered again, but the right compartment is

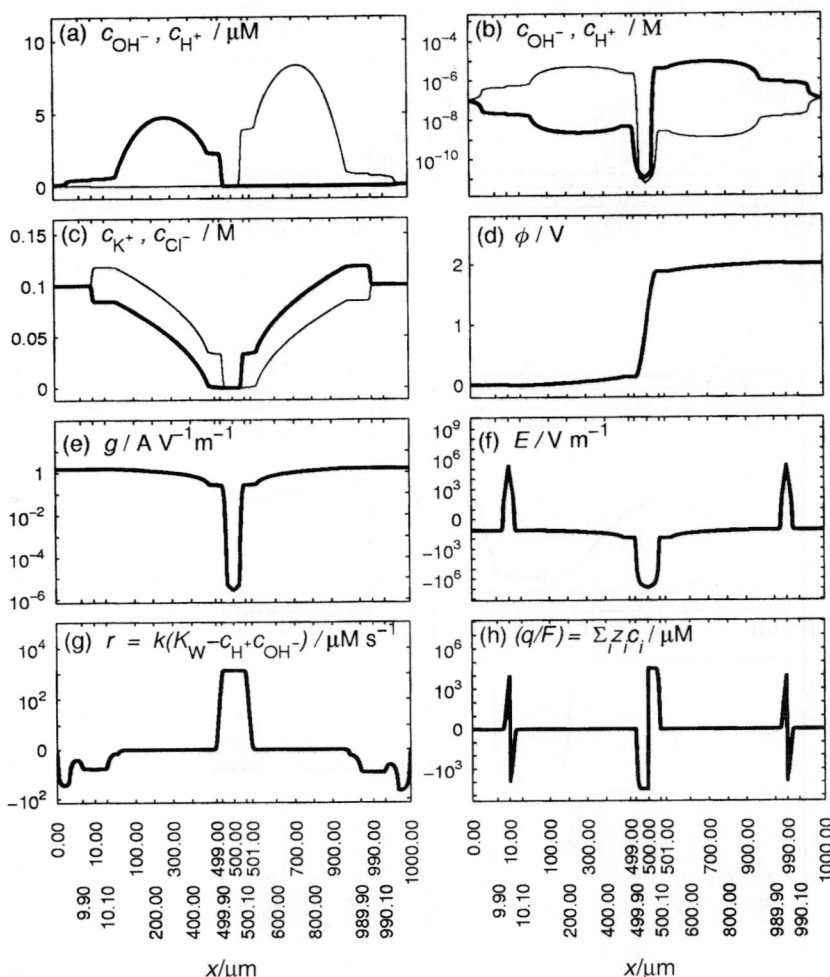

Fig. 5 The structure of the bipolar membrane diode, *cf.* point 3 in Fig. 2(c). Configuration of Fig. 1(c), left and right reservoirs contain 0.1 M KCl, left and right parts of the membrane have lengths $L_{CL} = L_{CR} = 0.49$ mm, left part has fixed anionic charge ($c_{FA^-} = 0.033$ M, $c_{FC^+} = 0$) and right part has fixed cationic charge ($c_{FC^+} = 0.033$ M, $c_{FA^-} = 0$), boundary layers $L_L = L_R = 0.01$ mm, applied potential difference $\Delta\phi = \phi_{left} - \phi_{right} = -2$ V. Spatial profiles of: (a) c_{H^+} (thick) and c_{OH^-} (thin line), (b) c_{H^+} (thin line) and c_{OH^-} (thick line) on a logarithmic scale, (c) c_{K^+} (thin) and c_{Cl^-} (thick line), (d) electric potential ϕ, (e) ionic conductivity $g = (F^2/RT) \sum z_i^2 D_i c_i$, (f) electric field intensity $E = -\partial\phi/\partial x$, (g) reaction rate of water dissociation reaction r, (h) scaled charge density q/F. The spatial axis x has a non-equidistant scale.

contaminated by the presence of KCl in addition to 0.1 M HCl. The diode-like action is preserved for 0.15 M KCl in the right compartment, *cf.* curve (A) in Fig. 2(b), but it ceases to exist for the level of contamination set to 0.225 M KCl, *cf.* curve (B) in Fig. 2(b). The structure of the contaminated acid–base diode corresponding to the point 2 from Fig. 2(b) (externally applied potential difference $\Delta\phi = \phi_{left} - \phi_{right} = -2$ V) is displayed in Fig. 4. Let us note the non-symmetrical shape of the electric double layer at the acid–base interface in Fig. 4(h). The migration of K^+ ions through the acid–base interface in the hydrogel (to the left) causes some increase in the magnitude of electric current intensity I passing through the diode when compared to the pure uncontaminated diode, compare curve (A) from Fig. 2(b) with Fig. 2(a).

At a high level of contamination the diode-like action ceases to exist, *i.e.*, the electric current passing through the system is directly proportional to the applied voltage both for $\Delta\phi > 0$ and for

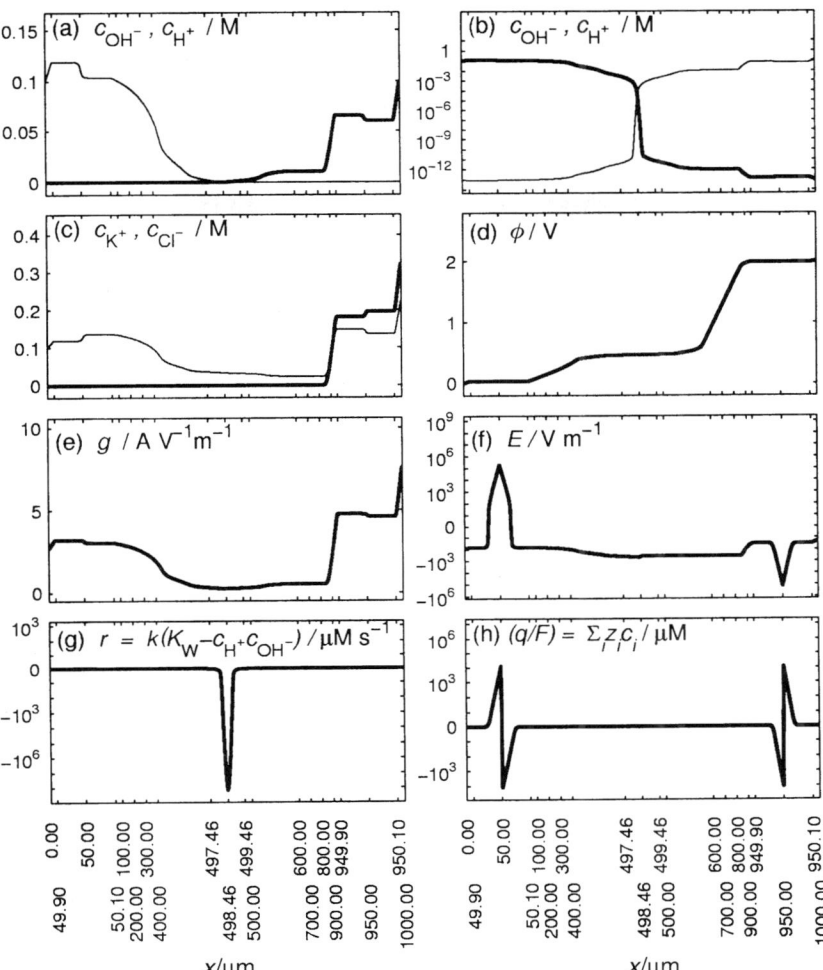

Fig. 6 The structure of the closed mode of the anionic hydrogel (c_{FA^-} = 0.033 M) "contaminated" by KCl in the right compartment in the backward voltage direction for $\Delta\phi = \phi_{left} - \phi_{right} = -2$ V, cf. point 4 in Fig. 2, 7(d), 7(a) and 7(b). Left compartment (0.1 M KOH), right compartment (0.1 M HCl contaminated by 0.225 M KCl), hydrogel length $L = 0.9$ mm, boundary layers $L_L = L_R = 0.05$ mm, cf. configuration in Fig. 1(b). Spatial profiles of: (a) c_{H^+} (thick) and c_{OH^-} (thin line), (b) c_{H^+} (thin) and c_{OH^-} (thick line) on a logarithmic scale, (c) c_{K^+} (thin) and c_{Cl^-} (thick line), (d) electric potential ϕ, (e) ionic conductivity $g = (F^2/RT)\Sigma z_i^2 D_i c_i$, (f) electric field intensity $E = -\partial\phi/\partial x$, (g) reaction rate of water dissociation reaction r, (h) scaled charge density q/F. The spatial axis x has a non-equidistant scale.

$\Delta\phi < 0$ because the diffusion and migration fluxes of K$^+$ ions across the acid–base boundary in the hydrogel "flood" the zone of the low conductivity, cf. curve (B) in Fig. 2(b).

4. Bipolar cation–anion exchange membrane diode

The current–voltage characteristic of the bipolar membrane consisting of the cation- and anion-exchange layers on the left and the right-hand side, respectively, and separated from the two mixed compartments containing 0.1 M KCl by a hydrodynamic boundary layer is shown in Fig. 2(c). This system has already been thoroughly studied semi-analytically.[15,16] Our numerical solution of this problem is free of a number of simplifying assumptions typically used in such semi-analytical studies.

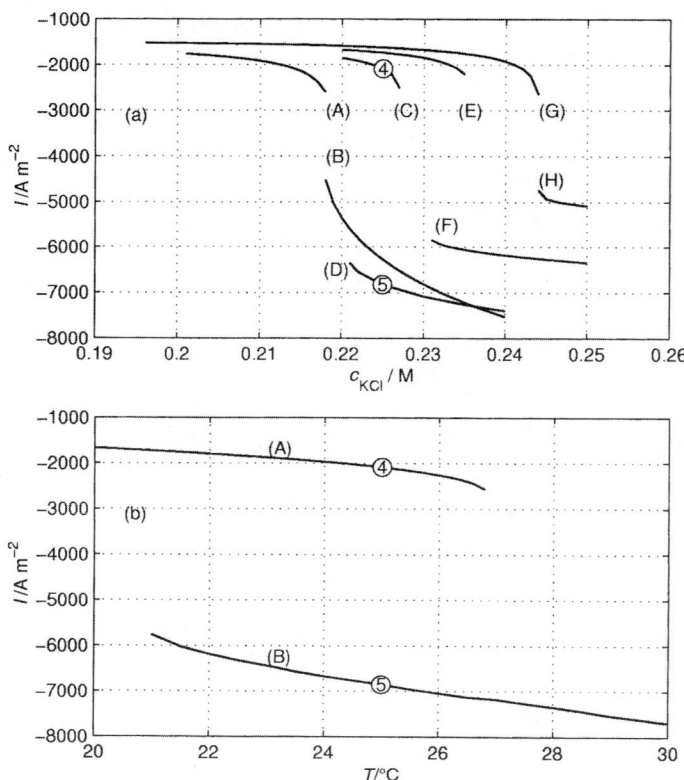

Fig. 7 The multiplicity of stationary solutions for the hydrogel with anionic fixed charge $c_{FA^-} = 0.033$ M, separated from the left reservoir (0.1 M KOH) and the right one (0.1 M KCl contaminated with KCl) by boundary layers, cf. Fig. 1(b). The applied voltage is $\Delta\phi = \phi_{left} - \phi_{right} = -2$ V. (a) Parametric dependence of electric current intensity I on the concentration of KCl contaminant in the right compartment, $T = 25\,°C$. The length of the membrane is $L_C = 1$ mm, $L_L = L_R$, where the thickness of the boundary layers is: (A), (B) $L_L = L_R = 0.1$ mm, (C), (D) $L_L = L_R = 0.05$ mm, (E), (F) $L_L = L_R = 0.02$ mm, (G), (H) $L_L = L_R = 0.001$ mm. (A), (C), (E), (G) are closed branches in the backward direction, (B), (D), (F), (H) are open branches in the backward direction. (b) Parametric dependence of electric current intensity I on the temperature, concentration of KCl in the right compartment is 0.225 M, membrane length $L_C = 0.9$ mm, boundary layers $L_L = L_R = 0.05$ mm. (A) closed branch in the backward direction, (B) open branch in the backward direction. Points 4 and 5 in Fig. 2(d), 7(a) and 7(b) are identical.

The anionic fixed charge $c_{FA^-} = 0.033$ M in the bulk of the left part of the membrane is compensated by H^+ and K^+ cations and the cationic fixed charge $c_{FC^+} = 0.033$ M in the right half is compensated by OH^- and Cl^- anions, cf. Fig. 5(a)–(c). A focused electric double layer of characteristic width 0.2 μm is formed in the centre of the membrane and two electric double layers appear on the outer membrane surfaces, cf. Fig. 5(h). The external potential difference $\Delta\phi = \phi_{left} - \phi_{right} = -2$ V is applied and a strong electric field $E = (-\partial\phi/\partial x)$ in the centre of the membrane causes fast migration of all ions and a zone of low conductivity electrolyte is created in the centre of the bipolar membrane, cf. Fig. 5(e). The product $(c_{H^+} c_{OH^-})$ in the middle of the bipolar membrane diode is much smaller than $K_w = 10^{-14}$ M^2, cf. Fig. 5(b), hence the water dissociation reaction $H_2O \rightarrow H^+ + OH^-$ takes place in the central part of the system, cf. Fig. 5(g). The small potential drops at the outer surfaces of the bipolar membrane predicted by numerical simulations correspond well to the equilibrium Donnan potentials.

Let us note that in the case of zero externally applied voltage $\Delta\phi = 0$ V, a zero electric current passes through the bipolar membrane but a small electro-diffusion current passes through the

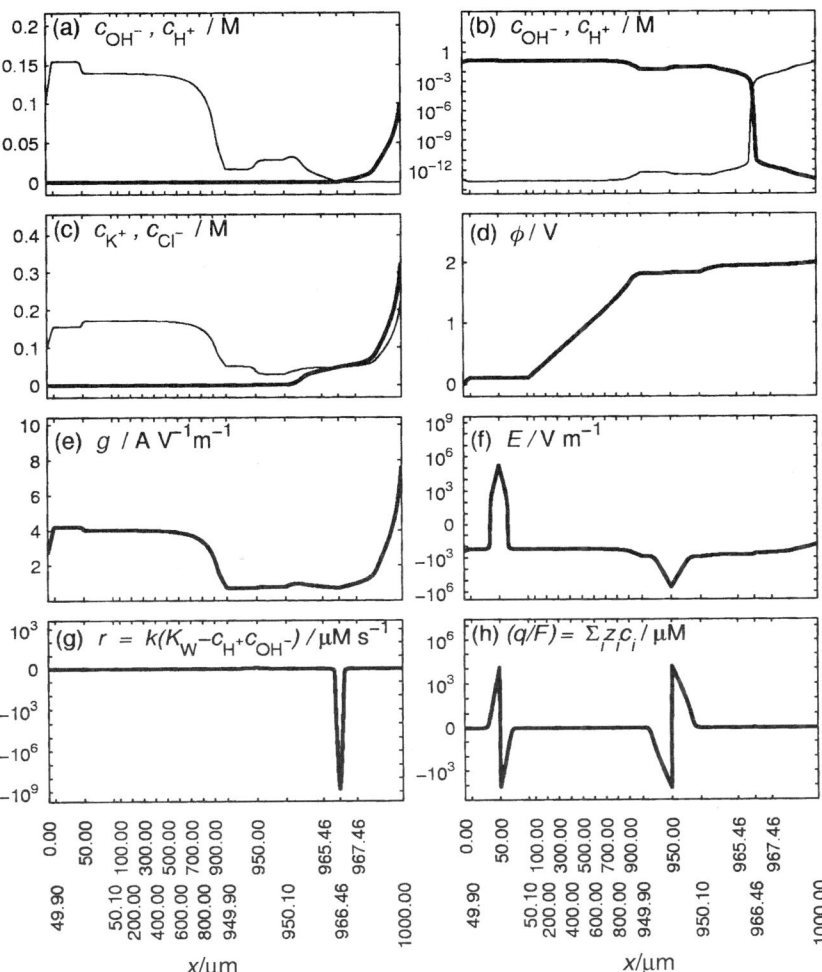

Fig. 8 The structure of the opened mode of the anionic hydrogel ($c_{FA^-} = 0.033$ M) "contaminated" by KCl in the right compartment in the backward voltage direction for $\Delta\phi = \phi_{left} - \phi_{right} = -2$ V, cf. point 5 in Fig. 2d, 7(a) and 7(b). Left compartment (0.1 M KOH), right compartment (0.1 M HCl contaminated by 0.225 M KCl), hydrogel length $L = 0.9$ mm, boundary layers $L_L = L_R = 0.05$ mm, cf. configuration in Fig. 1(b). Spatial profiles of: (a) c_{H^+} (thick) and c_{OH^-} (thin line), (b) c_{H^+} (thin) and c_{OH^-} (thick line) on a logarithmic scale, (c) c_{K^+} (thin) and c_{Cl^-} (thick line), (d) electric potential ϕ, (e) ionic conductivity $g = (F^2/RT)\Sigma z_i^2 D_i c_i$, (f) electric field intensity $E = -\partial\phi/\partial x$, (g) reaction rate of water dissociation reaction r, (h) scaled charge density q/F. The spatial axis x has a non-equidistant scale.

acid–base diode in the pure hydrogel because of the unequal values of the diffusion coefficients of the ionic species, compare characteristics in Fig. 2(a) and (c). It is obvious that the magnitude of the electro-diffusion electric current depends on the values of the diffusion coefficients and on the concentration gradients of particular ions.

5. Thyristor-like action of the cation-exchange hydrogel contaminated by KCl in the right compartment

Let us add the fixed anionic charge $c_{FA^-} = 0.033$ M to the hydrogel of length $L_C = 0.9$ mm separated from the two boundary reservoirs by $L_L = L_R = 0.05$ mm thick boundary layers, cf.

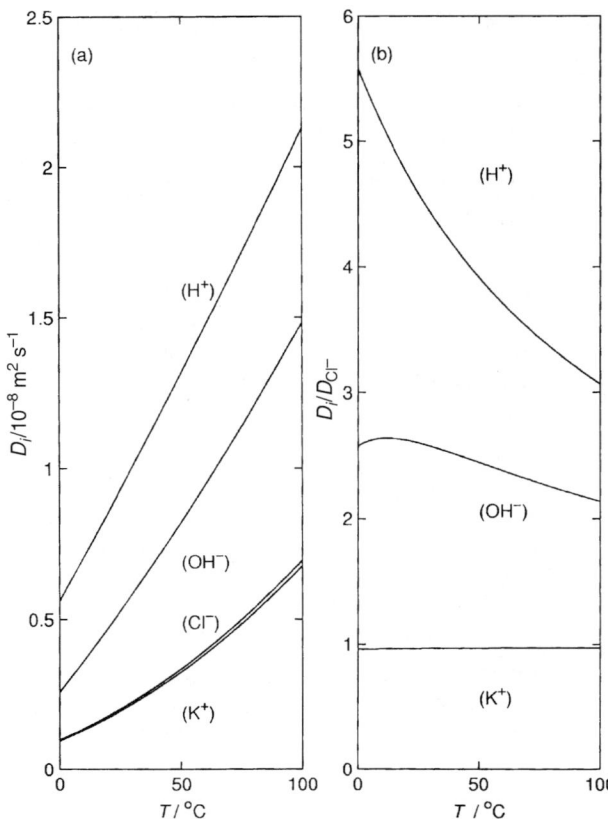

Fig. 9 (a) The dependence of the diffusion coefficients of H^+, OH^-, K^+ and Cl^- ions on temperature. (b) The dependence of the ratios of the diffusion coefficients D_{H^+}/D_{Cl^-}, D_{OH^-}/D_{Cl^-}, D_{K^+}/D_{Cl^-} on temperature.

configuration in Fig. 1(b). The left reservoir contains 0.1 M KOH and the right reservoir is 0.1 M HCl contaminated with 0.225 M KCl. Two different branches of stationary states in the current–voltage characteristic exist for this specification of the membrane system, cf. branches (A) and (B) in Fig. 2(d). The current–voltage response at branch (A) is characterised by different proportionality constants ("effective conductivities") for the forward ($\Delta\phi > 0$) and the backward ($\Delta\phi < 0$) applied voltage. The compensation of the anionic fixed charge by the H^+ and K^+ cations is the reason for the enhancement of the membrane conductivity, compare Fig. 2(b) and (d) and Fig. 6 demonstrating spatial profiles corresponding to point 4 in Fig. 2(d). The branch (A) in Fig. 2(d) ends at the limit point $\Delta\phi \approx -2.8$ V because of the global electric current-driven interactions resulting in flooding of the zone of the low conductivity by K^+ cations.

The parametric dependences of the passing electric current on the concentration of KCl contaminant in the right compartment for several values of the characteristic thickness of the hydrodynamic boundary layer are shown in Fig. 7(a). This figure illustrates the importance of reaction and transport processes in boundary layers for the explanation of hysteresis phenomenon observed experimentally by Nosticzius and coworkers.[2] In the case of a thick boundary layer, cf. branches (A) and (B) with $L_L = L_R = 0.1$ mm, as well as in the case of a very thin boundary layer, cf. branches (G) and (H) with $L_L = L_R = 0.001$ mm, the region of steady state multiplicity projected to the horizontal axis is very narrow. The thickness of the boundary layers for branches (C) and (D) is the same as in Fig. 2(d) and the points 4 and 5 in Fig. 2(d), 7(a) and 7(b) are identical (the concentration of contaminant in the right compartment is $c_{KCl} = 0.225$ M). The "upper" and "lower" stationary states represented, for example, by points 4 and 5, are characterized by different electric current densities I. The region of steady state multiplicities for branches (C) and (D) is relatively wide. Complex dynamic behavior could be expected in this parametric region in the case of externally controlled constant electric current passing through the system, set to values between the limit points of branches (C) and (D), that is $-6400 < I/\text{A m}^{-2} < -2600$.

The acid–base boundary where the water recombination takes place is for the point 5 from Fig. 2(d), 7(a) and 7(b) located in the right boundary layer of the membrane, *cf.* Fig. 8(a) and (b) (the externally applied voltage is $\Delta\phi = \phi_{\text{left}} - \phi_{\text{right}} = -2$ V). The current–voltage characteristic (B) in Fig. 2(d) ends at a limit point at $\Delta\phi \approx -1.4$ V because of the complex interaction of the acid–base boundary with the right-hand interface of the cation-exchange membrane. The acid–base boundary at the branch (B) in Fig. 2d is not accompanied by the zone of very low conductivity g and strong electric field intensity E, *cf.* Fig. 8(e) and (f).

Our hypothesis is that the limit points of branches (A) and (B) in Fig. 2(d) or the limit points of branches (C) and (D) in Fig. 7(a) are probably connected by the unstable branch of the stationary solutions. At present we can only speculate about the complex dynamic behavior of the system with a controlled constant electric current intensity I passing through the membrane. The investigation of this dynamic behavior is the subject of continuing research.

6. Temperature effects on the behavior of electrolyte diode

As follows from Fig. 9(b), the ratios of the diffusion coefficients vary with temperature as the values of the diffusion coefficients of H^+ and OH^- ions increase with temperature more slowly than the values of the diffusion coefficients of K^+ and Cl^- ions.[14,17]

The multiplicity of stationary states corresponding to points 4 and 5 in Fig. 2(d) and 7(a) is projected into the temperature parametric axis in Fig. 7(b). The multiplicity of stationary states characterized by different electric current densities I arises in the temperature interval 21–27 °C. The cause of this multiplicity is the temperature dependence of the diffusion coefficients shown in Fig. 9. A more detailed discussion of effective diffusion coefficients in the poly(vinyl alcohol) hydrogel was published recently.[18]

7. Conclusions

Numerically computed spatial profiles of the concentrations of ionic species, the reaction rate of the water dissociation reaction, electric potential, ionic conductivity, electric field intensity and charge density for several typical configurations of a thin layer of neutral and/or charged hydrogel with the corresponding boundary layers enable us to better understand some causes and effects of nonlinear behavior (*e.g.*, diode-like characteristics, multiple steady states) observed experimentally. Sharp variations of the above quantities often occur on the scale of 0.2 to 2 µm and present significant difficulties in the numerical solution of the corresponding balance equations. In the limited scope of this paper the protonisation–deprotonisation reversible reactions taking place in the system were not considered. At steady state local dissociation equilibria are established for immobile components in contrast to mobile components whose fluxes can somewhat shift species concentrations out of local equilibria. More complex behavior in the simulated hydrogel membranes was obtained when considering the dissociation reactions of fixed charged groups. These dissociations are certainly taking place in a number of hydrogel membranes containing different types of fixed groups. Systematic studies of the parametric dependences and dynamic properties of the observed spatial patterns are in progress.

Acknowledgements

Financial support from the Czech Ministry of Education (project MSM 2234000007/VZ) and from the Czech Grant Agency (project 104/99/1408) is acknowledged.

References

1 L. H. Hegedus, N. Kirschner, M. Wittmann and Z. Noszticzius, *J. Phys. Chem. A*, 1998, **102**(32), 6491.
2 L. H. Hegedus, N. Kirschner, M. Wittmann, P. Simon and Z. Noszticzius, *Chaos*, 1999, **9**(2), 283.
3 K. Krischer, in *Modern Aspects of Electrochemistry*, Kluwer, New York, 1999, vol. 32, p. 1.

4   B. Lovreček, A. Despić and J. Bockris, *J. Phys. Chem.*, 1959, **63**, 750.
5   C. Forgacs, *Nature*, 1961, **4773**, 339.
6   M. Pačes, J. Lindner, J. Havlica, J. Kosek, D. Šnita and M. Marek, in *Proceedings of the IMRET 4 conference*, AIChE, New York, 2000, p. 264.
7   M. Pačes, J. Kosek, D. Šnita and M. Marek, *J. Phys. Chem.*, in preparation.
8   I. Rubinstein, *Diffusion of Ions*, SIAM, Philadelphia, 1990.
9   I. Rubinstein, B. Zaltzman and O. Kedem, *J. Membr. Sci.*, 1997, **125**, 17.
10 I. Rubinstein and B. Zaltzmann, *Phys. Rev. E*, 2000, **62**, 2238.
11 D. Šnita, H. Ševčíková, M. Marek and J. H. Merkin, *J. Phys. Chem.*, 1996, **100**, 18740.
12 R. A. Mosher, D. A. Saville and W. Thormann, *The Dynamics of Electrophoresis*, VCH Verlag, Weinheim, 1992.
13 D. Šnita and M. Marek, *Physica D*, 1994, **75**, 521.
14 *CRC Handbook of Tables for Applied Engineering Science*, ed. R. E. Bolz and G. L. Tuve, CRC Press, Boca Raton, FL, 2nd edn., 1979.
15 S. Mafé, P. Ramírez and A. Alcaraz, *Chem. Phys. Lett.*, 1998, **294**, 406.
16 S. Mafé, J. A. Manzanares and P. Ramírez, *Phys. Rev. A*, 1990, **42**(10), 6245.
17 Y. Marcus, *Ion Properties*, Marcel Dekker, New York, 1997.
18 F. Fergg and F. J. Keil, *Chem. Eng. Sci.*, 2001, **56**, 1305.

Self-oscillating polymer chain in a laser field

Hiroyuki Mayama* and Kenichi Yoshikawa

Department of Physics, Graduate School of Science, Kyoto University, and CREST (Core Research for Evolutional and Scientific Technology) of JST (Japan Science and Technology Corporation), Kyoto 606-8502, Japan.
E-mail: mayama@chem.scphys.kyoto-u.ac.jp

Received 26th June 2001
First published as an Advance Article on the web 12th November 2001

We have investigated the dynamical behaviour of the rhythmic conformational change between the folded compact state and the unfolded state in a single polymer chain under thermodynamically open conditions. It is shown that the spontaneous rhythmic change in the conformation of a single polymer chain (T4DNA, 166 kbp, contour length: 56 µm) is generated using a focused continuous wave (CW) Nd : YAG laser beam (wavelength $\lambda = 1064$ nm), where the focused laser beam plays a dual role, both trapping a polymer chain at the focus and creating a temperature gradient around the focus. Furthermore, the whole process of the rhythmic conformational change: the course of melting, nucleation and growth between the folded and unfolded states has been clarified. The rhythmic change in the conformation is discussed in terms of the limit-cycle oscillation driven by the dissipation of the photon energy.

Introduction

Living matter is maintained under nonequilibrium conditions in an isothermal system.[1] Therefore, to understand the essence of life, it is essential to investigate a molecular system with energy dissipation.[2] Recently, numerical experimental studies have been performed on functional proteins, such as myosin, kinesin, dynein, and F_0F_1 adenosine triphosphate (ATP) synthase.[3-9] From these studies, it has been found that functional proteins generate rhythmic vectoral motion with energy dissipation. Myosin, kinesin and dynein move along polymer substrates with ATP hydrolysis: myosin along actin filaments in muscle and other cells, and kinesin and dynein along microtubules. F_0F_1 ATP synthase is the smallest functional protein found in mitochondria or bacterial membranes, in which synthesis and hydrolysis of ATP occur with an electrochemical proton gradient. It should also be noted that functional proteins generate vectoral motion under large thermal fluctuations in an isothermal system because even pollen floating on the water displays Brownian motion.

Contrary to this, in a Carnot cycle unidirectional motion is brought about by a temperature difference between high and low temperature heat reservoirs, where the Carnot-cycle contains hard bodies to minimize the effect of thermal fluctuations. In the Carnot cycle, the efficiency of energy transduction from thermal energy to vectoral motion is represented as $\eta_{Carnot} < (T_h - T_l)/T_h$, where T_h and T_l are the temperatures of the high and low temperature heat reservoirs, respectively. Thus vectoral motion is not generated in an isothermal system because $T_h = T_l$. It is expected therefore that there is an essential difference in the working mechanisms between natural machinery and currently available man-made systems.

DOI: 10.1039/b105573a

To explain how vectoral motions of functional proteins occur under isothermal conditions, theoretical and experimental studies based on the ratchet model have been performed.[10–18] The essentials are a periodic asymmetric potential and fluctuations. Theory predicts that the driving forces required to generate the unidirectional motion are fluctuating forces in the potential,[11,12] the fluctuating barrier of the potential,[13,14] the fluctuating state in the asymmetric potential with nonequilibrium chemical reaction[15] and the collective behaviour of large ensembles of a many-functional system,[16] where fluctuation means time-correlated noise. Though a few experimental studies have succeeded in inducing the drift of a particle with Brownian motion using an electric field[17] and a light field,[18] the experiment using a biopolymer has not yet been reported. On the other hand, in experimental study on a functional protein, it was proposed that the unidirectional motion in functional proteins could be explained by the ratchet model.[19] The theories based on the ratchet model are very useful to interpret the generation of vectoral motion in molecular machinery, but a relationship between the ratchet model and functional proteins is still unclear. Furthermore, the current experimental and theoretical studies on the ratchet model and a functional protein have not been based on a molecular system under thermodynamically open conditions.

Very recently, a hypothesis on the mechanism of molecular machinery in an isothermal system has been proposed.[20] The essence of the hypothesis is a limit-cycle oscillation in a single polymer chain with external energy supply under thermodynamically open conditions, where the single polymer chain exhibits a first-order phase transition between the unfolded state (the random coil state) and the folded compact state (the condensed state). Since the hysteresis in the first-order phase transition is concerned with the irreversibility of the path of conformational change, i.e., breaking of detailed balance,[21,22] it is expected that unidirectional motion will be generated rhythmically on its conformational change between the folded state and the unfolded state under thermodynamically open conditions, even if the polymer chain is exposed to large thermal fluctuations. Then, the state of the polymer chain describes a closed orbital in a phase space in one direction.[20] This causes vectoral motion in molecular machinery under thermodynamically open conditions. Considering that the hydrolysis of ATP with its associated vectoral motion in functional proteins corresponds to an external energy supply under thermodynamically open conditions in an isothermal system, it is expected that the hypothesis will help one to understand the essence of the mechanism of molecular machinery.

Based on our hypothesis, we have tried to realize the rhythmic conformational change in a single polymer chain under thermodynamically open conditions. There are three essential factors in the experiment: to create thermodynamically open conditions, to select a single polymer chain with a first-order phase transition, and to design a folding/unfolding transition sensitive to the degree of nonequilibrium under thermodynamically open conditions. To create thermodynamically open conditions, we need to induce local heating in aqueous solutions, i.e., a temperature gradient. It has been reported that a focused CW Nd:YAG laser beam ($\lambda = 1064$ nm) at suitable laser power induces local heating around the laser beam focus,[23] and also that the focused laser beam, known as optical tweezers, can trap a single DNA molecule in the folded state.[24–26] Thus, we decided to use a CW Nd:YAG laser beam to generate thermodynamically open conditions. As a single polymer chain with a first-order phase transition, we have selected individual T4DNA molecules. We have found that individual DNA molecules (T4DNA, 166 kbp, the contour length ca. 56 μm) exhibit a first-order phase transition, the folding/unfolding phase-transition, upon changes in concentration of the condensation agents used such as hydrophobic polymers,[27–29] multivalent cations,[30–32] alcohol,[33] surfactants,[34–35] and polyanions.[36–38] Furthermore, we have investigated the effect of temperature on the folding/unfolding transition, and have found that the folded DNA chain unfolds with increasing temperature in poly(ethylene glycol) (PEG) solutions,[27,28] whereas the unfolded DNA chain folds with increasing temperature in a spermidine^{3+} solution.[30] We have found that the folding/unfolding transition has a suitable sensitivity to temperature gradient to generate the rhythmic conformational change in PEG solutions. Based on the experimental results, we have succeeded in generating a rhythmic conformational change, or a limit-cycle oscillation in its conformation, of a single polymer chain (T4DNA) in a PEG solution under thermodynamically open conditions using a focused CW Nd:YAG laser beam.[39] We have observed that the folded DNA chain trapped by the focused laser beam unfolds rhythmically, where the period of the conformational change depends on suitable laser power between the upper and lower threshold values and it is of the order of several seconds.[39] Thus, we have found that a single polymer chain exhibits a rhythmic conformational change under thermodynamically open conditions. To establish the hypothesis, or the generality of molecular machinery, it is necessary to investigate the dynamical behaviour of the rhythmic conformational change in a single polymer chain and to improve the working hypothesis. Through further experiments, we have investigated the dynamical behaviour of the limit-cycle oscillation of a single polymer chain and have considered the model equations of rhythmic conformational change in a single polymer chain. In this article, we discuss the dynamical behaviour of the self-oscillating polymer chain under thermodynamically open conditions in terms of a limit-cycle oscillation driven by energy dissipation.

Experimental

Sample preparation

T4GT7 DNA (165.6 kbp), the fluorescent dye 4′,6-diamidino-2-phenylindole (DAPI) and potassium chloride (KCl) were obtained from Wako Pure Chemical Industries, Ltd. Poly(ethylene glycol) (PEG, average $M_r = 20\,000$) was purchased from Kishida Chemicals. 2-Mercaptoethanol was obtained from Nacalai Tesque. Heavy water (D_2O) was obtained from Aldrich. The sample solution was adjusted as follows: Tris–HCl buffer 0.1 mM (pH 7.75), DAPI 0.9 µM, T4 DNA 0.3 µM in nucleotides, KCl 0.3 M and different concentrations of PEG. The prepared samples were allowed to stand at ambient temperature, $20 \pm 1\,°C$, for 30 min before observation.

Temperature effect

Single-molecule observation was performed with a Nikon TE-300 microscope equipped with a SIT camera (Hamamatsu Photonics) and an Argus 50 image processor (Hamamatsu Photonics) at different temperatures. The fluorescence images of DNA obtained by single-molecule observation were recorded on videotape. The temperature of the sample was controlled by a Thermo Plate (TOKAI HIT, AXMS type). The accuracy of the temperature was $\pm 0.5\,°C$. To characterize the size of the DNA, we measured the long-axis length L, which was defined as the longest distance in the outline of the DNA image. We only counted the moving DNA in the bulk solution and omitted the adsorbed DNA on the glass surface. There is a blurring effect of $ca.$ 0.3 µm due to the high sensitivity of the SIT camera and the resolution limit associated with the wavelength of the observation light.

Rhythmic folding/unfolding transition

Single-molecule observation and optical trapping were carried out with a Nikon TE-300 microscope equipped with a SIT camera (Hamamatsu Photonics) and a continuous wave (CW) Nd : YAG laser (Spectron SL902T, wavelength $\lambda = 1064$ nm, beam mode: TEM_{00}) at room temperature and different laser powers. The axis of the laser beam was tilted for the optical axis of the fluorescence microscope by adjustment of the optical path, the angle between the optical axis and the laser beam axis was $ca.$ 0.174 rad. The fluorescence images of DNA obtained by single-molecule observation were recorded on videotape.

Results and discussion

We needed to select a suitable condensation agent to induce the folding/unfolding phase-transition, sensitive to thermodynamically open conditions, $i.e.$, temperature gradient. In particular, it was desired that the DNA should exhibit a discrete conformational transition from the folded state to the unfolded state with increasing temperature. The folded DNA trapped by a focused laser beam is exposed to the local heating and is unfolded. Then, the unfolded DNA chain escapes from the trapping point with Brownian motion and enters the ambient "cool" environment. Since the unfolded DNA is less stable at lower temperature, it becomes folded in the ambient environment. The folded DNA chain is pulled back towards the laser focus and is trapped again. The cycle is, thus, repeated.[39] We have found that folded DNA in PEG solution is unfolded with increasing temperature,[27,28] $i.e.$, it exhibits a suitable response to the temperature gradient by local heating. Thus, we decided to adopt PEG as the condensation agent to investigate the dynamical behavior of the self-oscillation polymer chain.

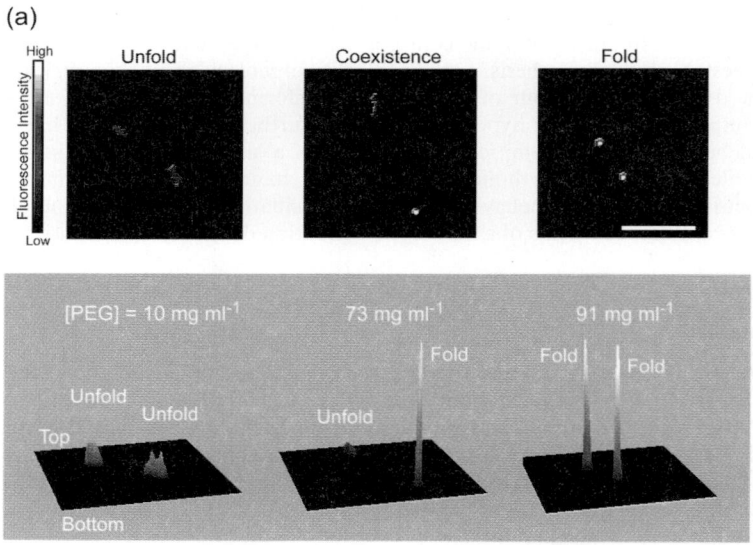

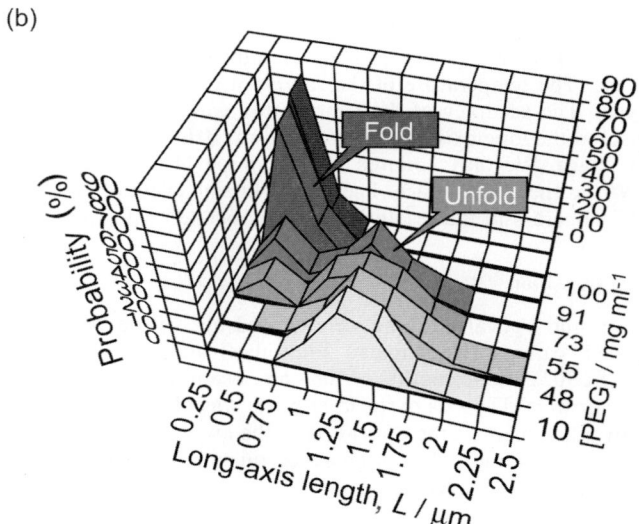

Fig. 1 (a) Typical fluorescence images (upper) and quasi-3-dimensional profile of typical fluorescence images of DNAs (lower) in PEG solution. The unfolded state (left), the coexistence region (middle), and the folded state (right) are at PEG concentrations of 10, 73, and 91 mg ml^{-1}, respectively. The scale bar represents 5 μm. (b) Distribution of the long-axis length L of T4DNA molecules at different concentrations of PEG (average M_r = 20 000) including a fixed concentration of KCl 0.3 M at 20 °C.

In order to investigate the dynamical behavior of the self-oscillating polymer chain two additional experimental conditions are helpful. First, it is expected that in a high viscosity solution the kinetic rate of the folding/unfolding process would be decreased, *i.e.*, the characteristic timescale of the rhythmic change would be extended. We can prepare a high viscosity solution using high molecular weight PEG. Secondly, since the unfolded part in the folded DNA is elongated along the axis of the optical tweezers (the focused laser beam) by the scattering force,[39] then if the optical tweezers tilt the optical axis we would be able to observe the projection image of the unfolded DNA chain in the rhythmic conformational change. Hence the axis of the focused laser beam and the optical axis of the fluorescence microscope are tilted.

We tried to find suitable experimental condition in PEG solution, *i.e.*, the coexistence region.[39] We adopted PEG (average $M_r = 20\,000$) to induce the folding/unfolding transition and to simultaneously prepare a high viscosity solution. Fig. 1(a) shows typical fluorescence images of individual DNAs within three regions: the unfolded coil state (left), the coexistence region (middle) and the folded compact state (right). The fluorescence images of the unfolded and folded DNAs are observed as blurred swollen images and bright spots, respectively. To characterize the folding transition, we measured the long-axis length L of the fluorescence images of at least 50 DNA chains at different concentrations of PEG including a fixed KCl concentration (0.3 M) at 20 °C. Fig. 1(b) shows the distribution of L for different PEG concentrations. The unfolded DNA chain is stable at lower PEG concentration, 10 and 48 mg ml^{-1}, in which a characteristic long-axis length L is *ca.* 1.25 µm. On the other hand, the folded DNA chain is stable at higher PEG concentration, 91 and 100 mg ml^{-1}, in which the characteristic L is *ca.* 0.5 µm. The unfolded and folded DNA chains coexist at intermediate PEG concentrations of 55 and 73 mg ml^{-1}, *i.e.*, the coexistence region, in which bimodal distribution is clearly presented. Thus, it was observed that DNA exhibits the discrete transition between the unfolded state and the folded state at the level of a single molecule. Fig. 2 shows the dependence of the long-axis length L on the PEG concentration. The unfolded region, the coexistence region and the folded region are clearly shown.

Next we examined the effect of temperature. We chose a PEG concentration of 73 mg ml^{-1}, since in the coexistence region individual T4DNA molecules are sensitive to temperature change. Fig. 3(a) shows the distribution of L at different temperatures. The portion corresponding to the folded state decreases with increasing temperature, while that of the unfolded state increases. The temperature dependence found is in agreement with that found by us previously.[27,28] Fig. 3(b) represents schematic free energy profiles of individual DNA chains within the lower and higher temperature ranges. Thus, it was confirmed that the individual DNAs in PEG solution shows a suitable response to the temperature change, generating the rhythmic conformational change in the presence of a temperature gradient.

To evaluate the sensitivity of the folding/unfolding transition to temperature change, or temperature gradient, we evaluated the Gibbs energy difference ΔG from the folded state to the unfolded state with increasing temperature. By evaluating the ratio of the folded and unfolded states, we can estimate ΔG between both states: $\Delta G = -\ln(P_{\text{fold}}/P_{\text{unfold}})$, where P_{fold} and P_{unfold} correspond to the probabilities of existence of the folded and unfolded states, respectively. Fig. 4 shows that $\ln(P_{\text{fold}}/P_{\text{unfold}})$ changes linearly with $1/T$. Here, P_{fold} and P_{unfold} were obtained from the bimodal distribution in Fig. 3(a). From the slope of the straight line, the values of the thermodynamic parameters, the transition temperature T_c, the entropy change ΔS and the enthalpy

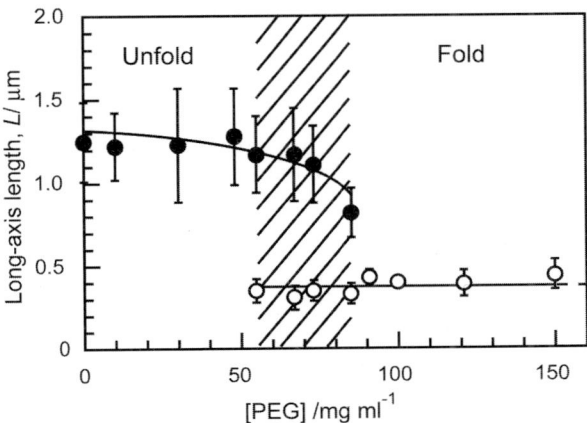

Fig. 2 Dependence of the long-axis length L of T4DNA molecules on the concentration of PEG at a fixed concentration of KCl 0.3 M. L changes in a discontinuous manner at the level of a single DNA chain in the coexistence region.

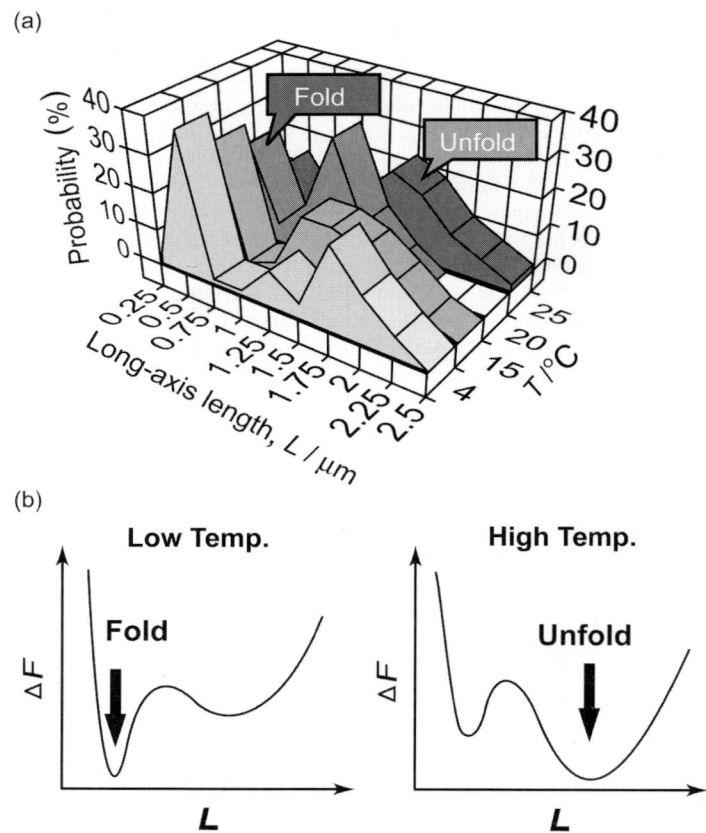

Fig. 3 (a) Distribution of the long-axis length L of T4DNA molecules in the coexistence region of PEG, 73 mg ml^{-1}, at different temperatures. (b) Schematic representations of the free energy difference ΔF vs. the long-axis length L in a single DNA chain at lower and higher temperatures. The folded state becomes less stable with increasing temperature, while the unfolded state becomes stable in the PEG solution.

change ΔH at T_c were estimated to be 272 K, 14.6 k_B and 5.5 × 10^{-20} J (T4DNA)$^{-1}$ respectively, where T_c is defined at $P_{fold} = P_{unfold}$ and k_B is Boltzmann's constant. Here, we defined that the Gibbs energy change depends on the entropy change within a narrow temperature range. The value of ΔS is less than that obtained previously ($\Delta S = 38\ k_B$).[28,39] It is expected that the folding/unfolding transition which shows mild temperature dependence will be preferable for investigation of the dynamical behaviour in the rhythmic conformational change of a single polymer chain from the viewpoint of kinetic rate.

Fig. 5(a) shows a schematic representation of the folding/unfolding transition in PEG solution containing a monovalent cation. We have shown that, in PEG solution, the folded DNA chain becomes stable at lower temperature, and less stable at higher temperature. Fig. 5(a) shows that the translational entropy of the counterions increases with increasing temperature and this is the driving force for transition from the folded state to the unfolded state. In spermidine solution, a trivalent polyamine, the opposite temperature effect occurs,[30] this means that the folded DNA chain is less stable at lower temperature, and more stable at higher temperature under the presence of multivalent cations. Fig. 5(b) shows a schematic representation of the folding transition in spermidine solution. Monovalent cations (native cations) and trivalent cations (added cations) act as counterion. At lower temperatures, the monovalent cations are the counterions, while the trivalent cations are the counterions at higher temperatures through ion-exchange. In this case, the entropy of a system consisting of a DNA chain and counterions increases with increasing temperature and the free energy of the system decreases. Thus, the temperature effect in the folding

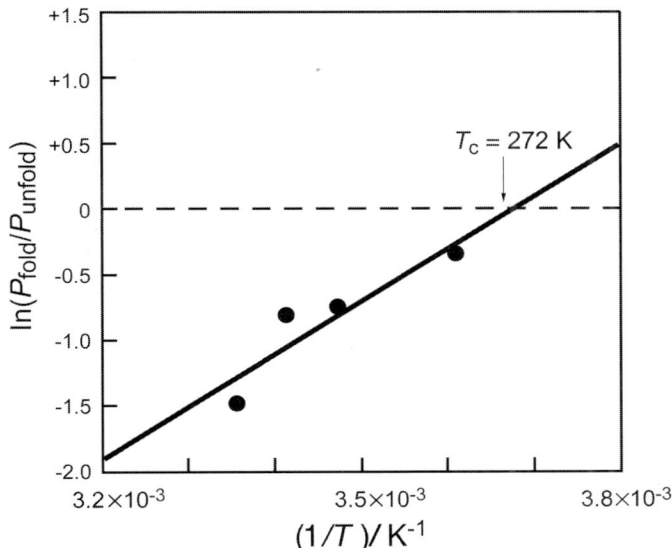

Fig. 4 Temperature dependence of the ratio of the probability of the folded compact state and the unfolded coil state in a plot of $\ln(P_{\text{fold}}/P_{\text{unfold}})$ vs. $1/T$ plot, where P_{fold} and P_{unfold} corresponds to the probabilities of occurrence of the folded compact state and the unfolded coil state, respectively. From the slope of this linear relationship and the transition temperature at $P_{\text{fold}} = P_{\text{unfold}}$, the thermodynamic parameters ΔS and ΔH are estimated to be 14.6 k_B and 5.5×10^{-20} J (T4DNA)$^{-1}$, at 272 K, respectively.

transition with different cations is accounted for by the translational entropy of the counterion although we omit detailed discussion based on the statistical mechanics.[27–30] Here, we emphasize that the opposite temperature effects in different solutions have not been explained by the previous theories.[40–43]

Based on the above experimental results, we tried to generate a rhythmic conformational change using a CW Nd : YAG laser beam. Fig. 6(a) shows a time trace of the fluorescence DNA image around the laser focus at a laser power of 700 mW, indicating the change in the fluorescence intensity in several seconds, although the fluorescence intensity of the image of the folded DNA chain is gradually decreased by fading in the time tracer. It should also be noted that the shape of each blurred swollen image is slender, or distorted, in contrast to that of the bright spots. In comparison with the fluorescence images of the DNA chain as shown in Fig. 1(a), it is clear that the bright spots correspond to the folded DNA chains at the laser focus, while the blurred images are the unfolding state around the focus. Thus, the rhythmic conformational change could be observed.

Here, it is helpful to consider the characteristic scale in this phenomenon in order to understand the rhythmic conformational change. The diameter of the focused laser beam is ca. 1 µm and the temperature rise is estimated to be at least $>10\,°C$.[23] The characteristic values of the long-axis length L in the folded and unfolded DNAs are 0.5 and 1.25 µm, respectively, as shown in Fig. 1(b) and 2. Furthermore, T_c is roughly evaluated to be 272 K, as shown in Fig. 4. In this situation, the folded DNA trapped by the focused laser beam is exposed to the local heating. This process corresponds to an external energy supply to the single DNA chain. Since the folded DNA chain becomes less stable in the area of local heating, it is unfolded. The unfolded DNA chain then enters the ambient "cool" temperature and becomes folded. This process is equal to energy dissipation from the single DNA chain to the external environment. The folded DNA chain is pulled back towards the trapping point. Thus, the rhythmic conformational change of a single DNA change is generated.

To confirm whether the local heating plays a significant role in the rhythmic phenomena, we performed the experiment using heavy water, D_2O, instead of H_2O, because it is well known that heavy water does not absorb the photon energy of a 1064 nm laser beam.[23] Fig. 6(b) shows the

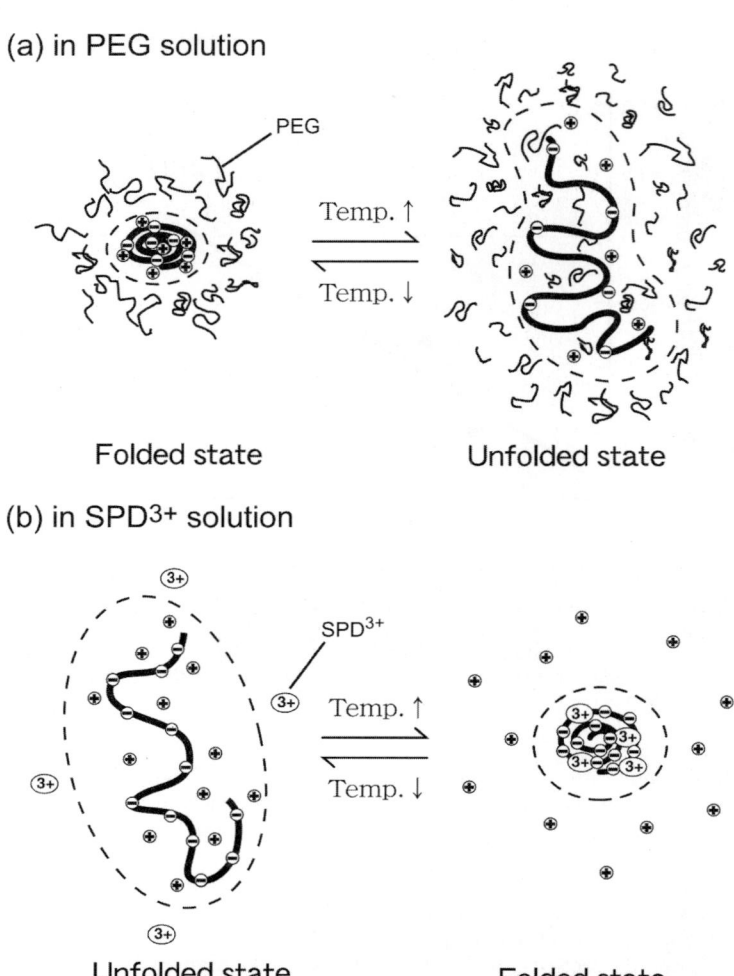

Fig. 5 Schematic representations of the folding/unfolding transition of individual DNAs driven by temperature change in PEG solution (a) and spermidine^{3+} solution (b).

time series of the fluorescence images in D_2O solution. It was found that the folded DNA chain is stationary at the laser focus and there is no rhythmic change. Thus, it was established that local heating by a focused CW Nd:YAG laser beam is an essential factor in causing the rhythmic change.

To discuss the dynamical behaviour of the rhythmic change in aqueous solution, we examined a more detailed time series. Fig. 7 exemplifies a time series of the fluorescence image with a time interval of 1/30 s. In the series, the fluorescence image of $\Delta t = 0.47$ s corresponds to that of $t = 10$ s in Fig. 6(a). The schematic representations of individual DNA images are also shown below their corresponding fluorescence images in Fig. 7. It is clearly observed that the shape and the fluorescence intensity are changed in the timescale of 1/30 s. Fig. 8 shows a time series of quasi-3-dimensional profiles of the fluorescence images of DNA corresponding to Fig. 7. The change in the distribution of the fluorescence intensity of the DNA images is clearly presented. In Fig. 7 and 8, the fluorescence images in $\Delta t = 0$–0.17 s are the bright spots, and their shapes and fluorescence intensity are changed in $\Delta t = 0.20$–0.77 s. In particular, the bright spot at $\Delta t = 0.33$ s is divided into two bright parts and elongated downward at $\Delta t = 0.37$–0.47 s. The elongated chain then collapses upward in $\Delta t = 0.50$–0.77 s. At $\Delta t = 0.80$ s, the image becomes a bright spot again.

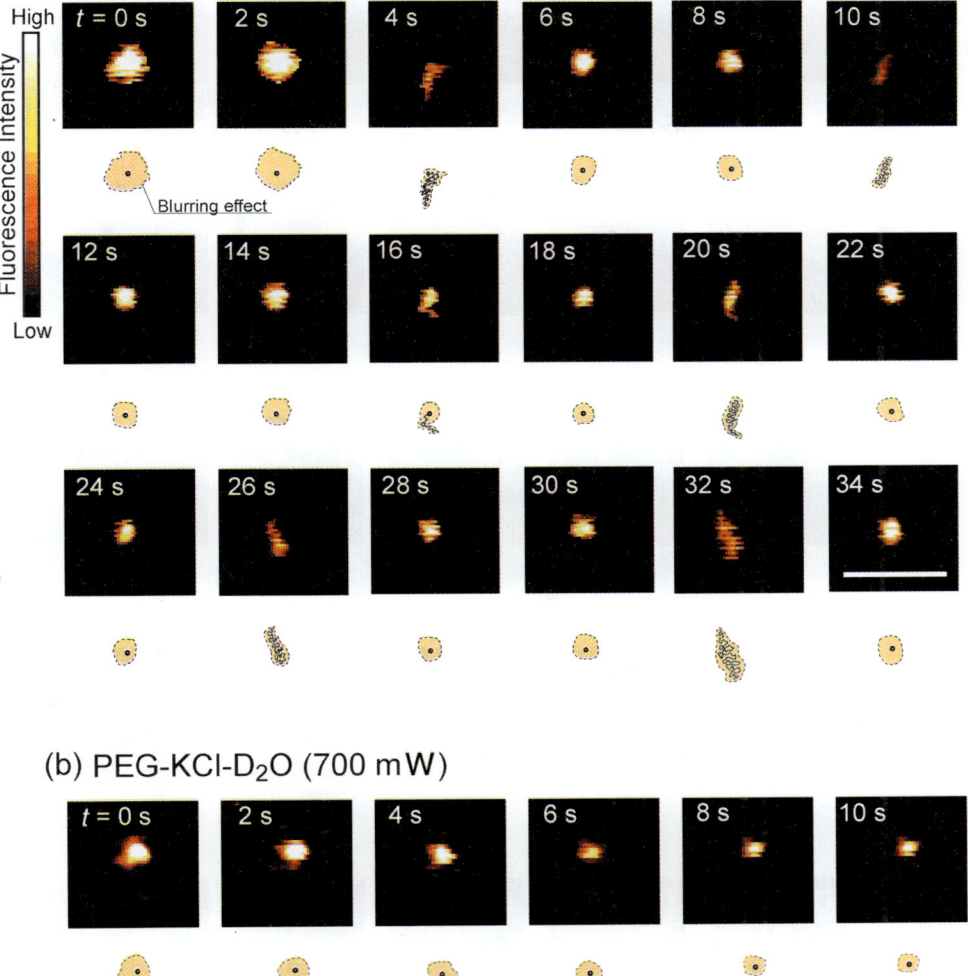

Fig. 6 Rhythmic folding/unfolding transition in a single T4DNA molecule. (a) Upper: time trace of fluorescence images of rhythmic conformational change of T4DNA (laser power: 700 mW, time interval: 2 s) in an aqueous solution. Lower: the schematic representations of corresponding fluorescence images. The scale bar represents 2 μm. (b) Upper: the time trace in D_2O (laser power: 700 mW, time interval: 2 s) in a D_2O solution under the same conditions as the aqueous solution. Lower: the schematic representations.

Through detailed investigation of the shape and distribution of the fluorescence intensity, the time series may be classified into several steps: (i) the folded state ($\Delta t = 0$–0.17 s), (ii) the unfolding process, or melting from the folded state to the unfolded state ($\Delta t = 0.20$–0.43 s), (iii) the unfolded state ($\Delta t = 0.47$–0.53 s), (iv) the nucleation and growth from the unfolded state to the folded state ($\Delta t = 0.57$–0.77 s) and (v) the folded state ($\Delta t = 0.80$ s). In Fig. 7 and 8, it is shown that melting, nucleation and growth, kinetics in a first order phase transition, occur during the rhythmic conformational change.[44] Thus, the dynamical behaviour has been clarified.

Fig. 9 shows a schematic representation of the dynamical behaviour of the self-oscillating DNA chain in a laser field. The axis of the optical tweezers tilts to the optical axis as shown. The laser beam passes downwards through the observed snapshots from the upper side to the lower one. In the optical tweezers, the trapping force appears around the laser focus and the scattering force

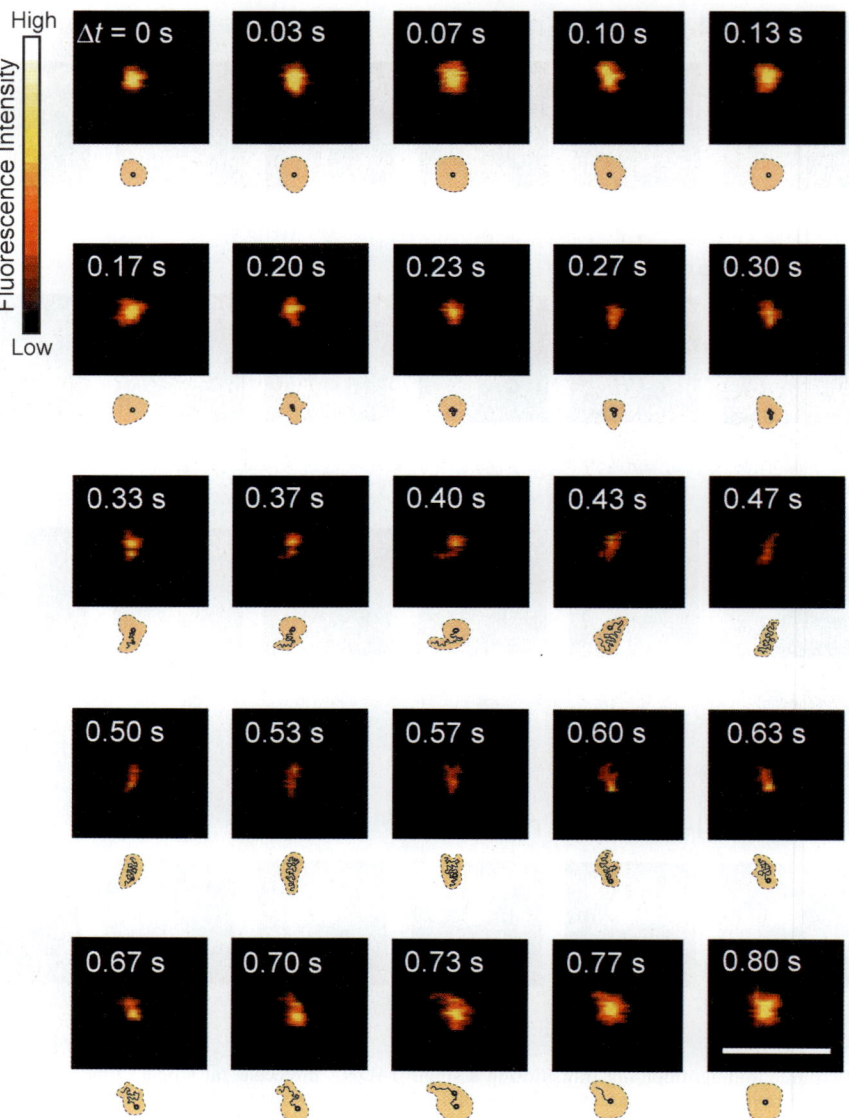

Fig. 7 Detailed time trace of the fluorescence images in the rhythmic folding/unfolding transition of a self-oscillating polymer chain (upper) and their schematic representations (lower). The image at $\Delta t = 0.47$ s corresponds to the image at $t = 10$ s in Fig. 6(a). The scale bar represents 2 µm.

exists along the laser beam. The folded DNA chain is trapped at the focus because the segment density is very high and the trapping force is rather stronger than the scattering force. On the other hand, the unfolded DNA chain is not trapped because the segment density is very low and the trapping force becomes relatively weaker in comparison with the scattering force. Therefore, the unfolded part in the folded DNA is scattered downwards. The projection images of the unfolded DNA chain elongated along the axis of the optical tweezers are then observed as the distorted images shown in Fig. 6 and 7. Furthermore, Fig. 9 is helpful in understanding the relationship between external energy supply and energy dissipation in the rhythmic conformational change. Fig. 9-1 and 9-2 show that the photon energy is supplied to the folded part in the

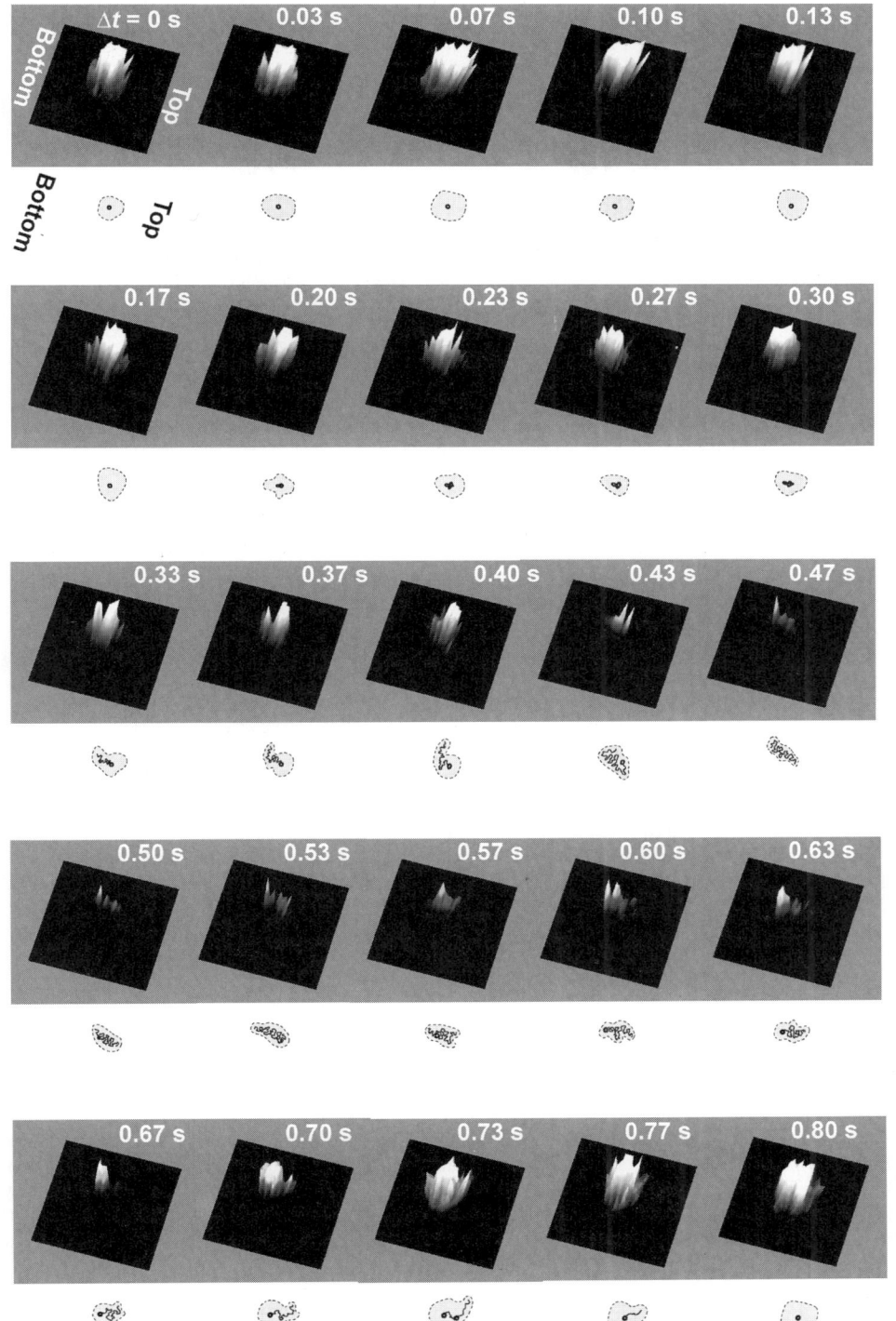

Fig. 8 Time trace of quasi-3-dimensional profiles of the fluorescence images of DNA corresponding to Fig. 7 and their schematic representations.

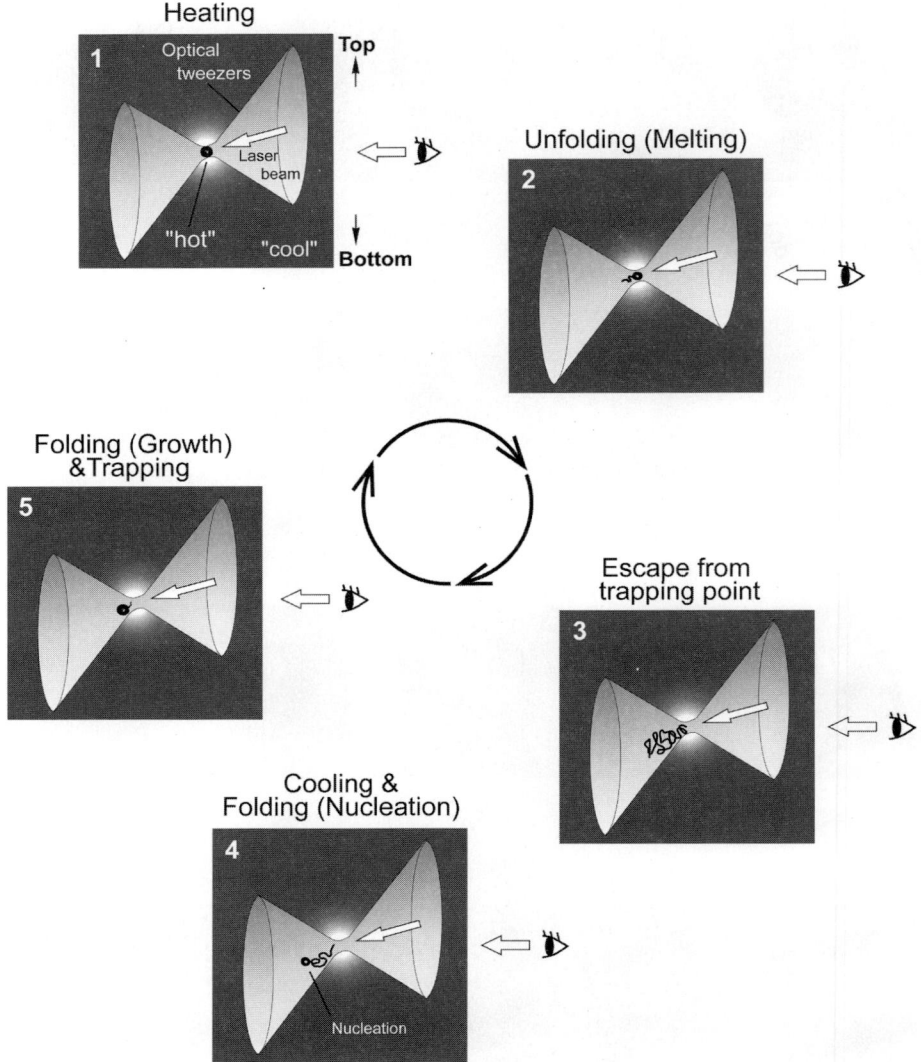

Fig. 9 Schematic representations of the rhythmic conformational change of a DNA molecule between the folded state and the unfolded state in a laser field. The cross angle between the laser beam and the optical axes is *ca.* 0.174 rad. The diameter of the laser focus is *ca.* 1 µm.

DNA chain through local heating. Fig. 9-3, 9-4 and 9-5 show that the energy is dissipated from the unfolded part in the DNA chain to the ambient environment. Thus, the limit-cycle oscillation of the single DNA chain is associated with the energy supply and dissipation under thermodynamically open conditions.

We estimated the size of the unfolded DNA chain in the rhythmic conformational change as its long-axis length, L. The unfolded DNA leans away from the optical axis, from the angle between the laser and optical axes, L of a fully elongated chain from Fig. 7 at $\Delta t = 0.57$ s is roughly estimated to be 4.8 µm. This value is larger than that of the folded DNA chain in Fig. 1 and is in agreement with that found in our previous experiments.[27,30–38] Since it is established that microphase separation is induced at temperatures of the order of several 10 °C higher than room temperature in the H_2O–PEG system,[45] it is expected that microphase separation would be induced in the local heating area, where the transition temperature of the phase separation depends on the concentration and the degree of polymerization of the PEG. Then the DNA chain would be incorporated in the H$_2$O domain since DNA is a highly negatively charged polymer chain. Further experiments on the folding/unfolding transition with a microphase separation should provide useful information on the L of the unfolded DNA chain in the present experiment. It should be noted that the characteristic L, as shown in Fig. 2 and 3, is measured in homogeneous solution.

Next, we tried to analyze the rhythmic conformational change. Fig. 10(a) shows the time traces of the fluorescence intensity of the trapped DNA chains at different laser powers (left) and their Fourier transformations (right). The appearance of a characteristic frequency is noted, and this frequency at the main peak increases with rising laser power. In the time traces, the light grayish area corresponds to the folded state with external photon-energy supply, while the dark grayish

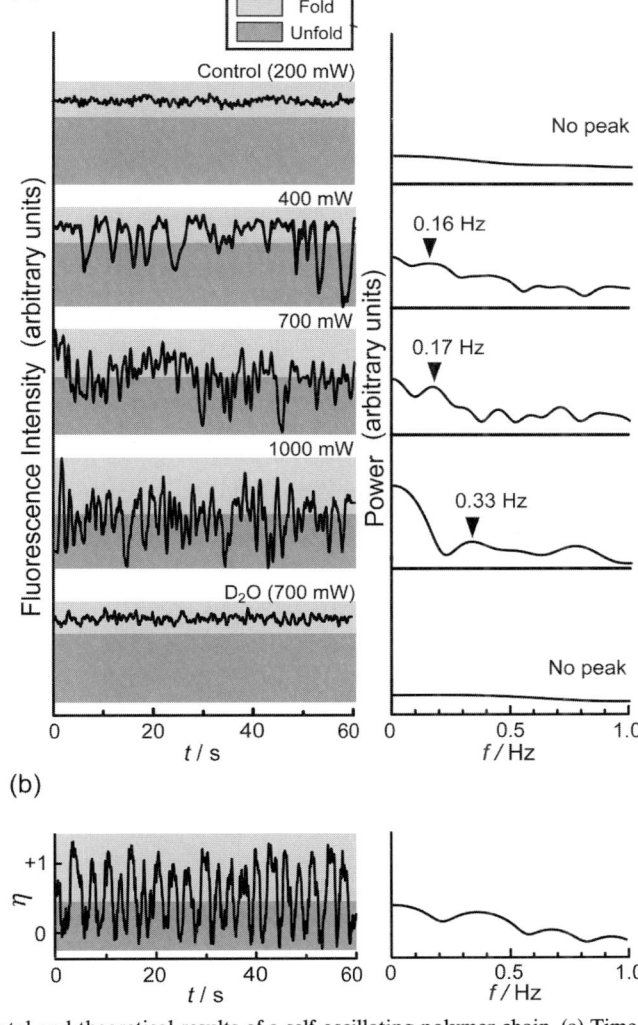

Fig. 10 Experimental and theoretical results of a self-oscillating polymer chain. (a) Time traces of the fluorescence intensity of the folding/unfolding DNA images at different laser powers (left), and their Fourier transformations (right), where f is frequency. Here, the controls are the time trace of the folded DNA chain at 200 mW (no oscillation) and its FT. (b) Results by numerical simulation based on eqns. (4) and (5), for $\eta(t=0) = 0.5$, $\tau(t=0) = -0.2$, $|\xi(t)|_{max} = 40$, $\varepsilon = 1$, $h = 2$, $\beta = 5$ and $\gamma = 2$.

area is an unfolded DNA chain with energy dissipation. Fig. 11 shows the laser power dependence of the characteristic frequency of the self-oscillating polymer chain. The frequency gradually increases with increasing laser power, but not proportionally to the laser power, and that at least a lower threshold of the laser power exists. We could not determine the higher threshold because of the durability of the fluorescence microscope. The laser power dependence of the characteristic frequency and the existence of a lower threshold show that the rhythmic conformational change in a single DNA chain is a limit-cycle oscillation. Based on these experimental findings, we have established that the rhythmic conformational change is a limit-cycle oscillation between the folded and unfolded states under thermodynamically open conditions. On the other hand, the appearance of the characteristic frequency was not observed in D_2O because of the absence of thermodynamically open conditions.

We describe briefly the mechanism in terms of the nonlinear dynamics of the rhythmic folding/ unfolding transition. For simplicity, we consider a double-minimum profile on free energy in respect of an order parameter,[46–48] where the order parameter is the segment density of a single polymer chain. By a symmetry argument in terms of the order parameter η, we interpret the first-order phase transition in a polymer chain without losing the essence of the first-order phase transition. The free energy profile F of a first-order phase transition is represented as[46–48]

$$F = a\eta^4 - b\eta^3 + c\eta^2 + \tau\eta \quad (1)$$

where η is the order parameter to change relative stable states between two stable points at 0 and +1, and the coefficient of each term is determined arbitrarily to describe a double-minimum free energy profile. We chose a condition of $a = 3$, $b = 6$ and $c = 3$. Here, τ is an environmental parameter in terms of the reduced temperature around a DNA molecule.

$$\tau = \alpha(T - T_c)/T_c \quad (2)$$

where α is a positive constant and T_c is the transition temperature of the folding/unfolding transition in individual DNA chains. The order parameter η accounts for the reduced segment density and is described by the radius of gyration, R, as follows.

$$\eta = (R^{-3} - R_{unfold}^{-3})/(R_{fold}^{-3} - R_{unfold}^{-3}) \quad (3)$$

where R_{fold} and R_{unfold} are the radii of gyration in the folded ($\eta \approx +1$) and unfolded ($\eta \approx 0$) states, respectively.[32] From the symmetry argument of the free energy profile, the essence of the temporal

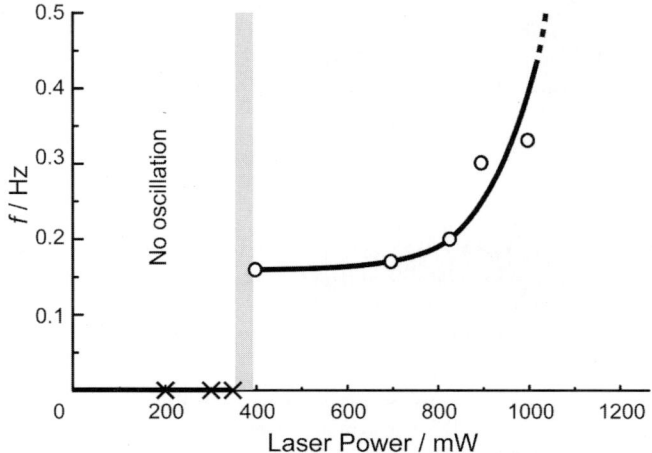

Fig. 11 Dependence of the frequency of the main peak on the laser power. The lower threshold of laser power is *ca.* 400 mW.

change in the order parameter is given as

$$\varepsilon \frac{d\eta}{dt} \cong -\frac{\partial F}{\partial \eta} = -12\eta^3 + 18\eta^2 - 6\eta - \tau + \xi(t) \quad (4)$$

where ε is a positive constant and $\xi(t)$ is white noise. In our experimental system, the DNA chain generates the rhythmic conformational change under large thermal fluctuations. Let us consider briefly the relationship of η and τ to describe the temporal change of τ. As shown in Fig. 9, the folded DNA chain trapped by the focused laser beam is exposed to a "hot" temperature and is unfolded. After that the unfolded chain is scattered by the laser beam and enters into the ambient "cool" temperature where it is folded. The folded chain is trapped at the laser focus again. This means that the temperature around the DNA chain is increased when the DNA chain is folded ($\eta \approx +1$), and decreased when it is unfolded ($\eta \approx 0$). Thus, based on the relationship of η and τ, the time dependence of temperature τ is given as

$$\frac{d\tau}{dt} = h - \beta(1-\eta)^\gamma \quad (5)$$

where h is the heating rate at the local heating centre (the laser focus), β is a positive constant and γ is an index to represent the profile of the heating rate. The right side in eqn. (5) represents the temperature gradient around the local heating area, or a profile of the degree of nonequilibrium, where the temperature gradient profile corresponds to the laser beam profile (TEM_{00}). Fig. 10(b) shows the time trace of the order parameter in the rhythmic conformational change and its Fourier transformation constructed by eqn. (4) and (5) for $\eta(t=0) = 0.5$, $\tau(t=0) = -0.2$, $|\xi(t)|_{max} = 40$, $\varepsilon = 1$, $h = 2$, $\beta = 5$ and $\gamma = 2$. Here, the magnitude of the order parameter corresponds to the fluorescence intensity of the DNA chain.[39] The time trace and its FT reproduce the characters of the experimental results very well. Thus, eqn. (4) and (5) describe the rhythmic conformational change in a single polymer chain under thermodynamically open conditions. In this condition the temporal change of the white noise is larger than that of the order parameter. Here, we emphasize that our experimental findings and our hypothesis point out that a polymer chain with a first-order phase transition generates a limit-cycle oscillation on its conformation under thermodynamically open conditions, even if the polymer chain is exposed to large thermal fluctuations.

Besides the oscillation of a single DNA molecule, we have found various kinds of rhythmic phenomena under steady irradiation with a laser beam (see Table 1). We have found that periodic growth and bursting are generated on a cluster of micrometre-sized plastic beads.[49] Firstly, the beads are pulled into the laser focus and a cluster is formed there by the attractive force acting between the beads. The cluster size gradually increases and reaches a critical size, where the attractive force is equal to the scattering force. The cluster is then smashed by the scattering force when it becomes larger than the attractive one, and the beads in the cluster are scattered along the laser beam. The cluster then reappears and grows again. Thus, this cycle is repeated. In the periodic growth and bursting of the cluster of sub-micrometre-sized beads, the competition of attractive

Table 1 Oscillatory phenomena under nonequilibrium field induced by a laser beam

System	Phenomenon	Period/s	Mechanism
Giant DNA molecule (ref. 39 and this work)	Rhythmic conformational change between the folded state and the unfolded state	3–6	First-order phase transition Temperature gradient
Micro-beads (ref. 49)	Periodic growth and bursting of cluster of beads	1–2	Attractive force Scattering force
Giant lipid tubule (ref. 50)	Rhythmic swinging motion	1–25	Bimodal trapping field Temperature gradient
Water droplet (ref. 51)	Periodic growth and disappearance of water droplet	1–3	Evaporation Condensation

and scattering forces plays a significant role. This cycle is described in terms of a limit-cycle oscillation with dissipation of sub-micrometre sized beads in a laser field.[49] We have also found that a giant lipid tubule shows a rhythmic swinging motion in a laser field having a bimodal trapping field, where the laser beam is split into two laser beams by adjustment of an optical system.[50] When an end of the tubule is trapped steadily at one stable point, the side of the tubule is trapped along one trapping field (one of the split laser beams). The tubule is thus exposed to asymmetric local heating by the trapping field. As a result, a surface tension gradient appears, causing tubule to tilt from one trapping field to another one in one direction (the other split laser beam). After tilting towards the other trapping field, the tube is affected by asymmetric heating and the tubule tilts to the initial trapping field in the opposite direction. Thus, the rhythmic swinging motion of a giant lipid tubule is generated in the bimodal trapping field, where the end is fixed throughout the process. In this case, the competition between the bimodal trapping field and local heating plays a significant role. This phenomenon is discussed in terms of a limit-cycle oscillation with dissipation of photon energy.[50] In a further experiment, the periodic growth and disappearance of a water droplet has been observed.[51] In the experiment, a large water source in a small closed space is heated by a focused laser beam. Liquid water evaporates gently and supersaturated vapor is generated. When a small particle of dust or a crack on the glass surface exists as a nucleus, vapor condenses to form a small droplet at the nucleus and the droplet grows. The growing droplet then disappears by absorption into the large water source on contact with it. However, since the air is still supersaturated, another small water droplet immediately appears at the nucleus. Thus, the growth and disappearance of the water droplet is repeated. In this case, the competition between evaporation and condensation at the nucleus play a significant role. This phenomenon is also discussed in terms of a limit-cycle with the dissipation of photon energy.[51] We have shown various oscillatory phenomena induced by laser irradiation. We emphasize that rhythmic phenomena in laser fields give a deep understanding of the generality of molecular systems under nonlinear nonequilibrium conditions.

In this article, we have shown novel rhythmic phenomena in a single polymer chain in a laser field and described our theory. The essence of the rhythmic phenomena is that the hysteresis of the first-order phase transition of a single polymer chain is concerned with the irreversibility on the path of the conformational change, or a breaking of the detailed balance in the statistical system.[20–22,39] The molecular machinery based on our hypothesis generates a unidirectional motion in nonequilibrium conditions, even under large thermal fluctuations. Since the temperature is equal to a chemical potential on free energy, the rhythmic change will also be generated in a chemical potential gradient under isothermal conditions. In relation to this, recently, it has become clear that vectoral motions of functional proteins are associated with their conformational changes under energy dissipation.[3,6,8,9,19] In muscle myosin, it has been suggested that an order-to-disorder transition induced by reaction with ATP in the myosin head, *i.e.*, a first-order phase transition, is essential for force generation and vectoral motion,[52–55] where myosin has two heads to attach to actin filaments, it moves along the filament in one direction using two heads with ATP hydrolysis. In the mobility cycle, it is expected that the order-to-disorder transition generates stroking of the two heads along the filament.[9] Even in kinesin, it is expected that a similar conformational change strategy is used.[9] Here, it should be noted that ATP hydrolysis and the stroking of two heads along the filament correspond to an energy dissipation structure and a rhythmic conformational change, respectively. Thus, the close relationship between the order-to-disorder transition and vectoral motion in molecular machinery would be, eventually, explained by our theory. It is rewarding to discuss the vectoral motions of functional proteins in terms of a limit-cycle oscillation with energy dissipation. Since a highly charged polymer chain is also expected to undergo a discrete conformational transition,[29,56,57] it would generate rhythmic conformational change under thermodynamically open conditions. The experimental and theoretical findings in this article give a deeper understanding of the mechanism of molecular machinery in living matter.

Conclusion

We have experimentally examined the rhythmic conformational change in a single polymer chain under the irradiation of a focused CW Nd : YAG laser beam. In a solution of PEG–KCl–H_2O, the rhythmic change of the conformation of a single DNA chain between the folded and the unfolded states is generated, the characteristic frequency depending on the laser power. Furthermore, we have investigated the dynamical behaviour of the rhythmic conformational change, *i.e.*, the unfolding process from the folded to unfolded states, and the process of nucleation and growth from the unfolded to folded states. In contrast, in a solution of PEG–KCl–D_2O, the rhythmic change was absent. This shows that the local heating induced by a focused laser beam plays a significant role in causing the rhythmic change. The rhythmic conformational change in a single polymer chain is discussed in terms of the limit-cycle oscillation under thermal fluctuations.

Acknowledgement

We thank Dr N. Magome, Graduate School of Environmental Studies, Nagoya University, and Mr S. M. Nomura, Graduate School of Science, Department of Physics, Kyoto University for useful advice on the experiments.

References

1 E. Schrödinger, *What is Life?*, Cambridge University Press, Cambridge, 1944, p. 71.
2 G. Nicolis and I. Prigogine, *Self-Organization in Nonequilibrium Systems*, Wiley, New York, 1977, p. 10.
3 M. W. Walker, S. A. Burgess, J. R. Sellers, F. Wang, J. A. Hammer, J. Trinick and P. J. Knight, *Nature (London)*, 2000, **405**, 804.
4 K. Svoboda, C. F. Schmidt, B. J. Schnapp and S. M. Block, *Nature (London)*, 1993, **365**, 721.
5 C. Shingyoji, H. Higuchi, M. Yoshimura, E. Katayama and T. Yanagida, *Nature (London)*, 1998, **393**, 711.
6 T. Tani and S. Kaminuma, *Biophys. J.*, 1999, **77**, 1518.
7 Y. Sambongi, Y. Iko, M. Tanabe, H. Omote, A. Iwamoto-Kihara, I. Ueda, T. Yanagida, Y. Wada and M. Futai, *Science*, 1999, **286**, 1722.
8 R. Yasuda, H. Noji, M. Yoshida, K. Kinosita, Jr. and H. Itoh, *Nature (London)*, 2001, **410**, 898.
9 R. D. Vale and R. A. Milligan, *Science*, 2000, **288**, 88.
10 R. P. Feynman, R. B. Leighton and M. Sands, *The Feynman Lectures in Physics*, Addison-Wesley, Reading, MA, 1963, vol. 1, ch. 46.
11 M. O. Magnasco, *Phys. Rev. Lett.*, 1993, **71**, 1477.
12 R. Bartussek, P. Hänggi and J. G. Kissner, *Europhys. Lett.*, 1994, **28**, 459.
13 R. Dean Astumian and M. Bier, *Phys. Rev. Lett.*, 1994, **72**, 1766.
14 J. F. Chauwin, A. Ajdari and J. Prost, *Europhys. Lett.*, 1994, **27**, 421.
15 H. X. Zhou and Y. Chen, *Phys. Rev. Lett.*, 1996, **77**, 194.
16 F. Jülicher, A. Ajdari and J. Prost, *Rev. Mod. Phys.*, 1997, **69**, 1269.
17 J. Rousselet, L. Salome, A. Ajdari and J. Prost, *Nature (London)*, 1994, **370**, 447.
18 L. P. Faucheux, L. S. Bourdieu, P. D. Kaplan and A. J. Libechaber, *Phys. Rev. Lett.*, 1995, **74**, 1504.
19 S. P. Tsunoda, A. J. Rodgers, R. Aggeler, M. C. J. Wilce, M. Yoshida and R. A. Capaldi, *Proc. Nat. Acad. Sci.*, 2001, **98**, 6560.
20 K. Yoshikawa and H. Noguchi, *Chem. Phys. Lett.*, 1999, **303**, 10.
21 D. D. Fitts, *Nonequilibrium Thermodynamics*, McGraw-Hill, New York, 1962, p. 135.
22 R. L. Stratonovich, *Nonlinear Nonequilibrium Thermodynamics I*, Sprinter-Verlag, Berlin, 1992, p. 72.
23 M. Ishikawa, H. Misawa, N. Kitamura, R. Fujisawa and H. Masuhara, *Bull. Chem. Soc. Jpn.*, 1996, **69**, 59.
24 Y. Matsuzawa, K. Hirano, K. Mori, S. Katsura, K. Yoshikawa and A. Mizuno, *J. Am. Chem. Soc.*, 1999, **121**, 11581.
25 D. T. Chiu and R. N. Zare, *J. Am. Chem. Soc.*, 1996, **118**, 6512.
26 S. M. Nomura, Y. Yoshikawa, K. Yoshikawa, O. Dannenmuller, S. Chasserot-Golaz, G. Ourisson and Y. Nakatani, *CHEMBIOCHEM*, 2001, no. 6, 457.
27 H. Mayama, T. Iwataki and K. Yoshikawa, *Chem. Phys. Lett.*, 2000, **318**, 113.
28 H. Mayama and K. Yoshikawa, *Macromol. Symp.*, 2000, **160**, 55.
29 V. V. Vasilevskaya, A. R. Khokhlov, Y. Matsuzawa and K. Yoshikawa, *J. Chem. Phys.*, 1995, **102**, 6595.
30 H. Murayama and K. Yoshikawa, *J. Phys. Chem. B*, 1999, **103**, 10517.
31 K. Yoshikawa, S. Kidoaki, M. Takahashi, V. V. Vasilevskaya and A. R. Khokhlov, *Ber. Bunsen-Ges. Phys. Chem.*, 1996, **100**, 876.
32 Y. Yamasaki and K. Yoshikawa, *J. Am. Chem. Soc.*, 1997, **119**, 10573.
33 M. Ueda and K. Yoshikawa, *Phys. Rev. Lett.*, 1996, **77**, 2133.
34 S. M. Mel'nikov and K. Yoshikawa, *Biochem. Biophys. Res. Commun.*, 1997, **230**, 514.
35 S. M. Mel'nikov, V. G. Sergeyev, K. Yoshikawa, H. Takahashi and I. Hatta, *J. Chem. Phys.*, 1997, **107**, 6917.
36 Y. Ichiba and K. Yoshikawa, *Biochem. Biophys. Res. Commun.*, 1998, **242**, 441.
37 N. Makita and K. Yoshikawa, *FEBS Lett.*, 1999, **460**, 333.

38 K. Tsumoto and K. Yoshikawa, *Biophys. Chem.*, 1999, **82**, 1.
39 H. Mayama, S. M. Nomura, H. Oana and K. Yoshikawa, *Chem. Phys. Lett.*, 2000, **330**, 361.
40 C. B. Post and B. H. Zimm, *Biopolymers*, 1986, **18**, 1487.
41 A. Yu. Grosberg, I. Ya. Erukhimovitch and E. I. Shakhnovitch, *Biopolymers*, 1982, **21**, 2413.
42 H. L. Frisch and S. Fesciyan, *J. Polym. Sci. Polym. Lett.*, 1979, **17**, 309.
43 I. M. Lifshitz, A. Yu. Grosberg and A. R. Khokhlov, *Rev. Mod. Phys.*, 1978, **50**, 683.
44 Y. Matsuzawa and K. Yoshikawa, *J. Am. Chem. Soc.*, 1996, **118**, 929.
45 G. N. Malcolm and J. S. Rowlinson, *Trans. Faraday Soc.*, 1957, **53**, 921.
46 L. D. Landau and E. M. Lifshitz, *Statistical Physics*, Pergamon, New York, 3rd edn., 1996, Part 1, p. 451.
47 P. M. Chaikin and T. C. Lubensky, *Principles of Condensed Matter Physics*, Cambridge University Press, London, 1st edn., 1995, p. 173.
48 C. Kittel and H. Kroemer, *Thermal Physics*, Freeman, New York, 2nd edn., 1982, p. 302.
49 N. Magome, H. Kitahata, M. Ichikawa, S. M. Nomura and K. Yoshikawa, *Phys. Rev. Lett.*, submitted.
50 S. M. Nomura and K. Yoshikawa, *Phys. Rev. Lett.*, submitted.
51 M. Ichikawa and N. Magome, *Meeting Abs. Phys. Soc. Jpn.*, 2000, **55**No. 2-2, 243.
52 I. Brust-Mascher, L. E. W. LaConte, J. E. Baker and D. D. Thomas, *Biochemistry*, 1999, **38**, 12607.
53 D. D. Thomas, S. Ramachandran, O. Rpppnarine, D. W. Hayden and E. M. Ostap, *Biophys. J.*, 1995, **68**, 135s.
54 S. Y. Bershitsky, A. K. Tsaturyan, O. N. Bershitskaya, G. I. Mashanov, P. Burns and M. A. Ferenczi, *Nature (London)*, 1997, **388**, 186.
55 M. Walker, X. Jinang Zhang, J. Trinick and H. White, *Proc. Natl. Acad. Sci. USA*, 1999, **96**, 465.
56 S. Takagi, K. Tsumoto and K. Yoshikawa, *J. Chem. Phys.*, 2001, **114**, 6942.
57 H. Noguchi and K. Yoshikawa, *Chem. Phys. Lett.*, 1997, **278**, 184.

General Discussion

Prof. Harrison opened the discussion of Prof. Nicolis' paper: I wish to question the birth date of 1960s given by Prof. Nicolis for nonlinear kinetics. W. H. Mills suggested in 1932 that spontaneous optical resolution would occur in a system with stereospecific autocatalysis of formation of a chiral product, if the catalysis went with a power of catalyst concentration greater than 1. This is not a well-controlled pattern-forming process; but the idea clearly and very simply illustrates the power of nonlinear dynamics.

When I wrote in a manuscript that reaction–diffusion theory of pattern formation started with Turing's 1952 paper, I was taken to task by a referee for failing to mention the suggestion of Rashevsky.[1]

Prof. Nicolis mentioned the Lotka–Volterra equations. I thought they arose out of an attempt to account for the failure of the anchovy population in the Adriatic to increase when fishing was interrupted in the First World War, and that the equations were well established by the 1930s.

1 N. Rashevsky, *Bull. Math. Biophys.*, 1940, **2**, 15.

Prof. Nicolis responded: Prof. Harrison is perfectly correct in the specific examples he provides. Actually, one can find even earlier examples of complex behavior in chemical kinetics: Fechner's electrochemical observations around 1839, Ostwald's discovery of periodic release of H_2 in the reaction of Cr in concentrated acid solutions around 1899, the Bray–Liebhafsky extensive work on catalytic decomposition of H_2O_2 in the 1920's among others.[1] These were, however, isolated attempts. Nonlinear kinetics as a branch of chemistry in its own right arose when the scientific community realized that results that were being derived independently at the level of experimental observations, irreversible thermodynamics, mathematical modeling and dynamical systems theory could all be tied up together in a well-defined body of knowledge. This provided then a unifying framework that made further progress possible at a very fast pace. This view is further supported by the "sociological" analysis of the field carried out some time ago by Burger and Bujdoso.[2]

1 See, *e.g.*, A. Pacault and J.-J. Perraud, *Rythmes et formes en chimie*, P.U.F., Paris, 1997.
2 M. Burger and E. Bujdoso, in *Oscillations and Traveling Waves in Chemical Systems*, Wiley, New York, 1985.

Prof. Harrison commented: Yes, I agree with Prof. Nicolis that very little notice was being taken of Turing's 1952 paper by the early 1960s. I show herewith a plot of number of citations per year of that paper from 1961 to 1999 (Fig. 1). It shows a striking increase from 1–2 citations per year to about 100. The increase, though noisy, is essentially continuous and not stepwise. There could be various opinions on whether the notice that has been taken over the years of this and other reaction–diffusion papers shows a phenomenon for which "birthday" is a good analogy. I am inclined to think that the date of publication is the birthday. But surely the kineticist must be intrigued as to what was going on in the "induction period" of 9 years between that date and the start of my graph.

Prof. Westerhoff commented: After this issue of the history of the field, I should like to return to what is important for its future. With the specific example of Turing patterns and developmental biology in mind, I think that the field should elaborate on nonlinear dynamics and self organization

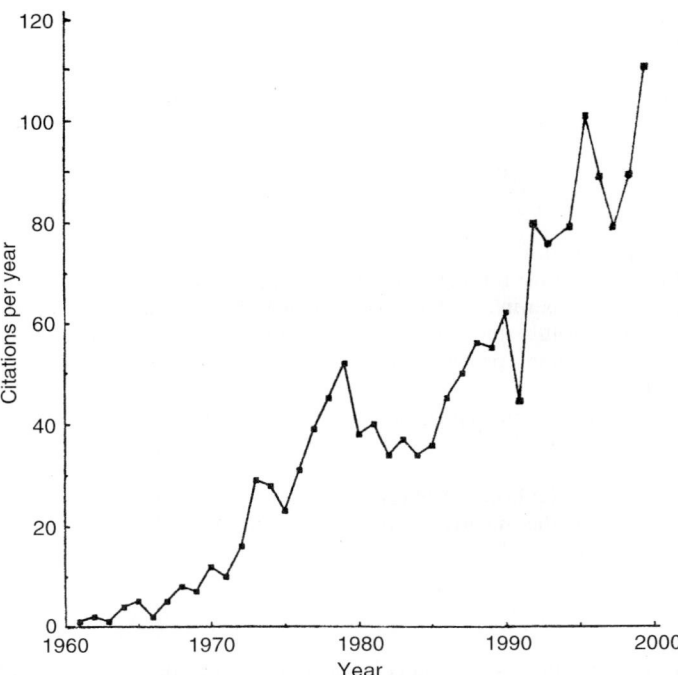

Fig. 1 Citations per year of A. M. Turing, *Philos. Trans. R. Soc. London, Ser. B*, 1952, **237**, 37.

in the context of a self perpetuating, pre-existing heterogeneous environment with partial pre-specification. Would you agree that this is the way to go, rather than to focus on simple nonspecific models focusing on principles only?

Prof. Nicolis replied: Simple, generic models have still a great deal to say when the principles and mechanisms underlying a hitherto unexplored system need to be unraveled. They also help one to identify some key questions that could be hidden in the details of a full-fledged model. However, I fully agree with you that future work in nonlinear kinetics as applied to biology should incorporate much more explicitly than in the past the characteristics of the environment, including its heterogeneous character (*cf.* also Section 4 and 5 of my paper).

Prof. Epstein communicated: The results described on the adsorption of diatomic molecules to surfaces have antecedents in work by Flory[1] on the reactions of side chains of bifunctional linear polymers and of Epstein[2] on the binding of multi-site ligands, *e.g.*, proteins, to one-dimensional chains such as nucleotides. Flory found analytical results for the numbers of reacted and unreacted sites at equilibrium, while Epstein treated the kinetics, obtaining double exponential solutions analogous to those presented by Prof. Nicolis in the one-dimensional surface adsorption problem.

1 P. J. Flory, *J. Am. Chem. Soc.*, 1939, **61**, 1518.
2 I. R. Epstein, *Biopolymers*, 1979, **18**, 765.

Prof. Nicolis replied: I appreciate this interesting comment. There are, indeed, analogies between the binding of ligands to a one-dimensional chain and the problem of cooperative desorption considered in Section 7 of my paper. It would be interesting to look at the analogy more closely and compare, in the two cases, the sources of cooperativity or the statistics of higher order clusters of reacted and unreacted sites.

Prof. Hanke opened the discussion of Prof. Orbán's paper: What is meant by anomalous dispersion relation of BZ under certain conditions?
Please add citation of first paper published according to ref. 1.

1 W. Hanke, *Int. J. Bifurcation Chaos*, 1999, **9**(10), 2099.

Prof. Orbán responded: In my short talk (but not in our paper) reference was given to the work of Steinbock et al.,[1] who reported on observing "anomalous dispersion relation" in the BrO_3^-–CHD–ferroin system. For further details, please read the cited papers or contact their authors.

1 N. Manz, S. C. Müller and O. Steinbock, *J. Phys. Chem. A*, 2000, **104**, 5895; C. T. Hamik, N. Manz and O. Steinbock, *J. Phys. Chem. A*, 2001, **105**, 6144.

Prof. Noszticzius asked: What is the bromide source in the suggested mechanism? There is one bromide source in Table 1 of the paper namely reaction (M5):

$$Br_2 + acetone \rightarrow Br-Ac + Br^- \tag{M5}$$

but the Br_2 necessary for this reaction is produced *e.g.* in reaction (M4)

$$Br(\text{I}) + Br^- \rightarrow Br_2 \tag{M4}$$

which process is a bromide sink. Thus there is no net bromide production in the combined process of reactions (M4) + (M5). The situation is similar with reaction (M7):

$$Ru(\text{III}) + Br^- \rightarrow Ru(\text{II}) + Br_2. \tag{M7}$$

Is it possible that the active $H_2PO_2^-$ can react with Br(I) in an oxygen atom transfer reaction to give bromide or it is the reduction of Ru(III) by $H_2PO_2^-$ which is the important initial step in the generation of bromide?

Prof. Orbán replied: The source of Br^- ions in the oscillatory mechanism is included in the reduction of BrO_3^- by $H_2PO_2^-$ [see eqn. (1) in the paper] through catalysis by Mn(II). BrO_3^- first reacts with Mn(II) in process (M1) forming Br(I) and Mn(III). The products of reaction (M1) are then reduced by the active form of $H_2PO_2^-$, restoring Mn(II) in eqn. (M3) and generating Br^- ions according to the step:

$$Br(\text{I}) + H_2PO_2^- \text{ (active)} + H_2O \rightarrow Br^- + H_3PO_3 + H^+$$

The second member of the dual catalyst, Ru(II) is also oxidized by BrO_3^- and Br(I) formed this way can also be reduced to Br^- with $H_2PO_2^-$. In the catalyst pair the concentration of Ru(II) is two order of magnitude lower than the concentration of Mn(II) therefore, participation of the Ru(III)–Ru(II) couple in the production of Br^- may not be significant. The reduction of Ru(III) by $H_2PO_2^-$ was proved to be slow compared to the reaction between Ru(III) and Br^-. Process (M7) is supposed to play an important role in the Br^- consumption part of the oscillatory cycle.

Prof. Showalter commented: The redox chemistries of Ce^{3+} and ferroin are affected by the high concentration of SO_4^{2-} in the classical BZ reaction. Is it possible that these catalysts are not effective in the bromate–hypophosphite–acetone catalyst system because SO_4^{2-} is not present in this system in such high concentration?

Prof. Orbán responded: In our experiments the concentration of H_2SO_4 varied between 1.0 and 2.5 M which is the regular range for $[H_2SO_4]$ in the classical BZ reaction. The effectiveness of Mn(II) and the ineffectiveness of Ce(IV) or ferroin as catalyst in the BrO_3^-–$H_2PO_2^-$–acetone system may be attributed to the differences in the rate of reduction of their oxidized form by $H_2PO_2^-$. The reaction between Mn(III) and $H_2PO_2^-$ is much faster than the reaction between Ce(IV) or ferriin and $H_2PO_2^-$. On the other hand complex formation between Mn(III) and hypophoshite may also be a factor which makes the Mn catalyst effective in the oscillatory oxidation of $H_2PO_2^-$ by BrO_3^-.

Prof. Noszticzius commented: I agree with Prof. Showalter that H_2SO_4 and somewhat surprisingly the sulfate ion concentration itself can be an important parameter of BZ systems. As a striking example I want to mention an unpublished result of Prof. H. D. Försterling. In the classical BZ reaction (bromate, malonic acid, cerium ion) no oscillations can be observed in perchloric acid medium in the absence of sulfate ions. Reported oscillations in perchloric acid medium appear only if the catalyst is cerium sulfate.

Prof. Orbán replied: This phenomenon may be connected with the stability of Ce^{4+} ions in different strong acids. The Ce^{4+} ions are most stable in H_2SO_4 where they form a sulfato complex. In $HClO_4$ and HNO_3 the Ce^{4+} ions undergo photoreduction (H_2O is oxidized to O_2), in HCl the Ce^{4+} ions oxidise Cl^- to Cl_2.

The redox potential of the Ce^{4+}/Ce^{3+} couple spans from 1.2 to 1.7 V in different acids due to the complex formation between cerium ions and the anion of the acids. The highest value of redox potential was measured in $HClO_4$ (1.70 V), medium in H_2SO_4 (1.44 V) and the lowest in HCl (1.28 V).

Dr Schreiber asked: Is the bromate–hypophosphite–acetone-dual catalyst reaction capable of multistable steady states and in particular, would a prospective experimentalist be able to observe front waves in addition to pulse waves?

Prof. Orbán answered: One of the most attractive features of the bromate–hypophosphite–acetone-dual catalyst system is that it exhibits long lasting oscillations in batch. All of our experiments have been carried out under batch condition. Multistable steady states can only be observed and maintained in a CSTR. This reaction, like some other bromate oscillators may be capable of showing bistability but we have not tested the system for this purpose. Once the existence of bistability will be established we may look for front waves.

Prof. Scott commented: You mention the presence of an air gap in the apparatus. Would it be possible to run this reaction in a completely closed reactor with no air gap?

Prof. Orbán responded: Yes, it would be possible if an air gap-free CSTR were filled and fed with the oscillatory mixture, and then the flow were stopped.

Unfortunately we have not carried out such an experiment. We plan certainly to run this reaction in a completely filled and closed reactor in order to gain information about the origin of stirring rate sensitivity of the oscillations.

Prof. Scott said: Oxygen is known to affect the response of both well-stirred BZ systems and wave propagation when using a ferroin catalyst.[1] Does oxygen have a similar influence on the spatial structure reported in the bromate–hypophosphite system and, if so, what do you think might be the mechanism for this interaction?

1   A. F. Taylor, B. R. Johnson and S. K. Scott, *Phys. Chem. Chem. Phys.*, 1991, **1**, 807.

Prof. Orbán replied: The ferroin alone does not induce oscillations in the bromate–hypophosphite–acetone system. However, it does if used in combination with Mn(II) catalyst. I have to admit no systematic efforts have been made so far to keep check on the effect of oxygen in the system where ferroin was used as second catalyst. Both batch oscillations and wave evolution were studied in the presence of air and no noticeable inhibitory effect was observed.

Prof. Dewel asked: Do you think that the new chemical system you have presented could be adapted to produce stationary patterns of the Turing type?

Prof. Orbán responded: The new system suggested by us to use for studying pattern formation belongs to the family of bromate oscillators where the oscillations and nonlinearity come from the chemistry of bromine. The diffusion coefficients of bromine species do not differ much, therefore one of the required conditions for Turing structures to form—significant difference in diffusion coefficient of the activator and the inhibitor—is not fulfilled in a bromate oscillator. Efforts in order to produce Turing structures using bromate oscillators have been unsuccessful so far. However, quite recently Vanag and Epstein[1] were able to observe Turing structures when the reagents of the classical BZ reaction were dispersed in water droplets of a reverse AOT microemulsion. Under such conditions the new system—in principle—may also be a candidate for producing stationary patterns.

1   V. K. Vanag and I. R. Epstein, *Phys. Rev. Lett.*, 2001, **87**, 228 301.

Prof. Menzinger asked: With reference to the preceding question about the suitability of the new reaction system for observing Turing structures: do you have any idea which species is the activator (which would have to be immobilized)?

Prof. Orbán replied: In order to answer this question I have to refer to Prof. Epstein's lecture presented on the ESF "Reactor" Workshop held in Leeds, September 7–9, 2001. He reported on observing Turing structures in a BZ mixture when it was run in an AOT microemulsion. In this medium some species (*e.g.* BrO_2) diffuse slower by 2–3 orders of magnitude in the dispersed phase than in the oil, and such a difference in diffusion coefficients made it possible to observe Turing structures in the BZ reaction. The oscillations in the new reaction are based on similar chemistry as in the BZ, therefore in order to make this reaction suitable for observing Turing structures, the diffusion of the species such as in BZ, should be "slowed down".

Prof. Gáspár commented: Before trying to point out the activator and inhibitor species in this system, I think it is important to establish and understand the mechanism of the system. Then we can decide if the system is capable of supporting Turing-type pattern formation besides travelling waves. On the other hand, we can not exclude this possibility right away. I would like to remind you of the BZ–AOT system reported by Epstein *et al.* at the ESF "Reactor" Workshop held in Leeds, September 7–9, 2001. In that system the authors observed wave-like behaviour and the formation of Turing patterns as well depending on the experimental conditions.

Prof. Epstein commented: We have recently shown[1] that it is possible to obtain stationary (Turing) patterns in the Belousov–Zhabotinsky reaction when it is run in a reverse microemulsion consisting of water, octane and the surfactant AOT. It is quite likely that similar patterns can be generated in the system described by Prof. Orbán if it is run in such a microemulsion.

1 V. K. Vanag and I. R. Epstein, *Phys. Rev. Lett.*, 2001, **87**, 228 301.

Prof. Gáspár said: You show in your paper that the Mn(II)–Ru(bpy)$_3$SO$_4$ dual catalyst system is sensitive to light. What do you think of the effect of light? Could it be that it can be explained by the same effect as that of the Ru(bpy)$_3^{2+}$ catalyst in the BZ system producing a Br$^-$ ion which then acts as an inhibitor?

Prof Orbán responded: My answer is yes. In the BrO$_3^-$–H$_2$PO$_2^-$–acetone–Mn(II)/Ru(bpy)$_3$SO$_4$ oscillator the only photosensitive species is the Ru-complex. Since the activator and inhibitor in this system are supposed to be the same as in BZ, the explanation of the light sensitivity must be similar in both cases.

Prof. Scott commented: Tóth *et al.* report that relatively intense illumination of ferroin-catalysed BZ systems can initiate wave behaviour through a positive stimulus on the reaction mechanism.[1] It would be interesting to examine this effect for the dual catalyst systems described here where inhibitory effects can also be seen.

1 R. Tóth, V. Gáspár, A. Belmonte, M. C. O'Connell, A. F. Taylor and S. K. Scott, *Phys. Chem. Chem. Phys.*, 2000, **2**, 413.

Prof. Orbán replied: The effect of light on the behavior of the BrO$_3^-$–H$_2$PO$_2^-$–acetone system in the presence of the Mn(II)–ferroin catalyst pair has not been tested so far, but we plan to examine and to compare the light sensitivity of the Mn(II)–Ru(bpy)$_3$SO$_4$ and Mn(II)–ferroin dual catalyst containing oscillators.

Prof. Noszticzius commented: It is possible that a part of the observed change in the amplitude of the Pt electrode potential oscillations is due to the light sensitivity of the Pt electrode. As the amplitude of the unperturbed oscillations can be very high (more than 300 mV, see Fig. 1 of the paper) it is possible that the potential determining redox pair is not the same at high and at low potentials and light can affect mostly one of them. In this respect it is interesting to observe that light seems to "cut down" the low potential part of the oscillations (Fig. 2 of the paper).

Prof. Orbán responded: We didn't notice a change in the response of the Pt electrode in a dilute H$_2$SO$_4$ solution when the system was illuminated. The effect shown in Fig. 2 of the paper should be attributed to the effect of light on the oscillatory chemistry. Waves are also affected by light where no Pt electrode is present. At this point we cannot suggest a detailed mechanism for the BrO$_3^-$–H$_2$PO$_2^-$–acetone–Mn(II)Ru(bpy)$_3$SO$_4$ oscillator, which is needed to answer your question about the potential determining redox couple in high and low potential states.

Prof. Sørensen opened the discussion of Prof. Noszticzius' paper: Almost ten years ago, complex oscillations in a closed BZ system were observed for the first time by Dr Jichang Wang in Copenhagen (Fig. 2).[1] Following a suggestion by Prof. Noyes at the Faraday Discussion meeting in 1974 that BrMA is the essential species for complex behaviour we modelled successfully this phenomenon by extending the simple Oregonator model with a variable representing (BrMA) and adding a new reaction for the removal of BrMA without producing Br$^-$ (Fig. 3). The mechanistic nature of this reaction has remained obscure. The HPLC measurements show in Fig. 6 of the paper clearly demonstrates a slow accumulation of BrEETRA, Br$_2$AcA and Br$_3$AcA. This accumulation

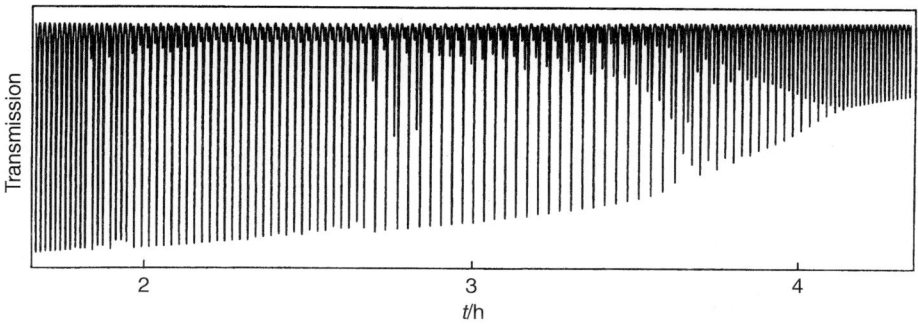

Fig. 2 Transient oscillations in the BZ system. The initial conditions are: [MA] = 0.44 M, [BrO$_3^-$] = 0.126 M, [Ce^{3+}] = 0.00133 M. The temperature was 25 °C. [H$_2$SO$_4$] = 1 M.

	Reactions			Rate constants
1	BrO$_3^-$ + Br$^-$	→	BrMA + HBrO$_2$	$k_1 = 1.8$ M^{-1}s^{-1}
2	HBrO$_2$ + Br$^-$	→	2BrMA	$k_2 = 3 \times 10^6$ M^{-1}s^{-1}
3	2HBrO$_2$	→	BrO$_3^-$ + BrMA	$k_3 = 3000$ M^{-1}s^{-1}
4	BrO$_3^-$ + HBrO$_2$	→	2HBrO$_2$ + 2Ce^{4+}	$k_4 = 42$ M^{-1}s^{-1}
5	BrMA + Ce^{4+}	→	Br$^-$	$k_5 = 32$ M^{-1}s^{-1}
6	MA + Ce^{4+}	→	P	$k_6 = 0.30$ M^{-1}s^{-1}
7	BrMA	→	Q	$k_7 = 0.0009$ s^{-1}

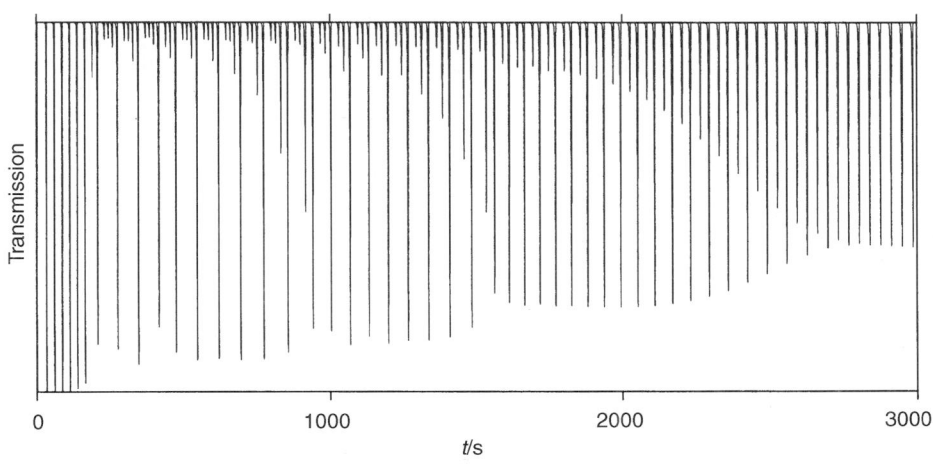

Fig. 3 Modified Oregonator model for transient complex oscillations in the Belousov–Zhabotinsky reaction together with a timetrace from the model obtained with the rate constants given above and initial conditions as in the caption to Fig. 2. The variations of the concentration of Ce^{4+} has been converted to transmittance values for easier comparison with Fig. 2. MA is malonic acid and P and Q are 'inert' reaction products.

of brominated species is equivalent to a removal of BrMA without generation of the equivalent amount of Br^-.

J. Wang, P. G. Sørensen and F. Hynne, Z. Phys. Chem., 1995, **192**, 63.

Prof. Gáspár asked: Have you tried to model the chaotic oscillations in the Ce(IV)-catalysed BZ system?

Prof. Noszticzius responded: We are planning to do this. In this respect, modelling of the transient complex oscillations mentioned in Prof. Sørensen's comment can be a good starting point as our present experiments and calculations were also performed in closed systems. Naturally it is also worthwhile to try to model CSTR experiments and real chaotic behavior applying the MBM mechanism.

Prof. Sørensen said: The most important remaining discrepancy between the experiments and the simulations with the extended Oregonator is the time of switching from periodic oscillations to complex behaviour. In the experiments this time is 5–10 times longer than in the simulations. I suggest that the inclusion of a correct mechanism for the production of Br_2AC, Br_3AC in the 'full scale model' will also allow for a quantitative modelling of this effect.

Prof. Westerhoff commented: 1. It was very nice to see this development of applying HPLC to the BZ reaction. This should enable much more complete analysis and understanding of the process. As a biologist I am a bit spoiled by the fact that in biological systems virtually all reactions are enzyme-catalysed and can hence be readily quenched by denaturing the enzymes (acid, heat). You stop the reaction by simple dilution in water. Aren't there more secure ways of freezing the chemical status quo?

2. I was actually asking this as a prelude to a more significant question: Table 1 of the paper shows a model that contains 48 reactions. In the 'Test for the model' section of the paper you state that you have tested the model for two cases. These turn out to be two different initial conditions. I agree that the correspondence between simulation and experiment are not bad in Figs. 7 and 8 of the paper. However, two such tests for 48 dynamic equations must be insufficient. It seems to me that your HPLC analysis should enable you to inject at any time point in the experiment any compound and then measure the responses of many of the concentration variables (perturbation; phase shifts). Are you planning such experiments (*cf.* the paper by Danø *et al.*, refs. 12–14).

Prof. Noszticzius replied: 1. The HPLC technique allows one to measure the concentration evolution of those components only which are relatively stable. As Figs. 4 and 6 of the paper show, these concentrations change mainly on a slow time scale (not on the time scale of a single oscillation). The applied hundred times dilution with water resulted in relatively stable samples: repeated HPLC analysis after several hours gave the same result for nearly all components. A notable exception is MAMA: it hydrolyses slowly to tartronic and malonic acids (*cf.* ref. 20 in our paper). It is true, however, that changes within a single cycle cannot be resolved with the present technique.

2. The suggested perturbation experiments would be certainly interesting. A limitation of the HPLC is, however, that the oscillatory intermediates like bromous acid, bromine dioxide and the various organic free radicals cannot be followed directly with this technique. The components measured by HPLC play rather the role of the slowly changing parameters in the oscillatory dynamics. Nevertheless I agree that stepwise modification of these parameters could generate significant changes in the dynamical behavior and it would be very informative to compare the results of such perturbation experiments with the predictions of the MBM model.

Prof. Müller commented: I consider the analyses of organic products by HPLC a very valuable progress in understanding the dynamics of the BZ reaction and wish to point out that in a previous study[1] on spiral wave dynamics in a continuously fed gel reactor, mesoxalic acid was systematically added to the ferro-catalyzed BZ system in order to demonstrate that this supports transition from rigid to meandering spiral tip rotation. In fact, such transitions are frequently observed in ageing BZ solution in which organic products accumulate. The previous assumption about the significant role of mesoxalic acid in the influence on the dynamical evolution has now been nicely corroborated. My question is why is the concentration of this product orders of magnitude higher in the ferroin-catalysed system as compared to the cerium-catalysed one?

1   F. Krüger, Zs. Nagy-Ungvárai and S. C. Müller, *Physica D*, 1995, **84**, 95.

Prof. Noszticzius responded: This interesting comment is in accordance with Prof. Westerhoff's suggestion that it is worthwhile to study the effect of the various organic intermediates (which are now identified by HPLC) on the dynamics of the BZ reaction. Answering your question; most probably mesoxalic acid is an intermediate in both cerium- and ferroin-catalysed BZ systems but its extremely fast oxidation by Ce^{4+} drives the mesoxalic acid concentration below the detection limit when the catalyst is cerium.

Prof. Nicolis said: Your model is composed of a large number of steps. Furthermore, for its quantitative study one has also to specify the values of the numerous rate constants involved. I presume that there are uncertainties on some of these steps and values. Sensitivity analysis along, for instance, the lines of ref. 1 allows one to assess how the predictions may be affected by such uncertainties. Have you performed or do you plan to perform this study, and which parts of the mechanism are the most vulnerable from this points of view?

1   D. Edelson and H. Rabitz, in *Oscillations and Traveling Waves in Chemical Systems*, Wiley, New York, 1985.

Prof. Noszticzius answered: I agree that sensitivity analysis could help to clarify further the mechanism. We have not yet carried out such calculations. Most of the rate constants in the MBM mechanism are known values as they are measured experimentally. Nevertheless there are uncertainties even regarding the possible intermediates and reaction routes when the inorganic bromine dioxide radicals react with the organic malonyl and bromomalonyl radicals. These reactions play an important role in the negative feedback loops. Thus I would expect that this part of the mechanism would be the most vulnerable.

Prof. Orbán commented: In Figs. 4 and 6 of the paper the concentration evolution of 19 species is shown during the Ce-catalysed and ferroin-catalyzed BZ reaction. I have two questions related to these concentration *vs.* time curves:

1. Each curve runs smooth, not a single outlying point is found along the curves. Neither curve shows periodic or stepwise changes in time which are expected for the concentration of intermediates or reagents and products during the oscillatory regime. I think the uncertainty in the concentration measurements by HPLC and/or small number of samples taken in the time scale for analysis may be the reasons for not noticing the characteristic stepwise or oscillatory changes in some curves. Can you comment on this remark?

2. In Figs. 4d and 6d of the paper the accumulation of Br_2AcA and Br_3AcA is presented, which—at the end of the oscillatory regime—reaches a value of about 0.02 M. The formation of 0.02 M Br_2AcA (and Br_3AcA) is equivalent to the consumption of 0.04 M bromate which means that the bromine content of about 40% of the initial bromate ends up as Br_2AcA and Br_3AcA. If this value is realistic, Br_2AcA and Br_3AcA are seen to be an important intermediate or product of the BZ reaction. Table 1 of the paper summarises all essential chemical processes (48 reactions) and species which are claimed to participate in the mechanism but Br_2AcA and Br_3AcA are not listed. Why are these species not considered in the mechanism and left out from Table 1?

Prof. Noszticzius replied: 1. I agree with your remark. During the oscillatory regime the intermediate concentrations should show periodic stepwise or oscillatory changes. For example the oxalic acid concentration calculated by the MBM model (see ref. 39 of the paper) displays periodic concentration changes but the measured curve in Fig. 4c of the paper does not show this feature. This is partly due to the fact that HPLC samples were not collected frequently enough to resolve oscillations. But even if the samples were collected with a high enough frequency probably the present quenching method would not be fast enough to freeze immediately "the chemical status quo" as Prof. Westerhoff said. Thus we measure not an instantaneous but a kind of average concentration within one oscillatory period. To see details within one period a quicker dilution method should be developed. For example a fast mixing as in a stopped flow apparatus or a simultaneous application of a quenching agent can help if we want to achieve a better resolution of the concentration *vs.* time diagrams.

2. Dibromoacetic acid (Br_2AcA) is a decarboxylation product of dibromomalonic acid (Br_2MA). By the time of the HPLC analysis most of the Br_2MA decomposes and it is measured as Br_2AcA. The decarboxylation reaction is discussed as reaction (NR4) in our paper but it is not included in Table 1 as the exact value of its rate constant is not known presently. This process, however, will not affect the calculated dynamics, as long as Br_2MA does not take part in other reactions. The amount of tribromoacetic acid (Br_3AcA) is small but it is well measurable by HPLC because of its high UV molar absorption coefficient. Its source, however, is not known presently.

Dr Hauser commented: Since we studied the effect of O_2 on the reaction of malonic acid with Ce^{4+} in acidic medium,[1] we have also performed preliminary EPR experiments using ferriin instead of Ce^{4+}. In experiments under N_2-atmosphere, we observed intensive EPR spectra from malonyl (MA˙) radicals when using Ce^{4+}, whilst such EPR signals could not be seen in the case of ferriin. Does this observation of difference in reactivity of MA towards ferriin and Ce^{4+} fit into the picture of the differences in reactivity you observed between Ce^{4+}- and the ferroin-catalysed versions of the BZ reaction?

1 B. Neumann, S. C. Müller, M. J. B. Hauser, O. Steinbock, R. H. Simoyi and N. S. Dalal, *J. Am. Chem. Soc.*, 1995, **117**, 6372.

Prof. Noszticzius answered: Your observations are in complete agreement with our results: ferriin does not react with malonic acid to give malonyl radicals. These radicals would form recombination products such as ETA and MAMA but these compounds were not found by HPLC in the ferriin–malonic acid reaction. It is interesting to remark, however, that ferriin does react with bromomalonic acid and in that case the main product is BrEETRA.

Prof. Gáspár commented: Indeed, the ferroin/ferriin chemistry in the BZ system is still a mystery after so many years of using the BZ reaction in our field. Let me give you just one example. If you dilute the stock solution of ferriin with pure water, the color of the blue solution turns to red. What is happening exactly and how is it that, even today, nobody as yet knows.

I really would encourage younger chemists who are interested in the chemistry of the BZ system to try to study the mechanistic details of reactions involving ferroin/ferriin. There is a lot to be discovered.

Dr Wang said: The Belousov–Zhabotinsky (BZ) reaction exhibits quite different dynamical behavior when a metal catalyst, for example, cerium, is replaced by a ferroin complex, even though the concentration of the metal catalysts remains constant. These phenomena have been understood largely on the basis of theft differences in redox potential, which in turn affects reaction rates. Other factors have also been discussed by Richard Field.

To throw light on whether the redox potential is the major reason for causing such dramatic differences in dynamics, I suggest conducting a series of experiments with the same metal atom which forms complexes with different ligands. By controlling the size of the ligand, the redox potential of the metal complexes can be manipulated systematically. In this way, the contribution from other factors may be minimized. This study may also be helpful in understanding the mechanism of the BZ reaction.

Prof. Noszticzius replied: This is an interesting idea. It would be especially interesting to study with HPLC how the oxidized form of these complexes react with malonic, oxalic and mesoxalic acids because these substrates react with Ce^{4+} (in the case of oxalic and mesoxalic acid very rapidly) but they do not react with ferriin.

Prof. Nicolis opened the discussion of Dr Turco Liveri's paper: In addition to modifying the physical properties of the medium could it be that the micelles interfere also with the reaction mechanism by, for instance, sequestering some of the reactants inside the micelle or by fixing them momentarily on its surface?

Dr Turco Liveri responded: Several reasons urged us to ascribe the effects of the micelles to the increase in the medium viscosity instead of partition of reactants. The $C_{14}E_6$ micelles have a more hydrophobic core than the $C_{10}E_6$ ones. Thus, if partition of the BZ components occurs, they would affect the dynamics of the BZ system to a different extent and this does not happen. In fact, it has been found that both the induction period and the kinetics of the reaction between Ce(IV) and MA are not significantly affected by the presence of micelles.

Prof. Westerhoff commented: You explain the effects of the added non-ionic micelles solely by their effect on viscosity. Because of the provision of additional, hydrophobic surfaces, explanations based on adsorption of compounds to the micelles should perhaps also be considered. Have you considered adding other, partly hydrophobic, compounds (*e.g.* liposomes) and other agents that affect viscosity (such as glycerol, DNA)?

Dr Turco Liveri answered: It is not possible to use either vesicles or liposomes because these systems are not stable in acidic media. Moreover, it is very difficult to find additives that do not react with the BZ components. We have used both ethylene glycol and polyethylene glycol and they react with the BZ components. Thus, the presence of these additives not only influences the viscosity but also the kinetics of the reactions. We have not tried with either glycerol or DNA, but we think that, analogously to the non-ionic polymer, they will react.

Prof. Epstein commented: I am delighted that the authors have begun to investigate complex reaction–diffusion behavior in micelles, one of the media suggested by Prof. Nicolis in his Introductory Lecture as affording an opportunity to bridge the gap between the macroscopic and mesoscopic or microscopic scales. I would like to propose alternative explanations for two of the phenomena reported by the authors.

It seems likely to me that the origin of the change in oscillation frequency with surfactant concentration results, not from viscosity effects, but rather from the fact that at higher surfactant concentrations more of some key non-polar species, bromine and perhaps bromine dioxide, will migrate to the micelles, thereby decreasing their concentrations in the aqueous phase, where the reaction takes place.

The chaotic behavior that is reported may reflect convective motion and/or spatial pattern formation rather than the local chemical kinetics. As shown in Fig. 4, reverse microemulsions of the BZ reaction with the ionic surfactant AOT yield a rich variety of spatio–temporal patterns.[1] It is possible that the present system, even with a non-ionic surfactant, gives rise to at least some similarly complex behavior.

Also, in an experiment such as the one described in an unstirred system monitored spectrophotometrically, even simple concentration gradients combined with convective motion generated

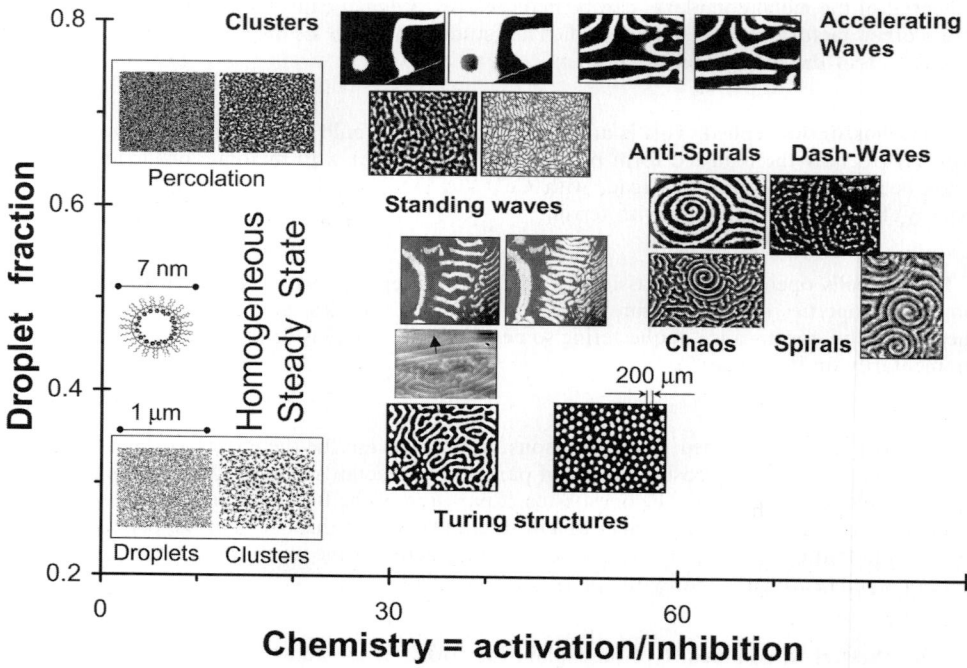

Fig. 4 A schematic phase diagram summarizing complex spatio-temporal behaviour in the BZ–AOT system.

by evaporation and/or heating from the apparatus can lead to complex temporal patterns. Such behavior has been demonstrated[2] in the biacetyl–oxygen system.

1 V. K. Vanag and I. R. Epstein, *Phys. Rev. Lett.*, 2001, **87**, 228 301; V. K. Vanag and I. R. Epstein, *Science*, 2001, **294**, 835.
2 I. R. Epstein, M. Morgan, C. Steel and O. Valdés–Aguilera, *J. Phys. Chem.*, 1983, **87**, 3955.

Dr Turco Liveri replied: As we said to Prof. Nicolis, the experimental trends can be mainly ascribed to the medium viscosity enhancement. We agree with Prof. Epstein that complex behaviour can also arise in our system, analogously to that observed with the AOT microemulsions.

Prof. Hanke said: I think it would be useful to test the hypothesis of changed viscosity of the medium by doing comparable experiment in gels of the BZ reaction, where viscosity is surely higher than in fluid systems, and can be controlled up to a certain degree.

Dr Turco Liveri responded: We planned to perform experiments in the presence of gels formed by the non-ionic surfactants used in the present work.

Dr Satnoianu commented: I wanted to discuss the methodology of the analysis of the time series. Essentially the results of the paper are based on the use of FFT techniques. I note that this is not enough in order to assess the eventual chaotic behaviour of a signal. This idea is clearly explained in many books, *e.g.*, ref. 1. In this book, for example, other methods are explained, with examples, to analyse complex spatiotemporal series such as: Lyapunov exponents, dimension calculations, singular value decompositions and Poincare maps.

Of course, the difficulty of the work resides in the short length of the time series. However, similar problems have been encountered in biology already where usually measurements cannot be made for a reasonable long time without interfering fundamentally with the system. There are now published methods (in *Physica D*) to analyse these short time series.

1 T. Mullin, *The Nature of Chaos*, Clarendon Press, Oxford, 1995.

Dr Turco Liveri replied: We have employed the FFT analysis in order to understand the route to chaos and not to characterize it. It has been observed[1] that the system shows sensitivity to the initial conditions and this can be considered as evidence of chaotic transient behaviour. Nevertheless, we are still working on quantifying chaos.

1 M. Rustici, M. Branca, C. Caravati and N. Marchettini, *Chem. Phys. Lett.*, 1996, **263**, 429.

Dr Wittmann commented: I want to show that in similar experiments, where instead of your surfactants, ethylene glycol/polyethylene glycol was added to the Ce-catalysed BZ system, the dependence of the time period on the polymer concentration is similar as shown on Figs. 4 and 6 of your paper (see Fig. 5 below).

For this system we have made simulations with the MBM model (see the paper by Noszticzius *et al.*) to explain it on a chemical basis introducing a reaction between bromate and the alcoholic group of the surfactant generating $HBrO_2$ (see Fig. 6 overleaf). We present a poster (see poster list) about the results which show a qualitatively good agreement (k is the pseudo first order rate constant of the reaction).

This indicates that beside the viscosity change, chemical reactions could account for this effect also in your system as your surfactants also contain alcoholic groups.

Dr Turco Liveri responded: As far as we know the non-ionic surfactants we have used do not react with the $HBrO_3$. We have measured the surface tension during the time course of the reaction and no change occurs. This indirectly demonstrates that the kinetics of the reaction you have suggested can be either very slow or not take place. The lack of the reaction can be considered a consequence of the lower availability of the OH groups in the micelles with respect to those in the polymer chain.

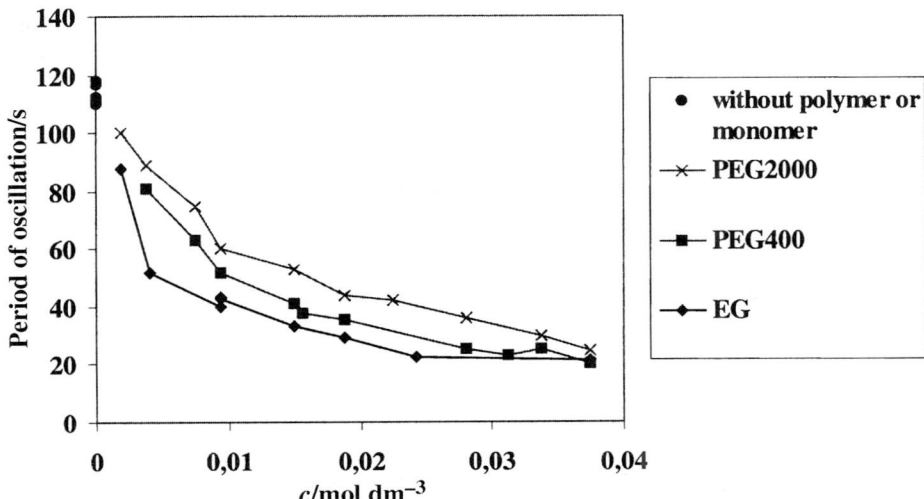

Fig. 5 Period of the oscillations as a function of the polymer/monomer concentration.

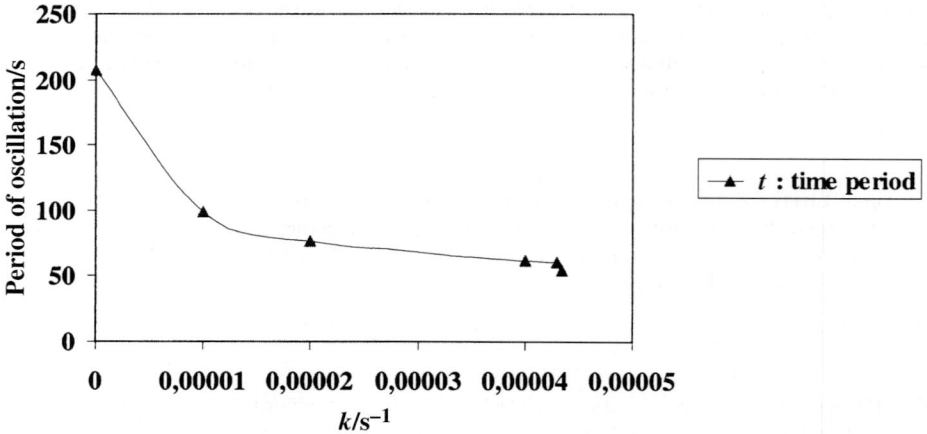

Fig. 6 Period of the oscillations as a function of the pseudo first order rate constant of the reaction R–CH$_2$–OH + HBrO$_3$ → R–CH=O + HBrO$_2$ + H$_2$O.

Prof. Sørensen said: I have a general comment regarding your use of an unstirred cuvette for the study of oscillatory systems such as the BZ reaction. Traditionally, this procedure would have been considered inappropriate because of the occurrence of unavoidable and uncontrollable hydrodynamic convection. But as remarked by Prof. Nicolis in his Introductory Lecture, oscillations under non-ideal conditions are ubiquitous in nature and such systems should also be studied in the laboratory. To be useful it is however necessary to characterize the non-ideality either by measuring the spatial correlations of the cell oscillations by monitoring at several positions simultaneously and calculating the correlations or by simultaneous anemometric measurements of the fluid flow.

Dr Turco Liveri responded: We have previously measured the difference of potential time series with couples of electrodes dipped in different parts of the solution. It has been observed that the position of the electrodes did not affect the results. Obviously, different chaotic series were detected for every couple of electrodes. It would be useful to calculate the spatial correlations of these cell oscillations.

Dr Münster said: 1. In recent years techniques have been developed which provide insight to the microviscosity and—even more important—the micropolarity at the surface of micelles. In particular, the solvent relaxation technique has been used to characterize the membrane/solution interface. Did you consider applying methods of that kind in order to obtain more detailed information about the interplay between the chemistry of the BZ reaction and the micelles?

2. Furthermore, a large part of the data analysis you presented was based on the power spectra computed from your measured time series. These time series, however, are relatively short. Did you take measures to avoid alias-lines to emerge in your power spectra which might easily lead to a misinterpretation of the spectra?

Dr Turco Liveri answered: 1. We have not determined the micelle microviscosity, but we planned to perform EPR measurements in order to obtain useful information. This technique is one of the most powerful for detecting microviscosity of the interior of surfactant micelles.

2. We know that the time series are relatively short and in order to improve the spectra analysis we have applied the method given in the experimental part of the paper.

Dr Schreiber commented: A transition from quasiperiodicity to chaos may essentially be one of the following: (1) periodicity → two-torus → three-torus, which is however, structurally unstable and is superseded by chaos, or (2) periodicity → two-torus → fractalization of the two-torus → chaos. Is it possible to identify one of these in your experiments?

Dr Turco Liveri replied: It is not simple to discriminate between these two routes. We did not succeed in observing three irrationally related frequencies that would confirm the first hypothesis and it is quite difficult to find in the literature a structurally unstable three-torus. Until now we are not able to choose between these two roads to chaos.

Prof. Orbán commented: I would like to draw your attention to consider the effect of convection when you explain the observed irregular oscillations in an unstirred BZ system.

We have measured about 0.1 °C increase in temperature during one oscillatory cycle of the BZ reaction and in a lengthy experiment, such as yours, 3–4 °C rise in temperature may occur. Even if the BZ solution is thermostatted, but not stirred, the solution near the wall of the reaction vessel is colder than the middle part of the bulk, inducing significant convection what may turn the regular oscillatory response of the Pt electrode to chaotic.

Dr Turco Liveri and Mr Lombardo replied: We do not think that the chaotic behaviour is due to a temperature gradient. Recently, we have monitored the behaviour of the unstirred BZ system by using cuvettes of different path length, *i.e.* from 1 cm to 0.5 mm.

It has been found that the dynamics of the system does not change with decreasing path length. This result suggests that the observed patterns cannot be attributed to the temperature difference between the wall and the middle part of the reactor.

Prof. Noszticzius opened the discussion of Dr Šnita's paper: In our laboratory, electrolyte diodes were connected to construct so-called electrolyte transistors and complex oscillations were observed in these devices under specific conditions.[1] It was suggested that the experimentally found bistability of the salt-contaminated electrolyte diodes can explain the oscillations. Nevertheless our simple model[2] did not show bistability. Thus, it is important to emphasize that the new model of the Prague laboratory is capable in describing the bistable behaviour.

In the conclusion it is mentioned that "more complex behavior in the simulated hydrogel membranes was obtained when considering the dissociation reactions of fixed charge groups." Can you describe briefly those complex phenomena? I ask this question because recently we found[3] that the hydrogel of the electrolyte diode contains chemically bound carboxylic acid groups in low concentration and it is the dissociation of these groups which is responsible for the fixed negative charges in the gel.

1 L. Hegedüs, N. Kirschner, M. Wittmann and Z. Noszticzius, *J. Phys. Chem. A*, 1998, **102**, 6491.
2 L. Hegedüs, N. Kirschner, M. Wittmann, Z. Noszticzius, T. Amemiya, T. Ohmori and T. Yamaguchi, *Chaos*, 1999, **9**, 283.
3 K. Iván, N. Kirschner, M. Wittmann, P. L. Simon, V. Jakab, Z. Noszticzius, J. H. Merkin and S. K. Scott, *Phys. Chem. Chem. Phys.*, submitted.

Dr Šnita responded: We have studied the effects of electric field imposed on the gel with fixed groups of weak acid, weak base and ampholyte. The degree of dissociation of the fixed groups varies depending on local pH. For example, the gel layer with the fixed ampholyte between the acidic and the alkaline boundaries can partially exhibit similar properties as the bipolar membrane. The hypothesis on the type and the concentration of the fixed groups in the gel can be verified comparing the relevant experiments with the results of simulations based on our modeling methodology.

Prof. Hanke commented: In the gels discussed, fixed negatively charged anions of high mass are surrounded by smaller positively charged cations. This gives rise to a situation in which oscillations, for example, of the cations relative to the anions, can occur in a manner which has similarly been described for example in classical plasma physics (hydrodynamics). A detailed discussion is given in ref. 1.

The impedance of the gels should be measured as a function of frequency to find out more about the detailed electrodynamical behaviour.

1 J. J. Brandstetter, *An Introduction to Waves, Rays and Radiation in Plasma Media*, McGraw-Hill, New York, USA, 1963.

Dr Šnita replied: Some nonlinear effects studied in the plasma physics can serve as a motivation and one can try to find analogies in the liquid electrolyte systems. But the collisions of ions in liquid electrolytes with electroneutral molecules of solvent (*e.g.* water) suppress classical oscillations following from the competition of electrostatic and inertial forces, which takes place in the plasma. In the case of local macroscopic deviation from electroneutrality in a fluid medium (both plasma and electrolyte) the electromagnetic forces are important for fluid dynamics. This phenomenon probably causes the instability in the case of overlimiting current passing through an ion exchange membrane.

The observation of gels in transient and ac electric fields is important. We plan to follow this type of experiment in the cooperation with the group of Prof. Nosztíczius.

Dr Feigin said: In the pre-printed paper you have analysed a stationary situation and revealed, for some configurations of the system, very significant separation of charges and correspondingly extremely high intensities of electric fields. I agree with Prof. Hanke that your systems are similar qualitatively to the plasmas placed into an external electric field, but in my opinion, the closest example is isotropic (non-magnetised) plasma. Based on this analogy I can assume that your systems have to demonstrate non-trivial time evolution. In particular the systems can possess intrinsic frequencies and demonstrate, as a reaction on external effect (*e.g.* turning on of external electric field), oscillations which are similar to Langmuir oscillations of isotropic plasmas. It seems to me that you can observe a resonance effect applying a periodic external field and changing the frequency of this forcing. Note that under similar forcing, Langmuir turbulence (non-regular oscillations of both electric field, and charge density) may be observed in plasmas.

Prof. Menzinger commented: Over more than a decade, primarily your group has published several experimental and modeling studies on the effects of external electric fields on real and model nonlinear reactions. With the benefit of hindsight, could you comment on the impact of your present findings, particularly of the nonuniform field gradients that you find in simple electrolyte solutions, on that prior work. Even if the gross features of your earlier conclusions did not change significantly, can you anticipate some changes in the details—and if so, what are they?

Dr Šnita responded: Systems with high ionic strength and approximately uniform conductivity were studied in most of our previously published studies, *cf.* refs. 1–10. Migration of ions in an external electric field can be the cause of a number of nonlinear phenomena in reaction–diffusion systems (wave acceleration, deceleration, reversal, splitting, annihilation, change of formal stoichiometry of reacting mixture, and convective instability). Nonuniform spatial distribution of ions with different mobility is connected with internal electric field (diffusion potential) which modifies effective diffusivities of ions. Evolution of concentration fields in the above mentioned systems does not influence significantly the field of electric current, because the solution conductivity is determined mainly by the background electrolyte (*e.g.* the sulfuric acid in the BZ reaction mixture). The assumption of local electroneutrality can be then mostly accepted in such systems, as relatively high local conductivity is reflected in the relatively low value of the local Debye length.

The systems studied in this contribution can exhibit significant change of spatial field of electric conductivity. Particularly the local conductivity of the solution can by very low and hence the corresponding value of the local Debye length can be relatively high and significant deviations from the local electroneutrality can be observed.

As it has been illustrated in the paper, this can be the cause of the existence of multiple steady states and probably of other nonlinear effects. The description of previously published experimental results will not be affected qualitatively. The differences in spatiotemporal patterns of concentrations and electric fields quantities can be found by comparing the results of modeling with and without the local electroneutrality assumption. Some such simulations are the subjects of current work.

1 H. Ševčíková and M. Marek, *Physica D*, 1983, **9**, 140.
2 H. Ševčíková and M. Marek, *Physica D*, 1984, **13**, 379.
3 H. Ševčíková and M. Marek, *J. Phys. Chem.*, 1984, **88**, 2181.
4 H. Ševčíková, M. Marek and S. C. Müller, *Science*, 1992, **257**, 951.
5 F. Münster, P. Hasal, D. Šnita and M. Marek, *Phys. Rev. E*, 1994, **50**, 546.
6 D. Šnita and M. Marek, *Physica D*, 1994, **75**, 531.
7 P. Kaštánek, J. Kosek, D. Šnita, I. Schreiber and M. Marek, *Physica D*, 1995, **84**, 79.
8 H. Ševčíková, J. Kosek and M. Marek, *J. Phys. Chem.*, 1996, **100**, 1666.
9 D. Šnita, H. Ševčíková, J. Lindner, M. Marek and J. H. Merkin, *J. Chem. Soc., Faraday Trans.*, 1998, **94**, 213.
10 L. Forštová, H. Ševčíková, M. Marek and J. H. Merkin, *J. Phys. Chem. A*, 2000, **104**, 9136.

Prof. Gáspár said: I think that you are mixing two pictures here. In the case of the earlier experiments, the voltage drop occurred on the whole ionic medium resulting in a small value of the potential gradient. However, in the case of the electrode diode and the other systems discussed in this paper, the chemical reaction takes place only in a very narrow part of the system (few micrometers), and the potential drop within this narrow region results in an enormous gradient. This makes this system very different from the previous ones in which the effect of electric field on chemical wave behaviour was studied.

Prof. Noszticzius commented: As I see one of the major differences between the previous and present work of Marek and coworkers is that presently the electric field is strongly non-homogeneous. This was not the case in the previous experimental and theoretical investigations.

Dr Šnita responded: As discussed previously the earlier experimental and modeling studies dealt with the systems with relatively high local conductivity and hence approximately homogeneous electric field.

Local electric field is given as sum of two parts. The first one is given by the ratio of the value of local electric current and the value of local conductivity. This part was approximately homogeneous in the earlier studies and strongly non-homogeneous in this paper. The second part is given by diffusion potential and represents the cause of modification of transport of ions with the different mobilities and can be non-homogeneous even if the conductivity field is approximately homogeneous.

Very strong simplification (not accepted by us) considers constant electric field and does not take into account the presence of diffusion potential.

Prof. Müller opened the discussion of Dr Mayama's paper: As I understand, the folded DNA trapped in the focus of a laser beam is exposed to local heating of more than 10 °C, then it escapes from this hot area and unfolds in a cooler environment. Can one imagine a conformational change associated with such a high temperature change to have any significance in real biological tissue?

Nevertheless, the reported oscillations are certainly of high interest for the dynamical behaviour in nonequilibrium polymer/biopolymer systems, in general.

Dr Mayama responded: In a real biological system, various environmental parameters, such as the concentration of ATP, RNA and small ions, takes on the role of generating the conformational transition of giant DNA molecules. Here, the perspective based on free energy is very important, *e.g.*, that temperature corresponds to chemical potential in this work. As for the actual experimental examples on the regulation by chemical parameters, see refs. 37 and 38 of the paper.

Prof. Dewel said: Your model (eqns. (4) and (5)) predicts that the local temperature, τ, would vary periodically with time (limit cycle in the η,τ phase). I rather think that the motion of the DNA chain takes place in the presence of a stationary temperature gradient created by the laser. The main dissipative process occurring in your system is heat diffusion in the solvent (no heating in D_2O!)

Dr Mayama replied: We have already mentioned that the temperature gradient is the driving force for the oscillation, as you suggest. The most important subject concerning the oscillation is as follow. If the conformational transition of DNA is interpreted with a linear kinetic equation, the DNA molecule would only show thermal fluctuation (not oscillation) under a stationary temperature gradient. The necessary condition to generate the oscillation is the nonlinearity in the kinetics of the conformational transition of DNA.

Prof. Sørensen asked: Will it be possible from your experimental data to see if the melting of a single DNA molecule is a single step process or whether it occurs in successive steps?

Can your method be used for studying other molecules, for example, proteins?

Dr Mayama answered: We have reported the kinetic process of the unfolding, as well as the folding of single DNA, indicating that the process is described with a single exponent with time (see ref. 44 of the paper).

For the second question, in principle, our method is applicable for other molecules, as far as the size of the molecule is over the limitation of optical resolution.

Prof. Menzinger commented: Referring to Fig. 9 of the paper, the observed rhythmical conformational change of DNA is the result of two effects. Since the coiling/uncoiling at the low/high temperatures outside/inside the laser focus, respectively, are relatively well understood, the explanation of the periodic coiling/uncoiling reduces to that of the rhythmic trapping of coiled DNA by and escape of melted DNA from the laser focus. To invoke a trapping and a scattering force and to describe the rhythmic phenomenon phenomenologically by eqns. (4) and (5) does not constitute a physical explanation and only pushes it only one level back.

Could Dr Mayama give us a physical picture of the trapping and untrapping mechanisms? Which part of the literature on optical traps is relevant to this issue?

Dr Mayama replied: We have reported the physico-chemical studies on the trapping mechanism of giant DNA molecules. Concerning the physical picture of the trapping, these are reviewed in refs. 1–3.

1   Y. Matsuzawa, Y. Koyama, K. Hirano, T. Kanbe, S. Katsura, A. Mizuno and K. Yoshikawa, *J. Am. Chem. Soc.*, 2000, **122**, 2200.
2   S. Katsura, K. Hirano, Y. Matsuzawa, K. Yoshikawa and A. Mizuno, *Nucleic Acids Res.*, 1998, **26**, 4943.
3   Y. Yamasaki, K. Hirano, K. Morishima, A. Mizuno, F. Arai and K. Yoshikawa, *Forma*, 1998, **13**, 397.

Dr Satnoianu commented: The oscillatory behaviour consisting in the folding/unfolding of the DNA raises questions of this process in relation to the dynamics of the cell cycle especially around the division time. I suggest that the authors clarify that the conditions used in their experiments are not usually met in a real cell nucleus (for example, the temperature cannot be above 10 °C). Also it is of interest for the authors to discuss the robustness of the oscillatory regime in raport to variations in the dynamical parameters.

Dr Mayama responded: We don't consider that the temperature is the regulatory parameter in a real cell nucleus. Instead of the case in a living system, a non-specific environmental parameter, such as the concentration of ATP, RNA, H^+, *etc.*, should exhibit the essential role in controlling the conformation of DNA and other macromolecules. It is important that you understand our argument based on free energy, that temperature corresponds to chemical potential. The importance of the chemical parameters on life phenomenon are discussed in ref. 1.

As for the robustness of oscillation, we have confirmed that the oscillation is recovered after a short time erruption of the laser irradiation.

1 K. Yoshikawa, *J. Biol. Phys.*, in press.

Prof. Fernandes de Lima commented: This is a very interesting and elegant experiment. I have not a question but two short comments. The first is about the biological significance of volume phase transitions in an isolated DNA molecule. One isolated DNA is very different from the DNA within the cell nucleus, where it is associated with histones and non-histone proteins that control conformational changes and transcription.

The second comment is about the biological significance of volume phase transitions of polyanionic gels in general (DNA is such a polymer). An important role for these transitions has been proposed by Ichigi Tasaki (of NIH) since the early nineties.[1]

He proposes that the transition from quiescent to excited state at the axonal and synaptic membranes are just such phase transitions at the external surfaces of biological membranes. These external surfaces are made of polyanionic gels due to fixed charges in the glycoproteins and glycolipids in the hydrophilic part of the amphipatic bilayer.

1 I. Tasaki, *Ferroelectrics*, 1999, **220**, 305; I. Tasaki and Byrne, *Biopolymers*, 1992, **32**, 1019; I. Tasaki and Byrne, *Biopolymers*, 1994, **34**, 209.

Dr Mayama replied: As for the first comment on the effect of histone and non-histone proteins, we have reported structuring in the conformational transition on the scale of several tens and hundred base pairs for a giant DNA complex with histone (see, *e.g.*, ref. 1). In this article, we have given a discussion on the conformational transition in relation to the genetic activity. Here, it is important to note that a single DNA chain and histone molecules correspond to a negative charged chain and multivalent counterions, respectively, and therefore the conformational change of a DNA–histone complex in living cells is briefly interpreted as the folding phase-transition in individual DNA molecules as shown in Fig. 5(b) of the paper.

As for the second comment, we also agree with the idea of Tasaki on the important role of phase transition in dynamical aspects of life including membrane phenomena. In connection to the importance of phase transitions in a small system, we have reported our hypothesis in the literature (see, ref. 20 of the paper and ref. 2).

1 Y. Yoshikawa, Y. S. Velichko, Y. Ichiba and K. Yoshikawa, *Eur. J. Biochem.*, 2001, **268**, 2593.
2 K. Yoshikawa, *Complexity in a Molecular String: Hierarchical Structure as is Exemplified in a DNA Chain*, in *Complexity and Diversity*, Springer-Verlag, Tokyo, 1997.

Prof. Westerhoff commented: I should like to get this straight: you are proposing a large conformational change in the DNA due to heating by a number of degrees Kelvin. Clearly this does not happen in biology at time scales of seconds (too large heat conductances, too limited local heat sources at the DNA). The only possible direct relevance for biology I see is that the thermodynamic force provided by the heating in your experiment is in reality provided by other factors, such as (the removal of) histones or nucleosomes, or the modulation of supercoiling by DNA gyrase. In this vein, have you added spermine/spermidine to the DNA in your experiments? These compounds should exert counteracting compacting forces that are understood in the literature, and this may allow you to estimate the thermodynamic forces.

Dr Mayama responded: We think that "temperature" is one of the important intensive variables of a thermodynamic system, as your suggest. We think that other thermodynamic parameters beside temperature, such as chemical potential, play the essential role in a biological system. We have performed a systematic research on the effect of various kinds of chemical species. For example, the dynamic structure of the complexes of DNA with histone is reported in ref. 1.

As for the effect of spermidine/spermine, we have performed the detailed experiment on the level of a single molecule. To be honest, spermidine/spermine are not suitable condensation agents to generate oscillatory phenomena under the presence of a temperature gradient because of the opposite temperature effect to that in PEG solution as shown in Fig. 5 of the paper. As can be predicted from the theoretical consideration in the present paper, any oscillation can never be generated in individual DNA molecules under the presence of spermidine/spermine (see ref. 2). In these studies, we have given a full interpretation of statistical thermodynamics on the conformational transition, including the opposite temperature effect to the case in PEG solution, *i.e.*, DNA shows an unfolding transition from the compact state to the elongated coil state accompanied by the increase in temperature.

1 Y. Yoshikawa, Y. S. Velichko, Y. Ichiba and K. Yoshikawa, *Eur. J. Biochem.*, 2001, **268**, 2593; T. Sakaue and K. Yoshikawa, *Phys. Rev. Lett.*, 2001, **87**, 078105-1.
2 M. Takahashi, K. Yoshikawa, V. V. Vasilevskaya and A. R. Khokhlov, *J. Phys. Chem.*, 1997, **101**, 9396; H. Murayama and K. Yoshikawa, *J. Phys. Chem. B*, 1999, **103**, 10517.

Prof. Müller said: I see from the forgoing discussion how important it is to look not only at the absolute values of temperature changes, but to consider also the induced spatial gradients associated with the heating. Locally, these temperature gradients could be quite the same in your experiment and in biological tissue.

Investigation of nonlinear dynamical properties by the observed complex behaviour as a basis for construction of dynamical models of atmospheric photochemical systems

Alexander M. Feigin, Yaroslav I. Molkov, Dmitrii N. Mukhin and Eugenii M. Loskutov

Institute of Applied Physics, Russian Academy of Sciences, 46 Ulyanov Street, Nizhny Novgorod 603950, Russia

Received 2nd April 2001
First published as an Advance Article on the web 12th November 2001

The importance of the investigation of nonlinear dynamical properties (NDPs) of the atmospheric photochemical systems (PCSs) was demonstrated in ref. 1 and 2 (A. M. Feigin and I. B. Konovalov, *J. Geophys. Res.*, 1996, **101** (D20), 26038; 1. B. Konovalov, A. M. Feigin and A. Y. Mukhina, *J. Geophys. Res.*, 1999, **104** (D3), 3669). The only known way to study NDPs of any natural dynamical system (including atmospheric PCSs) is to construct a mathematical model of the system. The key point here is adequacy of the NDPs of the constructed model to the system observed. We propose a new approach to construction of such an adequate model for systems manifesting nonstationary chaotic behaviour and describe an algorithm based exclusively on nonlinear dynamical analysis of the observed time series (TS) without invoking any *a priori* knowledge about the properties of the system observed. Potentialities of the algorithm are demonstrated with the aid of a computer model of the mesospheric PCS. The duration of the "observed" TS is limited so that the system demonstrates only one—chaotic—type of behaviour, without any bifurcations throughout the observed TS. The proposed algorithm enabled us to make a correct prognosis of bifurcation sequences and calculate probabilities to reveal, at the time instant of interest, predicted regimes of the system's behaviour for times much greater than the length of the initial TS.

1. Introduction

1.1.

Atmospheric photochemical systems (PCSs) are ensembles of interrelated chemical processes, including photolysis processes, that occur in the Earth's atmosphere and affect dynamics of minor gaseous constituents of the atmosphere (*i.e.*, all chemical compounds present in the atmosphere, excluding molecular nitrogen and oxygen). The processes and their characteristics differ appreciably in different regions of the atmosphere. The following PCSs are typically distinguished: PCS of the boundary layer, of the free troposphere, of the polar lower stratosphere, of the mesosphere, and others. Photochemical processes are among the principal elements in the chain of global atmospheric processes defining the thermal structure of the Earth's atmosphere, its radiation balance and global circulation. Analysis of PCS evolution is of key importance for investigation of changes in the state of the ozone layer, climatic consequences of civilization (in particular, the greenhouse effect), control of the chemical composition of the air in densely populated regions and in regions of intensive plant cultivation, *etc.*

1.2.

One of the principal goals in investigating the problems mentioned above and many others is prognosis of the expected changes in the characteristics of the processes and phenomena observed in the atmosphere today. Nonlinearity is the inherent property of atmospheric photochemistry. Therefore, knowledge of NDPs of atmospheric PCSs, namely, types of possible regimes of behaviour for different combinations of parameter values, is of principal importance for construction of an adequate prognosis of phenomena in which photochemistry plays a significant role. Firstly, it allows for prognosis of possible (due to parameter trends) switchings of the system between regimes of behaviour co-existing in its phase space under current parameter values. Secondly, knowledge of NDPs enables one to foresee bifurcations (disappearance of the existing regimes and birth of the new ones) and evaluate their consequences.

A general approach to the study of NDPs is well known and consists in investigation of the structure of the phase space of the system and analysis of changes occurring in this structure when the values of control parameters vary, in other words investigation of the structure of the space of the parameters of the system. Apparently, this approach implies that there is a definite mathematical model of the system under consideration.

A traditional method of modelling—construction of a model on the basis of "first principles"—demands general understanding of the origin of the processes playing the main role in the studied phenomenon, as well as specific knowledge about the composition of the "ensemble" of processes of "required" origin. The generally adopted method of verification of the "first principles" models is comparison of results of modelling with the observed ones (and earlier observed results) by *quantitative* characteristics of the phenomenon under consideration. Clearly, in terms of *prognosis* of the evolution of a *nonlinear* dynamic system (DS), such verification is sound only under the condition of revealing the bifurcations that occurred in the course of observations and were reproduced by the model. However, for "natural" DSs another situation is typical, when, in spite of the "intrinsic" nonlinearity of such systems and *nonstationarity* of the processes running in them, no bifurcations occur during the observation time. In this case, the traditional method of verification can only testify to adequate reproduction by the model of the current type of behaviour of the system, giving no guarantees as regards prognosis of possible future bifurcations. Indeed, satisfactory quantitative adequacy of the model of the observed evolution can be attained by either taking into account or neglecting processes that introduce into the model *qualitatively different* NDPs. A "qualitative" error in creating the "first principles" model of such complex and insufficiently studied DSs as various natural systems is highly probable.

1.3.

The above is absolutely true for atmospheric PCSs. A classical mathematical model of an atmospheric PCS is a set of equations of chemical kinetics, complicated to a greater or lesser extent by contributions from processes of a non-chemical nature. In the simplest case, such a model is a box in which "correctly" selected chemical reactions proceed. This model can be written in the form of a system of first-order ordinary differential equations for concentrations of minor gaseous constituents of the atmosphere:

$$\frac{dX_i}{dt} = f_i(\vec{X}, \vec{A}), \quad \vec{X} = \{X_i\}_{i=1}^N, \quad \vec{A} = \{A_j\}_{j=1}^M. \tag{1}$$

Here, X_i is the concentration of the chemical constituents which are part of the PCS, A_j are the parameters determining the character of the evolution of the system, and t stands for time. The functions $f_i(\vec{X}, \vec{A})$ on the right-hand side of eqn. (1) describe the sources and sinks of the corresponding chemical components; and f_i are the nonlinear functions of the arguments $\vec{X}$. The local character of the function $f_i(\vec{X})$ means that model (1) includes only chemical (and photochemical) reactions proceeding in the atmosphere. Such box models are usually part of complex models describing more or less adequately the interrelated atmospheric processes (chemical processes, heat and mass transfer, radiation transfer) characterised by definite temporal and spatial scales as a whole.

One of the main features of atmospheric PCSs is their nonstationarity. Conditions, that are external relative to photochemical processes, vary in the course of time for different reasons, both of natural and anthropogenic nature. Correspondingly, mathematical models of atmospheric PCSs of the form (1) are nonautonomous; the control parameters entering these models are functions of time.

1.4.

The essential influence exerted by processes of a non-chemical nature upon variations of concentrations of chemical components significantly impedes clarification of the role of NDPs of "chemical" origin in the observed phenomena. Nevertheless, today we are in a position to speak about possible manifestations of a multistable box model of a *boundary layer* PCS and multistability induced bifurcations of the birth/disappearance of a pair of "additional" equilibrium states in the change of the so-called regimes with high and low concentration of NO_x (a family of odd nitrogen) in the boundary layer of the troposphere.[3,4]† Another phenomenon that may be caused by NDSs of a *mesospheric* PCS is multiple amplification of the so-called quasi-2-day oscillations of different characteristics (zonal and meridional wind amplitudes, temperature, *etc.*) that occur in the upper mesosphere and thermosphere in the summer season.‡ The corresponding bifurcation of period doubling of a limit cycle, that corresponds to "conventional" oscillations of concentrations of chemical components with a period of one day, occurs with variation of control parameters, such as water vapour concentration, coefficient of vertical eddy diffusion and temperature.[6–9]

Of special interest is manifestation of NDPs of a *polar lower stratospheric* PCS in formation of an Antarctic ozone hole. This phenomenon first revealed in 1985[10] could not be explained within the scope of the known models and, consequently, had not been predicted by them (*cf. e.g.*, ref. 11). It did not take much time to understand that the cause of the error was neglect of heterogeneous chemical reactions with participation of particles of polar stratospheric clouds.[12,13] One of the properties of these reactions, that is most important in terms of construction of an adequate prognosis, is a special type of nonlinearity which they introduce into equations of chemical kinetics. It was shown recently[1,2] that the influence of heterogeneous reactions on the phase space structure of a polar lower stratospheric PCS increased as one of the control parameters (concentration of inorganic chlorine) increased and resulted in a sequence of unpredicted bifurcations in the behaviour of Antarctic PCS in the 1980s. Note that, in the course of formation of an ozone hole, the evolution of a given PCS depends weakly on processes of non-chemical origin, so that the box model of the form (1) reproduces the key quantitative characteristics of the phenomenon to a satisfactory accuracy. Therefore, we have strong arguments to believe that NDSs of the polar lower stratospheric PCS played a decisive role in the appearance of the Antarctic ozone hole and can cardinally affect its future evolution.

1.5.

The above allows us, firstly, to speak about the need to investigate NDPs of atmospheric PCSs in constructing a prognosis of evolution of different atmospheric phenomena. Secondly, the use for this purpose of "first principle" models is fraught with prognostic error that is possible even in modelling future behaviour of "long-lived" systems in nature that demonstrate stable evolution.

We propose a basically new method of constructing mathematical models of different systems existing in nature, intended for investigation of their NDPs and long-range prognosis of their

---

† The rate of emission into the atmosphere of free radicals is the control parameter whose seasonal variations may lead to these bifurcations. Study of this phenomenon is of practical importance: the two mentioned regimes correspond to essentially different (by two orders of magnitude and more) boundary ozone concentrations, increased values of which are the cause of exacerbation of many diseases in humans, lead to decreased productivity of crop and forest reproductivity, and so on.
‡ A fairly detailed analysis of experimental data was given in ref. 5 for more than 20 years of observations.

qualitative behaviour. The method is based on analysis of the observed time series and is most effective for systems demonstrating nonstationary chaotic dynamics. This behaviour is typical, in particular, for various systems determining the occurrence of paramount processes in the Earth's atmosphere and hydrosphere (evolution of the ozone layer,[14] behaviour of the concentrations of chemical components within the boundary layer,[15] large-scale variations of surface temperature of the tropical Pacific Ocean (the El Niño phenomenon)[16,17]). Note that even today the scope of real systems demonstrating chaotic dynamics ranges from the above enumerated atmospheric and atmospheric–oceanic processes to diverse systems in living organisms[18,19] and tectonic activity[20] and has a tendency to expand as modern methods of nonlinear dynamical analysis are applied to new data bases of different nature.

We will address the case when a chaotic time series (TS) *only* is available, and there is no additional information about the dynamic system that generated it. As was said above, our goal is to construct a model that would be adequate in its NDPs to the observed system. In other words, it must not only reproduce the observed evolution of the system but also allow us to predict qualitative changes in the system's behaviour (bifurcations). We will refer to such models as *prognostic* ones. We will also suppose that no bifurcations occur in the course of the observed time series and the system demonstrates only one—chaotic—type of behaviour.

Because the change in the type of behaviour of the system is a consequence of the variation of control parameters, the time series generated by such a system is nonstationary. Obviously, nonstationarity of the observed TS can be revealed only if the corresponding process is characterised by at least two, strongly differing timescales.§ This lays the basis for two hypotheses that are natural in terms of physics.

Hypothesis 1. The observed nonstationary time series is generated by a *weakly nonautonomous* "*fast*" dynamical subsystem, the parameters of which vary slowly on characteristic timescales of the evolution of its dynamical variables.

Hypothesis 2. Evolution of a "slow" subsystem, interaction with which is the cause of nonautonomy of the "fast" subsystem, remains unchanged in the not too remote future of the "slow" subsystem. By the "not too remote future" of the slow subsystem we mean a time interval of the order of several lengths of the observed TS during which changes of the parameters of the fast subsystem are relatively small.

Interpretation of nonstationarity of TS as nonautonomy of the observed DS (hypothesis 1) is meaningful when the time dependence of parameters of the system is known. The second hypothesis allows one to reconstruct the trends of parameters of a fast subsystem over a sufficiently extended time interval.

Productivity of this approach was first demonstrated in ref. 24 and 25, where an algorithm based on the hypotheses formulated above was proposed for construction of prognostic models of systems demonstrating low-dimensional chaotic dynamics. The potential of this algorithm was illustrated by means of computer models. A computer modelled nonstationary chaotic series was used to construct a prognostic model that was employed for constructing prognosis of bifurcations of the "observed" system. The algorithm allowed us to predict correctly a sequence of bifurcations and determine other characteristics of prognosis for times much greater than the length of the initial series.

In Section 2 of this paper we propose a much more universal algorithm that enables one to create prognostic models for both low- and high-dimensional dynamical systems. In Section 3 the new algorithm is used for construction of the prognostic model of a mesospheric PCS employing a nonstationary chaotic time series generated by a computer model of the system. We also consider criteria for choosing technical parameters of a prognostic model. In the concluding section, results of the present research are formulated and possible further development of the proposed approach and its application to analysis of real time series is discussed.

---

§ Several methods for revealing nonstationarity of TS were proposed (see, *e.g.*, ref. 21–23). Still another is part of the algorithm that will be described in this paper.

2. Algorithm of constructing a prognostic model by observed time series

2.1.

The proposed algorithm can be represented as the following step-by-step procedure:
(1) The phase space of the observed system is reconstructed.
(2) A prognostic model is elaborated in the form of a discrete time map written in an analytic form.
(3) A covariance matrix of parameters of the prognostic model is found, and parameter trends and their covariance matrix are extrapolated to the "future" (outside the scope of the initial TS).
(4) A bifurcation prognosis is constructed: a bifurcation sequence is predicted and the time dependence of the probability to reveal the predicted regimes of the system's behaviour is determined.

Methods of phase space reconstruction have been discussed in many papers (see, for instance, ref. 26 and the literature cited therein). Therefore, in Section 2.2 we will restrict ourselves to a brief consideration of the first step of the procedure and to some comments on its features within the framework of the proposed algorithm. The next step will be quite novel, which distinguishes the proposed algorithm from that proposed earlier[24,25] for elaboration of prognostic models of low-dimensional DSs. This step will be addressed in more detail in Section 2.3. Steps 3 and 4 do not differ "ideologically" from the corresponding steps of the "low-dimensional" algorithm. Nevertheless, bearing in mind the key importance of these steps in construction of bifurcation prognosis and aiming at giving the reader a complete picture of the proposed approach we will address in more detail the most significant aspects of the corresponding operations in Sections 2.4 and 2.5.

2.2.

In this work, as in ref. 24 and 25, the model of the observed DS is constructed in the form of a discrete time map. In conformity with the choice of a discrete model, the vector of state in the reconstructed phase space at the time instant t_k is specified by co-ordinates with delays:[27]

$$\vec{Y}(t_k) = \{y(t_k), y(t_k + \Delta t), \ldots, y(t_k + (d_E - 1)\Delta t)\}, \qquad (2)$$

where $y(t)$ is the initial time series, d_E is the dimension of phase space (minimal embedding dimension), and Δt is the delay time determined, for example, from the condition of the first minimum of the mutual information function.[28] The value of d_E was calculated by the "false neighbours" method.[26,29]

Further, in the reconstructed phase space we chose the Poincare section for which we used the zero of the second co-ordinate $y_k^{(2)} = y(t_k + \Delta t) = 0$. As a result, a new data series $\{\vec{X}_j = \vec{Y}(t_j)\}_{j=1}^{J}$ (where t_j are the instants at which the reconstructed phase trajectory intersects the Poincare section in a definite direction) was obtained from the initial TS. It is worthy of note that $\{\vec{X}_j\}$ can be determined to a finite accuracy that depends on the discretization scale of the initial TS.

2.3.

The model was constructed in the form of a certain function $f(.)$ approximating the discrete time map:

$$x_j = f(x_{j-1}, \ldots, x_{j-N}; t_{j-1}; \vec{\mu}) + \xi_j, \qquad (3)$$

$$f(x_{j-1}, \ldots, x_{j-N}; t_{j-1}; \vec{\mu}) = f_0(x_{j-1}, \ldots, x_{j-N}; \vec{\mu}_0) + \vec{\beta} \cdot t_{j-1} \cdot \vec{f}_1(x_{j-1}, \ldots, x_{j-N}; \vec{\mu}). \qquad (4)$$

Here, N is the order of mapping, ξ_j is the error of approximation, and $\vec{\mu} = \{\vec{\mu}_0, \vec{\mu}_1, \vec{\beta}\} \in \mathcal{R}^{\{M_0 + M_1 + B\}}$ is a set of parameters of the model. The evolution operator was written in the form (4) employing the hypotheses 1 and 2 formulated in Section 1.5. Namely, the slow dependence on time t of parameters of the observed DS and small relative variation of their values over the time interval of about several lengths of the initial TS allowed us to describe nonautonomy as a time-dependent correction $[\vec{\beta} \cdot t \cdot \vec{f}_1(.)]$ to the autonomous part of the model $f_0(.)$. In other words, slowness of the dependence of evolution operator on time means the possibility of its Taylor series expansion in time with a restricted number of terms. In this work we took into account only the linear terms of the expansion. Apparently, this approximation is equivalent to reconstruction of nonautonomy of the observed DS in the form of a linear time dependence only of those parameters that enter the model linearly, which is clearly seen in eqn. (4).¶

A nontrivial issue is a choice of the form of functions $f_0(.)$ and $\vec{f}_1(.)$. For the algorithm to allow for construction of prognostic models of various *high-dimensional* DSs, these functions must enable one to approximate nearly any single-valued function of an arbitrary number of variables with an arbitrary preset accuracy. Such universality is inherent in artificial neural networks[30] that allow one to increase approximation accuracy by a simple increase in the number of neurons, while the form of the function remains unchanged. Still another property of a neural network (NN) is its dissipativity outside the "learning region" (the region of variation of the arguments of the function within the scope of the initial TS), thus providing global stability of the obtained models.∥

In this research we chose as $f_0(.)$ and $\vec{f}_1(.)$ functions a perceptron, *i.e.*, the simplest three-layer NN,[30] in which the linearly entering parameters were sought in the form of linear functions of time:

$$f(x_{j-1}, \ldots, x_{j-N}; t_{j-1}; \vec{\mu}) = \sum_{i=1}^{m} (\alpha_i + t_{j-1}\beta_i)\tanh\left(\sum_{k=1}^{N} w_{ki} x_{j-k} + \gamma_i\right). \tag{5}$$

Here, m is the number of neurons in the hidden layer of the network and $\mu = \{\alpha, \beta, w, \gamma\} \in \mathcal{R}^{(N+3)m}$ is the complete set of model parameters. In eqn. (5) the number of NN inputs is chosen to be equal to the order of the mapping.

As soon as the specific form of the discrete time map has been chosen, construction of a model reduces to finding optimal values of its parameters. The slowness of the time dependence of the evolution operator, that was used in writing a prognostic model in the forms (3) and (4), enables us to employ the corresponding perturbation theory and optimise parameter values in two stages. At the first stage, we seek the values of the parameters $\tilde{v}, \tilde{w}, \tilde{\gamma}$ that would provide the minimum of "autonomous" r.m.s error.**

$$\chi_a^2 = \left\langle \left[x_j - \sum_{i=1}^{m} v_i \tanh\left(\sum_{k=1}^{N} w_{ki} x_{j-k} + \gamma_i\right) \right]^2 \right\rangle_j, \tag{6}$$

where $\langle . \rangle_j$ has the meaning of an average for the corresponding subscript. At the second stage, the values of parameters $w = \tilde{w}, \gamma = \tilde{\gamma}$ are fixed and the values of parameters $\bar{\alpha}, \bar{\beta}$ corresponding to the minimum of "nonautonomous" r.m.s. error are found:

$$\chi_n^2 = \left\langle \left[x_j - \sum_{i=1}^{m} (\alpha_i + t_j \beta_i)\tanh\left(\sum_{k=1}^{N} \tilde{w}_{ki} x_{j-k} + \tilde{\gamma}_i\right) \right]^2 \right\rangle_j. \tag{7}$$

The resulting model may be interpreted as an autonomous DS

$$x_j = \sum_{i=1}^{m} \bar{v}_i(t)\tanh\left(\sum_{k=1}^{N} \tilde{w}_{ki} x_{j-k} + \tilde{\gamma}_i\right) + \xi_j, \quad \bar{v}_i(t) = \bar{\alpha}_i + \bar{\beta}_i t, \tag{8}$$

the qualitative behaviour of which is determined by a *single* control parameter, namely, time t.

2.4.

The model constructed in Section 2.3 reproduces not only the topological structure of the observed attractor, but slow variations of this structure also. This makes it possible to investigate *future* NDSs of the observed system by constructing bifurcation prognosis, *i.e.* analyse the depen-

---

¶ Linear approximation is insufficient when quantitative characteristics of the observed attractor, that vary due to the system's nonautonomy, reach their extreme values *within* the initial TS. Such a situation occurs, for example, in construction of a prognostic model of a mesospheric PCS by a TS corresponding to the so-called "external" chaotic attractor of the given system (see ref. 24 and 25, as well as Section 3.1).
∥ Problems encountered when using NNs will be considered in Sections 3.1 and 3.4.
** Note that the search of the "best" approximation of the data series $\{x_j = y(t_j)\}_{j=1}^{J}$ by minimising the errors (6) and (7) corresponds to the use of the method of least squares, which imposes restrictions on the application of the algorithm to analysis of noisy TS. This issue will be addressed in Section 4.3.

dence of qualitative behaviour of model (8) with time. It is quite obvious that results of such analysis depend significantly on errors inevitable in constructing a prognostic model that are characterised by a random quantity ξ_j entering into eqn. (8). The cause of such errors is, firstly, non-zero accuracy of reconstruction of the phase trajectory and the corresponding point map with respect to *finite* and *discrete* TS. Secondly, error may be due to incomplete correspondence of the operator, describing evolution of the prognostic model, to the analogous operator of the DS that generated the initial TS. The most dangerous consequence of the above-mentioned errors in constructing the prognosis of qualitative behaviour of the DS by the prognostic model (8) is the error in seeking values of the parameters α and β controlling changes in the behaviour of the model in the course of time. Available information about the observed DS (*i.e.*, the initial time series) does not allow one to determine the magnitude of this error accurately. Nevertheless, reasonable physical assumptions can give an upper estimate of this error.

Assume that there exists a set of parameters (α^*, β^*) for which a prognostic model describes the observed DS *exactly*, and the differences between the model and the modelled system, characterised in eqn. (8) by a random quantity ξ_j, are generated exclusively by error in determining these parameters. Assume also that ξ_j is normally distributed white noise with zero mean. Then, parameters $\{\alpha, \beta\}$ are normally distributed random quantities with average values $\{\bar{\alpha}, \bar{\beta}\}$ minimising the "nonautonomous" error (7), with $\{\bar{\alpha}, \bar{\beta}\} = \{\alpha^*, \beta^*\}$, and with the covariance matrices $C_{\alpha\alpha}$, $C_{\alpha\beta}$, $C_{\beta\alpha}$, $C_{\beta\beta}$ determined from expansion of this error near the minimum:

$$\chi_n^2(\alpha, \beta) \approx \chi_n^2(\bar{\alpha}, \bar{\beta})(1 + J^{-1}((\alpha - \bar{\alpha})^T(C_{\alpha\alpha}^{-1})(\alpha - \bar{\alpha}) + (\alpha - \bar{\alpha})^T(C_{\alpha\beta}^{-1})(\beta - \bar{\beta})$$
$$+ (\beta - \bar{\beta})^T(C_{\beta\alpha}^{-1})(\alpha - \bar{\alpha}) + (\beta - \bar{\beta})^T(C_{\beta\beta}^{-1})(\beta - \bar{\beta}))). \tag{9}$$

2.5.

The last step is investigation of future NDPs of the observed DS, *i.e.*, construction of the prognosis of bifurcations of its qualitative behaviour. Within the framework of the proposed algorithm it is sufficient to study the dependence of qualitative behaviour of the prognostic model (8) on the single control parameter t. This dependence is specified by a random, normally distributed set of parameters $v(t)$ with an average

$$\bar{v}(t) = \bar{\alpha} + \bar{\beta}t, \tag{10}$$

and a covariance matrix

$$C_{vv}(t) = C_{\alpha\alpha} + t(C_{\alpha\beta} + C_{\beta\alpha}) + t^2 C_{\beta\beta}. \tag{11}$$

The random spread of parameter trend $v(t)$ means that the predicted bifurcations must be considered as random events, and the instants of the bifurcation transitions as random values. Consequently, prognosis of bifurcations is expected to give answers to the following questions: What is the probability with which the predicted bifurcations will occur up to a definite instant of time? What is the mathematical expectation of the instant of a specific bifurcation transition? What is the accuracy of prediction of an average instant of a specific bifurcation? What is the probability to reveal the type of the system's behaviour of interest to us at a given instant of time?

We use the Monte Carlo method to find answers to the formulated questions. Consider a set of "Π" supplementary to the $\bar{v}(t)$ trends constructed in the parameter space of the model so that the sets of parameters specified by them at an arbitrary time instant t should be described by a normal probability distribution characterised by the found dependences $\bar{v}(t)$ and $C_{vv}(t)$. On having constructed a bifurcation diagram (BD) corresponding to each supplementary trend we obtain an ensemble of bifurcation instants for each predicted bifurcation transition. To answer the questions posed above, it is sufficient to find the time dependence of the probability density of the instants of all the predicted bifurcations. For this we divide the entire timescale into sufficiently small intervals δt_i and record how often elements of each of the obtained ensembles fall within these intervals; this quantity is denoted by v_b^i. The time dependence of the probability density of the b-bifurcation, $\varphi_b(t)$, is defined as

$$\varphi_b = \frac{v_b^i}{\delta t_i \Pi}, \quad b = 1, \ldots, B. \tag{12}$$

Here, B is the total number of predicted bifurcations.

Knowledge of $\varphi_b(t)$ gives answers to the first three of the four questions formulated above if we determine in a standard manner the dependence on slow time of the integral probability Φ_b of the corresponding bifurcation:

$$\Phi_b(t^*) = \int_T^{t^*} \varphi_b(t) dt, \qquad t^* > T \tag{13}$$

and calculate the mathematical expectation of the bifurcation instant and the corresponding dispersion.

In addition, analysis of the whole set of functions $\varphi_b(t_s)$ allows us to determine the probability P_k with which the "k"-th type of behaviour of the system can be revealed at a given instant of time t^*:

$$P_k(t^*) = P_k^+(t^*)(1 - P_k^-(t^*)). \tag{14}$$

Here,

$$P_k^{+,-}(t^*) = \int_T^{t^*} \varphi_k^{+,-}(t) dt, \qquad t^* > T \tag{15}$$

and the functions $\varphi_k^+(t)$ and φ_k^- are the sums of probability densities of all possible bifurcations of the transition to and from the regime (type of behaviour), respectively. In eqn. (13) and (15) T denotes the boundary of the observed TS.

3. Construction of prognostic model of a mesospheric PCS

3.1.

In this section we demonstrate the proposed algorithm taking as a data source a system describing the behaviour of minor chemical constituents of the Earth's atmosphere in the mesopause region (heights of 70 to 90 km).[6,7] For the convenience of the reader, we present below a brief description of this system and its NDPs described earlier in ref. 7 and 31.

The system includes five first-order nonlinear differential equations under periodic external forcing:

$$\frac{dx_1}{dt} = -(a_9 + 2a_{11}x_1 + a_{10}x_3 + a_4 x_4 + a_5 x_5)x_1 + a_1 x_2 x_5 + a_{15} x_4^2 + a_{16} s(t) x_3 + 2a_8 s(t)$$

$$\frac{dx_2}{dt} = -(a_6 + a_{12} x_3 + (a_1 + a_2 + a_{14}) x_5) x_2 + a_4 x_1 x_4 + a_7 s(t) r$$

$$\frac{dx_3}{dt} = -(a_{10} x_1 + a_{12} x_2 + a_{13} x_4 + a_{16} s(t)) x_3 + a_9 x_1 \tag{16}$$

$$\frac{dx_4}{dt} = -(a_4 x_1 + 2a_{15} x_4 + a_3 x_5 + a_{13} x_3) x_4 + a_5 x_1 x_5 + a_{12} x_2 x_3 + 2a_{14} x_2 x_5 + a_7 s(t) r$$

$$\frac{dx_5}{dt} = -(a_5 x_1 + a_3 x_4 + (a_1 + a_2 + a_{14}) x_2) x_5 + a_6 x_2 + a_{13} x_3 x_4.$$

These are equations of chemical kinetics for 16 chemical reactions (including photolysis reactions) running in the mesosphere. The system contains five chemical species (O, H, O_3, HO, HO_2) whose concentrations are dynamical variables denoted, respectively, by $x_1, \ldots, x_5$. The dynamics of these species is determined by the magnitudes of the coefficients $a_1 \cdots a_{16}$ (constants of the reactions) and has a characteristic timescale of one day or less under mesospheric conditions. The concentrations of the other reagents participating in the reaction are much higher than those of the above five species and are supposed to be constant. Periodic external forcing is due to daily variations of exposure to light as a result of sunrise and sunset and is manifested by periodic modulation of photodissociation rates of the molecules of ozone, one of the dynamical variables of system (16), oxygen, the concentration of which is assumed to be constant, and water vapour, whose relative concentration "r"†† is one of the control parameters of the system. This modulation enters into system (16) in the form of the periodic function $s(t)$ that reaches its maximum $s_{max} = 1$ at midday and minimum $s_{min} = 0$ at midnight.‡‡ By virtue of such nonautonomy, the phase space of system (16) has dimension six.

The variable x_1 (the concentration of atomic oxygen [O]) is the dynamical variable whose time series "was observed", although this choice is not of fundamental importance. Fig. 1(a) displays a BD where the concentration of atomic oxygen at the instant of sunrise is on the vertical axis, and the magnitude of relative concentration of water vapour "r", that is the control parameter of the system, is on the horizontal axis. Clearly, the system possesses a wide set of dynamic regimes that are realised depending on the value of the control parameter. If several values of [O] correspond to some value of r, then a regime with a period of the same number of days is realised. The "dark" regions in the BD are for those intervals of parameter values in which the system behaves chaotically.

One can see that the system has three regions of chaotic behaviour in the considered range of variation of the control parameter. Of particular interest are two of them that are realised at relatively small values of $r \in [1.5; 2.1]$. The corresponding fragment is magnified to scale on the abscissa in Fig. 1(b). It is clearly seen that the two regions of chaotic behaviour partially overlap. This feature is a consequence of the bistable behaviour of the system accompanied by hysteresis with the variation of the control parameter. As the control parameter changes from the left to the right within the limits of Fig. 1(b), a sequence of bifurcations takes place. First, the motion with a

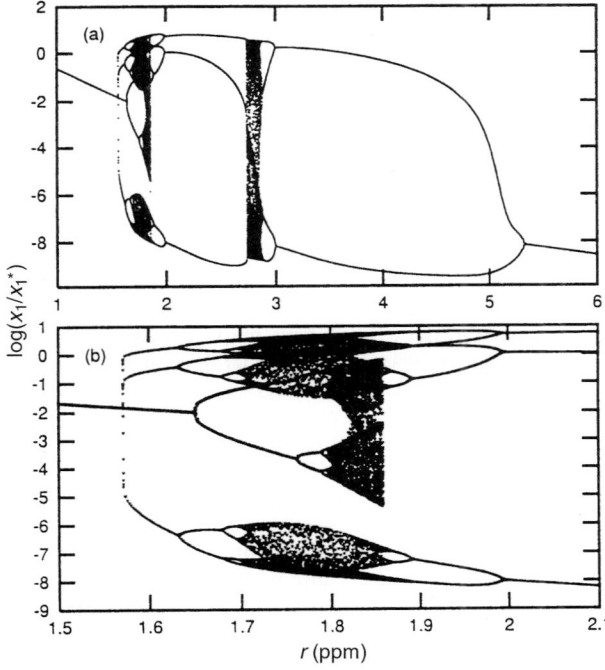

Fig. 1 (a) Bifurcation diagram of system (16); (b) fragment of bifurcation diagram with the scale extended along the abscissa axis. Logarithm of atomic oxygen concentration at sunrise, normalised to $x_1^* = 1.13 \times 10^9$ cm^{-3} along the vertical axis; value of relative concentration of water vapour r along the horizontal axis.

---

†† Given below in ppm units (particles per million) defined as the number of molecules of water vapour per million molecules of air.
‡‡ Detailed consideration of system (16) (values of coefficients a_1–a_{16}, ranges of variation of control parameter "r", form of function $s(t)$), and its nonlinear-dynamical properties can be found in ref. 31 and 32.

period of one day loses stability (at $r \approx 1.65$) and period doubling occurs (a stable regime with a period of two days is born). Then follows a cascade of such doublings, and the dynamics becomes chaotic at $r \approx 1.8$. We will call the resulting chaotic set an "internal attractor". Finally, for $r \approx 1.87$, this motion disappears and the system jumps into an absolutely different regime of behaviour.

With a reverse change of parameter, we start at $r = 2.1$ from stable motion with a period of three days which doubles at $r \approx 2$, quadruples at $r \approx 1.9$, and then a transition to a chaotic regime through a cascade of doublings occurs. The obtained chaotic set will be referred to as an "external attractor". As is clear from Fig. 6(b), this attractor consists of three regions and, with a period of external forcing of one day, the map jumps successively from one region into another. With a further decrease of the control parameter, an inverse cascade of doublings is initiated at $r \approx 1.7$, a regime with a three-day period is, again, realised at $r \approx 1.63$ and vanishes at $r \approx 1.58$, and the system passes to a stable one-day regime. Thus, catastrophic bifurcations occur in the system both with an increase and a decrease of parameter r.

3.2.

In this paper we use chaotic evolution of a mesospheric photochemical system (MPCS) "observed" within the basin of internal attractor for construction of prognostic models for the given dynamical system.§§

So, we generate a TS of variable x_1 having duration $T = 1500$ days within the limits of which parameter r decreases (linearly in time) from 1.855 to 1.82 (Fig. 2(a)), and make use of the algorithm described in Section 2.

First, we reconstruct the attractor in phase space and establish that (i) the minimal embedding dimension of system (16) is equal to 3; and (ii) the correlation dimension of the attractor is 1.9.¶¶

We take as the Poincare section the surface $\dot{x}_1 = 0$ and record the intersections of this section by the phase trajectory from the side of $\dot{x}_1 < 0$ (local minima of variable x_1). Such a section in the expanded phase space corresponds to the instant of sunrise. A first-order discrete time map corresponding to the chosen Poincare section is given in Fig. 2(b). This discrete time map is multi-valued (Fig. 2(b)) due to the presence of an "additional" branch in the left-hand side of the map. Since in ref. 24 and 25 we used the low-dimensional algorithm, the constructed prognostic model neglected the multivalued nature of the map. The high-dimensional algorithm described in Section 2 permits one to construct a model in the form of a higher-order discrete time map, thus avoiding rough approximation of the observed data.

The prognostic model was constructed in the form of the NN (8) (number of inputs being $N = 2$) approximating the second-order discrete time map extracted from the initial TS. The total number of model parameters was $\mu = \{\alpha, \beta, w, \gamma\} = (N + 3)\, m = 5m$, of which $2m$ parameters $\{\alpha_i, \beta_i\}_{i=1}^{m}$ determine the behaviour of the model outside the scope of the initial TS. The number of neurons m is a technical parameter of the model; its influence on the characteristics of prognosis of qualitative behaviour of the system will be discussed in Section 3.3.

In line with the general algorithm, the NN was learnt in two stages. At the first stage, the values of the parameters $\{\tilde{v}, \tilde{w}, \tilde{\gamma}\}$ minimising the autonomous error (6) were sought; at the second stage $\{\tilde{w}, \tilde{\gamma}\}$ were fixed, and the values of the parameters $\{\tilde{\alpha}, \tilde{\beta}\}$ minimising the nonautonomous error (7) were found. The minimum of the first error was sought by *Variable Metric Methods in Multidimensions* (also called *quasi-Newton methods*),[33] and the minimum of the second one by the *Singular Value Decomposition (SVD) Method*.[33] At the first stage we had to take into account that a complicated nonlinear dependence of the NN (8) on parameters $\{w, \gamma\}$ may result in the existence of many local minima of the error (6). For construction of a model that would be maximally sensitive to a change of dynamical variables within the range of values corresponding to the initial TS, the "observed" series $\{x_j\}$ was normalised to provide $\langle x_j \rangle_j = 0$, $\langle x_j^2 \rangle_j = 1$, after which the right-hand side of (6) was supplemented by the factor $(1 + \lambda(4m)^{-1} \sum_{i=1}^{m} (v_i^2 + w_{1i}^2 + w_{2i}^2 + \gamma_i^2))$,

---

§§ No specific problems arise when a prognostic model is constructed by the time series generated by two other chaotic attractors given in Fig. 1(a) and (b).
¶¶ The cause of such low values of the dimensions of system (16) was discussed in detail in ref. 31 and 32.

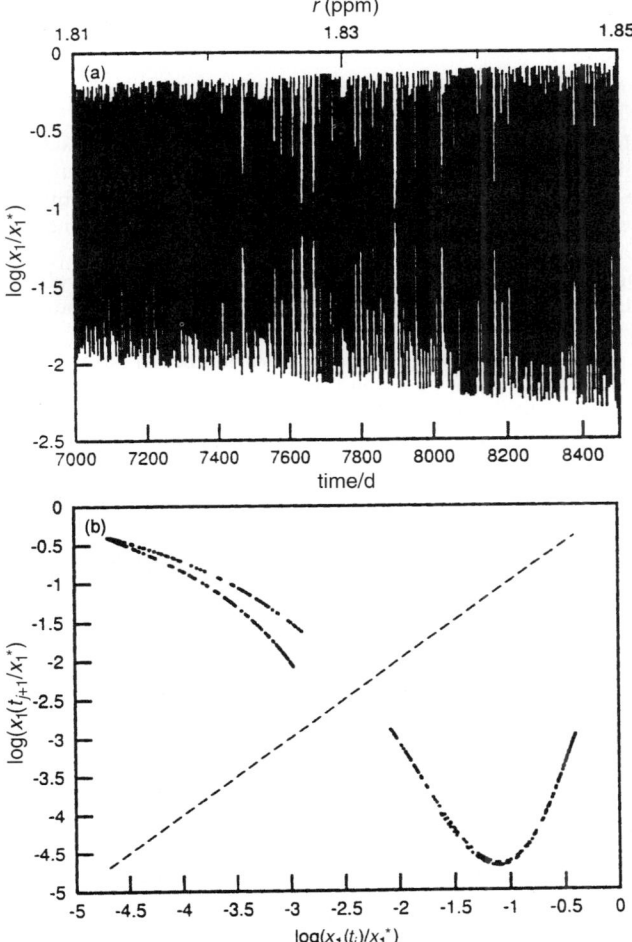

Fig. 2 (a) "Observed" TS employed for construction of prognostic model. Logarithm of atomic oxygen concentration, normalised to $x_1^* = 1.13 \times 10^9$ cm^{-3} along the vertical axis; time along the lower horizontal axis and current value of r along the upper horizontal axis; (b) first-order discrete time map reconstructed by the given TS.

where λ is the technical parameter of order unity, and the minimum of the autonomous error corrected in this manner was sought.‖‖

Further, steps 3 and 4 of the general algorithm were accomplished. Namely, the error (7) was expanded into series (9) and the covariance matrix $\begin{pmatrix} C_{\alpha\alpha} & C_{\alpha\beta} \\ C_{\beta\alpha} & C_{\beta\beta} \end{pmatrix}$ of parameters $\{\alpha, \beta\}$ characterising nonautonomy of the prognostic model was calculated.

Still further, the time dependence of qualitative behaviour of the prognostic model was investigated. For this, 1000 trends of parameters $v(t) = \alpha + \beta \, t$ were specified so that, at each time instant, distribution of these parameters should be normal, with the average (10) and the covariance matrix (11). Analysis of changes in the behaviour of the model in the "future"*** determined by each trend, allows one to set up a statistical ensemble of bifurcation instants for each of the

‖‖ Note that for $\lambda \ll 1$ this procedure means minimisation of (6) with minimal norm of the values of parameters $\{v_i, w_{1i}, w_{2i}, \gamma_i\}_{i=1}^m$.
*** Apparently, the "future" is conventional here because the proposed alogirithm allows for analysis of the changes in the model's behaviour in both directions of the time axis.

bifurcation transitions found and calculate the probability characteristics (12)–(14) of the predicted bifurcations and regimes of behaviour.

3.3.

Before we pass to discussion of the results of the considered research, consider an important issue of choosing a technical parameter of model (8), namely, the number of neurons m. Since the order of the discrete time map N (the number of inputs of the NN) is defined by the dimension of the reconstructed phase space of the observed system (the minimal embedding dimension), m determines completely the total number of prognostic model parameters $\mu = \{\alpha, \beta, w, \gamma\} = (N + 3) m$. Therefore, we will actually speak about the optimal number of parameters of the model.

On the one hand, for the model to guarantee sufficiently accurate reproduction of the observed evolution of the system, the number of parameters must not be too small: the reproduction error is obviously a monotonically decreasing function of the number of parameters. This statement is illustrated in Fig. 3(a) and (b), where the monotonically decreasing autonomous (6) and nonautonomous (7) errors, normalised to dispersion of the observed signal, are plotted vs. the number of neurons m.

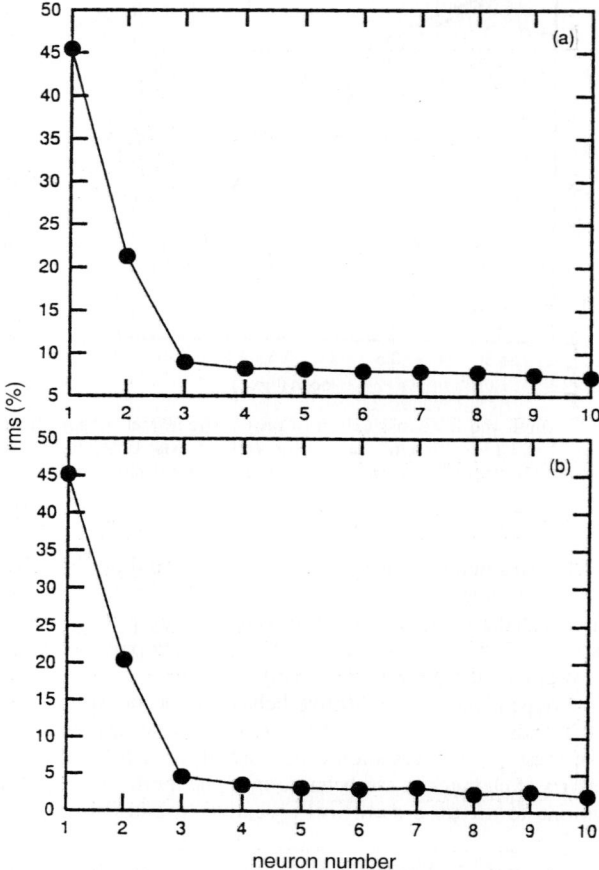

Fig. 3 Least r.m.s. error plotted vs. the number of neurons in the hidden layer of NN (8): (a) for autonomous error (6); (b) for nonautonomous error (7).

Our task, however, is not pinpoint accuracy of reproducing the observed evolution, but *prognosis of changes in the qualitative behaviour* of the observed system. The informativity of such a prognosis within the scope of the algorithm under consideration is determined by the probability characteristics (12)–(14). Clearly, the prognosis is the more informative the more accurately the bifurcation instants are predicted and the greater the probability of revealing the most probable regime of behaviour at the time instant of interest. Both these characteristics depend significantly on "arrangement" of m-dimensional space of nonautonomous parameters of the model $\{v_i\}_{i=1}^m = \{\alpha_i + t\beta_i\}_{i=1}^m$. An increase in m means an increase in the dimension of this space and, generally speaking, the complication of its structure. The resulting number of qualitatively different regimes of behaviour co-existing in time for different trends $v(t)$, forming a statistical ensemble of bifurcation instants, can increase too. Therefore, it should be expected that a too large number of parameters will result in deterioration of the characteristics of prognosis.

So, we can conjecture that there exists an optimal range of the number of model parameters that provides the most informative prognosis of qualitative behaviour of observed DS with the aid

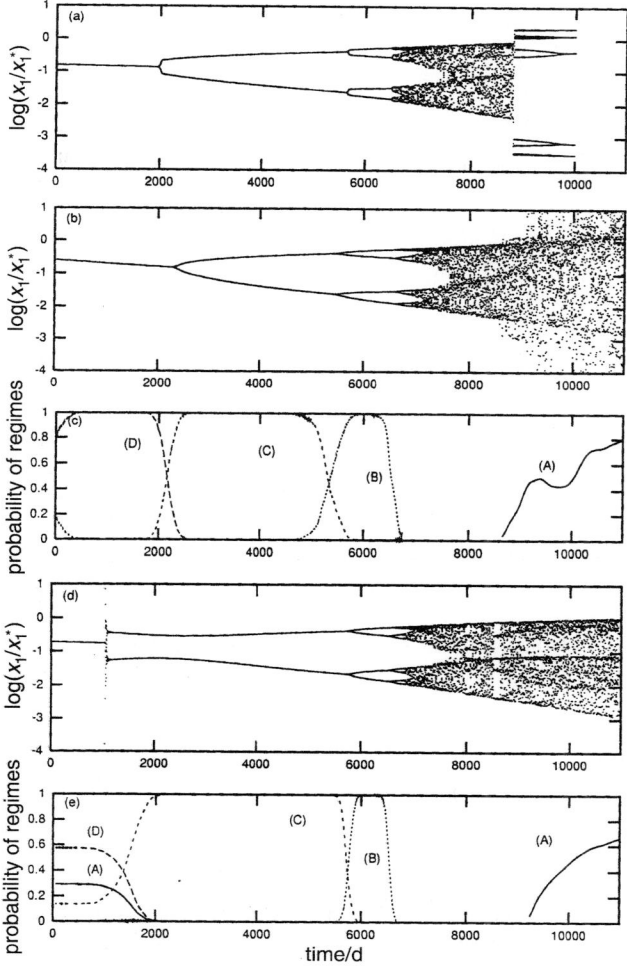

Fig. 4 (a) Fragment of bifurcation diagram of system (16) from Fig. 1. Prognosis of qualitative behaviour of mesospheric PCS constructed using model (8): (b), (d) bifurcation diagrams corresponding to average trend (10) with the number of neurons $m = 3$ and $m = 4$, respectively; (c), (e) time dependence of the probability to reveal different regimes: "regime" of catastrophe (A); oscillatory regimes with a period of four (B), two (C) and one (D) days for $m = 3$ and $m = 4$, respectively.

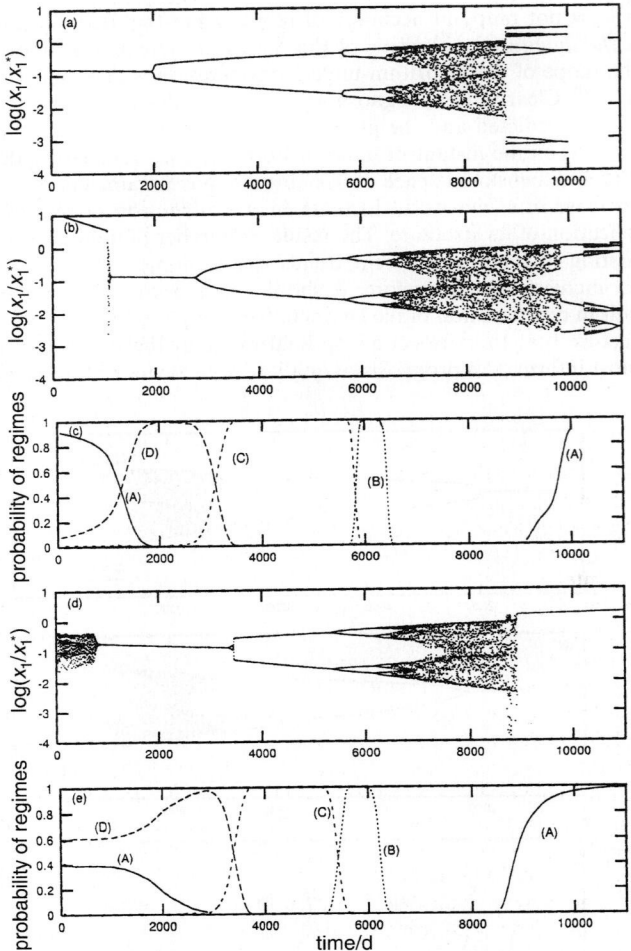

Fig. 5 The same as in Fig. 4 but for $m = 5$ (b) and (c) and $m = 6$ (d) and (e).

of a prognostic model of the given form. The lower limit of this range (*i.e.*, the minimal needed number of parameters) can be found in a standard manner by analysing the dependence of the quality of reproduction of the *observed* behaviour of a DS on the number of parameters of the model. For example, for the situation under consideration, analysis of Fig. 3(a) and (b) leads straightforwardly to the conclusion that the number of neurons in the NN (8) must be not less than three: $m \geqslant 3$. As to the upper limit of the range, we do not find it possible to make any conclusions *before* we construct the prognosis; moreover, it is necessary to compare the characteristics of the prognoses constructed using a different number of parameters (different number of neurons in the hidden layer of the NN in the model (8) of interest to us). It is also worthy of note that the informativity of the prognosis using a "good" prognostic model must depend sufficiently weakly on the number of parameters, if this number falls within the optimal range.

To conclude this section we would like to note the following circumstance. Representation of the evolution operator in the form of the expansion (4) in time explicitly entering this expression means that the proposed algorithm can be employed as long as the nonautonomous part of the operator is of the order of its autonomous part or less. Our calculations verify that, for models with $m \in [3; 10]$, the applicability limit of the algorithm is a time interval having duration of about four lengths of the initial TS $T \in [7000; 8500]$ calculated from its centre (moment $t = 7750$). Thus, we can expect to obtain a correct prognosis over the time interval $1750 \approx < t < \approx 13\,750$.†††

3.4.

We investigated prognoses of bifurcations constructed using model (8) with the number of neurons $m \in [1; 10]$. As was to be expected, for $m = 1, 2$ the model does not reproduce even the observed evolution of the system. The most informative was prognosis for $m = 5, 6$. For a greater number of neurons, prognosis informativity decreases.

In Fig. 4(b) and (d), 5(b) and (d) and 6(b) we present the bifurcation diagrams corresponding to the average trend (10) for $m = 3$–7, respectively. Time dependences of the probability to reveal different predicted regimes of behaviour, determined by the expression (14), are shown on the panel below each diagram (Fig. 4(c) and (e), 5(c) and (e) and 6(c)). This characteristic allows one to judge the accuracy of prediction of bifurcation transition instants as well as the maximal probability of the regime most probable at the current time instant. We remind the reader that all the results given here were obtained *exclusively* on the basis of analysis of the TS that represented variations of one dynamical variable (concentration of atomic oxygen) over the time interval $T \in [7000; 8500]$ days (see Fig. 2(a)) *without* any other information about the system (16) that generated it. For estimation of the quality of prognosis, correct bifurcation diagrams obtained by direct computation of the system (16) and describing variations of qualitative behaviour of this system over the entire time interval are shown on the upper panel of each figure (Fig. 4(a), 5(a) and 6(a)). Finally, the bifurcation diagram corresponding to the average trend and time dependences of the probability of predicted regimes plotted for the same initial TS (see Fig. 2(a)) but using a low-dimensional algorithm[24,25] are given for comparison in Fig. 6(d) and (e).

Prognoses for different values of parameter m will be compared by analysing the time dependence of the probabilities of the predicted regimes, $P_k(t)$. Being the most typical "representatives" of the corresponding statistical ensembles, the bifurcation diagrams for the average trend (10) may, however, reproduce local features of the structure of the space of parameters of prognostic model, that are not typical of the statistical ensemble as a whole. For example, in Fig. 4(d) one can see that no bifurcations are predicted in the "future" $t > 8500$ prognosticated within the limits of this panel for a model with $m = 4$, whereas statistical analysis shows (see Fig. 4(e)) that a catastrophic bifurcation will occur by the time instant $t = 11\,000$ with probability $P_\infty \simeq 0.62$, as a result of which the chaotic attractor that generated the initial TS will disappear.

As noted above, comparison of the time dependences of the regime probabilities depicted in Fig. 4(c) and (e), 5(c) and (e) and 6(c) shows that the most informative are prognoses made employing models with $m = 5$ and $m = 6$. Indeed, one can see in Fig. 5(c) and (e) that the maximal probabilities of all predicted regimes are equal to unity in this case. An exception is a catastrophic "regime", the most remote into the "past" ($t < 7500$), to which the model passes when a single-periodic regime breaks up. Comparison with the correct bifurcation diagram (Fig. 5(a)) reveals that this regime, unlike all the others, was predicted erroneously; the observed system has no such regime. However, it should be taken into consideration that the probability of this "false" regime is rather small at the boundary of the prognosis interval (for $t \approx 1750$): $P_\infty \approx 0.2$ for $m = 5$ and $P_\infty \approx 0.3$ for $m = 6$.

Models with a larger or smaller number of neurons predict much worse the catastrophic regime that actually occurs in the future for $t > 8800$. Note that the corresponding catastrophic bifurcation is a natural restriction of the interval of prognosis into the "future". Moreover, in good agreement with what was said in Section 3.3, models with a large number of neurons provide a better accuracy of prediction of the instants of nearest bifurcations in the "past", and a model with $m = 3$ gives the most accurate prognosis of the bifurcation of the birth of a single-periodic regime, that is the most remote into the past, which is not prognosticated at all by models with

---

††† It is the upper estimate of the prognosis interval. The interval of prognosis may be, for example, shortened as a result of a catastrophic bifurcation, when the system passes to the earlier unattainable region of phase space.

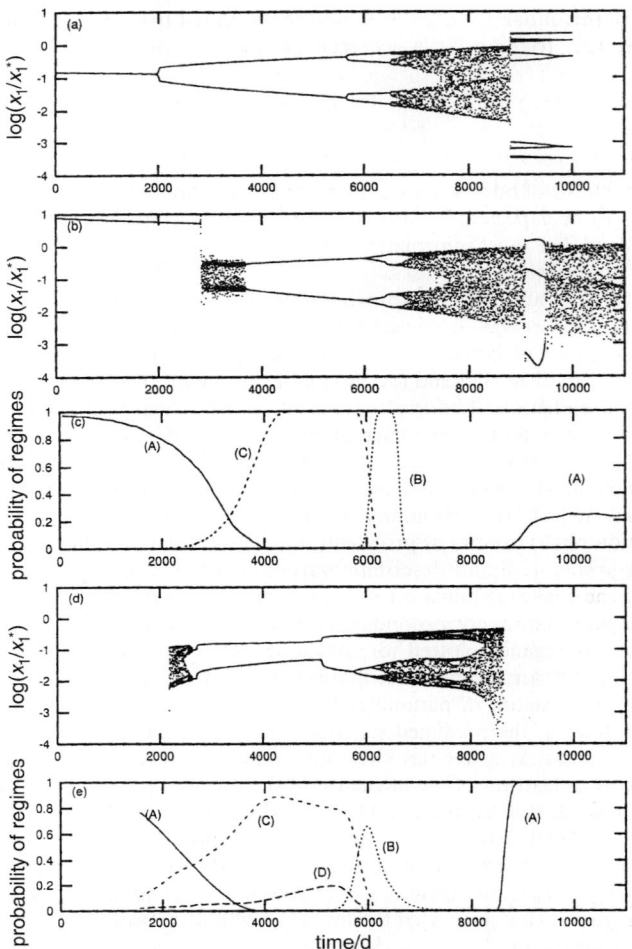

Fig. 6 The same as in Fig. 4 but for $m = 7$ (b) and (c) and calculated using low-dimensional algorithm[24,25] (d) and (e).

$m \geqslant 7$. Nevertheless, we can conclude that the informativity of the prognosis constructed with the aid of model (8) depends weakly on the number of parameters for $m \in [3; 7]$.

Let us make some conclusions from the results presented in Fig. 4–6:

(1) The prognostic model (8) with the optimal number of neurons $m = 5$ and $m = 6$ predicted correctly all the bifurcation transitions within the prognosis interval, with the maximal probabilities of all regimes of behaviour predicted over this interval being equal to unity.

(2) The "future" catastrophe—the nearest bifurcation to the boundaries of the initial TS—is predicted worse than other bifurcations that are more remote in time. This result has two causes. Firstly, dissipativity of a NN that provides global stability of the prognostic model decreases sensitivity of the model *outside* the "learning region" (the region of variation of the arguments of a discrete time map *within* the initial TS). Secondly, it is exactly the situation that occurs for the internal attractor of a mesospheric PCS: the characteristic size of the attractor grows with approaching catastrophe. Consequently, model (8) prognosticates the future worse than the past, recession into which results in a decreased range of variation of dynamical variables.

(3) The prognosis constructed employing a low-dimensional algorithm[24,25] shown in Fig. 6(d) and (e) predicts bifurcations and regimes of behaviour in the past much worse than the NN (8),

but it is better in predicting the future catastrophe. The cause of a poor prognosis of the "past" is insufficiently accurate reproduction of the observed behaviour of the DS, which is due primarily to the use of evolution operator in the form of a first-order discrete time map. However, the form of this function—the fifth degree polynomial—provides a more informative prognosis of the "future" bifurcation that is very near in time. Clearly, the use of polynomial functions of many variables in a "pure" form for construction of prognostic models of high-dimensional DSs is hardly promising because of the inevitable global instability of such models. However, hybrid models that combine, for instance, a NN as an autonomous component of the prognostic model (3),(4) and a system of multidimensional polynomials orthogonalised on the observed attractor as a nonautonomous component may be an optimal solution for modelling some high-dimensional DSs.

4. Conclusion

4.1.

We proposed an algorithm for constructing prognostic models of systems demonstrating complex (high-dimensional) dynamic behaviour. The basis for this algorithm is an investigation of a NDPs of the system exclusively by analysing the TS that represents variations of one dynamical variable, *without* any additional information about the system that generated this TS. Prognostic models are intended for prediction of qualitative behaviour of observed DS and its bifurcations. The algorithm essentially broadens the scope of the general approach to analysis of nonstationary TS recently proposed by the authors.[24,25]

The algorithm was used for analysis of the TS generated by a computer model of a mesospheric photochemical system. We employed the TS calculated for slow change in the control parameter of the system, namely, relative concentration of water vapour. The duration of the "observed" series was restricted so that the system demonstrated *only one*—chaotic—type of behaviour, *without* any bifurcations.

The constructed prognostic model enabled us to make a correct prognosis of bifurcation sequences and calculate probabilities to reveal at the time instant of interest predicted regimes of the system's behaviour for times much greater than the length of the initial TS.

4.2.

Of basic importance for application of the developed approach is *prediction of bifurcations of the more complex behaviour to the simpler one*: the dimension of attractors arising in the phase space of a DS as a result of bifurcations must not exceed the dimension of the attractor corresponding to the initial TS. It is quite obvious, for example, that the described general procedure does not allow for prediction of a cascade of period doubling bifurcations for the case when the initial TS contains information about the type of behaviour that is simplest for such a cascade. We emphasise that the proposed algorithm enables one to overcome windows of regular behaviour between chaotic attractors, but use of the most complex type of behaviour is decisive for construction of a prognostic model.

4.3.

The assumption that the noise is additive and dynamical, to which corresponds additive allowance for the random component in eqn. (3), imposes significant restrictions in terms of the applicability of the algorithm proposed in this research to analysis of real data bases. The assumption of additivity imposes purely quantitative restrictions on the degree of noise of the analysed TS. Whereas the dynamical noise enabled us to use the simplest way for optimising parameters of the model—the method of least squares. This permitted us to write cost functions (errors) in the form (6), (7), to find covariance matrices of the parameters characterising nonautonomy of the system from eqn. (9), and to use for these parameters a normal distribution function. It is clear, however, that a stochastic component of the actual TS nearly always contains the so-called measurement noise. Under these conditions, the method of least squares gives a systematic error in seeking optimal values of parameters.[34] Then, the procedure of finding a distribution function of the parameters of a prognostic model based on this method also becomes incorrect. One of the ways to solve this problem is to employ the method of total least squares.[35] The corresponding modification of the algorithm is being undertaken at present, and we expect that it will enable us to use the elaborated approach for construction of prognostic models of real atmospheric processes on the basis of observed TS.

4.4.

Consider, in conclusion, possible practical applications of the prognostic models. We take as an example the construction of a prognostic model of a system determining ozone layer evolution in middle and low latitudes. The analysis of satellite measurements of ozone abundance in an atmosphere in ref. 14 showed that the corresponding time series is generated by a chaotic (strange) attractor, which makes this dynamic system an appropriate candidate for construction of a prognostic model.

The prognostic model is expected to provide the following opportunities:

First, reconstruction of nonautonomy of the considered system and, consequently, finding nonstationary characteristics of the ozone layer.

Second, prediction of bifurcations, *i.e.*, of qualitative changes in ozone evolution caused by nonstationarity.

Third, verification of the available "first principles" models by which the future state of the ozone layer is estimated at present. We mean comparison of NDPs of the prognostic model and of the "first principles" model. Note that the system under consideration is typical for the Earth's atmosphere in the sense that direct analysis of the NDPs of the "first principles" model is nearly impossible because of a high order of this system. Therefore, for verification we will first have to construct the so-called essential (or basic) dynamical model possessing the minimal possible number of degrees of freedom and, at the same time, retaining the NDPs of the complete "first principles" model. The general algorithm of constructing dynamical models was first described in ref. 31 and was used successfully for construction of essential dynamical models of the polar lower stratospheric[1] and mesospheric[31] PCSs.‡‡‡ Discrepancy between the NDPs of the essential and prognostic models will indicate that the essential and, hence, the complete "first principles" model must be corrected.

Finally, comparison of the verified "first principles" model with the prognostic model will allow one to set a correspondence between dynamical variables and nonautonomous parameters of the prognostic model and real physical and chemical characteristics of the ozone layer. This will enable one (i) to identify the "principal" dynamical variables that determine ozone layer evolution; (ii) to ascertain the trends of which parameters may lead the system to the revealed bifurcations and (iii) estimate the bifurcation instants and their quantitative consequences.

Acknowledgements

The research was done under support of the RFBR (project 99-02-16162).

References

1 A. M. Feigin and I. B. Konovalov, *J. Geophys. Res.*, 1996, **101**(D20), 26023.
2 I. B. Konovalov, A. M. Feigin and A. Y. Mukhina, *J. Geophys. Res.*, 1999, **104**(D3), 3669.
3 L. I. Kleinman, *J. Geophys. Res.*, 1991, **96**, 721.
4 L. I. Kleinman, *J. Geophys. Res.*, 1994, **99**, 16831.
5 T. Thayaparan, W. K. Hocking and J. MacDougall, *J. Geophys. Res.*, 1997, **102**, 9461.
6 B. Fichtelmann and G. Sonnemann, *Ann. Geophys.*, 1992, **10**, 719.
7 G. Sonnemann and B. Fichtelmann, *J. Geophys. Res.*, 1997, **102**, 1193.
8 G. Sonnemann and A. M. Feigin, *Phys. Rev. E*, 1999, **59**(2-A), 1719.
9 G. R. Sonnemann, A. M. Feigin and Y. I. Molkov, *J. Geophys. Res.*, 1999, **104**(D23), 30591.
10 J. C. Farman, B. G. Gardiner and J. D. Shanklin, *Nature*, 1985, **315**, 207.
11 *National Research Council. Causes and Effects of Changes in Stratospheric Ozone: Update 1983*, National Academy Press, Washington, DC, 1984.

---

‡‡‡ A particular version of this algorithm was quite recently published in ref. 36.

12 S. Solomon, *Rev. Geophys.*, 1988, **26**(1), 131.
13 S. Solomon, *Nature*, 1990, **347**, 347.
14 P. Yang, G. P. Brasseur and J. C. Gille *et al.*, *Physica D*, 1994, **76**, 331.
15 I.-F. Li, P. Biswas and S. Islam, *Atmos. Environ.*, 1994, **28**, 1707.
16 J. D. Neelin and M. Latif, *Phys. Today*, 1998, **N12**, 32.
17 B. Wang, A. Barcilon and Z. Fang, *J. Atmos. Sci.*, 1999, **56**, 5.
18 H. D. I. Abarbanel, R. Huerta and M. I. Rabinovich *et al.*, *Neural Comput.*, 1996, **8**(8), 1567.
19 G. W. Frank, T. Lookman and M. A. H. Nerenberg *et al.*, *Physica D*, 1990, **46**, 427.
20 H. N. Srivastava, S. N. Bhattacharya and K. C. Sinha Ray, *Geophys. Res. Lett.*, 1996, **23**(24), 3519.
21 R. Manuka and R. Savit, *Physica D*, 1996, **99**, 134.
22 T. Schreiber, *Phys. Rev. Lett.*, 1997, **78**, 843.
23 A. Witt, J. Kurths and A. Pikovsky, *Phys. Rev. E*, 1998, **58**, 1800.
24 A. M. Feigin, Y. I. Molkov, D. N. Mukhin and E. M. Loskutov, *Preprint No. 508, Institute of Applied Physics, RAS*, 1999, pp. 1–53.
25 A. M. Feigin, Y. I. Molkov, D. N. Mukhin and E. M. Loskutov, *Radiophys. Quantum Electron.*, 2001, **44**, 348.
26 H. D. I. Abarbanel, *Analysis of Observed Chaotic Data*, Springer-Verlag, New York, 1997.
27 F. Takens, in: *Dynamical Systems and Turbulence*, ed. D. A. Rand and L.-S. Young, 1981, Springer, Berlin, vol. 898, p. 366.
28 A. M. Fraser and H. L. Swinney, *Phys. Rev. A*, 1986, **33**, 1134.
29 M. B. Kennel, R. Brown and H. D. I. Abarbanel, *Phys. Rev. A*, 1992, **45**, 3403.
30 *The Handbook of Brain Theory and Neural Networks*, ed. M. A. Arbib, The MIT Press, 1995.
31 A. M. Feigin, I. B. Konovalov and Y. I. Molkov, *J. Geophys. Res.*, 1998, **103**(D19), 25447.
32 I. B. Konovalov and A. M. Feigin, *Nonlinear Proc. Geophys.*, 2000, **7**, 87.
33 H. W. Press, B. P. Flannery and S. A. Teukolskyl, *Numerical Recipes*, Cambridge University Press, Cambridge, 1990.
34 P. E. McSharry and L. A. Smith, *Phys. Rev. Lett.*, 1999, **83**, 4285.
35 S. Van Huffel and J. Vandewalle, *The Total Least Squares Problem*, SIAM, Philadelphia, 1991.
36 L. V. Kalachev and R. J. Field, *J. Atmos. Chem.*, 2001, **39**, 65.

Low-dimensional manifolds in tropospheric chemical systems

Alison S. Tomlin,*[a] Louise Whitehouse,[a] Richard Lowe[a] and Michael J. Pilling[b]

[a] *Department of Fuel and Energy, University of Leeds, Leeds, UK LS2 9JT*
[b] *School of Chemistry, University of Leeds, Leeds, UK LS2 9JT*

Received 4th April 2001
First published as an Advance Article on the web 13th November 2001

Ordinary differential equations derived from large nonlinear systems of chemical reactions are computationally expensive to solve because of the large number of coupled species and the large range of time-scales present. The range of time-scales often spans several orders of magnitude leading to stiff systems of equations requiring implicit numerical techniques. The use of slow manifolds for the description of long time-scale chemical processes has two advantages in that it reduces the number of variables required and also the stiffness of the chemical system by assuming that the fast time-scales are in local equilibrium with respect to the slower ones. The method exploits the existence of a low-dimensional manifold within a large-dimensional species phase space onto which the system quickly collapses. This paper investigates the existence of slow manifolds for nonlinear tropospheric chemical systems and presents a simple method for estimating the local dimension of the manifold using linear perturbation theory. The method is demonstrated for several tropospheric mechanisms over diurnal simulations including a subset of the Master Chemical Mechanism describing butane oxidation and the formation of ozone in the troposphere. It is shown that the intrinsic dimension of the slow manifold varies diurnally and depends on photolytic processes and the relative concentrations of major pollutants.

1 Introduction

Time dependent chemical reaction systems are often important subcomponents of larger reactive flow models with applications in combustion, chemical reactor design and atmospheric modelling. If detailed reaction mechanisms are used then a large number of coupled chemical species need to be represented and the chemical model often consumes a considerable part of available computational resources. For example, the Leeds Master Chemical Mechanism (MCM)[1] describing tropospheric gas phase chemistry, contains 3500 coupled species. Further difficulties arise because of the large range of time-scales present in detailed mechanisms. Intermediate radical species may have relaxation times of at least ten orders of magnitude less than stable species, leading to stiff systems of equations that must be solved using special numerical techniques such as implicit methods. Although certain simplifications such as approximate Jacobian methods can be used for stiff systems, the elimination of fast time-scale processes offers significant advantages in that it reduces both the number of variables and the stiffness of the system. It is often possible to assume a local equilibrium with respect to the fastest time-scales leading to a reduced set of equations describing the long term behaviour of the system on a slow manifold of lower dimension than the original phase space. Using such techniques for determining reduced chemical models is often essential for two- and three-dimensional reactive flow calculations where even the use of large parallel computers does not provide the resources required for the inclusion of detailed chemical reaction schemes.

Several issues arise relating to the use of slow manifold techniques. The first is the dimension of the slow manifold required to give an accurate representation of the full system over the chosen time period. Ideally the lowest dimension capable of accurately representing the important system dynamics should be determined so that only a small number of equations is required in the reduced model. The second issue is that qualitative changes in the long time dynamics of the system should not occur as a result of describing the system on a slow manifold. For example, if oscillatory behaviour exists in the full system it should be retained in the reduced system.[2] Finally practical methods for model description on the slow manifold need to be implemented as efficiently as possible. Several techniques have been used previously.[3] The quasi steady-state approximation (QSSA)[4] is the oldest technique and is used to define algebraic equations for the concentration of fast chemical species. In combustion models the intrinsic low-dimensional manifold (ILDM) method[5] is often used. The ILDM adopts a geometric approach by assuming that after the fast time-scales have collapsed the system can be described by look-up tables of rates of change on a lower-dimensional hyperspace or slow manifold. Application of the ILDM concept has also been made through the use of fitted polynomial representations of changes in concentration over a fixed time-step for a reduced number of variables on the slow manifold (sometimes called repro-modelling).[6–9] The latter has been applied more extensively than traditional look-up tables in atmospheric chemistry systems, perhaps because of the difficulties of searching in high-dimensional tables which may arise in diurnally varying atmospheric models.

Usually, application of the ILDM takes no account of the dimension changing with time/space during a simulation and a constant dimension is used for the look-up table or repro-model. In principle however, varying dimensions could be used and therefore methods for calculating the changing local intrinsic dimension of the manifold over time are required. Section 2 of this paper will present a linear perturbation method for the evaluation of system time-scales. Section 3 will discuss couplings between species through the left and right eigenspaces of the linearised system and how species may be related to time-scale modes. Section 4 presents a simple method for the estimation of local intrinsic manifold dimension and Section 5 its evaluation for a simple 3-variable system. Section 6 demonstrates the method for a nonlinear 6-variable system and a tropospheric butane oxidation scheme.

2 The evaluation of system time-scales

Given that the ILDM relates to the collapse of fast equilibrium processes it follows that the local intrinsic manifold dimension will depend on the time-scales of the system of equations. For equation systems of low dimension the investigation of such time-scales can be carried out through a non-dimensionalisation process. Small parameters can then often be identified indicating fast variables. For larger systems such as those typically found in tropospheric chemical mechanisms, non-dimensionalisation may be impractical and hence perturbation methods are generally used to investigate system dynamics. By studying the evolution of a small disturbance or perturbation to the nonlinear system it is possible to reduce the problem to a locally linear one. The resulting set of linear equations is easier to solve, and information can be obtained about the local time-scales and stability of the nonlinear system. Usually such a stability analysis is applied to a fixed point, for example to establish the stability of a stationary state of the system. In this work it is applied to a more general case of systems with time-scale separation *i.e.* it is used to study the relaxation of a fast variable onto a manifold representing slower processes. Useful information can therefore be derived from the analysis since the concentrations of the slow species will not change significantly over the time-scale of relaxation of the perturbed species.

The rate of change of concentration in a homogeneous reaction system can be described by the following system of ordinary differential equations (ODES):

$$\frac{d\boldsymbol{c}}{dt} = \boldsymbol{f}(\boldsymbol{c}, \boldsymbol{k}), \quad \boldsymbol{c}(0) = \boldsymbol{c}^0, \tag{1}$$

where c is an n-dimensional concentration vector and k the vector of reaction rate coefficients. Where operator splitting is used in a reactive flow problem this formulation describes the chemical reaction step and $f(c, k)$ can be represented as a function of the reaction rates:

$$f_i(c, k) = \sum_j v_{ij} R_j. \tag{2}$$

Here i is the species number, v_{ij} is the stoichiometric coefficient of the ith species in the jth reaction and R_j is the jth reaction rate.

If we disturb the system of eqn. (1) at a particular point c^{fx} with a *small* perturbation Δc_i so that:

$$c_i = c_i^{fx} + \Delta c_i \tag{3}$$

and expand as a Taylor series, we get:

$$\frac{\partial c_i}{\partial t} = f(c_i^{fx}) + \left(\frac{\partial f_i}{\partial c_1}\right)_{fx} \Delta c_1 + \left(\frac{\partial f_i}{\partial c_2}\right)_{fx} \Delta c_2 + \cdots + \left(\frac{\partial f_i}{\partial c_n}\right)_{fx} \Delta c_n + \cdots \tag{4}$$

Substituting

$$\frac{\partial c_i}{\partial t} = \frac{\partial c_i^{fx}}{\partial t} + \frac{\partial \Delta c_i}{\partial t} = f(c_i^{fx}) + \frac{\partial \Delta c_i}{\partial t} \tag{5}$$

then to leading order, we are left with a set of linear equations describing the motion of the perturbation:

$$\frac{\partial \Delta c_i}{\partial t} = \left(\frac{\partial f_i}{\partial c_1}\right) \Delta c_1 + \left(\frac{\partial f_i}{\partial c_2}\right) \Delta c_2 + \cdots + \left(\frac{\partial f_i}{\partial c_n}\right) \Delta c_n. \tag{6}$$

The general solution to this equation can be written as:

$$\Delta c(t) = e^{\int_0^t J(s)\,ds}, \tag{7}$$

where J is the Jacobian matrix given by $J = \partial f / \partial c$. For a constant Jacobian eqn. (7) reduces to

$$\Delta c(t) = e^{Jt} A, \tag{8}$$

where A is a constant matrix. In the following analysis we assume that the Jacobian does not change significantly over the time-scale of relaxation of a small perturbation so that the solution to eqn. (7) can be approximated using:

$$\Delta c(t) \approx e^{Jt} A, \tag{9}$$

or in the form:

$$\Delta c_i(t) \approx C_1 e^{\lambda_1 t} + C_2 e^{\lambda_2 t} + \cdots + C_n e^{\lambda_n t}, \tag{10}$$

where C_i depend on the size and direction of the perturbation.

The exponents λ_i are the eigenvalues of the Jacobian matrix calculated at the point c^{fx}. Each eigenvalue of the Jacobian matrix is therefore associated with a different time-scale of the locally linear solution to the full equations. The eigenvalue with the largest negative real part corresponds to a perturbation which decays very quickly and is therefore associated with the fastest time-scale. In the analysis that follows we assume that the eigenvalues have zero imaginary components since relaxation according to complex eigenvalues would not normally be assumed to be in equilibrium. Removing complex relaxation processes may possibly change the dynamics of the reduced system and would need to be considered as a special case.[2] This approach is essentially local and if we are studying nonlinear trajectories then the Jacobian will be time dependent. Thus the ordering of time-scales may change as the reaction proceeds.

3 Species couplings and time-scale modes

If $\lambda_1, \ldots, \lambda_n$ are distinct eigenvalues of J and X_r and X_l the matrices of corresponding right and left eigenvectors then:

$$\text{diag}(\lambda_1, \ldots, \lambda_n) = X_l^T J X_r, \tag{11}$$

where $\text{diag}(\lambda_1, \ldots, \lambda_n)$ is a matrix with the n eigenvalues as diagonal elements and all other elements are zero.

Therefore:

$$Jt = X_r \text{diag}(\lambda_1 \ldots, \lambda_n) t X_l^T, \quad (12)$$

$$e^{Jt} = X_r e^{\text{diag}(\lambda_1, \ldots, \lambda_n)t} X_l^T \quad (13)$$

and

$$\Delta c(t) \approx X_r e^{\text{diag}(\lambda_1, \ldots, \lambda_n)t} X_l^T A. \quad (14)$$

If at $t = 0$ $\Delta c = \Delta c_0$ i.e. the size of the initial perturbation then:

$$\Delta c_0 \approx X_r I X_l^T A = X_r X_l^T A = A. \quad (15)$$

where I is the identity matrix, and therefore A is equal to the matrix of initial perturbation. The linearised solution for a small perturbation from the point c^{fx} is therefore given by:

$$\Delta c(t) \approx X_r e^{\text{diag}(\lambda_1, \ldots, \lambda_n)t} X_l^T \Delta c_0 \quad (16)$$

The time dependence of a small perturbation of a species concentration from a point on the slow manifold is governed by the size of the initial perturbation, the time-scales of the system as determined by the eigenvalues of the Jacobian, and the coupling between the species as determined by the left and right eigenvectors of the Jacobian. For an uncoupled system the Jacobian matrix is diagonal therefore:

$$\Delta c(t) = e^{\text{diag}(\lambda_1, \ldots, \lambda_n)t} \Delta c_0. \quad (17)$$

Each species perturbation decays or grows to a single time-scale i.e. eigenvalue. It is therefore possible for an uncoupled or a very weakly coupled system, to relate the eigenvalues to the lifetimes τ_i of equivalent chemical species where $1/\tau_i = |\lambda_i|$.

For nonlinear systems with species coupling, several different time-scales may contribute to the decay or growth of each species perturbation and conversely several different species may contribute to each time-scale. The calculation of eigenvectors is useful since the off-diagonal terms tell us about the couplings between species and the contributions of individual species to different time-scale modes.

3.1 The relationship between species concentrations and time-scale modes

It is useful at this point to construct a new variable set from linear combinations of the original variables so that each variable in the new set behaves according to a single time-scale. In this way we can easily divide the system into slow and fast time-scales according to the eigenvalues. We use a set of basis vectors as the transformation matrices and in this case we use the left and right eigenvectors. For systems with degenerate eigenvalues Maas and Pope suggest the use of Schur vectors.[5]

We therefore define a new set of variables z where

$$z = X_l^T c \quad (18)$$

and

$$\frac{dz}{dt} = X_l^T f. \quad (19)$$

Conversely it follows that the inverse transformation matrix X_r exists and transforms the new variables back to the original species concentrations:

$$c = X_r z. \quad (20)$$

Using eqn. (16) it can be shown that a perturbation in z is given by:

$$\Delta z(t) = X_l^T \Delta c(t) = X_l^T X_r e^{\text{diag}(\lambda_1, \ldots, \lambda_n)t} \Delta c_0 = e^{\text{diag}(\lambda_1, \ldots, \lambda_n)t} X_l^T \Delta c_0 = e^{\text{diag}(\lambda_1, \ldots, \lambda_n)t} \Delta(z_0) \quad (21)$$

The system will therefore respond to a perturbation in z according to single time-scales, each perturbation growing or decaying exponentially depending on the size and sign of the eigenvalues.

The variables z can therefore be considered as the modes of the system. What this means physically is that the transformation matrix X_1^T shows us how each species contributes to the modes associated with each eigenvalue. By ordering the eigenvalues we can see which species are associated with the slow and fast modes of the system. This can allow us to identify species contributing to the fast decaying modes which can potentially be removed from the system without a loss in accuracy in representing the slow chemical system.

Assuming we can order the eigenvalues according to their values from large and negative to large and positive we can separate the system into slow and fast modes. This can usually be achieved since large gaps appear in the spectrum of eigenvalues. We therefore partition the transformation matrix X_1^T into fast and slow parts:

$$X_1^T = \begin{bmatrix} X_{1f}^T \\ X_{1s}^T \end{bmatrix}, \quad (22)$$

allowing the modes to be separated into fast:

$$z_f = X_{1f}^T c \quad (23)$$

and slow modes:

$$z_s = X_{1s}^T c. \quad (24)$$

4 Estimating the distance from the slow manifold

By ordering the n eigenvalues as above we can consider individual modes i and investigate their collapse onto an $(n-1)$-dimensional manifold in an n-dimensional phase space. Treating the point c^{fx} as a point on the slow manifold we can use the method described above to investigate the behaviour of a perturbation from the manifold in the region close to it. Qualitatively we know that modes with large negative eigenvalues will decay back to the manifold quickly.

The equation describing the decay of each fast mode is given by:

$$\frac{dz_i}{dt} = X_{1,\lambda_i}^T \frac{dc}{dt} = X_{1,\lambda_i}^T f, \quad (25)$$

where X_{1,λ_i}^T is the transpose of the left eigenvector associated with eigenvalue λ_i. If the mode has collapsed then $X_{1,\lambda_i}^T f = 0$ giving the equivalent of the Maas and Pope equation defining an $(n-1)$-dimensional manifold following from the collapse of an individual mode z_i.

Following from eqn. (25) we can also show that:

$$\frac{dz_i}{dt} = X_{1,\lambda_i}^T \frac{dc}{dt} = X_{1,\lambda_i}^T f(c^{fx}) + X_{1,\lambda_i}^T \frac{d\Delta c}{dt} = \frac{d\Delta z_i}{dt}. \quad (26)$$

since we know that $X_{1,\lambda_i}^T f(c^{fx}) = 0$ at the point c^{fx} which lies on the manifold. The equation $X_{1,\lambda_i}^T f = 0$ corresponds to a mode having collapsed onto the manifold but a mode which has decayed close to the manifold may also be assumed to be in equilibrium with a small tolerated error. Estimating the distance of each mode from the manifold and comparing this distance with a tolerance parameter provides some way of estimating the dimension of a manifold at a particular point in phase space or along a reaction trajectory.

Close to the manifold (*i.e.* in the region of the fixed point about which the original expansion was made) we can estimate the distance of an individual mode i from an $(n-1)$-dimensional manifold using the equation:

$$\frac{d\Delta z_i}{dt} = \lambda_i \Delta z_i, \quad (27)$$

where Δz_i gives the size of the perturbation or distance from the manifold. It also follows from eqn. (26) that

$$\frac{d\Delta z_i}{dt} = X_{1,\lambda_i}^T f. \quad (28)$$

Equating the right hand sides of eqn. (27) and (28) gives an estimation for Δz as:

$$\Delta z_i = \frac{X_{1,\lambda_i}^T f}{\lambda_i} \qquad (29)$$

This only gives a relative distance since the choice of eigenvectors will affect the absolute value. By normalising, we can obtain a measure of the relative distance of each mode from the manifold. In this work we have normalised according to the size of each mode although it is necessary to add a small parameter to avoid division by zero. The normalised distance from the manifold is therefore given by:

$$\Delta \tilde{z}_i = \frac{X_{1,\lambda_i}^T f}{\lambda_i} \frac{1}{(\kappa + |X_{1,\lambda_i}^T c|)} \qquad (30)$$

By defining a suitable tolerance parameter we therefore have an approximate system for judging when a mode is approaching close to an $(n-1)$-dimensional manifold relating to its collapse.

Following from the ideas of Roussel and Fraser[10] we can imagine the system collapsing onto a cascade of manifolds of decreasing dimension with the fastest modes collapsing first and the slowest last. By estimating the distance of each mode from each $(n-1)$-dimensional manifold and comparing against a tolerance parameter we can therefore determine at each time-point along the trajectory how many modes have effectively collapsed. It also follows that once the fastest mode has collapsed then the error of assuming an $(n-2)$-dimensional manifold can be estimated by the distance of the next slowest mode from the manifold. If we assume n_f fast modes, the error associated with the $(n-n_f)$-dimensional slow manifold will be largely determined by the distance of the z_{n_f} mode from the manifold although there may be some contribution from the faster modes where time-scale separation is weak.

5 Application to a simple three-variable system

The example chemical system used in this section is taken as an extremely small subset of the MCM developed by Jenkins *et al.*[1,11] It is presented not as a realistic tropospheric model but as a suitable 3-variable linear test case for the first application of the method presented above. The linearised system is formed by taking a subset of the MCM with 8 reactions and 7 species as shown in Table 1 and assuming that the major species have reached steady state conditions after

Table 1 Reaction mechanism for the simple example tropospheric system

Number	Reaction	Rate coefficient
1a	$O^1D \xrightarrow{k_{O_2}} O^3P$	2.101×10^8 s^{-1}
1b	$O^1D \xrightarrow{k_{N_2}} O^3P$	5.060×10^8 s^{-1}
1c	$O^1D \xrightarrow{k_{H_2O}} 2OH$	5.412×10^7 s^{-1}
2	$OH + CO \xrightarrow{k_2} HO_2 + $ products	2.384×10^{-13} molecule^{-1} cm^{-3} s^{-1}
3	$HO_2 + NO \xrightarrow{k_3} OH + NO_2$	8.941×10^{-12} molecule^{-1} cm^{-3} s^{-1}
4	$OH + NO_2 \xrightarrow{k_4} HNO_3$	1.408×10^{-11} molecule^{-1} cm^{-3} s^{-1}
5	$HO_2 + HO_2 \xrightarrow{k_5} H_2O_2 + $ products	2.67×10^{-12} molecule^{-1} cm^{-3} s^{-1}
J1	$O_3 \xrightarrow{j_1} O^1D + $ products	2.11×10^{-5} s^{-1}

reacting for a given time interval, due to the photostationary state. We consider fixed photolysis conditions corresponding to mid-day under clear skies with a solar declination of 23.79° and a zenith angle of 16.2° at a latitude of 40°. From the reaction set in Table 1 we can form the rate equations of the system:

$$\frac{d[O^1D]}{dt} = j_1[O_3] - k'_1[O^1D] \tag{31}$$

$$\frac{d[OH]}{dt} = 2k'_{H_2O}[O^1D] - k_2[CO][OH] + k_3[HO_2][NO] - k_4[OH][NO_2] \tag{32}$$

$$\frac{d[HO_2]}{dt} = k_2[OH][CO] - k_3[HO_2][NO] - 2k_5[HO_2]^2 \tag{33}$$

$$\frac{d[NO_2]}{dt} = k_3[HO_2][NO] - k_4[OH][NO_2] \tag{34}$$

$$\frac{d[NO]}{dt} = -k_3[HO_2][NO] \tag{35}$$

$$\frac{d[O_3]}{dt} = -j_1[O_3] \tag{36}$$

where O^3P is not considered since it is a product only in this reduced system. The system is reduced to one of three equations by taking CO, O_3, NO, NO_2 to be in steady state and by assuming a pseudo-linear form for the decay of HO_2. We compare the relaxation of the fast variables of the system for both low and high NOx (NO + NO_2) with general initial conditions for other species given in Table 2.

5.1 Eigenspaces for low NOx conditions

For low NOx conditions the concentrations for the NOx species are [NO] = 1.00×10^6 molecules cm^{-3} and $[NO_2] = 5.00 \times 10^6$ molecules cm^{-3}. Pseudo-first-order rate coefficients are defined as in Table 3 leading to the following system of three equations for $[O^1D]$, [OH] and $[HO_2]$:

$$\frac{d[O^1D]}{dt} = j_1[O_3] - k'_1[O^1D] = f_1 \tag{37}$$

$$\frac{d[OH]}{dt} = 2k'_{H_2O}[O^1D] - k'_2[OH] + k'_3[HO_2] - k'_4[OH] = f_2 \tag{38}$$

$$\frac{d[HO_2]}{dt} = k'_2[OH] - k'_3[HO_2] - 2k'_5[HO_2] = f_3 \tag{39}$$

with the Jacobian given by:

$$J = \frac{\partial f_i}{\partial c_j} = \begin{pmatrix} -k'_1 & 0 & 0 \\ 2k'_{H_2O} & -k'_2 - k'_4 & k'_3 \\ 0 & k'_2 & -k'_3 - 2k'_5 \end{pmatrix}$$

Table 2 Initial conditions used for all scenarios for the simple example tropospheric system

Species	Initial concentration/molecules cm^{-3}
$[O^1D]$	3.06×10^5
[OH]	5.66×10^6
$[HO_2]$	5.57×10^8
$[O_3]$	7.38×10^{11}
[CO]	2.458×10^{12}

Table 3 Pseudo-first-order rate constants for low NOx scenario

$k'_1 = k_{O_2}[O_2] + k_{N_2}[N_2] + k_{H_2O}[H_2O]$	$= 7.70 \times 10^8 \text{ s}^{-1}$
$k'_2 = k_2[CO]$	$= 5.87 \times 10^{-1} \text{ s}^{-1}$
$k'_3 = k_3[NO]$	$= 8.49 \times 10^{-6} \text{ s}^{-1}$
$k'_4 = k_4[NO_2]$	$= 7.04 \times 10^{-5} \text{ s}^{-1}$
$k'_5 = k_5[HO_2]$	$= 1.49 \times 10^{-3} \text{ s}^{-1}$
k'_{H_2O}	$= 5.41 \times 10^7 \text{ s}^{-1}$

The eigenvalues are therefore $\lambda_1 = -7.7 \times 10^8$, i.e. large and negative corresponding to the relaxation of $[O^1D]$, and $\lambda_2 = -0.586$, and $\lambda_3 = -2.97 \times 10^{-3}$, with corresponding right eigenvectors:

$$X_r = \begin{pmatrix} 1 & 0 & 0 \\ -1.41 \times 10^{-1} & -9.95 \times 10^{-1} & 1.45 \times 10^{-5} \\ 1.07 \times 10^{-10} & 1 & 1 \end{pmatrix}$$

The first eigenvector shows us that $[O^1D]$ is largely decoupled from the other species.

The corresponding left eigenvectors are found from the transpose of the Jacobian or by taking the inverse of the transpose of the right eigenvectors and are given by:

$$X_l = \begin{pmatrix} 1 & 1 & 1 \\ 0 & 7.11 & 7.11 \\ 0 & -1.03 \times 10^{-4} & 0.708 \end{pmatrix}.$$

This now provides information on the relationship between species and time-scale modes in the system. The first column of the matrix of left eigenvectors shows us that only O^1D contributes to the first mode which collapses on the time-scale of 10^{-8} s. Once O^1D has collapsed the third time-scale has contributions from both OH and HO_2 demonstrating strong coupling between these species.

5.2 Eigenspaces for high NOx conditions

NOx concentrations for the high NOx scenario are given by $[NO] = 1.00 \times 10^{12}$ molecules cm^{-3} and $[NO_2] = 5.00 \times 10^{12}$ molecules cm^{-3}. The corresponding pseudo-first-order rate constants are given in Table 4. The eigenvalues are now given by $\lambda_1 = -7.7 \times 10^8$, $\lambda_2 = -71.7$ and $\lambda_3 = -8.39$ with corresponding matrix of right eigenvectors:

$$\begin{pmatrix} 1 & 0 & 0 \\ -1.41 \times 10^{-1} & -1.06 \times 10^2 & 1.35 \times 10^{-1} \\ 1.07 \times 10^{-10} & 1 & 1 \end{pmatrix},$$

and left eigenvectors:

$$\begin{pmatrix} 1 & 1 & 1 \\ 0 & 7.11 & 7.11 \\ 0 & -0.962 & 760 \end{pmatrix}.$$

Table 4 Pseudo-first-order rate constants for high NOx scenario

$k'_1 = k_{O_2}[O_2] + k_{N_2}[N_2] + k_{H_2O}[H_2O]$	$= 7.70 \times 10^8 \text{ s}^{-1}$
$k'_2 = k_2[CO]$	$= 5.87 \times 10^{-1} \text{ s}^{-1}$
$k'_3 = k_3[NO]$	$= 8.49 \text{ s}^{-1}$
$k'_4 = k_4[NO_2]$	$= 7.04 \times 10^1 \text{ s}^{-1}$
$k'_5 = k_5[HO_2]$	$= 1.49 \times 10^{-3} \text{ s}^{-1}$
k'_{H_2O}	$= 5.41 \times 10^7 \text{ s}^{-1}$

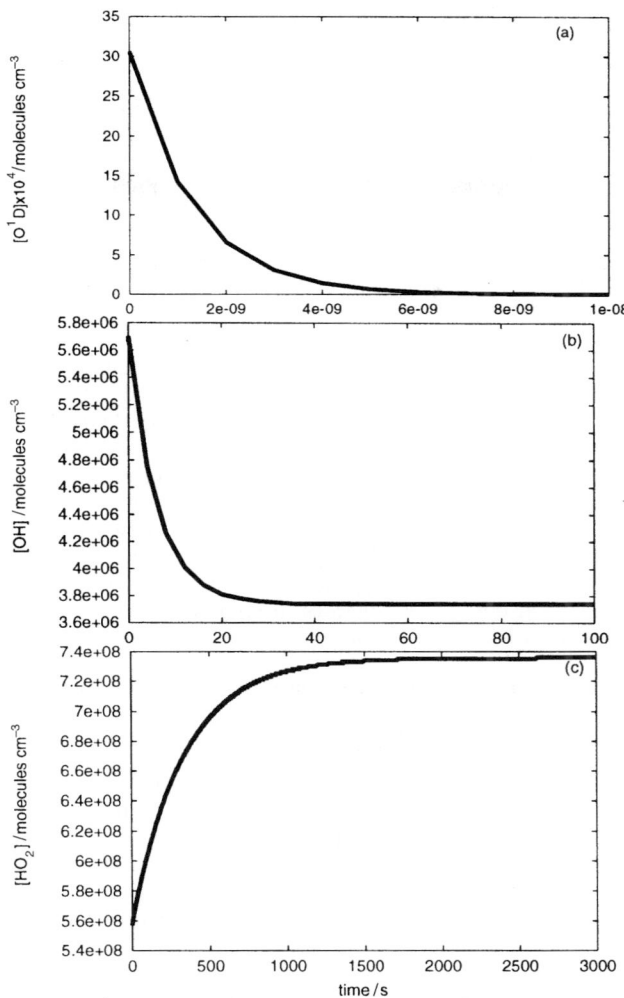

Fig. 1 Species concentration against time for three variable system. (a) [O^{1}D], (b) [OH], (c) [HO$_2$].

From this we can see that the time-scales of the system are faster than for the low NOx case and that the coupling between the radical species OH and HO$_2$ is weaker at high NOx. Under high NOx conditions [HO$_2$] dominates the third mode and the contribution from [OH] will be small at longer times since it will already have collapsed.

5.3 Collapse to the steady state of the system for low NOx conditions

Low NOx conditions have been chosen for an illustration of the methods since they represent the largest degree of coupling between the radical species. For constant photolysis conditions the radicals in the system will eventually reach a steady state with respect to the major species. This we can think of as a zero-dimensional manifold in species phase since it corresponds to a single point.

The steady state values to which the concentrations collapse are given by setting the right-hand sides of eqn. (37)–(39) to zero:

$$[O^1D] = \frac{j_1[O_3]}{k'_1} = 2.02 \times 10^{-2} \text{ molecules cm}^{-3}$$

$$[OH] = \frac{1}{k'_2 + k'_4}\left(k'_3[HO_2] + \frac{2k_{H_2O}j_1[O_3]}{k'_1}\right)$$

$$0 = [HO_2]^2 + \frac{k'_3 k'_4}{2k_5(k'_2 + k'_4)} - \frac{k'_2 k_{H_2O} j_1 O_3}{k'_1 k_5(k'_2 + k'_4)}$$

giving steady state concentrations $[OH]_{ss} = 3.74 \times 10^6$ molecules cm^{-3}, $[HO_2]_{ss} = 7.35 \times 10^8$ molecules cm^{-3}. By computing the trajectories over time we can determine the relative speeds with which each species collapses onto the zero-dimensional manifold as shown in Fig. 1.

Fig. 1(a) shows that $[O^1D]$ collapses very quickly, with a time-scale of the order of 1×10^{-8} s as expected from the large eigenvalues associated with its time-scale. From the right and left eigenspaces we saw that the other species were to a certain extent coupled. If we study the right eigenvectors we can see that the collapse of $[OH]$ has contributions mainly from the first and second modes. However since the first mode collapses almost immediately we expect $[OH]$ to collapse according to the second mode on a time-scale of $1/\lambda_2$ i.e. of the order of seconds. This is indeed demonstrated in Fig. 1(b). For $[HO_2]$ there are contributions from both the second and third modes but since the second model collapses in seconds we expect $[HO_2]$ to collapse the most slowly on a time-scale of $1/\lambda_3$ i.e. of the order of 1000 s as demonstrated in Fig. 1(c).

5.4 Investigating the distance from the low-dimensional manifolds for low NOx conditions

For such a simple 3-variable system calculating the distance of each species from the zero-dimensional manifold along a reaction trajectory is trivial since it only involves computing the difference between the trajectory value and the steady state concentrations. However, we wish to investigate a more general method for determining the distance of an individual mode from successively lower-dimensional manifolds that can be applied in higher-dimensional models. A simple estimation of this distance of a single mode from an $(n-1)$-dimensional manifold in a phase space of dimension n was presented in Section 4. In this section we compare this method to two other methods for calculating the distance of a mode from its corresponding low-dimensional manifold to assess its applicability.

Comparison with a geometric approach. Using a geometric approach we first calculate the distance of a mode along the left eigenvector corresponding to each eigenvalue of the trajectory, from an $(n-1)$ i.e. 2-dimensional manifold given by $X^T_{1,\lambda_i} f = 0$ for $i = 1, 3$. This effectively gives us the distance of a fast mode from the manifold in the direction of the eigenvector associated with its fast time-scale. We know that when this distance is zero the trajectory lies on the $(n-1)$-dimensional manifold associated with the collapse of the fast mode.

The first mode corresponding to O^1D has not been considered since its distance from the 2-dimensional manifold corresponding to its collapse is trivially given by $[O^1D]-[O^1D]_{ss}$ as shown in Fig. 1(a). The second mode, as shown by the left eigenspace, corresponds mainly to the collapse of $[OH]$ and the third to both $[OH]$ and $[HO_2]$. The distance calculated by the geometric method described above is compared in Fig. 2(a) with the use of the simplified expression for normalised distance derived in Section 4 eqn. (30), with κ in this simple case set to zero. The distances calculated geometrically for the second and third modes have been scaled by dividing by a factor of 4.3×10^6 and 9.2×10^8 respectively in order to compare against the normalised values calculated according to eqn. (30). These scalings are somewhat arbitrary but are used to demonstrate that the relative time-scales of collapse are well represented by eqn. (30) when compared to the exact computation of the geometric distance of a mode from the manifold.

It can be seen in Fig. 2 that the simple expression for normalised distance predicts the collapse onto the manifolds almost concurrently with the geometric method after the first 150 s. It is expected that the simplified expression becomes more accurate for later times as the perturbation

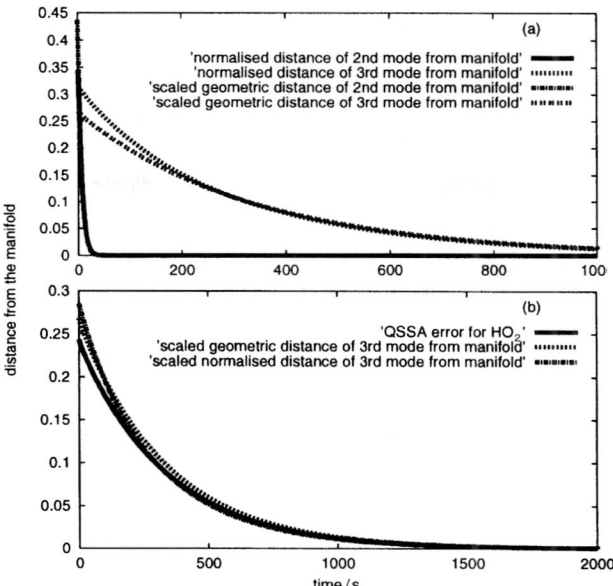

Fig. 2 (a) Geometric distance of second and third modes from manifolds scaled onto corresponding normalised distance calculated using $(X^T_{1,\lambda_i}f)/[(\kappa + |X^T_{1,\lambda_i}c|)\lambda_i]$. (b) Relative distance of third mode from manifold with two approximate distances scaled onto it.

analysis used in the derivation of the expression was essentially a local technique applicable to small perturbations. However, since we intend to use a small threshold value when deciding whether a mode has collapsed, the early part of the decay is not important. By choosing a threshold for the normalised distance we can see that the method may be used to show how the intrinsic manifold dimension changes with time.

Comparison with distance from the zero dimensional manifold—the steady state. We return to the idea that for this system the trajectory will relax onto a hierarchy of manifolds of decreasing dimension before finally reaching a dimension of zero at the steady state. If we assume the first two modes have collapsed after 50 s then from Fig. 1(a) and (b), the distance of the third mode from the manifold should represent the error in assuming steady state for the slowest species [HO_2] given by:

$$\text{relative steady state error} = \left(\frac{[HO_2]_{ss} - [HO_2]}{[HO_2]_{ss}}\right),$$

where $[HO_2]_{ss} = 7.36 \times 10^8$ molecules cm^{-3}.

Fig. 2(b) shows the relative error in assuming steady state compared with the scaled distances using the previous two methods. The simple expression for normalised distance compares well with both the relative steady state error for the slowest mode and with the geometric method, and for this simple 3-variable system provides an easy way of estimating the distance of a trajectory from lower and lower dimensional manifolds. By choosing an acceptable level of error and identifying the time at which it occurs we can choose a suitable tolerance parameter for identifying collapsed modes in the system. Selecting a 5% error gives a tolerance parameter of 0.05 for the normalised distance from the manifold. Fig. 2 therefore indicates that a 1-dimensional manifold would be sufficient to represent the system after approximately 20 s and a zero-dimensional manifold after approximately 600 s for the initial conditions used here.

6 Application to nonlinear systems

6.1 Six-variable system

In this section we relax the steady state constraints imposed on the major species in Section 5 giving a non-linear set of rate equations. The reaction system is given in Table 1 and the rate equations for the system described by eqn. (31) to (36). We compare the system time-scales for both low and high NOx scenarios with initial conditions for each as in Section 5, although the NOx levels no longer remain constant throughout the simulation.

As before we wish to calculate the normalised distance of each time-scale mode from the manifold and therefore it is necessary to calculate the eigenvalues and left eigenvectors from the Jacobian where:

$$J = \frac{\partial f_i}{\partial c_j} = \begin{pmatrix} -k_1 & 0 & 0 & 0 & 0 & j_1 \\ 2k'_{H_2O} & -k'_2 - k_4[NO_2] & k_3[NO] & -k_4[OH] & k_3[HO_2] & 0 \\ 0 & k'_2 & -k_3[NO] - 4k_5[HO_2] & 0 & -k_3[HO_2] & 0 \\ 0 & -k_4[NO_2] & k_3[NO] & -k_4[OH] & k_3[HO_2] & 0 \\ 0 & 0 & -k_3[NO] & 0 & -k_3[HO_2] & 0 \\ 0 & 0 & 0 & 0 & 0 & -j_1 \end{pmatrix}$$

The Jacobian is time dependent and hence so are the eigenvalues and eigenvectors of the matrix. They are therefore recalculated at each time-step.

The normalised distances for the two second fastest modes calculated from eqn. (30) with κ chosen to be 10^5 in the low NOx scenario are shown in Fig. 3(a). The use of a non-zero value for κ stems from the fact that the mode by which we normalise crosses the zero axis and therefore division by zero would otherwise occur. Since concentration units are in molecules cm^{-3}, 10^5 represents a small value. Similar behaviour is shown as in the 3-variable linear system with the second fastest mode collapsing of the order of 10 s and the third fastest slightly quicker than in the previous system of the order of 100 s. The distances for each of the six modes are also presented for low NOx conditions along with their associated eigenvalues at selected time-points in Table 5. After 200 s the normalised distances of the fastest three modes from the manifold are very small with much larger values for the slower modes. The distance of the slowest mode is always 1 since it relates exclusively to [O$_3$] and $X^T_{1,\lambda_6} f = -j_1[O_3]$ with λ_6 equal to j_1 and $X^T_{1,\lambda_6} c = [O_3]$. The table also demonstrates that the eigenvalues λ_i do not change significantly with time, allowing the assumption of a slowly varying Jacobian to be used. The time-scale separation under low NOx conditions is however small, with the third eigenvalue relating strongly to [HO$_2$] being only slightly larger than the fourth which is strongly linked to [NO$_2$]. In contrast, for high NOx

Table 5 Dimension analysis for low NOx scenario

Time/s	Eigenvalue	Normalised distance	Description
200	-7.70×10^8	6.480×10^{-12}	Fast mode relaxed
200	-5.87×10^{-1}	1.05×10^{-2}	Relaxed
200	-6.60×10^{-3}	8.06×10^{-2}	Off manifold
200	-5.24×10^{-3}	4.44	Off manifold
200	-5.27×10^{-5}	6.31×10^{-1}	Off manifold
200	-2.11×10^{-5}	1.0	Off manifold
500	-7.70×10^8	0	Fast mode relaxed
500	-5.87×10^{-1}	3.42×10^{-3}	Relaxed
500	-6.84×10^{-3}	1.00×10^{-2}	Relaxed
500	-5.43×10^{-3}	2.37	Off manifold
500	-5.25×10^{-5}	0.628	Off manifold
500	-2.11×10^{-5}	1.0	Off manifold

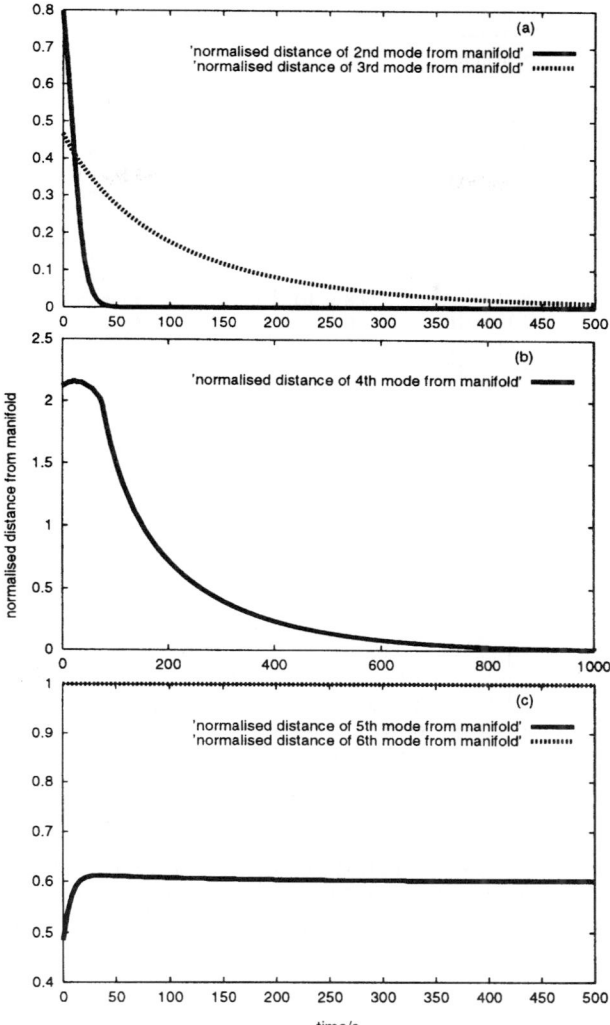

Fig. 3 The distance of the modes from the manifold found using $(X_{1,\lambda_i}^T f)/[(\kappa + |X_{1,\lambda_i}^T c|)\lambda_i]$. (a) second and third modes, (b) fourth mode, (c) fifth and sixth modes.

conditions as shown in Table 6, the time-scale separation between the 3 fast modes and the 3 slow modes is very pronounced. As with the 3-variable simple case the low NOx scenario provides a more stringent test of the methods developed here since it shows more inter-species coupling and lower time-scale separation between the collapsed and non-collapsed modes.

Comparison of normalised distance with QSSA errors for low NOx scenario. Choosing a similar tolerance parameter of 0.05 as in the previous section it follows that after approximately 200 s the full 6-variable system should collapse onto a 3-variable slow manifold. Using the QSSA method for the 3-radical species it is possible to define a reduced 3-variable model. We first check that the 3 fastest modes are really related to the 3-radical species which raises an interesting point about the application of the QSSA. Often the application of the QSSA is applied *a priori* to the radical species in a system. More quantitatively, other criteria related to the species life-time, the relative rates of production and destruction of a species or the QSSA error[4] can be employed. All of these techniques ignore the fact that the time-scales are not governed necessarily by individual species

Table 6 Dimension analysis for high NOx scenario

Time/s	Eigenvalue	Normalised distance	Description
0.15	-7.70×10^8	0	Fast mode relaxed
0.15	-7.11×10^1	9.0×10^{-3}	Relaxed
0.15	-8.41	0.498	Off manifold
0.15	-2.11×10^{-5}	1.0	Off manifold
0.15	-2.51×10^{-6}	2.4×10^{-2}	Off manifold
0.15	-2.65×10^{-7}	5.9×10^{-2}	Off manifold
1.94	-7.70×10^8	0	Fast mode relaxed
1.94	-7.11×10^1	3.0×10^{-7}	Relaxed
1.94	-8.41	9.8×10^{-3}	Relaxed
1.94	-2.11×10^{-5}	1.0	Off manifold
1.94	-3.65×10^{-9}	1.6×10^{-2}	Off manifold
1.94	-5.14×10^{-17}	2.95×10^5	Off manifold

but by combinations of species as shown by eqn. (23). The reason why the QSSA often works well is that often the radical species to which the QSSA is applied are effectively decoupled from the major species as a group even if they are coupled to each other. This is demonstrated by considering the left eigenvectors shown below for the low NOx case at $t = 200$ s:

$$X_l = \begin{pmatrix} \lambda_1 & \lambda_2 & \lambda_3 & \lambda_4 & \lambda_5 & \lambda_6 \\ 1 & 0 & 0 & 0 & 0 & 0 \\ 0.139 & 0.99 & 0 & 0 & 0 & 0 \\ 0.099 & 0.707 & 0.699 & 0 & -0.05 & 0 \\ 0 & 0 & 0 & 0 & 1 & 0 \\ 0 & 0 & 0 & 0.704 & 0.710 & 0 \\ 0 & 0 & 0 & 0 & 0 & 1 \end{pmatrix}$$

Here all values below 10^{-2} have been omitted for clarity. A schematic version of the relationship between modes and species is also shown in Fig. 4 where mode 1 is the fastest mode and mode 6 the slowest. It is clear that the matrix is block triangular and while the radical species are coupled to each other they do not couple to the major species.

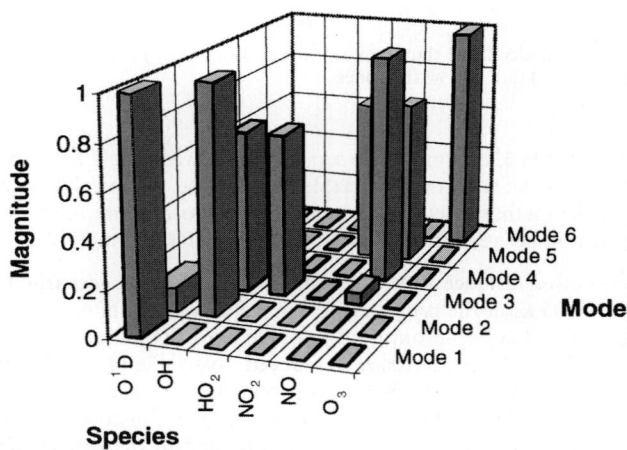

Fig. 4 A schematic diagram showing the relative relationships between species and modes for the six-variable system.

The same is true for initial high NOx conditions with an example at $t = 1.94$ given below:

$$X_1 = \begin{pmatrix} \lambda_1 & \lambda_2 & \lambda_3 & \lambda_4 & \lambda_5 & \lambda_6 \\ 1 & 0 & 0 & 0 & 0 & 0 \\ 0.138 & 0.981 & -0.133 & 0 & 0 & 0 \\ 0.0013 & 0.009 & 1 & 0 & 0 & 0 \\ 0 & 0 & 0 & 0 & 0 & 1 \\ -0.09 & -0.697 & 0.0063 & 0.703 & 0 & -0.106 \\ -00.7 & -0.514 & -0.48 & 0.514 & 0.48 & -0.07 \end{pmatrix}$$

For the high NOx scenario shown in Table 6 the three fast modes collapse to the manifold in less than 2 s, and again the three slow modes remain off the manifold. Following consideration of several time-points the application of the QSSA for all scenarios seems to be applicable in terms of the long time behaviour of the system. The QSSA solution on the 3-variable slow manifold therefore becomes:

$$0 = j_1[O_3] - k'_1[O^1D] \qquad (40)$$

$$0 = 2k'_{H_2O}[O^1D] - k_2[CO][OH] + k_3[HO_2][NO] - k_4[OH][NO_2] \qquad (41)$$

$$0 = k_2[OH][CO] - k_3[HO_2][NO] - 2k_5[HO_2]^2 \qquad (42)$$

$$\frac{d[NO_2]}{dt} = k_3[HO_2][NO] - k_4[OH][NO_2] \qquad (43)$$

$$\frac{d[NO]}{dt} = -k_3[HO_2][NO] \qquad (44)$$

$$\frac{d[O_3]}{dt} = -j_1[O_3]. \qquad (45)$$

We compare the difference between the quasi-steady state values of the fast species and their concentrations when calculated from the full 6-variable system using eqn. (31) to (36) for low NOx conditions in order to evaluate the error in assuming a 3-dimensional manifold. When compared against the scaled normalised distance of the slowest of the fast modes from its low-dimensional manifold in Fig. 5, we can see that the distance estimate represents well the time behaviour of the steady state error.

Diurnal variation of manifold dimension. In this section we allow a variable photolysis rate in order to simulate typical diurnal variations in species profiles and to assess the changes in the

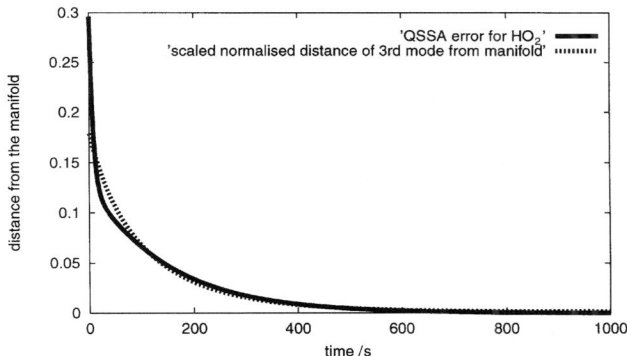

Fig. 5 Comparison of relative steady state error for [HO$_2$] and normalised distance from manifold of 3rd fastest mode.

dimension of the slow manifold. The variation in photolysis rate is taken from Derwent and Hov[12] and the reaction rate coefficient j_1 drops to almost zero at night. Simulations run from midnight to midnight with photolysis rates very low at the start of each simulation and peaking in the middle of the day.

In the calculation of slow manifold dimension we consider the normalised distance of the collapsing modes from the manifold and also the presence of conservation relations signified by small eigenvalues. For this purpose we define two tolerances: *mtol*, defining the normalised distance from the manifold below which the mode is said to have relaxed to the manifold, and *ctol* defining the value of the eigenvalue below which the mode is said to be a conservation relation. A very small eigenvalue for a mode implies that some combination of species variables changes extremely slowly over time so that the mode can be considered to be constant. It does not necessarily imply a constant species concentration as indicated by the QSSA where a conservation relation is applied to a group of variables while each individual concentration may change with time. The tolerance parameters used were *mtol* = 0.1, and *ctol* = 1.6×10^{-6} relating to conservation over the time-scale of approximately one week. The manifold dimension is then calculated as $n_s - n_m - n_c$, where n_s is the number of original variables, n_m the number of relaxed modes and n_c the number of conservation relations.

Fig. 6(a) shows the diurnal variation of dimension for the high NOx scenario with initial conditions as described in Section 5. In this case the manifold drops to zero after a few hundred seconds since the three fastest modes relax to the manifold and the three slow modes form conservation relations. The onset of photolytic reactions at around 5 h causes a short increase of dimension to 3 settling down to 1 since the slow mode relating to predominantly O_3 is no longer conserved. It should be stressed that no transport terms or emissions are included in this simple model and the collapse of the system to a zero-dimensional equilibrium at the end of the simulation in this case would be affected by the resulting changes in concentrations from these processes.

Fig. 6(b) shows the diurnal variation of dimension for the low NOx scenario with initial concentrations of the major species as described in Section 5. After the first few hundred seconds of the simulation the dimension has dropped to 2 since the two fastest modes have relaxed to the manifold and the two slowest modes form conservation relations as a consequence of zero photolysis rates. At the onset of daylight conditions the dimension increases to 4 since conservation relations are now broken and O_3 and NO_2 now vary significantly. After about 7 h simulation the modes

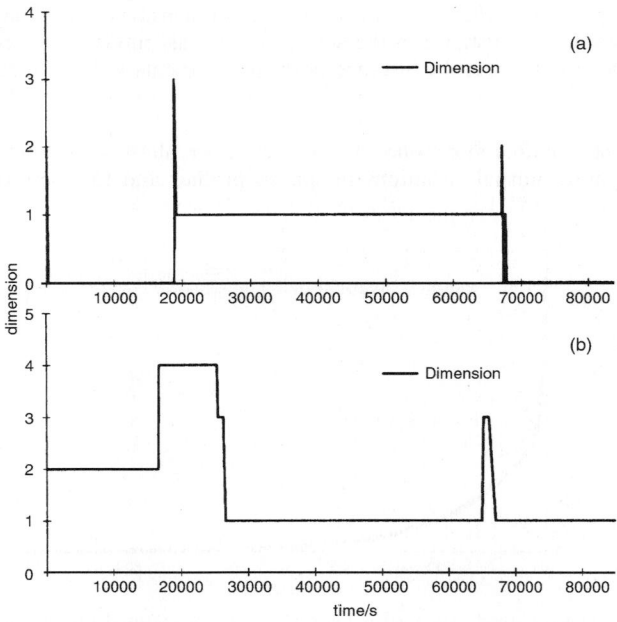

Fig. 6 The dimension of the six-variable system over time for (a) polluted and (b) clean air scenarios.

relating to the concentrations of NO_2 and HO_2 have now relaxed and the concentration of NO has dropped to zero giving a dimension of 1. At the onset of night-time conditions the change in photolysis rates forces the modes relating to the concentrations of NO_2 and HO_2 off the manifold for a short time so that a peak of 3 appears until night-time conditions have been fully established. Already we begin to see the difficulties in assuming a constant dimension for the slow manifold since variable photolysis rates change the nature of the underlying dynamics.

6.2 Application to butane oxidation scheme

In this section we apply the techniques described above to a mechanism describing butane oxidation in the troposphere. The mechanism is derived from a detailed subset of the MCM[1,13,14] consisting of 512 reactions and 185 species. This extended butane oxidation mechanism has previously been reduced[15] to 53 reactions and 42 species for a scenario representing polluted conditions. The reduction was based on local rate sensitivities with principal component analysis and rate of production analyses for the identification of redundant species and reactions.[15] Here we seek to determine whether a lower dimensional slow manifold exists that could accurately represent the long time dynamics of this reduced chemical scheme. The 5 day simulations represent initial high NOx levels of 150 ppb with data taken from Bowman and Seinfeld.[16] The initial butane concentration was scaled to 118 ppb to represent the sum of all alkane concentrations. Fig. 7 shows the diurnal concentrations of O_3, NO, NO_2 and PAN over 5 days and Fig. 8 the diurnal variation in manifold dimension using $mtol = 0.01$ and $ctol = 5 \times 10^{-7}$. Complex modes are retained despite having large negative real parts if their imaginary part is greater than 0.1. This prevents the removal of modes relating to stable focus type dynamics. The maximum dimension using the above tolerances is 13, significantly lower than the original 42 variables. This demonstrates that although sensitivity analysis resulted in a reasonable reduction of the mechanism, the dynamics of the system is confined to a much lower dimensional manifold after only a short reaction time. There is significant variation in the dimension of the slow manifold throughout the day with a much lower dimension at night and small periods of high dimension at the onset of photolysis.

Table 7 presents eigenvalues and normalised errors for each mode after 12.2 h simulation time when the dimension of the slow manifold is at its highest *i.e.* 13. The table shows that the real components of the eigenvalues span 24 orders of magnitude from 10^9 to 10^{-15} and fast modes are defined as having an eigenvalue of less than -10^3. Two complex eigenvalues are present and despite showing a small normalised distance these are not considered to have collapsed since they may affect the system dynamics. Using the chosen tolerances 4 conservation modes are defined relating to very slow time-scale processes. With one exception the fastest 26 modes are defined as having collapsed.

Analysis of the left and right eigenspaces can give us useful information relating to species couplings and the contribution of each species to the live and collapsed modes. In terms of deriving a reduced model it is useful to examine those modes remaining on or off the manifold and the species with major contributions to them. An automatic method has been developed whereby these major species can be identified from the left eigenspaces. An example is given below for a time-point of 12.2 h in the butane simulation where the dimension is 13. Redundant species are determined by analysing the collapsed modes. Starting with the fastest collapsed mode the major eigenvector components are examined for each mode. If a single eigenvector has a large contribution to a dead mode then the species to which it relates is removed from the set of remaining species and is discounted in all subsequent modes, including live ones. This process is relatively straightforward for those modes which are dominated by a single species. Other modes may have large contributions from several species. However, if a group of dead modes relate to the same group of species then it is possible to remove those species as a group. Following this process for each of the dead modes we are left with a group of species contributing to the live modes remaining off the manifold at each point on the trajectory.

Table 8 lists the species remaining for the butane scheme at 12.2 simulation time. The modes lying off the manifold are dominated by the nitrogen-containing species, NO_2, NO_3, HONO, HNO_3, by PAN ($CH_3CO_3NO_2$) and by several carbonyl species. In fact this group of species or subsets of the group dominate the live modes throughout most of the simulation. It is interesting

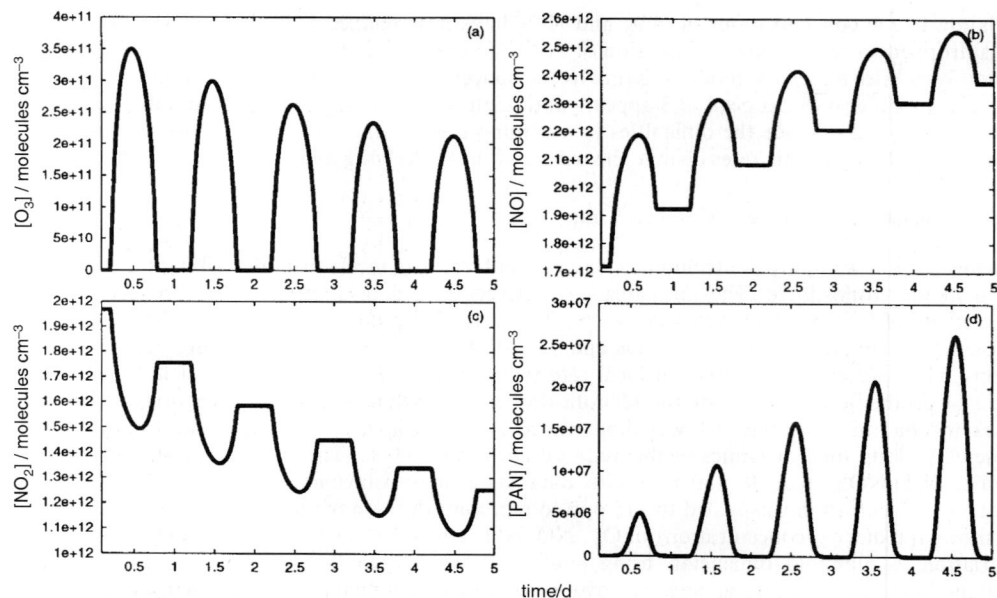

Fig. 7 Species concentration against time for the butane system. (a) $[O_3]$, (b) $[NO]$, (c) $[NO_2]$, (d) $[PAN]$.

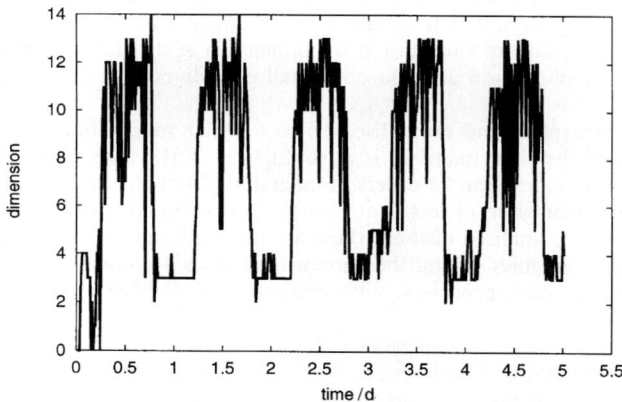

Fig. 8 The dimension of the butane system over time for a trajectory with high initial NOx concentrations.

to compare the species contributing to the live modes with the non QSSA species identified in Zeng et al.[15] Ozone and NO are absent from the set of live modes because of the strong coupling between NO, NO_2, NO_3 and O_3 via the reactions:

$$NO + O_3 \rightarrow NO_2 + O_2 \tag{6}$$

$$NO_2 + h\nu(+O_2) \rightarrow NO + O_3 \tag{J2}$$

and

$$NO_2 + O_3 \rightarrow NO_3 + O_2 \tag{7}$$

$$NO_3 + NO \rightarrow 2NO_2, \tag{58}$$

so that $[O_3] = j_2[NO_2]/(k_6[NO])$ corresponding to the photostationary state, and $[NO_3] = k_7[NO_2][O_3]/(k_8[NO])$. It follows that only two of the four species are needed to define the

Table 7 Status of modes after 12 h simulation for the butane system under high NOx conditions

Eigenvalue				
Real part	Imaginary part	Speed of mode	Dist. from manifold	State of mode
-0.7151×10^9	0	Fast	0.0000	Collapsed
-0.6664×10^7	0	Fast	0.0000	Collapsed
-0.6165×10^6	0	Fast	0.0000	Collapsed
-0.1930×10^6	0	Fast	0.0000	Collapsed
-0.1664×10^6	0	Fast	0.0000	Collapsed
-0.7274×10^5	0	Fast	0.0000	Collapsed
-0.4323×10^5	0	Fast	0.1627×10^{-1}	Collapsed
-0.4883×10^5	0	Fast	0.0000	Collapsed
-0.4323×10^5	0	Fast	0.0000	Collapsed
-0.4073×10^5	0	Fast	0.0000	Collapsed
-0.4073×10^5	0	Fast	0.0000	Collapsed
-0.9896×10^4	0	Fast	0.0000	Collapsed
-0.6186×10^2	0	Medium	0.2858×10^{-4}	Collapsed
-0.5985×10^2	0	Medium	0.1285×10^{-4}	Collapsed
-0.5673×10^2	0	Medium	0.9862×10^{-1}	Off manifold
-0.4364×10^2	0	Medium	0.4985×10^{-4}	Collapsed
-0.1772×10^2	1.10	Medium	0.1140×10^{-3}	Complex
-0.1772×10^2	-1.10	Medium	0.1140×10^{-3}	Complex
-0.1634×10^2	0	Medium	0.1894×10^{-4}	Collapsed
-0.1634×10^2	0	Medium	0.1713×10^{-4}	Collapsed
-0.1159×10^2	0	Medium	0.2591×10^{-4}	Collapsed
-0.1160×10^2	0	Medium	0.2184×10^{-4}	Collapsed
-0.1159×10^2	0	Medium	0.2591×10^{-4}	Collapsed
-0.1069×10^2	0	Medium	0.7713×10^{-5}	Collapsed
-0.9806×10^1	0	Medium	0.0000	Collapsed
-0.6753×10^1	0	Medium	0.1205×10^{-4}	Collapsed
-0.8634×10^{-1}	0	Medium	0.1748×10^{-2}	Collapsed
-0.5499×10^{-1}	0	Medium	0.1020×10^{-3}	Collapsed
-0.1017×10^{-2}	0	Medium	0.2090×10^{-1}	Off manifold
-0.3901×10^{-3}	0	Medium	0.2346×10^{-1}	Off manifold
-0.8452×10^{-4}	0	Medium	0.7072	Off manifold
-0.2786×10^{-4}	0	Medium	0.5003×10^1	Off manifold
-0.2066×10^{-4}	0	Medium	0.1507	Off manifold
-0.1383×10^{-4}	0	Medium	0.2010	Off manifold
-0.8004×10^{-5}	0	Medium	0.5559×10^{-1}	Off manifold
-0.5988×10^{-5}	0	Medium	0.1266×10^{-1}	Off manifold
0.4076×10^{-5}	0	Medium	0.1270	Off manifold
-0.1300×10^{-5}	0	Medium	0.5166×10^{-1}	Off manifold
-0.1166×10^{-7}	0	Slow	0.2122×10^{-1}	Conservation rel
-0.4388×10^{-8}	0	Slow	0.2148×10^{-1}	Conservation rel
-0.1582×10^{-9}	0	Slow	0.2171×10^{-1}	Conservation rel
0.2246×10^{-15}	0	Slow	0.8879×10^2	Conservation rel

system. It is perhaps surprising that the analysis returns NO_3 rather than NO as the largest contribution to a live mode given its much shorter lifetime. It is also important to note that ozone becomes a live species for a short period at dawn as the photostationary state is established and at dusk when the photolysis rates are rapidly dropping. In the analysis of Zeng et al. NO_3 was identified as a QSSA species, while NO and O_3 were not.

HONO and HNO_3 act as both sinks and as photolysis sources of OH. The mechanism used does not include deposition of HNO_3, so that its role as a source may be overemphasized. PAN relates almost entirely to the only non-complex mode remaining off the manifold at night, reflecting the fact that it is an important reservoir species. The other two species which occur briefly as contributing to live modes are $HO(CH_2)_3CO_3$ and CH_3CO_3 although they appear only during the night at the very early stages of simulation. The absence of CH_3CO_3 from the main set of live species derives from its strong reversible coupling to PAN. Once a significant concentration of

Table 8 The main species contributing to remaining live modes for the butane simulation at $t = 12.2$ h

Main species contributing to remaining alive modes
NO_2
NO_3
HONO
HNO_3
CH_3O_2
HCHO
$C_2H_5COCH_3$
C_3H_7CHO
CH_3CHO
$C_2H_5O_2$
$CH_3CO_3NO_2$
$HO(CH_2)_3CHO$
$HO(CH_2)_2CHO$

PAN has been formed the mode dominated by CH_3CO_3 remains collapsed. CH_3CO_3 was not identified as a QSSA species by Zeng et al.

It is at first sight surprising that CH_3O_2 and $C_2H_5O_2$ remain in the live modes. This arises because they are the main contributors to the complex modes, retained in order to preserve any inherent dynamical complexity. It is unlikely that these modes are significant because the real part of the eigenvalue that relates to their modes is large and negative and their distance from the appropriate low-dimensional manifold is very small. The carbonyl compounds are retained because they are the major relatively stable intermediates in the oxidation process, with lifetimes of approximately 5×10^4–10^7 s. The major peroxy radical formed from butane is that derived from the 2-butyl radical leading to formation of $C_2H_5COCH_3$ and CH_3CHO. The former is photolysed to form the ethyl peroxy and peroxyacetyl (PA) radicals, while acetaldehyde reacts with OH to form PA and is photolysed to form methyl peroxy, which generates HCHO. Zeng et al. showed that $C_2H_5COCH_3$ and CH3CHO dominate the secondary chemistry. The hydroxy aldehydes listed in Table 8 are formed in the minor oxidation route via the 1-butyl radical which accounts for only 15% of the loss of butane at 298 K. The main reactants, C_4H_{10}, CH_4 and CO are excluded from Table 8 because they contribute to the very slow conservation modes.

Determining such a group of species provides useful information for the generation of reduced models based on slow manifold techniques. For example, a repro-model can be generated for a suitable group of major species as demonstrated by previous work relating to a carbon bond mechanism for tropospheric chemistry.[9] In this work a 9-species repro-model was generated that accurately represented the changes in concentration over time of the original 29 species scheme. The reduced model was generated by fitting polynomial difference equations describing the change in concentration of the 9 key species over fixed time-steps to a large data base of simulated trajectories using the full scheme. The generation of the data to be fitted is quite time consuming since a wide range of concentration and photolysis conditions must be represented and the integration of all coupled species carried out. Typically several thousand data points representing the time dependence of the group of key species must be generated from the detailed model. This calculation needs to be carried out only once however, and once fitted polynomials have been generated, their use in reactive flow models in place of integration techniques can lead to speed-ups of up to a factor of 1000. Previous applications of repro-modelling[9] have been based on the use of high-order orthonormal polynomials and the method determines the non-effective parameters, whilst fitting the effective parameters using the method of least squares. Such fitted models therefore, although not providing detailed information on reaction kinetics, are very useful for reactive flow simulations where many repeated calculations over similar ranges of conditions are required. Air pollution forecasting is one area where such speed-ups may be crucial in achieving high resolution simulations for important species.

7 Discussion and conclusions

A simplified perturbation method has been presented for the calculation of intrinsic slow manifold dimension for chemical reactions systems. The method is based on the calculation of the eigenvalues and eigenvectors of the local Jacobian during trajectory simulations and on the definition of certain tolerance parameters related to the normalized distance from the manifold and the local eigenvalues. The method has been demonstrated *via* a simple linear 3-variable tropospheric system where the time dependence of the simple normalised distance parameter was shown to be almost identical to the geometric distance of the trajectory from each 2-dimensional manifold. The normalised distance of the slowest mode from the zero-dimensional steady state was also shown to behave similarly to the steady state error. This was also the case for a simple nonlinear 6-variable system where the normalised distance from the manifold accurately represented the error in application of the QSSA for the 3 fastest species. The method seems to provide a simple way to calculate the dimension of the slow manifold and may find uses in the application of low-dimensional tabulation and repro-modelling techniques. In general it is desirable to tabulate or fit polynomials to the lowest number of species possible without the introduction of model inaccuracies. The method presented here provides a first step in identifying the lowest number of variables which need to be included. It should be stated however that model inaccuracies can only be calculated exactly *via* application of the chosen method such as repro-modelling in comparison with the full solution.

The simulation of variable photolysis conditions for the 6-variable system and the butane mechanism demonstrated that the dimension of the slow manifold can both increase and decrease throughout a simulation. This diurnal variation in dimension poses problems for the practical application of the ILDM. The simplest method of application is the use of a constant dimensional look-up table or a constant number of polynomial functions and has been predominantly used to date. In this case the maximum dimension present in a single calculation is usually used. For atmospheric problems however, the maximum dimension may be quite large in some cases and searching within high-dimensional tables is computationally expensive. Fitting polynomials for higher-dimensional manifolds is possible as has been shown in previous work by the authors[9] and is more efficient than searching in high-dimensional tables. The number of polynomial terms and therefore the fitting and simulation time will however increase with the number of variables required and therefore methods to identify the minimum dimensionality of the problem are very useful.

In principle it may be possible to use either variable dimensional look-up tables or a variable number of polynomial functions during a trajectory simulation or in different parts of the computational domain for reactive flow calculations. This method has been shown to be possible for the application of repro-modelling. For example, fitted polynomial functions can be determined which describe the concentration of certain fast species as functions of the slow species at the same time point[9] (a similar method to the QSSA). In this way if the manifold dimension increases over a time-step the concentrations of the fast species can be estimated using these fitted functions and the fast species replaced into the coupled repro-model. The manifold dimension would need to be stored as a function of the input variables since the full Jacobian would not be available during the simulation using a reduced variable set.

References

1 M. E. Jenkin, S. M. Saunders and M. J. Pilling, *Atmos. Environ.*, 1997, **31**, 81.
2 J. Toth, G. Li, H. Rabitz and A. S. Tomlin, *SIAM J. Appl. Math.*, 1997, **57**, 1531.
3 A. S. Tomlin, T. Turányi and M. J. Pilling, in *Autoignition and Low Temperature Combustion of Hydrocarbons*, ed. M. J. Pilling, Elsevier, Amsterdam, 1998.
4 T. Turányi, A. S. Tomlin and M. J. Pilling, *J. Phys. Chem.*, 1993, **97**, 163.
5 U. Maas and S. B. Pope, *Combust. Flame*, 1992, **88**, 239.
6 T. Turányi, *Comput. Chem.*, 1994, **18**, 45.
7 A. M. Dunker, *Atmos. Environ*, 1986, **20**, 479.

8 C. M. Spivakovsky, R. Yevich, J. A. Logan, S. C. Wofsy, M. B. McElroy and M. J. Prather, *J. Geophys. Res. D*, 1990, **95**, 18441.
9 R. Lowe and A. S. Tomlin, *Atmos. Environ.*, 2000, **34**, 2425.
10 M. R. Roussel and S. J. Fraser, *J. Chem. Phys.*, 1991, **94**, 7106.
11 N. Bell, M. J. Pilling and A. S. Tomlin, in preparation.
12 R. G. Derwent and ØHov, Report R9434, UK Atomic Energy Research Establishment, Harwell, UK, 1979.
13 http://www.chem.leeds.ac.uk/Atmospheric/MCM/mcmproj.html.
14 G. Zeng, PhD Thesis, University of Leeds, 1997.
15 G. Zeng, M. J. Pilling and S. M. Saunders, *J. Chem. Soc., Faraday Trans.*, 1997, **93**, 2337.
16 F. M. Bowman and J. H. Seinfeld, *J. Geophys. Res.*, 1994, **99**, 5309.

A numerical study of spatial structure during oscillatory combustion in closed vessels in microgravity

Roger Fairlie[a] and John F. Griffiths[b]

[a] *School of Computing, The University, Leeds, UK LS2 9JT.*
  E-mail: rogerf@comp.leeds.ac.uk
[b] *School of Chemistry, The University, Leeds, UK LS2 9JT.*
  E-mail: johng@chem.leeds.ac.uk

Received 10th April 2001
First published as an Advance Article on the web 12th November 2001

The existence and spatial development of gas-phase, thermokinetic oscillations under the influence of mass and thermal diffusion have been investigated by numerical methods in a 1-dimensional system. The conditions correspond to those that would be experienced under microgravity. The interest arises because there have been recent experimental investigations of oscillatory reactions, involving cool flames during butane oxidation, as part of the NASA, KC135 microgravity flight programme. The Sal'nikov, thermokinetic scheme, which is a two-variable model representing an intermediate chemical species and reactant temperature (taking the form P → A → B), forms the basis of the present work. In this model, thermal feedback occurs through the exothermicity of the second step and the non-linearity is derived from its temperature dependence. There are no known chemical examples that satisfy Sal'nikov's formal structure but Griffiths and co-workers conceived an experimental analogue under terrestrial conditions whereby a gaseous reactant was allowed to flow from an external reservoir into a closed, heated reactor at a controlled rate *via* a capillary tube which fed the reactant to the centre of the vessel. The exothermic reaction that occurred in the vessel satisfied the necessary conditions for the second step and the inflow, with no temperature dependence, represented a physical analogue to the first step of the Sal'nikov scheme. Thermokinetic oscillations were observed and the range of conditions for their existences was investigated. One of the experimental systems was the exothermic reaction between hydrogen and chlorine. To represent the Sal'nikov conditions hydrogen was fed slowly into the reactor, which already contained chlorine. We have exploited this chemical system and its experimental implementation in the present paper to investigate the behaviour when no convection or bulk gas motion occurs and when heat and mass transport is driven solely by diffusion. We study the response of alternative numerical approaches to the way in which the first step of the scheme is simulated. In the first, the precursor (P) is supplied at the same rate simultaneously throughout the cells representing the reactor. This is close to the concept of the Sal'nikov model. In the second method, a fixed rate of supply is applied at the inner boundary of the axisymmetric, 1-dimensional system. This is analogous to the experimental procedure. The numerical results show how oscillatory states can be sustained as a result of heat and mass transport by diffusion. The temporal and spatial evolution of reaction in a range of circumstances is discussed.

Introduction

Thermokinetic oscillations in gaseous reactions are most familiar as the repetitive cool flames that are associated with the low temperature combustion of hydrocarbons.[1] Other examples are also known.[2] These non-isothermal phenomena arise under non-adiabatic conditions as a result of the interaction between exothermic, chemical processes with non-linear feedback and heat loss to the surroundings. Under terrestrial conditions, the major mode of heat transfer in closed systems is by natural convection to the vessel surface, and this has justified the basis for theoretical interpretation in terms of an idealised, spatial uniformity of temperature and species concentrations.[3] In such cases the thermal losses are defined to be of Newtonian form, and they are quantified by a surface heat transfer coefficient or a characteristic Newtonian cooling time.[4]

Experimental procedures have followed suit in order to seek quantitative links between theory and experiment.[3,5,6] Some of the first experiments in which a mechanical stirrer was used to smooth out temperature and concentration gradients during oscillatory cool flames of propane were reported at Faraday Discussions 9, in 1974 (Fig. 1).[7] These were the first experiments to confirm the damped oscillatory nature of cool flame phenomena in closed vessels, a realisation that had eluded experimentalists until then because natural convection perturbed the system to a significant extent,[8] as shown in Fig. 1.

Thirty years on there is an incentive to consider another idealised extreme, whereby a diffusive flux is the only heat loss mode. The experiments cannot be implemented in a terrestrial environment but the condition is reasonably well satisfied in microgravity, at which convection is almost entirely suppressed. An experimental programme, conducted by NASA, exploits the parabolic trajectory of an aircraft to achieve a microgravity condition, typically over a time interval of 22 s. This duration has proved to be sufficiently long for butane + oxygen cool flames in microgravity to be investigated in a closed vessel.[10] An important, distinguishing feature of these studies from the terrestrial experiments was that multiple cool flames did not occur in mixtures that contained only butane and oxygen. This limitation was attributed to the much reduced heat transport rate in the absence of convection. The expedient was to include helium in the reaction mixture in order to enhance the thermal conduction. This proved to be successful in promoting the occurrence of multiple cool flames. The result and its interpretation were vindicated in numerical analyses based on a formal mathematical structure to represent the kinetics in simplified form.[11]

These novel developments involve a spatial as well as a temporal evolution, which adds a supplementary dimension concerning the nature of the propagation of combustion fronts when oscillatory reaction occurs and the way in which diffusion of reactive intermediates and thermal conductivity control events. Some progress has been made recently on the basis of a numerical interpretation of the cool flames in microgravity.[11] That species and thermal diffusion may play a part in the spatial evolution of the cool flames of butane seems reasonable because both non-linear kinetic and thermal feedback are known to be involved. This thermokinetic interaction was elucidated by Gray and Yang,[12–14] in the period 1965–9.

There is an interesting alternative, conceived by Sal'nikov in the 1940s,[15,16] which relates to a more rudimentary analysis of thermokinetic oscillations. Sal'nikov's model, based on the two-step kinetic representation $P \rightarrow A \rightarrow B$, incorporates only the non-linear thermal feedback component through the exothermicity and activation energy associated with the second step. The first step is regarded to be thermoneutral and to have no temperature dependence. As shown by Sal'nikov[15,16] and elaborated more recently by P. Gray et al.,[17] and B. F. Gray and Roberts,[18] this simple scheme is capable of exhibiting oscillatory reaction in spatially uniform, non-adiabatic conditions.

There is no known combustion system that matches the Sal'nikov model but Griffiths and co-workers[19–21] were able to create analogues to it by devising an experiment in which the first kinetic step was represented in a physical way by the controlled inflow of reactant from a cold reservoir to a heated reaction vessel. The chemistry that then occurred in the reaction vessel was identified in a global sense with the second step. In one case [19,20] the reactant was di-*tert*-butyl peroxide, which decomposes exothermically. Although the free radical mechanism involves several elementary reactions, and rather more reactions (with an attendant gain in exothermicity) if oxygen is also present,[22] the overall reaction rate is governed by the initial, first order decomposition of the peroxide linkage. There is no chain branching. The experiments involving di-*tert*-butyl

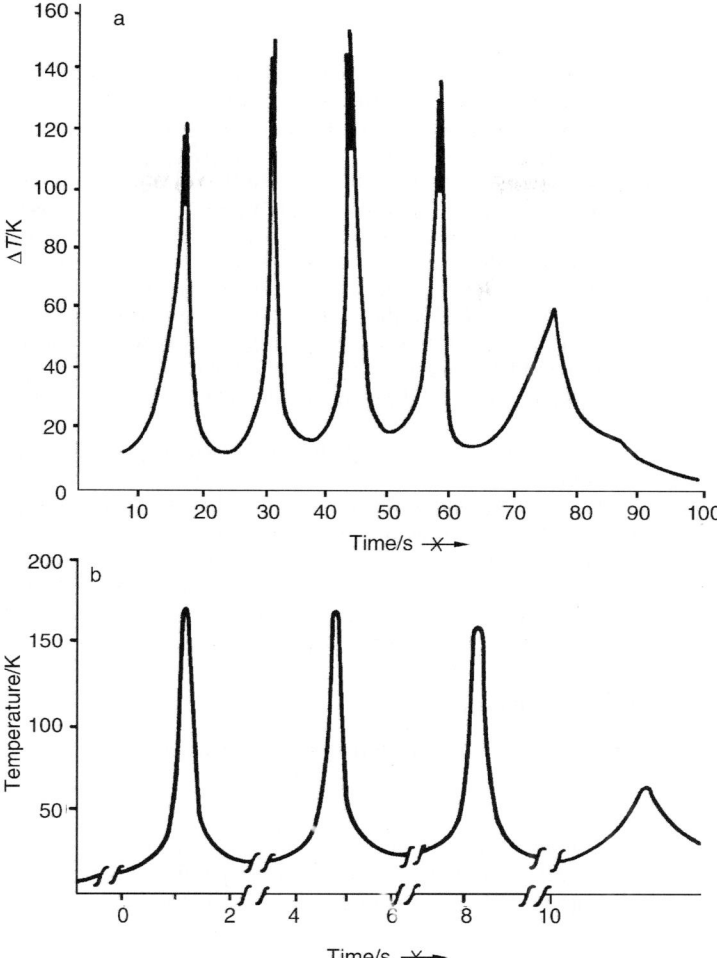

Fig. 1 Temperature–time profiles of cool flame phenomena obtained during the combustion of propane + oxygen in closed vessels under similar conditions at approximately $T_a = 600$ K and $p = 60$ kPa. The upper trace includes the effect of natural convection induced by the non-isothermal reaction. The lower trace, showing the sustained oscillatory reaction perturbed eventually by fuel consumption, was obtained with mechanical stirring to create a spatially uniform environment.[9]

peroxide were performed in a reaction vessel containing a mechanical rotor to smooth out the concentration and temperature gradients.[19,20]

In another case the exothermic reaction between $H_2 + Cl_2$ was investigated, the experiment being performed by allowing a controlled flow of hydrogen into the reactor with chlorine already present.[21] This experimental analogue provided a test for the robustness of the Sal'nikov thermokinetic oscillations with respect to exothermic reaction of order different from unity. Natural diffusion and convection remained a possibility in the $H_2 + Cl_2$ reaction, but there was no forced mixing in these particular experiments. Although a chain reaction propagated by free radicals, in the early stages of reaction between H_2 and Cl_2 the overall rate of formation of HCl can be expressed in the form

$$v = k[H_2][Cl_2]^{1/2}, \tag{1}$$

where k represents a composite of rate constants for the elementary steps involved. Later the retardation by HCl gives more complex concentration dependences than those expressed in eqn. (1). The overall exothermicity is 186 kJ mol^{-1}.

Although experiments of these kinds have not been studied in microgravity, there may be some interest to do so if it is perceived that there are fundamental properties of interest from the Sal'nikov structure when species diffusion and thermal conductivity are the controlling factors. The purpose of the present paper is to explore how the $H_2 + Cl_2$ reaction might behave in circumstances where there is no opportunity for convection to play a part. Following from eqn. (1), this numerical study is restricted to a global representation of the reaction in its relatively early stages, based on an assumption of a sufficient excess of Cl_2 in the reaction vessel, such that the consumption of Cl_2 may be ignored with respect to the rate and extent of H_2 flowing into it. This gives the early time rate expression (eqn. (1)) in a form that is a pseudo-first order reaction with respect to [H_2]. The numerical treatment is then restricted to a two-variable model representing the reactant temperature and the concentration of H_2 within the reaction vessel, which is identified with A in the Sal'nikov model. Thus we seek to determine how, using a one-dimensional, numerical code, the system behaves when it is exposed to mass and thermal diffusion, and the spatial structure that may develop.

A sphere is the most practical vessel consistent with a 1-dimensional representation, as was used experimentally.[21] The "infinite cylinder" may also be reasonably approximated in an experimental test by a cylinder of suitably high length/diameter ratio. Although less easily matched experimentally, the "infinite slab" has the distinction of offering no variation in cross-section as the reaction progresses to the outer boundary. We examine the predicted differences in behaviour that would arise in a planar system from that in a sphere and also how these are affected when an asymmetric source of the pre-cursor is introduced.

Theoretical background

Formulation of the problem with diffusive fluxes

The Sal'nikov model comprises a pair of consecutive, first order reactions

$$P \xrightarrow{1} A \xrightarrow{2} B \qquad (2)$$

which are subject to the conditions $E_1 = 0$, $q_1 = 0$, $E_2 > 0$, $q_2 > 0$, where E and q represent the activation energy and the exothermicity for the respective steps. P represents a pre-cursor to the active intermediate A. In purely diffusive conditions, the conservation equations for the intermediate species A and energy take the form

$$\frac{\partial a}{\partial t} = k_1 p_0 e^{(-k_1 t)} - k_2(T)a + D_A \nabla^2 a \qquad (3)$$

$$\frac{\partial T}{\partial t} = \frac{q_2 k_2(T) a}{C_V \sigma} + D_T \nabla^2 T \qquad (4)$$

where lower case letters represent species concentrations and k represents a rate constant. The heat capacity at constant volume is C_V and σ is the (constant) reactant density. D_A and D_T are species and thermal diffusion coefficients respectively and ∇^2 is the Laplacian operator. The first term in eqn. (3) represents the decay of precursor in the external reservoir. If a "pool chemical approximation" is invoked this term is reduced to $k_1 p_0$.[23] The boundary conditions were, in each case,

$$\frac{\partial a}{\partial x} = 0 \quad \text{and} \quad \frac{\partial T}{\partial x} = 0 \quad \text{at } x = 0, \qquad (5)$$

$$(T - T_a) = 0 \quad \text{at } x = r, \qquad (6)$$

This formulation implies that there is an identical supply of material from the reservoir to all points in the reaction volume (P → A).

For the purpose of numerical analysis it is more convenient to cast these equations in a dimensionless form. We base the following summary on the analysis by Gray et al.[17] and the more recent work by Scott *et al.*[24] to which diffusion fluxes were an accompaniment. The non-dimensionalisation includes a reference concentration (c_{ref}) and length (l_{ref}) as well a characteristic time, chosen here to be the chemical timescale of reaction (2) defined at the ambient temperature for the system (t_{ch}). Thus let

$$c_{ref} = \frac{C_V \sigma R T_a^2}{q_2 E_2}, \qquad l_{ref} = (D_A t_{ch})^{1/2}, \qquad k_2 = Z_2 e^{-(E_2/RT)},$$

$$t_{ch} = \frac{1}{k_2(T_a)}, \qquad \tau = \frac{t}{t_{ch}}, \qquad \alpha = \frac{a}{c_{ref}},$$

$$\mu = \frac{k_1 p_0 q_2 t_{ch}}{C_V \sigma} \frac{E}{R T_a^2}, \qquad \varepsilon = \frac{R T_a}{E_2}, \qquad \theta = \frac{E(T - T_a)}{R T_a^2}$$

$$f(\theta) = e^{\theta/(1+\varepsilon\theta)} \qquad Le = \frac{D_T}{D_A} \qquad D_T = \frac{\lambda}{C_V \sigma}$$

Le is the Lewis number and λ is thermal conductivity. Eqn. (2) and (3) thus take the non-dimensionalised form:

$$\frac{\partial \alpha}{\partial \tau} = \mu^{-(k_1 t_{ch})\tau} - \alpha f(\theta) + \nabla^2 \alpha \qquad (7)$$

$$\frac{\partial \theta}{\partial \tau} = \alpha f(\theta) + Le \nabla^2 \theta \qquad (8)$$

Formulation allowing for an asymmetric supply of reactant

In the $H_2 + Cl_2$ experiment the hydrogen was admitted to the vessel from a capillary tube at the centre. This constitutes an asymmetric, pressure-driven feed that can be simulated as a source at the left-hand boundary of an axisymmetric model, as has been used similarly elsewhere to describe the effect of a constant heat flux into a reactive solid.[25] Here we explore the behaviour of the pair of eqn. (7) and (8) in an axisymmetric, planar system, with the modified eqn.

$$\frac{\partial a}{\partial t} = P_0 - k_2(T)a + D_A \nabla^2 a \qquad (9)$$

$$\frac{\partial T}{\partial t} = \frac{q_2 k_2(T) a}{C_V \sigma} + D_T \nabla^2 T \qquad (10)$$

The first term on the right-hand side of eqn. (9) is invoked only at the interface adjacent to the left-hand boundary. Its implication is that the precursor, P, is held in a reservoir at the left-hand boundary but external to the system and that the reactant, A, is supplied at a constant rate, P_0, to the vessel at that interface. A pool chemical approximation is applied to the external supply source. The reactant is able to spread through the reaction volume as a result of the diffusion term in eqn. (9). The behaviour of A in the reaction volume is determined only by the second and third terms of eqn. (9). The equations were non-dimensionalised in a similar way to eqn. (7) and (8).

Stability analysis based on spatially uniform conditions

To anticipate what type of behaviour might be expected, we summarise the results of a Hopf bifurcation analysis.[26] The analysis has been developed only for spatially uniform conditions and when the pool chemical approximation is assumed with respect to the pre-cursor P. That is $p = p_0$ throughout, such that μ represents a constant supply rate given by $\mu e^{-(k_1 t_{ch})\tau} = 1$. Including $f(\theta)$ in its approximate form,[27] $f(\theta) = e^\theta$, which is reasonable at $\varepsilon\theta \ll 1$, neglecting the species and thermal diffusion terms and including a term to represent a Newtonian heat loss as a function of θ, eqn. (7)

and (8) then become

$$\frac{\partial \alpha}{\partial \tau} = \mu - \alpha e^{\theta} \qquad (11)$$

$$\frac{\partial \theta}{\partial \tau} = \alpha e^{\theta} - \kappa \theta \qquad (12)$$

where κ relates to the Newtonian heat transfer coefficient χ through

$$\kappa = \frac{t_{ch}\chi S}{C_V \sigma V} \qquad (13)$$

The stationary state solutions to eqn. (11) and (12) are single valued and they exist at

$$\alpha_s = \frac{\mu}{e^{\theta}}, \qquad \theta_s = \frac{\mu}{\kappa} \qquad (14)$$

The Hopf curve, which distinguishes the region of instability from that of stability in the (α–θ) plane, is defined by the equations

$$\frac{d\alpha}{d\tau} = \frac{d\theta}{d\tau} = 0 \qquad (15)$$

and

$$\text{tr } J = 0$$

$$\det J > 0 \qquad (16)$$

where tr J and det J are the trace and the determinant of the Jacobian matrix from the conservation eqn. (11) and (12). Instability exists when tr $J < 0$. The system is stable when tr $J > 0$, but there is a distinction between nodal and focal behaviour in the stable region at the condition

$$\Delta = (\text{Tr } J)^2 + 4(\det J) \qquad (17)$$

The behaviour of the Sal'nikov model has been investigated comprehensively both analytically and numerically by Gray and Scott[23] under a range of circumstances.

From eqn. (11) and (12)

$$J = \begin{vmatrix} -e^{\theta_s} & -\alpha e^{\theta_s} \\ e^{\theta_s} & \alpha_s e^{\theta_s} - \kappa \end{vmatrix} \qquad (18)$$

From eqn. (18), and eliminating α by eqn. (14), at the limit of stability

$$\text{tr } J = \kappa\theta_s - \kappa - e^{\theta_s} = 0$$

$$\det = \kappa e^{\theta_s} \qquad (19)$$

from which

$$\kappa = e^{\theta_s}(\theta_s - 1)^{-1} \qquad (20)$$

There is a minimum in the locus $\kappa(\theta)$ below which the system is always stable. The minimum exits at $\theta = 2$, at which $\kappa = e^2 (=7.389)$. For values of $\kappa > e^2$, there is a range of θ within which unstable states exist. As κ increases, the range of instability broadens from the minimum value at $\kappa = 7.389$, $\theta = 2$ (Fig. 2). The dimensionless temperature excess (θ) is, itself, governed by the supply

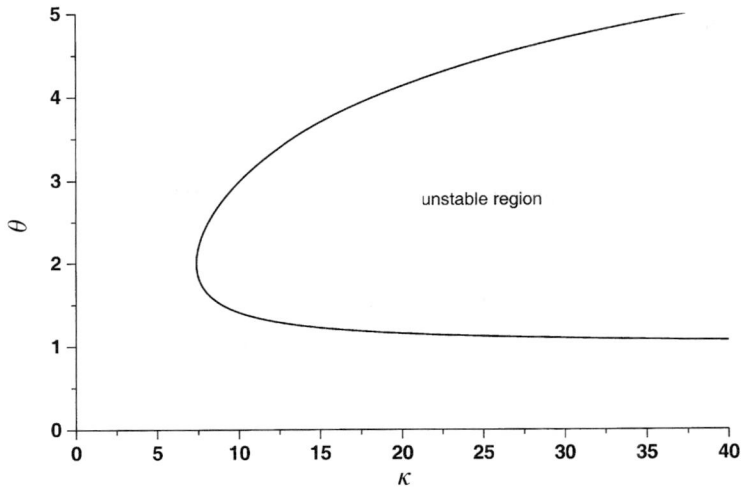

Fig. 2 Analytical solutions for stable and unstable behaviour in the Sal'nikov scheme when the pool chemical approximation is applied in spatially uniform conditions. θ is the non-dimensionalised temperature excess and κ is the ratio of the chemical time to the Newtonian cooling time, as expressed in eqn. (13).

rate parameter μ, such that eqn. (20) is solved for a given supply rate utilising the stationary state solutions (14). For a given value of κ the unstable (oscillatory) states can exist only within a certain range of θ.

Physicochemical parameters and numerical methods

For ease of computation certain parameters were assumed to be constant. This would have a limited quantitative rather than qualitative effect on the results. Thus, heat capacity for the mixture of bimolecular reactants and product and reactant density were taken to be 30 J mol^{-1} K^{-1} and 1.0×10^{-6} mol cm^{-3} respectively, from which the volumetric heat capacity was 3×10^{-5} J cm^{-3} K^{-1}. The numerical results are not sensitive to the reactant density in a direct way but its value determines the magnitude of k_2 and of the volumetric heat capacity, which is relevant to the interpretation of thermal diffusivity (D_T) and, through Le, the magnitude of the diffusion coefficient, D_A. The reactant density corresponds to that used experimentally for the initial partial pressure of chlorine in the reactor at 625 K (40 Torr).[21]

The detailed kinetic mechanism appropriate to early stages of the $H_2 + Cl_2$ reaction is set out in Table 1. The data are those used by Coppersthwaite et al.[21] to represent the oscillatory reaction under spatially uniform conditions within an appropriate temperature range (600–650 K). The most significant departures from published data relate to reactions (i) and (iv), for which surface activity is believed to play a part.[21] The formalised rate constant for the overall reaction expressed

Table 1 Reactions and kinetic parameters

	Reaction	Rate constant
(i)	$Cl_2 + M \rightarrow 2Cl + M$	$k_i = 1 \times 10^{17}\ e^{(-182\,900/RT)}$ mol^{-1} cm^3 s^{-1}
(ii)	$H_2 + Cl \rightarrow HCl + H$	$k_{ii} = 1 \times 10^{12}\ e^{(-23\,000/RT)}$ mol^{-1} cm^3 s^{-1}
(iii)	$Cl_2 + H \rightarrow HCl + Cl$	$k_{iii} = 8 \times 10^{13}\ e^{(-600/RT)}$ mol^{-1} cm^3 s^{-1}
(iv)	$Cl + Cl + M \rightarrow Cl_2 + M$	$k_{iv} = 1 \times 10^{17}$ mol^{-2} cm^6 s^{-1}

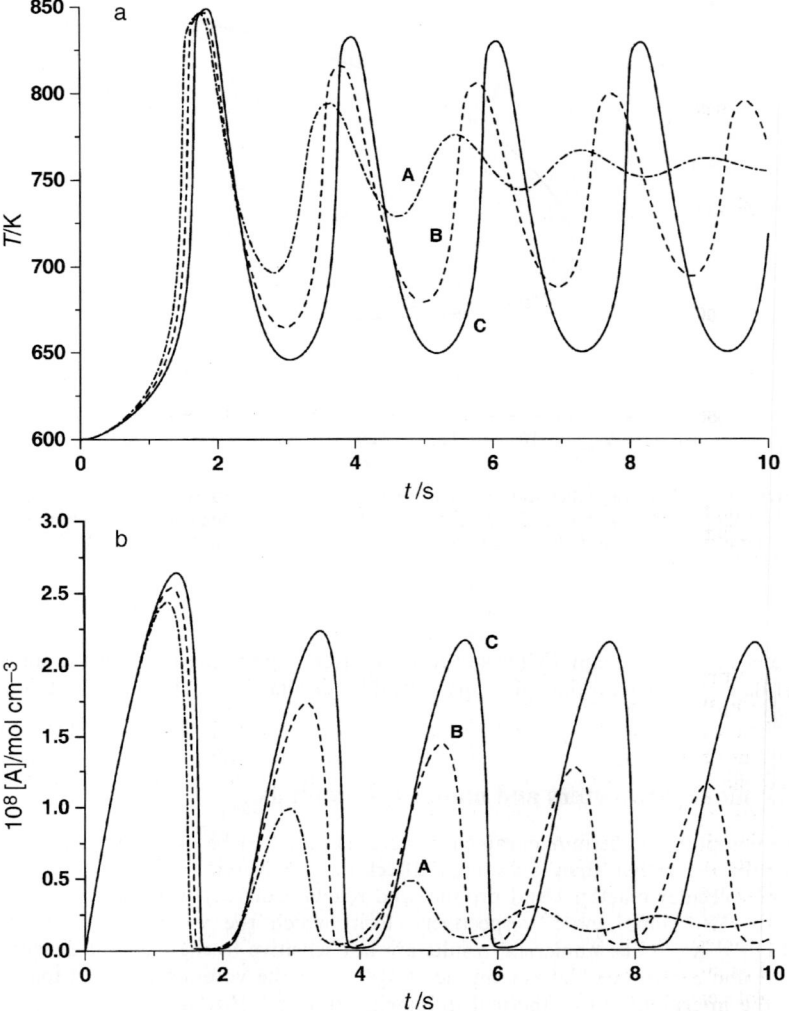

Fig. 3 (a) Centre temperature (T)–time and (b) intermediate concentration ([A])–time profiles at $T_a = 600$ K under the effect of mass and thermal diffusion in pool chemical conditions at the lower bound between damped and sustained oscillations. A is at $D_a = D_T = 10$ cm² s⁻¹, B is at $D_a = D_T = 11$ cm² s⁻¹ and C is at $D_a = D_T = 12$ cm² s⁻¹.

as a pseudo-first order reaction is given by $k_2 = 2k_{ii}(k_i/k_{iv})^{1/2}[Cl_2]_0^{1/2}$, giving a value $k_2 = 2 \times 10^9$ $e^{(-113\,450/RT)}$ s⁻¹. The magnitudes selected for D_T and D_A in order to investigate behaviour under the influence of diffusive fluxes are discussed below.

Eqn. (7) and (8), subject to the boundary conditions (5) and (6) cast in non-dimensional form, and the initial conditions $\theta = 0$ and $\alpha = 0$, were integrated using the NAG routine D03PSF,[28] which can be utilised to represent the 1-dimensional shapes (infinite slab, infinite cylinder or sphere). This routine is set up for the integration of a system of non-linear convection—diffusion equations in one dimension, using the Method of Lines to reduce the PDEs to ODEs. The ODE system is then solved by a BDF method.[29] The BDF algorithm makes use of formulae that are varied automatically up to order 5. The scheme is an implicit, variable step-size method, the time step and order being selected automatically to satisfy a user-specified tolerance. This allows temperature spikes, which may develop on very short timescales, to be accurately modelled while not

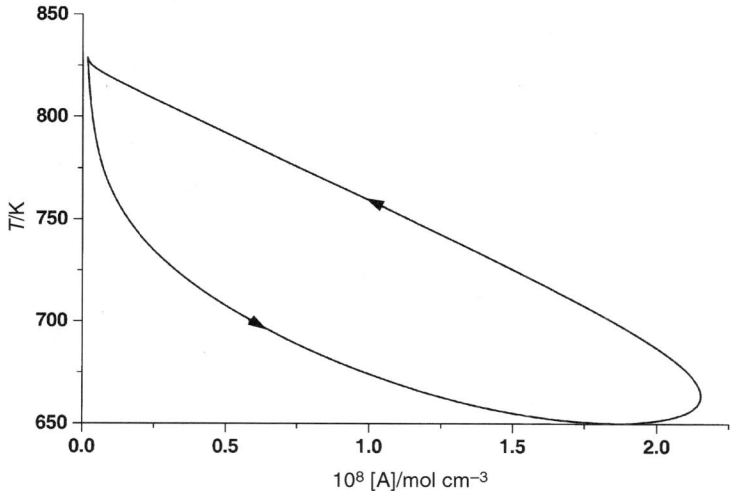

Fig. 4 A phase plot showing the relationship between the intermediate concentration, [A], and temperature, T, during an oscillatory cycle at $T_a = 600$ K and $D_A = D_T = 12$ cm^2 s^{-1}. Between the time intervals 7.3 and 9.4 s. The direction of motion round the cycle is marked by arrows.

restricting the time step in other phases of the simulation. A modified Newton iteration scheme is used to solve the system of non-linear equations at each time step.

The problem was discretised using a finite volume approach.[28] The source terms were calculated at each node and the diffusive terms were evaluated at each nodal mid-point using values derived from the nodes at either side.[30] These values were calculated by assuming that the solution varied within a cell, and employing standard upwind techniques combined with a slope limiter.[31] In the present application, the domain was meshed uniformly with 126 points. Neumann conditions were specified at the symmetry boundary, and Dirichlet conditions were specified at the outer boundary, as noted above.

For eqn. (9) and (10) the interior boundary is completely remodelled to allow a constant mass flux into the domain. A "ghost cell" is introduced as the first cell. This represents a cell within the reservoir and as such is assumed to have a constant value for the concentration of the pre-cursor, and a constant temperature. The boundary to this cell is set to the same fixed conditions. The interface between this first cell and the second cell thus becomes the boundary between the reservoir and the reaction vessel. An advective flux is imposed across this, which represents the forced inflow of A into the domain. While this representation implies an advection term at only one interface, the bleed rate into the domain is sufficiently slow for diffusion to realistically dominate transport thereafter.

Results

Analyses for a spherical system

Although not truly representative of the experimental study, calculations based on a pool chemical approximation permit the distinction between sustained and damped oscillatory behaviour to be established. When the pool chemical approximation was applied to the pre-cursor concentration in eqn. (3), sustained oscillations were generated at 600 K under the effect of diffusive fluxes for mass and energy within the range of values $D_A = D_T = 12$–23 cm^2 s^{-1}. Damped oscillatory reaction was obtained at values for the diffusive fluxes just outside this range, as might be expected from eqn. (20). As shown in Fig. 3, when the mass and thermal diffusion coefficients were

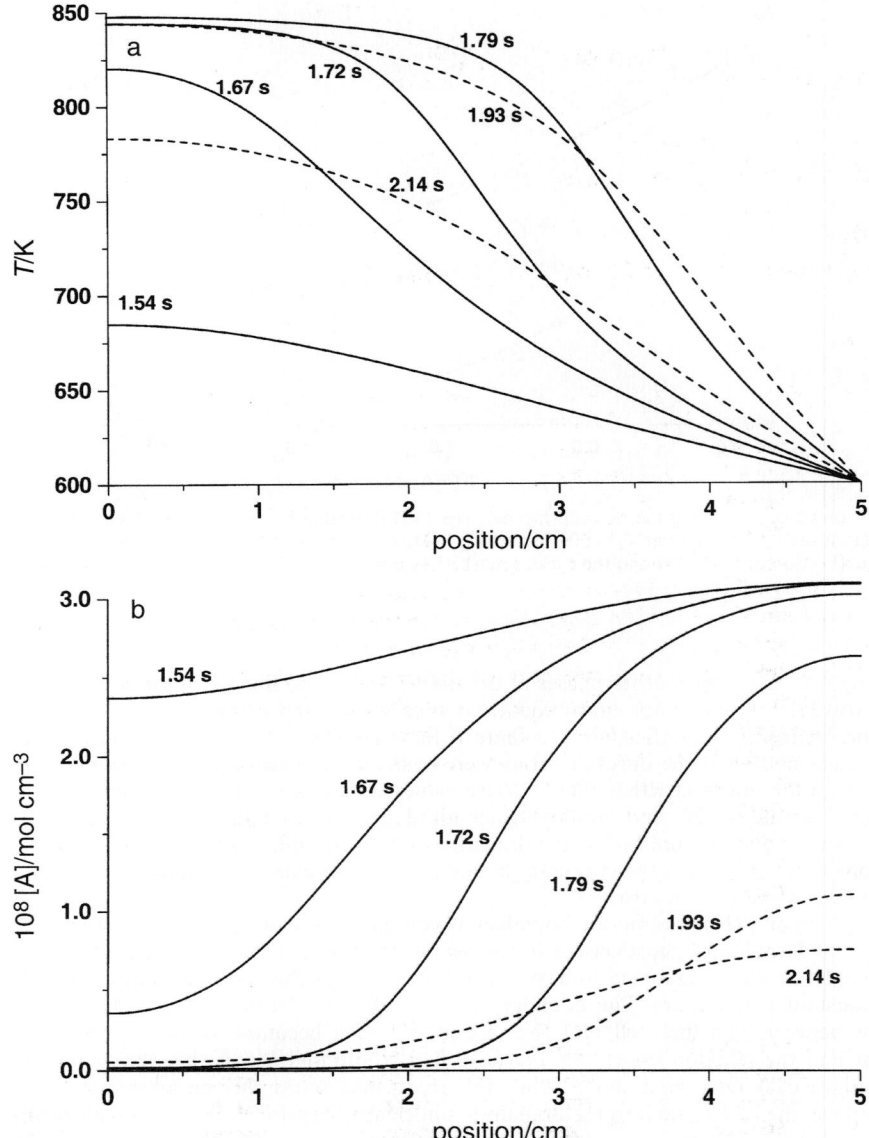

Fig. 5 Spatial profiles at selected times showing (a) temperature and (b) intermediate concentration, [A], under the effect of mass and thermal diffusion when $T_a = 600$ K. The solid lines represent the period during which the temperature is increasing and the broken lines represent the period during which the temperature is decreasing. $D_A = D_T = 12$ cm^2 s^{-1}.

varied over the range 10–12 cm^2 s^{-1} ($Le = 1$) there is a marked transition from damped oscillatory behaviour to sustained oscillations. The maximum amplitude approaches 200 K at the centre of the reactant mass (Fig. 3a), and the concentration of the intermediate, [A], decays almost to zero in each successive cycle. The phase relationship between intermediate concentration and temperature throughout the cycle of sustained oscillations shown in Fig. 3 ($D_A = D_T = 12$ cm^2 s^{-1}) is displayed in Fig. 4. As the initial peak temperature is approached, there is a distortion in the shape of the temperature maximum. This is connected with the incipient emergence of a combustion front from the centre of the reacting mixture. However, this is not fully sustained and the development fails at about half distance to the wall (Fig. 5). This failure is probably connected with a dramatic reduction in the concentration of the intermediate at the periphery of the reactor as a result of its diffusion from the wall region into the highly reactive, central zone.

There is a very high sensitivity of the amplitude where the transition from sustained to damped oscillations exists at the upper bound of the diffusion coefficients, as shown in curves A and B in Fig. 6, for which $D_A = D_T = 23$ cm^2 s^{-1} and $D_A = D_T = 24$ cm^2 s^{-1} respectively. If we revert to spatially uniform conditions by incorporating a Newtonian heat transfer at the surface ($t_N = C_v \sigma V / \chi S = 0.3$ s) and adopting Neumann conditions at the external boundary, the system exhibits damped oscillations of similar but slightly smaller period and amplitude as those under the effect of mass and thermal diffusion close to the upper limit at the same inlet flow rate of H$_2$ (Fig. 6, curve C).

Spatial temperature and intermediate concentration profiles throughout the first temperature maximum, corresponding to conditions in Fig. 6 curves A and B, are given in Fig. 7 and 8. In a qualitative sense the spatial development during sustained oscillation (Fig. 7) is remarkably similar to that at the lower mass and thermal diffusion coefficients (Fig. 5). There is a faster development of the combustion front from the centre when higher diffusion coefficients prevail.

Analyses for a planar system

Sustained oscillations are still possible if the Salnikov model derived from eqn. (3) and (4) is transposed to the infinite slab, but at $T_a = 600$ K they exist only at mass and thermal diffusion coefficients in excess of 35 cm^2 s^{-1} (Fig. 9a, curve A), when there is the same inflow condition as that for the results given in Fig. 3. The corresponding concentration profile of the intermediate A at the centre of the axisymmetric system is shown in Fig. 9b, curve A. Spatial variations in temperature and concentration are shown in Fig. 10. There is no predicted expansion of a combustion front in planar symmetry, and the concentration of the intermediate is depleted virtually to zero at the centre of the system when the temperature is close to its maximum. As is consistent with the phase plot in Fig. 4, the maximum concentration at the centre precedes the maximum in temperature in an oscillatory cycle.

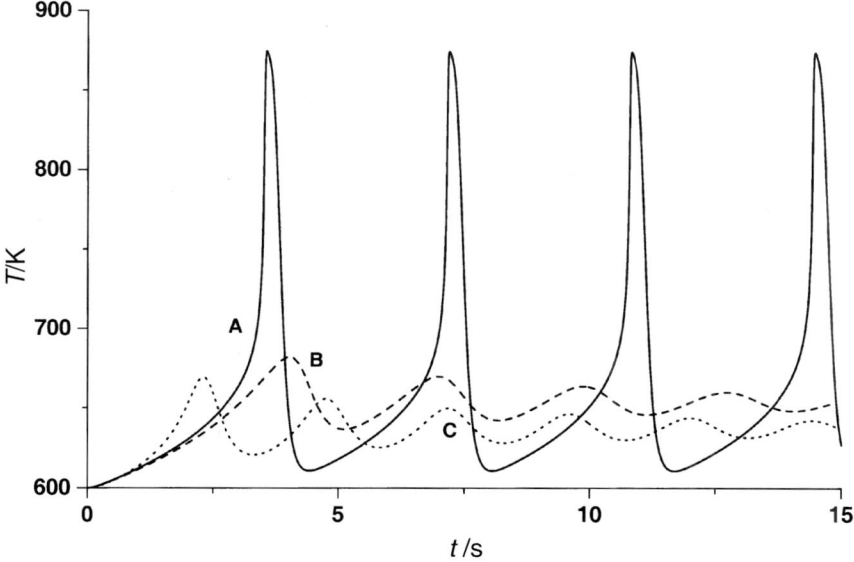

Fig. 6 Centre temperature–time profiles at $T_a = 600$ K under the effect of mass and thermal diffusion in pool chemical conditions at the upper bound between sustained and damped oscillations. A is at $D_A = D_T = 23$ cm^2 s^{-1} and B is at $D_A = D_T = 24$ cm^2 s^{-1}. C represents the behaviour when the system is transposed to spatially uniform conditions and the surface heat transfer has a Newtonian cooling time of 0.3 s, which is typical for a low pressure gas-phase reaction in a small-scale stirred reactor.

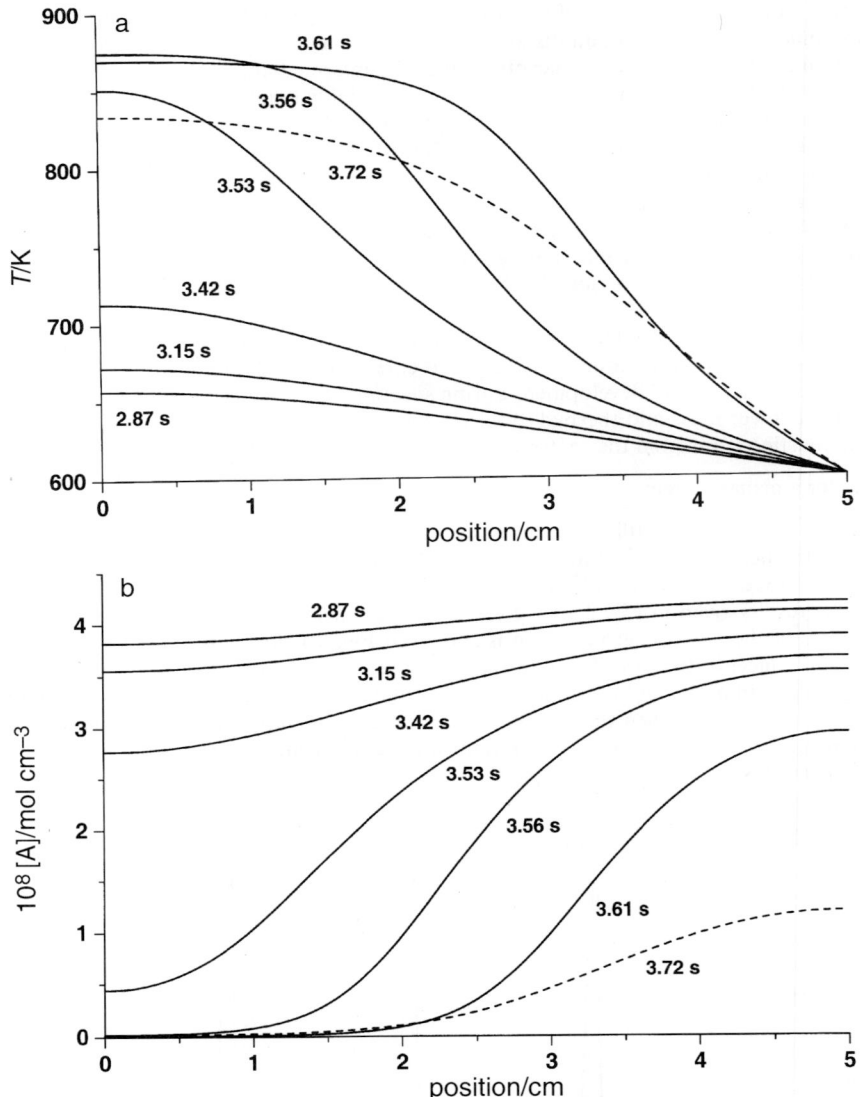

Fig. 7 Spatial profiles at selected times showing (a) temperature and (b) intermediate concentration, [A], under the effect of mass and thermal diffusion when $T_a = 600$ K. The solid lines represent the period during which the temperature is increasing and the broken line represents the period during which the temperature is decreasing. $D_A = D_T = 23$ cm^2 s^{-1}.

Oscillations were also computed for inflow of precursor at the central plane of the axisymmetric slab (eqn. (9) and (10)), assuming a pool chemical approximation at the source and with the same values for D_A and D_T as in Fig. 10. The centre temperature–time and concentration–time profiles from this calculation are shown in Fig. 9, curves B. The molar density inflow rate was chosen to give a period of oscillations that is comparable to that of curves A in Fig. 9. The spatial temperature and concentration profiles through an oscillatory cycle are shown in Fig. 11. Considerably higher concentrations were created throughout the reactor than in the former case (*cf.* Fig 10b and 11b). There is a constant gradient for the supply rate at the central plane, which creates the highest concentration in that vicinity when the temperature in an oscillatory cycle is at its

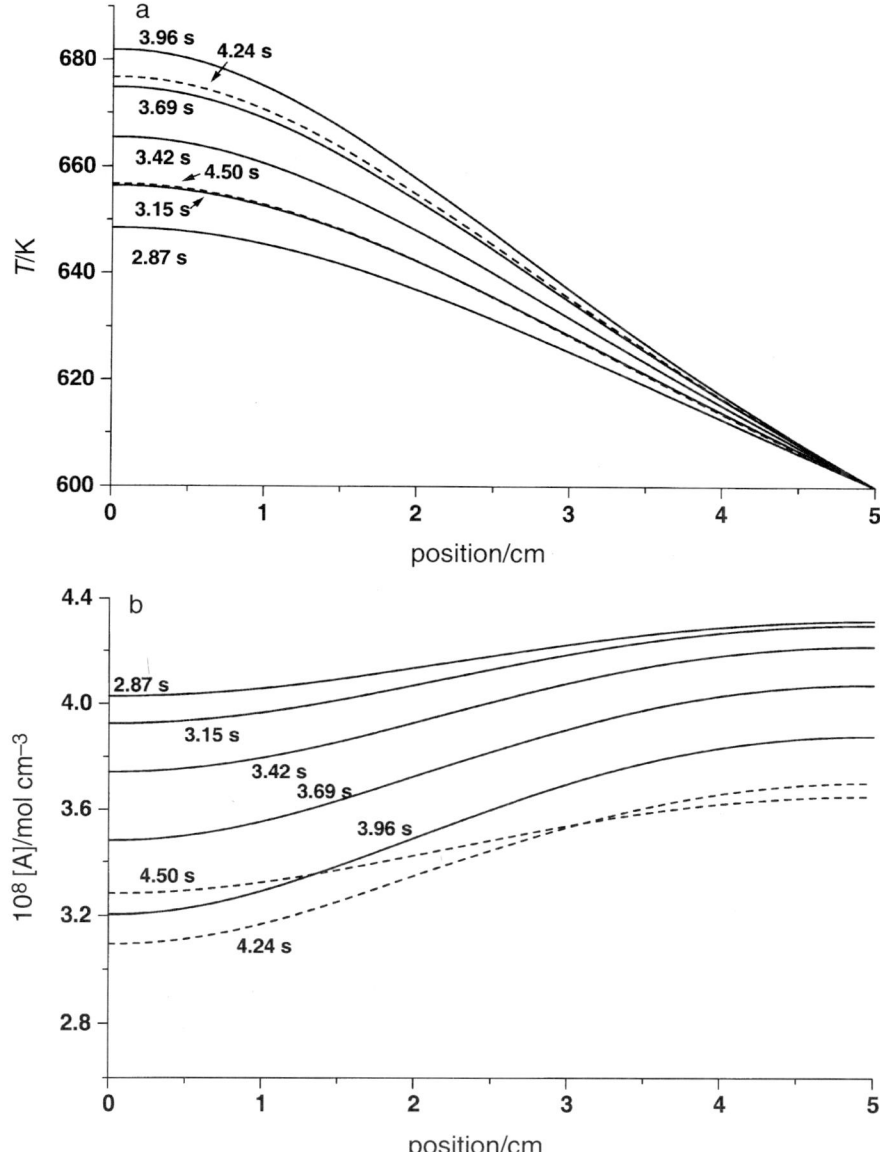

Fig. 8 Spatial profiles at selected times showing (a) temperature and (b) intermediate concentration, [A], under the effect of mass and thermal diffusion when $T_a = 600$ K. The solid lines represent the period during which the temperature is increasing and the broken line represents the period during which the temperature is decreasing. $D_A = D_T = 24$ cm^2 s^{-1}.

lowest (Fig. 11b). Complex concentration profiles are created as the temperature varies through its cycle and at no time or place does the concentration approach zero. By contrast, when the precursor supply is distributed simultaneously to all cells, the concentration of A is always highest at the external surface of the reaction volume and lowest at the central plane at any given time (Fig. 10b).

Discussion

The application of a global kinetic approach to the Sal'nikov representation of the $H_2 + Cl_2$

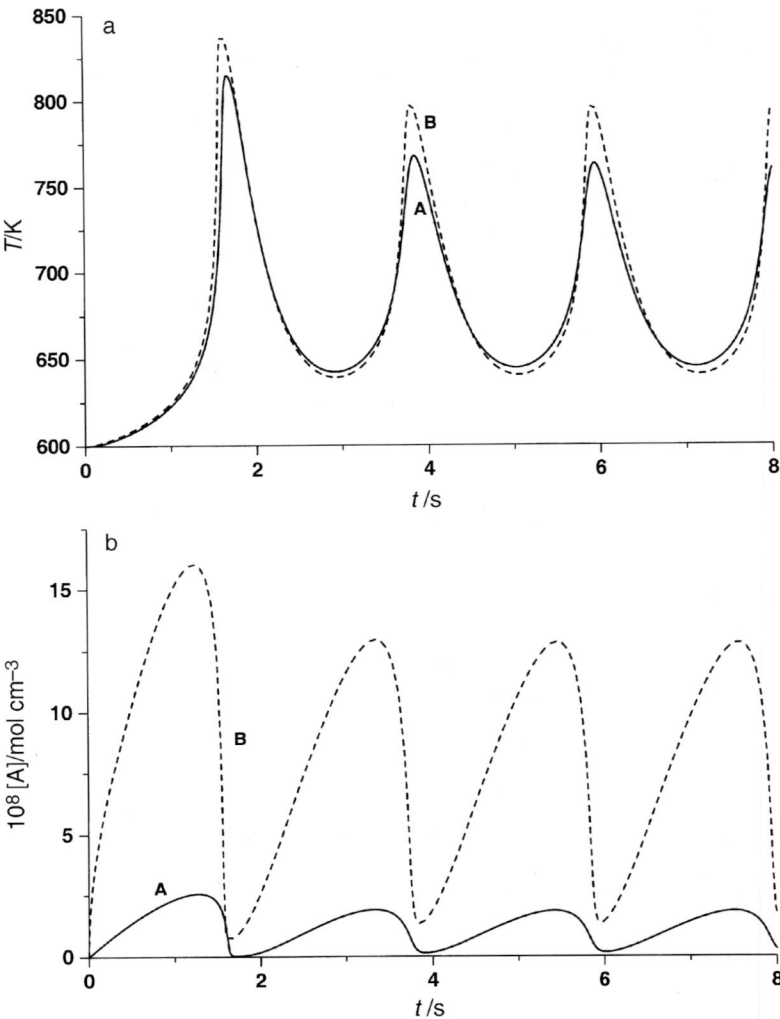

Fig. 9 (a) Centre temperature (T)–time and (b) intermediate concentration ([A])–time profiles at $T_a = 600$ K under the effect of mass and thermal diffusion at $D_A = D_T = 37$ cm^2 s^{-1} in pool chemical conditions close to the lower bound between damped and sustained oscillations in an infinite slab. Curves A correspond to the computation with the pre-cursor admitted simultaneously to each cell. Curves B correspond to the asymmetric supply of pre-cursor from the left hand side. The supply rates have been selected to give a period which corresponds to that in Fig. 3. There is a different timescale on the abscissa from that shown in Fig. 3.

reaction, as used here, was vindicated by Coppersthwaite et al.[21] They showed that, under spatially uniform conditions, the change in concentration of Cl atoms during oscillations was tied directly to the temperature change. The change of H atom concentration was found to be slightly out of step with that of Cl atoms, but the phase shift was of an order of magnitude smaller than that seen here between the temperature and the concentration of the intermediate species A, identified as [H$_2$] in the reaction vessel. However, in reducing the kinetic representation of the chemistry to a pseudo first order reaction with respect to [H$_2$], the opportunity to explore the potential effect of the selective diffusion of reactive intermediates, and especially that of H atoms, has been sacrificed.

The present calculations are restricted to an investigation of the effect of variations in the thermal diffusivity (D_T) and the overall mass diffusion coefficient (D_A). Normally these parameters

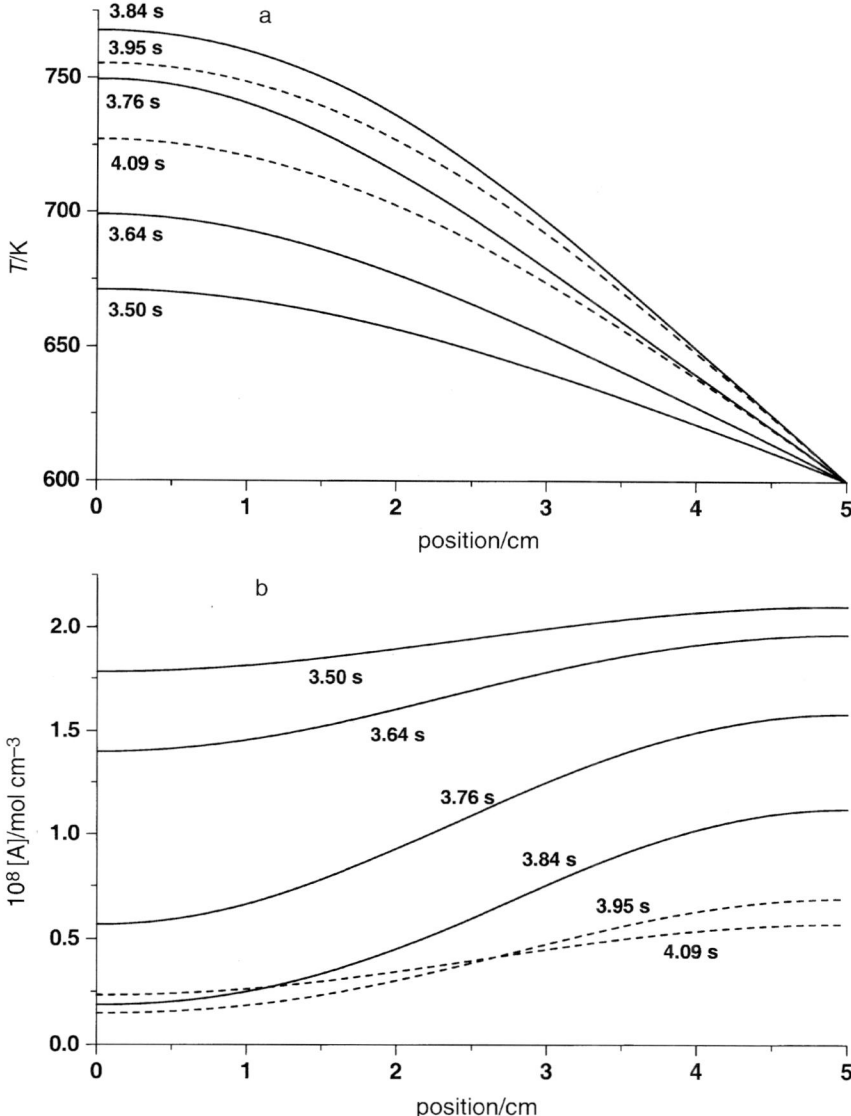

Fig. 10 Spatial profiles showing (a) temperature and (b) intermediate concentration, [A], at selected times in an infinite slab under the effect of mass and thermal diffusion when $T_a = 600$ K, with precursor supply to every cell. The solid lines represent the period during which the temperature is increasing and the broken line represents the period during which the temperature is decreasing. $D_A = D_T = 37$ cm^2 s^{-1}. The results correspond to curves A in Fig. 9.

are expected to be linked at $Le = 1$. An empirical approach has been adopted to establish the magnitude of the diffusion coefficients that make oscillations possible at a reaction vessel temperature and supply rate that are representative of the experimental conditions. Values for thermal diffusivity or the mass diffusion coefficient of $H_2 + Cl_2/HCl$ mixtures are not available. However, thermal conductivities of H_2 and N_2 are known at temperatures up to 373 K,[32] and these can be extrapolated to higher temperatures. The thermal conductivity of H_2 is about 7 times that of N_2, but the variation in binary mixtures varies non-linearly, there being a relatively weak dependence for compositions containing less than 20% by volume of H_2. Since the hydrogen reacts rapidly on

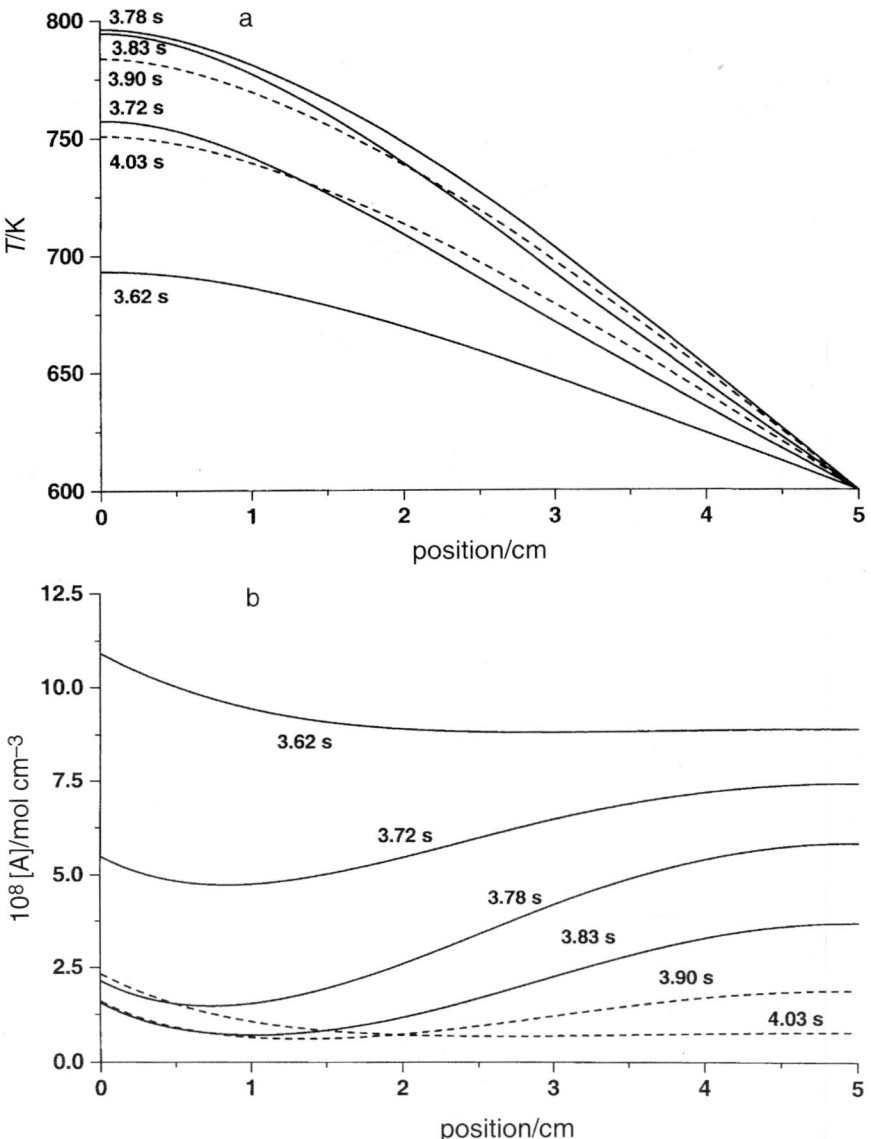

Fig. 11 Spatial profiles showing (a) temperature and (b) intermediate concentration, [A], at selected times in an infinite slab under the effect of mass and thermal diffusion when $T_a = 600$ K, with an asymmetric precursor supply to the left-hand side. The solid lines represent the period during which the temperature is increasing and the broken line represents the period during which the temperature is decreasing. $D_A = D_T = 37$ cm^2 s^{-1}. The results correspond to curves B in Fig. 9.

entry to the hot vessel, we may presume that a typical thermal conductivity for the reactive mixture may not differ greatly from that of N_2. Taking the measured values of λ (N_2) = 5.4 (6.1) × 10^{-4} J cm^{-1} K^{-1} s^{-1} at 333 (373) K and extrapolating to 600 K,[33] giving $\lambda = 8.8 \times 10^{-4}$ J cm^{-1} K^{-1} s^{-1}, the thermal diffusivity of N_2 at 40 Torr is 29.3 cm^2 s^{-1}. This indicates that the empirically derived diffusivities at which oscillatory behaviour is predicted to occur during the $H_2 + Cl_2$ reaction in a sphere under purely diffusive conditions appear to be physically acceptable. The required increase in mass and thermal diffusivity in order for oscillations to exist in a planar reaction vessel (the "infinite slab") is to be expected since there is no longer a surface area dependence on the distance from the origin, which strongly affects the transport processes in a sphere.

The three timescales that determine the existence of oscillations in the formal Sal'nikov scheme relate to the rate constant for step 1 ($t_1 = 1/k_1$), the rate constant for step 2 ($t_2 = 1/k_2$) and the heat loss term (expressed as a Newtonian cooling time in spatially uniform conditions, where $t_N = \chi S/C_v \rho V$). Non-linear feedback is encapsulated in the Arrhenius dependence of the rate constant for step 2. In the experimental analogue these timescales relate to the precursor supply rate, the rate constant for step 2 and the heat loss rate. For di-*tert*-butyl peroxide decomposition in a well-stirred reactor the heat loss is characterised by the Newtonian cooling time, t_N. For the hydrogen + chlorine in terrestrial conditions the heat loss would be characterised by convection dependent behaviour. Purely diffusive heat transport would be approached only at microgravity.

Numerical analyses permit various interpretations of the critical timescales. The two extremes of spatially uniform and distributed temperature cases for heat transport are readily simulated, as illustrated in Fig. 6 for spherical symmetry. The alternative interpretations of precursor supply are subtle and the diffusion of species A has different consequences in each case, applied here to the infinite slab. It might be supposed that, in a system that relies exclusively on thermal feedback, species diffusion would not be significant. When the precursor supply is distributed simultaneously to all cells, the diffusion of A plays a subsidiary, but not negligible part. As can be seen in Fig. 10, substantial concentration gradients of [A] develop, and the gradients are more prominent in a sphere (Fig. 5, 7 and 8). If D_A is increased (D_T being held constant) the amplitude of oscillations is increased by the enhanced transport of the intermediate species to the centre of the reacting mass. Conversely, a decrease in D_A causes the amplitude of oscillations to be reduced because the transport of A to the most reactive zone is suppressed. The sensitivity of the amplitude to changes in the magnitude of D_A is low and there is no qualitative change associated. The period of oscillations is not affected. The inflow of precursor at a controlled rate at the central plane of the axisymmetric slab relies entirely on diffusion for A to be distributed through the reaction volume. Consequently, D_A is involved in the timescale for reactant supply throughout the vessel and there is both a quantitative and qualitative sensitivity to its variation. Both amplitude and period are affected by small changes of D_A, which reflects movement around the parameter space within which unstable states exist, analogous to that shown in Fig. 2. Sustained oscillations exist only within a finite range of D_A.

The representation of precursor supply as a simultaneous distribution to all cells resembles the Sal'nikov, hypothetical model and permits numerical exploration of its behaviour under the effect of mass and heat transport by diffusion. The representation of precursor supply locally at the centre of the axisymmetric system resembles more closely the way in which experiments have been implemented. The next step is to extend this particular investigation from the planar symmetry to the cylinder and sphere. Further developments to explore the response in two dimensions are also in prospect, since there is then scope for spatial patterns to evolve in a cylinder under the influence of mass and thermal diffusion in different spatial dimensions.

References

1. J. F. Griffiths, in *Oscillations and Traveling Waves in Chemical Systems*, ed. R. J. Field and M. Burger, Wiley, New York, 1984, p. 529.
2. P. Gray and S. K. Scott, in *Oscillations and Traveling Waves in Chemical Systems*, ed. R. J. Field and M. Burger, Wiley, New York, 1984, p. 493.
3. J. F. Griffiths and S. K. Scott, *Prog. Energy Combust. Sci.*, 1987, **13**, 161.
4. P. Gray, J. F. Griffiths and K. Kishore, *Combust. Flame*, 1974, **22**, 197.
5. J. F. Griffiths, *Prog. Energy Combust. Sci.*, 1995, **21**, 25.
6. J. F. Griffiths and C. Mohamed, in *Comprehensive Chemical Kinetics*, ed. M. J. Pilling, Elsevier, Amsterdam, 1997, vol. 35, p. 545.
7. P. Gray, J. F. Griffiths and R. J. Moule, *Faraday Symp. Chem. Soc.*, 1974, **9**, 103.
8. J. F. Griffiths, P. Gray and B. F. Gray, *Proc. Combust. Inst.*, 1971, **13**, 239.
9. R. J. Moule, PhD Thesis, University of Leeds, 1979.
10. H. Pearlman, *Combust. Flame*, 2000, **121**, 390.
11. R. Fairlie, J. F. Griffiths and H. Pearlman, *Proc. Combust. Inst.*, 2000, **28**, 1693.
12. B. F. Gray and C. H. Yang, *J. Phys. Chem.*, 1965, **69**, 2747.

13 C. H. Yang and B. F. Gray, *J. Phys. Chem.*, 1969, **73**, 3395.
14 B. F. Gray and C. H. Yang, *Trans. Faraday Soc.*, 1969, **65**, 1614.
15 I. E. Sal'nikov, *Dokl. Akad. Nauk. SSSR*, 1949, **60**, 405 and 611.
16 I. E. Sal'nikov, *Zh. Fiz. Khim.*, 1949, **2**, 258.
17 P. Gray, S. R. Kay and S. K. Scott, *Proc. R. Soc. London, Ser. A*, 1988, **416**, 321.
18 B. F. Gray and M. J. Roberts, *Proc. R. Soc. London, Ser. A*, 1988, **416**, 391.
19 P. Gray, J. F. Griffiths, S. R. Kay and S. K. Scott, *Proc. Combust. Inst.*, 1989, **22**, 1597.
20 P. Gray and J. F. Griffiths, *Combust. Flame*, 1989, **78**, 87.
21 D. P. Coppersthwaite, J. F. Griffiths and B. F. Gray, *J. Phys. Chem.*, 1991, **95**, 6961.
22 J. F. Griffiths and C. H. Phillips, *Combust. Flame*, 1990, **81**, 304.
23 P. Gray and S. K. Scott, *Chemical Oscillations and Instabilities*, Clarendon Press, Oxford, 1990.
24 S. K. Scott, J. Wang and K. Showalter, *J. Chem. Soc., Faraday Trans.*, 1997, **93**, 1733.
25 J. Brindley, J. F. Griffiths and A. C. McIntosh, *Chem. Eng. Sci.*, 2001, **56**, 2037.
26 B. D. Hazzard, N. D. Kazarinov and Y.-H. Wan, *Theory and Applications of Hopf Bifurcation*, Cambridge University Press, Cambridge, 1981.
27 D. A. Frank-Kamenetskii, *Diffusion and Heat Transfer in Chemical Kinetics*, trans. J. P. Appleton, Plenum Press, New York, 2nd edn., 1969.
28 S. V. Pennington and M. Berzins, *ACM Trans. Math. Software*, 1994, **20**, 63.
29 M. Berzins, *Appl. Numer. Anal.*, 1986, **2**, 109.
30 R. D. Skeel and M. Berzin, *SIAM J. Sci. Stat. Comput.*, 1990, **11**, 1.
31 B. Van Leer, *J. Comput. Phys.*, 1990, **14**, 361.
32 J. O. Hirschfelder, C. F. Curtis and R. B. Bird, *Molecular Theory of Gases and Liquids*, Wiley, New York, 1964.
33 D. T. Jones, PhD Thesis, University of Leeds, 1969.

Spatial bifurcations of fixed points and limit cycles during the electrochemical oxidation of H_2 on Pt ring-electrodes

Peter Grauel, Hamilton Varela and Katharina Krischer*

Fritz-Haber-Institut der Max-Planck-Gesellschaft, Faradayweg 4-6, 14195 Berlin, Germany. E-mail: krischer@fhi-berlin.mpg.de

Received 12th April 2001
First published as an Advance Article on the web 12th November 2001

Pattern formation during the oscillatory oxidation of H_2 on Pt ring-electrodes in the presence of electrosorbing ions was studied under potentiostatic control for three different positions of the reference electrode (RE). The position of the RE crucially affects the degree of the global feedback which is imposed by the potentiostatic operation mode, and the three configurations selected corresponded to zero, maximum and intermediate global coupling. In the absence of global coupling, 'communication' among different positions occurs exclusively through migration coupling (the electrochemical counterpart to diffusion in reaction–diffusion systems). In this case, spatially inhomogeneous oscillations that were attributed to a spatial bifurcation of the homogeneous limit cycle were observed throughout. This implies that the system is Benjamin–Feir unstable. For the strongest global coupling adjustable, travelling pulses were found that emerged in a wave bifurcation with $n = 1$ from the homogeneous steady state. The pulses exhibited modulations in velocity and width that most likely resulted from the interaction between inhomogeneities of the catalytic surface and the nonlinear reaction dynamics. In the case of an intermediate global coupling strength, a diversity of spatio-temporal motions was observed. The dynamics ranged from pulses over target patterns and so-called asymmetric target patterns to mixed states where two or three of these states alternate. For some parameters these mixed states were in addition separated by bursts of the system to a nearly homogeneous unreactive state.

1. Introduction

The vast majority of electrochemical systems undergo dynamic instabilities resulting, *e.g.*, in sustained temporal oscillations, bistability or spatio-temporal patterns (for recent review articles see ref. 1–3). In general, these phenomena are caused by the interaction between nonlinear electrode kinetics and electric properties of the entire system, most importantly the potential drop through the electrolyte. This second property makes pattern formation in electrochemical systems extremely sensitive to the relative arrangement of working electrode (WE), counter electrode (CE) and reference electrode (RE). In particular, it facilitates experimental studies of pattern formation in the presence of various spatial couplings.

To make this clear, let us review the mechanism of coupling among reacting sites in electrochemical systems. The central variable for pattern formation in electrode reactions is the potential drop across the double layer, ϕ_{DL}, and different locations of the electrode are coupled through the electric field in the electrolyte. Thus potential changes at some location of the WE are felt rapidly

DOI: 10.1039/b103345j

at other locations. The coupling is non-local because its effect is felt by a whole neighbouring range (in some geometries by the whole electrode) but with decreasing strength for increasing distance.

The electric field or, equivalently, the potential distribution in the electrolyte is determined mainly by the potential of the WE and CE and the relative locations of these electrodes. Since the CE constitutes an equipotential plane, a pattern in ϕ_{DL} that establishes at the WE exists all over the electrolyte up to the CE, though the pattern smooths and its amplitude diminishes with increasing distance from the WE. In the following, we only consider situations in which the WE and the CE are parallel to each other. Then, the distance between the two electrodes determines the range of the coupling. It is maximum if this distance is much larger than the lateral extension of the working electrode and becomes more localised for smaller distances. This coupling through the electric field is always effective in an electrochemical experiment, and is referred to as migration coupling. Thus, in electrochemical systems migration coupling takes the role of diffusion coupling in reaction–diffusion systems.

In a potentiostatic experiment, the voltage between the WE and the RE is kept constant. In general, the RE 'sees' the potential at a certain location in the electrolyte. If this location is behind the CE (see *e.g.* the external RE in Fig. 1), then it is located in an equipotential plane parallel to the electrode, and the coupling remains unaffected by the RE. If, on the other hand, it is located between the WE and the CE, *e.g.* to minimize the voltage drop through the electrolyte (see RE(2) in Fig. 1), the potential at the position of the RE is altered by any *local* change of the double layer potential (because the latter affects the potential distribution in the entire electrolyte up to the CE). As a consequence, the actual voltage between the WE and the RE differs from the set voltage, and the potentiostat changes the potential of the WE. In other words, charge is pumped into the entire double layer. If the distance between every position of the WE and the RE is equal, the potential at the position of the RE is a function of the average double layer potential, and the feedback by the potentiostat imposes a global coupling on the system. The strength of the coup-

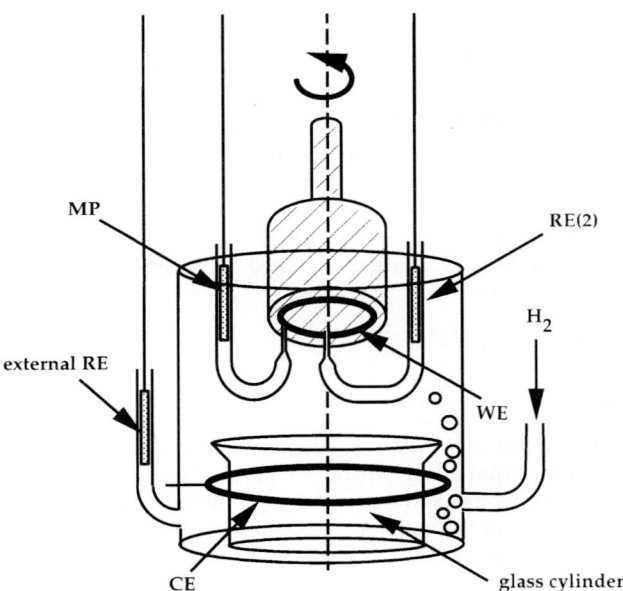

Fig. 1 Schematic view of the experimental set-up. WE: working electrode, a rotating Pt ring-electrode embedded into a Teflon body, RE: reference electrode. Two types of RE were used: (1) a Hg/Hg$_2$SO$_4$ electrode in a separate compartment (external RE) and (2) a Ag/AgCl electrode (RE(2)) put on the axis of the ring and close to the WE with the help of a glass (Haber–Luggin) capillary. In one set of experiments the distance between RE(2) and the plane of the WE was 3 mm, in another set it was 13 mm. CE: counter electrode, a Pt ring. MP: potential micro-probe.

ling is a function of the distance between the WE and the RE: The closer the WE and the RE are, the stronger is the global coupling.

In summary, the distance between the WE and the CE determines the nonlocality of the migration coupling. The distance between the WE and the RE determines the strength of an additional global coupling.

Global constraints exist in a variety of physical and chemical systems. Among them are catalytic reactors, in which the external constraint stems from the electric control of the temperature[4–6] or in which the global coupling is due to fast mixing of the gas phase,[7,8] gas-discharge devices,[9] semiconductor systems[10] or ferromagnetic systems.[11] Experimental and theoretical studies of these systems revealed a rich plethora of spatiotemporal patterns (see, *e.g.* ref. 6, 9 and 12–24).

Concerning electrochemical systems, there are some examples demonstrating the impact of global coupling on pattern formation in the bistable and the oscillatory region. The pioneering experiment goes back to Otterstedt *et al.* who observed pulses travelling around Co electrodes during electrodissolution of the electrode.[25] In later studies these authors also report more complex motions in the same system.[26,27] Stationary domains were found during the reduction of persulfate[28] and the electrooxidation of H_2.[3,29] Standing waves formed during formic acid oxidation when a close reference electrode was used.[30] In the bistable region of this reaction, fronts could be remotely triggered,[31] another peculiar manifestation of global coupling. The difference between the impact of global coupling in electrochemical systems with an N- and an S-shaped current potential curve is discussed in ref. 32.

In this contribution we extend these studies on the impact of global coupling on patterns in electrochemical systems. The system under consideration is the oscillatory oxidation of H_2 on Pt ring-electrodes. By adjusting three different positions of the RE, the strength of the global coupling is varied. In the first set of experiments the RE is located behind the CE. Thus, this system can be viewed as a 'reference system' without global coupling. In the second set of experiments, the RE is placed on the axis of the WE at approximately the closest distance that can be adjusted. Finally, an intermediate distance between the WE and the RE, and hence also an intermediate strength of the global coupling is realised. Besides some of the already known manifestations of global coupling, other novel wave types are reported.

2. Experimental

A schematic view of the experimental set-up is shown in Fig. 1. A rotating Pt ring (outer diameter 30 mm, width 1 mm) embedded into a Teflon cylinder served as the working electrode (WE). Two different configurations were used for the reference electrode (RE). Either a Ag/AgCl electrode (RE(2)) was put into a J-shaped glass capillary (Haber–Luggin capillary) whose tip was located on the axis of the WE and adjusted at two different distances (namely 3 and 13 mm $\pm$ 0.2 mm) to the plane of the WE, or a Hg/Hg_2SO_4 reference electrode in a separate compartment (external RE) that was connected to the main compartment below the plane of the CE was used. The CE, a 1 mm thick Pt wire bent to a ring of 65 mm in diameter, was located parallel and at a distance of 45 mm to the WE.

To monitor the angular potential distribution in front of the WE, the tip of a second glass capillary equipped with an Ag/AgCl electrode (the potential micro-probe, MP) was placed 1 mm $\pm$ 0.2 mm below the Pt ring (WE). During the experiments the WE was rotated at 20 Hz, and the voltage between the MP and the WE was measured with an acquisition rate of 1 kHz (*i.e.* 50 points/rotation), which allowed us to construct a spatiotemporal picture of the potential in front of the WE. As the resistance between the MP and the WE was negligible, the measured voltage represents the local potential drop across the double layer (called double layer potential, ϕ_{DL}, below) to a good approximation. For further experimental details see ref. 29 and 33.

The chemicals, H_2 (5N, Linde), H_2SO_4 (p.a., Merck), HCl (p.a., Merck) and $CuSO_4$ (p.a., Merck) were used as received and solutions were prepared from millipore water (suprapure system, Milli-Q). The electrolyte used in all experiments consisted of 0.5 mM H_2SO_4, 0.1 mM HCl and 0.025 mM $CuSO_4$ saturated with H_2, which was continuously bubbled through the cell during the experiments.

Prior to each experiment the base electrolyte (0.5 mM H_2SO_4) was purged with N_2 in order to remove dissolved oxygen. Then, the Pt working electrode was electrochemically cleaned through fast voltammetric scans for 1 h between the potentials at which hydrogen and oxygen evolution set in. Afterwards, the electrolyte was saturated with H_2, and HCl and $CuSO_4$ were added. The dynamic behavior was investigated under potentiostatic conditions, *i.e.* during one experiment the voltage between the WE and the RE (either the external RE or RE(2), *vide supra*), U, was kept constant by means of a potentiostat (FHI electronic laboratory).

Concomitant to the H_2 oxidation at the working electrode, Cu was deposited on the CE. Consequently, the Cu concentration changed slightly on the timescale of hours, being somewhat lower at the end of the experiment than in the beginning. However, the change in Cu concentration was slow compared to the characteristic time (*e.g.* the oscillation periods) of the phenomena studied. Thus, in the time window in which a certain time series was recorded the parameters of the system were constant to a very good approximation. The long-time drift in the Cu concentration resulted in the same dynamic behavior being found at somewhat shifted values of the external potential U when repeating the experiment after an hour. Also, on different days, the quantitative values of U at which a certain behavior was found varied somewhat, most likely due to slight differences in the roughness of the surface. But the qualitative dynamics, *i.e.* the types and sequences of dynamical states observed, were well reproducible.

3. Results

In the following we describe pattern formation during the oscillatory H_2 oxidation as a function of the external potential for the three different positions of the RE used. Firstly, the external RE is considered, and thus the case without global coupling. Then, we investigate the behavior for the closest distance between the WE and the RE, and thus the strongest coupling that can be adjusted. Finally we look at the dynamics in the case of intermediate coupling strength.

3.1 External reference electrode

In Fig. 2 a current–voltage (I/U) curve recorded during a potential scan in the positive direction is displayed. The curve represents a rough picture of the global behavior that is also found under stationary conditions. At low values of the applied potential a stable stationary fixed point exists which undergoes an instability that gives way to large amplitude oscillations. Directly beyond the bifurcation, the oscillations possess a very long period whereby the current remains quasi-stationary close to the former fixed point for most of the oscillation period. From this quasi-stationary state the trajectory spirals out exhibiting high frequency oscillations with small but slowly increasing amplitude (*cf.* Fig. 3a) until the system undergoes a large excursion to high values of the current density from where it returns back to the quasistationary state. This behavior suggests that the oscillations are born in a homoclinic bifurcation. For increasing external voltage

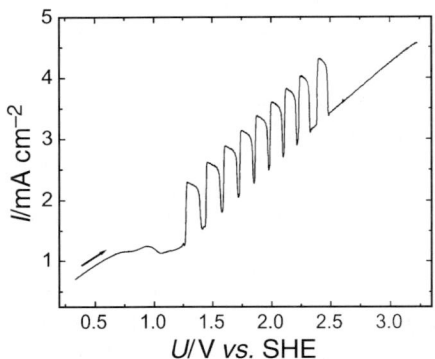

Fig. 2 I/U potentiodynamic profile of a rotating Pt ring-electrode during the H_2 oxidation reaction in 0.5 mM H_2SO_4 solution in the presence of 0.025 mM $CuSO_4$ and 0.1 mM HCl; rotation speed of the electrode: 20 Hz; scan rate: 50 mV s^{-1}. The external reference electrode was used, scan region: 0.33 to 2.23 V.

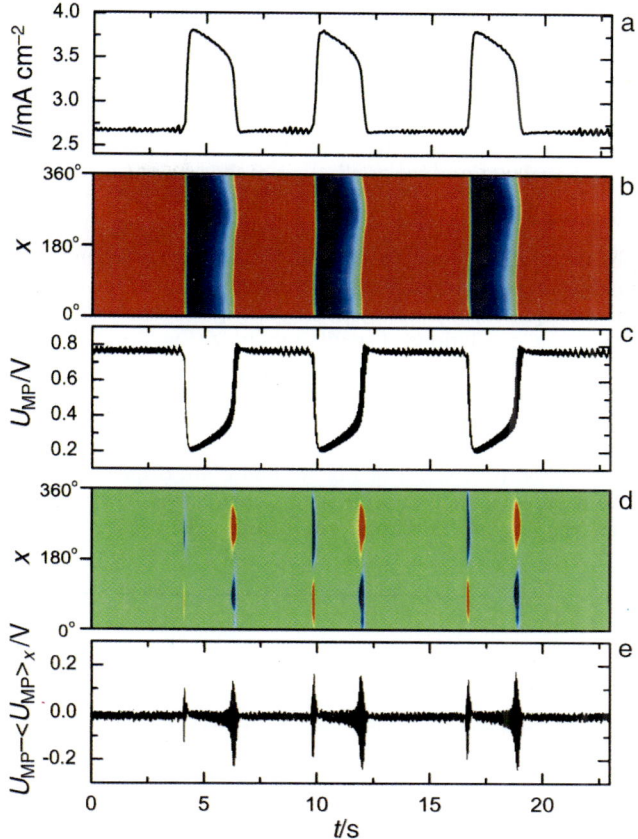

Fig. 3 Spatio-temporal dynamics in the oscillatory region under potentiostatic conditions with the external RE. External voltage U: 2.05 V vs. SHE, other conditions as in Fig. 2. (a) Time series of the global current. (b) Position–time plot of the local double layer potential as measured with the potential micro-probe ($U_{MP}(x,t)$). In the rainbow colour scale red denotes the highest potential value and blue the lowest one. (c) Plot of the micro-probe potential ($U_{MP}(x,t)$) as a function of time. (d) Position–time plot of the inhomogeneously oscillating part of (b) obtained by subtraction of the homogeneous part from the micro-probe potential ($U_{MP}(x,t) - \langle U_{MP}(t)\rangle_x$). (e) Plot of the inhomogeneously oscillating part of the micro-probe potential as a function of time.

the oscillation period becomes shorter, while the amplitude stays approximately constant up to the high potential limit of the oscillations where the limit cycle disappears in a second hard bifurcation and the behavior becomes stationary again.

A typical space–time measurement in the oscillatory region at a constant value of the external potential U is displayed in Fig. 3. Fig. 3a shows the time series of the global current and Fig. 3b the double layer potential as a function of space and time. At first glance, the oscillations appear to be homogeneous. However, when investigating the plot more carefully, in the transition region from high to low current a wavy structure of the double layer potential is clearly discernible. The inhomogeneous character of the oscillations becomes much better visible when subtracting at each instant in time the spatial average from the data (Fig. 3d). In this representation it is obvious that both flanks of the relaxation-type global oscillation are accompanied by a spatial symmetry breaking. The inhomogeneous part of the oscillation can be approximated as an oscillating sinusoidal structure of wavenumber one whereby the sign of the amplitude is different on the two flanks of the oscillation.

Fig. 3c and e show the time series of the micro-probe signal, U_{MP} (c) and the micro-probe signal after subtracting the spatial average, $U_{MP} - \langle U_{MP}\rangle_x$ (e), *i.e.* the same data of Fig. 3b and d in a different representation. In these plots the absolute values of the double layer potential as well as the amplitudes of the homogeneous and inhomogeneous oscillations can be easily read off. The homogeneous oscillation possesses an amplitude of about 600 mV, the inhomogeneous structure having a maximal amplitude of about 300 mV.

Qualitatively the same spatio-temporal behavior was observed over the entire oscillatory region; an inhomogeneous structure, close to a sinusoidal mode with wavenumber one was always excited on the flanks of the relaxation type oscillations of the total current.

3.2 Small distance between RE and WE

When minimizing the distance between the RE and the WE, *i.e.* when using a Haber–Luggin capillary (RE(2)) and adjusting a distance of 3 mm between the tip of the capillary and the plane of the working electrode, the spatio-temporal dynamics was qualitatively different. The total current exhibited relatively harmonic oscillations of small amplitude (Fig. 4). At the low potential end, the oscillations looked as if they were associated with a period-doubled limit cycle (Fig. 4a–c). For increasingly positive potential the difference between successive minima became smaller so that the current eventually exhibited period-1 oscillations (Fig. 4d–f).

Space–time plots of the local double layer potential, $U_{MP}(x,t)$ (without subtraction of the spatial average) corresponding to the time series of Fig. 4a and e are displayed in Fig. 5a and c. Obviously, this time we are dealing in the first place with a spatial symmetry breaking, namely a pulse that propagates around the ring in a fairly uniform manner. The oscillations in the current density

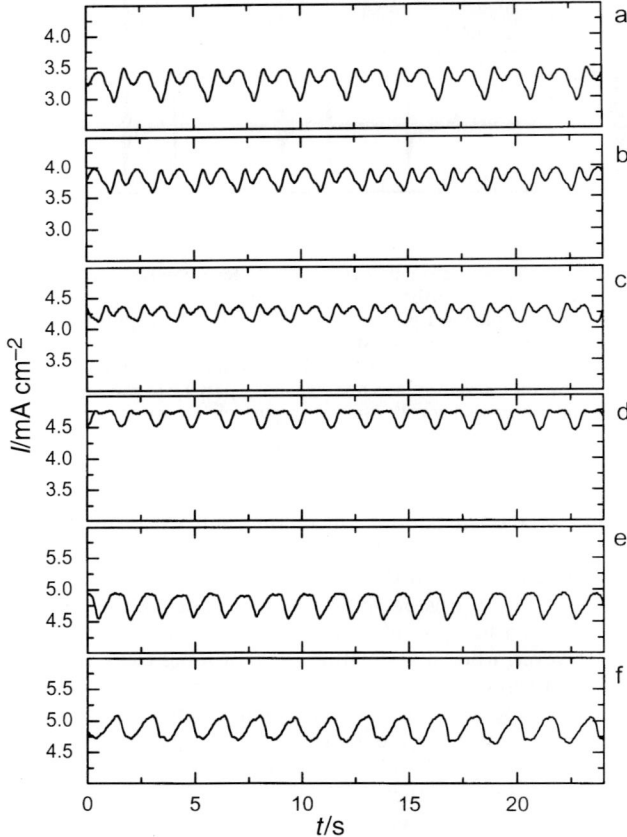

Fig. 4 Global current oscillations under potentiostatic conditions with the reference electrode (RE(2)) placed at a distance of 3 mm from the plane of the WE for different values of the external voltage U: (a) 0.96, (b) 1.06, (c) 1.16, (d) 1.26, (e) 1.36 and (f) 1.46 V *vs.* SHE. Other conditions as in Fig. 2.

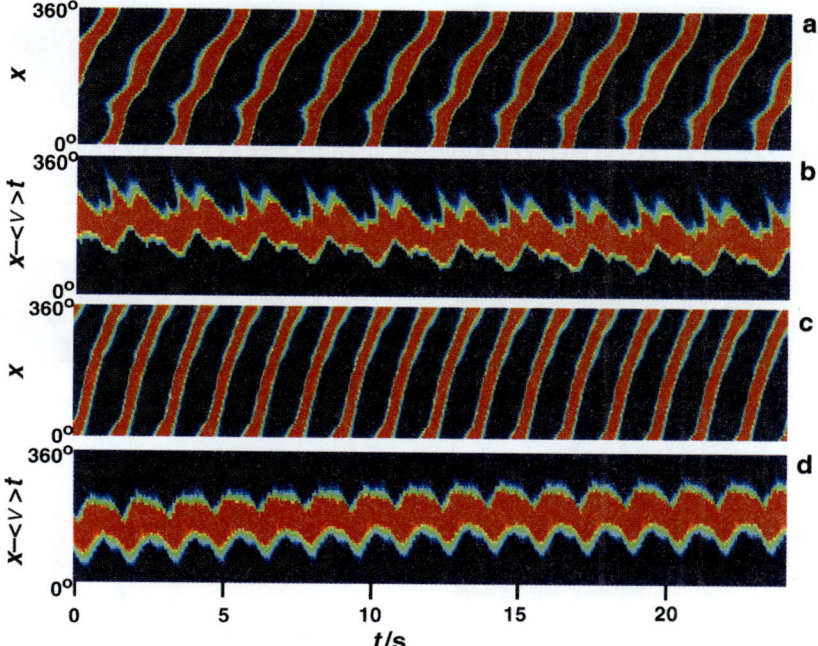

Fig. 5 Position–time plots of the local electrode potential (U_{MP}) in the laboratory coordinate system ((a) and (c)) and in a coordinate system moving with the mean velocity of the pulse ((b) and (d)) for the time series shown in Fig. 4(a) and (e).

arise as a result of changes in the width of the pulse during one rotation around the ring. This can be seen more clearly in Fig. 5b and d where the spatio-temporal behavior that one would observe in a coordinate system that moves with the mean velocity of the pulse is plotted. In the case of the simple periodic time series (Fig. 5d) the resulting picture is a 'zigzag' motion of the pulse, which exhibits some minor variations of the width. The latter cause the oscillations seen in the total current density (Fig. 4e), which possess the same period as the rotation period of the pulse, while the zigzag pattern indicates that the velocity of the pulse changes during one rotation. Not only is the back-and-forth motion of the pulse in the moving frame more pronounced in the first case (Fig. 5b), but the width of the pulse also oscillates more strongly and with twice the rotation period. In addition, a second active area is periodically excited at a certain position of the ring. (*cf.* the light blue-greenish offshoots in the upper part of the plot). This satellite pulse propagates quickly toward the main pulse and merges with the latter soon after its birth.

At this close distance between working and reference electrode, we never observed homogeneous oscillations. Rather, one of the two types of pulses emerged whenever the system did not attain a homogeneous stationary state. The mean pulse width and velocity increased with increasing U, the mean velocity ranging from *ca.* 4 cm s^{-1} (Fig. 4a) to 6.2 cm s^{-1} (Fig. 4e). Only very close to the end of the existence range of the pulses did the mean velocity decrease again slightly, *e.g.* in the case of Fig. 4f it amounts to *ca.* 5.5 cm s^{-1}.

3.3 Intermediate distance between RE and WE

The dynamics were by far richer for the intermediate distance between working and reference electrodes. An overview of the behavior is displayed in Fig. 6 where space–time plots of the double layer potential are shown together with the time series of the global current. The measurements in Fig. 6a–f were obtained for increasing values of the external voltage. At low values of the external voltage the most complex spatio-temporal patterns were observed (Fig. 6a). A closer view suggests that it is made up of a sequence of behaviors that appear 'isolated' at larger values of U, which we therefore describe first.

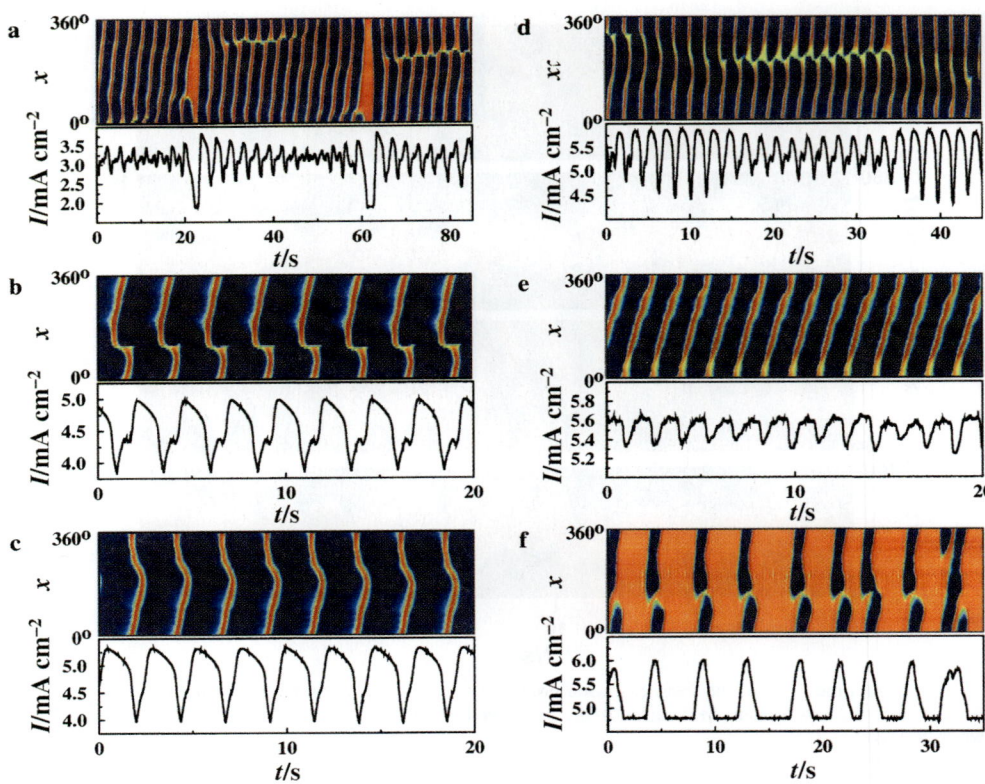

Fig. 6 Position–time plots of the local double layer potential and corresponding time series of the global current under potentiostatic conditions with the reference electrode (RE(2)) placed at a distance of 13 mm from the plane of the WE for different values of the external voltage U: (a) 1.06, (b) 1.37, (c) 1.5, (d) 1.67, (e) 1.76 and (f) 1.86 V *vs.* SHE.

In Fig. 6e we recognize again a pulse travelling around the electrode with slightly changing velocity, similar to that shown in Fig. 5c.

The next simplest behavior appears to be that displayed in Fig. 6c: Position 0° (or equivalently 360°) seems to be a wave source which periodically sends out two pulses that travel in opposite directions and meet in the center of the ring where they annihilate each other. From a phenomenological point of view, the pattern looks like the one-dimensional analogue of a 'classical' two-dimensional target pattern. Following Zhabotinsky *et al.*,[34] who obtained similar dynamics in simulations of a three-component reaction–diffusion system, we also refer to it as target pattern. Note, however, that neither in those simulations nor in our experiments are the patterns linked to a pacemaker or to an excitable medium.

The spatio-temporal plot shown in Fig. 6b resembles that of Fig. 6c in so far as there is also a source point that periodically emits waves. However, in this case the 'excitation' propagates only in one direction. The pulse that is about to travel into the opposite direction soon comes to a halt. The position at which the propagation of this second front is stopped remains in a more unreactive 'red state' until the propagating wave hits it. Upon this collision, both front and 'point excitation' are annihilated and the entire domain relaxes back to the 'blue' ground state until the cycle starts anew. In the following we call this behavior, which to our knowledge has not been described in the literature, an asymmetric target pattern.

In Fig. 6d we come across a mixture of the last two discussed behaviors: The system seems to switch back-and-forth between a more or less deformed variant of the target pattern, and the asymmetric target pattern. The complex overall behavior is also reflected in the time series of the total current density, which is apparently quasiperiodic. However, to determine whether we are indeed dealing with a quasiperiodic state requires much longer time series than we have recorded so far. The further characterization of the dynamics will be done in future work.

In Fig. 6a we meet a still higher degree of complexity. Four types of behavior take turns: Slightly imperfect target patterns are followed firstly by asymmetric target patterns, then by pulses and finally by the expansion of the less reactive portion of the pulse over the entire electrode before the scenario starts again. In the time series of the total current the global deactivation manifests itself in a large excursion toward smaller values. These large amplitude oscillations are separated by apparently quasiperiodic sections of the time series with smaller amplitudes that resemble the time series of Fig. 6d. Again, due to the complexity of the behavior which requires much longer measurements, that we have not yet obtained for technical reasons, we are currently not able to classify the behavior further, *e.g.* as periodic, quasi-periodic or chaotic. Note that the point-oscillation at which the front is extinguished in the asymmetric target pattern appears each time at a different position. The same is true for the measurement of Fig. 6d. This makes it very unlikely that structural defects of the electrode surface are responsible for the emergence of the asymmetric target patterns.

At the highest value of the applied potential a behavior similar to an asymmetric target pattern is observed (Fig. 6f): A front-like excitation travels once around the ring before it is extinguished again at a certain position. However, unlike in Fig. 6b the behavior is no longer strictly periodic, and once in a while the wave emerges at a different position, survives the first rotation, and is extinguished only when it approaches the 'critical position' for the second time. In Fig. 6f this is, *e.g.*, the case for the last pulse displayed.

4. Discussion

The patterns observed at large and small distance between the WE and the RE emerge in fundamentally different spatial instabilities, which can be attributed to the fact that in the first case the different positions of the electrode are coupled exclusively by migration currents, while for a close reference electrode, there is an additional global constraint present.

To make the different bifurcations involved most transparent, consider eqn. (1) and (2). They represent a general, dimensionless model for the spatio-temporal dynamics of an electrochemical oscillator under potentiostatic control in the case where the RE is far away from the WE such that we can exclude any global coupling.[35,36]

$$\frac{\partial \phi_{DL}(x)}{\partial t} = f(\phi_{DL}(x), c(x)) - \frac{\sigma}{\beta}\left(\frac{\partial \phi(x)}{\partial z}\bigg|_{z=WE} + (U - \phi_{DL}(x))\right) \quad (1)$$

$$\frac{\partial c(x)}{\partial t} = g(\phi_{DL}(x), c(x)) \quad (2)$$

U is the externally applied voltage, σ the specific conductivity of the electrolyte, which together with the geometric factor β determines the uncompensated cell resistance per unit electrode area. x and z are the coordinates parallel to the (one-dimensional) electrode and normal to the electrode, respectively, ϕ the potential in the electrolyte and $z = $ WE, a position at the working electrode.

The two variables ϕ_{DL} and c are the potential drop across the double layer, the central variable for the description of electrode reactions, and the concentration (or coverage) of a chemical species, respectively. The functions f and g define the homogeneous dynamics of the system, which is of the activator–inhibitor type, whereby ϕ_{DL} takes on the role of the activator and c that of the inhibitor. Spatial coupling due to diffusion of c is much smaller than migration coupling, and can thus be neglected.

The second term on the right-hand side of eqn. (1) represents migration currents that are induced by an inhomogeneous distribution of the double layer potential.[2,37] Thus, it describes the spatial coupling between different locations of the electrode owing to an inhomogeneous potential distribution, and it is the counterpart of diffusion in reaction–diffusion systems. Just as does diffusion coupling, migration coupling acts in a synchronizing manner, *i.e.* in the absence of a reaction term it smoothes any spatial inhomogeneities. Note that in order to solve eqn. (1) it is necessary to determine the electric field normal to the electrode, which requires the solution of Laplace's equation $\Delta\phi = 0$. This equation brings the directions parallel to the electrode into play, and it becomes clear why eqn. (1), which does not explicitly contain a spatial operator that depends on x, is capable of describing pattern formation along the electrode: the spatial coupling among different sites of the electrode is mediated through the bulk electrolyte. An inhomogeneous distribution of $\phi_{DL}(x)$ changes the potential distribution, $\phi(x, y, z)$, in the entire electrolyte, and thus also the electric field at the WE, $[\partial\phi(x)/\partial z]|_{z=WE}$.

In Fig. 3, i.e. for a large distance between the WE and the RE, the dynamics is dominated by the homogeneous mode which oscillates in time. Spatial structures pop up predominantly on the flanks of the oscillations when the homogeneous mode changes quickly, the amplitude of the inhomogeneous part of the oscillation being at most 50% of the amplitude of the uniform oscillation. This is a strong hint that the oscillations also exist in an infinitely small, 'point-like system'. In other words, we can view our system as being composed of (infinitely many) oscillatory elements coupled by migration currents.

It is apparent that the homogeneous oscillation is unstable with respect to large wavelength perturbations. This observation allows us to deduce the qualitative run of the real part of the largest Floquet multiplier m as a function of the wavenumber of a spatial perturbation, n (Fig. 7a). The homogeneous limit cycle is stable with respect to homogeneous perturbations and neutrally stable upon phase displacements. Thus, $m = 1$ for $n = 0$. In contrast, m lies outside the unit circle for perturbations with a large wavelength, i.e. within an interval $n \in [0, n_c]$ (see also ref. 38). For our later discussion it is useful to recall how the growth rate of perturbations of the *homogeneous fixed point* changes with the wavenumber of the perturbations (Fig. 7b). It is largest and positive for $n = 0$, and decreases monotonically with increasing n, reflecting the oscillating nature of the local elements and the synchronizing spatial coupling on the activator variable.

Pattern formation in oscillating media coupled by diffusion has been studied extensively.[39–42] Many studies were performed with the complex Ginzburg–Landau equation (CGLE) which is valid close to a Hopf bifurcation. The CGLE has the particular advantage that the stability of the homogeneous oscillation that is born in the Hopf bifurcation can be directly read off from the coefficients entering the equation. It was shown that if the so-called 'phase diffusion coefficient' is negative, the homogeneous oscillation is unstable with respect to spatial perturbations. For large systems, the long-term behavior is then typically an irregular turbulent motion. Such oscillating media have been called Benjamin–Feir unstable systems.[41,43]

According to the discussion in the second last paragraph, it appears obvious that the spatial structure in Fig. 3 arises because of a negative phase diffusion coefficient. Thus, we can classify our electrochemical oscillator as being Benjamin–Feir unstable, as was also done by Christoph for solutions of a reaction–migration system describing Ni dissolution.[38] In contrast to most studies of Benjamin–Feir unstable systems, in our experiment the system length is of the same order of magnitude as the characteristic length of the pattern, and instead of 'electrochemical turbulence', regular limit cycle oscillations established. One should also keep in mind that (1) the oscillations are relaxation-like and (2) that they are of the mixed mode type. The fact that the small amplitude

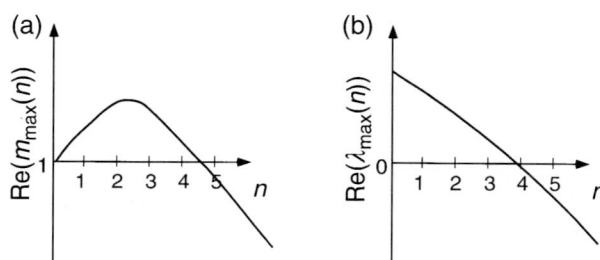

Fig. 7 (a) Schematic of the real part of the largest Floquet multiplier $Re(m_{max}(n))$ as a function of the wavenumber n of a perturbation of the *homogeneous limit cycle*. (b) Dispersion relation displaying the growth rate $Re(\lambda_{max}(n))$ of perturbations of the *homogeneous stationary state* vs. their wavenumber n. (a) and (b) apply to the case of the limit cycle oscillations of Fig. 3. Note that the Floquet multiplier is real, i.e. $Im(m_{max}(n)) = 0$ for the interesting range of n, whereas the largest eigenvalue $\lambda_{max}(n)$ is complex for $0 \leqslant n \leqslant n_c$.

spiraling out of the trajectory is not accompanied by spatial structures (*cf.* Fig. 3) strongly suggests that the mixed-mode character arises from an instability of the homogeneous system and not from spatial instabilities. Hence, the proper description of the data of Fig. 3 requires a 3-component reaction–migration system, and the oscillations are far from a supercritical Hopf bifurcation. Both instances restrict further comparison of the experimental behavior with solutions of the CGLE and call for additional theoretical studies that take into account these complications. We are currently developing an extended ordinary differential equation (ODE) system in order to describe the more complex temporal behavior. However, besides this realistic modeling, a more general, 'normal form-type' approach would also be desirable.

Benjamin–Feir unstable states were observed in hydrodynamical systems.[43–46] Numerical studies found that most reaction–diffusion systems also possess a Benjamin–Feir unstable region. However, corresponding experimental examples in reaction–diffusion type systems are rare. Turbulent states that might be associated with a negative effective phase diffusion coefficient were reported for CO oxidation on Pt(110)[47,48] and for the NO + NH_3 reaction on Pt(100),[49] both under low pressures. Further analysis of these turbulent states that would allow an unambiguous assignment of the dynamical state of the system has not yet been reported.

Now, let us turn to a discussion of the dynamics found for the closest distance between the WE and the RE. This geometry introduces a negative global coupling into the system which manifests itself in an additional term in eqn. (1) that depends on the average double layer potential, $\langle \phi_{DL} \rangle_x$:

$$\frac{\partial \phi_{DL}(x)}{\partial t} = f(\phi_{DL}(x), c) + \alpha(\phi_{DL}(x) - \langle \phi_{DL} \rangle_x) - \frac{\sigma}{\beta}\left(\frac{\partial \phi}{\partial z}\bigg|_{z=WE} + (U - \phi_{DL}(x))\right) \quad (3)$$

where $\alpha > 0$ is a parameter that depends on the difference of the resistances between the WE and the RE, on the one hand, and the WE and the CE, on the other. It determines the strength of the global coupling.[28] Qualitatively, the global coupling in eqn. (3) acts in a destabilizing manner: If at a certain position x_1, $\phi_{DL}(x_1)$ is larger than the average double layer potential, the global coupling term (the second term on the right-hand side in eqn. (3)) is positive and thus causes a *further increase* of $\phi_{DL}(x_1)$. If, on the other hand, $\phi_{DL}(x_1)$ is smaller than $\langle \phi_{DL} \rangle_x$, the term is negative and again the negative perturbation at x_1 is enhanced. Such type of global coupling has been coined negative global coupling (because of the negative sign in front of $\langle \phi_{DL} \rangle_x$). A different sign in front of the global coupling term turns the action of the coupling into a synchronizing one, which is also called a positive global coupling.

To recall the influence of the global coupling on the stability of a stationary state, consider two systems that possess identical homogeneous states, but one system experiences a global constraint, the other one not. In the case of our electrochemical system, for example, we can prepare these states by changing the position of the CE. If the CE is at the same height as the RE, α in eqn. (3) is 0 and the system is only coupled by migration coupling. If we put the CE further away than the RE, the stationary states remain identical but global coupling comes into play. Since the global coupling affects only the homogeneous mode (through the average double layer potential), the two systems exhibit the same dependence of the largest eigenvalue of the linearized system λ_{max}, on the wavenumber of the perturbation, n, for $n > 0$. But the respective eigenvalues differ for $n = 0$. If the average double layer potential enters into the activator equation with a negative sign, as in eqn. (3), it is straightforward to deduce that the homogeneous mode is stabilized and the dispersion relation possesses the qualitative shape shown in Fig. 8. In particular, for strong negative coupling or large systems $\lambda_{max}(n = 1)$ will be larger than $\lambda_{max}(n = 0)$. This means that when starting from a stable stationary state with only negative real parts of all eigenvalues and changing a parameter such that the system is driven into an unstable region, the first eigenvalue to become unstable is the one with $n = 1$. If the eigenvalue is real, this bifurcation gives rise to stationary domains, as were observed in nonisothermic catalytic reactions,[4,50] during the electrochemical reduction of peroxodisulfate[28] or also during the oxidation of H_2 on Pt for different electrolyte concentrations and compositions than used here.[3,29] For complex eigenvalues, a wave bifurcation occurs. Owing to the rotational invariance of our 1-dimensional spatial domain, the eigenfunctions of the system are Fourier modes whereby sine and cosine modes are degenerate, and become unstable simultaneously. Hence, for initial perturbations with predominantly even or odd symmetry, standing waves with wavenumber one might develop (*i.e.* only the sine, or equivalently the cosine is

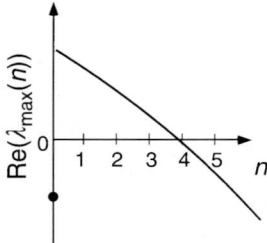

Fig. 8 Schematic dispersion relation displaying the growth rate $Re(\lambda_{max}(n))$ of perturbations of the homogeneous stationary state *vs.* the wavenumber n of the perturbation in the case of negative (desynchronizing) global coupling. This situation applies to the data shown in Fig. 5 and 6. Note that $\lambda_{max}(n)$ is complex for $0 \leq n \leq n_c$.

excited). Such waves were observed, *e.g.*, during formic acid oxidation on a Pt ring-electrode when the RE was put sufficiently close to the WE.[30] For all other perturbations, in which $\sin(x)$ and $\cos(x)$ are both excited sufficiently, travelling waves or phase pulses prevail.

We thus conclude that the pulses we observed for a close reference electrode (*cf.* Fig. 5) are associated with a wave bifurcation of the stationary state with $n = 1$. In contrast to the patterns observed with the external RE, that arose due to a spatial instability of the homogeneous *limit cycle*, here we are dealing with a non-trivial Hopf bifurcation of the homogeneous *stationary state*.

The experimental pulses were not at any time 'perfect' in the sense that they possessed a strictly constant shape or velocity (*cf.* Fig. 5). This can already be deduced from the time series shown in Fig. 4, which is oscillatory with a period that corresponds to the time in which the pulse travels once around the ring. A constant shape would result in a stationary current, a constant velocity in addition in a stationary domain in the moving frame. Instead, in the moving frame the experimental pulses displayed a zigzag motion, reflecting a variation in the speed of the pulse, superimposed by a minor alteration of the width. This behavior might be the manifestation of secondary bifurcations of the pulse, such as an oscillatory instability that leads to variations in width.

However another interpretation appears to be more likely: An electrode surface, like any catalytic surface, is never an ideally homogeneous medium. Rather, its reactivity is likely to differ somewhat from position to position, and it might also change with time. Thus, the pulse might experience different local properties while propagating around the electrode. These inhomogeneities will in general affect pulse width and velocity. The fact that the periods of the oscillations in width in Fig. 5b or d coincide with the rotation period of the pulse supports this conjecture (although a 1 : 1 locking of the two frequencies also cannot be excluded *a priori*).

Provided that the fine structure of the pulses stems from nonhomogeneities of the electrode surface, the sequence of time series shown in Fig. 4 suggests that the interaction between the intrinsic dynamics and the nonhomogeneous distribution of parameters is very intricate. This points to a possible control of the dynamics by deliberately designing catalysts with spatially modulating properties. In the electrochemical context, *e.g.*, the poisoning of the electrode by the reaction intermediate CO is the major problem in all oxidation reactions of small organic molecules. Their most important application is in fuel cells, such as the direct methanol fuel cell.[51] All these oxidation reactions exhibit oscillatory behavior and they belong to the same class of electrochemical oscillators as the system studied here. If it could be shown that the interaction between a nonuniform activity of the electrode and the nonlinear reaction dynamics can prevent complete poisoning of the catalytic surface at low overvoltage, this would be a major breakthrough in fuel cell research. Progress along these lines will not be possible without theoretical foundations. Work in this direction has begun,[52–56] and much more is needed to reach the goal of controlling patterns and overall reaction rates by the nonuniformity of active media.

In the case of intermediate distance between the WE and the RE, the strength of the global coupling is smaller than in the case of the close RE which apparently enables more diverse pattern formation.

Target patterns were found in simulations of an electrochemical oscillator with Benjamin–Feir stable homogeneous limit cycles and negative global coupling.[38] Because of the similarity between target patterns and asymmetric target patterns we believe that the patterns observed in Fig. 6b–f emerge owing to the global constraint and do not require a Benjamin–Feir unstable limit cycle. On the other hand, the spatio-temporal behavior of Fig. 6a appears to be of a different quality and we speculate that the large amplitude bursts in Fig. 6a might be linked to the fact that the homogeneous oscillation without global coupling is Benjamin–Feir unstable. Hence, in this case, the spatio-temporal pattern would exist only in Benjamin–Feir unstable systems that are subject to a negative global coupling. To substantiate this conjecture, model calculations are required, that extend those on pattern formation in the CGLE with global coupling,[14,22,57,58] in which the corresponding behavior has not been found. Furthermore, we have discussed that in the absence of global coupling, the homogeneous dynamics can only be described by a set of three essential variables (a three-component system). As long as the dynamics of Fig. 6a (and also 6d) is not further characterized, we cannot exclude that these complex motions are the result of the interaction of three variables.

5. Conclusions

We presented experiments on an electrochemical oscillator subject to global, desynchronizing coupling with varying strength. The experiments revealed several novel aspects that broaden our view on pattern formation in the presence of nonlocal constraints, and they point to directions in which further theoretical foundations are needed. Among them are (1) Pattern formation in Benjamin–Feir unstable systems: (a) in which the system size and characteristic size of the patterns are of the same order of magnitude, (b) close to a subcritical or homoclinic bifurcation, (c) in three-component reaction–transport models and (d) with an additional desynchronizing global constraint. (2) A more detailed understanding of how the strength of the global coupling affects possible bifurcation scenarios and the diversity of dynamical states. (3) Pattern formation in systems with distributed parameters, in particular with emphasis on tailoring inhomogeneous active media to obtain catalytic surfaces with improved performance.

Acknowledgements

We are grateful to G. Ertl for stimulating interest in this work and for generous support and thank A. Bonnefont, F. Plenge, J. Christoph and M. Eiswirth for fruitful discussions. A grant from the DFG in the framework of the Sfb555 is gratefully acknowledged.

References

1 M. T. M. Koper, in *Advances in Chemical Physics*, ed. I. Prigogine and S. A. Rice, Wiley, New York, 1996, vol. 92, p. 161.
2 K. Krischer, in *Modern Aspects of Electrochemistry*, ed. B. E. Conway, J. O. M. Bockris and R. White, Kluwer Academic/Plenum, New York, 1999, vol. 32, p. 1.
3 K. Krischer, N. Mazouz and P. Grauel, *Angew. Chem. Int. Ed.*, 2001, **40**, 851.
4 Y. E. Volodin, V. V. Barelko and A. G. Merzhanov, *Sov. J. Chem. Phys.*, 1982, **5**, 1146.
5 G. Philippou and D. Luss, *Chem. Eng. Sci.*, 1993, **48**, 2313.
6 U. Middya, M. D. Graham, D. Luss and M. Sheintuch, *Physica D*, 1993, **63**, 393.
7 M. D. Graham, S. L. Lane and D. Luss, *J. Phys. Chem.*, 1993, **97**, 7564.
8 R. Imbihl and G. Ertl, *Chem. Rev.*, 1995, **95**, 697.
9 H. Willebrand, T. Hüntler, F. J. Niedernostheide, R. Dohmen and H.-G. Purwins, *Phys. Rev. A*, 1992, **45**, 8766.
10 E. Schöll, *Nonlinear Spatio-Temporal Dynamics and Chaos in Semiconductors*, Cambridge University Press, Cambridge, 2001.
11 F. J. Elmer, *Physica D*, 1988, **30**, 321.
12 M. Sheintuch, *AIChE J.*, 1989, **42**, 1081.
13 U. Middya, M. D. Graham, D. Luss and M. Sheintuch, *J. Chem. Phys.*, 1993, **98**, 2823.
14 M. Falcke, H. Engel and M. Neufeld, *Phys. Rev. E*, 1995, **52**, 763.
15 M. Falcke and H. Engel, *Phys. Rev. E*, 1997, **56**, 635.
16 A. Alekseev, S. Bose, P. Rodin and E. Schöll, *Phys. Rev. E*, 1998, **57**, 2640.
17 S. Bose, P. Rodin and E. Schöll, *Phys. Rev. E*, 2000, **62**, 1778.
18 M. Meixner, P. Rodin and E. Schöll, *Phys. Rev. E*, 1998, **58**, 5586.
19 M. Meixner, P. Rodin and E. Schöll, *Phys. Rev. E*, 1998, **58**, 2796.

20 F.-J. Niederostheide, R. Dohmen, H. Willebrand, B. S. Kerner and H.-G. Purwins, *Physica D*, 1993, **69**, 425.
21 D. Battogtokh, M. Hildebrand, K. Krischer and A. S. Mikhailov, *Phys. Rep.*, 1997, **288**, 435.
22 F. Mertens, R. Imbihl and A. Mikhailov, *J. Chem. Phys.*, 1994, **101**, 9903.
23 K. C. Rose, D. Battogtokh, A. S. Mikhailov, R. Imbihl, W. Engel and A. M. Bradshaw, *Phys. Rev. Lett.*, 1996, **76**, 3582.
24 I. Z. Kiss, W. Wang and J. L. Hudson, *Phys. Chem. Chem. Phys.*, 2000, **2**, 3847.
25 R. D. Otterstedt, P. J. Plath, N. I. Jaeger and J. L. Hudson, *J. Chem. Soc., Faraday Trans.*, 1996, **92**, 2933.
26 R. D. Otterstedt, N. I. Jaeger, P. J. Plath and J. L. Hudson, *Chem. Eng. Sci.*, 1999, **54**, 1221.
27 J. Christoph, R. D. Otterstedt, M. Eiswirth, N. I. Jaeger and J. L. Hudson, *J. Chem. Phys.*, 1999, **110**, 8614.
28 P. Grauel, J. Christoph, G. Flätgen and K. Krischer, *J. Phys. Chem. B*, 1998, **102**, 10264.
29 P. Grauel and K. Krischer, *Phys. Chem. Chem. Phys.*, 2001, **3**, 2497.
30 P. Strasser, J. Christoph, W.-F. Lin, M. Eiswirth and J. L. Hudson, *J. Phys. Chem. B*, 2000, **104**, 1854.
31 J. Christoph, P. Strasser, M. Eiswirth and G. Ertl, *Science*, 1999, **284**, 291.
32 K. Krischer, N. Mazouz and G. Flätgen, *J. Phys. Chem.*, 2000, **104**, 7545.
33 H. Varela and K. Krischer, *Catal. Today*, 2001, in press.
34 A. Zhabotinsky, M. Dolnik and I. Epstein, *J. Chem. Phys.*, 1995, **103**, 10306.
35 G. Flätgen and K. Krischer, *J. Chem. Phys.*, 1995, **103**, 5428.
36 N. Mazouz, K. Krischer, G. Flätgen and G. Ertl, *J. Phys. Chem.*, 1997, **101**, 2403.
37 N. Mazouz, G. Flätgen and K. Krischer, *Phys. Rev. E*, 1997, **55**, 2260.
38 J. Christoph, PhD Thesis, FU Berlin, Berlin, Germany, 1999.
39 M. C. Cross and P. C. Hohenberg, *Rev. Mod. Phys.*, 1993, **65**, 851.
40 Y. Kuramoto, *Chemical Oscillations, Waves and Turbulence*, Springer, Berlin, 1984.
41 A. S. Mikhailov, *Foundations of Synergetics I*, Springer, Berlin, 1994.
42 A. S. Mikhailov and A. Y. Loskutov, *Foundations of Synergetics II*, Springer, Berlin, 1996.
43 T. Benjamin and J. Feir, *J. Fluid Mech.*, 1967, **27**, 417.
44 B. Lake and H. Yuen, *J. Fluid Mech.*, 1977, **83**, 75.
45 Y. Liu and R. E. Ecke, *Phys. Rev. Lett.*, 1997, **78**, 4391.
46 W. K. Melville, *J. Fluid Mech.*, 1982, **115**, 165.
47 S. Jakubith, H. H. Rotermund, W. Engel, A. v. Oertzen and G. Ertl, *Phys. Rev. Lett.*, 1990, **65**, 3013.
48 M. Kim, M. Bertram, M. Pollmann, A. v. Oertzen, A. S. Mikhailov, H. H. Rotermund and G. Ertl, *Science*, 2001, **292**, 1357.
49 G. Veser, F. Esch and R. Imbihl, *Catal. Lett.*, 1992, **13**, 371.
50 L. Lobban, G. Philippou and D. Luss, *J. Chem. Phys.*, 1991, **93**, 733.
51 L. Carrette, K. A. Friedrich and U. Stimming, *ChemPhysChem*, 2000, **1**, 162.
52 A. K. Bangia, M. Bär, I. G. Kevrekidis, M. D. Graham, H.-H. Rotermund and G. Ertl, *Chem. Eng. Sci.*, 1996, **51**, 1757.
53 A. Kulka, M. Bode and H.-G. Purwins, *Phys. Lett. A*, 1995, **203**, 33.
54 A. Hagberg, E. Meron, I. Rubinstein and B. Zaltsman, *Phys. Rev. Lett.*, 1996, **76**, 427.
55 M. A. Liauw, J. Ning and D. Luss, *J. Chem. Phys.*, 1996, **104**, 5657.
56 S. Y. Shvartsman, E. Schütz, R. Imbihl and I. G. Kevrekidis, *Phys. Rev. Lett.*, 1999, **83**, 2857.
57 D. Battogtokh, A. Preusser and A. S. Mikhailov, *Physica D*, 1997, **106**, 327.
58 D. Lima, D. Battogtokh, A. S. Mikhailov, P. Borckmans and G. Dewel, *Europhys. Lett.*, 1998, **42**, 631.

Identification of the intermittency-I route to chaos in oscillating CO oxidation on zeolite-supported Pd

Marina M. Slin'ko,[a] **Anatolii A. Ukharskii,**[a] **Nikolai V. Peskov**[b] **and Nils I. Jaeger**[*c]

[a] *Institute of Chemical Physics, RAS, Kosygina Str., 4, 117334 Moscow, Russia*
[b] *Department of Computational Mathematics and Cybernetics, Moscow State University, 119899 Moscow, Russia*
[c] *Institut für Angewandte und Physikalische Chemie, FB 2, Universität Bremen, PF 330440, 28334 Bremen, Germany*

Received 5th April 2001
First published as an Advance Article on the web 7th November 2001

For the oscillating oxidation of CO on a zeolite-supported palladium catalyst the transition to chaos could be observed in a very narrow region of the CO concentration in the feed. The reaction was carried out under the conditions of a continuous stirred tank reactor. A careful choice of the method for time series analysis led to the unambiguous identification of the intermittency-I route to chaos in the catalytic system despite the rather limited number of data points which can be acquired under normal pressure conditions. The route to chaos could be derived from the variation of the Fourier spectrum and the Poincare section as a function of the CO concentration in the feed. The embedding dimensions for the observed chaotic attractors of $d_E \geq 10$ are much higher than the embedding dimensions obtained during UHV single crystal studies. High embedding dimensions indicate that the dynamic behaviour of the system has to be simulated with a distributed model which describes the collective behaviour of many Pd particles in the zeolite crystallite.

Introduction

The analysis of temporal and spatiotemporal order observed experimentally under low and high pressure conditions in many heterogeneous catalytic systems has contributed enormously to the understanding of the details of the reaction mechanisms and to the construction of corresponding mathematical models.[1–3] The observed oscillatory behaviour includes regular periodic, quasi-periodic, complex and chaotic time series. While many time series reported in the literature for heterogeneous catalytic reactions appear to be complex, a distinction between periodic and chaotic oscillations was often only based on a visual inspection of the time series. Deterministic chaos has to be identified and distinguished from random noise and, *e.g.*, quasiperiodic behaviour or mixed mode oscillations. The required rigorous proof of the existence of chaos in heterogeneous catalytic systems has been reported in only a few studies.[4–6] Although the data have been shown to display unambiguous features of deterministic chaos, the understanding of the origin of the observed behaviour is still incomplete.

The identification of the scenario of transition to chaos in a dynamical system is an important and attractive problem in nonlinear dynamics. Its analysis produces additional information about the dynamic behaviour of the system. Moreover the identification of the route to chaos may be helpful in the discrimination of models used for the simulation of experimental data.

There are three main routes of transition to chaos in dynamical systems: *via* period-doubling bifurcations (Feigenbaum scenario), *via* the breaking of a torus (Ruelle and Takens scenario) and *via* intermittency (Pomeau–Manneville scenario).[7–9] For homogeneous catalytic systems all these scenarios could be identified in the intensively studied Belousov–Zhabotinsky reaction.[10–12]

In the case of heterogeneous catalytic systems it is more difficult to observe and examine the transition to chaos experimentally. One reason is the drift of the catalyst activity, which is common for heterogeneous catalytic systems and which makes it difficult to follow and analyse the phenomenon unambiguously. Another reason could be the very small region of experimental parameters in which the transition to chaotic behaviour can be observed. The first, and up to now the only, identification of a transition from periodic to chaotic behaviour has been demonstrated for kinetic oscillations in the catalytic CO oxidation on a well-defined Pt(110) surface under UHV conditions.[5] It could be shown that upon variation of the CO partial pressure in well-defined steps the system followed the period doubling route to deterministic chaos. The identification of the scenario was performed by evaluating the power spectra, the correlation integrals, and the Lyapunov exponents.

In this paper we present the results of the study of the transition to chaos *via* the Pomeau–Manneville type-I intermittency scenario in the case of CO oxidation on a zeolite-supported Pd catalyst. To our knowledge this is the first qualitative evidence for such a transition during reaction rate kinetic oscillations on a supported catalyst under normal pressure conditions.

Experimental

1. Catalytic measurements

The dynamic behaviour of the system has been studied in a continuous flow glass reactor under atmospheric pressure.[13] The catalyst was applied under shallow bed conditions on a glass frit in the reactor. The calculation of a small axial Peclet number and characteristics of the residence time spectra indicate, that at the given experimental conditions the behaviour of the flow reactor is similar to a continuous stirred tank reactor (CSTR). As was demonstrated in ref. 1 and 2 this type of a reactor is the best for the study of kinetic oscillations. A certified mixture of CO and N_2 (10 vol.% CO, 99.97% pure, in N_2, Messer–Griesheim) was passed through a glass tube with quartz pellets held at 500 K to remove traces of volatile metal carbonyls and then mixed with oxygen (99.995% pure, Messer–Griesheim). Nitrogen (99.995% pure, Messer–Griesheim) was added as a balance.

The reaction mixture was fed into the reactor with a flow rate of 150 cm^3 min^{-1} corresponding to a linear flow velocity of 1.89 cm s^{-1}. The flow rates were controlled by thermal mass flow controllers (HI-TEC, ~1% precision). The gas composition was changed by varying the CO concentration up to 3.5 vol.%. The outlet CO and CO_2 concentrations were measured with an IR analyser URAS 10E. The data for CO, CO_2 concentrations and temperature were digitised with a sampling time of 0.1 s.

The catalyst consisted of palladium dispersed within the cavities of a Faujasite X type zeolite. The average diameter of the Pd particles was 10 nm. The Pd loading was 0.05 wt.%. Prior to each experimental run the catalyst was oxidised for 12 h in streaming synthetic air at 633 K. Details of catalyst preparation and electron microscopy results can be found in ref. 14 and 15.

2. Numerical aspects

Since the discovery of deterministic chaos many sophisticated algorithms have been developed in order to extract information on the primary dynamics from experimental time series. A description of the most effective and frequently used methods of nonlinear time series analysis can be found in 3 reviews.[16–18] While processing experimental data it is important to have the possibility to apply several different methods to calculate the key dynamic and metric characteristics of the time series. The efficiency of each method depends on many factors, *e.g.* the length of the time series, the value of the sampling interval, the signal-to-noise ratio *etc.* Hence, one should be able to choose the most effective and suitable method for the given experimental time series to obtain the most reliable evaluation of its dynamic and metric characteristics. Moreover it is often necessary to apply a multitude of methods of time series analysis such as the Fourier or wavelet analysis, the construction of Poincare sections, various maps and histograms. For this purpose we have developed the new computer code "CDP—Chaotic Data Processing", which collects all methods described below and is based on all modern algorithms widely used in nonlinear time series analysis. More information about the program "CDP" and its possibilities can be obtained from ref. 19.

Experimental results and digital processing of experimental data

Regular oscillations have been observed at 503 K in the case of the oxidised catalyst, when the reaction mixture contained 0.3 vol.% CO, 20% O_2 and N_2 as a balance. The transition to complex oscillatory behaviour was detected while increasing the CO inlet concentration, which played the role of a bifurcation parameter in very small steps (0.02–0.04%). A list of experimental conditions together with the length of the corresponding time series is given in Table 1. Some selected sections of experimental time series for CO concentrations between 0.3% and 1.1% are shown in Fig. 1. As the CO concentration slowly increases the system undergoes a sequence of transitions from regular to more complex temporal behaviour. The observation of an aperiodic time series alone does not mean that the signal is chaotic and the analysis has to exclude multi-periodic or random signals. Fourier power spectral analysis allows the oscillatory behaviour to be resolved into the component frequencies that constitute the time series. Periodic and quasi-periodic oscillations produce sharp peaks at a few representative frequencies, while chaotic behaviour is characterised by the presence of broadened peaks at a large number of frequencies. Therefore power spectral analysis can distinguish between quasi-periodic and chaotic oscillations.

Fourier power spectra of the corresponding time series shown in Fig. 1 are presented in Fig. 2. For regular oscillations the fundamental frequency and two more harmonics can be detected. The presence of harmonics in the spectra may indicate deviation from the sinusoidal oscillations due to the relaxation character of the observed oscillations, which includes successive alternation of slow

Table 1 CO concentrations, length of the corresponding time series, maximal Lyapunov exponents and embedding dimensions which were obtained as the result of the nonlinear time series analysis

CO concentration (%)	Length of time series	Embedding dimension		Max. Lyapunov exponent		Rosenstein's algorithm
		d_{svd}	d_R	Wolf's algorithm		
0.34	12 751	5	9	0.028	0.019	0.049
0.36	14 902	6	9	0.036	0.034	0.065
0.38	14 501	6	8	0.066	0.035	0.077
0.40	24 772	7	9	0.042	0.034	0.076
0.42	26 802	7	8	0.064	0.050	0.075
0.44	22 001	8	10	0.071	0.070	0.084
0.48	35 000	9	9	0.074		0.084
0.52	21 001	9	12	0.052	0.062	0.087
0.56	20 502	9	10	0.075	0.068	0.082
0.60	23 501	9	12	0.078	0.075	0.084
0.64	22 001	9	12	0.081	0.065	0.075
0.68	19 501	9	12	0.066	0.067	0.078
0.72	360 000	9	11	0.079	0.077	0.070
0.76	33 701	9	9	0.048		0.060
0.80	25 801	8	11	0.062	0.060	0.068
0.85	30 001	8	8	0.070		0.067
0.90	33 001	7	8	0.081	0.082	0.071
0.95	40 001	7	7	0.080		0.070
1.00	44 558	6	7	0.083	0.084	0.072
1.10	38 661	6	6	0.077		0.067

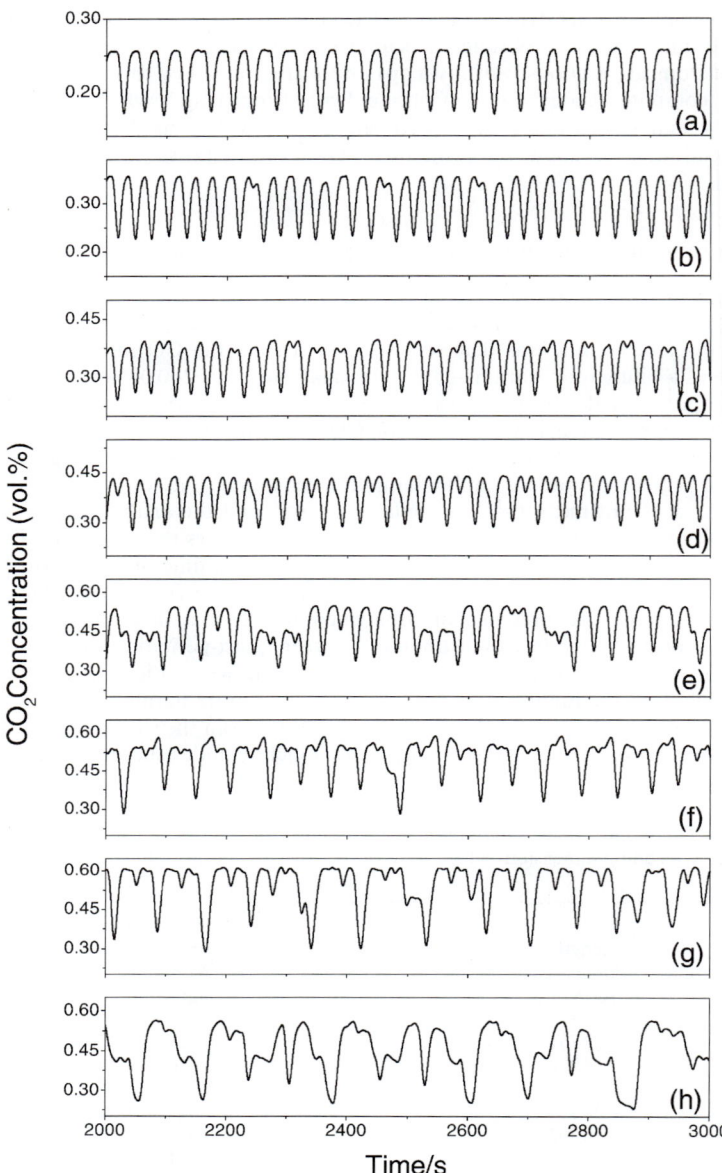

Fig. 1 Time series obtained during the increase in the CO inlet concentration: (a) 0.3%, (b) 0.4%, (c) 0.44%, (d) 0.52%, (e) 0.64%, (f) 0.76%, (g) 0.9%, (h) 1.1%.

and fast motions. With increasing CO concentration the power spectrum changes to a broadened peak, which is one of the main indications of both chaos and random noise.

Further analysis of the experimental time series has been carried out using the common tools of nonlinear dynamics. The experimental data represent a single-variable time series of the CO_2 concentration $s_k = s(k\Delta t)$, where $k = 0, 1, 2, \ldots, N$ and N is the number of points separated by the same time step $\Delta t = 0.1$ s. The data analysis is carried out as follows: (i) construction of the time series of d-dimensional vectors X_k of phase (or state) variables from a scalar time series of experimental data s_k; (ii) estimation of the minimal dimension d of X (embedding dimension) for phase

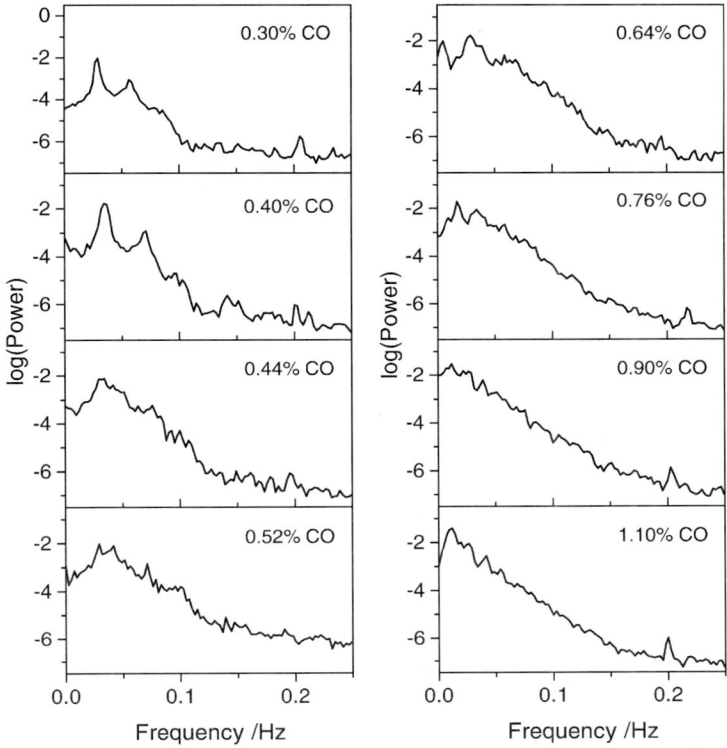

Fig. 2 Fourier power spectra for the corresponding time series, presented in Fig. 1.

space reconstruction of the system; (iii) estimation of the metric and the dynamical properties of the reconstructed attractor.

During the past two decades many sophisticated methods have been proposed for the solution of these problems. The most widely recognised methods were described in ref. 16–18. We will not review the known tools of data processing in this study, but rather shall discuss all the problems, difficulties and peculiarities of the treatment of experimental data presented in this work.

1. Noise reduction—high-frequency filter

Noise in the measurement of the CO_2 concentration was caused by the operation of the IR analyser URAS 10E. To estimate the noise component in the data, the device readings without reaction were written. Analysis of the measurements of the noise demonstrated that its magnitude was about 3–6% of the signal range. The Fourier spectrum of the noise has a main frequency of 2.5 Hz, which is more than an order of magnitude higher than the main frequency of the signal. This may be the consequence of the existence of a characteristic device frequency. Thus the noise is not white and could be effectively removed from the data by high-frequency filtering. For noise reduction all frequencies ten times higher than the main signal frequency were cut off from the signal Fourier spectrum. The results of filtering are shown in Fig. 1.

2. Choosing the time delay

There are several methods for phase space reconstruction from the signal of scalar sequence of measurements. In this work the *time-delay* method and the *differential* method were used.

In the differential method the phase variables are the successive time derivatives of $s(t)$. That is $x_1(t) = s(t)$, $x_2(t) = \dot{s}(t)$, $x_3(t) = \ddot{s}(t)$, and so on. The equations representing the dynamics have the following simple structure:

$$\dot{x}_1 = x_2,$$
$$\dot{x}_2 = x_3,$$
$$\ldots$$
$$\dot{x}_d = F(x_1, x_2, \ldots, x_d).$$

From the discrete time series the derivatives may be approximated by the finite differences, for example

$$\dot{x}_1(t_k) \approx (s_{k+1} - s_k)/\Delta t,$$
$$\dot{x}_2(t_k) = \ddot{x}_1(t_k) \approx (s_{k-1} - 2s_k + s_{k+1})/(\Delta t)^2, \ldots$$

The drawback of this method is that these equations are a poor representation of the derivatives for finite Δt. Moreover with the increase in the order of the derivative the noise-to-signal ratio also increases. Therefore the differential method is usually not applied for systems with a dimension larger than three. On the other hand the advantage of the derivative coordinates is their clear physical meaning.

From the definition of finite difference derivatives it can be concluded that each variable x_i is a function of the values of the measurements s_k at time t_k and at other times lagging by multiples of a fixed sampling time Δt. This introduces the idea of using time-delay coordinates to reconstruct the phase space of an observed dynamical system.[20] To reconstruct the structure of orbits in a phase space, the time delay variables $x_n(t) = s(t + (n-1)\tau \Delta t)$, where the lag or delay time τ is some integer number, can be used. If d is sufficiently large then the d-dimensional vector $X(t) = \{s(t), s(t+T), s(t+2T), \ldots, s(t+(d-1)T)\}$, $T = \tau \Delta t$, reconstructed from the time series $s(t)$ has the same topological properties as the vector consisting of the n relevant dynamic variables $u_1, u_2, \ldots, u_n$ from which the dynamic behaviour of $s(t)$ originates. The advantage of the time-delay method is that the noise-to-signal ratio is the same for all coordinates.

The embedding theorem of Mañé and Takens[21,22] also states that almost any time lag will be acceptable for the phase space reconstruction. However, for finite time series the proper choice of τ is important. If it is too small, there is no difference between the different components of the delay vector and the coordinates x_n and x_{n+1} will be correlated. Similarly, if τ is too large, then x_n and x_{n+1} are completely independent of each other in a statistical sense due to the intrinsic instability of a chaotic system. Although the validity of the time-delay method depends greatly on a proper choice of the time delay τ, there exists no rigorous way of determining its optimal value.[23]

One of the possible ways for choosing τ is to compute the autocorrelation function $C(\tau)$ and to look for that τ where $C(\tau)$ first passes through zero. In some studies including ref. 5, this method works well. However in our case the first zero of the autocorrelation function produces too large a value of τ ($\tau > 100$). For such large values of τ the reconstructed attractor includes strongly disordered orbits, which indicates that the coordinates obtained are statistically independent, in disagreement with the postulated deterministic nature of the system.

Another method for the determination of τ is to look for the first minimum of the average mutual information, as was recommended in ref. 24. For the time series s_k the average mutual information $I(\tau)$ between observations separated by a time lag τ is defined as follows:

$$I(\tau) = \sum_m P(s_m, s_{m+\tau}) \log_2 \left| \frac{P(s_m, s_{m+\tau})}{P(s_m) P(s_{m+\tau})} \right|,$$

where $P(s_m)$ is the probability of observing s_m at the time moment m, and $P(s_m, s_{m+\tau})$ is the joint probability of observing s_m and $s_{m+\tau}$ at the time moments m and $m + \tau$, respectively. During computations these probabilities are estimated directly from the histograms.

A typical view of the $I(\tau)$ graph is shown in Fig. 3(a). Here and everywhere in the text all the illustrations of the applications of the methods of nonlinear analysis will depict the results of the treatment of the longest time series, which was measured for an inlet CO concentration of 0.72%. Fig. 3(a) shows that, unfortunately, $I(\tau)$ does not exhibit a distinct minimum, i.e., we have to look

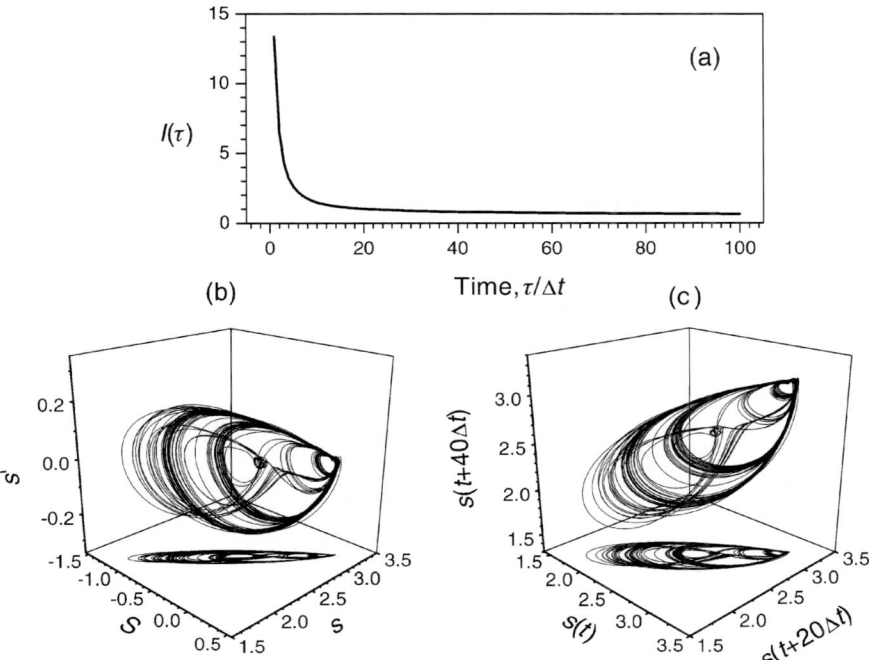

Fig. 3 (a) Average mutual information; (b) 3D projection of the reconstructed attractor in differential coordinates $\{S = \int_0^t (s(\tau) - \bar{s}) d\tau, s, \dot{s}\}$; (c) 3D projection of the reconstructed attractor in time-delayed coordinates with $\tau = 20\Delta t$.

for some other criterion. We assume that a suitable τ may be chosen in the interval, where the fast descent of $I(\tau)$ ends and the slow decrease begins. As can be seen from Fig. 3(a) this interval is approximately $15 < \tau < 30$. A more precise τ selection is based on a geometrical argument. The value of τ was adjusted by comparing the shapes of obtained orbits in 3D projection for both the differential and the time-delay reconstruction (in this case the derivative coordinates are $x_1(t) = \int_0^t (s(\xi) - \bar{s}) d\xi$, $x_2(t) = s(t)$, $x_3(t) = \dot{s}(t)$, ...). Fig. 3(b) and (c) show the best agreement between the three-dimensional reconstructions in both cases; this was achieved for a value of $\tau = 20$. This was also the best time delay for other measured time series. It must be noted that the method of the determination of τ from the first minimum of the information entropy as was recommended in ref. 25 was also unsuccessful for the treatment of experimental data presented in this work. The graphs of the information entropy for embedding dimensions between 2 and 15 have no distinct minima and show qualitatively similar behaviour to $I(\tau)$.

3. Choosing the embedding dimension

The embedding theorem of Mañé and Takens[21,22] states that for full unfolding of the reconstructed strange attractor, whose dimension is d_A, one should in general embed it into Euclidean space of (embedding) dimension $d_E \geq 2d_A + 1$. This means that the vector of phase variables $X(t)$ in the time-delay reconstruction method must consist of at least d_E components. The estimation of a low bound of d_E is very important for the performance of many algorithms, especially for the proper calculation of Lyapunov exponents. In the present work three well-known methods for d_E evaluation were used: saturation of the correlation dimension, false nearest neighbours, and singular-value decomposition.[17] Reliable results were obtained only with the simplest one of these three methods—the method based on singular-value decomposition of the covariance matrix.

The first method we used to estimate the embedding dimension is the widely used method based on the calculation of correlation integrals.[26] For time series the d-dimensional correlation integral

$C_2(\varepsilon)$ is defined by the following equation:

$$C_2(\varepsilon) = [N(N-1)]^{-1} \sum_{i \neq j} \Theta(\varepsilon - |X_i - X_j|_d),$$

where $\Theta(x)$ is the Heaviside step function, $\Theta(x) = 0$, if $x \leq 0$ and $\Theta(x) = 1$ for $x > 0$. The sum counts the pairs of data points separated in d-dimensional space by a distance less than ε. In the limit of an infinite amount of data points and for small ε the correlation integral scales as $C_2(\varepsilon) \approx \varepsilon^\nu$, where ν is the correlation dimension. Hence, one can evaluate $C_2(\varepsilon)$ as a function of dimension d and determine the conditions where the slope of its logarithm as a function of $\log \varepsilon$ becomes independent of d. Thus one can estimate the embedding dimension d_E, as well as the correlation dimension of the attractor.

However, the processing of experimental data does not display any indications of the local slope convergence. Fig. 4 demonstrates the results of correlation integral calculations. Fig. 4(a) shows the log–log graphs of $C_2(r)$ in embedding spaces of dimension from 2 to 20. Fig. 4(b) shows the corresponding graphs of $\log_2 C_2(r)$ local slopes. Although some convergence of local slopes at high dimensions can be seen in Fig. 4(b), a true scaling range cannot be detected. Thus we could hardly determine the embedding dimension and, therefore, the correlation dimension (as well as the Kol-

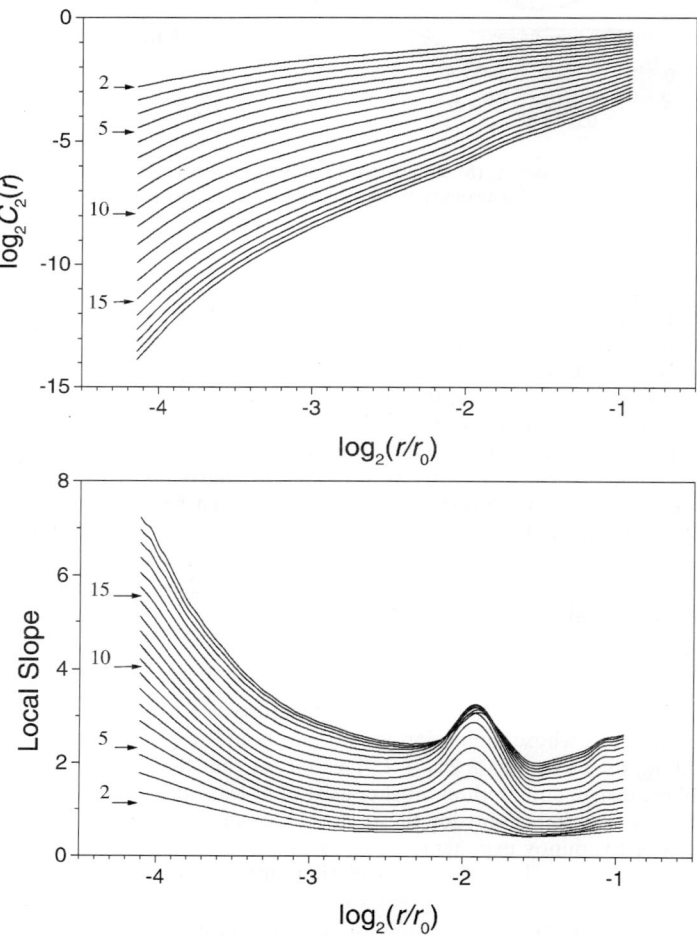

Fig. 4 (a) Correlation integrals for various embedding dimensions. (b) Local slope for various embedding dimensions as derived from the correlation integrals.

mogorov entropy and other dimensional features) with the help of the correlation algorithm. For the other time series with fewer data points the results obtained are even less reliable.

The second method we have used to estimate an optimal value of d_E is the false nearest neighbours (FNN) method.[27] The general idea of the method is to increase the dimension d of embedding space till the orbit projection onto this space will have no self-crossing point. Unfortunately, in the case of our data this method did not display robust behaviour relative to variations of the threshold parameter R_T (see ref. 17 for details) and did not give any definite value for d_E.

Finally the estimation of the embedding dimension of the system from the eigenvalues of the covariance matrix M_d was used:

$$M_d = N^{-1} \sum_n (X_n - \bar{X})(X_n - \bar{X})^T, \bar{X} = N^{-1} \sum_n X_n,$$

where X_d is a d-dimensional vector of time-delayed variables.[28] If d_A is the number of active degrees of freedom of the system and we choose $d > d_A$, then generally, in a downward ordered series of eigenvalues of M_d, the first number d_A of the eigenvalues arising from the real signal will gradually diminish, and the last number $d - d_A$ of the eigenvalues produced by the noise will be nearly equal. If this is the case, then by looking at the eigenvalues we may hope to find a "noise floor" at which the eigenvalue spectrum turns over and becomes flat. The number of eigenvalues where the floor is reached may be used to estimate the embedding dimension. In the present study the experimental data after filtering have a very low level of noise and the eigenvalue spectrum of the covariance matrix with sufficiently large d diminishes almost to zero. Therefore the estimations of the embedding dimension from singular-value decomposition of the covariance matrix (d_{SVD}) were obtained using the condition $\lambda_k/\lambda_1 > 0.01$. An example of the diagram of normalised eigenvalues for the longest time series obtained at 0.72% inlet CO concentration is shown in Fig. 5. The results of the analysis for the other time series are summarised in Table 1. From Fig. 5 an embedding dimension of $d_{SVD} = 10$ can be estimated which is a rather large value. As the estimation of embedding dimension produced by the singular-value decomposition method is usually lower then the real embedding dimension,[11] we can suggest that the space, which is necessary for the full unfolding of the attractor has a dimension >10. It is known that the higher the dimension of the system, the more data are necessary for reliable computations of correlation integrals. Therefore, the failure to determine the embedding dimension with the false nearest neighbours method and with the help of the correlation integral may be connected with the high dimension of the underlying dynamics and/or with the long term variation of the catalytic activity.

4. Estimating the maximum Lyapunov exponent

The instability of the motion on a strange attractor is characterised with the help of Lyapunov exponents, each of which measures the average spreading of trajectories within the attractor in a specific direction (eigendirection). There are as many different Lyapunov exponents for a dynamical system as there are phase space dimensions. The main indication of chaotic behaviour is the

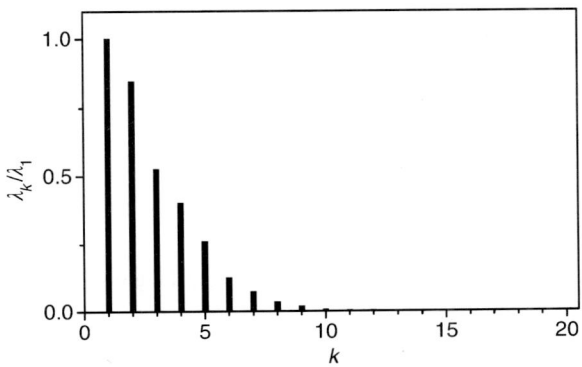

Fig. 5 Relative (λ_k/λ_1) eigenvalues of covariance matrix M_{20}.

existence of at least one positive value in the system's spectrum of Lyapunov exponents. The evaluation of a complete Lyapunov spectrum from experimental time series is, in principle, a solvable task, but very cumbersome. However, the main question, that we want to answer is whether the time series is chaotic or not. In this case only the largest Lyapunov exponent λ_{max} has to be estimated. A positive value for λ_{max} means an exponential divergence of nearby trajectories on the attractor or sensitive dependence on initial conditions, which is the main characteristic of chaos.

For the calculation of the largest Lyapunov exponent we have used two different methods: (i) the well-known method of Wolf et al.[29] and (ii) the method of Rosenstein et al.[30] The basic idea of Wolf's algorithm is to consider two different segments from the reconstructed trajectory with close initial points as the segments belonging to different trajectories, and to evaluate the divergence of these segments. The reconstruction trajectory follows in n-dimensional space a frequently wound and tortuous path. The procedure is to choose an initial point on the reconstructed trajectory and to find the neighbouring point, which is situated at the minimal distance from the initial point, but lagged in time (usually larger than one average period) from the initial point. This new point is considered as the starting point of the new trajectory, which forms a counterpart segment with the fiducial trajectory. The divergence of the trajectories from $r(t_0)$ to $r'(t_1)$ is monitored over a certain evolvent time. If the final separation is too large a new point of the reconstructed trajectory is chosen closer to the fiducial trajectory and approximately in the direction where the last point lies. The previous segments are replaced by the new ones and the increase in the distance from $r(t_1)$ to $r'(t_2)$ is monitored again. The procedure is repeated along the whole reconstructed trajectory and the Lyapunov exponent is determined as:

$$\lambda_{max} = \frac{1}{t_m - t_0} \sum_{i=1}^{m} \ln \frac{r'(t_j)}{r'(t_{i-1})},$$

where m is the total number of replacements along the complete reconstruction trajectory.

Typical results obtained with Wolf's algorithm are shown in Fig. 6. The period T_0 corresponding to the main frequency in the Fourier spectra of the time series at an inlet CO concentration equal to 0.72% can be evaluated as $T_0 \approx 50$ s. Fig. 6(a) displays the temporal convergence of the largest Lyapunov exponent as a function of time for embedding dimension equal to 10 and various evolvent times ($t_e = 8,10,12$ s). From these graphs the largest Lyapunov exponent could be estimated as $\lambda_{max} \approx 0.08$ bit s^{-1}. Fig. 6(b) shows the results of λ_{max} calculations in d_e-dimensional phase space with $d_e = 10,15,20$ and $t_e = 12$ s. Although the value of λ_{max} remains positive, its value decreases essentially with the increase in the value of the dimension. Therefore, as in the case of the calculation of correlation integral, the Lyapunov exponent does not converge properly with increasing embedding dimension. Hence it is not possible to estimate λ_{max} with the help of Wolf's algorithm without knowing the dimension of the system.

The method of Rosenstein et al.[30] is based on the direct evaluation of the average rate of divergence of neighbouring trajectories in phase space. In order to do this one should choose any point X_k in the reconstructed phase space and the sphere $S_k(r) = \{X_m : \|X_m - X_k\| < r\}$ with the centre at X_k. Now every point in $S_k(r)$ is considered as the initial point for the trajectory starting from this point and one can calculate the distance $R_{km}(t_i) = \|X_{m+i} - X_{k+i}\|$, $t_i = i\Delta t$, between the fiducial trajectory starting from the central point X_k and any other trajectory starting from a point $X_m \in S_k(r)$ as a function of the relative time i. Then one can compute the average distance $\bar{R}_k(t_i)$ over the initial sphere $S_k(r)$: $\bar{R}_k(t_i) = N_k^{-1}\Sigma_{X_m \in S_k(r)}R_{km}(t_i)$, where N_k is the number of points X_m in the sphere $S_k(r)$. Finally the global average distance $\bar{R}(t_i)$ should be calculated over the sufficiently large set of fiducial points X_k:

$$\bar{R}(t_i) = K^{-1} \sum_k \bar{R}_k(t_i),$$

where K is the number of fiducial points. The logarithm of the global average distance at time t_i is some effective expansion rate over the time span t_i. Therefore $\bar{R}(t_i)$ is plotted in semi-logarithmic coordinates vs. time (t_i, $\log_2 \bar{R}(t_i)$) and a sufficiently long, approximately linear, increasing segment indicates the exponential divergence of trajectories. The average slope of the linear segment of the curve $\log_2 \bar{R}(t)$ defines the largest Lyapunov exponent.

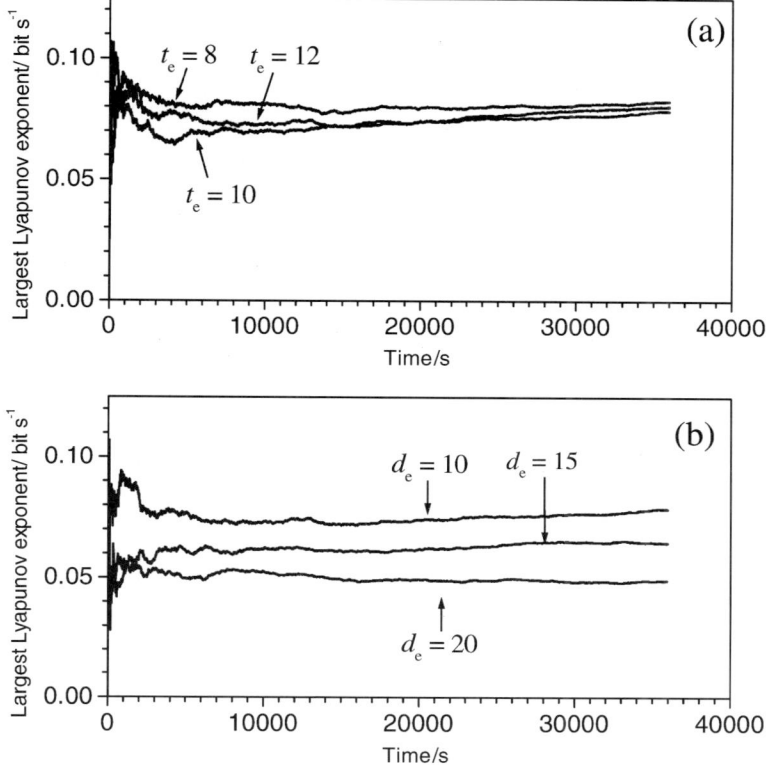

Fig. 6 Estimations of the largest Lyapunov exponent *via* Wolf's algorithm: (a) calculations for various time delays; (b) calculations for various embedding dimensions.

The results evaluated by this method from the longest time series measured for 0.72% CO inlet concentration are displayed in Fig. 7. Fig. 7(a) shows the average distance between neighbouring trajectories $\bar{R}(t_i)$ as a function of time for embedding dimensions ranging from 2 to 20. Averaging is carried out over more than 2500 spheres of radius equal to 10% of the amplitude of the time

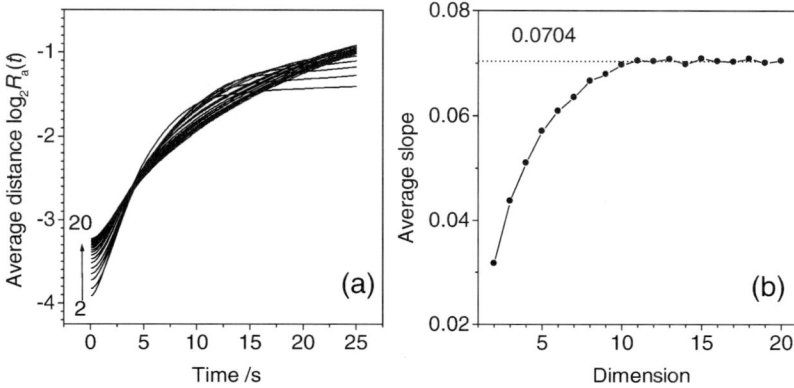

Fig. 7 Estimation of the largest Lyapunov exponent *via* the Rosenstein algorithm: (a) Logarithm of the average distance between the fiducial trajectory starting from the centre of the sphere with $r = 0.1R$ and all others trajectories starting from initial points, which belong to this sphere; (b) average slope of $\log_2 \bar{R}(t)$ in the interval of evolvent time $5 < t/s < 25$ in dependence on the dimension of the phase space.

series. The choice of the size of the sphere was based on the following arguments: (i) for small sizes each sphere contains only a few points and the fluctuations of $\bar{R}_k(t_i)$ are significant; (ii) for large sizes the divergence of trajectories is bounded by the size of the attractor in phase space. The best results were obtained with a medium-sized sphere, for which both factors could be neglected.

Fig. 7(b) displays the dependence of the average slope of $\log_2 \bar{R}(t)$ upon the embedding dimension in the interval $5 \leqslant t/s \leqslant 25$. The curves $\log_2 \bar{R}(t)$ display a clear tendency to converge with increase in the embedding dimension. Therefore it is possible to estimate not only the Lyapunov exponent, but also the embedding dimension d_R as the value, where the Lyapunov coefficient reaches the plateau. It can be seen that the average slope oscillates weakly around the value ≈ 0.07 s^{-1} beginning from the dimension equal to 11. The largest Lyapunov exponent was calculated as the average value over dimensions between the value of d_R obtained and the largest value for which the Lyapunov exponent was calculated, which was equal to 20. The values of embedding dimension d_R and the largest Lyapunov exponents for all considered time series are presented in Table 1. For comparison, the largest Lyapunov exponents, calculated by Wolf's algorithm for two dimensions (d_{svd} obtained by the method of singular-value decomposition of the covariance matrix and d_R obtained by Rosenstein's algorithm) are also shown in Table 1 (left and right numbers respectively). The difference between estimations obtained by Wolf and Rosenstein algorithms may be due to the relatively short length of the time series. We believe that the Rosenstein estimations are more reliable for the given experimental time series.

Route to chaos

The positive values of the largest Lyapunov exponents, together with the finite dimension of the attractor, indicate that the reaction dynamics is chaotic for CO inlet concentration larger than 0.4%. In addition to establishing the existence of deterministic chaos in a dynamical system it is also important to identify a route to chaos, which the system undergoes with the variation of the control parameter.

The time series in the region of the transition to chaos are shown in Fig. 8. Visual inspection of the presented data excludes period doubling. Furthermore no sign of the destruction of a torus can be detected. The regular oscillations destabilise at inlet concentration $C_{CO} = 0.34\%$, when the periodic state seems to be randomly disrupted by short perturbations. With increasing CO concentration the irregular perturbations become more and more frequent until fully developed chaos is eventually reached at a CO concentration equal to 0.4–0.42%. This suggests intermittency as a candidate for the route to chaos. The intermittency scenario is characterised by the existence of regular (laminar) phases along the evolution of a system variable, interrupted by bursts of irregular behaviour. Pomeau and Manneville established the existence of three types of intermittency depending on how a periodic orbit can lose its stability.[31] The classification is made according to the way the limit cycle Floquet multipliers cross the unit circle at the point of instability.[32,33]

For type I intermittency the limit cycle Floquet multiplier crosses the unit circle at the point 1. In this case a periodic state is destroyed by a cyclic fold and succeeded by chaos. The cyclic fold is the bifurcation, where stable and unstable limit cycles coalesce. For type III intermittency the Floquet multiplier crosses the unit circle at -1. During this bifurcation the limit cycle loses its stability *via* a flip subcritical or subcritical period doubling bifurcation. Type II intermittency is associated with a pair of complex conjugate Floquet multipliers crossing the unit circle at $e^{\pm i\omega}$ and introducing an additional incommensurate frequency with respect to the unstable periodic orbit at the onset of instability.

Since the paper of Pomeau and Manneville[31] the catalogue of intermittencies has been extended, *e.g.*, by type X, which extends type I to cover cases, with intermittency near hysteretic transitions, by type V for discontinuous maps and by on–off, which is induced by an irregular variation of a control parameter around the bifurcation point.[34] In all cases a reinjection mechanism must exist in order to approach the trajectory to the unstable periodic orbit after every irregular burst or perturbation.

In order to ascertain that the experimental time series obtained are consistent with an intermittency route to chaos, as well as to identify the intermittency type the Fourier power spectra and the return maps will be analysed.

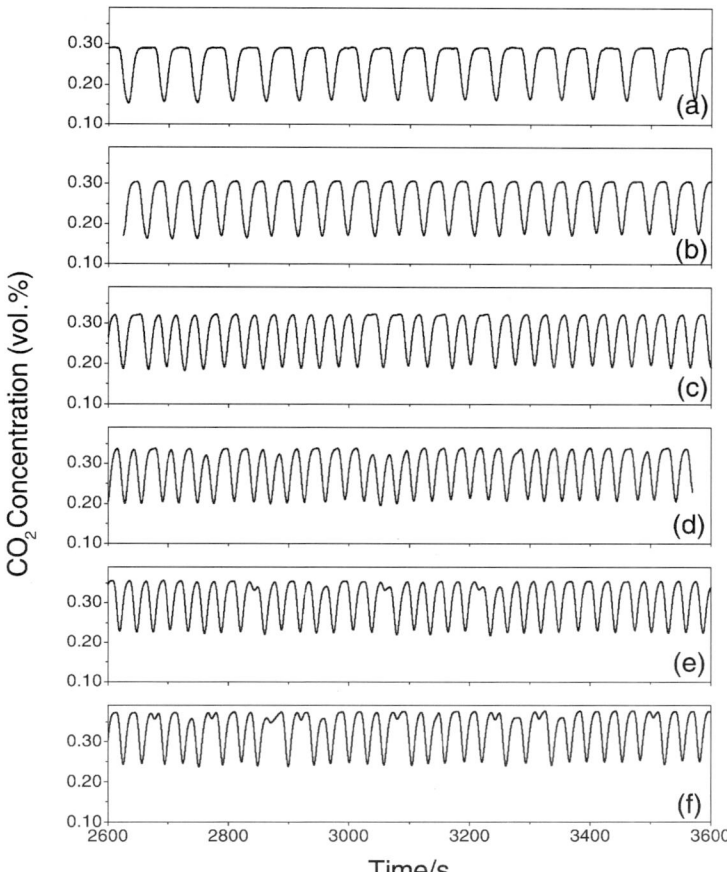

Fig. 8 Time series in the region of the transition to chaos for increasing CO concentration: (a) 0.3%, (b) 0.32%, (c) 0.34%, (d) 0.36%, (e) 0.38%, (f) 0.4%.

1. The Fourier power spectra

The Fourier power spectra presented in Fig. 9 demonstrate, that during the transition from regular to chaotic oscillations with increasing CO inlet concentration, no new peaks appear in the spectrum. This may be an indication of the type I intermittency route to chaos, during which the power spectrum changes continuously from a delta function to a broadened peak.[35] This is quite different from the behaviour during the transition *via* types II or type III intermittency, where new intrinsic frequencies arise in the power spectra.

Another signature of type I intermittency is provided by the specific characteristic of its corresponding power spectrum. One of the remarkable statistical properties of intermittency is the spectral power scaling law with frequency in the low-frequency region.[36] Fig. 10(a) shows the plot of $\log(P)$ in dependence on $\log(f/f_0)$ for 0.4% CO and 0.42% CO, where $P(f)$ is the spectral power and f_0 is the main frequency of the oscillations. It can be seen that in the region before the main peak the power spectrum may be approximated by the inverse power law $P(f) \approx 1/f^\kappa$, where κ is the scaling parameter, which becomes larger with increasing CO concentration towards the bifurcation point. Unfortunately, the time series for 0.36% and 0.38%CO inlet concentrations are too short and the values of κ for these time series are determined with a large error. However the tendency of κ to unity with decreasing CO inlet concentration from 0.42% to 0.36% was clearly identified, which represents one of the main signs of type I intermittency.

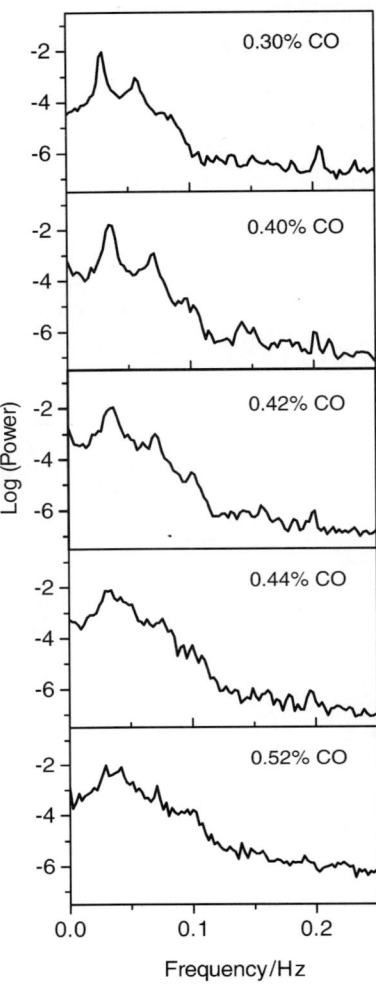

Fig. 9 Fourier power spectra for the corresponding time series in the region of the transition to chaos, presented in Fig. 8.

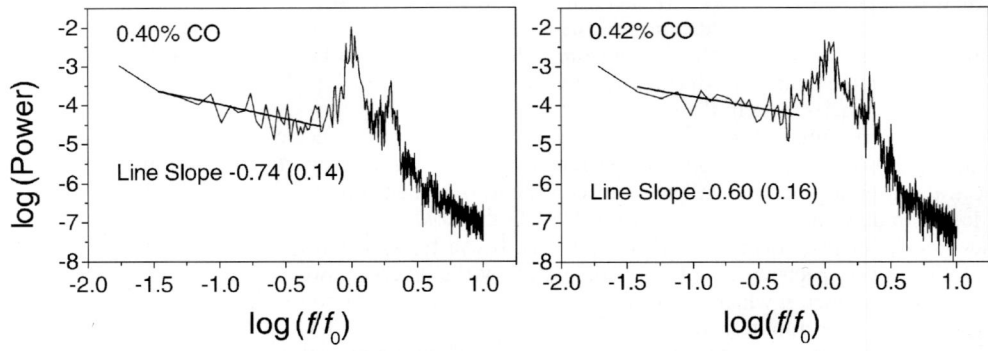

Fig. 10 Power spectra for 0.40% and 0.42% CO inlet concentration in log–log coordinates. The straight line obtained by linear regression is indicative of an inverse power law at low frequencies. The number in brackets is the square root deviation of the slope of the line.

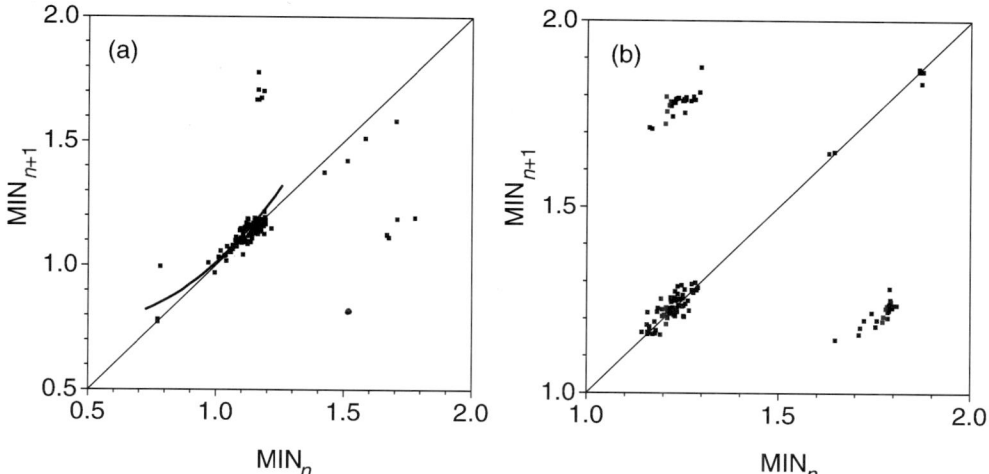

Fig. 11 Next minimum maps for two time series. The curved line in the left panel depicts the fitting polynomial $X_{n+1} = 1.03 - 0.99 X_n + 0.97 X_n^2$. (a) 0.40% CO; (b) 0.42% CO.

2. Next return map

The identification of type I intermittency can be derived from analysis of a one-dimensional reduced Poincare map. The theory of type I intermittency was developed on the quadratic map. The return map can be approximated by:

$$x_{n+1} = x_n + x_n^2 + \varepsilon$$

where ε is a control parameter which passes through zero at the onset of intermittency.

Fig. 11 demonstrates the next return map for 0.40% (a) and 0.42% (b) CO inlet concentrations. Data for the return map are obtained by taking the successive minima of the corresponding time series. Although some scatter of the data points can be seen, most points for the 0.4% CO time series are located around the fitting curve, which is obtained by the least-square-fitting procedure. The overall shape of the reconstructed map is almost the same as the theoretical model map for type I intermittency. This is a typical representation of a tangent bifurcation with a narrow channel between the reduced one-dimensional map and the diagonal corresponding to laminar motion. The groups of points far removed from the diagonal correspond to perturbations and reinjections. For 0.42% inlet CO concentration a similar picture is obtained. However, in this case the sets of points are less extended along the diagonal and more spread out far from the diagonal due to shorter intervals of the laminar motion. Unfortunately, owing to the insufficient length of the experimental time series for 0.36% and 0.38% CO near the tangent bifurcation it was not possible to work with the statistics for laminar intervals.

Discussion and conclusions

There are several conclusions that may be drawn from the present study. First the intermittency I type route to chaos has been identified for oscillating CO oxidation over a zeolite-supported Pd catalyst. The region of the CO inlet concentrations where this transition has been observed is very narrow, *i.e.*, only 0.04% CO. Nevertheless it was possible to determine the route to chaos from the variation of the Poincare section and the Fourier power spectra with CO concentration. The main indicator for the intermittency I type route to chaos is the tangent bifurcation due to which the system switches back and forth between a "ghost" periodic orbit and sudden bursts of chaotic behaviour. Another identification of the intermittency I route to chaos was obtained using the analysis of the power spectra, which are of the $1/f^\alpha$ type at low frequencies, where $\alpha \approx 1$ near the tangent bifurcation point.

Secondly, with increasing CO inlet concentration a variation of the value of the embedding dimension was determined. Table 1 demonstrates that for all time series $d_R \geqslant d_{SVD}$ and the change in these values with variation in the CO concentration indicates the evolution of the properties of the attractor. The study of intermittent chaos has been carried out using the power spectra, the shapes of the reconstructed attractors and the Lyapunov exponents. The results obtained demonstrate that we deal with a very complicated structure of chaos in a high dimensional system. The calculations of the embedding dimension gave a value $d_E \geqslant 10$. This value is much larger than the values which were obtained for chaotic oscillations during the CO oxidation over massive Pt foil by Razon et al.[4] ($d_E = 4$) and over a Pt(110) single crystal surface by Eiswirth et al.[5] ($d_E = 5$). The embedding dimension indicates an upper limit of state variables necessary to model the system. The large value, obtained in this study means that the supported catalyst has to be considered as a distributed system. This is an extension of the idea that internal diffusion limitations may play a role in the origin of the chaotic behaviour during the CO oxidation over Pd zeolite catalysts.[37]

It must be noted that the high embedding dimension and the relatively short time series available are the reason why not all the methods available for the non-linear analysis of chaotic time series could operate. It was not possible to obtain reasonable results by calculating the correlation dimension even for 360 000 data points. In this study the main periods of the oscillations are observed in the region between 30 and 100 s and in the case of a high embedding dimension the number of orbits is probably not large enough even for time series as long as 10 h. Another limitation of the analysis is the long term drift of the activity of the catalyst during such long measurements. In the case of nonstationary data one must be extremely careful when applying methods of nonlinear analysis. This problem is very important for the study of chaotic oscillations in heterogeneous catalytic systems, where the variation in the catalyst activity always occurs during long term measurements. Even in the case of single crystal studies it was quite difficult to choose the experimental conditions where the effect of long term drifts of the state of the catalyst was negligible.[5] This is even more difficult to do under high pressure conditions with supported catalysts. There are only few data available where the nonlinear analysis could be applied with confidence in the high pressure regime in the case of supported catalysts. Herzel et al.[38] could not follow the transition to chaos during the oxidation of methanol over Pd zeolite catalysts, but identified type-II intermittency chaos from the study of the histogram of the length of laminar phases. The authors could analyse the statistics of the laminar regions for 45 episodes due to the absence of any long term drift of the system. No visible changes in the dynamics occurred for up to 10 h. It must be noted that during the oxidation of methanol over a Pd zeolite catalyst the authors dealt with thermokinetic oscillations, which can be of much higher stability. These are nonisothermal oscillations, where the variations of the temperature together with the oxidation–reduction processes play an important role in their origin.

The mechanism of the kinetic oscillations observed in this study is closely connected with the periodic oxidation and reduction of Pd clusters embedded into a zeolite matrix. The properties of the oscillations are very sensitive to the preliminary pretreatment and the oxidation state of Pd. The point model, which was developed for the simulations of regular oscillations can be found in ref. 39. The results presented in this paper concerning the identification of the route from regular to chaotic oscillations and the determination of the embedding dimension of the strange attractor give additional important information for the improvement of the point model and for the development of a distributed model, describing the collective behaviour of many Pd particles in the zeolite crystallite. The results obtained will also be used for the discrimination of models. Work in this direction is currently in progress.

Acknowledgements

The authors acknowledge financial support from INTAS (grant 99-1882), Russian Fund of Fundamental Researches (grant 00-03-32125) and DFG (grant 436 RUS 113/435).

References

1 F. Schüth, B. E. Henry and L. D. Schmidt, *Adv. Catal.*, 1993, **39**, 51.
2 M. M. Slinko, N. I. Jaeger, "Oscillatory Heterogeneous Catalytic Systems", *Stud. Surf. Sci. Catal.*, 1994, **86**.

3   R. Imbihl and G. Ertl, *Chem. Rev.*, 1995, **95**, 697.
4   L. F. Razon, S.-M. Chang and R. A. Schmitz, *Chem. Eng. Sci.*, 1986, **41**, 1561.
5   M. Eiswirth, K. Krischer and G. Ertl, *Surf. Sci.*, 1988, **202**, 565.
6   M. Eiswirth, Th.-M. Kruel, G. Ertl and F. W. Schneider, *Chem. Phys. Lett.*, 1992, **193**, 305.
7   Hao Bai-Lin, *Chaos*, World Scientific, Singapore, 1984.
8   H. G. Schuster, *Deterministic Chaos*, VCH Verlag, Weinheim, 1988.
9   S. K. Scott, *Chemical Chaos*, Clarendon Press, Oxford, 1992.
10 R. H. Simoyi, A. Wolf and H. L. Swinney, *Phys. Rev. Lett.*, 1982, **49**, 245.
11 Y. Pomeau, J. C. Roux, A. Rossi, S. Bachelart and C. Vidal, *J. Phys., Lett.*, 1981, **42**, 271.
12 F. Argoul, A. Arneodo, P. Richetti and J. C. Roux, *J. Chem. Phys.*, 1987, **86**, 3325.
13 M. M. Slinko, N. I. Jaeger and P. Svensson, *J. Catal.*, 1989, **118**, 349.
14 D. Exner, N. I. Jaeger, K. Möller and G. Schulz-Ekloff, *J. Chem. Soc., Faraday Trans. 1*, 1982, **78**, 3537.
15 M. Liauw, PhD Thesis, Bremen University, 1994.
16 J.-P. Eckmann and D. Ruelle, *Rev. Mod. Phys.*, 1985, **57**, 617.
17 H. D. I. Abarbanel, R. Brown, J. J. Sidorowich and L. S. Tsimring, *Rev. Mod. Phys.*, 1993, **65**, 1331.
18 R. Gilmore, *Rev. Mod. Phys.*, 1998, **70**, 1455.
19 http://cmc.cs.msu.su/labs/mmp/CDP.html
20 N. H. Packard, J. P. Crutchfield, J. D. Farmer and R. S. Shaw, *Phys. Rev. Lett.*, 1980, **45**, 712.
21 F. Takens, in *Dynamical Systems and Turbulence*, ed. D. A. Rand and L. S. Young, Warwick, Springer, Heidelberg, 1981, p. 366.
22 R. Mañé and F. Takens, in *Dynamical Systems and Turbulence*, ed. D. A. Rand and L. S. Young, Springer, Heidelberg, 1981, p. 230.
23 H. Kantz and T. Schreiber, in *Cambridge Nonlinear Science Series*, Cambridge University Press, 1997, vol. 7.
24 A. M. Fraser and H. L. Swinney, *Phys. Rev. A*, 1986, **33**, 1134.
25 W. Liebert and H. G. Schuster, *Phys. Lett. A*, 1989, **142**, 107.
26 P. Grassberger and I. Procaccia, *Phys. Rev. Lett.*, 1983, **50**, 346.
27 M. B. Kennel, R. Brown and H. D. I. Abarbanel, *Phys. Rev. A*, 1992, **45**, 3403.
28 D. S. Broomhead and G. P. King, *Physica D*, 1986, **20**, 217.
29 A. Wolf, J. B. Swift, H. L. Swinney and J. A. Vastano, *Physica D*, 1985, **16**, 285.
30 M. T. Rosenstein, J. J. Collins and C. J. DeLuca, *Physica D*, 1993, **65**, 117.
31 Y. Pomeau and P. Manneville, *Commun. Math. Phys.*, 1980, **74**, 189.
32 P. Berge, Y. Pomeau and C. Vidal, *Order Within Chaos*, Wiley/Hermann, New York/Paris, 1986.
33 J. M. T. Thompson and H. B. Stewart, *Nonlinear Dynamics and Chaos*, Wiley, New York, 1986, p. 376.
34 G. J. de Vajcarcel, E. Roldan, V. Espinosa and R. Vilaseca, *Phys. Lett. A*, 1995, **206**, 359.
35 J. E. Hirsch, B. A. Huberman and D. J. Scalapino, *Phys. Rev. A*, 1982, **25**, 519.
36 H. Mori, B. C. So and S. Kuroki, *Physica D*, 1986, **21**, 355.
37 M. M. Slinko, N. I. Jaeger and P. Svensson, in *Proceedings of the International Conference on Unsteady State Processes in Catalysis*, ed. Yu. Matros, VSP, Utrecht, 1990, p. 415.
38 H. Herzel, P. Plath and P. Svensson, *Physica D*, 1991, **48**, 340.
39 M. M. Slin'ko, A. A. Ukharskii, N. V. Peskov and N. I. Jaeger, *Catal. Today*, 2001, **70**, in press.

General Discussion

Prof. Nicolis opened the discussion of Dr Feigin's paper: Your time series is non-stationary. A dynamical reconstruction using different slices of the series is therefore likely to yield different attractor dimensions and different lower bounds for the associated embedding dimensions. Under these conditions, how can you justify the use of a single, five variable model (eqn. (16)) throughout the analysis?

Dr Feigin responded: I would like to answer in three parts.
First, in this paper, we consider the situation which is typical for a natural system: the time series is *weakly* non-stationary and contains *no* bifurcations. Correspondingly, we have the same dimensions (correlation dimension, minimal embedding dimension, *etc.*) for all the slices of the initial time series. Note that throughout the analysis we know *nothing* about the system generating the analysed time series and use for reconstruction the dimensions calculated by the time series. More concretely calculated and used minimal embedding dimension equals *three*, and we have constructed a prognostic model as a two-order discrete time map.
Second, if the observed system has bifurcated *within* the time series analysed there is no problem to extract the bifurcations. This additional knowledge increases information about phase space structure as well about its modification and enlarges our possibilities in relation to both reconstruction, and prognosis.
Third, I would like to draw your attention to a fundamental limitation for prognosis of bifurcation: we can predict bifurcations of the more complex behaviour to the simpler one. In other words, the dimension of the attractor arising in the phase space as a result of bifurcation must not exceed the dimension of the attractor corresponding to the initial time series.

Prof. Westerhoff commented: Thank you for your fascinating presentation of your new method for qualitatively predicting the type of nonlinear dynamics a system will engage in. Intuitively, however, I think that there must be a limitation to the method. If the dynamics of the system during the sampling of the time series is insufficiently representative of its future dynamics, then it seems to me that your method must fail. This should even be so if the future of the system has lower dimensional dynamics than at present, when those dynamics are in entirely different degrees of freedom, or at different external parameter values. Can you, from a time series recorded in the summer, predict the weather in winter?

Dr Feigin replied: Thank you for the question. You are right, the limitations are connected not only with the number of degrees of freedom. The limitations may be formulated in a more general manner: we cannot predict system behaviour in a basin or subspace (of the phase space), which was not visited by the system *before* bifurcation. An example of the bifurcation after which the system "jumps" into a new basin/subspace is any catastrophic bifurcation, and every similar bifurcation is a fundamental (non-technical) horizon of our prognosis. I would like to underline that we cannot predict a type of system's behaviour *after* similar bifurcation, but we are able to predict this bifurcation and we demonstrate this in our paper.
As for *qualitative* (not *quantitative!*—this is beyond our approach) weather prediction, I cannot answer you unequivocally at this moment: I don't know whether season transition is a bifurcation or not. It seems to me, that the character of this transition depends on the system observed: it seems to be a quantitative change only if we measure, *e.g.*, air temperature, but probably is bifurcation for plant evolution.

Dr Slin'ko said: You report in the presented paper, that an advantage of the given work is the development of the universal algorithm, which makes it possible to create prognostic models for both low and high dimensional systems. Can you clarify why for the demonstration of the proposed algorithm you take as a data source the mathematical model eqn. (16), which produces low dimension chaos with correlation dimension 1.9 and embedding dimension 3?

Dr Feigin answered: Yes I can. The algorithm suggested is really much more universal than our first algorithm described earlier.[1] The choice of model for presentation of algorithm abilities within the scope of Faraday Discussion was dictated by our wish to demonstrate also a model of an atmospheric photochemical system. The only model having a relation to real atmospheric photochemistry and demonstrating chaotic behaviour is the model of a mesospheric photochemical system. Another advantage of this model from the point of view of this presentation is the possibility to compare a new algorithm with a previous one, and we specially paid attention to this in the paper. Naturally, we applied the new algorithm to more complex systems but limitation of the paper length made it impossible to include other examples in the paper.

For instant, we constructed a prognosis of bifurcation for the well-known system of Mackey and Glass[2] that models the process of leukocyte regeneration during chronic leukaemia illness. We used a time series generated by a chaotic attractor with a correlation dimension close to 3.5, and correctly predicated a sequence of the bifurcations and calculated their probability characteristics at a time interval having a duration of about three and half lengths the initial time series.

Note that computing time depends exponentially on dimensionality of the prognostic model, and so one can expect that the prognosis of bifurcations of very high dimensional systems will be computationally very expensive.

1 A. M. Feigin, Ya. I. Molkov, D. N. Mukhin and E. M. Loskutov, *Radiophys. Quantum Electron.*, 2001, **44**, 348.
2 M. C. Mackey and L. Glass, *Science*, 1977, **197**, 287.

Prof. Gáspár commented: You have demonstrated that the Earth's atmosphere in the mesospheric region, high up at 70–90 km, may show chaotic behaviour. I wonder if this chaos in Heaven may have an effect on the weather closer to Earth. I have also found it remarkable that by changing the relative concentration of water vapour in the ppm range brings about the chaotic behaviour. This means that by small perturbations of water we could control chaos in the mesosphere. Do you think it would be possible?

Dr Feigin responded: First of all I would like to say several words about the model of the mesospheric photochemical system which was used in the paper. This is the so-called box model, *i.e.*, it involves photochemical processes and neglects transport processes. The nonlinear dynamical properties of the box model are investigated in details in several papers (ref. 6, 7, 31, and 32 of the paper). Of course, water vapour mixing ratio is only one of the control parameters of the model. Others are altitude, air temperature, ratio of day to night times, duration of sunrise and sunset, and rates of photochemical reactions, which, for their part, depend on intensity of sun radiation in appropriate wavelength bands. Correspondingly, for control of the model behaviour we need to control all of these parameters.

Second, concerning the relation of this model to actual atmospheric processes. For applying the model to explain actual phenomenon we have to take into account transport processes. First of all we need to include in the model vertical eddy (turbulent) diffusion, since the characteristic time scale of this process in the mesopause region equals approximately 1 day and is close to the characteristic time of evolution of chemical species concentrations. Corresponding investigations were undertaken in our works with Dr Sonnemann from Leibniz-Institute for Atmospheric Physics (ref. 8 and 9 of the paper). We demonstrated that under real values of vertical eddy diffusion rate, more complex types of behaviour demonstrated by the box model are removed, and the single nonlinear regime may influence the actual mesospheric processes. This regime is two-day oscillations of chemical species concentrations, which represent a nonlinear response of mesospheric photochemistry on external periodic forcing—daily variations of sun radiation. So in respect to chaotic dynamics in the mesosphere, our conclusion is the following: there is no "chemical" chaos in the mesopause region. At the same time we have very sufficient grounds to suggest that namely nonlinear "chemical" oscillations are responsible for so-called summer quasi two-day oscillations of wind velocity and temperature. These oscillations have been observed during the last three decades (beginning from the first observations of mesopause processes) and are not explained satisfactory up to now.

Dr Tomlin said: The example you show in your paper has an embedding dimension of 3 and uses a Poincare' section which is a 2D surface. How can the method described in your paper be extended to higher embedding dimensions?

Dr Feigin answered: The universality of the described algorithm in respect to the system dimensionality is based on the type of functions used for reconstruction of the evolution operator of the observed system. These are artificial neural networks. First of all these functions are "ideal approximators", *i.e.*, they enable one to approximate nearly any single-valued function of an arbitrary number of variables with an arbitrary preset accuracy. We can increase approximation accuracy by a simple increase in the number of neurons, while the form of the function remains unchanged. Second, for increasing dimensionality of the prognostic model we need to increase the number of network inputs without changing the form of the function again. Of course, as I have mentioned, increasing dimensionality leads to an exponential rise of computing time but this is a technical, non-"ideological" problem.

Prof. Nicolis commented: The system under consideration is subjected to systematic variations of some of its parameters like, *e.g.*, those related to the seasonal cycle. Furthermore, it is most likely contaminated to a considerable extent by noise. How do you resolve these time scales and how do you separate the effects due to the noise? Did you construct correlation functions, Fourier spectra and similar indicators designed specifically for this purpose?

Dr Feigin replied: Let me separate your question in two parts: one is about different time scales and another concerning the noise.

I don't see any problem if system parameters oscillate periodically. If the corresponding period is close to the characteristic time scale of interest, these variations will be reflected in the evolution of prognostic model variables. If the period is significantly larger than the time scale of interest, it will be manifested in slow trends of prognostic model parameters. If, finally, the period is shorter than the characteristic time scale of interest, the corresponding fast variations of system parameters will influence the level of high frequency noises. It may be useful to analyse Fourier spectra for information about time scales manifesting themselves in the time series observed. First we can get a *preliminary* conclusion about the type of evolution (periodic, multiperiodic, chaotic or stochastic) corresponding to the time scales of our interest. Sometimes we can estimate from Fourier spectra a noise level, *etc.*, but of course, when we aim to construct a prognostic model and predict bifurcations of a dynamical system demonstrating chaotic behaviour, we have to use more adequate methods. Namely, we need to calculate dimensionality of an attractor and the minimal embedding dimension. We can also calculate a correlation function or mutual information function for determining appropriate time delay, *etc.*

As for the effect of noise, this is a serious problem. The algorithm suggested in the paper works well when the time series is practically noiseless and the main source of errors is discrepancy between the model and modelled system. This discrepancy is inevitable in our situation when all the information about dynamical properties of the system is concluded in the observed time series. The discrepancy leads to random errors of reconstruction of the evolution operator and therefore is equivalent to so-called dynamical noise. In reality, we almost always have the other situation when the main sources of noise are the different random effects on the measured signal. This is so-called noise of measurements, and in this situation the algorithm has to be generalised with respect to construction of distribution functions of the prognostic model parameters. This generalisation is now under consideration. I think that for "not too large" noises we have decided on "ideological" problems *via* the approach based on Bayes's theorems. Now we are trying to fabricate a fast enough numerical code generating any preset multidimensional distribution function. I hope that the generalised algorithm will enable one to predict bifurcations by noisy time series if the noise is not too large, *i.e.*, is of the order of 20–30% or less.

Dr Tomlin commented: In section 4.4 of your paper you discuss the possible practical application of prognostic models to a system determining ozone layer evolution at mid and low latitudes. Could you describe how your technique would be used in such models where other physical processes such as photolysis and transport processes are also present.

Dr Feigin replied: The principal distinctive feature of our approach is the following. We don't use any *a priori* information about the nature (chemistry, transport,...) of the process under investigation. We need to have at our disposal a long enough and not too noisy chaotic time series.

Prof. Scott said: The use of the chaotic response during parameter drift to provide information on the attractor in the system is very interesting. In other work, Showalter and colleagues[1] have used similar information to devise methods for controlling unstable periodic states coexisting with the chaotic attractor. In particular, they have in some cases imposed random perturbations on the system to explore the phase space more widely. I wonder if your techniques will also provide information that can be used to design control algorithms?

1   V. Petrov, E. Mihalink, S. K. Scott and K. Showalter, *Phys. Rev. E*, 1995, **51**, 3988.

Dr Feigin responded: Thank you for the question. Indeed, if we can predict qualitative behaviour of the system we can try to control it. There arises, however, a non-trivial problem. By predicting future bifurcations *exclusively* by observing the time series *without* any *a priori* knowledge about the nature and dynamical properties of the investigated dynamical system, we are unable to conclude any trends of what parameters of the system are responsible for the bifurcations predicted. This shortcoming of the prognostic model is retribution for its main advantage, that is, correct reproducing of both the structure of the system phase space, and the modification of this structure. As I see it, the only way to overcome the shortcoming is to fulfil the following: (i) To construct a "*first principles*" model of the investigated system; (ii) To verify this model *via* comparison with a corresponding *prognostic* model and correct the "first principal" model if its nonlinear dynamical properties differ from these properties of the prognostic model; (iii) To compare the prognostic model with the corrected "first principal" one and ascertain the trends of which parameters may lead the system to the predicted bifurcations.

After this we can think about controlling the future behaviour of the system.

Mr Li commented: In your paper, you reconstructed the phase space from the time series by the embedding method. In this method, you determined the delay time from the first minimum of the mutual information function and the embedding dimension by the 'false nearest neighbors' method. Certainly, there are many ways to determine delay time and embedding dimension, such as: the first zero of the auto-correlation function, average mutual information, singular value decomposition, saturation of system variants, and box counting methods, *etc.*

I would like to draw your attention to another method to determine delay time and embedding dimension. This method was proposed by Lipton and Dabke[1] and is based on a power spectra decay exponentially at high frequency when at least one Lyapunov exponent is positive, as presented by Sigeti.[2] In short, the power spectra of time series can be divided into three parts: the exponential fall-off asymptote, the polynomial fall-off asymptote, and the flat response asymptote. The asymptotes can have functions fitted to them using the method of least squares that enables the Kolmogorov–Sinai entropy and a sufficient embedding dimension to be estimated. The KS entropy is used to estimate an appropriate delay time.

From my experience,[3] the above method is a fast, reasonable and reliable method compared with other methods for reconstructing the phase space, even when the time series is noise corrupted and the number of samples is small.

1   J. M. Lipton, K. P. Dabke, *Phys. Lett. A*, 1996, **210**, 290.
2   D. E. Sigeti, *Physica D*, 1995, **82**, 136.
3   Y.-J. Li, Z.-S. Cai, Y.-N. Li, D.-H. Song, H. Song, B.-M. Xi, K.-Q. Ma, B.-X. Wu and X.-Z. Zhao, *Chin. Sci. Bull.* (int. edn.), 1998, **43**(17), 1447.

Dr Feigin answered: Thank you for this information. Our algorithm includes the reconstruction of the system evolution operator since a fast and reliable method for calculation of reconstruction parameters should help us to promote the algorithm to a more noisy time series generated by higher dimensional dynamic systems.

Prof. Sørensen opened the discussion of Dr Tomlin's paper: Has the method you describe in the paper been implemented as a fully automatic program which takes care of on the fly changes of the eigenvalues of the Jacobian by adjustments of the dimension of the slow manifold and the corresponding slow mode equations? When trajectories come close to bifurcation points for the fast reactions, the values of the eigenvalues may change rapidly.

Dr Tomlin responded: The method described determines the local dimension of the underlying slow manifold from the full set of ordinary differential equations. A secondary technique such as repro-modelling, the use of lookup tables for low dimensional manifolds or QSSA techniques must then be used to define the dynamic model for the slow modes. In the determination of the dimension the full Jacobian is calculated as part of the integration method and can therefore be easily determined for a range of trajectories in the full phase space. The extra calculation of the eigenvalues then allows the determination of the normalized distance coefficients in eqn. (30) of the paper. The integration method automatically performs extra Jacobian evaluations in regions where rapid changes in the Jacobian entries occur. The results clearly show that the local slow manifold dimension can change very quickly in such regions.

Prof. Scott asked: Presumably, any bifurcation phenomena of interest are likely to occur on the slow manifold rather than be associated with the fast relaxing modes?

Dr Tomlin replied: Complex behaviour such as oscillations are usually embedded in low dimensional manifolds. If only real eigenvlaues are considered to be associated with relaxation processes then the danger of losing such complex behaviour when confining the dynamics to a low dimensional manifold should not arise. In the present example we have chosen to retain the two complex modes in the system although they have negative real parts (see Table 7 of the paper). All modes which were assumed to have relaxed to the slow manifold were associated with eigenvalues with negative real and zero imaginary parts. The effect of reducing and/or lumping sets of chemical rate equations on the dynamical behaviour of the system is discussed in more detail in ref. 2 of the paper.

Multiple steady states pose a more interesting question. The method described in the paper will determine the local dimension of the slow manifold in a particular region of phase space through studying individual trajectories. The dimension of the slow manifold may be quite different in separated regions of phase space. If multiple stationary states exist then the definition of a single manifold containing more than one stationary state becomes more complex. This could provide an interesting area for further study although the situation is not common in typical tropospheric models.

Prof. Nicolis commented: It is not clear to me how nonlinear effects and, particularly, potential instabilities are accommodated in your analysis. In general the "elimination" of the fast modes is not tantamount to dropping them completely. One, rather, expresses them as a function of the slow modes through some sort of "equation of state" as, *e.g.*, in refs. 1 and 2. This allows one to close the system of equations of the slow modes, but the form of resulting evolution laws (including, possibly, their linear part) keeps track of this "elimination" procedure. As a result there may be a non-trivial contribution of nonlinearity to the time scales, entailing that the z variables are not clearly separated, except in the immediate vicinity of the steady state.

1   P. Coullet and E. Spiegel, *SIAM J. Appl. Math.*, 1983, **43**, 776.
2   W. Wasow, *Asymptotic Expansions for Ordinary Differential Equations*, Interscience, New York, 1965, see especially part devoted to Tikhonov's theorem.

Dr Tomlin responded: The method described in the paper does not actually remove the fast modes at this stage but merely defines which modes could either be removed or could be described by some other method such as an equation of state as you have suggested. Once the dimension of the slow manifold has been determined it is then possible to use several methods to describe the dynamics on the slow manifold and of the fast modes. The most common method of application is the use of the quasi-steady state approximation (ref. 4 of the paper) or QSSA where the concentrations of the species relating to the fast modes are described as functions of the slow species using the equation: $f(c,k) = 0$. Since these equations are often nonlinear and therefore difficult to solve algebraically, several alternative methods have been developed. The repro-modelling technique (ref. 9 of the paper) uses fitted polynomial difference equations to describe the dynamics of the slow variables over a fixed time interval. It is also possible using this method to use fitted polynomials that express the fast variables as functions of the slow ones, so that it is possible to retain the fast variables in the model at low computational cost. The standard ILDM method described by Maas and Pope (ref. 5 of the paper) uses lookup tables to describe quantities on the slow manifold so that the concentrations of the fast variables can be determined as functions of the look-up table quantities (essentially relating the fast variables to the slow ones). It is possible (as shown by the diurnal variation in slow manifold dimension for the butane system here) that modes may change from being collapsed to being off the manifold over time and therefore that either a conservative estimate should be made for the manifold dimension or that manifolds of varying dimension should be used in the model (a more difficult task). For this reason we have suggested the use of 11 variables (*i.e.* the maximum number given by the diurnal simulations) to describe the butane oxidation system here. It can be shown that where the dimension drops to 3 that the three slow variables are contained in the set of 11 shown in Table 8 of the paper.

It could be possible that for larger perturbations from the slow manifold nonlinearities may be important. However, the method is essentially testing when a trajectory has relaxed close to the manifold and therefore inaccuracies in the distance measure at larger distances will not significantly affect the method.

Dr Boissonade asked: How do you apply this approach to the case where the manifold is strongly folded? How can you define the "distance" to the manifold in this case where there are several branches?

Dr Tomlin answered: The distance defined is a "local" measure, *i.e.*, at a point on an individual trajectory in phase space of dimension n it defines the distance of the trajectory from an $(n-1)$ dimensional manifold where mode i is assumed to have collapsed. The distance is defined in the direction of the left eigenvector associated with mode i. There is no reason why this distance measure should not be applied to a folded manifold, and indeed, examples of 1 and 2 dimensional folded slow manifolds are shown in Tomlin *et al.* (ref. 3 of the paper) and Maas and Pope (ref. 5 of the paper). It should be noted however that in the region of the fold the range of applicability of the local method may be less than for more simple regions. The method may become more inaccurate at larger distances from the manifold. However, if a small threshold value is chosen for the collapse of a mode then this does not pose significant problems.

Dr Feigin commented: As far as I have understood the goal of this work is estimation of "current dimensionality" of the slow manifolds. In other words the authors study *quantitatively* transient behaviour of stiff systems. In this case, separation of variables according to their characteristic times is, generally speaking, impossible since at least a part of the variables is characterised by several time scales. In particular, for this reason authors use a new set of variables for estimation of a distance in the phase space between current state of the system and "current" slow manifolds. The new variables are "non-natural" but every new variable relaxes with the single time scale.

Dr Tomlin replied: This is correct. Traditional methods such as the QSSA have tended to identify time scales with chemical species when they are not strictly speaking equivalent. In section 6.1 we have explored this point and have shown that for the 6 variable system, of the three radical species O^1D, OH and HO_2, only O^1D reacts according to a single time scale. OH and HO_2 are coupled to each other by differing degrees depending on the concentrations of the other variables. In this case however, the identification of a "group" of QSSA species is appropriate, since although they are coupled to each other they do not couple to the "group" of species associated with the slow time scales. Other methods such as the CSP (computational singular perturbation)[1] method have also used a locally linear change of basis variables to explore system time scales.

1   S. H. Lam and D. A. Goussis, *Int. J. Chem. Kinet.*, 1994, **26**, 461.

Dr Satnoianu said: It is of interest to compare the present approach with a rigorous method called "the inertial manifold theory". The comparison could be helpful in trying to extend the presented method to more general systems. Another outcome of this will be an estimation of the dimensionality of the slow manifold.

Dr Tomlin responded: Fully nonlinear perturbation techniques such as those used in inertial manifold methods are certainly more robust than the method presented here which is based on locally linear techniques. The drawback of such approaches however is that they often require complex algebraic manipulation in order to determine inertial forms for the reduced equation set. Li *et al.*[1] and Tomlin *et al.*[2] applied such methods to a nonlinear equation system describing hydrogen combustion where a second order inertial form for the reduced equation set was determined. The resulting reduced equation system was of a higher accuracy than that achieved using a simple first order method. Such methods would, however, be extremely difficult to apply to a high dimensional nonlinear system such as the butane oxidation scheme described here, even with the use of computer based algebraic manipulation packages. The approach suggested here is therefore to use approximate methods for the determination of the dimension of the reduced system based on a small tolerated error, followed by the use of repro-modelling methods or look-up tables to describe the reduced chemical system for use in engineering applications. Although perhaps not robust for all general types of ordinary differential equation systems, the approach provides a method that can be applied to a wide class of chemical sub-models found in large-scale reactive flow applications.

1   G. Li, A. S. Tomlin, H. Rabitz and J. Toth, *J. Chem. Phys.*, 1994, **101**(2), 1171.
2   A. S. Tomlin, G. Li, H. Rabitz and J. Toth, *J. Chem. Phys.*, 1994, **101**(2), 1188.

Prof. Scott asked: Is it possible to imagine a computational algorithm in which the dimensional of the system is varied dynamically during the computations, switching between different size look-up tables or fitted polynomials, in response to the evolution of the system in the phase plane?

Dr Tomlin replied: It is possible in practice to apply reduced models of varying dimension throughout a large scale simulation. For example, using the repro-modelling approach for the butane oxidation system described in section 6.2, it would be possible to switch between models of dimension 3 and 11. For the 3 dimensional model used during night-time simulations fitted polynomial equations would be determined describing the 3 slow variables as functions of the concentrations of these 3 variables and solar angle at the previous time point. In addition, fitted equations could be derived for the remaining 8 variables as functions of the 3 slow variables at corresponding time points (approximate equations of state). Such additional equations could therefore be used to derive the concentrations of these 8 variables before they are restored to the full 11 variable model. The fitting of such functions could be separated easily into regions of low and high photolysis in this case. If approximate mass conservation is required from the model then such additional fitted functions should be described for all fast and conserved variables.

The difficulty with such an approach is that the dimension of the slow manifold is determined from the full Jacobian and therefore it would be necessary to store the local dimension for use in the reduced model. For example in the butane oxidation case, the dimension would be stored in an 11 dimensional look-up table representing possible concentrations of the 11 key variables. Searching in such a table would have a computational overhead which might out-weigh the use of a more highly reduced model for certain parts of the run.

Prof. Scott asked: In exothermic reactions, temperature inevitably arises with a strong nonlinear dependence. Is it still possible for the eigenvector approach to lead to only linear combinations of the dynamical variables (including temperature)?

Dr Tomlin answered: For non-isothermal problems, such as those which arise in combustion, the energy equation becomes supplementary to the set of species rate equations in the problem. If one treats temperature as an extra variable equivalent to the concentrations, then the Jacobian will contain an extra column containing the terms $\partial f_{(j+1)}/\partial T$ and an extra row relating to partial derivative of the energy equation with respect to each species and temperature. The use of approximate linear perturbation methods should not be affected by the different form of the nonlinearities. If the Jacobian is supplied analytically then the presence of temperature as an extra variable poses some extra work in forming the Jacobian elements. Many standard integration packages, e.g., CHEMKIN-II[1] however, use approximate finite difference methods to form the Jacobian and for such methods there is therefore no need to distinguish between nonlinearities in reaction terms and those arising from Arrhenius parameters. The use of finite difference methods will however affect the accuracy of the method in comparison to the use of an analytical Jacobian. Because of these extra terms it is possible that the energy equation may influence the normalised distance coefficients depending on the size of the relevant eigenvector elements in eqn. (30) of the paper. It may be interesting to study the influence of nonlinear temperature effects in future applications.

A slightly different approach, which is that adopted by Maas and Pop (ref. 5 of the paper), is to consider phase space as (enthalpy + concentrations) instead of (temparature + concentrations). These are equivalent representations. This has the advantage that for adiabatic systems enthalpy does not change and changes little near, for example, adiabatic explosions. It is interesting that in this case enthalpy change is described by the heat loss equations instead of the chemical kinetic equations. Therefore, the timescale of enthalpy change is dicated by physical processes only.

The author would like to note helpful discussions with Dr Tamás Turányi with regards to this comment.

1   R. J. Kee, F. M. Rupley and J. A. Miller, *A FORTRAN Chemical Kinetics Package for the Analysis of Gas Phase Chemical Kinetics*, SANDIA Report SAND-89-8009B, 1991.

Dr Schreiber commented: A remark on the selection of the dimension of the submanifold: it should be determined by a characteristic time scale of a process coupled to the original system, for example, by a period of a sinusoidal forcing term.

Dr Tomlin responded: It is true than when selecting suitable dimension of a slow manifold for a chemical sub-model within a larger coupled model one should take account of the relative timescales of the chemical *vs.* coupled processes. For typical reactive flow problems such as those found in atmospheric chemistry and combustion, the coupled processes are likely to be transport processes such as molecular diffusion, advection or turbulent mixing. To a certain extent therefore the threshold chosen for the determination of the slow manifold will depend on how such transport processes are described and their characteristic timescales. The current method shows however that it is not just the characteristic timescale as determined by λ_i that determines the accuracy of a model based on a slow manifold but also the right hand sides of the equations who's contributions are determined by $X^T_{1,\lambda_i} f$ as shown in eqn. (30) of the paper.

Dr Boissonade asked: What can be the relation of your technique with the standard numerical solutions for stiff ODE's which eliminate the fast variables at every time step (with calculation of the Jacobian) and automatically adapt themselves to the situation at the given point in the evolution?

Dr Tomlin replied: There are several numerical techniques for stiff systems that make use of the existence of fast variables in the system. Hesstvedt *et al.*,[1] for example, use the quasi-steady-state approximation (QSSA) for fast species within solvers for atmospheric chemistry simulations. The speed-up provided by such methods is however marginal when compared to other stiff solvers such as the Rosenbrock scheme with often significant loss of accuracy.[2-4] Repro-modelling methods on the other hand replace repeated integration of the chemical sub-model within reactive flow models by the simple algebraic calculation of fitted polynomial difference equations. Although many simulations must be carried out in order to provide the fitting data, once the polynomials have been determined they can show speed ups of up to a factor of 100–1000 over even stiff integrators (refs. 6 and 9 of the paper). They therefore provide highly computationally efficient representations of reduced chemical models that can be used in complex 3D models where many repeated calculations are required, *e.g.*, air quality models.

1 E. Hesstvedt, O. Hov and I. S. A. Isaksen, *Int. J. Chem. Kinet.*, 1978, **10**, 971.
2 J. G. Verwer and M. van Loon, *J. Comput. Phys.*, 1994, **113**, 347.
3 A. Sandu, J. G. Verwer, M. van Loon, G. R. Carmichael, F. A. Potra, D. Dabdub and J. H. Seinfeld, 1997, **31**, 3151.
4 A. Sandu, J. G. Verwer, J. G. Blom, E. J. Spee and G. R. Carmichael, *Atmos. Environ.*, 1997, **31**, 3459.

Prof. Scott opened the discussion of Prof. Griffiths' paper: For the "asymmetric case" the "fuel supply" p_0 term does not occur in the main domain of the reaction zone. The resulting eqns. (9) and (10) then have the form of the classic mass and heat balance in a porous catalytic particle, albeit with non-standard boundary conditions in the present case. In the catalytic particle case, oscillatory and other instabilities arise separately in the case that the mass and heat diffusivities are unequal (*i.e.* non-unit Lewis number). That could seem to be a case of interest in your system as the significant species is presumably the H-atom whilst the thermal diffusivity will be determined by Cl_2. Thus we might expect $D_H \gg D_T$. Have you extended your calculations into these regimes?

Prof. Griffiths replied: We are interested to establish whether or not other instabilities can exist. However, the calculations performed to date on the asymmetric supply case have been limited to the two variable model using a global rate expression for reactant consumption. This was derived from a stationary state approximation applied to the H and Cl atom concentrations. In developments from this work we propose to investigate the behaviour of the system with more detailed chemistry included, specifically taking the intermediates into account. For the reasons that you suggest, variations in Lewis number will then have relevance in the calculations. There is one facet of variations in Lewis number of which we are already aware, albeit not connected with the asymmetric supply case. That is, although the feedback in the Sal'nikov system is strictly thermal in origin, species diffusion in a spherical system can play a part in controlling the onset of oscillations through a "preferential" rate of diffusion of species A from the relatively unreactive periphery to the highly reactive central zone.[1]

1 R. Fairlie and J. F. Griffiths, *Math. Comput. Modell.*, 2001, in press.

Prof. Showalter asked: Have experimentalists observed cellular flames in microgravity conditions?

Prof. Griffiths responded: I do not believe so, at least not in the context of stabilised flames. However, Prof. Ronney, in a review of microgravity combustion experiments[1] remarks on the observation of "splitting cellular flames" associated with experiments to establish stable flame balls.

1 P. D. Ronney, *Twenty-Seventh Symposium (International) on Combustion*, The Combustion Institute, Pittsburgh, 1998, p. 2485.

Prof. Scott commented: The butane–oxygen flame can show cellular instabilities for compositions close to the flammability limits. Are the spherical fronts you describe obtained for compositions well away from this limit?

Prof. Griffiths answered: Our initial approach to this problem was guided by your theoretical and numerical study of the cellular instabilities of propagating flames that was based on the Sal'nikov model,[1] so we share a common interest. There is also some connection to your previous comment, insofar that these instabilities were linked to mixtures with Lewis number greater than unity ($Le = D_T/D_A$). As noted elsewhere in this discussion, we have yet to investigate the potential for the existence of interesting or different structures when Le is varied and, of course, it requires multi-dimensional modelling (which is in the planning stage). One fundamental difference from your theoretical investigation[1] is that, once one moves to conditions for spontaneous combustion as opposed to induced ignition, flammability limits as such (the composition range within which a propagating flame can be sustained) ceases to have significance. Even so, combustion fronts which develop spontaneously in mixtures for which Le is varied (*e.g.* by including inert diluents, such as helium, or by changing the reactant proportions) may show different sorts of structure/instability. To refer to the induced flame problem, a hydrogen flame can be supported in chlorine,[2] which may yield different sorts of structures, but I do not know the range of compositions over which this flame can be sustained.

1 S. K. Scott, J. Wang and K. Showalter, *J. Chem. Soc., Faraday Trans.*, 1997, **93**, 1733.
2 A. G. Gaydon and H. G. Wolfhard, *Flames, their Structure, Radiation and Temperature*, Chapman and Hall, London, 3rd edn., 1970.

Prof. Merkin said: A question about the spatially uniform model: were the numerical calculations done assuming that $k_1 t_{ch} = 0$ in eqn. (7) as they show uniform amplitudes? Previous calculations and analysis on a related model,[1] essentially taking $k_1 t_{ch}$ small but non-zero, show a slow change in period and amplitude of the oscillations. Similar behaviour might be seen here. In the spatially distributed model with concentration input on one boundary and heat loss in other, are there any steady, spatially inhomogeneous solutions? Do the oscillations arise from instabilities in these steady solutions?

1 J. H. Merkin, D. J. Needham and S. K. Scott, *Proc. R. Soc. London, Ser. A*, 1986, **406**, 299.

Prof. Griffiths replied: We assumed the pool chemical approximation in the calculations presented here so that we could distinguish the region of sustained oscillations from damping to a stable focus. We observe behaviour just as you describe it when the assumption is not made, as shown elsewhere.[1] Allowance for reactant consumption is more in keeping with the experiment because, in the experiment, the external reservoir was of finite volume and so the pressure of the reactant fell as it flowed into the reactor. This diminished the supply rate (to a good approximation as a first order decay), and the flow stopped when the pressures in the two volumes was equalised. We do not know what additional complexities of the inhomogeneous solutions might arise with these boundary conditions, and it is possible that the scope within which we have explored the asymmetric boundary conditions so far is too restrictive. We propose to explore the problem in two spatial dimensions. I am assuming that your emphasis is on "instability" in a physicochemical context. Nevertheless, this is an opportunity to confirm that the numerical procedure itself is robust. Tests were made using different tolerances, time steps and mesh points to ensure the precision of the computed solutions.

1   R. Fairlie and J. F. Griffiths, *Math. Comput. Model.*, 2001, in press.

Prof. Nicolis commented: The possibility to realize a purely diffusive mass and heat transport by eliminating bulk flow effects offers the perspective to observe in combustion systems some of the behaviors discovered in the last years on reaction–diffusion systems in isothermal, open non-stirred reactors (ref. 1 and the papers by Boissonade *et al.* and Dewel *et al.*). In addition to the oscillations reported in your work one could inquire about Turing patterns and interference effects leading, *e.g.*, to localized structures and to complex (non-planar or non-cylindrically symmetric) fronts. Do you believe that such effects can indeed be expected and are they accessible to observation by the present experimental methods?

1   See, *e.g.*, A. De Wit, *Adv. Chem. Phys.*, 1999, **109**, 435.

Prof. Griffiths responded: The features to which you refer are of considerable interest to workers within the fields of interest covered by this Faraday Discussion meeting and it would be our wish to maintain some involvement through developments of our present work. It is premature to speculate on the types of response that might be possible in gaseous thermokinetic systems, but the opportunity to perform "numerical experiments" certainly gives us the chance to explore whether or not the criteria for Turing patterns and related phenomena might be satisfied by reaction–diffusion experiments. Whether or not they could acquire a high enough priority and the necessary funding for study in the International Space Station is another matter altogether!

Prof. Hanke commented: The minimum gravity in parabolic flight is $10^{-2} g$, thus a zero gravity situation is not really given.

Prof. Griffiths answered: You are correct. However, one reason for us becoming engaged in these types of calculations is that there are proposals within NASA for the parabolic flight experiments on non-isothermal gas-phase oscillatory reactions to be upgraded to the International Space Station (ISS) combustion programme. Consequently it is important to establish how the response of these systems is likely to be affected by changing gravitational effects, following from variations in buoyancy and convection. Apart from a fundamental interest in "spatial structure" arising in these combustion systems, one of our major goals is to model the response to variations in convection so that we may give some guidance in the design of the new experiments. The "zero gravity" case is one extreme, and is representative of conditions in the ISS.

Prof. Scott commented: In combustion systems under microgravity, stable flameballs exist. In simple autocatalytical reactions the analogies may also exist. In a spherical coordinate system, autocatalytic reactions with removal of the autocatalyst stable structures similar to flameballs appear.

Dr Tóth replied: We are currently studying the existence and stability of steady front profiles numerically for simple autocatalytic reactions with various orders of autocatalysis. The resultant structure indeed can be stabilized by including a slow decomposition of the autocatalyst. These isothermal analogues of stable flameballs act as spherical sinks for the reactant and sources for the product.

Dr Münster opened the discussion of Dr Krischer's paper: In your contribution you showed space–time plots of the electric potential displaying an impressive variety of complex phenomena.

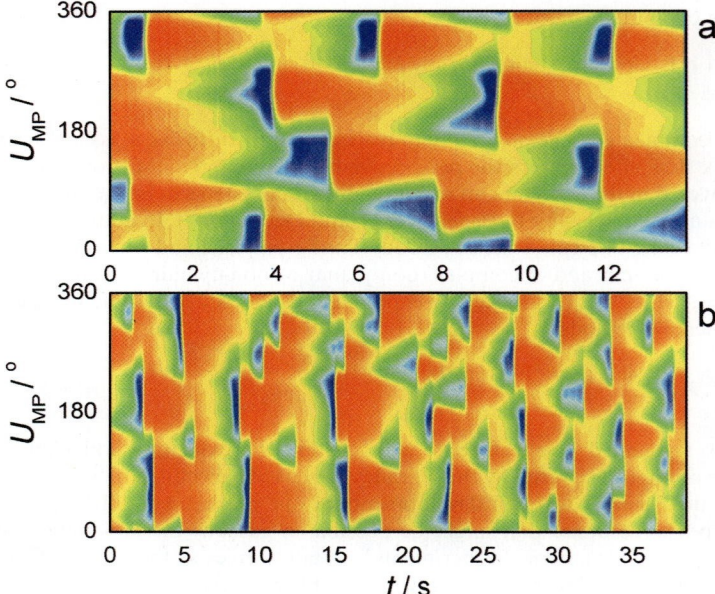

Fig. 1 Position–time plots of the local double layer potential as measured with the potential micro-probe ($U_{MP}(x, t)$). The external applied potential was increased from (a) to (b). Experimental conditions identical to that shown in the paper by Grauel *et al.* (this Discussion, pages 165–178), but with copper concentration of 0.01 mM.

In my opinion the Karhunen–Loeve decomposition technique might be a useful tool to quantify the effect of the coupling in your experiments. Have you applied the latter or any related techniques to your data and did you find it to be useful?

Dr Krischer responded: Yes, we indeed did a Karhunen–Loeve (KL) decomposition of the data shown in the paper. However, if you look at Fig. 6 of the paper, there are some data sets *e.g.*, Fig. 6b, c and e that have comparatively simple dynamics. In these cases we found that the dynamics can be described by the superposition of 2 or 3 dominating modes, which one would have expected beforehand. In the case of the more complicated data of Fig. 6a or d, the measured spatio-temporal time series turned out to be too short to extract the minimal number of spatial modes involved in the dynamics. However, we have other sets of data where the KL decomposition was extremely useful. I can show you two such spatio-temporal time series (Fig. 1). Successive images were observed for an increasing value of the external voltage. Performing a KL decomposition with these data reveals nicely how more and more spatial modes with higher wave numbers become excited when increasing U.

Prof. Noszticzius commented: In some chemical systems,[1,2] while changing a parameter, excitability of a stable steady state can be observed before the appearance of finite amplitude oscillations. Such experiments can help to determine the type of the bifurcation leading to limit cycle oscillations. In your electrochemical system a main parameter was the potential of the rotating Pt ring-electrode. Could you observe an excitable behaviour in a potential range close to the oscillatory state?

1 Z. Noszticzius, P. Stirling and M. Wittmann, *J. Phys. Chem.*, 1985, **89**, 4914.
2 Z. Noszticzius, M. Wittmann and P. Stirling, *J. Chem. Phys.*, 1987, **86**, 1922.

Dr Krischer replied: We have not yet performed these experiments, but they would be certainly helpful for a further characterisation of the bifurcation leading to oscillations at low values of the potential. We hope to be able to perform such experiments soon. Thank you for pointing out this aspect.

Prof. Dewel said: The temporal behaviours you present in Fig. 3 of the paper are very regular despite the fact that your limit cycle is Benjamin–Feir (BF) unstable. BF instability indeed generally generates a turbulent behaviour (phase or amplitude turbulence) as illustrated by the numerous integrations of the complex Ginzburg Landau equation (CGLE). CGLE does not properly describe the relaxation oscillations you observe. A better approach would be the recent theory developed in Nice[1] to study the effects of spatial inhomogenities on almost homoclinic limit cycles.

1   M. Argentina, P. Coullet and E. Risler, *Phys. Rev. Lett.*, 2001, **86**, 807.

Dr Krischer answered: We attribute the regular limit cycle oscillations we observe in Fig. 3 of the paper to a finite size effect. We are aware that the CGLE does not appropriately describe the dynamics in our case. We classified the limit-cycles as BF unstable because the point-like (or homogeneous) system possesses a stable limit cycle that is destabilised by the spatial coupling, just as it is the case in BF-unstable systems. However, I do not know the recent work by Argentina *et al.*, and I am grateful to you for this hint. If a differentiation between the destabilisation of a homogeneous limit cycle by spatial coupling should be made depending on, *e.g.*, the nature of the bifurcation that might be close to the limit cycle or other characteristics of the system, we might have to revise our classification.

Prof. Marek commented: You have presented examples of spatiotemporal patterns recorded within relatively short time intervals. How robust are these patterns, did you follow repeatedly the evolution of such patterns starting from different initial conditions and their transient behaviour after perturbations?

Dr Krischer responded: We performed the experiments with two ring-electrodes and observed the same succession of patterns. The absolute values of the applied voltage for which a certain pattern was observed differed slightly, though. The patterns were stable for a longer time than reproduced in the paper, however, any catalytic surface is subject to slow variations. Thus, on a time scale which is much longer than the time intervals shown in this work, we observed a drift, *i.e.*, a change in amplitude, for example, or even the evolution from one pattern to another one.

Prof. Gáspár commented: You mention in your paper that numerical studies show that reaction–diffusion systems also possess a Benjamin–Feir unstable region. You have shown the existence of such instability in an electrochemical system using a ring electrode. Does it imply that we should use a ring reactor in a chemical setting as well? What characteristics should a chemical system have to show this instability?

Dr Krischer replied: The geometry of the reactor is irrelevant as long as the reactor is larger than the intrinsic wavelength of the system. Whether a chemical system is Benjamin–Feir unstable or not depends purely on the local dynamics. However, there is no easy way to determine for which parameter values the system is Benjamin–Feir unstable. The only possibilities I am aware of are either to perform a simulation of the RD-system or to transform the equations into amplitude equations, which might be very tedious. Then, however, the parameter combinations that correspond to a negative phase diffusion coefficient can be easily determined.

Prof. Müller commented: It is known that in global feedback studies of some reaction–diffusion patterns the sign of the feedback signal influences the dynamic behaviour significantly. For instance, both for positive and for negative sign, stabilizing and destabilizing effects have been recently demonstrated in the case of light-sensitive Belousov–Zhabotinsky waves (see ref. 1). What is the role of the feedback sign in your experiments?

1   O. Kheowan, C. K. Chan, V. S. Zykov, O. Rangsiman and S. C. Müller, *Phys. Rev. E*, 2001, **64**, 035201-1.

Dr Krischer responded: The pulses, target patterns and most likely also the asymmetric target patterns we observe in our experiments are linked to a bifurcation of the homogeneous *fixed point* to spatio-temporal patterns. Since the global coupling acts on the double layer potential, *i.e.*, the activator variable, such a bifurcation is only possible with a negative, *i.e.*, destabilizing feedback. (If the feedback acts on the inhibitor variable, things are slightly more complicated.[1])

1   K. Krischer, N. Mazouz and G. Flätgen, *J. Phys. Chem. B*, 2000, **104**, 7545.

Dr Schreiber said: One of the measured irregular spatiotemporal patterns with distinct triangular structure resembles those known to occur in a transition zone between front and pulse waves in reaction–diffusion systems, such as the Gray–Scott model, and involves a heteroclinic connection between multiple steady states. Could your observation not be explained in this way?

Dr Krischer replied: In our system we can exclude that a second stationary state is involved in the dynamics; thus the system does not exhibit fronts in parameter regions close by. This seems to be a strong hint that the patterns we observed emerge in a different scenario.

Prof. Sørensen commented: Your Fig. 4f shows small amplitude simple oscillations which indicates that the wave possibly is described by amplitude equations such as the complex Ginzburg–Landau equation. This hypothesis might be tested by monitoring the response to small perturbations of the waves for example *via* the global feedback and comparing the measured response with the solutions to the Ginzburg–Landau equation.

Dr Krischer answered: This is a nice suggestion. In the parameter region in which the data shown in Fig. 4 were measured, it should be possible to describe the dynamics by amplitude equations and it is certainly worth verifying this by the method you propose.

Dr Krischer commented further: As you bring the issue up, I wonder whether you also have a suggestion how to proceed in the case of the data shown in Fig. 3 in order to obtain a deeper insight in the underlying dynamics. The limit cycle shown in Fig. 3 was most likely born in a homoclinic bifurcation and it is spatially inhomogeneous. Since we are dealing with locally oscillatory dynamics that are spatially destabilized by the spatial coupling, we classified the system as Benjamin–Feir unstable. However, we do not know how to determine from the data a quantity that would correspond, for example, to a formal phase diffusion coefficient (if this is still a good measure in this case).

Prof. Sørensen said: The amplitude of oscillations in a limit cycle arising from a homoclinic bifurcation is generally large and the convergence of any state to the limit cycle fast. When spatial dependence is taken into account the system corresponds to a continuous field of oscillators where all local states remain close to the limit cycle. Any scalar function on the limit cycle having a range in the interval of 0 to 2π can be used as a phase. A universal equation for the phase which is analogous to the complex Ginzburg–Landau equation for Hopf systems is called the nonlinear phase diffusion equation.[1] The computation of the parameters in this equation from a kinetic model involves evaluations of integrals along the limit cycle. This is much more difficult than computing parameters in the complex Ginzburg–Landau equation and reduces the practical usefulness of the nonlinear phase diffusion equation.

1   Y. Kuramoto, *Chemical Oscillations Waves and Turbulence*, Springer-Verlag, Berlin, 1984.

Prof. Nicolis opened the discussion of Dr Slin'ko's paper: The time series reported in your work do not fit to the idea of a succession of long quiescent periods interrupted by short-living bursts usually associated to intermittency. They look, rather, like weakly chaotic multi-modal oscillations which one tends to associate to homoclinic chaos. The evidence coming from the first return map is not conclusive, since one needs to extrapolate it to the whole range of values of the relevant variable and to complete the branch near the bissectrix by another branch accounting for reinjection.

Dr Slin'ko responded: The small amplitude oscillations, which appear on the top of some waves and are visible in Figs. 8e and 8f of the paper, can be considered from two points of view: (i) as the interruptions between laminar motions and (ii) as close approaches to the saddle point. The first case is related to the intermittency type behaviour and the second case can be associated with the homoclinic chaos. We could not obtain any other features of the homoclinic chaos from the available experimental data. On the other side, in spite of the non-standard picture of the experimental time series, the intermittency was confirmed by specific features of the Fourier spectra and Poincare sections. We agree that the evidence coming from the first return map, presented in Fig. 11 of the paper is not conclusive. Due to the short length of the time series the number of reinjection points is not large and hence the corresponding reinjection branch in Fig. 11a is weakly visible. However in Fig. 11b two sets of points in the left-upper and right-lower corners of the graph are evident and may be associated with the reinjection mechanism.

Dr Münster said: The nearest neighbour method to estimate the embedding dimension has been applied to relatively high dimensional attractors in the study of hydrodynamic phenomena. In the paper you demonstrated that this method could not be applied to the data obtained in your work. Is there an explanation for the failure of the nearest neighbour method in the case you reported?

Dr Slin'ko replied: The general idea of the *False Nearest Neighbours* (FNN) method is to increase the dimension d of embedding space till the orbit projection onto this space will have no self-crossing point.[1] If by passing from dimension d to dimension $d+1$, two points move away from each other, than in dimension d we have false nearest neighbours. When increasing d, starting with small values, one can estimate the minimal embedding dimension by finding no more false neighbours. One of the problems which make the method difficult to apply is that the instability of the chaotic dynamics sometimes makes it difficult to decide whether neighbours are true or false.[2] Some threshold size R_T is required to decide whether neighbours are false or not. The FNN method works well when the distribution of false neighbours over dimensions only slightly depends upon the threshold value in sufficiently wide range of R_T (for example, $5 < R_T < 50$ as in ref. 3).

Unfortunately in the present study during the treatment of the experimental time series we find out a considerable dependence of this distribution upon R_T. For example, during the treatment of the time series for the CO concentration equal to 0.72%, at low threshold ($R_T < 8$) the FNN method estimates the embedding dimension as small as $d_E \leq 5$. The situation at $R_T \geq 8$ is shown in Fig. 2. As can be seen from this figure at $R_T \geq 8$ the false neighbours arise at higher dimensions, up to 20. Therefore, the FNN method apparently cannot give the rigorous estimation of the system dimension in our case. The reason may be the complex structure of the strange attractor and the nonstationarity of the obtained experimental time series.

1 M. B. Kennel, R. Brown and H. D. I. Abarbanel, *Phys. Rev. A*, 1992, **45**, 3403 (ref. 27 from the paper).
2 H. Kantz and T. Schreiber, *Nonlinear Time Series Analysis*, in *Cambridge Nonlinear Science Series*, Cambridge University Press, Cambridge, 1997, vol. 7 (ref. 23 from the paper).
3 H. D. I. Abrarbanel, R. Brown, J. J. Sidorowich and L. S. Tsimring, *Rev. Mod. Phys.*, 1993, **65**, 1331 (ref. 17 from the paper).

Dr Feigin commented: First of all I should like to say that this is my first knowledge of work in which different techniques of time series analysis are compared by application to the real time series (TS).

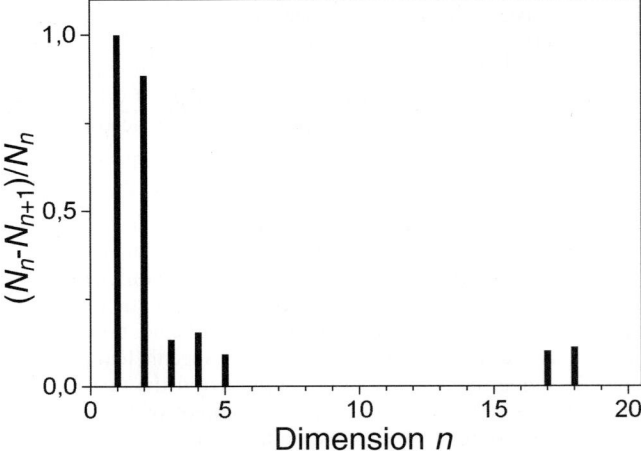

Fig. 2 The distribution of false nearest neighbours over dimensions at $R_T = 8$, N_n is the number of nearest neighbours in n-dimensional space.

My question is related to calculations of minimal embedding dimension. It seems to me that the only TS generated for CO concentration of 0.72%, of length 360 000, makes it possible to calculate correctly high (up to 12) dimensionality. As for all the other TS, their lengths are non-sufficient and dimensionality equalling 9 or 12 seems to be in contradiction with fundamental limitations obtained by Eckmann and Ruelle.[1]

1 J. P. Eckmann and D. Ruelle, *Physica D*, 1992, **56**, 185.

Dr Slin'ko responded: Fundamental limitations, obtained by Eckmann and Ruelle deal with the estimation of dimension with the help of correlation integrals and the estimation of Lyapunov exponents by the Wolf method. In agreement with the conclusions, which have been done by Eckmann and Ruelle, the application of these methods for the given experimental time series was not successful. The values of embedding dimensions shown in Table 1 of the paper were obtained from singular-value decomposition of the covariance matrix (d_{SVD}) and from calculations of the maximum Lyapunov exponent by the Rosenstein method (d_R). Both methods are more crude and were developed particularly for the analysis of the short time series.

Dr Schreiber said: Type I intermittency in 1D maps and low dimensional systems relates to a reappearance of a chaotic attractor that has been generated at some earlier point when scanning parameter space, by a simple mechanism such as repeated period doublings. Do you have any experimental indication or an idea what this original generation of chaos might be?

Dr Slin'ko replied: No, we have no experimental indication of the presence of the period doubling route for the generation of the chaos in the present system. Moreover, we cannot detect any windows with periodic oscillations inside the chaotic region. The reason may be a very small region of CO concentration, where chaos is generated *via* period doublings, which cannot be identified in the experiment. Moreover, the complex reaction dynamics on zeolite-supported Pd catalyst depends not only upon one parameter, namely the CO inlet concentration, but also upon many other parameters, such as catalyst temperature, the size of Pd particles, the size of zeolite crystallites and so on. It can be assumed that the development of the chaotic attractor can be detected during the variation of some other parameter. To clarify this problem we need a rigorous mathematical model of the reaction of CO oxidation in a zeolite crystallite. The development of such a model is now in progress.

Dr Krischer commented: You stressed in your talk that period doubling cascades were found in studies of the CO oxidation on single crystal surfaces whereas the intermittency-I route to chaos was observed in the catalytic CO oxidation on zeolith-supported catalysts. Moreover in the former studies the minimal embedding dimension extracted from the data was much smaller than in the latter one.

If I understood you correctly you related these different findings to the more complicated nature of the zeolith-supported catalyst. Can you explain this statement in some more detail and comment also on the fact that the high embedding dimension which you determine is in contrast to the fact that intermittency-I routes to chaos do not require a high-dimensional phase space (they are already found in 1D maps).

Dr Slin'ko responded: Yes, you are right in your understanding that from the comparison of the results, obtained in the present study and over single crystal surfaces under UHV conditions, the conclusion has been made about the more complicated dynamics over zeolite supported catalysts. The most serious arguments for this conclusion are the higher minimal embedding dimension and much greater region of chaotic oscillations, obtained in the present study in comparison with the UHV single crystal studies. We do not correlate the observation of the intermittency with the obtained high embedding dimension. You are right, that this phenomenon can arise even in 1D maps. However, the high value of the embedding dimension may indicate that the zeolite supported catalyst has to be considered as a distribution system. In this case the intermittency chaos, which has been identified may represent the spatiotemporal intermittency. This type of chaos can be observed only in distributed systems, for example in the array of coupled oscillators.

In comparison with the single crystal surface in the case of Pd zeolite catalyst we have an array of Pd particles, embedded into the zeolite matrix and coupled through the gas phase. The internal diffusion limitations in the zeolite crystallite makes the coupling between Pd particles through the gas phase in pores to be local and puts the particles into different conditions depending upon their positions inside the zeolite crystallite. As is shown in ref. 1 such a system could produce very complicated dynamic behavior including regular, quasiperiodic, and chaotic oscillations.

1   E. S. Kurkina, N. V. Peskov and M. M. Slin'ko, *Physica D*, 1998, **118**, 103.

Prof. Marek said: You have used infrared analysis for the recording of the CO_2 concentration. What would be the time delay of the measured signal to actual value?

Dr Slin'ko replied: The time delay between the signal of CO_2 concentration, measured by infrared analyses and the measurement of the small amplitude (2–3 °C) variations of the catalyst temperature was 20 s. It may indicate the time delay of the measured signal to the actual value. The main period of oscillations is about 50 s. During this time at liner flow rate equal to 1.89 cm s^{-1} one oscillatory peak passes the distance of about 95 cm. From the other side during the time delay of 20 s the transformation of the peak due to CO diffusion equal to 0.2 cm^2 s^{-1} can be calculated as 2 cm. Therefore the effect of diffusion-like stirring is about 2%, which as we suppose is not essential for the qualitative conclusions, which were obtained during the time series analysis. Moreover the comparison of regular oscillations of temperature variation and CO_2 concentration demonstrates that the waveforms and periods of both time series are approximately the same. Therefore the diffusion-like and hydrodynamic stirring does not affect significantly the CO_2 concentration measurements. To understand the affect of stirring upon the dimension of the strange attractor we shall try to treat the chaotic time series of the temperature measurement, which is a much more difficult task due to the much smaller amplitude and slow drift of the signal.

Oscillatory dynamics protect enzymes and possibly cells against toxic substances

Marcus J. B. Hauser,[a] Ursula Kummer,[b] Ann Z. Larsen[c] and Lars F. Olsen[c]

[a] *Institut für Experimentelle Physik, Abteilung Biophysik, Otto-von-Guericke Universität, Postfach 4120, D-39016 Magdeburg, Germany*
[b] *European Media Laboratory, Schloss-Wolfsbrunnenweg 33, D-69118 Heidelberg, Germany*
[c] *Celcom, Department of Biochemistry and Molecular Biology, SDU Odense University, Campusvej 55, DK-5230 Odense M, Denmark. E-mail: lfo@dou.dk*

Received 4th April 2001
First published as an Advance Article on the web 7th November 2001

We have used the oscillating peroxidase–oxidase (PO) reaction as a model system to study how oscillatory dynamics may affect the influence of toxic reaction intermediates on enzyme stability. In the peroxidase–oxidase reaction reactive intermediates, such as hydrogen peroxide, superoxide, and hydroxyl radical are formed. Such intermediates inactivate many cellular macromolecules such as proteins and nucleic acids. These reaction intermediates also react with peroxidase itself to form an inactive enzyme. The fact that the PO reaction shows bistability between an oscillatory and a steady state gives us a unique possibility to compare such inactivation when the system is in one of these two states. We show that inactivation of peroxidase is slower when the system is in an oscillatory state, and using numerical simulations we provide evidence that oscillatory dynamics lower the average concentration of the reactive intermediates.

Introduction

With the increased ability to make continuous measurements of metabolites in cells the number of reports of oscillating biochemical processes has grown considerably.[1] The first metabolic process shown to oscillate *in vivo* was glycolysis in yeast and muscle cells.[2,3] More recently oscillations in secondary messengers such as cyclic AMP[4] and Ca^{2+} (ref. 5) have attracted the interest of biochemists and cell biologists. Also other types of oscillations, such as oscillations in intracellular NAD(P)H, pH, hydrogen peroxide, and superoxide have been measured in various cells *in vivo*.[6,7]

In spite of numerous experimental investigations and theoretical studies, the purpose of metabolic oscillations in cells is still not completely understood. It has not yet been established whether oscillations occur as a harmless side effect of the nonlinear properties of metabolic enzymes, or if they serve one or more important functions. It has been shown experimentally that oscillations may provide metabolism with an increased thermodynamic efficiency,[8] and that oscillations of second messengers, such as calcium ions, encode information in their frequency.[9] Roles as biological time-keepers[1] and encoders of transmembrane signalling have also been proposed.[10] Presumably, oscillations serve many functions in cell metabolism. Here we wish to point to another potential function of oscillating biochemical processes, namely the protection of proteins against otherwise harmful substances, such as reactive oxygen species, that are produced during cell

DOI: 10.1039/b103076k

metabolism or cell signalling. A similar idea has been presented earlier in the context of calcium oscillations: oscillations in cytosolic calcium were postulated to prevent the precipitation of calcium phosphates in cytoplasm,[11] but this assumption has to our knowledge not yet been verified experimentally.

In the present article we study the oscillating PO reaction as a model for the interaction of toxic reaction intermediates with cellular components such as nucleic acids, proteins and lipid membranes. The PO reaction involves the oxidation of an organic electron donor (typically NADH) by molecular oxygen:[12,13]

$$2NADH + O_2 + H^+ \rightarrow 2NAD^+ + 2H_2O \qquad (1)$$

catalyzed by peroxidase. When NADH and O_2 are supplied continuously to a stirred aqueous solution with a pH between 5 and 6.5 containing peroxidase, a suitable phenolic compound and Methylene Blue, the reaction shows both simple and complex oscillations.[12,14,15] Furthermore, such oscillations have been demonstrated using peroxidases from both plants and animals.[16] Oscillations have also been observed in extracts of plant material rich in peroxidase.[17] Thus, there are reasons to believe that oscillations do occur *in vivo*. During the PO reaction hydrogen peroxide and superoxide are formed as intermediates.[13] These reactive oxygen species were previously considered as undesired in cellular metabolism, because of their ability to oxidize a number of biomolecular substances, such as nucleic acids, enzymes and membrane lipids, resulting in a breakdown of these components. However, recently it has been shown that reactive oxygen species also seem to function as secondary messengers in certain cell signalling processes.[7] Thus, the cell faces the problem of handling a substance with both a beneficial and a harmful effect. The same problem arises in a number of other contexts, *e.g.* in a liver cell when calcium ions are liberated into the cytoplasm following the binding of certain hormones to receptors in the cell membrane. In the present study we provide evidence that one way to minimize the harmful side effects of a toxic substance is to keep the corresponding reaction in an oscillatory state.

Experimental

Experiments with the peroxidase–oxidase reaction were conducted[18,19] at 28 (± 0.1) °C in a $2.0 \times 2.0 \times 4.3$ cm^3 quartz cuvette fitted with a thermostating jacket. The cuvette was connected to a Zeiss S 10 diode array spectrophotometer through optical fibers. Oxygen in the solution was measured with a Clark-type oxygen electrode (Microelectrodes Inc.). The reaction mixture consisted of an 8 ml well-stirred homogenous aqueous solution containing 0.1 M sodium acetate, pH 5.1, 0.75–1.3 µM of peroxidase from either horseradish (Boehringer Mannheim) or cow's milk (lactoperoxidase, Sigma), 0.05–0.2 µM Methylene Blue (Merck), and 300–600 µM 4-hydroxybenzoic acid. O_2 enters the reaction mixture from a 1.05% (v/v) O_2/N_2 gas mixture supplied to the approximately 9 ml gas head space above the liquid. The rate of oxygen diffusion v_{O_2} into the liquid is given by the equation:

$$v_{O_2} = K([O_2]_{eq} - [O_2]) \qquad (2)$$

where $[O_2]$ and $[O_2]_{eq}$ are the actual oxygen concentration in the liquid and the oxygen concentration at equilibrium between the gas and the liquid, respectively. The oxygen transfer constant K depends on the surface area and hence on the stirring rate. K was 4.4×10^{-3} ($\pm 0.1 \times 10^{-3}$) s^{-1} corresponding to a stirring rate of 900 (± 10) rpm. NADH (Boehringer Mannheim) was supplied by infusion of a 0.1 M NADH solution into the reaction mixture through a capillary whose tip was below the surface of the liquid. The infusion was mediated by a Harvard Apparatus, model 22, syringe pump, and the infusion rate was typically 35 µl h^{-1}.

We recorded the time series of the absorbences in the range 350 to 600 nm (1 nm resolution) and the O_2 concentration every 2 s and stored the data on a computer for later analysis. Specifically the absorbences at wavelengths corresponding to NADH, ferric peroxidase, compound III (oxyferrous peroxidase), and ferrous peroxidase were used for spectral deconvolution of the absorbance measurements to concentrations of these four species.[20] Their concentrations were determined by solving the system of linear equations:

$$A = l \times \varepsilon \times c \qquad (3)$$

Table 1 Horseradish peroxidase molar absorption coefficients in M^{-1} cm^{-1}

Species	$\varepsilon_{360\,nm}$	$\varepsilon_{403\,nm}$	$\varepsilon_{418\,nm}$	$\varepsilon_{439\,nm}$
NADH	4.3×10^3	50	0	0
Per^{3+}	4.3×10^4	1.0×10^5	5.4×10^4	1.3×10^4
Compound III	2.7×10^4	7.1×10^4	1.16×10^5	3.2×10^4
Per^{2+}	3.8×10^4	5.1×10^4	6.6×10^4	8.9×10^4

where A is a vector containing the absorbances at 4 different wavelengths, l is the length of the light path through the sample, ε is a 4×4 matrix containing the molar absorption coefficients of NADH, ferric peroxidase, ferrous peroxidase, and compound III at the four wavelengths while c is the vector of the concentrations of these four species. The molar absorption coefficients ε used in the calculations of c for NADH and three intermediates of horseradish peroxidase were measured as described previously[20] and are listed in Table 1. The molar absorption coefficients for NADH and lactoperoxidase were measured as for horseradish peroxidase[20] and are listed in Table 2.

Experimental results

The PO reaction involves a number of reactive oxygen species, several of which will react unspecifically with different cell components and inactivate those. These reactive oxygen species also inactivate peroxidase itself. We show in Fig. 1 time series of NADH, ferric peroxidase, compound III, ferrous peroxidase and oxygen. The concentrations of the four species were computed from experimental data as described in the Experimental section. The experiment is started by infusion of NADH into a solution equilibrated with O_2 in the gas phase and containing the enzyme and the two modifiers, 4-hydroxybenzoic acid and Methylene Blue. Other enzyme species participating in the reaction, such as compound I and compound II, (see Table 3, later) could not be detected, *i.e.* their concentrations must be lower than the detection limit of the apparatus which corresponds to 0.05 µM. Thus, the sum of concentrations of the three enzyme species shown in Fig. 1 represents the total enzyme concentration at any time t. The time evolution of this sum is shown in the bottom part of the figure. At time $t = 17\,000$ s (indicated by the arrow) 6 µM hydroquinone, which is a potent inhibitor of the reaction,[21] is added to the reaction mixture. The reaction stops immediately as seen by the abrupt increase in the concentration of NADH and the return of the oxygen concentration to the equilibrium value of 12 µM. As for the total enzyme concentration, there is an almost linear decline in the time interval $1000 < t/s < 17\,000$, due to reaction of the enzyme with oxygen free radicals, *e.g.* superoxide or hydroxyl radicals.[22,23] The radicals probably react with the protein part of the enzyme rather than with the heme group[22,23] as we could not detect degradation products such as P670.[24] When the reaction is inhibited the inactivation stops immediately.

The PO reaction catalyzed by peroxidase shows a variety of dynamic behaviours depending on the reaction conditions (for a review see ref. 25). The dynamics include stationary (non-oscillatory) and oscillatory states. In addition, the PO system is known to display bistability, that is, two different coexisting dynamic states are simultaneously stable for the same experimental parameters. Which of these dynamic states is approached depends on the "history" of the reaction system. Experimentally this means that for exactly the same experimental parameters and very

Table 2 Lactoperoxidase molar absorption coefficients in M^{-1} cm^{-1}

Species	$\varepsilon_{360\,nm}$	$\varepsilon_{412\,nm}$	$\varepsilon_{430\,nm}$	$\varepsilon_{439\,nm}$
NADH	4.3×10^3	25	0	0
Per^{3+}	3.66×10^4	1.12×10^5	4.0×10^4	1.98×10^4
Compound III	3.0×10^4	7.36×10^4	8.32×10^4	5.5×10^4
Per^{2+}	3.0×10^4	5.5×10^4	9.3×10^4	9.18×10^4

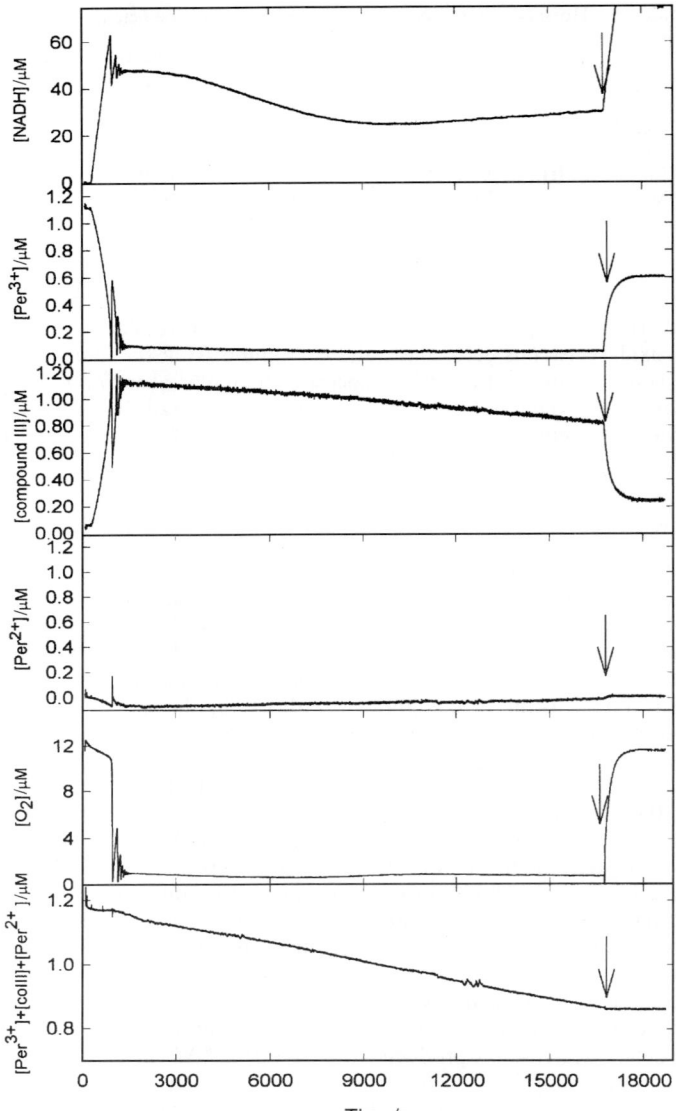

Fig. 1 Time series of the concentrations of NADH, ferric peroxidase (Per^{3+}), compound III, ferrous peroxidase (Per^{3+}), oxygen and the total enzyme concentration in an experiment showing non-oscillatory behaviour. At time $t = 17\,000$ s the reaction is inhibited by addition of 6 µM hydroquinone. Note that the inactivation of the enzyme stops when the reaction is blocked. The reaction mixture contained 1.2 µM lactoperoxidase, 300 µM 4-hydroxybenzoic acid, and 0.2 µM Methylene Blue in 8 ml of a 0.1 M sodium acetate buffer, pH 5.1. The experiment was started by infusion of 0.1 M NADH into the reaction mixture at a rate of 35 µl h^{-1}. The concentrations of NADH, ferric peroxidase (Per^{3+}), ferrous peroxidase (Per^{2+}) and compound III were determined by spectral deconvolution of the absorbences at wavelengths 360, 412, 430 and 439 nm using the molar absorption coefficients listed in Table 2.

similar initial conditions the reaction may converge on either one of the two stable dynamic regimes. Examples are (i) two coexisting steady states[26] and (ii) a steady state coexisting with periodic oscillations,[27] also known as hard excitation.[1] The latter type of behaviour provides us with a unique possibility to study the inactivation of the enzyme while the reaction is either in a non-oscillatory state or in an oscillatory state under the exact same experimental conditions. Fig. 2a shows time series of the concentrations of NADH, ferric peroxidase (Per^{3+}), compound **III**, ferrous peroxidase (Per^{2+}), and O_2 for a typical experiment where the PO reaction is in a non-oscillatory (stationary) state. About 1500 s after the start of the NADH infusion the oxygen concentration in the liquid reaches a stationary value of approximately 2–3 µM corresponding to a constant rate of oxidation of NADH. This rate remains essentially the same throughout the experiment, *i.e.* for more than 10 000 s. Fig. 2b shows the corresponding time series for an experiment where the PO reaction converges on an oscillatory state. It is worth emphasizing that the

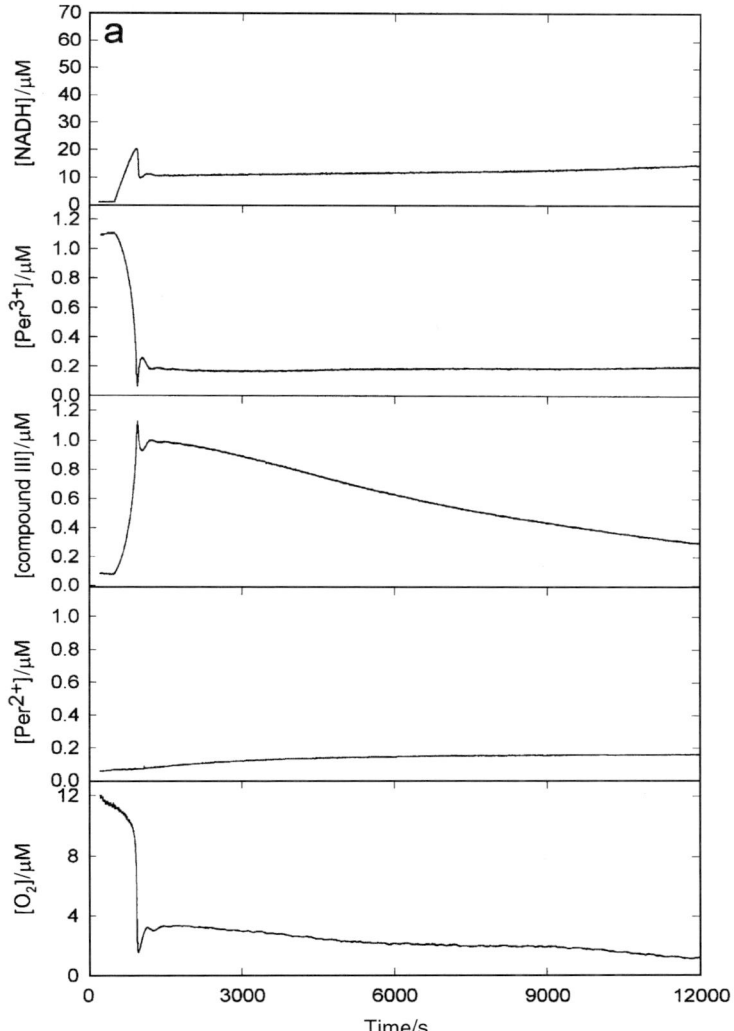

Fig. 2 Bistability between a steady state and oscillations in the peroxidase–oxidase reaction. Time series of the concentrations of NADH, ferric peroxidase (Per^{3+}), compound **III**, ferrous peroxidase (Per^{2+}) and oxygen during (a) a steady state and (b) an oscillatory state. The reaction mixture contained 1.3 µM horseradish peroxidase, 600 µM 4-hydroxybenzoic acid and 0.2 µM Methylene Blue in 8 ml of a 0.1 M sodium acetate buffer, pH 5.1. The experiment was started by infusion of 0.1 M NADH into the reaction mixture at a rate of 35 µl h^{-1}. The concentrations of NADH, ferric peroxidase (Per^{3+}), ferrous peroxidase (Per^{3+}) and compound **III** were determined by spectral deconvolution of the absorbences at wavelengths 360, 403, 418 and 439 nm using the molar absorption coefficients listed in Table 1.

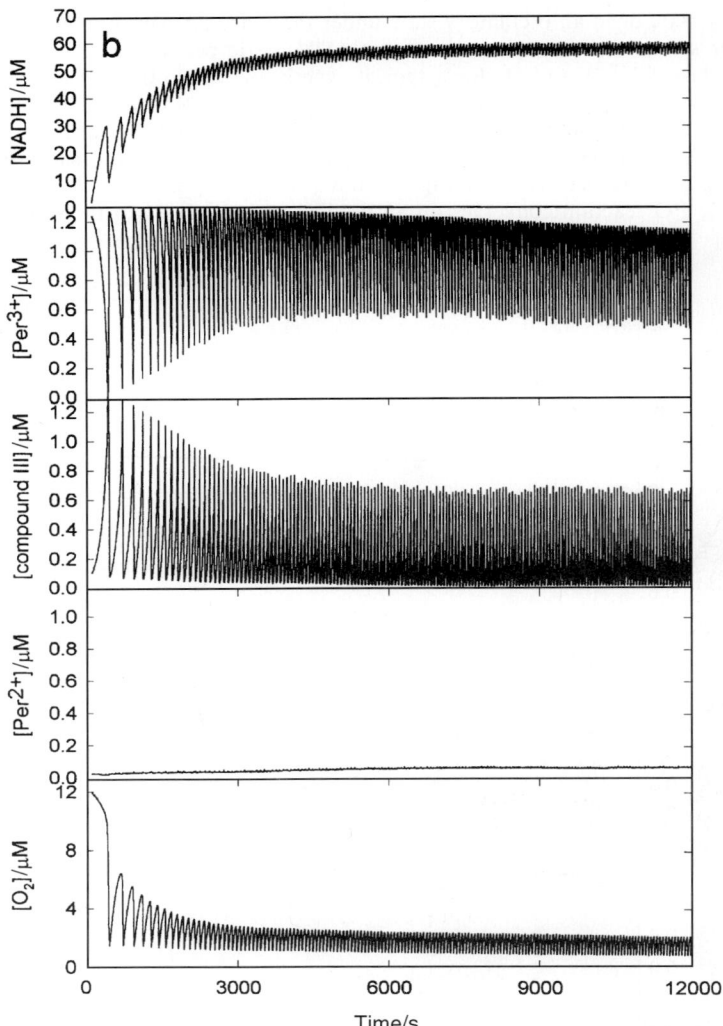

Fig. 2 Continued.

two experiments only differ in the dynamics shown by the PO reaction. The average rates of oxidation of NADH are the same. The argument for this is as follows: The rate of oxidation of NADH is stoichiometrically linked to the reduction of O_2 through reaction (1). Even though intermediates such as NAD_2 are formed, these can be further oxidized in the reaction, such that the overall stoichiometry of NADH consumption to oxygen consumption is 2 : 1.[28] Furthermore, if the concentration of O_2 is constant then the rate of consumption of O_2 is equal to the rate of oxygen diffusion into the liquid (eqn. (2)) which again is half the rate of oxidation of NADH. Thus, if the average concentration of O_2 is the same during an oscillatory and a non-oscillatory state then the average rate of oxidation of NADH must be the same. This is indeed the case (within experimental error) for the experiments shown in Fig. 2a and b.

As mentioned above the sum of concentrations of the enzyme intermediates ferric peroxidase, ferrous peroxidase and compound **III** is essentially identical to the total concentration of enzyme present in the reaction mixture. In Fig. 3 we have plotted the sum of the concentrations of the three enzyme intermediates calculated from Fig. 2a and b. The sum of concentrations from the

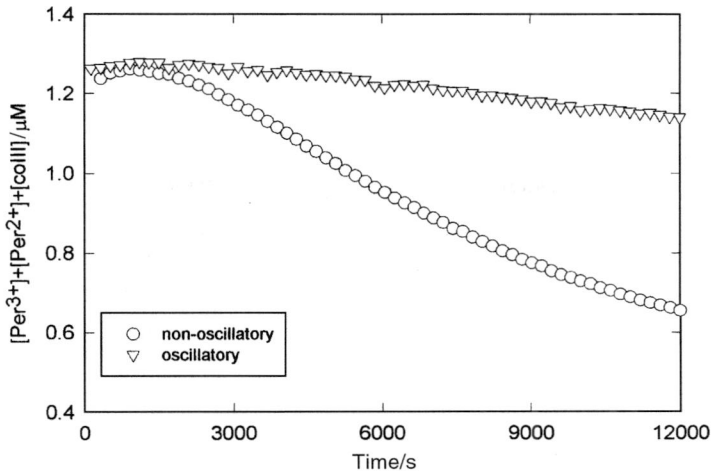

Fig. 3 Total enzyme concentration is plotted against time. The sum of the concentrations of ferric peroxidase (Per^{3+}), ferrous peroxidase (Per^{2+}) and compound **III** from the experiments in Fig. 2 are plotted against time. (▽) oscillatory state (Fig. 2b); (○) non-oscillatory state (Fig. 2a).

experiment showing oscillatory kinetics decreases at a much lower rate compared to the steady state experiment.

We have conducted a large number of experiments showing oscillations and non-oscillatory behaviour, respectively, to compare the rates of inactivation of the enzyme. In all cases we found that the rate of inactivation of the enzyme is always significantly lower in an oscillatory state than in the corresponding non-oscillatory state. To illustrate this further we show the time series of the total concentration of enzyme from cow's milk (lactoperoxidase) from an experiment (Fig. 4) similar to that in Fig. 1, but here the PO reaction starts in an oscillatory state and following a

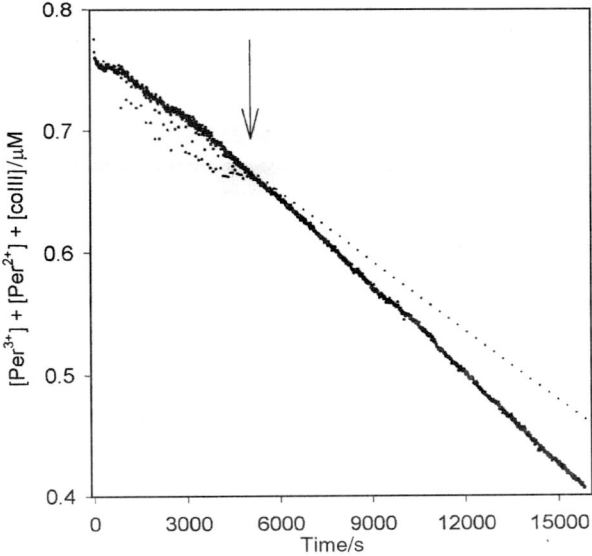

Fig. 4 Time series of the total concentration of lactoperoxidase. At $t \sim 5000$ s (indicated by the arrow) the dynamics change from a periodic oscillation to a stationary state. The experimental conditions are the same as in Fig. 1, except for the concentration of Methylene Blue which is 0.05 μM. The dotted line represents the linear regression of the rate of enzyme inactivation during the oscillatory state.

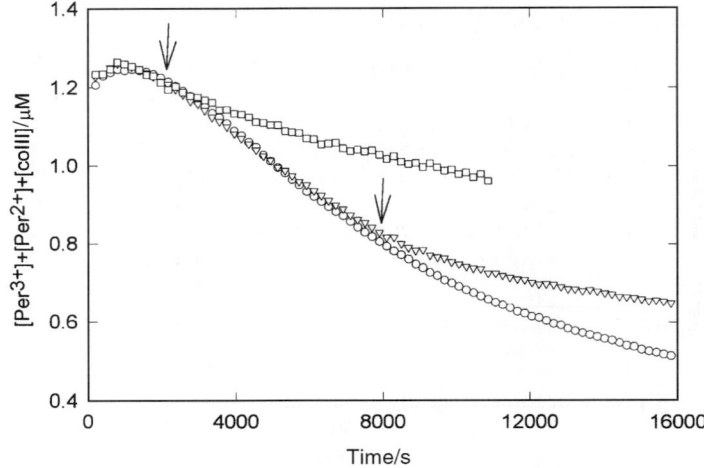

Fig. 5 Total enzyme concentration plotted against time. The total enzyme concentrations of two experiments where a transition from a non-oscillatory state to an oscillatory state is induced at 2300 (□) or at 7900 s (▽) (indicated by arrows), and an experiment where the system is always in a non-oscillatory state (○). Experimental conditions as in Fig. 2. Note that the rate of enzyme deactivation during a steady state operation is much faster than that during oscillations.

small random perturbation at about 5000 s (indicated by the arrow) switches to a stationary state. The dotted line represents the linear regression of the decline in total enzyme concentration during the oscillatory state. We note that as the reaction switches from the oscillatory state to the steady state the rate of enzyme decay increases from 18.5 to 24.4 pM s^{-1}. In other experiments the PO reaction starts in the steady state and then later switches to an oscillatory state. In those cases—using peroxidase from horseradish—(Fig. 5) the rate of decay of the enzyme slows from 69 to 16–23 pM s^{-1} when the reaction switches from the steady state to the oscillatory state. Thus, oscillatory kinetics seems to protect the enzyme against degradation. The degradation of the enzyme can be ascribed to the presence of reactive intermediates like superoxide radical, hydrogen peroxide, and hydroxyl radical which are generated during the reaction.[13,22–24]

Table 3 Detailed (BFSO) model of the peroxidase–oxidase reaction

Reaction[a]	R_i	Constant
$NADH + O_2 + H^+ \rightarrow NAD^+ + H_2O_2$ (1)	$k_1[NADH][O_2]$	10.0^b
$H_2O_2 + Per^{3+} \rightarrow$ compound I (2)	$k_2[H_2O_2][Per^{3+}]$	$1.8 \times 10^{7\,b}$
Compound I + NADH $\rightarrow$ compound II + NAD· (3)	$k_3[\text{compound I}][NADH]$	$4.0 \times 10^{5\,b}$
Compound II + NADH $\rightarrow Per^{3+} + NAD·$ (4)	$k_4[\text{compound II}][NADH]$	$2.6 \times 10^{5\,b}$
$NAD· + O_2 \rightarrow NAD^+ + O_2^-$ (5)	$k_5[NAD·][O_2]$	$2.0 \times 10^{7\,b}$
$O_2^- + Per^{3+} \rightarrow$ compound III (6)	$k_6[O_2^-][Per^{3+}]$	$1.7 \times 10^{7\,b}$
$2O_2^- + 2H^+ \rightarrow H_2O_2 + O_2$ (7)	$k_7[O_2^-]^2$	$2.0 \times 10^{7\,b}$
Compound III + NAD· $\rightarrow$ compound I + NAD$^+$ (8)	$k_8[\text{compound III}][NAD·]$	$6.0 \times 10^{7\,b}$
$2NAD· \rightarrow NAD_2$ (9)	$k_9[NAD·]^2$	$5.6 \times 10^{7\,b}$
$Per^{3+} + NAD· \rightarrow Per^{2+} + NAD·$ (10)	$k_{10}[Per^{3+}][NAD·]$	$1.8 \times 10^{6\,b}$
$Per^{2+} + O_2 \rightarrow$ compound III (11)	$k_{11}[Per^{2+}][O_2]$	$1.0 \times 10^{5\,b}$
$\rightarrow$ NADH (12)	k_{12}	variable^c
O_2(gas) $\rightarrow O_2$(liquid) (13)	$k_{13}[O_2]_{eq}$	$4.4 \times 10^{-3\,d,e}$
$(-13)\ O_2$(liquid) $\rightarrow O_2$(gas)	$k_{-13}[O_2]$	$4.4 \times 10^{-3\,d}$
$O_2^- +$ compound III $\rightarrow E'$ (14)	$k_{14}[O_2^-][\text{compound III}]$	1.0×10^3

[a] Per^{3+} and Per^{2+} indicate iron(III) and iron(II) peroxidase respectively. E' indicates inactivated enzyme. [b] In M^{-1} s^{-1}. [c] Between 4.0×10^{-8} and 1.0×10^{-7} M s^{-1}. [d] In s^{-1}. [e] The value of $[O_2]_{eq}$ is 1.2×10^{-5} M.

Numerical simulations of the PO reaction

In order to study further the effect of oscillatory *vs.* non-oscillatory states we performed numerical simulations using the so-called BFSO model[29] which was shown to describe the PO reaction reasonably well. This model involves 5 enzyme intermediates and 5 different chemical species, including hydrogen peroxide and superoxide. Thus the complete model yields 10 nonlinear first order differential equations. The elementary reactions of the model are listed in Table 3. To the list of reactions of the original model[29] we have added a reaction (reaction (14)) to account for the inactivation of the enzyme. Although the exact mechanism for the inactivation of peroxidase is not known we have in our model proposed that this inactivation is due to the reaction of superoxide with compound III:

$$O_2^- + \text{compound III} \rightarrow E'$$

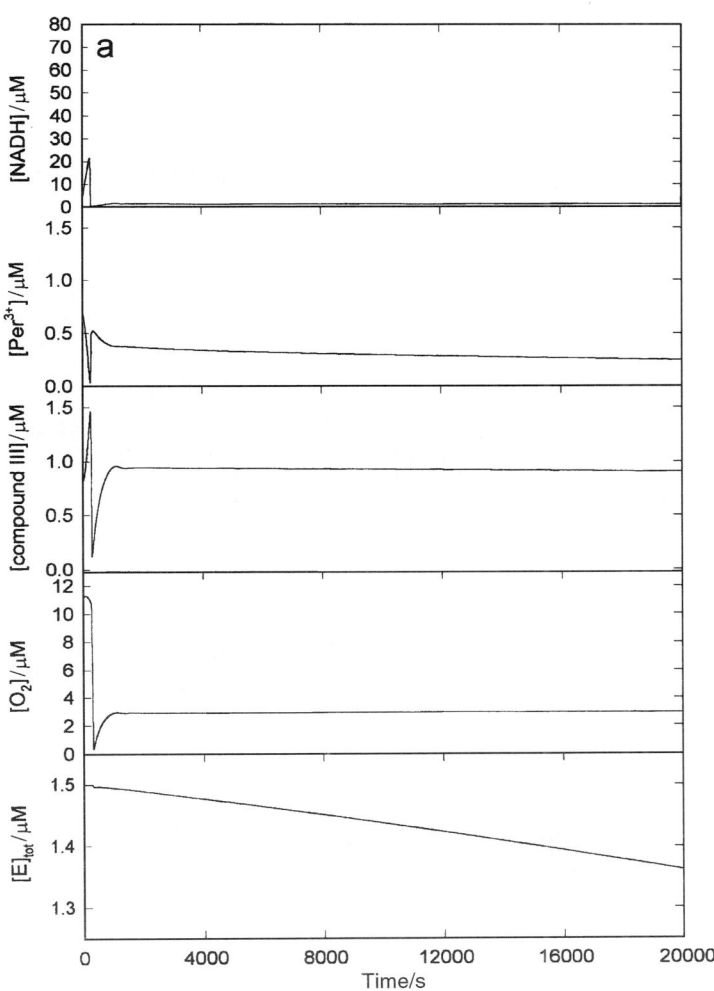

Fig. 6 Simulations of the peroxidase–oxidase reaction. Time series of the concentrations of NADH, ferric peroxidase, compound III, superoxide and O_2 generated by the model listed in Table 2 and Table 3. The graphs in (a) represent a steady state while the graphs in (b) represent the coexisting oscillatory state. The NADH flow rate was 0.07 µM s^{-1}. The initial concentration of O_2 was 12 µM, that of ferric peroxidase was 1.5 µM, while the initial concentration of H_2O_2 was 0.4 µM in (a) and 0 µM in (b). All other initial concentrations were 0.

where E′ is the inactivated enzyme. This need in fact not be the case in reality.[22,23] However, the choice of radical species and enzyme intermediate is not critical. We obtain essentially the same result if we assume that the inactivation is due to the reaction of any of the other enzyme species with superoxide or hydrogen peroxide. Most of the rate constants listed in Table 3 have been determined experimentally.[13] However, the model assumes that NAD_2 formed in reaction (R9) in Table 3 is not oxidized further as opposed to the recent experimental observation that essentially all NAD_2 formed is reoxidized to NAD^+.[28]

For a proper set of rate constants, which correspond to the present experimental conditions, the BFSO model shows bistability similar to that of the experimental system, *i.e.* depending on the initial conditions the system either settles on a steady state (as in Fig. 2a) or on a periodic oscillation (as in Fig. 2b). Time series of the corresponding simulations of the concentration of NADH, ferric peroxidase, compound III, oxygen and the total enzyme concentration are shown in Fig. 6. We note that while in the oscillatory state the inactivation of the enzyme is slower than in the corresponding stationary (non-oscillatory) state. To understand the effect of oscillations on the rate of enzyme inactivation we computed the concentrations of superoxide under oscillatory and non-oscillatory conditions. The simulations (Fig. 7) reveal that although the maximum concentra-

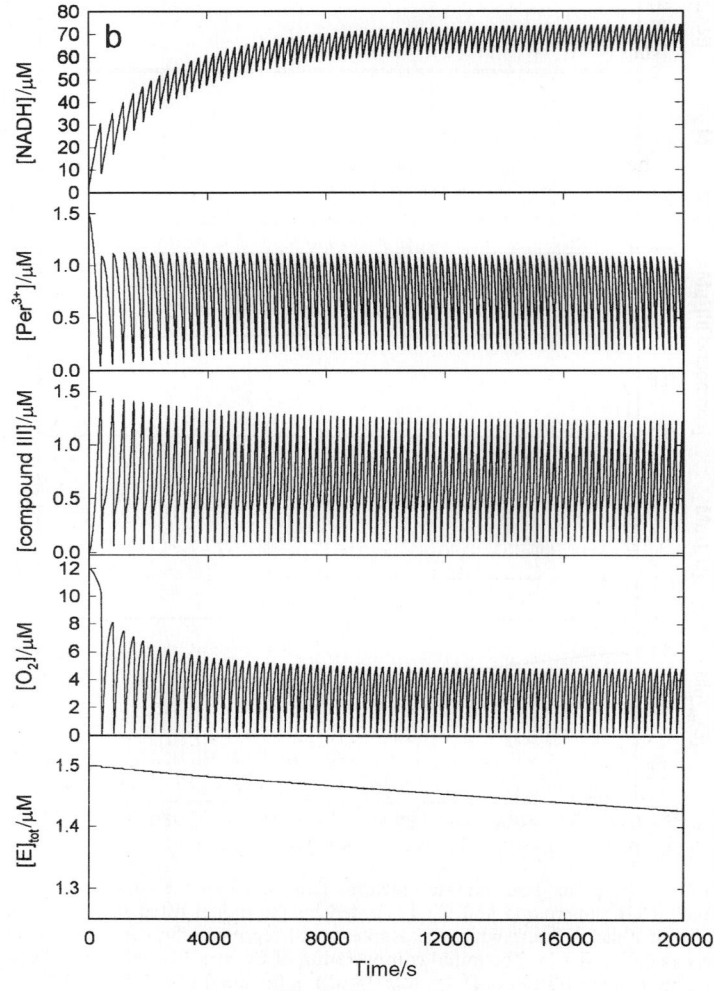

Fig. 6 Continued.

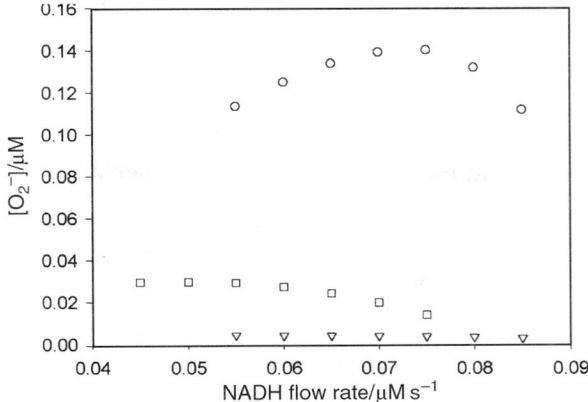

Fig. 7 Concentrations of superoxide under steady state and oscillatory conditions obtained from the BFSO model (Table 3). Steady state concentration (□) as well as maximum (○) and average (▽) concentrations of superoxide during oscillations plotted against NADH flow rate. In order for the reaction to go into a stable state the rate constant k_{14} is set to zero. The initial concentration of O_2 was 12 µM, that of ferric peroxidase was 1.5 µM, while the initial concentration of H_2O_2 was 0.6 µM under steady state conditions and 0 µM under oscillatory conditions. All other initial concentrations were 0.

tion of superoxide during oscillations is much higher than the values observed in the corresponding steady state, the average concentrations of this species are up to 8 times lower during oscillations than during steady state conditions. At the same time, the product (NAD^+) concentration increases with a rate that is only 5% lower in the oscillatory state compared to the steady state. This shows that the overall flux through the system is essentially the same in both cases, in accordance with the experimental observations (Fig. 2). Thus, we conclude from the simulations that a likely explanation for the increased degradation of the enzyme in the steady state is the higher average concentration of toxic reaction intermediates.

Discussion

In the previous sections we used the oscillation–steady state bistability (hard excitation) of the PO reaction to demonstrate that oscillatory dynamics slow down the inactivation of peroxidase. This occurs for peroxidases from both plant (horseradish) and animal (milk) sources. We have also indicated that the reason for the lower inactivation rate is that the average concentration of the inactivating species in the oscillatory state is lower than the concentration in the corresponding steady state. We believe that our finding that oscillatory dynamics seem to protect enzymes from inactivation by toxic reaction intermediates is not limited to the peroxidase–oxidase reaction, but is of much more general nature. There are other biochemical oscillators that involve reactive oxygen species.[7] High concentrations of H_2O_2, superoxide, and hydroxyl radicals can lead to the peroxidation of membrane lipids and structural changes in DNA.[30]

Furthermore, there are biochemical oscillators that involve species which are just as toxic to the cell as reactive oxygen species. An example is cytoplasmic calcium. On the one hand, prolonged high concentrations of calcium are lethal to the cell.[31] On the other hand, the cell relies on calcium as a second messenger and thus the concentration of calcium has to exceed a certain threshold from time to time. As we have shown here, oscillatory dynamics offer the possibility to employ rather harmful substances as messengers while maintaining them at very low average concentrations. At the same time they can reach fairly high concentrations for brief time intervals which facilitates the transmission of information.

A somewhat related problem arises in apoptosis, also referred to as programmed cell death. Apoptosis is important in normal development and formation of tissue patterns as well as physiological homeostasis, but also in certain pathological conditions such as Alzheimer's disease, AIDS and cancer.[9] The role of calcium in this process is that of an activator of several endonucleases which catalyze the breakdown of genomic DNA.[32–35] In rat hepatocytes the calcium concentration oscillates between 0.3 and 0.8 µM.[36–38] It has been shown that isolated rat liver nuclei contain an endonuclease which is activated by submicromolar concentrations of Ca^{2+} when physiological levels of ATP and NAD^+ are present.[34] In principle, the concentration range of Ca^{2+}-oscillations following stimulation with an agonist should trigger the endonuclease and lead to cell death. However, this has not been observed. This apparent inconsistency might well be due to the fact that the Ca^{2+} oscillations maintain the average Ca^{2+} concentration at a level well below the threshold that activates the endonuclease and hence indirectly protects the cell from destruction.

An interesting analogy to the protection of the enzyme from deactivation by toxic reaction intermediates is encountered in a large number of industrial syntheses, where chemicals are produced in reactions involving solid catalysts. Reaction products frequently tend to adsorb irreversibly onto the catalyst, thereby deactivating or "poisoning" it. To prevent such poisoning, many technical syntheses are performed under a "periodic (or oscillatory) operation mode": here, the reaction conditions are changed periodically by external manipulation to induce oscillations in the reaction conditions, which, in turn, prevent the irreversible deactivation of the catalyst. An example of such an industrial process is the synthesis of maleic anhydride.[39] Note, that technical processes involving "periodic reactor operation" do not oscillate spontaneously. Oscillations are instead imposed on the steady state processes by periodic variation of external parameters, in order to profit from the benefits associated with the oscillatory dynamics. By contrast, the PO system and other cellular oscillators are capable of operating under autonomous oscillatory conditions, with a similar function of protecting enzymes and other biological materials against poisoning and irreversible degradation. It is possible that nature invented this principle first for protective purposes, but then discovered that the oscillations in the concentrations of the reaction intermediates also provide a very good way of encoding information. Thus, employing these substances that exhibited oscillatory dynamics as second messengers enabled information to be stored not only in the mere presence of the substance, but also in the frequency or amplitude of the oscillation. Since, according to this speculation, mainly toxic intermediates displayed such behaviour, this possibly explains why nowadays, many toxic intermediates serve as second messengers in higher organisms.

Acknowledgements

MJBH and UK thank the Deutsche Forschungsgemeinschaft and the Klaus Tschira Foundation, respectively, for financial support, while LFO acknowledges the support of the Danish Natural Science Research Council. The authors wish to thank Anita Lunding, Line Planck Kongstad, Torben Christensen, and Mikkel Olrik for valuable technical assistance.

References

1. A. Goldbeter, *Biochemical Oscillations and Cellular Rhythms. The Molecular Bases of Periodic and Chaotic Behaviour*, Cambridge University Press, Cambridge, 1996.
2. L. N. M. Duysens and J. Amesz, *Biochim. Biophys. Acta*, 1957, **24**, 19.
3. R. Frenkel, *Arch. Biochem. Biophys.*, 1968, **125**, 151.
4. G. Gerisch and B. Hess, *Proc. Natl. Acad. Sci. USA*, 1974, **71**, 2118.
5. N. M. Woods, K. S. R. Cuthbertson and P. H. Cobbold, *Nature*, 1986, **319**, 600.
6. H. R. Petty, R. G. Worth and A. L. Kindzelskii, *Phys. Rev. Lett.*, 2000, **84**, 2754.
7. A. Amit, A. L. Kindzelskii, J. Zanoni, J. N. Jarvis and H. R. Petty, *Cell. Immunol.*, 1999, **194**, 47.
8. J. G. Lazar and J. Ross, *Science*, 1990, **247**, 189.
9. M. J. Berridge, M. D. Bootman and P. Lipp, *Nature*, 1998, **395**, 645.
10. A. L. Kindzelskii and H. R. Petty, *Biochim. Biophys. Acta*, 2000, **1495**, 90.
11. M. J. Berridge, in *Cell to Cell Signalling: From Experiment to Theoretical Models*, ed. A. Goldbeter, Academic Press, London, 1989, p. 449.
12. S. Nakamura, K. Yokota and I. Yamazaki, *Nature*, 1969, **222**, 794.
13. A. Scheeline, D. L. Olson, E. P. Williksen, G. A. Horras, M. L. Klein and R. Larter, *Chem. Rev.*, 1997, **97**, 739.
14. L. F. Olsen and H. Degn, *Biochim. Biophys. Acta*, 1978, **523**, 321.
15. L. F. Olsen and H. Degn, *Nature*, 1977, **267**, 177.

16   U. Kummer, K. R. Valeur, G. Baier, K. Wegmann and L. F. Olsen, *Biochim. Biophys. Acta*, 1996, **1289**, 397.
17   A. C. Møller, M. J. B. Hauser and L. F. Olsen, *Biophys. Chem.*, 1998, **72**, 63.
18   M. J. B. Hauser and L. F. Olsen, *J. Chem. Soc., Faraday Trans.*, 1996, **92**, 2857.
19   A. C. Møller and L. F. Olsen, *J. Phys. Chem. B*, 2000, **104**, 140.
20   M. J. B. Hauser, A. Lunding and L. F. Olsen, *Phys. Chem. Chem. Phys.*, 2000, **2**, 1685.
21   M. J. B. Hauser and L. F. Olsen, *Biochemistry*, 1998, **37**, 2458.
22   L. Gebicka and J. L. Gebicki, *Acta Biochim. Polon.*, 1996, **43**, 673.
23   L. Gebicka and J. L. Gebicki, *Biochimie*, 1996, **78**, 62.
24   S. N. Krylov and H. B. Dunford, *Biophys. Chem.*, 1996, **58**, 325.
25   M. J. B. Hauser and L. F. Olsen, in *Transport and Structure—Their Competitive Role in Biophysics and Chemistry*, ed. S. C. Müller, J. Parisi and W. Zimmermann, Springer, Berlin, 1999, p. 252.
26   H. Degn, *Nature*, 1968, **217**, 1047.
27   B. D. Aguda, L.-L. H. Frisch and L. F. Olsen, *J. Am. Chem. Soc.*, 1990, **112**, 6652.
28   E. S. Kirkor and A. Scheeline, *Eur. J. Biochem.*, 2000, **267**, 5014.
29   T. V. Bronnikova, V. R. Fed'kina, W. M. Schaffer and L. F. Olsen, *J. Phys. Chem.*, 1995, **99**, 9309.
30   H. Wiseman and B. Halliwell, *Biochem. J.*, 1996, **313**, 17.
31   P. Nicotera, G. Bellomo and S. Orrenius, *Annu. Rev. Pharmacol. Toxicol.*, 1992, **32**, 449.
32   A. G. Yakovlev, G. Wang, B. A. Stoica, C. M. Simbulan-Rosenthal, K. Yoshihara and M. E. Smulson, *Nucleic Acid Res.*, 1999, **29**, 1999.
33   A. G. Yakovlev, G. Wang, B. A. Stoica, H. A. Boulares, A. Y. Spoonde, K. Yoshihara and M. E. Smulson, *J. Biol. Chem.*, 2000, **275**, 21302.
34   D. P. Jones, D. J. McConkey, P. Nicotera and S. Orrenius, *J. Biol. Chem.*, 1989, **264**, 6398.
35   M. K. L. Collins, I. J. Furlong, P. Malde, R. Ascao, J. Oliver and A. L. Rivas, *J. Cell. Sci.*, 1996, **109**, 2393.
36   A. Sanches-Bueno, C. J. Dixon, N. M. Woods, K. S. R. Cuthbertson and P. Cobbold, *Biochem. J.*, 1990, **268**, 627.
37   T. A. Rooney, K. Suresh, C. Q. Joseph and A. P. Thomas, *J. Biol. Chem.*, 1996, **271**, 19817.
38   U. Kummer, L. F. Olsen, C. J. Dixon, A. K. Green, E. Bornberg-Bauer and G. Baier, *Biophys. J.*, 2000, **79**, 1188.
39   R. M. Contractor, *Chem. Eng. Sci.*, 1999, **54**, 5627.

pH oscillations in the hemin–hydrogen peroxide–sulfite reaction

Marcus J. B. Hauser,*[a] Anika Strich,[a] Rebeka Bakos,[ab] Zsuzsanna Nagy-Ungvarai[b] and Stefan C. Müller[a]

[a] *Abteilung Biophysik, Institut für Experimentelle Physik, Otto-von-Guericke-Universität, Universitätsplatz 2, 39106 Magdeburg, Germany.*
E-mail: marcus.hauser@physik.uni-magdeburg.de
[b] *Department of Inorganic and Analytical Chemistry, Eötvös University, Budapest, Hungary*

Received 29th June 2001
First published as an Advance Article on the web 7th November 2001

Oscillatory behaviour in the pH value has been observed during the oxidation of sulfite by hydrogen peroxide mediated by hemin, a well known enzyme model compound, in a continuous-flow stirred tank reactor. The dynamics of this reaction has been studied for a variety of flow rates of the reactants. As the flow rates increase, the oscillations evolve from relaxation oscillations to more complex shapes, displaying, among others, bursting behaviour. A reaction mechanism is proposed that involves the autocatalytic oxidation of HSO_3^- by H_2O_2, while slow equilibria between different pH-dependent forms of hemin account for the feedback loop which gives rise to oscillatory dynamics. It is shown in experiments that no participation of CO_2 is required for oscillations to occur.

Introduction

Oscillatory dynamics are of great importance in biochemical and biophysical reaction systems.[1,2] Over the last decades, numerous examples of oscillating biochemical reactions have been reported, most of which involve a vast number of components and rather complicated reaction pathways. The complexity of such mechanisms allows for a precise and fine tuned regulation of biological metabolism. However, the high degree of complexity hampers the fundamental understanding of the underlying mechanisms which lead to such nonlinear behaviour.

The increasing evidence for the importance of biochemical and biophysical oscillatory processes has brought new life to the study of simple biochemical oscillatory reactions. The most simple conceivable oscillators consist of reactions which comprise single enzymes and their substrates. So far, only a surprisingly scant number of such simple enzymatic systems are known, beyond any doubt, to oscillate. Simple biochemical oscillators include the well-studied peroxidase–oxidase reaction (see reviews, ref. 3–5), a catalase–ascorbate–oxygen system,[6] and a biomimetic cytochrome P450 reaction system.[7] However, the latter system is already more complex as it inherently involves transport across a phospholipid membrane.

In parallel to studies of oscillatory enzyme systems, considerable research efforts have been dedicated to the design and investigation of simple chemical reaction systems which show oscillations in the pH of the reaction medium. The first reaction showing this property was designed systematically a decade ago and involved hydrogen peroxide, sulfite and hexacyanoferrate.[8] Mechanistically, this system is composed of two subsystems, an autocatalytic reaction, as realised in the pH-dependent oxidation of HSO_3^- by H_2O_2, and a negative feedback which consumes

DOI: 10.1039/b105707n

H^+.[8,9] Later, modifications in the nature of the feedback reactions gave rise to a family of pH oscillators;[10–15] among these, it is important to mention a prototypical minimal system that involves HSO_3^-, H_2O_2 and hydrogen carbonate. It has recently been shown that even traces of hydrogen carbonate may be sufficient to establish an efficient negative feedback loop.[10,13]

The dynamics observed in this family of pH oscillators are very rich. When investigated as an open system (*i.e.* operated in continuous-flow stirred tank reactors, CSTR), coexisting stationary states,[16] as well as periodic[9,10,13,15] and chaotic[11,12] oscillations have been generated. The light-sensitivity of hexacyanoferrate has been exploited to manipulate the dynamics of this variant of the pH oscillator by applying suitable illumination.[12]

In biological systems, changes in the pH of the reaction medium play a prominent role, since many physiological parameters are affected by the pH value, *e.g.* the activity of enzymes or the permeability of membranes. Propagating waves of protons have also been observed in a cellular medium, as in the case of yeast cells during glycolysis[17] and in neutrophil cells.[18]

We are interested in studying a minimal pH oscillator which involves an enzyme or an enzyme model compound. To this purpose, we chose hemin, which is known to be part of the active centres of many enzymes that carry a heme as prosthetic group. Thus, hemin plays the role of a model substrate for a variety of enzymes which possess such a structural feature. In the present article, we study the dynamics displayed by a reaction system which combines the autocatalytic step of the pH oscillators with a feedback loop which is provided by reactions of the enzyme model compound hemin.

Experimental

Reactant solutions were prepared daily from reagent grade H_2O_2 (Baker), hemin (Fluka), Na_2SO_3 (Merck) and H_2SO_4 (Riedel-de Haën) without further purification. Water produced by an ion-exchanger and a membrane filtration purifying system was used to prepare the solutions. All solutions were stored in the dark in a refrigerator (at 8 °C). Stock solutions were discarded 36 h after preparation.

The H_2O_2 solution was prepared by dilution of a 30% stock solution. Its H_2O_2 content was checked spectrophotometrically using three wavelengths in the range 210–250 nm.[19] The 0.075 M sulfite solution was prepared by dissolving Na_2SO_3 in N_2 or argon-saturated aqueous solutions that contained 0.2 mM H_2SO_4. Finally, hemin was dissolved in a 40 μM NaOH solution, since this enzyme model compound is insoluble in acidic or neutral aqueous medium. To obtain complete dissolution of the hemin crystals, solutions of hemin in alkaline medium were shaken for 12 h prior to use.

For experiments under argon atmosphere, the water was bubbled with argon for 30 min prior to the preparation of the stock solutions. During the preparation of these solutions and while transferring them to the gas-tight syringes, an argon stream was passed over the solutions. Bubbling was avoided in order to prevent the loss of any volatile compounds. During the experiments a stream of argon was supplied to the head volume of the CSTR to prevent the uptake of CO_2 from the atmosphere.

The reactant solutions were stored in gas-tight syringes (50 ml) that were placed in the thermostatting block of the syringe pump. Immediately prior to the start of an experiment, the reactor was filled with 15 ml of each of the three reactant solutions. The experiment was subsequently started by activating the step-motor driven syringe pump, which supplied the reactants to the reactor. All experiments were carried out at 25.0 ± 0.3 °C.

The dynamics of the hemin system were studied in a water-jacketed glass CSTR with a liquid volume of 45 ml. The reactant solutions were supplied to the reactor through three Teflon tubes which enter the reactor through its lid. The ends of the tubes were immersed in the reaction solution in the CSTR. Excess liquid was removed from the reactor by a capillary connected to an aspirator. A glass pH electrode (Mettler-Toledo) was used to monitor the dynamics of the reaction. It is connected to a computer *via* an A/D converter and to a *x,t*-chart recorder. Data is stored on a computer for later analysis.

The dynamic behaviour of the hemin–H_2O_2–sulfite reaction system was investigated at different flow rates of the reactants. The flow rate k_f, which is the reciprocal residence time τ of the reactants in the CSTR, is used as the bifurcation parameter in all CSTR experiments. Since the reactor is fed by three reactant streams containing 50–100 µM hemin, 0.1 M H_2O_2, and 0.075 M Na_2SO_3, respectively, the reactor concentration of these reactants is one third that of the corresponding stock solutions.

Results

CSTRs allow a reaction system to settle on its asymptotic dynamic states. Therefore, the subsequent experiments were performed under CSTR operation mode. The flow rate of the reactants through the reactor was varied and the dynamic behaviour of the reaction system was monitored by a pH-sensitive electrode.

At low flow rates (*i.e.* $k_f < 7.0 \times 10^{-4}$ s^{-1}), the hemin–H_2O_2–sulfite reaction settles onto an acidic stationary state, the pH of which lies below ~6.5. When the flow rate exceeds the critical value of $k_{f, crit} = 7.0 \times 10^{-4}$ s^{-1}, the reaction system begins to show oscillatory behaviour (Fig. 1(a)). At flow rates slightly above $k_{f, crit}$, these periodic oscillations are sinusoidal. As the flow rates increase further, the amplitudes of the oscillations grow significantly (Fig. 2), and sharp peaks

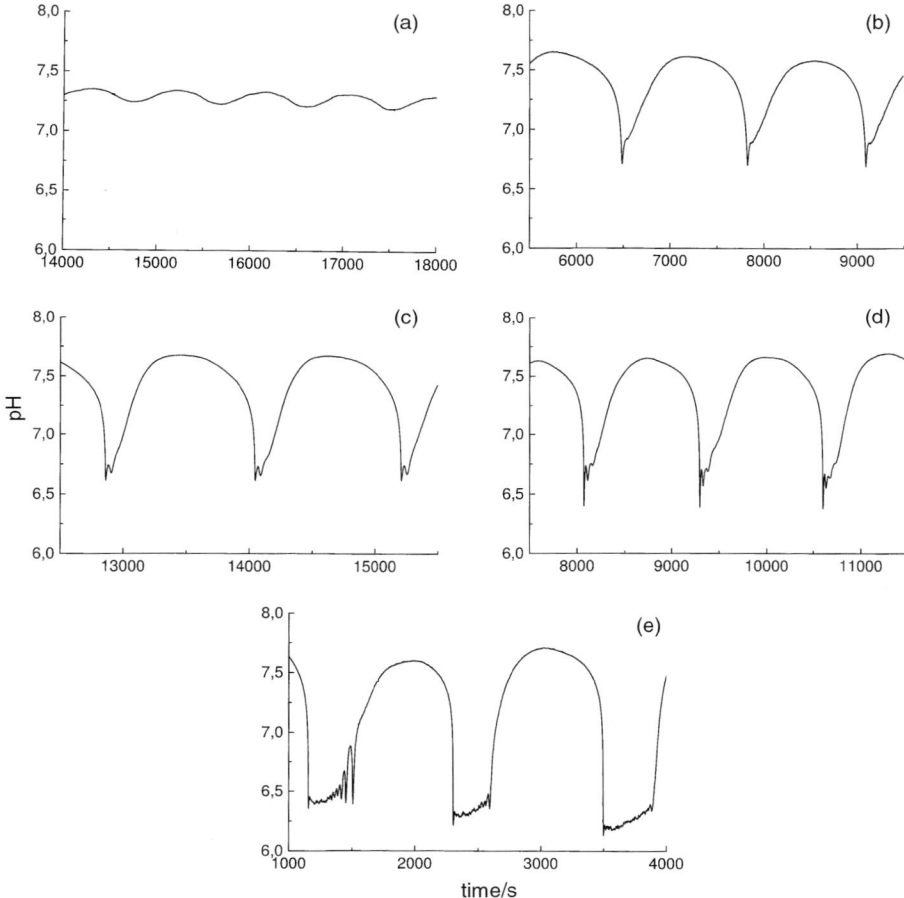

Fig. 1 Oscillations observed for increasing values of the flow rate k_f in the hemin–H_2O_2–SO_3^{2-} system under atmosphere. (a) At $k_f = 7.4 \times 10^{-4}$ s^{-1} the oscillations are sinusoidal and have low amplitudes. (b) At $k_f = 1.3 \times 10^{-3}$ s^{-1} the oscillations still have a simple periodicity, however, the peaks in the acidic domain become sharper. (c) At $k_f = 1.5 \times 10^{-3}$ s^{-1} a second minimum arises in the acidic region, thus yielding a bursting oscillation composed of one large-amplitude oscillation and one small-amplitude oscillation. (d) Bursting oscillations, consisting of one large-amplitude oscillation and two small-amplitude oscillations (at $k_f = 1.8 \times 10^{-3}$ s^{-1}) and (e) oscillations where bursting plays a prominent role, at $k_f = 2.2 \times 10^{-3}$ s^{-1}.

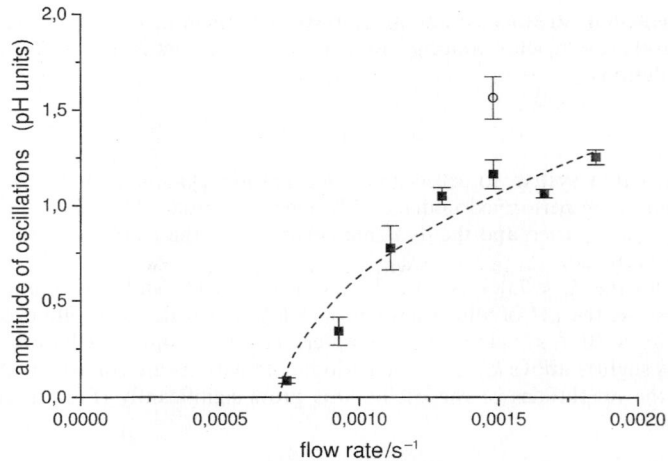

Fig. 2 Dependence of the amplitude of the oscillations on the flow rate k_f of reactants into the CSTR. Data obtained (■) under atmosphere, (○) under argon. The dashed line represents a fit of a square root dependence of the amplitude of oscillations A on the distance of the onset of oscillations to the data, according to $A = A_0(k_f - k_{f,\,crit})^{1/2}$, with $k_{f,\,crit} = 7.0 \times 10^{-4}$ s^{-1} and $A_0 = 38$.

develop at low pH region of the oscillations (Fig. 1(b)). The growth rate of the amplitudes is well described by a square root dependence of the amplitude A from the difference between the actual flow rate k_f and the critical flow rate $k_{f,\,crit}$, i.e. $A = A_0(k_f - k_{f,\,crit})^{1/2}$. By contrast, the periods of the oscillations are much less affected by the rise in k_f, as seen in Fig. 3.

Upon further increases in the flow rate, the simple periodicity of the oscillations no longer persists. Instead, we observe the emergence of bursting oscillations at $k_f \sim 1.4 \times 10^{-3}$ s^{-1}. The oscillations develop a second minimum in the pH, which is characterised by a much lower amplitude than the original minimum (Fig. 1(c)). The initial transition occurs from "simple" periodic oscillations with a single periodicity to periodic oscillations whose oscillatory cycle comprises one large-amplitude oscillation and one small-amplitude oscillation (Fig. 1(c)). As the flow rate is further increased, the number of small-amplitude oscillations augments by one, resulting in oscillatory cycles composed of one large-amplitude oscillation and two small amplitude oscillations

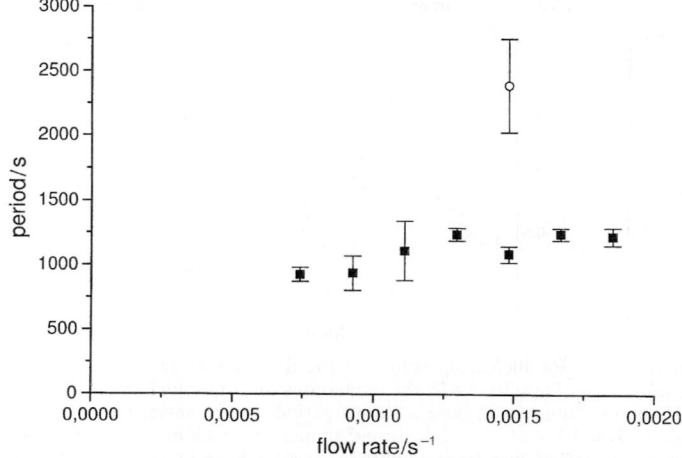

Fig. 3 Dependence of the period of the oscillations on the flow rate k_f of reactants into the CSTR. Data obtained (■) under atmosphere, (○) under argon.

(Fig. 1(d)). The number of such small bursts augments continuously as the flow rate increases. Such behaviour is known for mixed-mode oscillations which follow a so-called period-adding scenario.[20–22] At high flow rates these small-amplitude bursts become a quite prominent feature. One observes oscillations that perform a long series of such spikes when the oscillation is in the acidic part of the oscillatory cycle (Fig. 1(e)). At high flow rates, the reaction system spends almost as much time in the acidic part of the oscillatory cycle as in its alkaline part. Finally, at even higher values of the flow rate (*i.e.* at $k_f > 3.3 \times 10^{-3}$ s^{-1}) the oscillations cease and give way to an alkaline stationary state whose pH is above ~ 7.6.

The experiments described so far were conducted under aerobic conditions, *i.e.* the reactant solutions and the reactor were equilibrated with the atmosphere. However, Rábai and Hanazaki[10] and Frerichs and Thompson[13] reported that even trace amounts of hydrogen carbonate may generate a HCO_3^--driven feedback loop and hence induce oscillatory behaviour. In order to assess the contribution of CO_2 and O_2 to the oscillatory dynamics of the hemin–H_2O_2–sulfite reaction system, we repeated the CSTR experiments using an argon atmosphere instead of aerobic conditions. To eliminate traces of CO_2 and O_2 from participating in the reaction, all solutions were prepared under argon. The head space of the CSTR was flushed with argon throughout the course of the experiments.

As in the experiments under atmosphere, the reaction under argon also showed oscillatory dynamics, which comprised simple periodic as well as bursting oscillations (Fig. 4). However, for given values of the flow rates, the amplitudes of the oscillations were found to be larger than those obtained under aerobic conditions (Fig. 2). Similarly, the period of oscillations was observed to be almost twice as long under anaerobic conditions as the experiments run under aerobic conditions (Fig. 3). Due to the lengths of the oscillatory periods (*ca.* 40 min) only a few oscillations could be detected in the present reactor (of 45 ml volume). In addition, the dynamics depicted in Fig. 4 is seen to be still transient. At present, experiments are under way to determine the asymptotic dynamics of this system under argon.

Finally, we conducted a series of experiments to assess whether the hemin–H_2O_2–sulfite reaction may also show oscillatory behaviour under batch conditions, *i.e.* in a reactor without any inflow or outflow of chemicals. Under batch conditions, we did not detect oscillations, however, by operating the reactor under fed-batch conditions, damped oscillations of the pH value could be observed. Damped oscillations (not shown) lasted for up to 30 min, before the reaction settled to a stationary state at around pH 7.2 ± 0.5. It is worth noting that the window of reaction conditions leading to damped oscillations is narrow. If the reaction is performed under reaction conditions

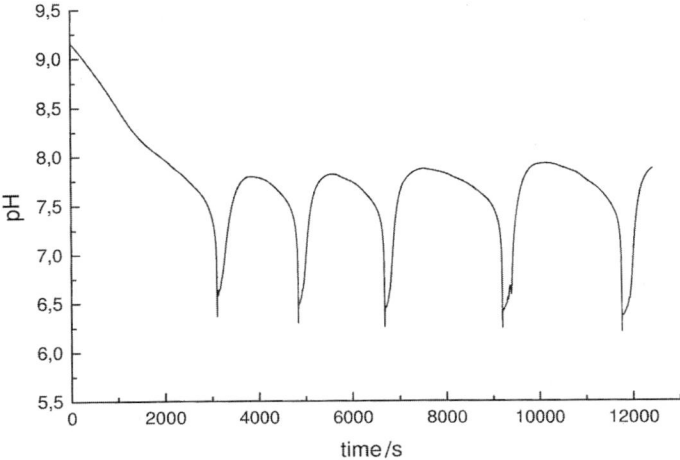

Fig. 4 Dynamical behaviour of the hemin–H_2O_2–SO_3^{2-} system under argon atmosphere at $k_f = 1.5 \times 10^{-3}$ s^{-1}. Note that the shapes of the oscillations are similar to those shown in Fig. 1. However, the dynamics are still in the transient regime, due to the long duration of the period of the oscillations.

that do not lead to damped oscillations, the reaction settles in stationary states which are characterized either by a very high pH (9–10) or a very low pH (3–4), respectively.

Mechanism

The hemin oscillator belongs to the family of pH oscillators based on the oxidation of sulfite by H_2O_2. These chemical oscillators share a common set of reactions that yield an autocatalytic production of H^+, while they differ in the H^+-consuming reactions that are involved in the feedback loop required for oscillatory dynamics.

The reactions involved in the autocatalytic oxidation of sulfite by H_2O_2 are well-established,[8–15] and their rate constants have been reinvestigated recently in dependence on the temperature.[16] The autocatalytic part of the mechanism comprises the protonation/deprotonation of SO_3^{2-} (eqn. (1)), and the reduction of H_2O_2 by SO_3^{2-} (eqn. (2)) and by HSO_3^- (eqn. (3) and (4)):

$$SO_3^{2-} + H^+ \rightleftharpoons HSO_3^- \tag{1}$$

$$H_2O_2 + SO_3^{2-} \rightarrow SO_4^{2-} + H_2O \tag{2}$$

$$H_2O_2 + HSO_3^- \rightarrow SO_4^{2-} + H^+ + H_2O \tag{3}$$

$$H_2O_2 + HSO_3^- + H^+ \rightarrow SO_4^{2-} + 2H^+ + H_2O \tag{4}$$

It is important to emphasise that the oxidation of HSO_3^- by H_2O_2 (eqn. (3) and (4)) is pH dependent and that it is autocatalytic in protons in an acidic medium (eqn. (4)).

The remaining task is to establish which reactions are involved in the negative feedback loop, i.e. which reactions slowly consume H^+. In the following, we discuss which reactions may take over this function in the hemin–H_2O_2–SO_3^{2-} system.

It has been shown recently that low concentrations of CO_2 and HCO_3^- may give rise to a negative feedback loop.[10,13] To test whether these compounds are responsible for the occurrence of oscillations in the hemin oscillator, we performed experiments under argon atmosphere, which exclude the participation of CO_2 and HCO_3^- in the reaction. As in the experiments run in contact with air, we also observed oscillations under argon atmosphere conditions (Fig. 4). Therefore, we conclude that participation of CO_2 and HCO_3^- is not essential for the emergence of a feedback mechanism. In other words, other H^+-consuming reactions provide the feedback steps of the hemin–H_2O_2–SO_3^{2-} system.

A plausible source for the negative feedback is the acid–base behaviour of aqueous hemin solutions. In an aqueous medium hemin may coordinate either one of two water molecules or hydroxy groups to the iron atom. Which and how many of these ligands are coordinated is strongly dependent on the pH.[23–26] In aqueous alkaline solutions, hemin shows acid–base equilibria between a porphyrin which coordinates with two hydroxy ligands $[Fe^{III}Por(OH)_2]^-$, a porphyrin coordinating to one hydroxy and one aquo ligand $[Fe^{III}Por(OH)(H_2O)]$ and a porphyrin coordinating to two aquo ligands $[Fe^{III}Por(H_2O)_2]^+$ (in this notation Fe and Por stand for the central iron atom and the porphyrin moiety of hemin, respectively):

$$[Fe^{III}Por(OH)_2]^- + H^+ \rightleftharpoons [Fe^{III}Por(OH)(H_2O)] \tag{5}$$

$$[Fe^{III}Por(OH)(H_2O)] + H^+ \rightleftharpoons [Fe^{III}Por(H_2O)_2]^+ \tag{6}$$

The pK_a value of equilibrium (5) has been determined as $pK_{(5)} \sim 13$ while the pK_a values reported for equilibrium (6) vary from 4.8^{23} to $\sim 6.5.^{24}$ Due to its high pK_a value, equilibrium (5) can be disregarded when modelling the hemin oscillator.

In addition to equilibria (5) and (6), hemin is also known to form μ-oxo-dimers, which in turn are in H^+-dependent equilibrium with monomer species. Of interest for the hemin oscillator is the equilibrium between the monomers $[Fe^{III}Por(OH)(H_2O)]$ and the corresponding μ-oxo-dimers:

$$2[Fe^{III}Por(OH)(H_2O)] \rightleftharpoons [Fe^{III}Por(OH)]_2(OH^-) + H^+ + H_2O \tag{7}$$

for which pK_a values of 7.1^{25}–7.24^{26} have been determined. These equilibria show that, depending on the pH of the reaction medium, hemin may exchange ligands, thus releasing or consuming H^+.

In our model, we include the autocatalytic reactions (1)–(4), as well as the acid–base equilibria (6) and (7) involving hemin species. The latter two acid–base equilibria form the feedback reactions in the hemin oscillator. This mechanism is able to model the essential dynamical features observed in the experiments.

Discussion

In the present paper we investigated oscillations of the pH value that occur during the hemin mediated oxidation of sulfite by hydrogen peroxide. Different types of oscillatory behaviour were detected as a function of the flow rates of the reactants through the CSTR. Oscillations begin at $k_{f, crit} = 7.0 \times 10^{-4}$ s^{-1} and their amplitudes grow with increasing flow rates. The exponent of the increase in amplitude is found to be 0.5, in good agreement with predictions for supercritical Hopf bifurcations.[27] At flow rates exceeding 1.4×10^{-3} s^{-1} the simple relaxation oscillations give way to mixed-mode or bursting oscillations. The shape of these oscillations is reminiscent of those observed in other physiological systems,[1,2,28] like in the oscillations of Ca^{2+} in hepatocytes stimulated by adenosine triphosphate (ATP).[29]

Upon a continuous increase in the flow rate the number of bursting oscillations per oscillatory cycle successively increases by one burst. Thus, the complexity of the periodic (bursting) state as a function of k_f increases by following a so-called period-adding sequence. Bursting behaviour has previously been reported for other, inorganic, pH oscillators which also pertain to the family of pH oscillators involving the oxidation of SO_3^{2-} by H_2O_2.[30,31] However, to our knowledge, a period-adding sequence in response to changes in a bifurcation parameter has not yet been observed in a member of the SO_3^{2-}–H_2O_2–familiy of pH oscillators.

Period-adding sequences have already been observed in an enzymatic reaction system, the peroxidase–oxidase (PO) reaction.[20,21] The major difference between the period-adding sequences of the PO system and the hemin–SO_3^{2-}–H_2O_2 system is that, so far, chaotic dynamics have not been detected in the hemin system. However, in both reaction systems, the period-adding sequences allow for a fine tuning of the shape of the bursting oscillations by small changes in the bifurcation parameter. Given that the shape of bursting oscillations may be the "carrier" of physiological information, e.g. as known for Ca^{2+} oscillations,[29] the possibility to tune the shape of the oscillations may be of physiological relevance for these enzyme and enzyme model systems.

The active centres of the peroxidases, that support the oscillating PO reaction, contain a heme group which structurally and chemically is very similar to hemin. Thus, hemin is predestined to serve as a model compound for mimicking processes occurring at the reactive centre of the peroxidases and other heme-containing enzymes.[32]

Interestingly, horseradish peroxidase has been found to catalyse the oxidation of sulfite by H_2O_2.[33–35] However, the peroxidase-catalysed reaction of SO_3^{2-} and H_2O_2 proceeds without production or consumption of H^+ by peroxidase. In addition, experiments with denatured horseradish peroxidase, i.e. an enzyme whose protein part is irreversibly altered, do not support the oxidation of sulfite by H_2O_2.[35] These findings indicate that there is a difference between the hemin–H_2O_2–sulfite reaction system and the peroxidase-catalysed oxidation of sulfite by H_2O_2: While the protein moiety of peroxidase is involved in the enzymatic catalysis, the hemin system neglects any contribution that does not stem directly from processes occurring at the hemoporphyrin. Thus, hemin can serve as a model compound for processes that occur at the active centres of heme-carrying enzymes.

In the hemin–H_2O_2–sulfite reaction system, hemin is actively involved in the feedback reactions which consume H^+. Preliminary experiments performed under argon atmosphere (Fig. 4) indicate that the hemin system displays the same dynamic behaviour as under air (Fig. 1). Thus, it is concluded that neither CO_2 nor HCO_3^- are essential to induce oscillatory and bursting behaviour in the hemin system. However, the difference in period and amplitude of oscillations observed for identical experimental conditions, but for different compositions of the gas-phase, show that—albeit not essential to generate oscillations—CO_2 and HCO_3^- contribute to the dynamics of the hemin system under atmospheric conditions.

To conclude, the hemin–H_2O_2–sulfite reaction system can be considered as a new member of the family of pH oscillators that have the reaction of H_2O_2 and sulfite reaction in common. In addition to its rich dynamics (which is presently under investigation), an attractive feature of the hemin–H_2O_2–sulfite system is the biological relevance of hemin as an enzyme model compound which can mimic processes that take place at the heme groups of enzymes. In this context, an interesting function is that of a model for catalases. These enzymes promote the disproportionation of H_2O_2 and they are known to be involved in oscillatory reactions.[6]

Acknowledgements

This study was supported by the Deutsche Forschungsgemeinschaft, the Stiftung Volkswagenwerk, Hannover, and the Hungarian Scientific Reasearch Fund (OTKA).

References

1. A. Goldbeter, *Biochemical Oscillations and Cellular Rhythms. The Molecular Bases of Periodic and Chaotic Behaviour*, Cambridge University Press, Cambridge, 1996.
2. L. Glass and M. C. Mackey, *From Clocks to Chaos. The Rhythms of Life*, Princeton University Press, Princeton, 1988.
3. R. Larter, L. F. Olsen, C. G. Steinmetz and T. Geest, in *Chaos in Chemistry and Biochemistry*, ed. R. J. Field and L. Györgyi, World Scientific, Singapore, 1993, pp. 175–224.
4. A. Scheeline, D. L. Olson, E. P. Williksen, G. A. Horras, M. L. Klein and R. Larter, *Chem. Rev.*, 1997, **97**, 739.
5. M. J. B. Hauser and L. F. Olsen, in *Transport and Structure—Their Competitive Roles in Biophysics and Chemistry*, ed. S. C. Müller, J. Parisi and W. Zimmermann, Springer, Heidelberg, 1999, pp. 252–272.
6. A. J. Davison, A. J. Kettle and D. J. Fatur, *J. Biol. Chem.*, 1986, **261**, 1193.
7. A. P. H. J. Schenning, J. H. Lutje Spelberg, M. C. P. F. Driessen, M. J. B. Hauser, M. C. Feiters and R. J. M. Nolte, *J. Am. Chem. Soc.*, 1995, **117**, 12655.
8. Gy. Rábai, K. Kustin and I. R. Epstein, *J. Am. Chem. Soc.*, 1989, **111**, 3870.
9. Gy. Rábai and I. Hanazaki, *J. Phys. Chem.*, 1994, **98**, 2592.
10. Gy. Rábai and I. Hanazaki, *J. Phys. Chem.*, 1996, **100**, 10615.
11. Gy. Rábai, *J. Phys. Chem. A*, 1997, **101**, 7085.
12. Gy. Rábai and I. Hanazaki, *J. Am. Chem. Soc.*, 1997, **119**, 1458.
13. G. A. Frerichs and R. C. Thompson, *J. Phys. Chem. A*, 1998, **102**, 8142.
14. V. Vanag, *J. Phys. Chem. A*, 1998, **102**, 601.
15. Gy. Rábai, N. Okazaki and I. Hanazaki, *J. Phys. Chem. A*, 1999, **103**, 7224.
16. I. Hanazaki, N. Ishibashi, H. Mori and Y. Tanimoto, *J. Phys. Chem. A*, 2000, **104**, 7695.
17. Th. Mair and S. C. Müller, *J. Biol. Chem.*, 1996, **271**, 627.
18. H. R. Petty, R. G. Worth and A. L. Kindzelskii, *Phys. Rev. Lett.*, 2000, **84**, 2754.
19. M. S. Morgan, P. F. van Trieste, S. M. Garlick, M. J. Mahon and A. L. Smith, *Anal. Chim. Acta*, 1988, **215**, 325.
20. M. J. B. Hauser and L. F. Olsen, *J. Chem. Soc., Faraday Trans.*, 1996, **92**, 2857.
21. M. J. B. Hauser, L. F. Olsen, T. V. Bronnikova and W. M. Schaffer, *J. Phys. Chem. B*, 1997, **101**, 5075.
22. K. P. Zeyer and F. W. Schneider, *J. Phys. Chem. A*, 1998, **102**, 9702.
23. O. S. Ksenzhek and S. A. Petrova, *Bioelectrochem. Bioenerg.*, 1978, **5**, 661.
24. C. Bartocci, F. Scandola, A. Ferri and V. Carassiti, *Inorg. Chim. Acta*, 1979, **37**, L473.
25. T. Uno, A. Takeda and S. Shimabayashi, *Inorg. Chem.*, 1995, **34**, 1599.
26. M. F. Zipplies, W. A. Lee and T. C. Bruice, *J. Am. Chem. Soc.*, 1986, **108**, 4433.
27. R. Seydel, *From Equilibrium to Chaos—Practical Bifurcation and Stability Analysis*, Elsevier, New York, 1988.
28. J. Keener and J. Sneyd, *Mathematical Physiology*, Springer, New York, 1998.
29. C. J. Dixon, N. M. Woods, K. S. R. Cuthbertson and P. H. Cobbold, *Biochem. J.*, 1990, **269**, 499.
30. Gy. Rábai and I. Hanazaki, *J. Phys. Chem. A*, 1999, **103**, 7268.
31. Gy. Rábai, A. Kaminaga and I. Hanazaki, *Chem. Commun.*, 1996, 2181.
32. Z. Genfa and P. K. Dasgupta, *Anal. Chem.*, 1992, **64**, 517.
33. T. Araiso, K. Miyoshi and I. Yamazaki, *Biochemistry*, 1976, **15**, 3059.
34. C. Mottley, R. P. Mason, C. F. Chignell, K. Sivarajah and T. E. Eling, *J. Biol. Chem.*, 1982, **257**, 5050.
35. C. Mottley, T. B. Trice and R. P. Mason, *Methods Enzymol.*, 1982, **22**, 732.

Control of the excitability of neuronal tissue by weak external forces

Wolfgang Hanke,[*a] **Meike Wiedemann**[a] **and Vera M. Fernandes de Lima**[b]

[a] *University of Hohenheim, Institute of Physiology, Garbenstrasse 30, 70599 Stuttgart, Germany. E-mail: hanke@uni-hohenheim.de*
[b] *University of Sao Paulo – POLI, SP, Brazil*

Received 23rd March 2001
First published as an Advance Article on the web 7th November 2001

The spreading depression (SD) is a pronounced example of excitation–depression waves in excitable media, to which neuronal tissue according to its structure and functions belongs. SD waves can especially easily be observed in the vertebrate retina which is neuronal tissue and a true part of the central nervous system (CNS). According to the high intrinsic optical signal (IOS) concomitant with the retinal spreading depression (rSD), it can be monitored with standard video imaging techniques, thus the retina has been used in our studies as a suitable model system for neuronal tissue in general. In particluar, the control of wave set-up and propagation in excitable media by weak external forces is of high interest. Accordingly, the interaction of rSD waves with DC and AC electromagnetic fields of low amplitude and frequency and with gravity has been investigated in this study. The dependence of rSD-wave propagation velocity on the given parameters as one important indication of excitability control has been investigated in detail. Our results with rSD waves are partially compared to another well known excitable medium, the Belousov–Zhabotinsky reaction, where some data about the effects of electrical fields and gravity have already been published.

Introduction

Neuronal tissue and thus also the vertebrate retina, which is a true part of the CNS, has all the properties of excitable media.[1] It shows a variety of behaviours known from such systems *e.g.* oscillations, pattern formation and propagating waves. The SD in the CNS, and especially that in the retina, is a very pronounced example of such behaviour, being an excitation–depression wave which travels at about 3 to 5 mm min^{-1} through the retinal tissue.[2,3] As an additional significant experimental advantage, due to its high IOS the SD can be seen in the retinal tissue with the naked eye and thus can easily be investigated with standard video imaging techniques.[4] Furthermore, the chicken retina used in our experiments has the advantage of being completely avascular. Thus plain neuronal effects can be studied in a very homogenous tissue.[5]

The properties of excitation–depression waves in different excitable media are known to be affected by a wide variety of chemical and physical parameters, including even weak external forces. The effects of low-amplitude electromagnetic fields, and to some extent also gravity, on processes in excitable media have thus been studied in more detail, but mainly in other excitable media than the retinal SD or neuronal tissue, especially in the well known Belousov–Zhabotinsky reaction (BZ).[6,7] Considering electrical fields, both theoretical[8] and experimental evidence has been presented to show that wave propagation parameters are affected by the direction and the amplitude of an applied electrical field, as well as by its frequency.[9,10] Additional studies have shown that the tip movement of spirals of BZ waves can be controlled by electrical fields.[11] A first study[7] has also shown that BZ-waves respond to gravity. From the theoretical point of view, the mechanism of the BZ reaction is quite well understood and the interaction of this system at least with electrical fields (currents) can thus basically be described by adequate theories. The waves in the BZ system are very similar in their behaviour to the SD, and consequently generalised theories of the BZ may also be used to describe the behaviour of the SD. Thus the BZ system can be used (partially) as a model system to gain understanding of the SD. The interactions of the BZ with electrical fields and gravity, as described in the literature, inspired us to do this study, and the BZ results are finally compared with our results concerning the retinal SD.

Concerning the interaction of spreading depression waves in neuronal tissue with electromagnetic fields or currents, not very much is known. Some basic theoretical considerations about SD waves have been published,[12] but the interaction of SD waves, and more generally of neuronal tissue, with electromagnetic fields has not yet been investigated in depth, although it would be of high interest. However, the electromagnetic fields induced by SD waves have been described[13,14] and partially measured by magnetoencephalographic (MEG) experiments.[15] Additionally, a variety of mostly epidemiological studies on the influence of electromagnetic field (DC and low frequency as well as high frequency) on neuronal tissue and especially the human brain have long existed.[16–18]

The basic units of neuronal tissue, neurons and glia cells, have been very carefully investigated and their electrical properties are well understood. The influence of currents and fields on the cells and on subunits and membrane components of such cells have been described extensively in the literature.[19] On the basis of these studies it also seems reasonable to expect effects of currents and fields on networks of such cells. Consequently, excitation–depression waves (for example rSD waves) in systems of coupled cells should also be affected by electrical fields. The same argumentation holds for the action of gravity, as it at least basically has been shown that gravity directly influences ion channels incorporated into membranes,[20,22] thus it again seems reasonable to postulate that gravity can influence SD waves.

In this study we have investigated the effects of low amplitude electrical fields on neuronal tissue, using the retinal spreading depression as a suitable model system. We have especially looked for the effects of low frequency AC as well as of DC currents and fields on the propagation parameters of rSD waves. In the AC studies 16 2/3 Hz and 50 Hz were used (high frequencies are not subject of this study), because these frequencies are most common in power lines and thus their effects are of some general interest. It is shown that the velocity of the rSD waves depends on direction, amplitude and frequency of the currents. Polarization effects, using isolated electrodes without currents, are very small, even at much higher fields (up to $\sim$kV cm^{-1}) and are excluded from our results. The effects of electrical fields (currents) were comparable in fully recovered and in partially refractory tissue.

It is furthermore shown in this study that gravity as another weak force, being permanently present under terrestrial conditions, directly modifies the velocity of SD waves propagating in retinal tissue.

Materials and methods

Experiments concerning electromagnetic fields

Retinal eye-cups including the intact retina were prepared from 7 to 14 day old chicken as described in detail previously.[5] These eye-cups were glued with cyan-acrylate glue into Petri dishes, which were equipped with an additional electrode system to apply the electrical field. The eye-cups were covered with a Ringer solution of the following ionic composition (all in mM): 100 NaCl, 6 KCl, 1 MgSO$_4$, 1 CaCl$_2 \cdot$ H$_2$O, 1 NaH$_2$PO$_4$, 30 NaHCO$_3$, 10 TRIS, 30 glucose. The pH was adjusted to 7.4 and the temperature of the system was kept at 30 °C for the complete experiment. All salts and solvents used were purchased from Sigma, Fluka or Merck and were at least of p.a. purity. The water used was double quartz distilled.

After preparation, the eye-cups were kept at rest for about 30 min to allow recovery from the preparation. During the complete experiment the eye-cups were perfused with the Ringer solution at a rate of 6 ml min^{-1} (the Petri dishes have a volume of 8 to 10 ml depending on the filling height). The diameter of the Petri dishes used is 35 mm, they were typically filled to about 10 mm height. The chicken retinas used have a thickness of about 0.3 mm and are about 10 mm in diameter. The specific electrical conductance of the aqueous solution used is 12.9 mΩ^{-1} cm^{-1} as directly measured. The electrical conductance of neuronal tissue (and thus the retina) is about 0.02 mΩ^{-1} cm^{-1}.[23]

In our experiments rSD waves were elicited mechanically by gently touching the retina with a fine glass needle. The procedure was observed through a binocular microscope to avoid unnecessary damage of the tissue. In all experiments, at the beginning, a number of waves without applied field were measured as controls. Following that, waves under the influence of an applied electrical field were elicited. The field was always switched on directly before eliciting the wave and switched off directly after the wave had passed the complete retina, to avoid long-term effects. At the end of each experiment, controls without field were again measured. Either waves were elicited every 20 min (fully recovered tissue) or every 5 min (partially refractory tissue). The parameters of the waves under the action of electrical fields were normalised to those of the control waves.

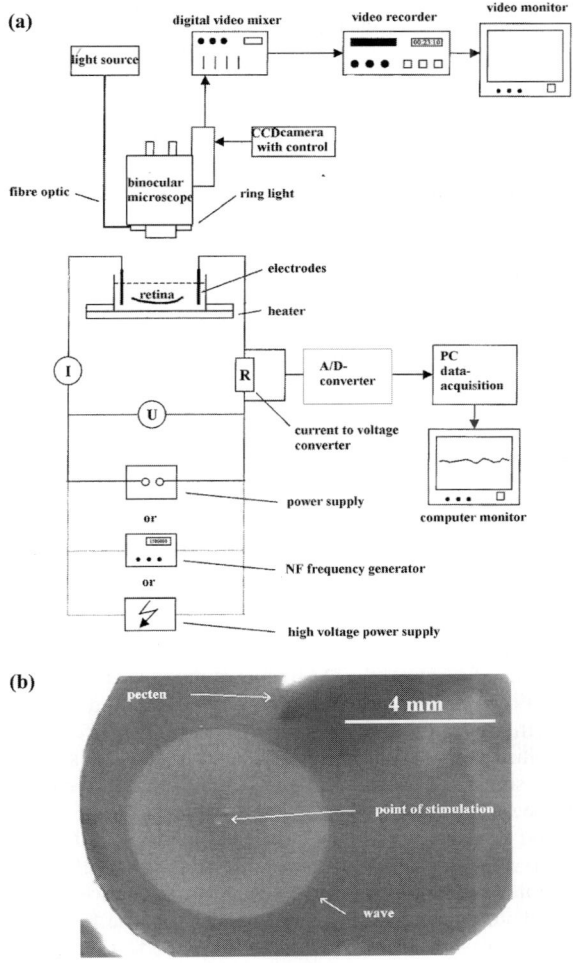

Fig. 1 (a) A cartoon of the complete set-up used for measurements of retinal spreading depression waves under the action of electrical fields. The mechanical stimulation of the retina is not explicitly shown in the cartoon. (b) Photo of a circular wave in a retina eye-cup. The pecten, the wave, and the place of stimulation, which in this case has a small lesion, are marked.

The electrical field was established by platinum–iridium electrodes immersed in the bathing solution; they did not touch the retinal tissue. The size of the electrodes was 3×5 mm^2. For current-free measurements the electrodes were isolated with a thin silicon coating. The AC or DC voltage was applied to the electrodes and the current induced in the system was measured (all AC measurements are true RMS). The Petri dishes with the eye-cups were mounted under a binocular microscope with a video-camera attached. The complete experiments were optically recorded on video tape. The electrical data were simultaneously stored on a computer-supported data-acquisition system. Wave propagation velocities were calculated online for control during the running experiments and later offline in detail. Each experiment was repeated at least 4 times. Averaged values are given as means $\pm$SD. A more detailed description of the electrical and optical systems used for rSD experiments can be found in the literature[5] together with additional information about the preparation, stimulus procedures, and the data evaluation.

In Fig. 1(a) a somewhat simplified cartoon of the complete set-up is given. The stimulation is not separately depicted in this figure. In Fig. 1(b) a photo of a wave in a retinal eye-cup is shown. The pecten marked separately in the figure is a pronounced structure in the avascular chicken retina which is necessary for metabolic support of the neuronal tissue. The wave can be seen as a concentric circle of changed brightness with a considerable sharp front. In this experiment the mechanical stimulation resulted in a small lesion (in the middle of the wave) which can also be seen in the photo.

In a smaller number of additional experiments DC and AC magnetic fields up to 100 Hz were applied to retinas by a set of coils (138 mm average diameter) in the Helmholtz configuration allowing homogenous fields up to 3 mT. The retina could be mounted in any direction relative to the applied field in a proper chamber, otherswise the set-up is identical to that for electrical current measurements as described above.

Experiments concerning gravity

To investigate the influence of gravity on the retinal spreading depression, in one set of experiments a complete rSD set-up slightly modified from that described above was used in parabolic flight campaigns (founded by the German Space Agency) for micro-gravity studies. Here the rSD could be investigated at g-values of 1 g (control), 1.8 g (acceleration phases) and 10^{-2} g (micro-g phase).

Hyper-gravity experiments were carried out in a similar set-up in a home-built low-speed centrifuge, being used at up to $5g$ in this study, applied perpendicular to the retina. In the centrifuge, the set-up was completely remote controlled.

Results

Electrical fields

Different series of experiments were carried out to investigate the influence of electrical fields and/or currents on propagating waves of the retinal spreading depression in chicken retinas. In Fig. 2 at the left side a series of photos of a rSD wave is given without electrical field applied. At the right side, under identical conditions, a series of photos with an electrical field (DC, wave is travelling towards the cathode) applied is shown. As can be seen from the photos, the average velocity of the wave travelling towards the cathode is smaller as the distance the wave proceeds in the same time interval is smaller. The wave propagation velocity is calculated from a series of photos as shown, using the stacking method.[4]

Waves were elicited and measured every 20 min in fully recovered tissue under the action of constant applied electrical fields (DC) of different amplitude. The field was switched on directly before each wave was elicited to avoid long term effects. After the wave had passed the complete retina the field was immediately switched off again. In controls the field was switched on directly after eliciting the wave, there was no difference found compared to the above protocol.

In Fig. 3 the wave propagation velocity is given as a function of the total current in the system under three conditions, first the wave travelling towards the cathode, second travelling towards the anode. As can be seen the velocity decreases with increasing current when travelling towards the cathode, whereas it increases slightly when travelling towards the anode. Finally the field was

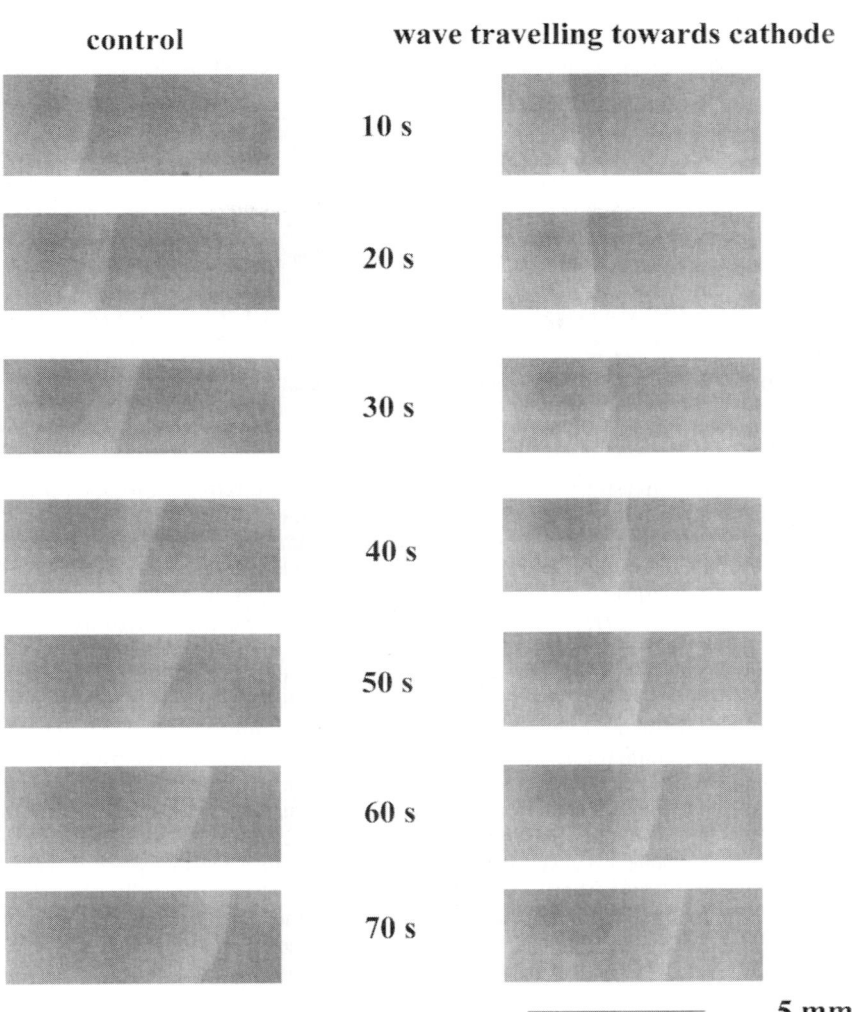

Fig. 2 Left: a series of photos from a propagating rSD wave under control condition given. The wave propagates with a homogenous velocity of about 4.5 mm min^{-1}. Right: a wave is displayed under the application of 80 mA in otherwise identical conditions, with the wave travelling towards the cathode. As can be seen, the IOS and the wave propagation are changed under these conditions, this wave is slower with about 3.5 mm min^{-1}. The time after stimulus is given between the photos.

set perpendicular to the wave propagation and, as can be seen, has no significant effect on the propagation velocity. When waves were induced every 5 min in partially refractory tissue the effects found were qualitatively identical to those in fully recovered tissue.

At higher currents ($I > 50$ mA) reproducibly, with increasing probability at increasing current, spontaneous waves were elicited at the anode side and sometimes also at the cathode side. Comparable effects have been reported previously, however, the waves were then preferentially induced at the cathode. Additionally, at even higher currents ($I > 75$ mA), again with increasing probability at higher currents, lesions were induced in the tissue as also is known from the literature.[24] In partially refractory tissue such lesions were induced already at smaller currents. Both, lesions and spontaneous waves, limit the current range accessible in our experiments to <80–100 mA, as in this study experiments with spontaneous waves or lesions were excluded from the data evaluation.

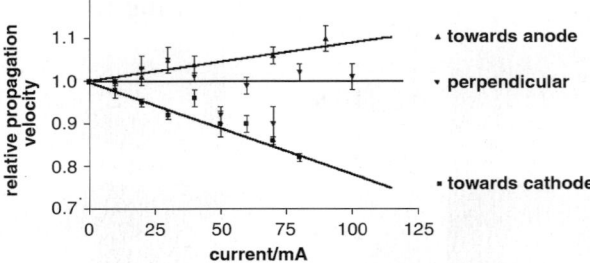

Fig. 3 Dependence of the rSD wave propagation velocity on the current in the system at DC fields. The wave is either travelling towards the cathode, when the velocity decreases with increasing current, or the wave is travelling towards the anode and the velocity slightly increases with increasing current. When the field is perpendicular to the wave propagation direction it has no significant effect on the velocity.

When rSD waves were investigated at higher currents, the wave velocities were frequently inhomogenous in time, being slightly faster at the beginning. In the data evaluation in such cases the wave propagation velocity in the later region was used, as after a short interval of higher velocity at the beginning the velocity typically was constant at a somewhat lower value.

Further effects on wave propagation velocity were observed when the system was pre-treated by a longer period of constant current. In such experiments, waves were measured at the beginning without current being present, then for 30 min 50 mA were applied and then another series of waves was elicited. As can be seen in Fig. 4, directly after the current application, the wave propagation velocity is reduced significantly, but recovers within about 100 min. This effect was independent of the direction of the current application. When 25 mA were applied, the effect was smaller but qualitatively comparable. At higher currents the effect was induced much faster.

To decide whether the effects measured were due to currents or to field induced polarization effects, another series of experiments was carried out with isolated electrodes. In this case, at DC-voltages up to 5000 V at an electrode distance of 3 cm, no reducible effects on the wave propagation parameters were observed (the biggest reduction in wave propagation velocity ever found in these experiments was about 6% at 5000 V applied). At voltages below 100 V no effects were found. At experiments with non-isolated electrodes, and thus current in the system, the voltages necessary to induce the given currents were never higher than about 10 V.

In addition to the effects electrical fields had on the wave propagation velocity, the IOS signal of the waves was affected. As shown in Fig. 5, at higher currents the IOS of rSD waves is partially suppressed. At currents >75 mA the optical profile of the wave changed significantly after a short time of field application, and the IOS of the waves sometimes disappeared completely. Under these conditions we also observed that previously circular waves with homogenous fronts on

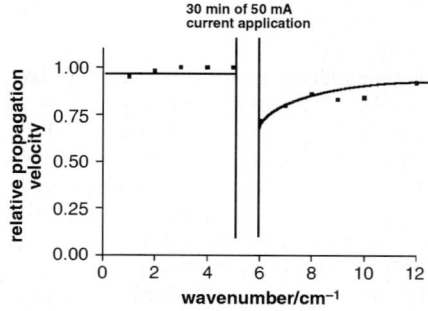

Fig. 4 Decrease of the rSD wave propagation velocity after a 30 min application of 50 mA. Within 100 min the system recovered. Waves before and after current application were elicited every 20 min.

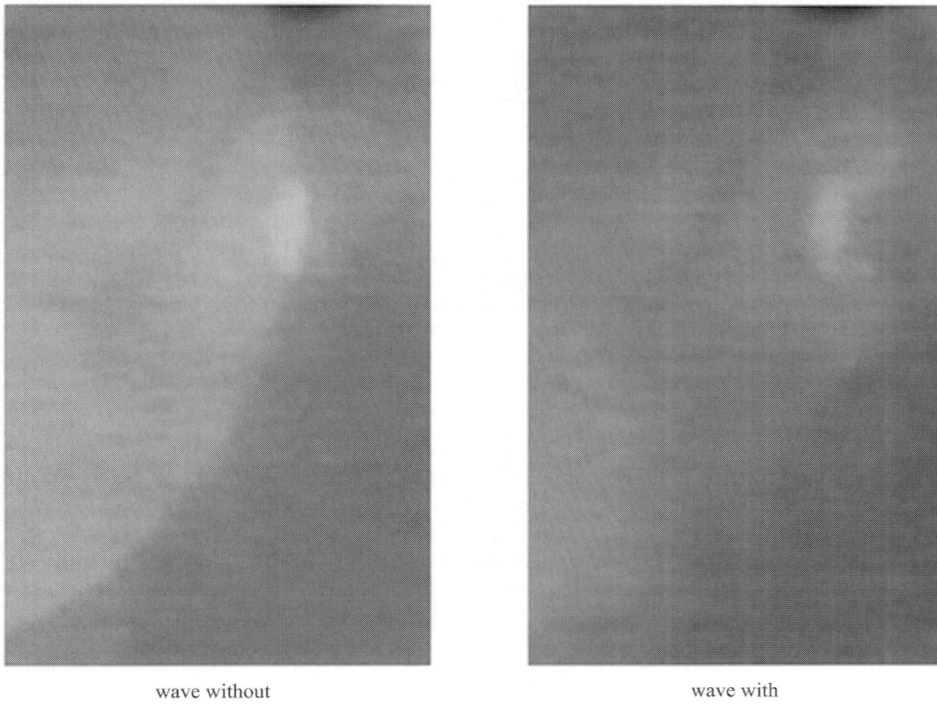

Fig. 5 Photo of a control wave (left) and the same wave under the action of 80 mA (right). The wave was travelling in the direction of the cathode. As can clearly be seen, the relative brightness change of the IOS accompanying the wave is partially reduced under the action of the high current. Additionally the wavefront becomes inhomogenous.

application of the current became inhomogenous or even broke down to fragments of waves with open ends, sometimes leading to spirals.

Finally, the effects of low frequency AC currents on the retinal spreading depression were investigated. At frequencies of 16 2/3 Hz and of 50 Hz currents up to 100 mA were applied to the system. The results for the wave velocity dependence were similar for both frequencies and are given in Fig. 6. Obviously, the wave propagation velocity is increasing with increasing AC-current at both frequencies, and, especially at higher currents, in a non-linear manner. This effect was independent of the direction of the field. At higher currents (>75 mA) spontaneous waves and lesions were again induced, however, under these conditions at both electrodes with about identical probability. The intrinsic optical signal was also affected similarly to the effect described for DC-currents.

Magnetic fields

In a smaller series of experiments homogenous magnetic fields, DC and AC up to 100 Hz, were applied to retinal SD-waves at a field strength up to 3 mT in a Helmholtz configuration coil-system in different orientations. No significant effects were found in these experiments on the velocity of retinal SD-waves propagating in completely recovered tissue. When the waves were initiated in the refractory state of the tissue after previous propagating waves, some long term effects of magnetic fields (min to h of field application) could be observed, which, however, were not very reproducible. A detailed study of these phenomena will be made in future.

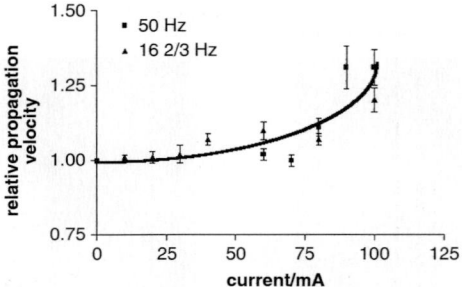

Fig. 6 Dependence of the rSD wave propagation velocity on the amplitude of an applied AC-current at 16 2/3 Hz and 50 Hz. As can be seen, the wave velocity increases at both frequencies with increasing current in a non-linear manner especially at higher currents. The behaviour is about identical for both frequencies, thus only one monotonically increasing curve is fitted to all the data.

Gravity

To get a first idea as to whether gravity interferes with retinal SD-waves, the velocity of such waves in completely recovered tissue was measured in a chamber with the retina being adjusted perpendicular to the surface, so that the g-vector is parallel to the retinal surface. In such a configuration, the normal g-vector (1 g) is in the plane of the propagating wave and there are three possible effects: the waves travel in the direction of the g-vector (down), in the opposite direction (up) or perpendicular to it. In Fig. 7(a) the results of these experiments are summarized in a bar

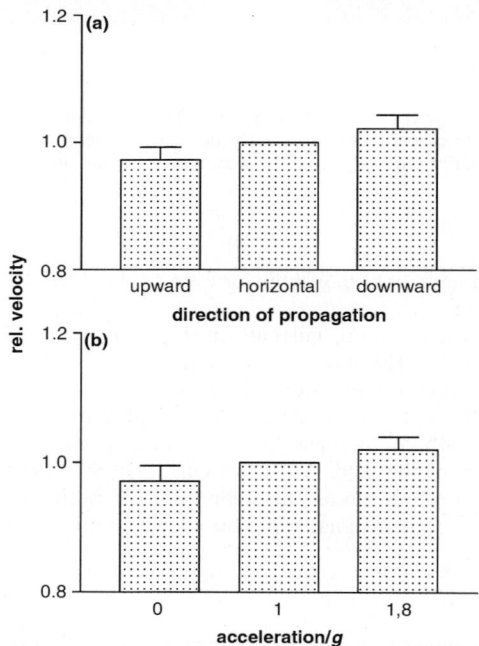

Fig. 7 (a) Bar graph of SD-wave velocity depending on the orientation of the wave relative to the g-vector under laboratory conditions (1 g). Waves travelling down in perpendicular adjusted retinas are faster than waves travelling upwards (against gravity). Waves travelling perpendicular to the g-vector have a medium speed, which is used as reference and set to 1. The other velocities are given relative to this value. (b) Bar graph of the velocity of retinal SD-waves measured in micro-g experiments in a parabolic flight campaign. For 1 g and 1.8 g the g-vector is perpendicular to the retinal surface. The wave velocity increases with increasing gravity. The velocities are normalized to the 1 g value.

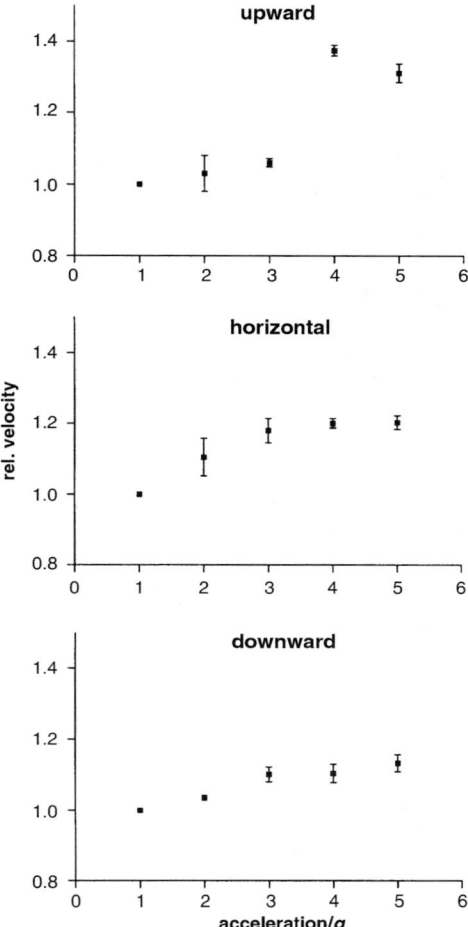

Fig. 8 Velocity of retinal SD-waves as function of gravity applied perpendicular to the retinal surface in a low-speed centrifuge. With increasing gravity the velocity of the waves increases in a nonlinear relation for all orientations. The difference between the waves travelling up and those travelling down may result from the additional terrestrial g-vector being present in these experiments (see Fig. 7(a)).

graph. Waves which are travelling in the direction of the g-vector (down) are faster than those propagating in the opposite direction (up), which are a little slower than the waves propagating perpendicular to the g-vector (right and left). These perpendicular waves are used as reference, their speed is normalized to one and the other values are given relative to this. Even from these simple experiments it is already obvious, that gravity directly influences the velocity of the propagating retinal SD-waves, depending on the orientation of the wave-front relative to the g-vector.

In a second approach SD experiments were carried out during parabolic flights, which deliver data for the propagation velocity of retinal SD-waves at micro-gravity ($<10^{-2}$ g). According to the given experimental conditions, in the control situation (1 g) the g-vector is perpendicular to the retinal surface, the same holds for the 1.8 acceleration phases of the airplane. In Fig. 7(b) a bar graph is given for the wave velocities under these three conditions, here using the value at 1g as reference. As can be seen, the retinal SD-wave velocity increases with increasing gravity (perpendicular to the retinal surface).

To further quantify the effects shown, we mounted retinas in a centrifuge and investigated SD-waves at different hyper-g values (up to 5 g). The retinas were mounted so that the g-vector given by the centrifuge is perpendicular to the retinal surface. We found that the velocity of propagating retinal SD-waves under these conditions increases with increasing gravity at all orientations. The results are shown in Fig. 8. It is obvious that there is a nonlinear relation between gravity and wave propagation velocity. The waves travelling up in the set-up, however, responded different from those travelling down. This difference can possibly be explained by the effect of the normal g-vector (1 g) in the plane of the retina which is permanently present on earth and has been already described above (Fig. 7(a)). A more detailed study about the effects of gravity on the retinal SD is in preparation.

Discussion

It has been known for a long time that the parameters of excitation–depression waves in excitable media can be controlled by weak external forces. In neuronal tissue and especially the CNS, spreading depression waves are the most pronounced example of such waves.[25] SD waves have been proposed to be the basis of classical migraine[26,27] and their induction and control is thus also of high medical interest. The control parameters of SD waves can be chemical and physical in nature. Whereas the effects of drugs on SD waves have been studied in detail,[5,28] much less is known about the effects of physical parameters, especially weak forces. In our study we have utilised the retinal SD as a specific versatile tool[4] to study the interaction of excitation–depression waves in neuronal tissue with electromagnetic fields (and currents), and with gravity as a permanently present weak force under terrestrial conditions.

As is obvious from the results, mainly, if not exclusively, the current flowing in the system and thus in the retinal tissue is the parameter affecting the wave, according to our experiments with electromagnetic fields. As experiments with isolated electrodes had only small effects, even at considerably high field strength (>1000 V cm^{-1}), polarisation effects most probably can be excluded at the lower field strength values (1–100 V cm^{-1}).

The effects of low amplitude magnetic fields also were small. In the case of AC magnetic fields the currents induced in the tissue are obviously too small to have effects. Long term effects of small magnetic fields on neuronal tissue might be possible, but are not very well reproducible and must be investigated in detail in a separate study.

The currents used in our experiments are comparatively high, up to 100 mA, however, the main part of the current is passing only through the bathing solution according to the following estimate. In the case of non-isolated electrodes, current is induced in the aqueous solution and in the tissue. The total current in the system is measured in our experiments. The maximum cross-section of the bathing solution in the Petri dishes used is about 350 mm^2, the retina has a maximum cross-section of about 3 mm^2. According to the data on the specific conductances of the tissue and the bathing solution, and furthermore due to inhomogeneities in the tissue, the current distribution in the complete system cannot be homogenous. At the middle of the Petri dish (cross-section 350 mm^2), where the eye-cup is placed, at a global current of 100 mA, the mean current density is about 0.29 mA mm^{-2} (equivalent to 290 A m^{-2}) in the bathing solution. With the given values for the conductances (12 mΩ^{-1} cm^{-1} and 0.02 mΩ^{-1} cm^{-1}) at a first rough estimate the current density in the tissue will be about 2–3 orders of magnitude lower than the total current measured in the system, this is, more accurately with the above numbers, about 500 mA m^{-2}. In the literature it has been stated that at current densities significantly higher than 100 mA m^{-2} severe damage can be expected in neuronal tissue.[29,30] The above argumentation is supported by the fact that at global currents of about 75 mA and higher, lesions are induced in the retinal tissue with increasing probability. The initiation of lesions and also of SD waves in retinal tissue by electrical currents has been described in detail previously.[24] In our experiments, the occurrence of lesions (and of spontaneous waves) is the natural limit for the current which can be applied to the system, as experiments with spontaneous waves and with lesions were excluded from the data evaluation. Also, inhomogeneities in the wavefronts showed up at higher currents and sometimes the splitting of waves was observed. Similar effects have been described for BZ-waves in the literature.[8]

At lower currents, the propagation velocity of the retinal SD waves was affected, but only when the field vector was adjusted in the direction of the wave propagation; perpendicular fields had no effect. More generally, wave propagation in excitable media obviously seems only to be affected by forces applied in the direction of the wave propagation, as has been shown for example with gravitational forces[7] and electrical fields[8,10] acting on BZ waves.

The effects of currents on the retinal SD were similar in fully recovered and in partially refractory tissue, thus most probably metabolic processes in the recovery period after a wave are not significantly affected by the current. However, long term effects of electromagnetic fields on processes in the recovery period of the neuronal tissue can be expected and will be investigated in more detail later.

AC-fields of 16 2/3 Hz and of 50 Hz at increasing currents increased the wave propagation velocity in a nonlinear manner. A possible frequency dependence, especially at higher frequencies, cannot be excluded (and will be investigated in future in more detailed frequency dependent measurements), although in our experiments the effects were similar at both frequencies. The effects of AC-currents were independent of the direction of the applied field, thus most probably they are based on another mechanism than the DC effects. This can also be concluded from the fact that DC-currents induced a linear effect, whereas AC-currents induced nonlinear effects.

At higher currents, independent of the direction of the field and for both AC and DC fields, long term effects are induced in the retinal tissue which are reversible on a timescale of hours, so long as no lesions are produced. Possibly changes on the cellular level, which adjust the tissue to a lower level of excitability induced by the current, might be responsible for this finding. These effects are more quickly induced and more pronounced at higher currents and might partially induce nonlinearities in the propagation of waves under the action of high currents.

According to a possible mechanism of the observed effects we have already excluded polarization effects and the interaction of currents with metabolic processes in the refractory period of the waves. However, as the propagation of SD waves is significantly defined by the liberation, diffusion and re-uptake of potassium and other ions[12,31] an interaction of electrical fields or currents with the local potassium concentration (but also with other ion concentrations) at the wavefront might well account for part of the effects observed. One mechanism, which might be involved in such an interaction can be due to the fact that current flowing through the tissue will also pass the cells, the current entering the cell being of the same amplitude as the current leaving the cell. Under our experimental conditions, at a global current of 100 mA in the system, equivalent to a current density of about 0.3 mA mm^{-2} (see above), an estimated current density of 0.5 µA mm^{-2} will be given in the tissue according to the above discussion. At a cross-section of a typical cell in the retina of about 100 µm^2 (bigger neuron) to about 10^4 µm^2 (Mueller cell), the corresponding currents through these cells will be 50 pA and 5 nA (rough estimate). The typical resistances for resting cells are in the range of MΩ to GΩ. Thus, a voltage drop in the range of some mV can be expected for one cell in our experiments. Such a voltage drop would be sufficient to shift at least slightly the global open state probability of voltage dependent channels in the cell membranes and thus influence the ion homeostasis and consequently the propagation parameters of spreading depression waves. One should also have in mind that due to the basic nature of excitable media, as the retinal tissue is, small changes in the system parameters can have severe consequences for the properties of propagating waves. Other mechanisms may (with high probability) also be involved, especially in the action of AC-currents, where they might induce the observed nonlinear effects, however, even the simple mechanistic discussion given here can be an interesting starting point for a theoretical approach to the problem in later studies.

At a more populist view, the effects found by us cannot be responsible for postulated effects of typical environmental electrical fields on the medical health of people, as the fields and currents used are much higher than typically found in the environment. Nevertheless, at least a partial control of the excitability of neuronal tissue seems to be possible with electrical fields (currents) of reasonable amplitude.

The effects of gravity on retinal SD-waves are multiple. In this study we have for the first time shown directly such effects, which are dependent on the absolute value of gravity as well as on the orientation of the tissue and the wave propagation relative to the g-vector. The finding that waves travelling in the direction of the g-vector are faster than those in the contrary direction is especially interesting, as it is different from the findings concerning the interaction of BZ-waves with gravity published by other groups.[7] The reason might be found in the different mechanisms underlying SD- and BZ-waves. Independent of this, even when the g-vector is perpendicular to the tissue in which the waves are travelling, gravity interacts with retinal SD-waves in a nonlinear relationship, giving an increase in velocity with increasing gravity amplitude.

According to a possible mechanism of the interaction of gravity with intact tissue, we have earlier shown that even ion-channels in (model) membranes can react to gravity in a nonlinear manner,[20-22] which might be one reason for the interaction of SD-waves with gravity as the propagation of SD-waves to a large extent is dependent on cell-membrane transport processes.

Acknowledgement

This work was partially supported by the DLR (German space agency) by grant No. 50WB9716.

References

1. W. Hanke, M. Goldermann, S. Brand and V. M. Fernandes de Lima, in: *A Perspective Look at Nonlinear Media: From Physics to Biology and Social Sciences*, ed. J. Paresi, S. C. Müller and W. Zimmermann, Springer-Verlag, Berlin, 1998, pp. 227–243.
2. M. Martins-Ferreira, in: *The Brain and the Behaviour of the Fowl*, ed. T. Okada, Japan Scientic Society Press, Tokyo, 1983, pp. 317–333.
3. V. M. Fernandes de Lima and W. Hanke, *Prog. Ret. Eye Res.*, 1997, **16**(4), 657.
4. W. Hanke, M. Goldermann, M. Wiedemann and V. M. Fernandes de Lima, *Microsc. Anal.*, Sept. 1996, 23.
5. V. M. Fernandes de Lima, M. Goldermann and W. Hanke, *The Retinal Spreading Depression*, Shaker Verlag, Aachen, 1999.
6. A. N. Zaitkin and A. M. Zhabotinsky, *Nature*, 1970, **225**, 535.
7. S. Fujeida, Y. Mogami, W. Zhang, H. Shinohara and S. Handa, *Anal. Sci.*, 1999, **15**, 159.
8. H. Sevcikova, J. Kosec and M. Marek, *J. Phys. Chem.*, 1996, **100**, 1666.
9. K. Miayakawa and M. Mizoguchi, *J. Chem. Phys.*, 1998, **109**, 7462.
10. H. Sevcikova, I. Schreiber and M. Marek, *J. Phys. Chem.*, 1996, **100**, 19153.
11. B. Schmidt and S. C. Müller, *Phys. Rev. E*, 1997, **55**, 4390.
12. H. C. Tuckwell and R. M. Miura, *Biophys. J.*, 1978, **23**, 257.
13. Y. C. Okada, M. Lauritzen and C. Nicholson, *Brain Res.*, 1998, **442**, 185.
14. R. S. Wijesenghe and N. Tepley, *Brain Topography*, 1996, **9**(3), 192.
15. K. M. A. Welch, G. Barkley, N. M. Ramadan and G. Dándrea, *Path. Biol.*, 1992, **40**(4), 349.
16. M. Persinger, *ELF and VLF Electromagnetic Field Effects*, Plenum Press, New York, 1974.
17. W. König, *Biological Effects of Environmental Electromagnetism*, Springer-Verlag, New York, 1981.
18. H. Röschke and K. Mann, *Neurophysiology*, 1996, **33**, 32.
19. T. F. Weiss, *Cellular Biophysics: Electrical Properties*, MIT Press, Cambridge, MA, 1997.
20. W. Hanke, *Adv. Space Res.*, 1996, **17**, 143.
21. N. Klinke, M. Goldermann and W. Hanke, *Acta Astronautica*, 2000, **47**, 771.
22. M. Goldermann and W. Hanke, *J. Microgravity Sci. Technol.*, 2001, in press.
23. P. L. Nunez, *Electrical Fields in the Brain: The Neurophysics of EEG*, Oxford University Press, Oxford, 1981.
24. M. Wiedemann and W. Hanke, *Neurosci. Lett.*, 1997, **232**, 99.
25. A. A. P. Leao, *J. Neurophysiol.*, 1944, **7**, 359.
26. P. M. Milner, *Electroenceph. Clin. Neurophysiol.*, 1958, **10**, 705.
27. M. Lauritzen, *Path. Biol.*, 1992, **4**, 332.
28. M. Wiedemann, V. M. Fernandes de Lima and W. Hanke, *Naunym Schmiedberg's Arch. Pharmacol.*, 1996, **353**, 1.
29. J. H. Bernhardt, in *Veröffentlichungen der Strahlenschutzkommission*, Gustav Fischer Verlag, Stuttgart, 1990, Band 6.
30. INIRC, *Health Phys.*, 1998, **74**, 494.
31. B. Grafstein, *J. Neurophysiol.*, 1956, **19**, 154.

Spatio-temporal dynamics in glycolysis

Thomas Mair, Christian Warnke and Stefan C. Müller

Otto-von-Guericke-Universität Magdeburg, Institute of Experimental Physics, Group of Biophysics, Universitätsplatz 2, 39106 Magdeburg, Germany.
E-mail: stefan.mueller@physik.uni-magdeburg.de

Received 10th May 2001
First published as an Advance Article on the web 12th November 2001

During the glycolytic degradation of sugar in a thin layer of yeast extract, travelling waves of NADH and protons can be generated that carry a state of high enzymatic activity through the system. The controlled initiation of such waves with an activator of the enzyme phosphofructokinase (PFK) and the influence of various salts and co-factors on the propagation dynamics are investigated. Furthermore a first study of the dispersion of waves is presented. The experimental characterisation of this *in vitro* system contributes to unravelling the possible role of glycolysis for biological information processing. In this context, the provision of chemically available energy in the absence of compartmentation by glycolysis is of primary importance.

1 Introduction

Rhythmicity is a common phenomenon of life, as we know from daily experience (*e.g.* our heart beat or breathing). It occurs on all levels of biological organisation and on a wide range of timescales. It is well known that cells and organs respond quite often to perturbations in their environment by rhythmic changes of cellular activity. Such a response requires the exchange of information between the cell and the environment and subsequent information processing within the cell. From this behaviour it has been concluded, that oscillatory reactions can have important impact on biological information processing, the oscillatory reaction being a measure of time and/or signal strength.[1–4]

Temporal oscillations in glycolysis were one of the first type of metabolic rhythms to have been intensively studied since the early sixties.[5–7] The degradation of sugar to pyruvate *via* glycolysis represents the primary pathway for the generation of energy in all living cells, except for a few bacteria. Moreover, the glycolytic intermediates are precursors for the synthesis of cellular components, such as amino acids or lipids. Due to this multitude of functions, glycolysis is connected to many different pathways which branch from or flow into glycolysis, thus representing a complex metabolic network. Accordingly, it plays a central role for the regulation and co-ordination of the cellular metabolism. The question has been raised whether the long known glycolytic oscillations have a particular biological function. Experimental results indicate an interaction between glycolytic oscillations and control of insulin secretion in pancreatic β-cells.[8] Furthermore, recent experiments on the activation and propagation of neutrophil cells have cast new light on the possible role of glycolytic dynamic spatial patterns. It was found that the activation of these cells gives rise to the generation of traveling waves of nicotinamide adenine dinucleotide reduced form (NADH) and protons, which result from oscillatory glycolysis.[9] The propagation direction of activated neutrophil cells coincides with that of the intracellular waves, indicating a functional coupling between cell movement and wave propagation. Similar patterns have been obtained with a yeast extract, where such travelling waves were observed for the first time.[10,11] Theoretical analyses in the early seventies, however, already predicted the formation of glycolytic waves.[12] Interestingly, the wave velocity in the two experimental systems was practically the same despite their different origin, indicating the existence of common properties of such patterns.

The detection of spatio-temporal dynamics in glycolysis opens a new and innovative way in which oscillatory reactions of the metabolism can be involved in biological information processing. It is suggested, that the propagation dynamics, *e.g.* wave velocity and direction, can translate the metabolic state of the cell into signals that induce co-ordinated cellular responses.

Most of our current knowledge about the temporal dynamics of glycolysis comes from experiments with yeast, both from cells and organelle-free extract. We have chosen the glycolytic degradation of sugar in a yeast extract as a model system for our investigations, in order to unravel the possible function of glycolytic waves for biological information processing. This *in vitro* system has the advantage of easy experimental handling. Moreover, the complexity of the metabolic network can be gradually changed by addition or removal of specific cellular components, thus supporting a detailed investigation of mutual interactions between the spatio-temporal patterns and particular metabolic pathways. This should give a basis for further investigations on the biological significance of reaction–diffusion waves.

2 Materials and methods

Preparation of yeast extract

Yeast extract was prepared from aerobically grown yeast *Saccharomyces carlsbergensis* (ATCC 9080) as described in ref. 6, except that the phosphate buffer was replaced by a 3-morpholino-propansulfonic acid (MOPS) buffer, 25 mM pH 6.5, containing 50 mM KCl.

Data acquisition

Temporal oscillations of NADH were recorded by means of its absorption at 340 nm with a spectrophotometer (Shimadzu UV-2501PC). The oscillations were initiated by the addition of trehalose and phosphate. Traveling waves of NADH were monitored with a home made 2D-spectrophotometer.[13] Briefly, the yeast extract (500 µl) was mixed with 100 mM trehalose, 50 mM KPO_4, pH 6.5 and 50 mM KCl (all final concentrations, final volume 550 µl) and transferred into a sealed reaction chamber. This reaction chamber was a petri dish made from quartz glass (diameter 35 mm covered with a cover glass, leaving a small air gap of 2 mm above the reactive layer (thickness 1 mm). The cover glass was sealed with grease. This chamber was placed into the light beam of the 2D-spectrophotometer ($\lambda = 340$ nm) and the spatial absorption changes of NADH were recorded with a UV-sensitive camera (Hamatsu C1000-13). The resulting movies were stored on a video tape (Sony video recorder EVT-801 CE).

Data evaluation

In order to analyse the propagation dynamics of the waves, the movies were digitised with a personal computer, equipped with a frame grabber (Matrox pulsar) using the image processing program WinPic (home-written). The digitised frames were further processed for improvement of image quality (background subtraction, smoothing) and for extraction of data sets (*e.g.* time–space plots) using image processing programs that were written with the interactive data language (IDL) software package.

3 Results and discussion

Effect of salts and co-factors on temporal dynamics

Preparation of an excitable medium from biological sources always includes the possibility of obtaining media of varying quality, since living systems are hard to maintain in a defined state. Therefore it is difficult to work with standardised material. As an example, the sugar-induced glycolytic oscillations for two different yeast extract preparations are shown in Fig. 1A and B. One

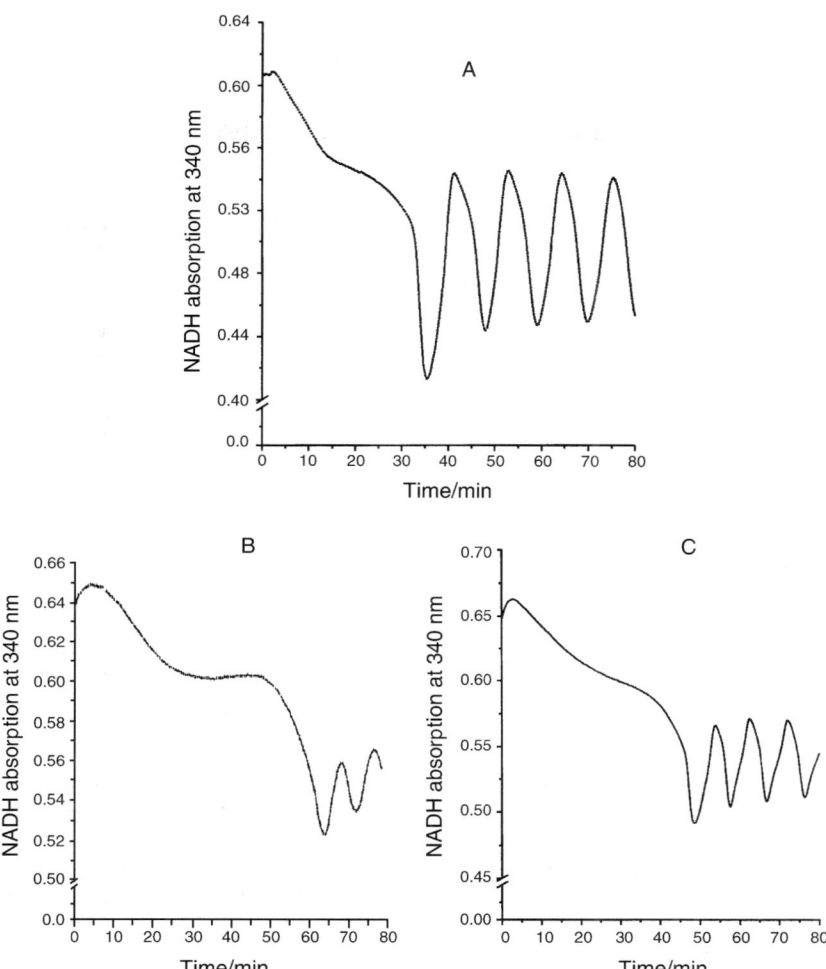

Fig. 1 Restoration of glycolytic oscillations by magnesium addition. Two different yeast extract preparations were tested for trehalose-induced oscillations in NADH: P05 (A) and P02 (B, C). P05 displays normal NADH oscillations whereas P02 has strongly reduced oscillatory activity (B). After addition of 2 mM magnesium to P02 (C), the amplitude is doubled and the onset of oscillations occurs earlier.

preparation (P05) exhibits significantly earlier onset of oscillations and larger amplitudes than the other preparation (P02).

In order to obtain improved standardisation of different preparations one has to control the basic reactions. It is well known that the allosteric enzyme PFK plays an important role in the generation of glycolytic oscillations *via* positive and negative feed-back regulation.[14] The adenine nucleotides serve as important effectors of this enzyme, with adenosin triphosphate (ATP) being an inhibitor (and substrate) and adenosin diphosphate (ADP) being an activator (and product) of the PFK. The enzyme needs several ions, *e.g.* magnesium and potassium, for optimal activity. The fact, that P02 has less oscillatory activity than P05 indicates that the enzyme PFK is impaired in P02. Therefore, we tried to restore this activity by means of magnesium addition. In fact, the onset of glycolytic oscillations occurs earlier and the amplitude is doubled when 2 mM magnesium is included in P02 (Fig. 1C). This effect is dose-dependent as can be seen from the results shown in Fig. 2, the optimum being around 5 mM.

Oscillatory control in glycolysis is subjected to complex interactions due to the multitude of regulatory functions of PFK. There are two feed-back loops which can directly or indirectly inter-

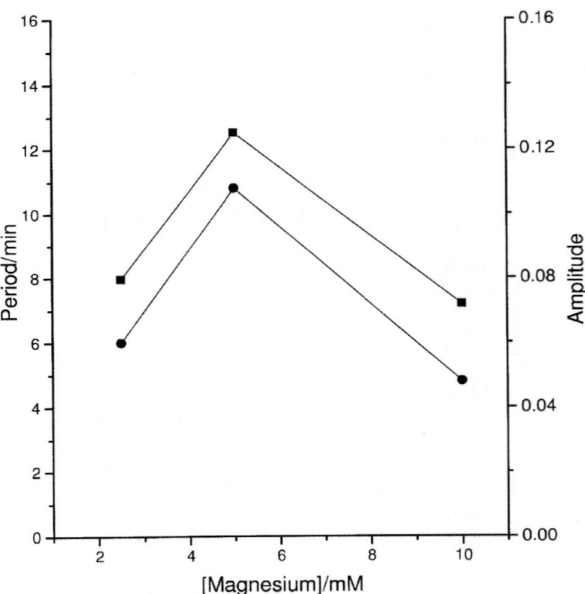

Fig. 2 Dose dependent effect of magnesium on amplitude and period of glycolytic oscillations. The yeast extract P02 was supplemented with trehalose, phosphate, KCl and varying amounts of magnesium. The NADH oscillations were monitored by absorption at 340 nm and their amplitudes (circles) and periods (squares) were evaluated at the different magnesium concentrations.

act with PFK activity, namely the adenine- and NAD/NADH cycles. The adenine nucleotide cycle couples the upper and lower part of glycolysis *via* the dephosphorylation and phosphorylation of ATP and ADP by means of the enzymes phosphofructokinase and pyruvate kinase, respectively. The NAD/NADH cycle is coupled to the phosphorylation of ADP, and hence the adenine nucleotide cycle, *via* the enzymes glyceraldehyde phosphate dehydrogenase (GAPDH)/phosphoglycerate kinase (PGK). Some preparations of yeast extract had a quite low concentration of NADH, as estimated from their absorbance at 340 nm. Addition of extra NAD to these preparations drastically reduced the time required for the onset of glycolytic oscillations and also increased the amplitude, indicating the impact of the NAD/NADH cycle for the dynamics of glycolytic oscillations.

These data demonstrate, that the preparation of a yeast extract as an excitable system requires careful control of those experimental parameters that can affect the allosteric regulation of PFK. Improved standardisation may be obtained by careful adjustment of magnesium, as well as adenine nucleotides and NAD in the yeast extract.

Spontaneous generation of traveling waves

Periodic reactions are not restricted to biological systems, but they also occur in chemical and physical systems and most of our current knowledge about the basic mechanisms of oscillatory reactions comes from experimental and theoretical investigations of such systems. It could be shown that the coupling of an autocatalytic reaction (*i.e.* an oscillatory one) with transport can lead to the formation of highly ordered spatio-temporal patterns.[15–17] Depending on the system properties, the patterns can be either dynamic (*e.g.* travelling waves) or stationary (*e.g.* Turing patterns, *cf.* ref. 18 and 19). Traveling waves occur as the result of a local perturbation of the autocatalytic reaction, which propagates by means of diffusive coupling through the medium, leading to a circular or spiral shaped reaction–diffusion wave (for overview see ref. 20). The process of wave generation and propagation has similarities with neuronal excitability, *e.g.* all-or-none reaction or threshold value, and that is why these waves are also called excitation waves.

The dynamics of glycolysis can be governed by autocatalytic reactions and hence pattern formation is possible. When there is reaction–diffusion coupling, addition of sugar spontaneously transforms the yeast extract into an excitable state, after an initial induction period.[10] Under these conditions, the oscillations are not homogeneous in space, but are characterised by the formation of travelling waves of glycolytic intermediates, composed of NADH or protons. As known from other excitable media, the shape of the waves can be either circular (Fig. 3A–D) or spiral shaped (Fig. 4). When looking at the intensity profile of NADH-waves, some overshoot in the wave back can be observed (Fig. 3E).

It is frequently seen, that the waves become narrower during the initial stage of wave formation (compare Fig. 3C and D). This wave sharpening is the result of different velocities of the wave front and back. In contrast to the wave front, the velocity of the wave back decreases with increasing radius until it coincides with the constant velocity of the wave front, which amounts to 5 µm s^{-1} (Fig. 5).

Controlled initiation of waves

The excitable nature of the yeast extract can be demonstrated by controlled initiation of waves. For this, the autocatalytic reaction, *i.e.* the PFK activity, has to be perturbed, since this reaction is part of the basic mechanism for excitability. One possible means to perform an excitation is to produce a local concentration increase of an activator of the autocatalytic reaction. Since fructose-2,6-bisphosphate (F-2,6-P_2) is a strong activator of PFK, we have applied this sugar phosphate *via* small glass capillaries by means of an injector.

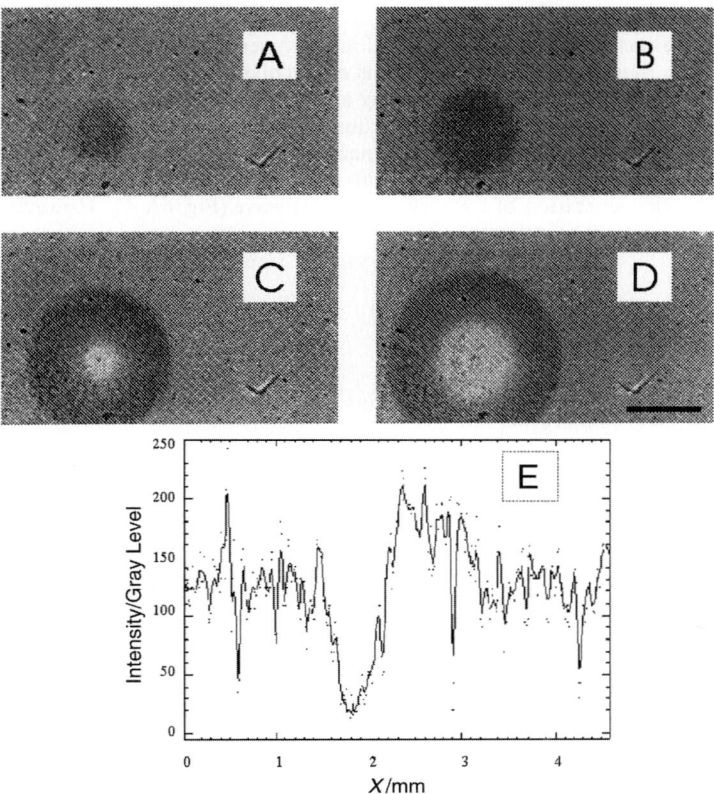

Fig. 3 Spontaneous formation of circular NADH waves. The spatial distribution of NADH in a yeast extract was recorded with a 2D-spectrophotometer (see Materials and methods). After an initial induction time, circular NADH waves propagated through the extract (A–D). Time intervals were 57 s (A, B) 150 s (B, C) and 58 s (C, D). Wave velocity was 5 µm s^{-1}. Scale bar: 1 mm. Note that the shape of the wave becomes narrow with time. The intensity profile of the wave shows some overshoot in the wave back (E).

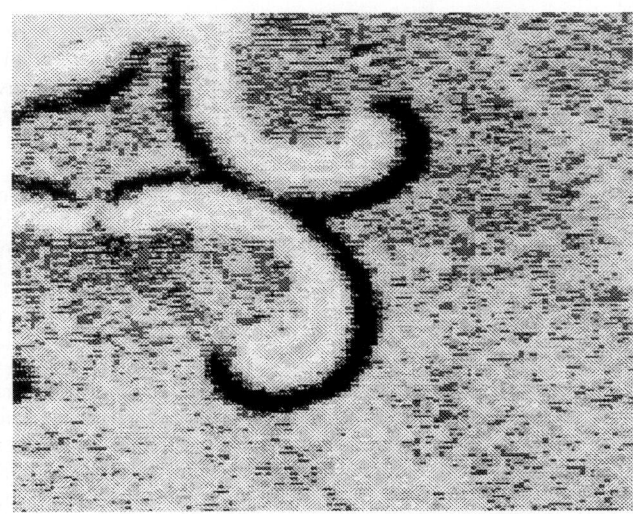

Fig. 4 Spiral-shaped NADH waves. Spontaneous break-up of circular shaped NADH waves (compare with Fig. 3) resulted in the formation of two open wave ends which subsequently generated a rotating double armed spiral (the image is contrast enhanced).

In particular, we have focused on the initiation of waves in the induction phase. During this phase of about one hour, only global oscillations of glycolysis, *i.e.* spatial homogeneous ones, can be observed. This induction phase is followed by an excitable phase which is characterised by the spontaneous formation of travelling waves. The question arises, whether travelling waves can form also in the induction phase. In the given experiment, the global homogeneous NADH oscillations had a period of about 12 min. In fact, a local injection of F-2,6-P_2 during a global NADH minimum leads to the generation of a travelling NADH wave (Fig. 6A–C). However, the wave can

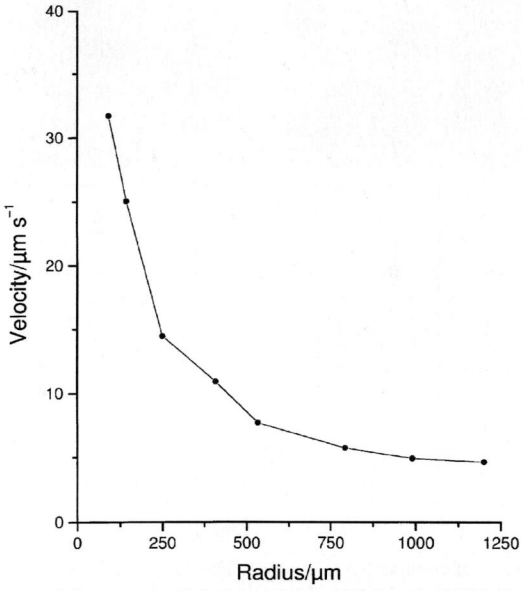

Fig. 5 Velocity of the wave back displays decreasing curvature dependence. During the initial stage of wave formation (see Fig. 3) the velocity of the wave back decreased with decreasing curvature until it matched the constant velocity of the wave front (5 µm s^{-1}).

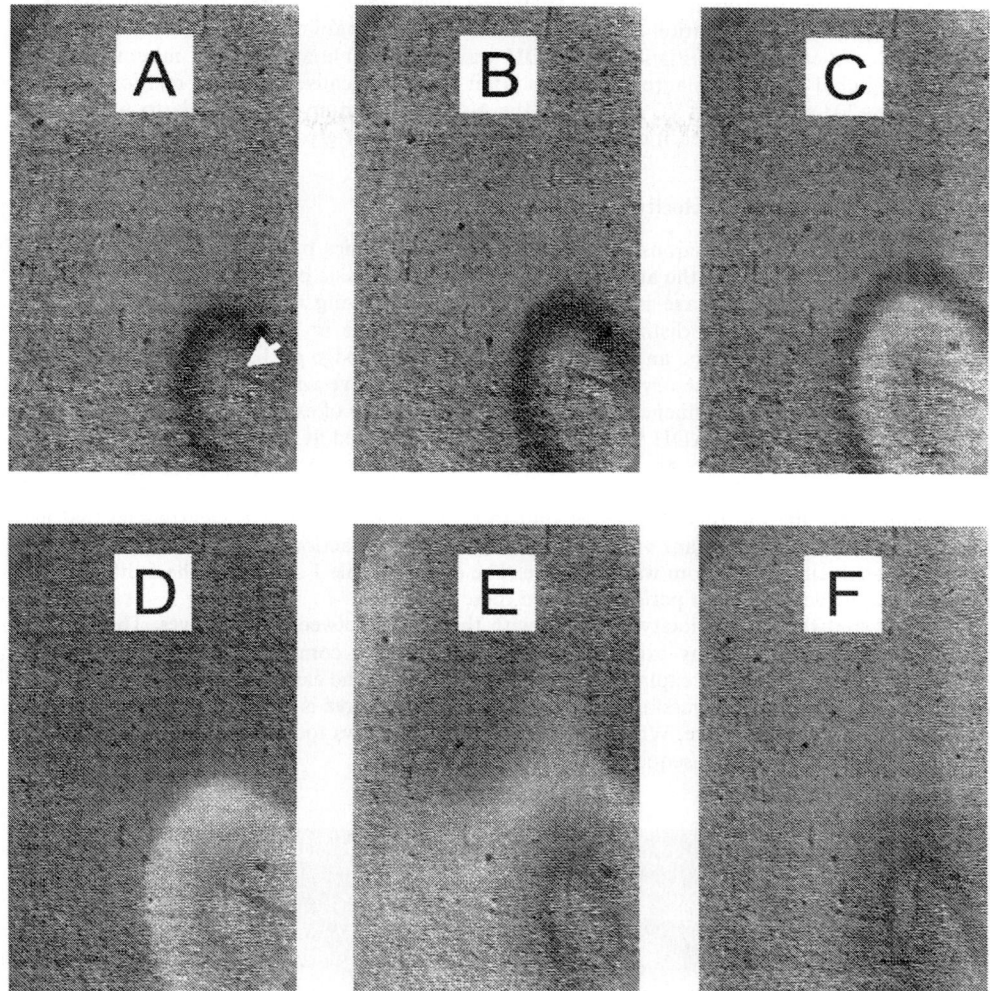

Fig. 6 Controlled initiation of NADH waves during the induction phase. The strong activator of phosphofructokinase, fructose-2,6-bisphosphate (23 nl of 0.5 mM dissolved in 1 mM KOH), was injected into the yeast extract at an oscillatory minimum of NADH during the induction phase (where no spontaneous wave formation occurs). The arrow in A marks the tip of the capillary used for the injection. A–C show the generation of a travelling NADH wave. D–E show the annihilation of the wave by a global NADH oscillation, leaving a small area of the yeast extract out of phase (F). Time intervals are 58 s (A, B), 63 s (B, C), 150 s (C, D), 69 s (D, E) and 57 s (E, F).

propagate only partially through the extract. After about 4.5 min the overall NADH concentration increases due to the global NADH oscillations and thereby stops the propagation of the wave (Fig. 6D–E). Since the back of an excitation wave has high inhibitor concentrations, the global oscillations cannot propagate across the back of the wave, leaving a small area that is out of phase with respect to the other parts of the yeast extract (Fig. 6F).

Similar patterns have been observed, when the yeast extract is supplied with purified ATPase.[11] The activity of this enzyme leads to a permanent break down of ATP into ADP, *i.e.* to a permanent increase in the activator concentration and decrease in the inhibitor concentration. In this case, the induction phase is followed by a phase that is still oscillatory but concomitantly shows the spontaneous formation of excitation waves. Obviously, the ATP/ADP ratio is an important determinant for the transition of the yeast extract from an oscillatory to an excitable state, whereby a high ATP concentration favours the transition to excitability.

The fact that a controlled initiation of NADH waves is possible at a NADH minimum during the phase of global oscillations agrees with this point of view, because the phase relations between NADH and ATP show a shift of 180°,[21] *i.e.* the NADH minimum corresponds to a maximum concentration of ATP. However, it cannot be excluded that other glycolytic intermediates can also influence this transition.

Dispersion of glycolytic wave velocity

Many of the autocatalytic reactions exert their rhythmic dynamics by feed-back regulation of the activator/inhibitor type. Since the autocatalytic reaction is the basic process for pattern formation, the profile of an excitation wave is determined by the underlying mechanisms of this reaction. Three different zones can be distinguished: An excitable zone in front of the wave with low concentrations of both effectors, an excited zone in the leading edge of the wave with high activator concentrations and a refractory zone in the back of the wave with high inhibitor concentrations. This profile markedly influences the propagation dynamics of excitation waves.

It is often observed, that NADH waves are frequently generated at some 'hot spots' in the yeast extract, which act as excitatory areas. As a result, we could observe the permanent generation of waves starting from a common centre. Fig. 7 shows the time–space plot of such an experiment. The diagonal dark lines in this plot correspond to propagating waves. It is clearly seen that each wave propagates with a constant velocity, as is expected for reaction–diffusion waves. However, this constant velocity differs from wave to wave. The data in Table 1 show that these differences in velocity correlate with different periods between these waves.

It is found that the wave velocity increases with the period between these waves. This relationship is representative for many excitable media and reflects a common form of the dispersion relation.[20,22,23] A mechanistic explanation for this is based on the existence of the refractory zone in the back of the waves. The medium in the back of the first wave is refractory and recovers only slowly to the full excitable state. When a subsequent wave follows too early, it will propagate into the refractory medium and consequently slows down.

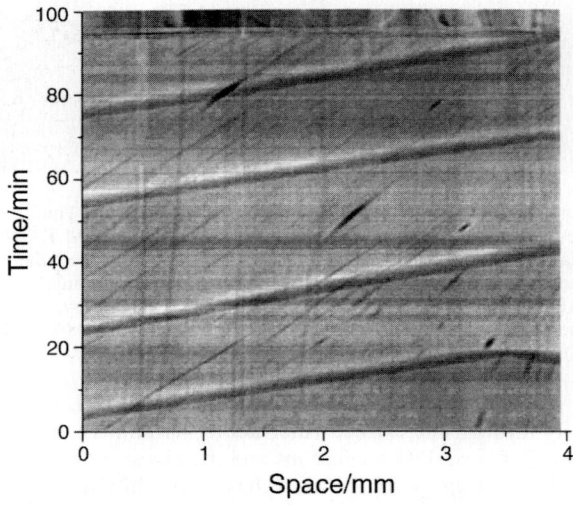

Fig. 7 Time–space plot of spontaneously generated wave trains with low frequency. The intensity profile of a horizontal line along the centre of a hot spot that permanently generated circular waves is plotted *vs.* time. The diagonal dark lines in the resulting time–space plot represent propagating NADH waves. (*cf.* legend to Fig. 8). The slope of these lines is a measure of wave velocity. At the right edge of the first diagonal line, mutual annihilation of colliding waves is visible.

Table 1 Dispersion of glycolytic wave velocity. Wave trains of spontaneous generated NADH waves were analysed for their propagation velocity and their wavelength

	Period/min	Velocity/$\mu m\ s^{-1}$
1. Wave	n.d.	3.8
2. Wave	21.9	3.3
4. Wave	23.3	3.3
3. Wave	33.6	4.0

Due to the concentration profile of the inhibitor (ATP in the case of PFK), there is an absolute refractory zone which is characterised by high inhibitor concentrations. The autocatalytic reaction is completely inhibited and consequently the generation or propagation of waves within this absolute refractory zone is not possible. This implies that the dispersion relation has a lower limit for the period as well as the velocity. The data in Table 1 do not indicate such a lower limit, probably because the waves were formed spontaneously. In order to test the lower limit, we have initiated the waves by injection of F-2,6-P$_2$ into the yeast extract during the excitable phase (Fig. 8A), thus controlling the period between the waves. The corresponding time–space plot (Fig. 8B) shows that a propagating wave can be initiated when the period is reduced to 12 min. Further reduction down to 10 min still allows the formation of a wave, but this wave propagates only a short distance through the extract and then exhibits self-annihilation. The same is true for shorter periods of 8 to 6 min. From these experimental results we can estimate the absolute refractory period to be around 12 min. It should be noted, that the wave velocity in this experiment was larger (period 12 min, velocity 4.2 $\mu m\ s^{-1}$) than for the experiment with controlled initiation of

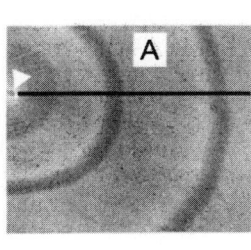
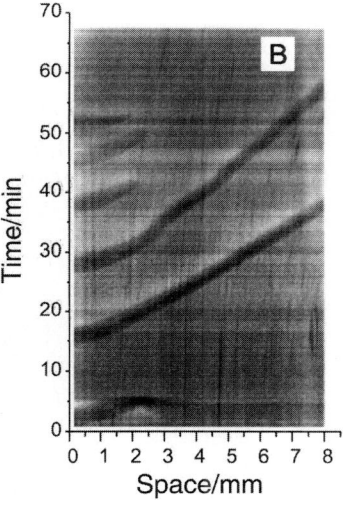

Fig. 8 Initiation of NADH waves at various time intervals. Controlled initiation of waves during the excitable state of the yeast extract was performed by fructose-2,6-bisphosphate injection (cf. legend to Fig. 6). A snapshot of three subsequently initiated waves is shown in A. The arrow head marks the position of the tip of the injection capillary. The horizontal line along the centre of the waves marks the image line that was used for the construction of a time–space plot. The intensity profile of this line from all images of the movie (i.e. from propagating waves) was plotted as a function of time (B). The diagonal dark lines in the time–space plot represent the propagating waves. Note, that the velocity of the waves across the first 0.5 mm seems to be accelerated. This is only an apparent acceleration which is due to some initial streaming phenomena caused by the injection procedure.

waves (see Table 1). This may be due to the fact that 2 different preparations of yeast extracts were used for these experiments (P02 for Table 1, P05 for Fig. 8).

Further confirmation for these results is obtained from experiments where a somewhat larger particle was present in the yeast extract. This particle acted as a strong excitatory signal and induced the permanent formation of waves with periods between 9 and 7 min. Only the first wave could propagate through the extract whereas all other waves displayed self-annihilation (Fig. 9).

Possible functions of glycolytic patterns

Glycolytic waves are of exceptional scientific interest with respect to other known biological wave phenomena for two reasons. Firstly, the generation of glycolytic patterns does not require compartmentation. They spontaneously form in a simple enzyme solution. Other known biological reaction–diffusion waves require compartmentation for structure formation, as for example calcium stores or amoeboid cells.[24–26] Secondly, glycolysis transforms chemically inert energy (sugar) into chemically available energy (ATP). Other known self-organised patterns in biology can only persist *via* the consumption of chemically available energy. In this sense, glycolytic waves provide energy instead of consuming it and therefore they are possibly involved in the genesis of biological structures, the formation of which requires spatial order and biological energy. Such properties might also have been of primary importance for the generation of a first primitive cell from an organic solution during the early days of evolution. Of course, the ancestor of the glycolytic pathway was surely constructed in a simpler way than the one we know from today's eukaryotic or prokaryotic organisms. However, this does not exclude that similar behaviour and properties were active in a proposed simple ancestor of glycolysis. Theoretical analyses of evolutionary optimisation indicate, that the overall concept of glycolysis can be realised with only a few proteins.[27]

Investigations of self-organisation in physical, chemical and biological systems greatly increased our knowledge about their fundamental properties. One main outcome of these studies is, that the spatio-temporal dynamics of reaction–diffusion waves are a universal property of dissipative systems. The presented results about the spatio-temporal dynamics of glycolysis give further experimental evidence for this view.

Very recently, investigations about NADH waves in neutrophil cells gave indications about one possible involvement of glycolytic waves for information processing. It was found, that polarised neutrophil cells generate travelling NADH waves and that the propagation direction of these waves coincides with the direction of cell migration. Upon receptor activation, the NADH waves change their propagation dynamics and so also do the cells.[28] Moreover, extracellular spatial gradients of *N*-formyl-methionyl-leucyl-phenylalanine induced reorientation of the waves. These results indicate that the extracellular signals are translated into ordered metabolic patterns, which in turn control cell migration.

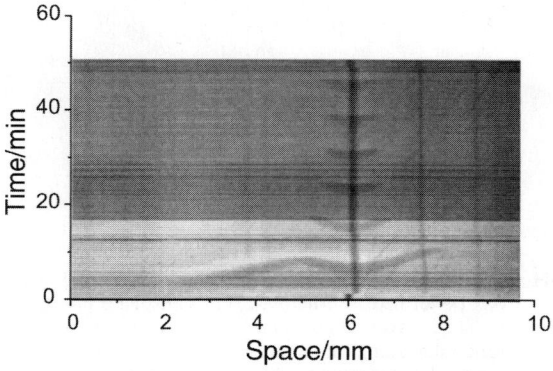

Fig. 9 Time–space plot of spontaneously generated NADH waves with high frequency. In this experiment, a small particle acted as a permanent source for excitation. This particle is visible as a dark vertical line in the time–space plot. Only the first wave can propagate for a larger distance. All other subsequent waves stop propagation shortly behind the wave centre. For construction of the time–space plot see legend to Fig. 8.

In general, the central meaning of glycolysis for the cellular metabolism opens a wide field of possible functions for glycolytic waves. The main goal behind this idea is that a marked part of the manifold metabolic interactions of a cell are translated into ordered glycolytic patterns which in turn participate in the control and co-ordination of cellular behaviour. The experimental model system yeast extract is of particular interest, because there it is possible to investigate the metabolic interactions on a subcellular level. This allows detailed studies of the basic mechanisms, which are hard to perform with living cells.

References

1. M. J. Berridge, P. H. Cobbold and K. S. R. Cuthbertson, *Philos. Trans. R. Soc. London*, 1988, **320**, 325.
2. P. De Koninck and H. Schulman, *Science*, 1998, **279**, 227.
3. Y. Tang and H. G. Othmer, *Proc. Natl. Acad. Sci. USA*, 1995, **92**, 7869.
4. G. Hajnoczky, L. D. Robb-Gaspars, M. B. Seitz and A. P. Thomas, *Cell*, 1995, **82**, 415.
5. A. Gosh and B. Chance, *Biochem. Biophys Res. Commun.*, 1964, **16**, 174.
6. B. Hess and A. Boiteux, *Hoppe-Seyler's Z. Physiol. Chem.*, 1968, **349**, 1567.
7. A. Betz and B. Chance, *Arch. Biochem. Biophys.*, 1965, **109**, 585.
8. B. E. Corkey, K. Tornheim, J. T. Deeney, M. C. Glennon, J. C. Parker, F. M. Matschinsky, N. B. Rudermann and M. Prentki, *J. Biol. Chem.*, 1988, **263**, 4254.
9. H. R. Petty, R. G. Worth and A. L. Kindzelskii, *Phys. Rev. Lett.*, 2000, **84**, 2754.
10. T. Mair and S. C. Müller, *J. Biol. Chem.*, 1996, **271**, 627.
11. S. C. Müller, T. Mair and O. Steinbock, *Biophys. Chem.*, 1998, **72**, 37.
12. A. Goldbeter, *Proc. Natl. Acad. Sci. USA*, 1973, **70**, 3255.
13. S. C. Müller, T. Plesser and B. Hess, *Anal. Biochem.*, 1985, **146**, 125.
14. A. Goldbeter and R. Lefever, *Biophys. J.*, 1972, **12**, 1302.
15. A. N. Zaikin and A. M. Zhabotinsky, *Nature*, 1970, **225**, 535.
16. A. T. Winfree, *Science*, 1972, **175**, 634.
17. S. C. Müller, Th. Plesser and B. Hess, *Science*, 1985, **230**, 661.
18. A. M. Turing, *Philos. Trans. R. Soc. London Ser. B*, 1952, **237**, 37.
19. V. Castets, E. Dulos, J. Boissonade and P. De Kepper, *Phys. Rev. Lett.*, 1990, **64**, 2953.
20. J. P. Keener and J. J. Tyson, *Physica D*, 1986, **32**, 327.
21. B. Hess, A. Boiteux and J. Krüger, *Adv. Enzym. Regul.*, 1968, **7**, 149.
22. J. M. Flesselles, A. Belmonte and V. J. Gaspar, *J. Chem. Soc., Faraday Trans.*, 1988, **94**, 851.
23. M. Wussling and T. Mair, in *Lecture Notes in Physics: Transport and Structure*, ed. S. C. Müller, J. Parisi and W. Zimmermann, Springer Verlag, Berlin, 1999, p. 151.
24. M. J. Berridge, *Nature*, 1993, **361**, 315.
25. J. D. Lechleiter and D. E. Clapham, *Cell*, 1992, **69**, 283.
26. G. Gerisch, *Naturwissenschaften*, 1971, **58**, 430.
27. E. Meléndez-Hevia, T. G. Waddell, R. Heinrich and F. Montero, *Eur. J. Biochem.*, 1997, **244**, 527.
28. H. R. Petty and A. L. Kindzelskii, *Proc. Natl. Acad. Sci. USA*, 2001, **98**, 3145.

Synchronization of glycolytic oscillations in a yeast cell population

Sune Danø,[a,b] Finn Hynne,[a,b] Silvia De Monte,[c] Francesco d'Ovidio,[c] Preben Graae Sørensen[a,b] and Hans Westerhoff[d]

[a] *Department of Chemistry, H. C. Ørsted Institute, University of Copenhagen, Universitetsparken 5, DK-2100 Ø, Denmark*
[b] *Center for Chaos and Turbulence Studies (CATS), H. C. Ørsted Institute, University of Copenhagen, Universitetsparken 5, DK-2100 Ø, Denmark*
[c] *Department of Physics, Technical University of Denmark, Building 309, DK-2800, Denmark*
[d] *Mathematical Biochemistry and Molecular Cell Physiology, BioCentum Faculty of Biology, Free University, 1081 HV Amsterdam, The Netherlands*

Received 10th April 2001
First published as an Advance Article on the web 23rd November 2001

The mechanism of active phase synchronization in a suspension of oscillatory yeast cells has remained a puzzle for almost half a century. The difficulty of the problem stems from the fact that the synchronization phenomenon involves the entire metabolic network of glycolysis and fermentation, and consequently it cannot be addressed at the level of a single enzyme or a single chemical species. In this paper it is shown how this system in a CSTR (continuous flow stirred tank reactor) can be modelled quantitatively as a population of Stuart–Landau oscillators interacting by exchange of metabolites through the extracellular medium, thus reducing the complexity of the problem without sacrificing the biochemical realism. The parameters of the model can be derived by a systematic expansion from any full-scale model of the yeast cell kinetics with a supercritical Hopf bifurcation. Some parameter values can also be obtained directly from analysis of perturbation experiments. In the mean-field limit, equations for the study of populations having a distribution of frequencies are used to simulate the effect of the inherent variations between cells.

Introduction

Synchronization phenomena are found in many physical and biological systems. A particularly well studied example is populations of yeast cells showing glycolytic oscillations with a half-minute period.[1–17] In these studies, suspensions with of the order of one billion yeast cells show macroscopic oscillations in the concentration of nicotinamide adenine dinucleotide (reduced) (NADH), which is monitored by fluorescence. At a very early stage the observed bulk oscillations were interpreted as the result of phase-synchronized oscillations of each individual cell. Phase coherence of independent oscillators will not persist unless the fundamental frequencies are equal. Obviously, the yeast cells are not identical. For example, variations in size and shape are easily seen in the microscope. If the cells were free-running independent oscillators, differences between the cells would eventually lead to a decrease in the macroscopic amplitude even if the cells were all in the same phase at the start of the experiment. For some experimental conditions, this is indeed the case.[18] However, if the cells were actively synchronizing their phases they could maintain a narrow phase distribution throughout the entire experiment. The existence of active phase synchronization was ingeniously demonstrated by the rapid reappearance of the oscillations after an instantaneous mixing of 2 populations 180° out of phase (Fig. 1). A comprehensive experimental determination of the final phase as a function of the inital phases is found in ref. 8 and discussed in ref. 19.

In this paper we will approach the synchronization problem using equations for a set of two-dimensional oscillators interacting through a common medium. These approximate equations are obtained directly from a single-cell model through a systematic derivation. The synchronization dynamics is determined by the value of some of the parameters of the approximate equations. The base model is assumed to give a realistic description of the yeast cell kinetics. The reliability of the result depends critically on the kinetic model, and thus on the biochemical data used for the optimization of this model. Finally, we show that the problem of non-identical cells can be addressed by describing the average behavior of a population of oscillators with a small dispersion in natural frequencies by just two complex differential equations for which the parameters can be calculated from a realistic base model.

A necessary condition for a chemical species to act as a synchronizer is that it can penetrate the cell membrane and that an instantaneous change in its concentration will influence the oscillations and give rise to an amplitude or phase change. Many different substances known to pass the cell membrane have been tested by perturbation experiments. Proper responses are observed with acetaldehyde,[6,13,16] O_2[9] and—when the glucose transporter is not saturated—glucose.[16] Substances which have been tested but without proper responses are H^+,[6] K^+,[6] pyruvate,[6,16] Mg^{2+},[8] ethanol,[8,16] phosphate,[8] cyanide[16] and glycerol.[20] Another necessary condition for a species to act as a synchronizer is that it oscillates in the extracellular medium. This can be the result of periodic excretion or absorption by the cells. This condition excludes O_2 as a synchronizer in anaerobic experiments or experiments where respiration is inhibited. Because of a limitation of the acetaldehyde assay it was concluded that no synchronizing species could be identified unambiguously.[8,19] A method for measuring the extracellular concentration of acetaldehyde in the presence of cyanide[13] showed, however, that the extracellular concentration of acetaldehyde exhibited large amplitude oscillations. This behavior is a strong indication that acetaldehyde is an important species for the synchronization.

The synchronization problem has previously been addressed using mathematical models. With a biochemical starting point, such studies have used models explicitly describing the "essential" metabolites while leaving out others for the sake of simplicity.[21–23] In the most recent study[23] a model with somewhat more realistic values for the parameters has been investigated by Wolf and Heinrich. It contains the main reactions of glycolysis and adjacent reactions producing ethanol and glycerol together with transmembrane transport of glucose and a coupling substance (pyruvate and/or acetaldehyde). For n cells it uses $6n + 1$ variables. The parameter values were

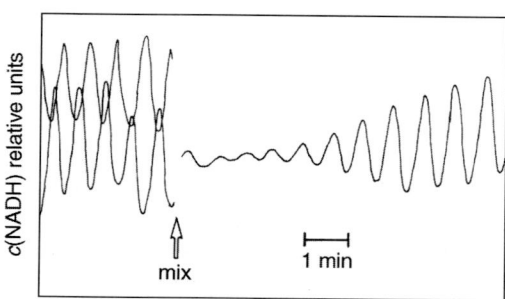

Fig. 1 Mixing experiment demonstrating the rapid reappearance of the oscillations following the mixing of two suspensions oscillating 180° out of phase. Reproduced from ref. 13.

selected such that the stationary concentrations are roughly in agreement with the experimental values. The input flux and the rate of an extracellular reaction removing the coupling substance were used as bifurcation parameters. For one cell their system has a subcritical Hopf bifurcation, and a large amplitude stable limit cycle. The phase relation for the oscillations of adenosine 5′-triphosphate [ATP] and [NADH] are well reproduced. For two cells the system has two subcritical Hopf bifurcations corresponding to inphase and antiphase synchronization and corresponding large amplitude stable limit cycles. Similar behavior is found for higher n. They show that the dynamics of addition of the coupling substance to the external medium is similar to the experimental behavior but the model does not account for the rapid synchronization seen in the experiments.

From a more abstract starting point, models of mutually coupled Stuart–Landau (SL) oscillators have been studied extensively, both numerically and theoretically. For a discussion of coupled identical SL oscillators with a linear mean-field interaction see ref. 24 and 25. For a discussion of the behavior of a system of coupled SL oscillators with a distribution in natural frequencies see ref. 26–28. However, there has been no explicit way to link these investigations to the synchronization phenomenon seen in yeast cells. Recent experiments have provided such a link by showing that each of the yeast cells is well described by the universal behaviour of a system close to a supercritical Hopf bifurcation.[16] Furthermore, some of the parameters needed for this description can be found experimentally. This result is the basis for the present work.

Experimental indications of Hopf behaviour

These experiments are performed in a CSTR with inflow of yeast cell suspension, glucose and cyanide and with outflow of surplus suspension (see Fig. 2). With this set-up, truly undamped sinusoidal oscillations can be observed by measuring the average fluorescence of NADH in a layer of cells at the front face of the cuvette (Fig. 3), and the exact operating point of the reactor can easily be controlled.

By changing the flow rate of glucose, steady as well as oscillatory signals can be observed. Fig. 4 shows the squared amplitude of the oscillations as a function of a measure of the glucose infusion rate, which is used as a bifurcation parameter. The straight line through the first four experimental points corresponding to the branch of emerging oscillations suggests a square root dependence of the amplitude on the bifurcation parameter. This behavior is consistent with a supercritical Hopf bifurcation. The deviation of the experimental points from the straight line at higher flow rates are

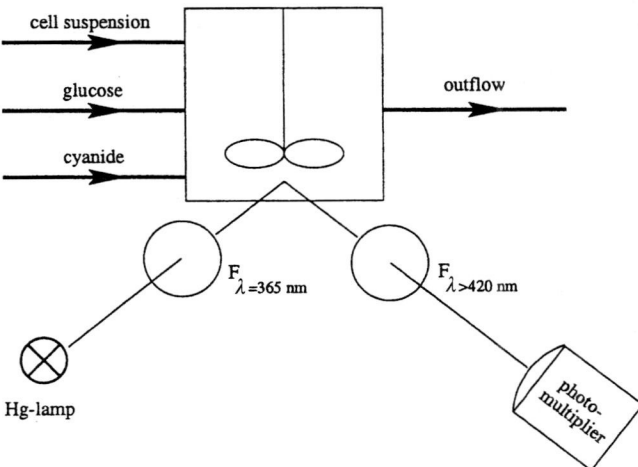

Fig. 2 Experimental set-up for the CSTR experiments. Cold starved cells, glucose, and cyanide flow into the reactor, which is a 1×1 cm^2 cuvette. The concentration of NADH is monitored by observing the fluorescence of NADH from cells close to the cuvette surface. F designates optical filters. Reproduced from ref. 16.

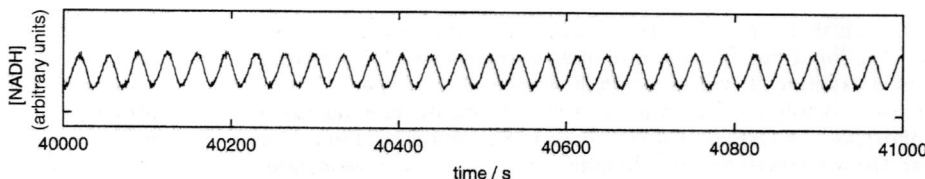

Fig. 3 Sustained oscillations in the CSTR. The cell density is 1.61×10^9 cells mL^{-1}, which corresponds to a dry weight of 30.5 mg mL^{-1}. The specific flow rate is 0.051 min^{-1}. The mixed flow concentrations of glucose and cyanide are 35.0 mM and 5.38 mM, respectively. The experiment was started at time 0.

probably caused by the saturation of the glucose transporter. This prevents the glucose concentration inside the cells from following a continued increase in the extracellular glucose concentration. This interpretation fits biochemical data on glucose transport well.[15,17,29,30] For changing cyanide concentrations an interval of oscillatory behavior is observed, in this case with a lower

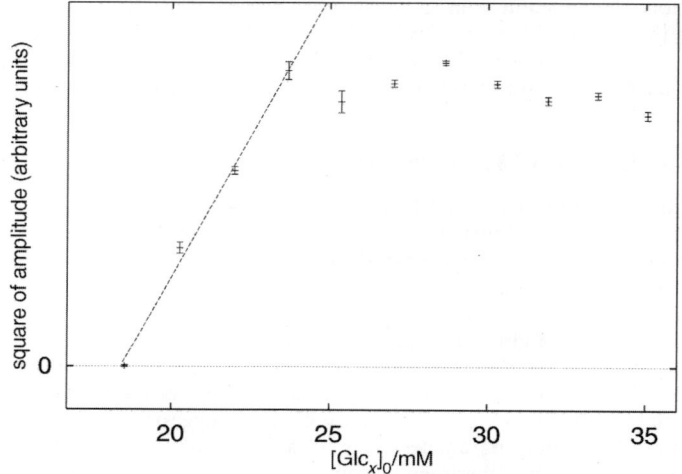

Fig. 4 Plot of the square of the amplitude A as a function of the mixed flow glucose concentration [Glc$_x$]$_0$. The bifurcation point is found at 18.5 mM, which corresponds to an extracellular glucose concentration of 1.6 mM. The fit of a straight line to the experimental points just after the bifurcation point demonstrates the square root dependence of A on the bifurcation parameter. This is characteristic of a supercritical Hopf bifurcation. Cell density, mixed flow cyanide concentration and specific flow rate were fixed at 1.61×10^9 cells mL^{-1}, 5.60 mM and 0.049 min^{-1}, respectively. Data from ref. 16.

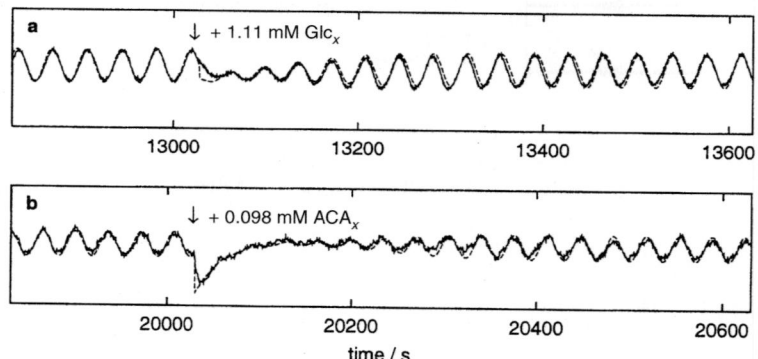

Fig. 5 Responses to instantaneous additions of extracellular glucose (a) and acetaldehyde (b) to yeast cells in a CSTR. The full line shows the experimental traces. The arrows indicate the time of the perturbations. The dashed curves show the fit of eqn. (9) to the experimental traces.

and an upper supercritical Hopf bifurcation,[31] see also ref. 12. The dynamics of a supercritical Hopf bifurcation has a characteristic behavior which can also be tested experimentally by perturbation experiments and analysis of the perturbation response.

Fig. 5 shows the result of perturbations with glucose and acetaldehyde together with the result of a fit to the experimental time traces of the universal form of a signal consistent with the dynamics of a supercritical Hopf bifurcation eqn. (9). The close correspondence between the curves demonstrates that the dynamics of the system follows closely the dynamics of a supercritical Hopf bifurcation.[31] The amount of the chemical species and the phase of oscillation at the perturbations in Fig. 5 have been chosen so that the oscillations just after the perturbation are suppressed. Such a perturbation is called a quenching perturbation. The phase of addition and the amount of chemical species used for a quenching is characteristic of each chemical species at the chosen operating point.

Modelling a population of identical oscillators in a CSTR

To model a population of oscillating yeast cells, we now consider N identical cells, each with volume ω, in an extracellular medium with volume Ω. Each cell is surrounded by a cell membrane able to transport certain chemical species between the interior of the cell and the extracellular medium. Enzyme concentrations are assumed to be constant. Concentrations of the M chemical species (metabolites) are $c_i(t)$ in cell i and $c(t)$ in the medium. By neglecting the changes in the amounts of metabolites caused by the state difference between cells flowing in and out of the reactor, the kinetic equations for this system in a CSTR with specific flow rate j and mixed flow concentrations c^f are

$$\dot{c}_i = F(c_i) + G(c_i, c)$$

$$\dot{c} = -\frac{\omega}{\Omega} \sum_i G(c_i, c) + H(c) + j(c^f - c) \qquad (1)$$

where $F(c_i)$, $G(c_i, c)$ and $H(c)$ describe the intracellular reactions, the transport across the cell membrane, and reactions in the extracellular medium, respectively.

G and H are expanded to first order from the stationary state (c_1^s, c^s) of a system where all c_i are equal to c_1^s. The stationary state is defined by

$$F(c_1^s) + G(c_1^s, c^s) = 0$$

$$-\frac{N\omega}{\Omega} G(c_1^s, c^s) + H(c^s) + j(c^f - c^s) = 0, \qquad (2)$$

and the expansion gives

$$G(c_i, c) = G(c_1^s, c^s) + \left.\frac{\partial G}{\partial c_i}\right|_{(c_1^s, c^s)} (c_i - c_1^s) + \left.\frac{\partial G}{\partial c}\right|_{(c_1^s, c^s)} (c - c^s)$$

$$= G(c_1^s, c^s) - D_I(c_i - c_1^s) + D_E(c - c^s)$$

and

$$H(c) = H(c^s) + \left.\frac{\partial H}{\partial c}\right|_{c^s} (c - c^s) = H(c^s) + H_E(c - c^s).$$

In this approximation the kinetic equations are

$$\dot{c}_i = F(c_i) + G(c_1^s, c^s) - D_I(c_i - c_1^s) + D_E(c - c^s)$$

$$\dot{c} = -\frac{\omega}{\Omega} \sum_i (G(c_1^s, c^s) - D_I(c_i - c_1^s)) - \frac{N\omega}{\Omega} D_E(c - c^s)$$

$$+ H(c^s) + H_E(c - c^s) + j(c^f - c). \qquad (3)$$

Defining $x_i = c_i - c_i^s$, $x = c - c^s$, $\tilde{F}(x_i) = F(c_i) - F(c_i^s)$ and using eqn. (2) we get

$$\dot{x}_i = \tilde{F}(x_i) - D_1 x_i + D_E x$$

$$\dot{x} = \frac{\omega}{\Omega}\sum_i (D_1 x_i) - \frac{N\omega}{\Omega} D_E x + (H_E - j)x$$

$$= \frac{N\omega}{\Omega}\left(D_1 \frac{1}{N}\sum_i x_i - D_E x\right) + (H_E - j)x. \tag{4}$$

In the somewhat unphysical limit of $\Omega \to 0$ we have $D_1(1/N)\Sigma_k x_k - D_E x \to 0$ which by substitution in eqn. (4) gives,

$$\dot{x}_i = \tilde{F}(x_i) + D_1\left(\frac{1}{N}\sum_k x_k - x_i\right) \tag{5}$$

which describes a population of cells with a linear mean-field coupling. The same approximation is obtained in the limit of D_1 and $D_E \to \infty$ at constant D_1/D_E (i.e. at infinite transport elasticities). It is seen that if $X(t)$ satisfies $\dot{X} = \tilde{F}(X)$ then $x_i(t) = X(t)$ for all i is a solution of eqn. (5). If this solution is stable the dynamics of the population is identical with the dynamics of a single cell in the CSTR. In the opposite limit of $(N\omega/\Omega) \to 0$ or D_1 and $D_E \to 0$ at constant D_1/D_E, c_i will evolve to a steady state set by the extracellular processes and the individual cells will become independent of each other.

Describing an oscillating yeast cell as a Hopf oscillator

The experimental evidence indicates that the dynamics of the population of yeast cells in the experiments can be considered as the dynamics of a single cell at a Hopf point, since the phase-synchronization is known to be fast (Fig. 1) and since, as shown above, the macroscopic system follows Hopf dynamics closely. Except for short transients the dynamics of the cell is then confined to a two-dimensional manifold embedded in the high-dimensional concentration space. On the manifold the original large system of kinetic equations can then be reduced to two differential equations describing the evolution of suitably chosen local coordinates on the manifold. The form of the Hopf kinetic equations is universal, and the parameters can be explicitly calculated from a kinetic model. The transformation from the local coordinates to actual concentrations also has a universal form with coefficients which can be calculated from the original kinetic model. For fast transients from any point in the concentration space, the evolution is essentially an instantaneous projection onto the slow manifold after which the state evolves in accordance with the local kinetic equations. If a complex variable z is chosen as local coordinate the local Hopf equations have a particularly simple form

$$\dot{z} = (i\omega_0 + \sigma\mu)z + gz|z|^2 \tag{6}$$

where ω_0 is the angular frequency at the bifurcation point, $\sigma = \sigma' + i\sigma''$ and $g = g' + ig''$ are system dependent complex parameters and μ is a real bifurcation parameter which is 0 at the bifurcation point.[38] This equation is known as the Stuart–Landau equation. To the lowest order the transformation from z to the metabolite concentration vector c is given by

$$c = c_s + zu + \overline{zu} \tag{7}$$

where c_s is the stationary metabolite concentration, and u is the right eigenvector of the Jacobian matrix at the bifurcation point corresponding to the eigenvalue $i\omega_0$. The overline denotes complex conjugation. The projection of a general state c onto the slow manifold is given by

$$z = u^*(c - c_s) \tag{8}$$

where u^* is the left eigenvector of the Jacobian matrix at the bifurcation point corresponding to the eigenvalue $i\omega_0$ and normalized such that $u^*u = 1$ and $u^*\bar{u} = 0$. For a Hopf bifurcation with slow hyperbolic modes an observed signal which is a linear function of the concentrations has the following universal form (up to second order):[41]

$$y(t) = y_0 + a_1 R \cos(\Theta + \Theta_0) + a_2 R^2 \cos(2\Theta + \Theta_d) + \sum_l b_l e^{\lambda_l t} \qquad (9)$$

where

$$R(t) = \frac{R_s}{\sqrt{1 + \left(\frac{R_s^2}{R_0^2} - 1\right) e^{2g' R_s^2 t}}},$$

$$\Theta(t) = (\omega_{ss} + g'' R_s^2)t - \frac{g''}{2g'} \ln\left(1 + \left(\frac{R_0^2}{R_s^2} - 1\right)(1 - e^{2g' R_s^2 t})\right).$$

Here, ω_{ss} is the imaginary part of the complex eigenvalue of the Jacobian matrix at the stationary state and R_s is the radius of the limit cycle in z space. The parameters g', g'', ω_{ss} and λ_l are charateristic of the system at the chosen distance from the bifurcation point. They should be the same for different perturbations. y_0 is the signal value of the stationary state. The parameters a_1, a_2, R_s, R_0, Θ_0, Θ_d and b_l are characteristic for the chemical species, the timing and size of the perturbation.

Estimates of the parameters σ and g may also be obtained directly by fitting eqn. (9) to experimental measurements of the response of small perturbations away from the limit cycle. In the present case the absolute amplitude of NADH cannot be obtained directly from the measurements. The best fit of the system parameters to the perturbation experiments shown in Fig. 5a gives $\omega_{ss} = 0.168$ s^{-1}, $R_s^2 g' = -0.014$ s^{-1}, $R_s^2 g'' = 0.0034$ s^{-1}, $\lambda_1 = -0.03$ s^{-1}. These values are consistent with the fit in Fig. 5b even though it is at a different operating point where the radius of the limit cycle is larger. This indicates that these parameters give a good general representation of the yeast dynamics. In Fig. 5b the hyperbolic exponential is different. We get $\lambda_2 = -0.012$ s^{-1}. This can be accounted for if different exponential modes are excited by perturbations with acetaldehyde and glucose, which is quite in line with biochemical expectations.

Hopf equation for a population of yeast cells

The orders of the system of kinetic eqn. (4) and (5) are $(N + 1)M$ and NM respectively where M is the number of metabolites. For a large number of cells these are huge numbers but they can be considerably reduced by taking advantage of the fact that the states of the individual cells are close to a supercritical Hopf bifurcation so that the state of each cell can be described by a local complex coordinate z_i. The external medium has no intrinsic oscillatory properties, however, and no similar description is possible. The contribution of the intracellular $\tilde{F}$ term is given by eqn. (6). The transformation of the linear terms is done by substituting eqn. (8) in eqn. (4), multiplying from the left by u^* and using the normalization conditions. We get

$$\dot{z}_i = (i\omega_0 + \sigma\mu)z_i + gz_i|z_i|^2 + u^* \cdot D_E \cdot x - u^* \cdot D_I \cdot x_i$$
$$= (i\omega_0 + \sigma\mu)z_i + gz_i|z_i|^2 + u^* \cdot D_E \cdot x - u^* \cdot D_I \cdot (uz_i + \overline{uz_i})$$

$$\dot{x} = \frac{N\omega}{\Omega}\left(D_I \cdot \frac{1}{N}\sum_i x_i - D_E \cdot x\right) + (H_E - j)x$$

$$= \frac{N\omega}{\Omega}\left(D_I \cdot \frac{1}{N}\sum_i (uz_i + \overline{uz_i}) - D_E \cdot x\right) + (H_E - j)x \qquad (10)$$

The dimension of this system is $2N + M$ which is considerably smaller than the dimension of eqn. (4). The elimination of the fast modes from the intracellular kinetics also has the advantage that the system is no longer stiff and can be integrated by explicit methods.

In the limit of $\Omega \to 0$ the variables of the external medium are eliminated and the dynamic equations are reduced to

$$\dot{z}_i = (i\omega_0 + \sigma\mu)z_i + gz_i|z_i|^2$$
$$+ \boldsymbol{u}^* \cdot \boldsymbol{D}_\mathrm{I} \cdot \left(u\left(\frac{1}{N}\sum_k z_k - z_i\right) + \bar{u}\left(\frac{1}{N}\sum_k \bar{z}_k - \bar{z}_i\right)\right)$$
$$= (i\omega_0 + \sigma\mu)z_i + gz_i|z_i|^2 + K_\mathrm{I}\left(\frac{1}{N}\sum_k z_k - z_i\right) + L_\mathrm{I}\left(\frac{1}{N}\sum_k \bar{z}_k - \bar{z}_i\right) \quad (11)$$

Except for the last term these equations are identical to equations used for coupled Stuart–Landau oscillators.[24,25,28] The dynamical properties of this equation depend on the actual values of the transport coefficients K_I and L_I.

From eqn. (11) it can be seen that the most important transport coefficient K_I is a complex number composed of a linearized transport matrix $\boldsymbol{D}_\mathrm{I}$ multiplied from the left by $\boldsymbol{u}^*$ and from the right by $\boldsymbol{u}$. This means that it depends on the transport of chemical species through the cell membrane as well as on the actual eigenvectors of the Jacobian matrix of a single cell.

This transport coefficient cannot be obtained directly from experiments. Therefore, a reliable estimate of the dynamics of synchronization depends critically on the reliability af the kinetic description of a single cell. For this reason we have constructed a complete model for the entire pathway which is active during the oscillations in the CSTR experiments.[31]

Modelling the dynamics of a single cell

Glycolysis is perhaps the best described biochemical pathway, and many detailed investigations of the system of oscillating yeast cells have been made. When seeking a model as realistic as possible, this information should be exploited in order to arrive at a reliable description despite the overwhelming amount of parameters which an explicit kinetic model will unavoidably contain.

In general, the glycolytic oscillations appear after addition of glucose when the oxygen is used up[2,3] or when aerobic respiration is inhibited by addition of cyanide which is an inhibitor of cytochrome c oxidase.[6] It has been demonstrated that all tested metabolites in the glycolytic and fermentative pathways are oscillating with the same frequency as NADH but with different amplitudes and phases.[4,14] By using the crossover theorem, phosphofructokinase was suggested to be a key enzyme in the mechanism.[2,7] The oscillations can be maintained for an extended period of time when the cells are harvested at the diauxic shift followed by a long period of aerated starvation.[10] Removal of extracellular acetaldehyde (either enzymatically or by addition of cyanide or other nucleophilic substances) also turned out to extend the duration of the oscillations.[8,12] Such dependences on experimental conditions make it prohibitively difficult to use data obtained under different conditions in the same quantitative optimization.

Fortunately, a series of compatible experiments (see below) have made it possible to use a new optimization method called "the direct method" to arrive at a full-scale model which gives a good fit to a host of dynamical and biochemical data.[31] The method has previously been used to find optimal values for the rate constants of a model for the Belousov–Zhabotinsky reaction where the optimized rate constants corresponded well to values obtained by other means.[32,33] The basis for the direct method is a description of the stationary velocity in terms of the nullspace of the stoichiometric matrix. The modelling is based on a network which has been imported from earlier extensive investigations of glycolytic oscillations in cell extracts,[34–36] supplemented with branching reactions for the production of glycerol and for glycogen build-up, together with a model for glucose transport[37] and diffusion across the cell membrane of acetaldehyde, ethanol and glycerol, see Fig. 6. The model comprises 22 chemical species, 24 net reactions and rate expressions with more than 60 parameters. The construction of a model with that many parameters will not give sensible results unless a sufficient amount of experimental data on the stationary state as well as on dynamic properties is available.

Such a solid experimental foundation is obtained by combining the CSTR experiments[16] and experiments done by the Amsterdam group.[12–14] All of these experiments were performed on the same yeast strain prepared by the same procedure.[11,14] Briefly, yeast cells (Sacchararomyces cere-

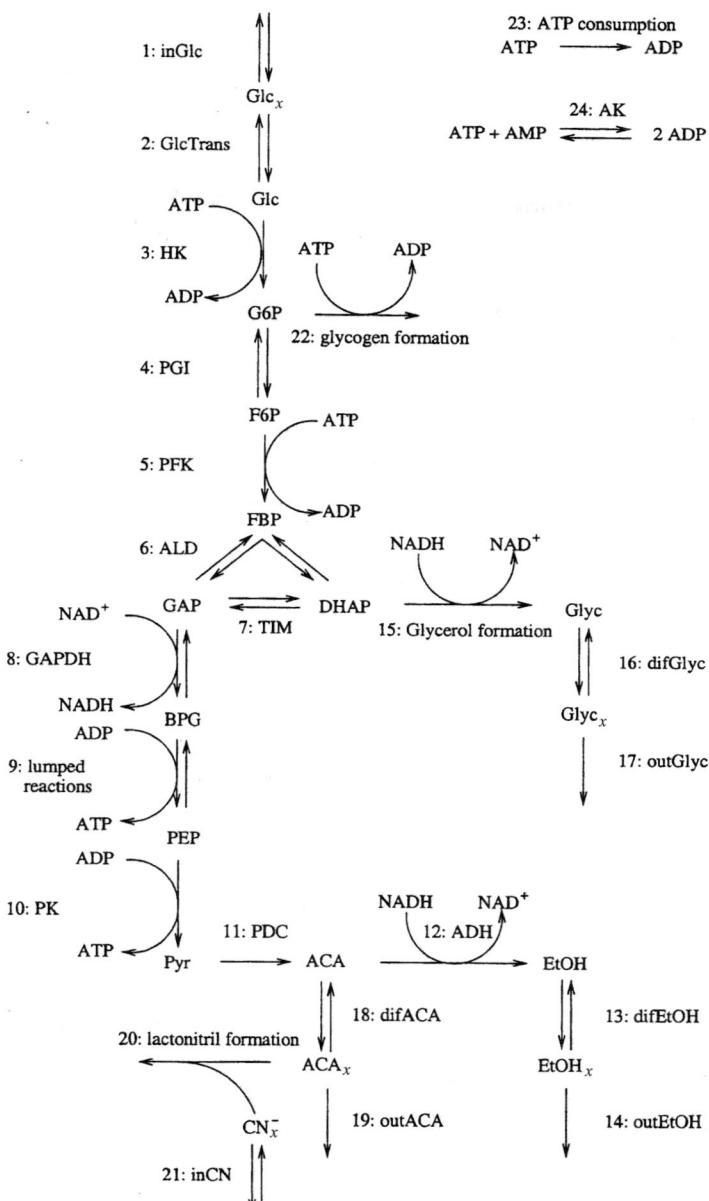

Fig. 6 Network for the full-scale model describing one cell in a CSTR. Reactions 1,14,17,19,21 represent CSTR inflows and outflows. Reactions 2,13,16,18 represent the transport reactions across the cell membrane which give rise to the non-zero elements of D_1. Reaction 22 is a one step model for a more complicated pathway. Inorganic phosphate and pH is considered constant. Subscript x refers to extracellular components of the respective chemical species. Abbreviations: ACA: acetaldehyde, ADH: alcohol dehydrogenase, ADP: adenosine 5'-diphosphate, AK: adenylate kinase, ALD: aldolase, AMP: adenosine 5'-monophosphate, ATP: adenosine 5'-triphosphate, BPG: 1,3-bisphosphoglycerate, DHAP: dihydroxyacetone phosphate, EtOH: ethanol, F6P: fructose 6-phosphate, FBP: fructose 1,6-biphosphate, G6P: glucose 6-phosphate, GAP: glyceraldehyde 3-phosphate, GAPDH: glyceraldehyd 3-phosphate dehydrogenase, Glc: glucose, Glyc: glycerol, HK: hexokinase, NAD$^+$: nicotineamide adenine dinucleotide(oxidised), NADH: nicotineamide adenine dinucleotide(reduced), PDC: pyruvate decarboxylase, PEP: phosphoenol pyruvate, PFK: phosphofructokinase-1, PGI: phosphoglucoisomerase, PK: pyruvate kinase, Pyr: pyruvate, TIM: triosephosphate isomerase.

Table 1 Eigenvectors u and u^* for the intracellular species. The last two columns display the absolute value of the components of u and the real part of $u_i^* u_i$, respectively. The eigenvectors have no NAD$^+$ or AMP components, since the concentrations of these metabolites are given implicitly by conservation equations. The units of u and u^* are concentration and reciprocal concentration, respectively. Only the relative magnitudes of the vector components have physical significance. Abbreviations: ACA: acetaldehyde, ADP: adenosine 5′-diphosphate, ATP: adenosine 5′-triphosphate, BPG: 1,3-bisphosphoglycerate, DHAP: dihydroxyacetone phosphate, EtOH: ethanol, F6P: fructose 6-phosphate, FBP: fructose 1,6-biphosphate, G6P: glucose 6-phosphate, GAP: glyceraldehyde 3-phosphate, Glc: glucose, Glyc: glycerol, NADH: nicotineamide adenine dinucleotide(reduced), Pyr: pyruvate

i	$\mathcal{R}u$	$\mathcal{F}u$	$\mathcal{R}u^*$	$\mathcal{F}u^*$	$\|u_i\|$	$\mathcal{R}(u_i^* u_i)$
Glc	5.39e−02	−1.16e−02	−1.15e−02	−7.37e−02	5.51e−02	−1.47e−03
G6P	−4.68e−01	7.93e−02	−3.16e−01	7.44e−01	4.75e−01	8.89e−02
ATP	−2.46e−01	−2.11e−01	−8.94e−01	2.52e+00	3.24e−01	7.53e−01
ADP	1.43e−01	1.24e−01	−4.63e−01	1.27e+00	1.90e−01	−2.25e−01
F6P	−6.50e−02	−2.07e−03	−2.53e−01	−7.65e−01	6.50e−02	1.48e−02
FBP	5.69e−01	−3.51e−01	2.44e−01	1.88e−01	6.69e−01	2.05e−01
GAP	7.70e−03	−4.41e−03	4.98e−02	1.85e−01	8.88e−03	1.20e−03
DHAP	1.65e−01	−1.29e−01	1.64e−01	4.25e−02	2.09e−01	3.25e−02
BPG	−4.34e−05	−4.25e−05	−7.52e−01	2.40e+00	6.08e−05	1.35e−04
NADH	3.00e−02	0	5.44e−01	−1.89e+00	3.00e−02	1.63e−02
PEP	6.62e−04	−2.14e−04	−3.43e−01	1.17e+00	6.96e−04	2.40e−05
Pyr	2.27e−02	−1.20e−01	1.07e−02	−7.99e−05	1.22e−01	2.33e−04
ACA	−2.58e−02	7.51e−03	1.48e−02	5.47e−01	2.68e−02	−4.49e−03
EtOH	3.30e−02	−1.62e−02	0	0	3.68e−02	0
Glyc	−7.22e−03	−5.00e−02	0	0	5.05e−02	0

visiae x-2180) are grown in a rotary shaker at 30 °C in a batch culture on glucose and yeast nitrogen base (YNB). The cells are harvested at the point of glucose depletion, washed and starved for 3 h at 30 °C. They can be stored for a couple of days at 5 °C. A basic requirement for the application of the direct method is that the set of states described by the oscillations and the response to the perturbations remain close to a stationary state, and it is advantageous if the dynamics of the system is the dynamics close to a bifurcation. If this is fulfilled, all stationary and dynamical properties can be quantitatively described by the behavior of the model at the stationary state. For the oscillatory yeast cell system these conditions are fulfilled, as all the experimental evidence indicate that the system behaves as a system close to a supercritical Hopf bifurcation. Furthermore, the saturation of the glucose transporter close to the Hopf point ensures that the system is never far from the bifurcation, even in batch experiments where glucose and cyanide are added in pulses at the beginning of the experiment. Therefore, it is reasonable to use the extensive data set on metabolite concentrations, phases and amplitudes made by the Amsterdam group[12–14] in conjunction with the data obtained from bifurcation analysis and perturbation experiments in the CSTR. When supplemented with data from an experiment performed under somewhat different experimental conditions,[5] the experiments by the Amsterdam group[12] also provide estimates of the metabolic fluxes. Enzyme kinetic parameters have been extracted from the literature where possible. *In vivo* enzyme activities sometimes differ significantly from *in vitro* values and are therefore calculated from a stationarity condition upon completion of the optimization. For details on the use of biochemical and dynamical data see ref. 31.

To avoid excessive use of computer time, the dynamic properties of the system are calculated without explicit integration of the kinetic equations. This is done by evaluating the local dynamics, in the simplest case by evaluating the eigenvalues of the Jacobian matrix. Briefly, the method involves the following steps:

(1) All stationary velocities are parametrized by utilizing that they are vectors in the nullspace of the stoichiometric matrix. As a result, all possible stationary velocities are described in terms of the total flux and a linear combination of four extreme stationary velocities.

(2) Known values of stationary concentrations and known parameters such as some of the enzyme kinetic parameters, experimentally controlled flows *etc.* are fixed at the desired values.

(3) Eigenvalues of the Jacobian matrix are calculated for as many different combinations of rates of remaining stationary concentrations, values of unknown enzyme kinetic parameters and relevant stationary velocities as possible. In this way we search for a pure imaginary eigenvalue of the Jacobian matrix of the model at which a Hopf bifurcation may occur.

(4) Right and left eigenvectors of the Jacobian matrix are evaluated at the Hopf points and dynamical features are calculated and compared with the experimental values.

(5) The parameter combination giving the "best match" between model calculations and experimental observations is chosen.

By searching through 100 millions of combinations of parameter values, a set of parameters is selected which largely agree with all available experimental data on the glycolysis of *Saccharomyces cerevisiae*. The resulting model has metabolite concentrations, metabolic fluxes, enzyme kinetic parameters, type and location of bifurcation, frequency of oscillations, perturbation responses and relative amplitudes of metabolic oscillations in good agreement with experiments, while the relative phases of some metabolites are somewhat off. Some higher-order dynamic properties are also inconsistent with the experiments: g'' has the wrong sign, and the value of σ' (which, together with μ, determines the rate of the return of the amplitude to the value for the limit cycle) is too small compared to the experimental value.

From the full-scale model the parameters in the single-cell Stuart–Landau equation can be calculated by the recipes in ref. 38. The frequency ω_0 is equal to the experimental value. The numerical values for the eigenvectors u and u^* at the chosen operating point (the Hopf bifurcation in Fig. 4) is given in Table 1 together with the absolute value of the complex components of u. The relative sizes of these indicate the relative amplitudes of the metabolites.

For the following modelling of a population of yeast cells we have used the Stuart–Landau parameters g', g'', $\mu\sigma'$ determined from the perturbation experiments (Fig. 5), and the eigenvectors determined from the full-scale model. This is a reasonable choice since the eigenvectors are first-order terms while the Stuart–Landau parameters depend on first-, second- and third-order terms which are more dependent on subtle details of the model.

Synchronization

Synchronization in a full model of two yeast cells with optimized parameters showed that they do synchronize but with a phase difference of 180° (anti-phase synchronization).[31] Simulation of eqn. (10) for 1000 coupled cells with kinetic parameters obtained by fitting to experimental response curves as explained in Fig. 5, and coupling parameters calculated from the eigenvectors in Table 1 is illustrated in Fig. 7 where the behavior of two cells and of the centroid $Z = (1/N)\Sigma_i z_i$ is shown.

The figure shows that the phases of the these two cells drift apart and that the amplitude of the centroid is decreasing so that essentially no persistent macroscopic oscillations are observed. This is consistent with results obtained by integrating two coupled full-scale models,[31] but is inconsistent with the experiments. To simplify the following arguments the discussion will be based on the

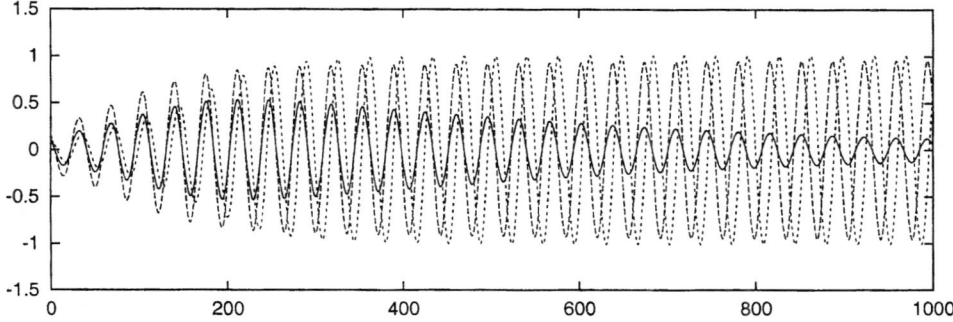

Fig. 7 Simulation of 1000 coupled cells. Timetrace of the real part of z_i for two different cells (dashed curves) and for the real part of the centroid (full curve). The Stuart–Landau parameters are obtained from perturbation experiments whereas the coupling parameters are calculated from the optimized full-scale model as described in the text.

approximate eqn. (11), for which an extensive literature exists and for which numerical studies indicate that the synchronization properties are similar to those of the more complete eqn. (10).

If the transport through the cell membrane is diffusive or mediated by a simple carrier, the matrix D_I is a diagonal matrix with positive elements. The sign of the contribution to the transport coefficient $K_I = u^* \cdot D_I \cdot u$ in eqn. (11) can then be determined by inspection of the components of u for the species, Table 1. For acetaldehyde it can be seen that the real part of this term is always negative, independent of the magnitude of the transport coefficient. This means that for the chosen operating point (i.e. the eigenvectors in Table 1), acetaldehyde works as an anti-phase synchronizing species. A similar conclusion can be made for glucose. The contributions from ethanol and glycerol are zero for this model and this is consistent with experiments. A possible solution to this puzzle is one feature of the full-scale model which can give the opposite effect. This is found in the glucose transport kinetics, for which we have chosen the expressions

$$v_\rightarrow = \frac{V_m \dfrac{[\text{Glc}_x]}{K_{\text{Glc}}}}{1 + \dfrac{[\text{Glc}_x]}{K_{\text{Glc}}} + \dfrac{P\dfrac{[\text{Glc}_x]}{K_{\text{Glc}}} + 1}{P\dfrac{[\text{Glc}]}{K_{\text{Glc}}} + 1}\left(1 + \dfrac{[\text{Glc}]}{K_{\text{Glc}}} + \dfrac{[\text{G6P}]}{K_{\text{IG6P}}} + \dfrac{[\text{Glc}][\text{G6P}]}{K_{\text{Glc}} K_{\text{IIG6P}}}\right)}$$

$$v_\leftarrow = \frac{V_m \dfrac{[\text{Glc}]}{K_{\text{Glc}}}}{1 + \dfrac{[\text{Glc}]}{K_{\text{Glc}}} + \dfrac{P\dfrac{[\text{Glc}]}{K_{\text{Glc}}} + 1}{P\dfrac{[\text{Glc}_x]}{K_{\text{Glc}}} + 1}\left(1 + \dfrac{[\text{Glc}_x]}{K_{\text{Glc}}}\right) + \dfrac{[\text{G6P}]}{K_{\text{IG6P}}} + \dfrac{[\text{Glc}][\text{G6P}]}{K_{\text{Glc}} K_{\text{IIG6P}}}}$$

modelling a reversible glucose (Glc) transporter regulated by glucose 6-phosphate (G6P). Note that the glucose transport kinetics in yeast cells is extremely complicated. Our motivation for choosing this particular set of equations is that it has been validated experimentally by perturbation experiments.[37] This expression gives rise to an off-diagonal term in D_I and the contribution of this term to K_I has a positive real part. The value of the off-diagonal term in D_I is 0.1 s^{-1} for the full-scale model and this is not sufficient to change the anti-phase synchronization behavior. The result of a simulation similar to the simulation in Fig. 7, but with an off-diagonal term of 0.35 s^{-1} is shown in Fig. 8. The phases of the two cells are now converging together with the phases of all the other cells, and the centroid shows persistent oscillations corresponding to the behavior observed in the experiments. Inspection of the value of K_I shows that the real part of this term is now positive. The dependence of the synchronization phenomenon on the parameters of the

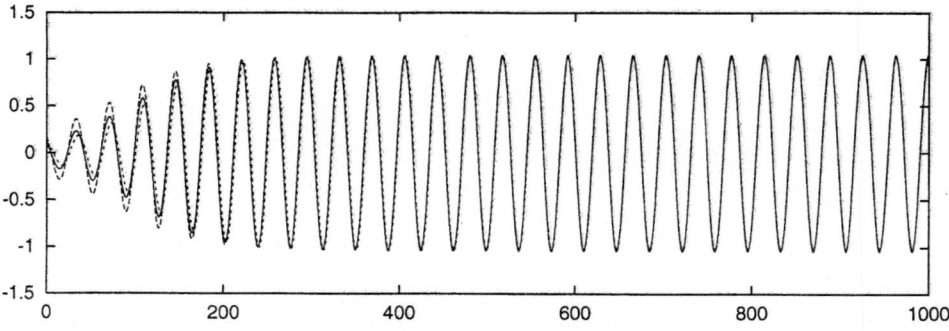

Fig. 8 Simulation of 1000 coupled cells. Timetraces similar to the traces in Fig. 7 but with the regulation of the flow of the glucose transporter by [G6P] increased from 0.1 to 0.35 s^{-1}.

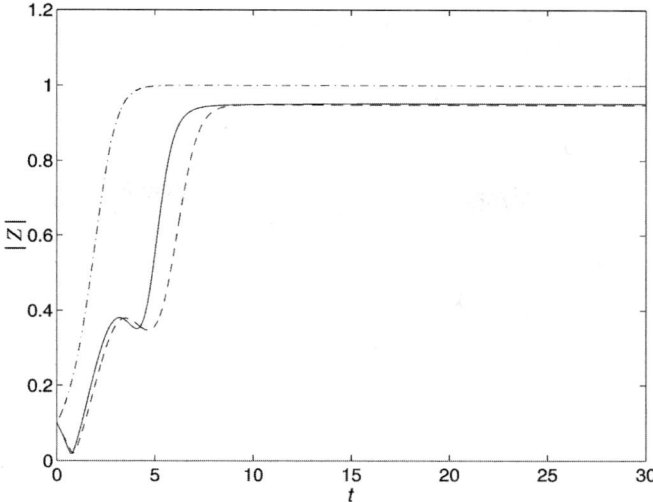

Fig. 9 Evolution of the centroid according to eqn. (14) (continuous line) from numerical integration of 1000 coupled oscillators (eqn. (13), dashed line) and evolution of a single uncoupled oscillator (dashed-dotted line). The parameters used are $K_1 = 6 + i2$, $L_1 = 4 + i3$, $\sigma\mu = 1$, and $g = 1$. The natural frequencies of the population are Gaussian distributed with standard deviation 1.

glucose transporter does not imply that glucose is "the synchronizer". Synchronization depends on the properties of the entire network and is observed experimentally even when the glucose transporter is saturated and when the amplitude of extracellular glucose oscillations is minimal.

Order parameter equations for a population of yeast cells

All macroscopic observable signals which are averages over single cell values are linear functions of the centroid of the state distribution $Z(t) = (1/N)\Sigma_i z_i(t) = \langle z_i(t) \rangle$. A kinetic equation for the centroid can be derived from eqn. (11) provided that the dispersion of cell state distribution remains small. If the cells are identical and in phase it can be shown that the kinetic equation for the centroid is equal to the kinetic equation for a single cell which is a simple Stuart–Landau equation.

$$\frac{dZ}{dt} = (i\omega_0 + \sigma\mu)Z + g|Z|^2 Z. \quad (12)$$

In nature, cells are never identical but have, for example, a distribution in age and cell sizes. A population of slightly different cells can be modelled by using a series expansion of the equations of motion, assuming a small distribution in the cell parameters. In the physics community ω_0 has usually been selected as the variable parameter,[26–28] but μ, g or K_1 may be more realistic choices for real yeast cells. If ω_i is the natural frequency of cell i, the kinetic equations in the strong coupling limit (defined above eqn. (11)) may be written

$$\frac{dz_i}{dt} = (i\omega_i + \sigma\mu)z_i + g|z_i|^2 z_i + K_1(Z - z_i) + L_1(\bar{Z} - \bar{z}_i) \quad (13)$$

Defining $z_i = Z + \varepsilon_i$ and assuming that $\langle \omega_i \rangle = \omega_0$ the kinetic equations for a system with a narrow distribution of ω_i can be approximated by two equations in the order parameters Z and $F = \langle \omega_i \varepsilon_i \rangle$, thus obtaining a macroscopic description of the system.[39]

$$\frac{dZ}{dt} = (i\omega_0 + \sigma\mu)Z + g|Z|^2 Z + iF$$

$$\frac{dF}{dt} = i(\langle \omega_j^2 \rangle - \omega_0^2)Z + (\sigma\mu - K_1 + 2g|Z|^2 + i\omega_0)F + (gZ^2 - L_1)\bar{F} \quad (14)$$

A comparison between the motion of the order parameter Z in a system with 1000 cells by eqn. (13) (2000 variables) and the reduced system eqn. (14) is given in Fig. 9. The behaviour of the population differs from that of an uncoupled oscillator. Even the complex transient of the population is captured by the macroscopic description in terms of the order parameters Z and F.

Discussion

The mechanism of synchroneous oscillations in a population of yeast cells is a problem which has remained elusive for almost half a century. The reason is that it depends critically on biochemical details of glycolysis and of the transport through the cell membrane. This means that a full-scale base model for glycolysis has to be used for each cell. The problem cannot be addressed on the level of a single enzyme or a single chemical species. For 1000 cells this implies a model with more than 20000 variables. For our full-scale model the ratio of timescales at the chosen operating point exceeds 10^7, implying that the model is also very stiff making the direct integration prohibitively time consuming.

Therefore, a method for reducing the dimensionality of the problem is needed. In this paper, we have described such a method. Another fundamental aspect is to model only the timescales relevant to the oscillations and the synchronization. All this is done in a systematic and therefore realistic way by considering only the dynamics in a small region around the stationary state. For the system of oscillating yeast cells, this is possible since it has been shown experimentally that the yeast cells are close to a supercritical Hopf bifurcation, where a normal form expansion can extract the relevant modes without sacrificing the accuracy. The essential kinetics of a single cell are then described by just two local coordinates.

The equations have a universal form which is the Stuart–Landau equation. The biochemical relevance of this equation is ensured by utilizing recipes[38] for calculating the parameters of the Stuart–Landau equations from the parameters of a base model and for calculating a transformation from the local coordinates to the concentration space and *vice versa*. In this way, all concentrations of the base model can be calculated at any time from the local coordinates.

Therefore, the changes in local coordinates can be calculated for any small exchange of a chemical species with the extracellular medium. If the coupling of the cells to the extracellular medium is small the transport between the cells and the extracellular medium can be described by linear terms with coefficients which can again be calculated from the transport terms of a base model. The result is that the number of variables in the kinetic equations for 1000 cells are reduced to 2000 and the kinetic equations are no longer stiff. A numerical integration modelling 30 min real time can then be performed on a standard PC in 15 s. A series expansion in order parameters allows further reductions of the complexity of the description of the system.

It turns out that the realistic transport coefficients for a Stuart–Landau mean-field description obtained by the procedure described here are generally complex numbers, and that they sometimes have a negative real part. This turns out to be of utmost importance for the synchronization dynamics (compare Fig. 7 and 8). Therefore, the existing work on mutually coupled Stuart–Landau oscillators[24,25,27,28] should be extended to include complex coupling constants, possibly with negative real parts. It must be emphasized that the biochemical value of our work depends critically on the faithfulness of the base model used, and conclusions on specific questions may change if the base model is changed. To make our base model as realistic as possible, the parameters were optimized for the actual operating point in the experiments using experimentally determined stationary state concentrations, measured amplitudes of the oscillations, perturbation experiments and realistic enzyme kinetic parameters. Our full-scale model reproduces almost all known biochemistry of the single-cell system, but still it cannot account for the experimentally observed phase synchronization.

In our full-scale model it is assumed that protein synthesis is negligible such that the concentrations of the enzymes remain constant. This is approximately true for all batch experiments because the cells are starved and the duration of the experiments does not exceed one hour. In the CSTR experiments the same is accomplished by the continuous replacement of old cells in the outflow with new cells from the inflow. In the full-scale model we also assume that cell metabolism can be described by homogeneous kinetic equations. This is not at all trivial. NADH waves have recently been observed in neutrophiles,[40] but similar complications are less likely for yeast cells because of their smaller size. For yeast cells the characteristic time of diffusion is 25 ms which is much shorter than the period of oscillation. One aspect which has not been taken into account in the Hopf kinetic equations is the presence of slow hyperbolic modes. In some cases such modes have a drastic influence on the kinetic behavior.[41]

The explicit link between our full-scale model and the reduced equations provides a biochemical interpretation of the synchronization problem for this particular model: The kinetic details of the glucose transporter play a crucial role for the phenomenon, but at the same time the relation $K_I = \boldsymbol{u}^* \cdot \boldsymbol{D}_I \cdot \boldsymbol{u}$ states explicitly that the entire glycolytic system contributes to the coupling term according to Table 1. This confirms the notion that cell synchronization is not a property of a single chemical species or enzyme but a dynamic property of the entire glycolytic and fermentative network as well as the transport of metabolites through the cell membrane. In agreement with these findings, recent experiments have shown that the glucose transporter exerts a substantial fraction of the control of both total flux and frequency during glycolytic oscillations.[17] Further progress on the understanding of active phase synchronization in yeast cells could be obtained by combining more biochemical data with the direct method of optimization and the method for obtaining low-dimensional descriptions of populations of oscillators described here.

Acknowledgement

The authors are grateful to Anders Bisgaard for valuable discussions.

References

1 L. N. M. Duysens and J. Amesz, *Biochim. Biophys. Acta*, 1957, **24**, 19.
2 A. Ghosh and B. Chance, *Biochem. Biophys. Res.*, 1964, **16**, 174.
3 B. Chance, R. W. Eastbrook and A. Ghosh, *Proc. Natl. Acad. Sci. USA*, 1964, **51**, 1244.
4 A. Betz and B. Chance, *Arch. Biochem. Biophys.*, 1965, **109**, 585.
5 A. Betz and R. Hinrichs, *Eur. J. Biochem.*, 1968, **5**, 154.
6 E. K. Pye, *Can. J. Bot.*, 1969, **47**, 271.
7 B. Hess, A. Boiteux and J. Krüger, *Adv. Enzyme Regul.*, 1969, **7**, 149.
8 A. K. Ghosh, B. Chance and E. K. Pye, *Arch. Biochem. Biophys.*, 1971, **145**, 319.
9 A. T. Winfree, *Arch. Biochem. Biophys.*, 1972, **149**, 388.
10 K. H. Kreuzberg and A. Betz, *J. Interdiscipl. Cycle Res.*, 1979, **10**, 41.
11 P. Richard, B. Teusink, H. V. Westerhoff and K. van Dam, *FEBS Lett.*, 1993, **318**, 80.
12 P. Richard, J. M. Diderich, B. M. Bakker, B. Teusink, K. van Dam and H. V. Westerhoff, *FEBS Lett.*, 1994, **341**, 223.
13 P. Richard, B. M. Bakker, B. Teusink, K. van Dam and H. V. Westerhoff, *Eur. J. Biochem.*, 1996, **235**, 238.
14 P. Richard, B. Teusink, M. B. Hemker, K. van Dam and H. V. Westerhoff, *Yeast*, 1996, **12**, 731.
15 B. Teusink, C. Larsson, J. Diderich, P. Richard, K. van Dam, L. Gustafsson and H. V. Westerhoff, *J. Biol. Chem.*, 1996, **271**, 24442.
16 S. Danø, P. G. Sørensen and F. Hynne, *Nature*, 1999, **402**, 320.
17 K. A. Reijenga, J. L. Snoep, J. A. Diderich, H. W. van Verseveld, H. V. Westerhoff and B. Teusink, *Biophys. J.*, 2001, **80**, 626.
18 B. Chance, G. Williamson, I. Y. Lee, L. Mela, D. DeVault, A. Ghosh and E. K. Pye in *Biological and Chemical Oscillators*, ed. B. Chance et al., Academic Press, New York, 1973.
19 A. T. Winfree, *The Geometry of Biological Time*, Springer, New York, 1980.
20 Nicolas Markadieu, personal communication.
21 J. Wolf and R. Heinrich, *BioSystems*, 1997, **43**, 1.
22 M. Bier, B. M. Bakker and H. V. Westerhoff, *Biophys. J.*, 2000, **78**, 1087.
23 J. Wolf and R. Heinrich, *Biochem. J.*, 2000, **345**, 321.
24 Y. Kuramoto, in *Lecture Notes in Physics*, ed. H. Araki, Springer, New York, 1975, vol. 39, pp. 420–422.
25 V. Hakim and W. L. Rappel, *Phys. Rev. A*, 1992, **46**, 7347.
26 A. T. Winfree, *J. Theor. Biol.*, 1967, **16**, 15.
27 P. C. Matthews, R. E. Mirollo and S. H. Strogatz, *Physica D*, 1991, **52**, 293.
28 Y. Kuramoto, *Chemical Oscillations, Waves and Turbulence*, Springer, New York, 1984.
29 J. A. Diderich, B. Teusink, J. Valkier, J. Anjos, I. Spencer-Martins, K. van Dam and M. C. Walsh, *Microbiology*, 1999, **145**, 3447.
30 B. Teusink, J. Diderich, H. V. Westerhoff, K. van Dam and M. C. Walsh, *J. Bacteriol.*, 1998, **180**, 556.
31 F. Hynne, S. Danø and P. G. Sørensen, *Biophys. Chem.*, in press.
32 F. Hynne, P. G. Sørensen and T. Møller, *J. Chem. Phys.*, 1993, **98**, 211.
33 F. Hynne, P. G. Sørensen and T. Møller, *J. Chem. Phys.*, 1993, **98**, 219.

34 K. Nielsen, P. G. Sørensen, F. Hynne and H.-G. Busse, *Biophys. Chem.*, 1998, **72**, 49.
35 O. Richter, A. Betz and C. Giersch, *BioSystems*, 1975, **7**, 137.
36 Y. Termonia and J. Ross, *Proc. Natl. Acad. Sci. USA*, 1981, **78**, 2952.
37 M. Rizzi, U. Theobald, E. Querfurth, T. Rohrhirsch, M. Baltes and M. Reuss, *Biotechnol. Bioeng.*, 1996, **49**, 316.
38 M. Ipsen, F. Hynne and P. G. Sørensen, *Chaos*, 1998, **8**, 834.
39 S. De Monte and F. d'Ovidio, *Europhys. Lett.*, submitted.
40 H. R. Petty, R. G. Worth and L. Kindzelskii, *Phys. Rev. Lett.*, 2000, **84**, 2754.
41 M. Ipsen, F. Hynne and P. G. Sørensen, *Physica D*, 2000, **136**, 66.

Complex morphogenesis of surfaces: theory and experiment on coupling of reaction–diffusion patterning to growth

Lionel G. Harrison,* Stephan Wehner and David M. Holloway†

University of British Columbia, Department of Chemistry, 2036 Main Mall, Vancouver B.C., Canada V6T 1Z1

Received 10th April 2001
First published as an Advance Article on the web 20th November 2001

Reaction–diffusion theory for pattern formation is considered in relation to processes of biological development in which there is continuous growth and shape change as each new pattern forms. This is particularly common in the plant kingdom, for both unicellular and multicellular organisms. In addition to the feedbacks in the chemical dynamics, there is then another loop linking size and shape changes with the reaction–diffusion patterning of growth controllers in the growing region. In studies by computation, the codes must incorporate, alongside the usual solvers of the partial differential dynamic equations, a versatile growth code, to express any kind of shape change. We have found that regulation of shape change in particular ways (*e.g.* to make narrow-angle branchings) demands new features in our chemical mechanisms. Our growth algorithm is for a surface growing tangentially, but moving outward and changing shape to accommodate the extra area. This is potentially applicable both to the tunica layer of multicellular plant meristems and to the growing tip of the cell surface, *e.g.* in the morphogenesis of single-celled chlorophyte algae which display branching processes: whorl formation in *Acetabularia* (Dasycladales) and repeated dichotomous branching in *Micrasterias* (Desmidiaceae). For computational studies, a hemispherical shell is a reasonable idealization of the initial shape. We describe results of two types of study: (1) Pattern formation by three reaction–diffusion models, with contrasted nonlinearities, on the hemispherical shell, particularly to find conditions for robust formation of annular pattern or pattern for dichotomous branching, both of which are common in plants. (2) Sequential dichotomous branchings in a system growing and changing in shape from the hemispherical start.

1 Introduction

1.1 Pattern formation by reaction–diffusion

The chemical basis of morphogenesis, as Turing[1] defined it in 1952, is the concept that new patterns in developing organisms appear first as nonuniform distributions in space of at least two chemical substances, established by the nonlinear dynamics of their chemical interactions together with transport by diffusion. Such dynamics can account for (or predict) the *de novo* appearance of multiple concentration maxima, spaced from each other in a quantitatively orderly manner. In

† Also at Mathematics Department, British Columbia Institute of Technology, Burnaby, B.C., Canada V5G 3H2.

DOI: 10.1039/b103246c

application to biological development, this property implies that Turing reaction–diffusion (R–D) can provide possible explanations of, for instance, branching processes in plants[2,3] or the initiation of single or multiple new organs in animals.[4–6] These symmetry-breaking events are fundamental to the establishment of plant architecture and the animal body plans. The same dynamics may be operating in and have been profitably studied in relation to more easily observed superficial patterns: mammalian coat patterns,[7] seashell pigmentation patterns[8] and tropical fish surface patterns.[9]

Some phenomena of biological development, such as the establishment of simple polarities defining head *vs.* tail or root *vs.* stalk, need only monotonically graded distributions of a chemical concentration. Some confusion can arise in that experimental biologists tend to use the word "morphogen" for such a simply-graded substance, while theorists use it, as Turing intended when he invented the word, for the substances with much greater patterning power in Turing dynamics. One of us[10] has suggested "type I or Wolpert morphogen" and "type II or Turing morphogen" to resolve this ambiguity. Here, we use "morphogen" in the latter sense.

Putative chemical mechanisms falling within the Turing class (as determined by linearization about the homogeneous steady state) are quite diverse, and can display substantially different behaviours when their full nonlinearities become important, at high pattern amplitudes. The range of mechanisms already proposed and studied theoretically by analysis and computation is quite broad[4–8,11–13] and often permits a choice from the existing literature of a suitable mechanism to explain a particular biological event. For instance, the Gierer–Meinhardt rate equations are very good at maintaining isolated single concentration peaks. In this group, we used a chemical mechanism giving an approximation to Gierer–Meinhardt dynamics to match theory to detailed experimental evidence on heart formation in a salamander.[5] The Brusselator,[11,12] on the other hand, is good at generating multiple peaks with regular spatial periodicity,[10] and has been used in most of our work on branching processes in plants.[2,10,14–17]

1.2 Coupling of R–D patterning to continuous growth with change of shape

Most theoretical work on R–D dynamics has been for domains of fixed size, shape and boundary conditions. Within these restrictions, there is plenty to study. Even the problems of what kinds of dynamics preferentially generate stripes or spots on a plane was not given even a partial solution, for restricted types of nonlinearities, until 1991,[18–20] and has continued to be an interesting and complex topic even for such well-studied models as the Brusselator.[21] The restrictions are often not serious for matching theory to many events of animal development. Apart from consumption of yolk, embryos cannot grow until they feed and cannot feed until they have undergone many patterning events.

Plants, however, continue to form new organs throughout their lives, with continuous growth and shape change as each pattern forms. The nature of plant development is remarkably similar in multicellular plants and some large unicellular ones that generate complex shapes. Fig. 1 and 2 illustrate morphogenesis for two unicellular examples, both green algae.

1.2.1 *Micrasterias*.
The genus *Micrasterias* of desmid algae has many species, common in freshwater ponds worldwide, which form diverse species-specific variants of a star-shaped cell by repeated dichotomous branchings at the tips of growing lobes. For this genus, it is fairly well established that the patterning is in the rate of addition of material to the cell surface,[2,22] with cell wall synthesis following in lock-step the extension of the cell membrane. Our previous publication on *Micrasterias*[2] described computations of simultaneous R–D patterning and growth matching diverse species of the genus, but only for 2-D spatial patterning. One can proceed a long way by considering the phenomenon as growth of an edge defining the cell outline, but eventually questions arise demanding 3-D computation. What controls successive dichotomous branchings so that they are all coplanar? The dynamics of a dome-shaped growing tip may be well represented by those of a section through it, because the essence of either is development of a convexity. But what about the processes at a cleft in the cell outline, which in 2-D is only concave, but in 3-D is a saddle-shaped region with two principal curvatures of opposite signs? In our 2-D modelling, the growth rate was zero at these points; but this is not experimentally correct. Fig. 1(c) shows the bottoms of clefts in the cell outline advancing outwards slowly, in fact at up to half the rate of advance of the fast-growing regions.[13] We seek to model such spatially differentiated growth

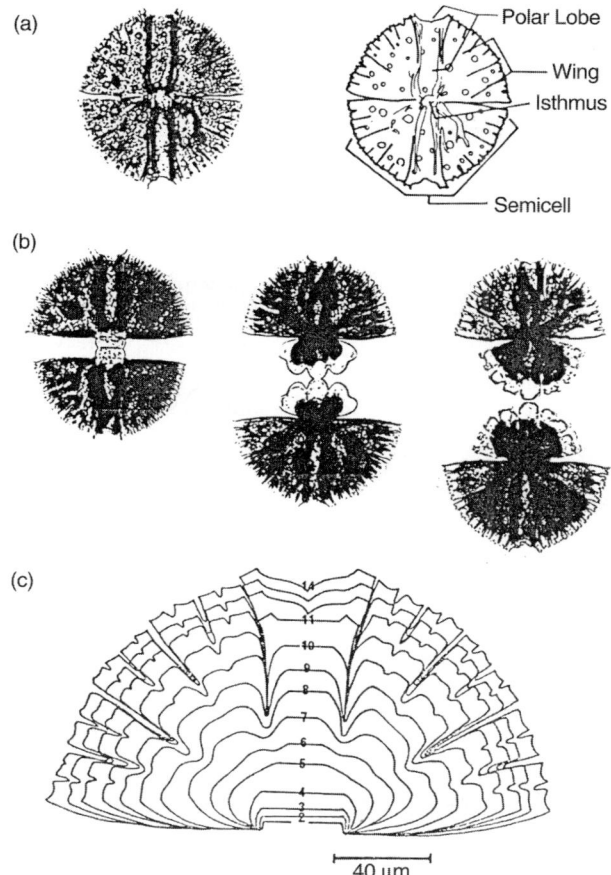

Fig. 1 Semicell morphogenesis in vegetative reproduction of the unicellular desmid alga *Micrasterias rotata*. (a) Fully-formed cell, photomicrograph and sketch with terminology; (b) Stages 3, 7 and 10 in the development of two daughter cells (60, 140 and 200 min after mitosis); (c) Stacked semicell outlines at 20 min intervals, showing the complete sequence of repeated dichotomous branchings. In the out-of-plane dimension, the cell is quite thin, *cf.* the shape of a common magnifying glass. All parts of the figure from Lacalli[32] with permission.

without being constrained to zero rates at saddle points, and to do this requires 3-D computations.

The above account carries the implication that we view such growth as that of *Micrasterias* as being shape change and volume increase driven by area increase of the cell surface. This tangential growth must be converted into outward advance of the surface, something quite different from accretive growth in which new material is layered on old, as in crystal growth. Our growth algorithms are described in sections 2.3 and 2.4.

1.2.2 *Acetabularia*. Fig. 2 illustrates the development of a unicellular seaweed, *Acetabularia acetabulum*, that grows as large and as complex in morphology as many a higher plant containing perhaps a million cells. The formation of vegetative whorls at the growing tip is always preceded by tip flattening (Fig. 2(b)), which suggests a succession of annular and periodic chemical patternings. These, postulated to be two-morphogen R–D processes, are called feedback loops I and II in Fig. 2(c). In addition, elongation of the main stalk and of each hair in a whorl is largely produced by rapid growth at the tip. To keep this confined to the tip, the growing region must perform a "skyhook" act of dragging its boundary up after it. A putative chemical mechanism for

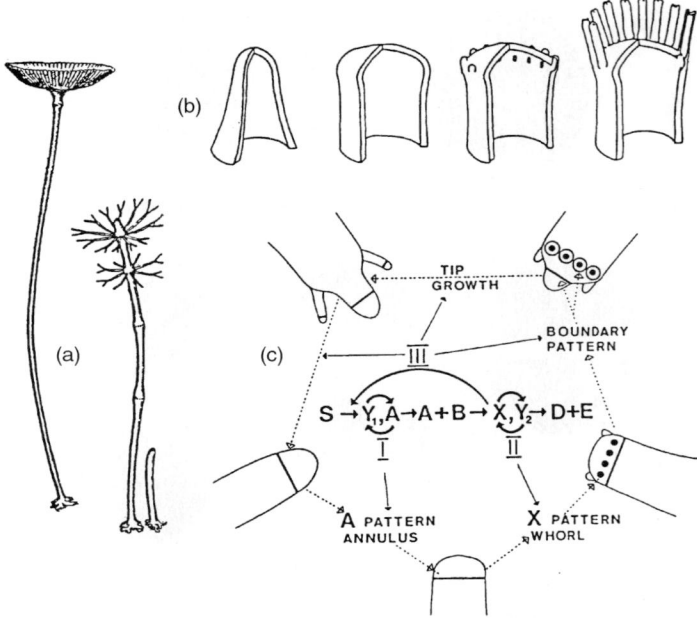

Fig. 2 Whorl formation in development of a very large unicellular marine alga, *Acetabularia acetabulum*. (a) Three stages in development. The single nucleus remains in the rhizoid, at bottom. Vegetative whorls form every few days and are deciduous (scars of two earlier whorls shown). The life cycle culminates in formation of the saucer-shaped reproductive whorl. Drawings are about ×2 actual size. Adapted with kind permission from ref. 33. (b) Four stages in vegetative whorl development, showing tip flattening before whorl formation. Reproduced with kind permission from ref. 3. (c) Feedback loops (Roman numerals) needed in a chemical mechanism to establish: I annular pattern leading to tip flattening; II whorl pattern; III redrawing of boundary conditions to confine developmental processes to the tip. Reproduced with kind permission from ref. 34.

this is control by the R–D systems of the supply of an initial input, substance A, from substrate S. This is feedback loop III in Fig. 2(c). It serves the purpose of confining morphogenetic activity to a number of regions which are progressively isolated from each other, allowing for good control of each in the succession of complex branching events. This is one of the most important characteristics of plant development, in all the higher plants as well as these lowly algae. Understanding of the nature of feedback loop III could be very significant in relation to the whole plant kingdom.

The occurrence of both annular and periodic patterning, and the need to explain both by processes which we number I and II, shows another need to do computational modelling in 3-D. Consider the idealization of a hemispherical growing tip, and the need to explain a dichotomous branching event, as in *Micrasterias*. In 2-D morphology, with an effectively 1-D domain of R–D along the growing edge, the refit of the tip must be from a one-wavelength to a two-wavelength pattern. But if this is supposed to represent a 3-D dome by a section through it, are the two new concentration maxima parts of a dichotomously branched pattern or of an annulus of high concentration? In linear Turing dynamics, these two patterns would begin to develop as the spherical surface harmonics $Y(l,m) = Y(3,0)$ (annular) and $Y(3,2)$ (dichotomous), and they would grow at equal exponential rates from whatever rudimentary amplitudes were provided by random noise or other antecedents. How does a plant choose one or the other pattern, robustly? In section 3.1, we describe our study of patterning on a non-growing hemispherical shell.

1.3 Chemistry *vs.* physics in our models

As physical chemists, we are chiefly interested in exploring the capabilities of nonlinear chemical kinetics, together with diffusive transport, for controlling plant morphogenesis, especially branching events. In the same spirit as Meinhardt's modelling of sea-shell pigmentation patterns,[8] we seek the minimal dynamic requirements, but Ockham's razor will not shave them down to the activities of only two morphogens. We need at least one more. The particular cases we have chosen to study involve also physics: a translation of tangential area increase into outward displacement of a surface. This is not uniquely defined without consideration of the mechanics of the surface, and how it may bend to relieve stresses generated by insertion of excess material. The possible role of mechanical stresses has been considered elsewhere,[3,23,24] but in entirely mechanical models without chemical R–D as the prime symmetry-breaker. In our work we have concentrated on chemistry by extreme simplification of the physics into algorithms to convert local area increase into sufficient outward movement to find places for the extra area and thereby relieve stresses.

1.4 Simplifying the physics: the algorithm for morphological expression of pattern

Fig. 3 illustrates the geometrical representations of shape used in our previous 2-D computations[2] and in the 3-D computations described here. In 2-D, the curved cell outline (initially a semicircle, Fig. 3(a)) was represented by nodes (*e.g.* A, B, C) initially equally spaced, joined by straight line segments. Increase in length in proportion to the concentration of an R–D patterned catalyst (morphogen X) was applied to the line segments (AB, BC) and the new position of the node B was found as an intersection of two circles centred on A and C and with radii as the new lengths of AB and BC. The intersection was always chosen for outward growth, as expected for a cell surface kept extended by turgor pressure. To simulate as well as possible the effect of simultaneous growth of the whole outline, nodes were visited in random order for such local computational steps. When the line segments became too long for good representation of the shape (*e.g.* twice original length) they were bisected with additional nodes.

Fig. 3(b) and (c) show the shape representation used in our 3-D work (by no means a unique choice). This mesh of triangles, its use in R–D and growth computations and its refinement by insertion of new triangles are described in section 2. Fig. 3(b) is the hemisphere used for the studies described in section 3.1. Fig. 3(c) illustrates a representation of a dichotomous branch; studies of formation of these are described in section 3.2.

2 Models and methods

2.1 Finite element representation

The mesh used to represent the initial shape (a hemisphere, for all computations so far undertaken, Fig. 3(b)) was one in which every node was joined by line segments to 6 neighbours (*i.e.* had coordination number 6), except for those on the equatorial boundary, each of which had 3 or 4 neighbours. The local environment of every node was not planar, but was as close as possible to a regular hexagon divided into 6 triangles. (Distortions from equilateral in these triangles will be greater in this arrangement than with the use of a mixture of hexagonal and pentagonal units, as in the fullerene or football arrangements. We felt it better to preserve the 6-coordinate geometry wherever possible.) In the R–D calculations and the growth algorithm (sections 2.2 and 2.3) this local geometry around each node was used as a finite element and was projected onto a plane so that this projection described the behaviour of the solutions at the nodes of the element. The areas of the 6 triangles appeared in all the calculations, and also their spatial orientations in the growth algorithm (section 2.3). Refinement of the mesh in regions that had grown substantially was done (section 2.4) in such a way that nodes on the boundaries of refined regions were given a coordination number 5 until it could later be restored to 6. (Only the nodes on the original equator had coordination number 3 or 4.)

2.2 Numerical solution of R–D equations

Numerical values of concentrations of morphogens and other reactants were defined at the nodes. For calculations of chemical kinetics, the functions of these concentrations appearing in the rate equations (section 2.5) were averaged over the three nodes of any triangle and multiplied by the area of the triangle. The diffusion terms in the equations were approximated assuming a constant concentration gradient down the triangle from outside edge to central node, and again taking into

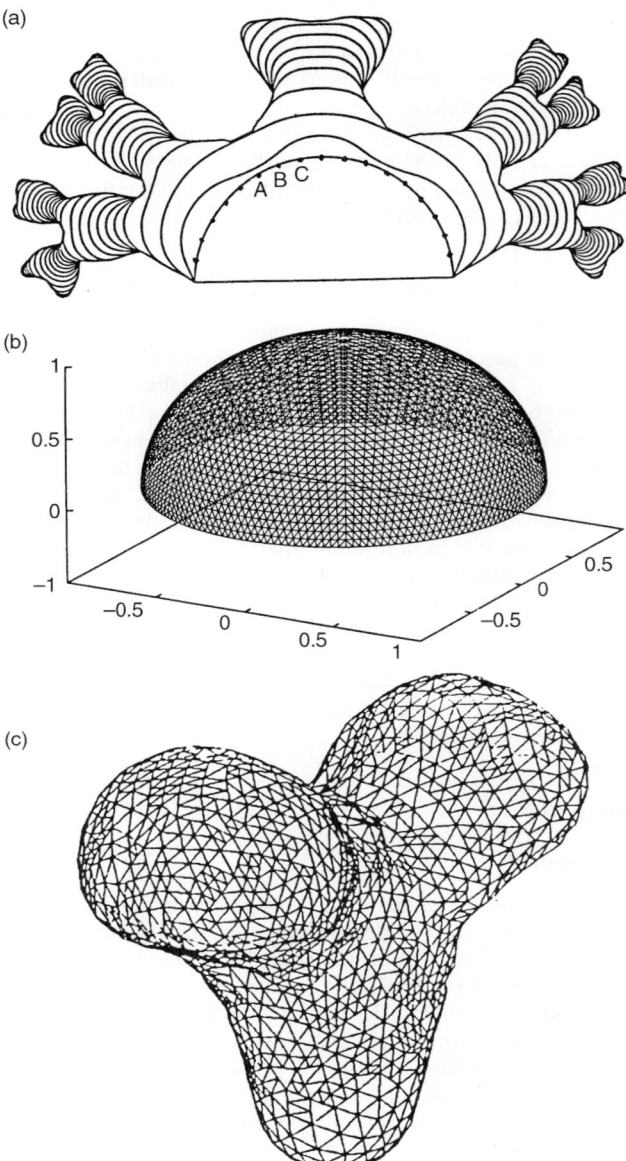

Fig. 3 From 2-D to 3-D computation. (a) Stacked stages (*cf.* Fig. 1(c)) in a 2-D computation of development of the outline of a semicell resembling *Micrasterias rotata*, starting from a semicircle. For a colour-coded version showing patterned distribution of morphogen X along each outline, see Holloway and Harrison,[2] Fig. 9. (b) For the start of a 3-D computation, a polygonal approximation to a semicircle is replaced by triangulation of a hemisphere. (c) How to refine the mesh as the surface grows? This example of dichotomous branching, reproduced with kind permission from ref. 26, is for a different growth process from the one we are studying, but shows mesh refinement by insertion of small triangles into large ones where needed. This account led us to our method.

account triangle area. Both reaction and diffusion were summed over all triangles of a finite element. The R–D partial differential equations (PDEs) were thus converted into a system of ordinary differential equations (ODEs) that were solved by a Runge–Kutta method, following a suggestion of Ascher *et al.*[25] The use in the diffusion calculation of finite elements rather than finite differences along line segments gives a better local averaging to cope with distortions of the mesh that may arise during the growth calculations.

2.3 The algorithm for morphological expression of pattern

In the finite element method, the growth algorithm is not at all closely analogous to the intersection-of-two-circles method used in our 2-D work. First, the direction of nodal movement is determined by averaging the normal vectors for all triangles of the finite element. Second, the magnitude of nodal movement is calculated. To do this, area increase is assumed to be a quadratic function of nodal movement in the specified direction. Three test movements are made to calculate the coefficients of the quadratic function, which is then used to calculate the nodal movement giving the area increase of the finite element, specified by the concentration of X morphogen at the node. (Fractional area at finite element i is $c_g X_i \Delta t$; growth is computed only every tenth time-step, and is zero at all others.) To simulate as closely as possible the result of simultaneous growth of the whole surface, nodes are visited in random order.

This method works well in convex regions of the surface, but can obviously give trouble in concavities. Our concept of outward movement to accommodate tangential area increase cannot be applied to an assembly of nodes in a concave region, because the overall movement would lead to shrinkage, not growth. This suggests that our neglect of bending stresses in such regions may be invalid.

This problem did not arise in our 2-D work, because clefts between growing lobes in our *Micrasterias* computations always had zero growth rate (regions colour-coded white in Holloway and Harrison[2]). In 3-D, these clefts are not completely concave pits, but are saddles, with concavity in one principal direction only. (In plant development in general, pits are much less common than saddles.) In our modelling, however, a region with any aspect of concavity is usually one of uniform unpatterned growth, which may be computed by simple enlargment, multiplying all the nodal coordinates by a small growth constant in each time step. Development much like that in the clefts of *Micrasterias* can thus be computed with the aid of a photocopier. In our computer codes where this procedure was needed, *all* nodes on the surface received this spatially homogeneous displacement, and regions of R–D patterning received the additional X-dependent displacement as computed by the finite element method.

2.4 Mesh refinement algorithm for a growing surface

The regions where morphogenetic patterning events happen are ones of greatly increasing surface area. Insertion from time to time of many new nodes is necessary to maintain good representation of the shape. In 2-D, the way to do this was trivially obvious (section 1.4) and the insertion of new nodes maintained the coordination number 2 for all new and previously-existing nodes.

To devise a method of mesh refinement in 3-D is a much more challenging problem, and does not appear to have a unique solution. The method described here, which limits coordination numbers to 5 or 6 (except on the original equator), does not eliminate mesh instabilities indefinitely, but postpones them until substantial growth has occurred, sufficient to model many of the biological events of interest.

Following Kaandorp,[26] who represented surfaces by triangulation for accretive growth (Fig. 3(c)), we insert a new triangle (3 new nodes) into an old triangle, with new vertices at the midpoints of the old sides. This is done once any side of an old triangle has exceeded twice its original length. The excessive elongation is discovered in the computation when one node (and the finite element it defines) is being visited. A new triangle is inserted into each of the triangles of the finite element, *e.g.* around node A, Fig. 4. For inhomogeneous growth, this mesh refinement creates boundaries between refined and unrefined (slower-growing) regions, at which some new nodes have only 4 neighbours, *e.g.* node B, Fig. 4. Our algorithm increases this to 5 without giving 7 neighbours to node C. When node B is visited, it recognizes C as a neighbour; when node C is visited, it does not recognize B as a neighbour.

2.5 Reaction–diffusion mechanistic schemes

For our study of repeated dichotomous branching, to approach a 3-D analogue of our 2-D work on *Micrasterias*,[2] only one model was used, namely, a Brusselator[11,12] with an additional feature

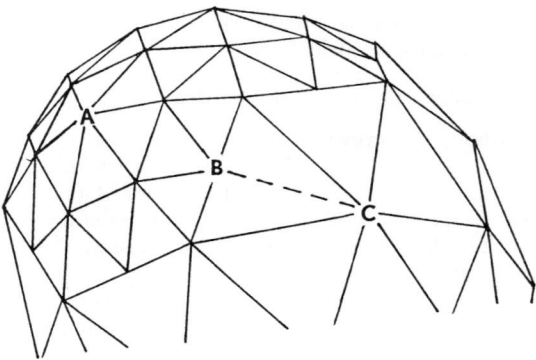

Fig. 4 Our treatment of the boundary between refined regions of the mesh and ones without new nodes: When one triangle with a vertex at node A has a new triangle inserted in it, the refinement is continued similarly for all 6 triangles around node A. New nodes on the boundary of the refined region, such as B, now have only 4 neighbours. To increase this to 5 without giving 7 neighbours to node C, our algorithm specifies that when node B is visited it recognizes C as a neighbour, but when C is visited it does not recognize B as a neighbour.

to provide "feedback loop III" (Fig. 2(c)), as described in our earlier work. For the study of competition between annular and dichotomously branched pattern on a non-growing hemisphere, we anticipated marked differences between different kinds of nonlinearity in the models, and therefore compared three R–D mechanisms with strongly contrasted nonlinearities: the Brusselator, the Gierer–Meinhardt model, and the Harrison–Lacalli hyperchirality model.

2.5.1 The Brusselator, with an additional feedback loop. The Brusselator two-morphogen activation-depletion (activator X, depleted substance Y) model is defined by the putative reaction mechanism:

$$A \xrightarrow{a} X \tag{1a}$$

$$B + X \xrightarrow{b} Y + D \tag{1b}$$

$$Y + 2X \xrightarrow{c} 3X \tag{1c}$$

$$X \xrightarrow{d} E \tag{1d}$$

X and Y are both transported by ideal Fickian diffusion, with diffusivities D_x and D_y. This is one of the three mechanisms compared in section 3.1, for patterning on a non-growing hemisphere. For some of the computations described in section 3.2, for patterning along with growth and shape change, there is an extra feedback from X to A:

$$S \xrightarrow{k_p X} A \tag{2a}$$

$$2A \xrightarrow{k_d} F \tag{2b}$$

With this addition, A is a non-diffusing morphogen. This additional control is intended to isolate growing tips chemically from each other, so that: (a) every new dichotomous branching event in a sequence will occur in a bounded region, independent of interference from other simultaneous events in other branches; (b) the boundaries will move so as to permit the generation of low-angle branching events, as discussed in the 2-D work.[2] There, together with the feedback in eqn. (2), a complete switch off of both X formation and growth was used below a threshold concentration X_{th}. Part of the purpose of the ongoing 3-D work is to avoid this arbitrary addition and explain how X_{th} arises in mechanistic terms. This has not yet been done. In the 3-D work described in section 3.2, X_{th} has not been used in the chemical dynamics, but has been used as a convenience to switch off growth in some regions while exploring the capabilities of the growth algorithm.

Parameter values were mostly taken from those found useful in previous work: for the unmodified Brusselator, the values for "point BR2" in our work on order in pattern;[27] for the modified Brusselator, the values quoted in the 2-D work.[2] For all 3-D computations, the diffusivity ratio was $D_y/D_x = 20$, but the absolute values of D_y and D_x varied from one computation to another of the results given in section 3.2. Changes in these values with no change in their ratio, and no changes in chemical-kinetic parameters can be used to change the wavelength (or for spherical harmonics, fit of l to radius) of an R–D model without moving the representative point in parameter space.[10]

2.5.2 The Gierer–Meinhardt model. The Gierer–Meinhardt model[4,6] has always been defined by its devisers in terms of rate equations, without connecting them back to a mechanistic scheme of chemical equations. Holloway *et al.*[5] showed how a mechanism based on enzyme kinetics, with allosteric control of an enzyme activity to produce autocatalysis, could give a close approximation to the Gierer–Meinhardt rate equations. Here, we use the original equations:

$$\partial X/\partial t = \rho_o \rho + c\rho X^2/Y - \mu X + D_x \nabla^2 X \quad (3a)$$

$$\partial Y/\partial t = c'\rho' X^2 - vY + D_y \nabla^2 Y \quad (3b)$$

Parameter values used in the computations described in section 3.1 were: $\rho_o = 0.018$, $\rho = 0.5$, $c = 0.045$, $\mu = 0.13$, $c' = 0.1$, $\rho' = 0.5$, $v = 0.135$.

2.5.3 The hyperchirality model. The hyperchirality model of Harrison and Lacalli[10,13] was based on a putative "Flatland" chirality arising from two modes of attachment of a tetrameric protein to a cell surface. The pattern development is then in the Flatland chiral asymmetries U and V of the two morphogens X and Y. The most significant feature of the model, however, is that, written in terms of U and V it contains only cubic nonlinearities; this gives it very strong stripe-forming power in a 2-D planar domain.[18–20] The rate equations are:

$$\partial U/\partial t = k_1 U + k_2 V - k_5 UV^2 - k_7 U^3 + D_x \nabla^2 U \quad (4a)$$

$$\partial V/\partial t = k_3 U + k_4 V - k_6 U^2 V + D_y \nabla^2 V \quad (4b)$$

Parameter values used in the computations described in section 3.1 were: $D_x = 4.4 \times 10^{-5}$, $D_y = 5.5 \times 10^{-4}$, $k_1 = 0.225$, $k_2 = 0.2$, $k_3 = 0.25-1$ (several values used), $k_4 = -0.125$ to -0.5 (several values), $k_5 = 0.1$, $k_6 = 0.125-0.5$ (several values) and $k_7 = 0.325$.

2.6 Terminology for concentrations

For morphogens X and Y, we use X and Y for their complete concentrations. The R–D equations have spatially homogeneous steady state concentrations X_0 and Y_0. For the departures of X and Y from this homogeneous state we use $U = X - X_0$ and $V = Y - Y_0$. In the case of the hyperchirality model, the rate equations are written in the first instance in terms of hyperchiral asymmetries U and V. In this case, the symmetry of the dynamics is such that for every solution in terms of spatially-patterned U and V there is another with the same numerical values but opposite sign for both U and V.

The initial state in all our computations on the constant-size hemispherical shell in section 3.1 is X_0, Y_0 plus low-amplitude concentration noise over the whole domain. For the hyperchirality model at all amplitudes and for all models at low enough amplitudes that only terms linear in U and V are significant, this starting condition implies that patterns with opposite signs for both U and V in all positions will grow with equal probability. This is not biologically realistic. Living patterns must be selected either by the properties of the nonlinear dynamic terms or by the growth of complex pattern out of pre-existing simpler pattern that determines the direction of the changes in U and V, *e.g.* dichotomous branching pattern arising out of simple tip growth pattern.

For a hemispherical shell, we use θ for co-latitude and φ for longitude.

2.7 Programming and computing facilities

The complete software package for combined R–D and growth computations consists of 15 linked C programs (each calculating, *e.g.* growth, refinement, triangulation, surface initialization, R–D, *etc.*), about 20 000 lines of code in total. The package compiles with the Gnu C compiler, and will run on Unix (AIX, HP/UX) and Linux. An advantage of C is the use of *structure* type to represent surface nodes, each with attendant coordinate positions, morphogen concentrations, identification numbers, element areas, *etc.* Attached to each node is a neighbour list, so that the whole surface is specified by a linked list of nodes. Iterative routines follow neighbour links (*while* there is a next neighbour), rather than predetermined numbers of iterations (*for* loop).

A graphical display program was written using the Silicon Graphics Graphics Library, to compile and run on Irix 6.x.

3 Computational projects and results

3.1 Pattern formation by R–D on a hemispherical shell of constant radius

3.1.1 Spherical harmonics as components of pattern. The hemispherical shell is an idealization of a meristematic region in a multicellular plant or a tip growth region in a unicellular plant, joined at its equator to a cylinder exhibiting slow unpatterned growth. In all computations in section 3.1, we use, at the equator, a Dirichlet boundary condition: X and Y are X_0 and Y_0 or $U = V = 0$. In the approximation of linearization about this spatially uniform state, the solutions of the dynamic equations are sums of spherical harmonics $Y(l,m)(\theta,\varphi)$, each (for U and V together reaching a fixed amplitude ratio) growing, decaying or oscillating in amplitude with an exponential time constant

$$k_g = (1/2)[k_1 + k_4 - \{l(l+1)/r^2\}(D_x + D_y) + (\beta^2 + 4k_2 k_3)^{1/2}] \tag{5a}$$

where

$$\beta = k_4 - k_1 - \{l(l+1)/r^2\}(D_y - D_x), \tag{5b}$$

and k_1–k_4 are the rate constants in linearized Turing equations, *e.g.* as in eqn. (4). (Eqn. (5) may be derived by analogy with eqn. (7.3) to (7.17) in Harrison,[10] with $\omega^2 = 4\pi^2/\lambda^2$ replaced by $\omega^2 = l(l+1)/r^2$.) For suitable choices of rate parameter values, k_g *vs.* r passes through a fairly sharp maximum at radius proportional to $l(l+1)^{1/2}$. The harmonic of any specific value of l is fastest-growing over only a narrow range of r. All modes of the same l will have the same value of k_g. This does not raise a problem in relation to the maintenance of simple nonbranching tip growth. For this, a radius appropriate to $l = 1$ will, with the boundary condition, select $Y(1,0) = \cos\theta$, the function which maintains a hemispherical growing tip, provided that it has a positive coefficient. At radius for $l = 2$, again there is no ambiguity. The boundary condition does not allow the circularly symmetric function (to chemists, the angular part of d_{z2}), and the other possibility $Y(2,1)$ gives the lop-sided branching appropriate to explain the monocotyledon embryo. (For the index m, we are using numbers 1, 2 without sign to represent the real functions.)

3.1.2 Competition between annular and dichotomously-branched pattern at $l = 3$. The problem requiring further study by computation and analysis arises for the range of radii selecting $l = 3$. Two quite different patterns, both common and important in plant morphogenesis, should grow at the same rate in the linear approximation:

$$Y(3,0) = 5\cos^3\theta - 3\cos\theta; \tag{6a}$$

$$Y(3,2) = \sin^2\theta \cos\theta \sin(2\varphi - \delta). \tag{6b}$$

$Y(3,0)$ is an annular pattern, with positive annulus if its coefficient is negative; $Y(3,2)$ is a dichotomous-branching pattern, for which change of δ or change of the sign of its coefficient serve only to rotate it. Fig. 5 illustrates the competition between these two kinds of pattern in typical computational results. Their equal exponential growth in the region of linear approximation are shown by the two parallel lines at earlier times in Fig. 5(b). It is not evident what is going to happen at later times, when the nonlinear terms become important. Table 1 summarizes the

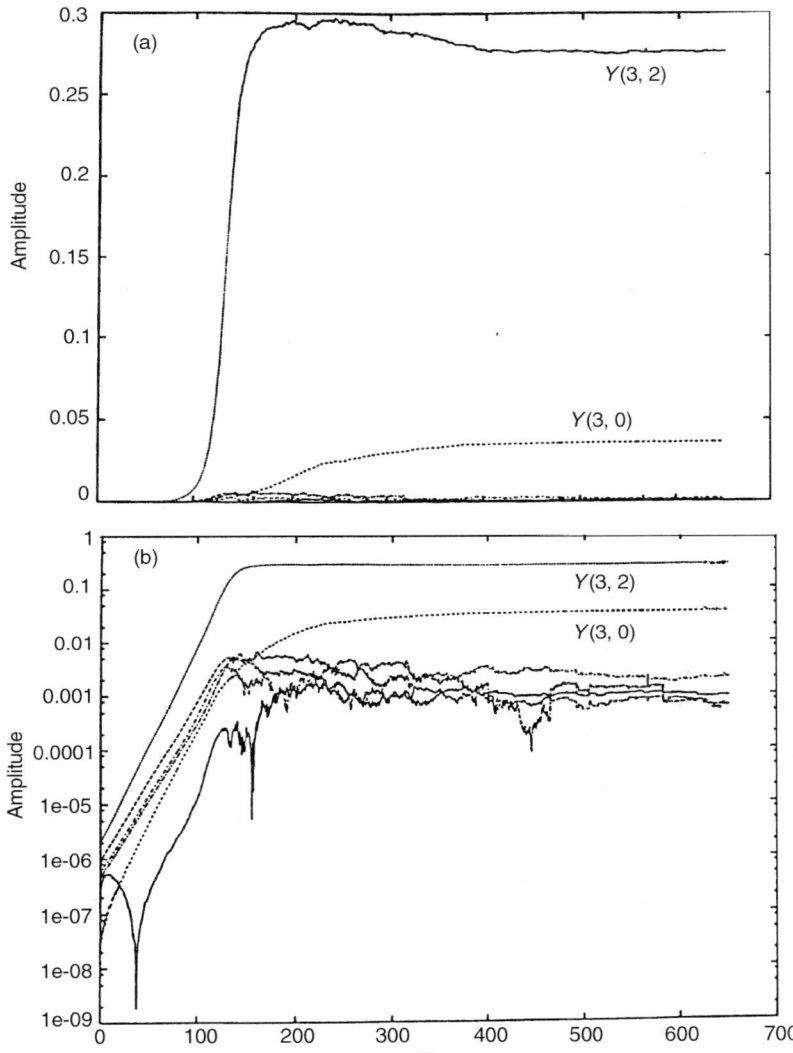

Fig. 5 Amplitude *vs.* time for several spherical surface harmonic components in a computation of pattern development on a hemispherical shell, computed by a Fourier-transform-like technique of convoluting the computed patterns with each harmonic. Unlabelled lines are for modes with $l = 1, 2$ and 4. (a) A hyperchirality model, linear scale of amplitude; (b) the same results as (a), logarithmic scale of amplitude. The time scale is in time-steps of the computation.

results of a statistically significant number of computations at optimal radius for $l = 3$, all continued well into the nonlinear region, for three R–D models. Dichotomous branching is strongly favoured, but not selected robustly enough by any of the models to account for the various instances in which it occurs in plant development. These results led us to three questions:

(1) They are for patterning on a domain of fixed size and shape. How are the pattern-forming capabilities of the dynamics affected by continuous feedback interaction with the morphological changes that chemical patterning stimulates? Our beginnings of 3-D modelling to address this question are described in section 3.2.

(2) Do the computational data of Table 1 correctly represent an intrinsic property of the R–D dynamics? This requires mathematical analysis, and led us to collaborate with a mathematician,

Table 1 Variability in computed pattern on a hemispherical shell, optimum r for $l = 3$

Mechanism or model	Numbers of computations which gave:		
	Dichotomous branching	Annular pattern	"Keyhole" (Fig. 7(e), 3)
Brusselator	84	19	0
Gierer–Meinhardt	29	12	0
Hyperchirality	303	59	53

Dr. Wayne Nagata. His bifurcation analysis of two of the models is to be described in full elsewhere.[28] In the most general terms, it supports the computational results. A summary of the present state of this work is given in section 4.

(3) Does the statistical split between branching and annular patterns arise from preservation through pattern development of a bias in the initial random input? We guessed that the input might have an initially larger amplitude for $Y(3,2)$ than for $Y(3,0)$ because the latter has, in a circular sense, a much longer correlation length, with uniform concentrations for distances up to $2\pi r$. Determination of the initial amplitudes for 1000 different random inputs gave the results plotted in Fig. 6, which support our guess.

Fig. 7 illustrates patterns for the two $l = 3$ harmonics, an equal mixture of the two, and computational results for R–D dynamics at $l = 3, 4, 5$ and 6. These pictures show that the equal mixture, and the commonest kind of pattern generated at $l = 3$, appear very good for the kind of patterning of a growth catalyst needed to stimulate dichotomous branching. But Fig. 7(e) shows that the hyperchirality model does not give good branching whorled patterns for $l > 3$ and indeed

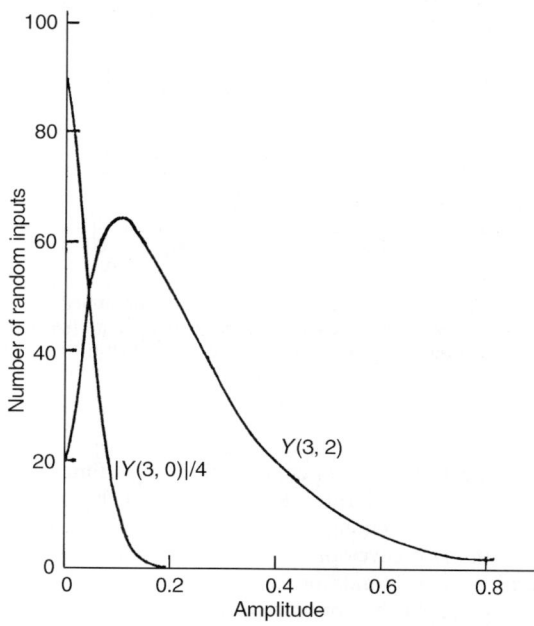

Fig. 6 Distributions of amplitudes of the two $l = 3$ harmonics in a set of 1000 random patterns on the hemisphere, as used for initiating pattern in our computations, drawn for increments of 0.01 in amplitude. Numbers on the scale of ordinates should be multiplied by 4 for the $|Y(3,0)|$ curve. This distribution contains both positive and negative annuli (for the variable U).

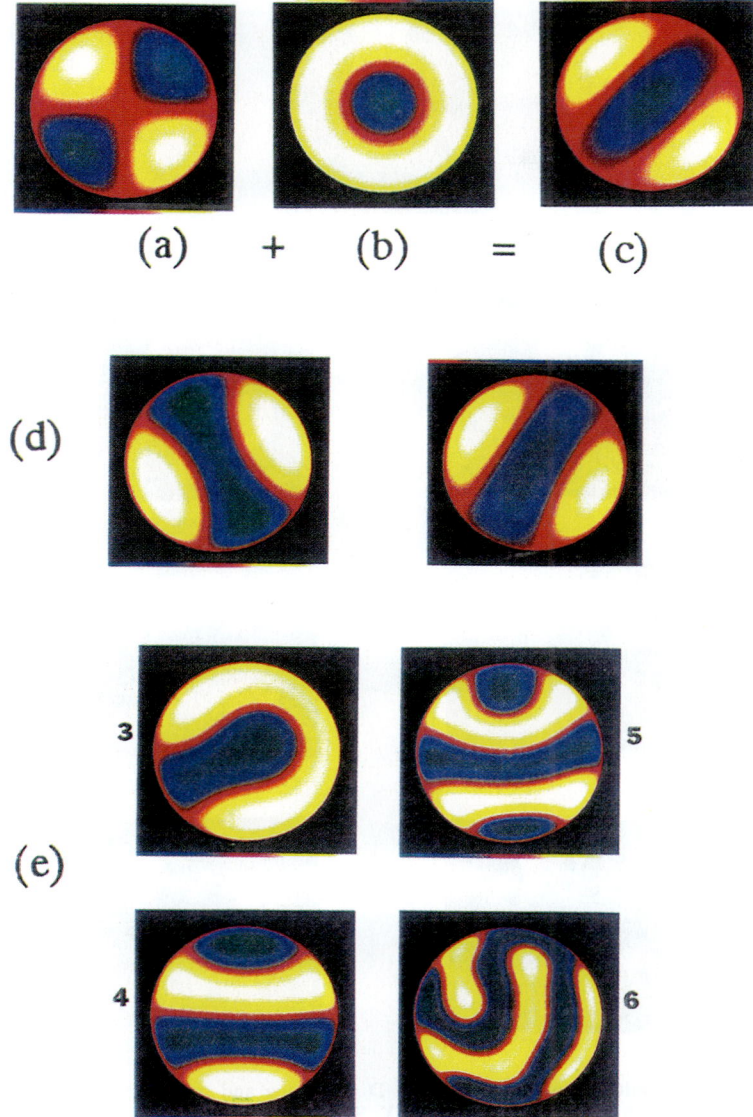

Fig. 7 Plan views (*i.e.* projections of the hemispherical shell on to its equator) of some concentration distribution patterns. Colour-coding is from maximum (white) to minimum (green) of the X or U concentration regardless of the quantitative range between these two. The circular boundary is at homogeneous steady state concentration $X = X_0$ or $U = 0$, and is not always the same colour because this value is not necessarily halfway between maximum and minimum. (a), (b) and (c) the spherical harmonics $Y(3,2)$, $-Y(3,0)$ and their sum; (d) Typical results of computations of pattern development by R–D at optimal radius for $l = 3$ and long times; (e) patterns not matching morphogenetic behaviour computed with the hyperchirality model at optimal radii for $l = 3, 4, 5$ and 6.

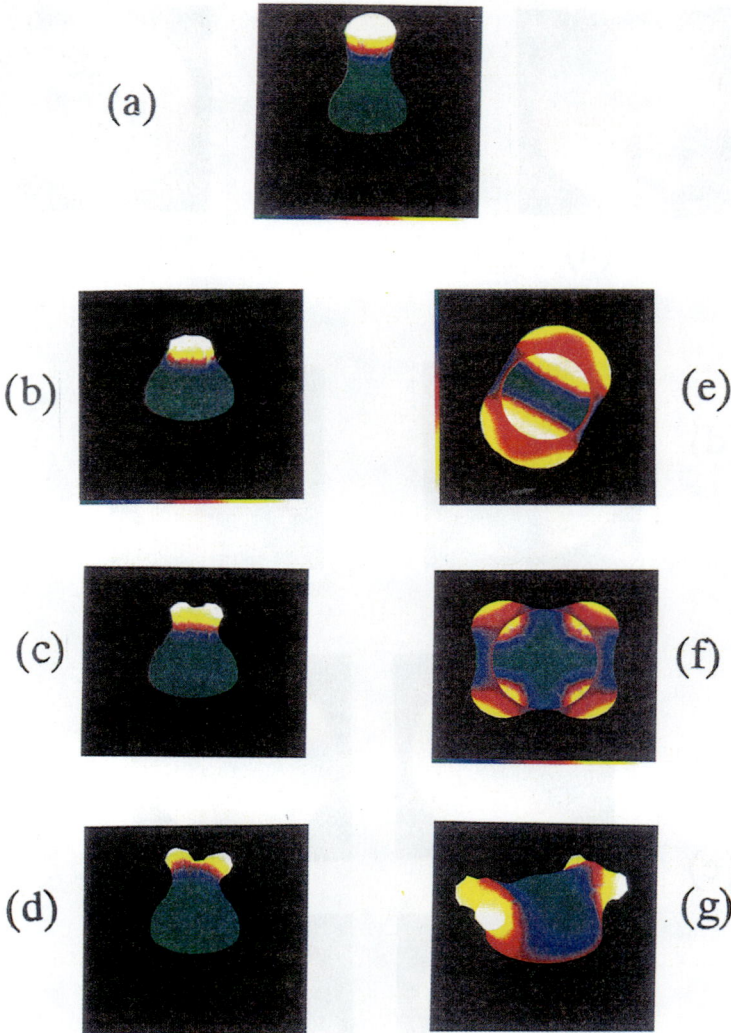

Fig. 8 Morphological results of computations of R–D patterning and simultaneous growth, colour-coded like Fig. 7 for concentration of the growth catalyst X. See text for details. (a) Tip-growing cylindrical outgrowth. (b)–(d) Time sequence of a dichotomous branching event. Perspective: the tip is inclined out of the vertical, 30° towards the viewer. (e)–(g) Three different computations of secondary dichotomous branching, from an initial mix of $Y(3,2)$ and $-Y(3,0)$ on a hemisphere. (e) and (f) are viewed looking upwards through the equator of the original hemisphere.

sometimes goes astray at 3. It seems to be beginning to show the stripe-forming tendency that is its dominant characteristic in planar domains.[18,19]

3.2 Three-dimensional computations on simultaneous reaction–diffusion patterning and morphological development

Fig. 8 shows the results of computations of simultaneous R–D patterning and morphological growth generating outgrowths from an initial hemispherical shell of unit radius. The regions colour-coded green for low X are essentially the regions in which growth has been cut off because $X < X_{th}$. The pictures illustrate tests of the ability of the model to generate: (1) a tip-growing cylindrical outgrowth (Fig. 8(a)); (2) a dichotomous branch, at an acute branching angle, from a tip-growing outgrowth (Fig. 8(b)–(d), a time sequence; (3) the formation of secondary dichotomous branching when the hemisphere was provided initially with a primary dichotomous branching pattern, a mixture of $Y(3,2)$ and $Y(3,0)$ (Fig. 8(e)–(g), results of three different computations). The most obvious feature of the result is that the second branching was in a plane at right-angles to that of the first. In *Micrasterias*, that is an unusual feature of just one branching event in one or two species. We have yet to model the common feature of that genus, that most successive branchings are all in exactly the same plane. We were able to treat growth in a saddle region (*e.g.* green region, Fig. 8(g)) differently from our 2-D work.[2] When the growth rate was reduced to zero, the saddle morphology thereby became stationary, while in the 2-D work the coordinates of the nodes in addition had to be "frozen" to avoid all parts of the growing edge still moving.

In these computations, X_{th} has been specified at $t = 0$ to restrict growth arbitrarily to a part of the hemisphere. It does not represent an intracellular timing event, as in the 2-D work.

Neither action on the whole hemisphere nor that in the growing region can be correlated precisely with the results for a non-growing hemisphere with boundary condition $U = V = 0$ reported in section 3.1. For the earlier period, the initial hemisphere may have different boundary conditions, because it may represent many different aspects of plant development. One of these is the initial state in *Micrasterias* morphogenesis, in which the equator of the hemisphere is the junction of the new half-cell, about to grow, to the old half-cell, morphogenetically "dead". A reasonable boundary condition to assume here is no-flux, and that has been used in Fig. 8(a)–(d). For the later period when, as in Fig. 8(a), a tip has started to advance and elongate a cylinder, the new boundary specification by "feedback loop III" (Fig. 2(c); eqn. (2)) has created something closer to a $U = V = 0$ boundary. The results of section 3.1 are then relevant to what will happen on the tip, but the boundary condition has more dynamic possibilities.

Fig. 8(a)–(d) were computed with the modified Brusselator, including feedback control from X to A. This provides the mechanism for continuous movement of the boundaries of the actively patterning region, "feedback loop III" in Fig. 2(c). This allows the tip to elongate the cylinder below it in Fig. 8(a), and controls the branching to an acute angle in Fig. 8(b)–(d). (Though it looks larger in the perspective shown, the branching angle is actually about 60°.) Fig. 8(e)–(g) were computed with the standard Brusselator only, and an X_{th} switching off growth in the low X regions. These computations are close to being 3-D analogues of some of the 2-D computations of Harrison and Kolář[17] which laid the foundation for both the 2-D work of Holloway and Harrison[2] and the present work.

4 Discussion

4.1 The robust selection of annular or dichotomously-branching pattern at $l = 3$

A pattern-forming event, in which R–D is the symmetry-breaker, -maker or -preserver, goes through three stages:

(1) The initial rudiment of possible patterns in concentration noise, together with any more orderly information from earlier developmental stages, which may specify, *e.g.*, polarity (*i.e.* the orientation to be adopted by the new pattern) or maxima *vs.* minima in growth rates (*e.g.* $Y(3,0)$ develops with a high-concentration or low-concentration annulus).

(2) The period of "linear" growth, which has linear dynamic equations, exponential growth of pattern amplitude and linear plots of log(concentration) *vs.* time, Fig. 5(b).

(3) The period of nonlinear growth, in which pattern amplitude eventually settles down to a steady state, but the pattern may change from what arose in the linear growth period.

In regard to expression of chemical pattern in morphological structures, one must ask how much the growth processes are reading from each of stages 1, 2 and 3. For the plant kingdom, the simultaneous continuity of size increase and shape changes suggests that all these stages must be considered *a priori* good candidates for items on the plant's library shelf of developmental information. Most studies of pattern formation by R–D have not assumed such a continuous linkage between patterning events and their expression. Rather, they have assumed that patterning goes into stage 3 before it is expressed. Usually, one must resort to computation for the greater part of such a study. At the start of the present work, we had anticipated that the selection of annular or branched pattern on a hemispherical shell would have some analogy to the selection of striped or spotted pattern on a plane. We expected that the pattern selection would be all-or-nothing for either $Y(3,0)$ or $Y(3,2)$, which one was selected being an idiosyncratic property of the nonlinearities of different models.

The computational results in section 3.1 show that the selection of one or the other mode was not robust for any of the types of nonlinearity we studied. All models favoured branching but gave annular pattern in something like 20% of the computations. Such developmental uncertainty is not entirely unknown in living systems. For instance, somatic embryos of a hybrid larch commonly progress *via* annular pattern to a whorl of several (average 6) cotyledons; but in about 20% of them, pattern arrests at the annular stage and no cotyledons form (von Aderkas;[29] this variability is in a clonal population of many genetically identical embryos). Our objective is to show how each of annular or dichotomously branching pattern can be generated robustly. For branched pattern, at the $l = 3$ fit to a hemispherical shell, stage 1 usually gives it an initial advantage (Fig. 6), and stage 2 linear dynamics simply maintain that advantage (Fig. 5). Even if the two modes have equal amplitudes, they will add to a very good pattern for catalysis of dichotomous branching (Fig. 7(c)). Thus dichotomous branching will start robustly if morphological growth is fast enough to read the chemical pattern during stage 2 and convert it to a broken symmetry which is quite unlikely to revert later to annular pattern. This much can be given as a solid conclusion without need of computations, since stage 2 involves only linear dynamic equations with well-known complete analytical solutions.

Analytical work by bifurcation analysis was undertaken by Nagata, and is in the course of publication,[28] for both the Brusselator and the hyperchirality models. This analysis was necessarily confined to regions of parameter space very close to the Turing instability boundary, while the computations for which statistics are given in Table 1 explore wider regions. The analysis, however, gave results quite close to those statistics. For the Brusselator, a positive amplitude for $Y(3,0)$ (low X-concentration annulus) was unstable, while a negative amplitude (high-X annulus) had a stable equilibrium point. The basins of attraction for $-Y(3,0)$ and $Y(3,2)$ had relative volumes of 16% and 84%, very close to the Table 1 statistics. The high symmetry of the hyperchirality model gives $Y(3,0)$ mirror-imaged basins of attraction for positive and negative amplitudes, with stable equilibrium points, and a ratio of volumes of 12% for annular pattern to 88% for dichotomous branching, a rather higher probability for the latter than our computations show.

If early expression of the developing pattern is the correct explanation for robust occurrence of dichotomous branching, the remaining problem is: how can annular pattern be formed robustly? The studies reported here have been entirely pattern arising out of nothing more than uniform initial distributions of chemical concentrations over the hemisphere, plus the inevitable noise that supplies small rudiments for all patterns, and does not favour the annular (Fig. 6). This may not represent fairly the morphogenetic sequence. A growing tip of dome shape already has chemical pattern of circular symmetry. This may link to the next patterning stage to give adequate advantage to the annular mode. The earliest computations on a hemisphere in this laboratory, by Zeiss,[14] in fact illustrated this for a hemispherical shell growing uniformly in time. The first pattern formed was $Y(1,0)$, the circularly-symmetric tip growth pattern which has no broken-symmetry competitors satisfying the boundary condition. As the hemisphere grew, this was replaced by a sequence of patterns, alternating annular pattern and resumption of apical tip growth. Earlier, in the first 1-D patterning work in this group, Lacalli and Harrison[30] found sequences leading them to remark: "In normal increase in size by growth . . . once a pattern has become established the possible later changes to it are those which conserve the symmetry of the morphogen distribution." Tip growth and annular pattern may be sequential products of the same mechanism, while branching needs *ab initio* operation of a new mechanism.

4.2 3-D computations of simultaneous R–D patterning and morphological change

This work is still at an early stage. Our computational methods have proved capable of handling, for the development of 3-D morphology, all the individual events that we were previously able to compute only in 2-D. In particular, dichotomous branching with the branching angle controlled by "feedback loop III" can be computed. In the saddle regions between branches, the artefact of movement of a saddle point after growth rate had been specified as zero there has been eliminated. In fact, this minor numerical instability in our 2-D work has been removed by a major conceptual improvement in representing in the computer code what a saddle point is. We have yet to fully test the computer codes with a form of this feedback that takes the slow-growing regions to homogeneous steady state, but our codes are now well enough developed that this work is for the immediate future. A promising line on which we have started work is feedback from morphogen X to reactant B in the Brusselator, because decrease of B more readily than decrease of A leads the dynamics to cross the Turing instability boundary into non-patterning behaviour.

A major challenge is to find mechanisms for the control of relative planes of successive branchings in a sequence. For all our previous success in modelling the development of diverse shapes of many species in the genus *Micrasterias*,[2] the coplanarity of sequential branchings is a fundamental feature of the morphogenesis that we were simply unable to address with 2-D work. Our first approach to this will use starting shapes without circular symmetry in place of the hemisphere. In relation to experimental data on *Micrasterias*, Lacalli[31] believed that the non-circular shape of the "isthmus" (corresponding to the equator of our initial hemisphere) was an important characteristic, non-genetic but yet heritable through generations of vegetative reproduction, and relevant to the specification of the plane of semicell development.

Acknowledgements

We wish to thank Thurston C. Lacalli, Jacques Dumais and Patrick von Aderkas for discussions and comments on this work and drafts of the manuscript; Wayne Nagata for active collaboration on the bifurcation analysis to be published in full elsewhere; Jesper S. Hansen, graduate student, University of Roskilde, Denmark, for work contrasting the Brusselator with and without our modification during a rotation in our group; and the Natural Sciences and Engineering Research Council of Canada and the British Columbia Institute of Technology Staff Applied R and D fund for financial support.

References

1 A. M. Turing, *Philos. Trans. R. Soc. London, Ser. B*, 1952, **237**, 37.
2 D. M. Holloway and L. G. Harrison, *Philos. Trans. R. Soc. London, Ser. B*, 1999, **354**, 417.
3 J. Dumais and L. G. Harrison, *Philos. Trans. R. Soc. London, Ser. B*, 2000, **355**, 281.
4 H. Meinhardt, *Models of Biological Pattern Formation*, Academic Press, London, 1982.
5 D. M. Holloway, L. G. Harrison and J. B. Armstrong, *Develop. Dynam.*, 1994, **200**, 242.
6 A. Gierer and H. Meinhardt, *Kybernetik*, 1972, **12**, 30.
7 J. D. Murray, *Philos. Trans. R. Soc. London, Ser. B*, 1981, **295**, 473.
8 H. Meinhardt, *The Algorithmic Beauty of Sea Shells*, Springer, Berlin, 1995.
9 S. Kondo and R. Asai, *Nature*, 1995, **376**, 765.
10 L. G. Harrison, *Kinetic Theory of Living Pattern*, Cambridge University Press, Cambridge, 1993.
11 I. Prigogine and R. Lefever, *J. Chem. Phys.*, 1968, **48**, 1695.
12 G. Nicolis and I. Prigogine, *Self-Organization in Non-Equilibrium Systems*, Wiley, New York, 1977.
13 L. G. Harrison and T. C. Lacalli, *Proc. R. Soc. London, Ser. B*, 1978, **202**, 361.
14 L. G. Harrison, J. Snell, R. Verdi, D. E. Vogt, G. D. Zeiss and B. R. Green, *Protoplasma*, 1981, **106**, 211.
15 L. G. Harrison and N. A. Hillier, *J. Theor. Biol.*, 1985, **114**, 177.
16 L. G. Harrison, G. Donaldson, W. Lau, M. Lee, B. P. Lin, S. Lohachitranont, I. Setyawati and J. Yue, *Protoplasma*, 1997, **196**, 190.
17 L. G. Harrison and M. Kolář, *J. Theor. Biol.*, 1988, **130**, 493.
18 M. J. Lyons and L. G. Harrison, *Chem. Phys. Lett.*, 1991, **183**, 158.
19 M. J. Lyons and L. G. Harrison, *Develop. Dynam.*, 1992, **195**, 201.
20 B. Ermentrout, *Proc. R. Soc. London, Ser. A*, 1991, **434**, 413.
21 P. Borckmans, G. Dewel, A. de Wit and D. Walgraef, in *Chemical Waves and Patterns*, Kluwer, Dordrecht, ed. R. Kapral and K. Showalter, 1995, p. 323.
22 O. Kiermayer, *Protoplasma*, 1964, **59**, 382.
23 P. B. Green, C. S. Steele and S. C. Rennich, *Ann. Bot.*, 1996, **77**, 515.
24 L. A. Martynov, *J. Theor. Biol.*, 1975, **52**, 471.
25 U. Ascher, S. Ruuth and R. Spiteri, *Appl. Numerical Math.*, 1997, **25**, 151.
26 J. A. Kaandorp, *Fractal Modelling*, Springer, Berlin, 1994, p. 151.

27 D. M. Holloway and L. G. Harrison, *Physica A*, 1995, **222**, 210.
28 W. Nagata, L. G. Harrison and S. Wehner, submitted.
29 P. von Aderkas, *Int. J. Plant Sci.*, 2001, acccepted.
30 T. C. Lacalli and L. G. Harrison, *J. Theor. Biol.*, 1978, **70**, 273.
31 T. C. Lacalli, *Protoplasma*, 1976, **88**, 133.
32 T. C. Lacalli, PhD Thesis, University of British Columbia, 1973.
33 A. Gibor, *Sci. Am.*, 1966, **215**, 118.
34 L. G. Harrison, *Int. J. Plant Sci.*, 1992, **153**, S76.

Chemical waves in open flows of active media: Their relevance to axial segmentation in biology

Mads Kærn,[*a] **Michael Menzinger,**[a] **Razvan Satnoianu**[b] **and Axel Hunding**[c]

[a] *Department of Chemistry, University of Toronto, 80 St. George Street, Toronto, Ontario M5S 3H6, Canada*
[b] *Centre of Mathematical Biology, Mathematical Institute, Oxford University, 24–29 St. Giles, Oxford, UK OX1 3LB*
[c] *Department of Chemistry, H. C. Ørsted Institute, University of Copenhagen, Universitetsparken 5, Copenhagen DK-2100 Ø, Denmark*

Received 10th April 2001
First published as an Advance Article on the web 12th November 2001

The boundary forcing of open flows of active media can lead to a variety of spatiotemporal structures, depending on the local kinetics of the medium and on the characteristics of the forcing. Here, we demonstrate that regardless of the local kinetics, the combination of flow and boundary forcing is a powerful method for replacing intrinsic modes with extrinsic ones. This entrainment of dynamics has important implications for biological morphogenesis. During early embryonic development it is frequently observed that stripes of gene expression and segments arise one after the other along a growth-axis. We show that axial growth can be viewed as an open flow of cells away from a growth zone. Based on this realisation, we demonstrate using three generic reaction–diffusion–advection schemes how a space-periodic structure is induced, one "segment" at a time along the growth/flow axis, by a segmental clock that is synchronised within the growth zone. The schemes are investigated in the context of an abrupt and a gradual change in the properties of the segmental clock. Experimental observations provide evidence that the latter is involved in the early development of many vertebrates.

1 Introduction

In past decades, pattern formation in far-from-equilibrium systems has been studied primarily in the context of reaction–diffusion systems with a rapidly diffusing inhibitor. The realisation[1] that the coupling of local kinetics to long-range inhibition may cause spontaneous symmetry-breaking has had a great impact on modern understanding of pattern formation.[2–11] On the other hand, symmetry breaking in open-flow reaction–diffusion–advection (RDA) systems has been considered only recently.[12–22]

The key element of an open flow[12] is that it begins at an upstream boundary where fresh medium is injected. This influx represents a boundary forcing that may entrain the temporally evolving medium into a variety of wave-patterns. The resulting spatio-temporal dynamics depends on the local kinetics of the medium (monostable, bistable, oscillatory) and on the character of the boundary forcing (constant, periodic, stochastic).

We argued recently[21] that axially growing embryos are actually equivalent to open flow systems: the movement of an organising centre or growth zone relative to the stationary cells is

DOI: 10.1039/b103244p

equivalent to a relative flow of cells away from this upstream boundary. In this paper, we extend our earlier analysis[22] to open-flow RDA systems whose local kinetics is either monostable, bistable or oscillatory. The idea is that cell differentiation during early development is regulated, at least in part, by a space-period morphogen pattern imposed by the combination of a growth-induced flow and a periodic process, a segmental clock, at the growth/flow boundary. Our investigation is also relevant to pattern formation in periodically boundary-forced plug-flow reactors and in light-sensitive media whose properties change at a moving boundary of illumination.

The open flows investigated here are "plug-flows" modelled by the RDA equations of the type:

$$\frac{\partial \boldsymbol{u}}{\partial t} = \boldsymbol{f}(\boldsymbol{u}) - \phi \frac{\partial \boldsymbol{u}}{\partial x} + \boldsymbol{D} \frac{\partial^2 \boldsymbol{u}}{\partial x^2}, \quad \boldsymbol{u}(x=0, t) = \boldsymbol{b}(t), \tag{1}$$

where $\boldsymbol{u}$ is a vector of interacting species, $\boldsymbol{f}(\boldsymbol{u})$ describes their local interactions and $\boldsymbol{D}$ is a diagonal matrix containing the diffusion coefficients. The flow coefficient, ϕ, is equal for all the species. No-flux boundary conditions are imposed at the outflow, $x = L$, while $\boldsymbol{b}(t)$ describes the characteristics of the temporal evolution at the inflow boundary, $x = 0$. The flow carries elements away from the boundary at velocity ϕ and an element is located at $x = a\phi$ after it has spent the time-period (or age) a in the flow.

In a chemical system, eqn. (1) describes an ideal plug-flow reactor illustrated in Fig. 1(a). In this context, Kuznetsov et al.[14] and later Andrésen et al.[16] studied eqn. (1) with oscillatory local kinetics, equal diffusion coefficients and constant boundary forcing. Based on linear analysis they predicted the formation of waves that are stationary relative to the inflow boundary. This finding was startling at that time since long-range inhibition is not involved. We have since confirmed their existence experimentally[17] and generalised the linear analysis.[19] We have also investigated periodic boundary forcing[18] which is of key relevance to biological morphogenesis.[21,22]

Space-periodic structure in the form of repetitive segments or bands of gene expression arises in many insects, worms and vertebrates one after the other along a growth-axis. A relevant case is the one where axial growth occurs by the divisions of stem cells at the tip of the growing embryo.

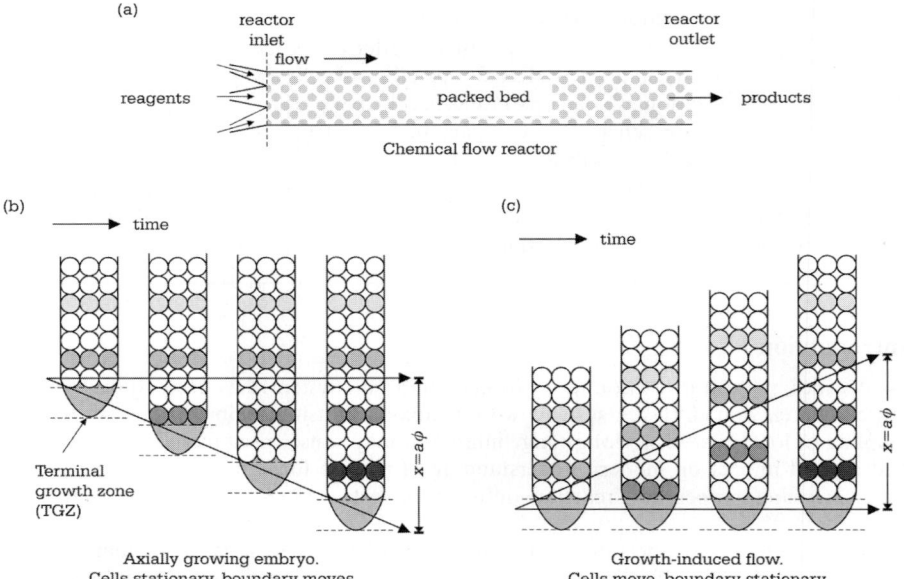

Fig. 1 (a) Open flow packed-bed chemical reactor. Reagents are injected at one end and products removed at the other. The constant influx imposes a boundary forcing that may be constant, periodic or stochastic in time. (b) The typical reference frame of axially growing embryos. Daughter cells are stationary and the terminal growth zone is pushed downwards. (c) The growth zone is taken to be stationary. The continued shedding of progeny cells now amounts to a steady flow of cells away from the growth zone.

This is illustrated in Fig. 1(b). While the terminal growth zone (TGZ) of proliferating stem cells advances downward (slanted arrow), the daughter cells remain stationary with respect to their immediate environment and the embryo as a whole (horizontal arrow). Thus, the distance between the TGZ and the daughter cells increases as the embryo continues to grow.

Fig. 1(c) shows the same process in the reference frame where the TGZ is stationary (horizontal arrow). The continuous shedding of daughter cells from the TGZ now amounts to a steady flow of cells away from it (slanted arrow). As in the plug-flow described by eqn. (1), the distance between the TGZ and a cell with age a is given $x = a\phi$ when the growth rate, ϕ, is constant. The net result of a growth-induced flow is thus the same as in a flow system of the plug-flow type and axially growing biological systems can be approximated by eqn. (1) when cell–cell communication is modelled as a diffusion-like process.

As indicated by Fig. 1(b) and (c), there are two ways to model axial (*i.e.* anisotropic) growth. One is to keep the cells fixed while allowing the growth-boundary to move, as in Fig. 1(b). The flow is absent in this reference frame and the length of the system increases at a rate given by the growth velocity. The other is to keep the growth-boundary stationary and have cells move, as in Fig. 1(c). We have chosen to adapt this latter reference frame since the theory of open flow systems is well developed. Since we are modelling a growth-process, the boundary at $x = L$ in eqn. (1) should strictly speaking move at a velocity given by ϕ. However, the outflow boundary does not have significant influence on the dynamics of the system provided that its length is sufficiently large at all times. It suffices for the present purposes to have a constant value of L.

To induce cell differentiation, a space-periodic morphogen pattern must be stationary relative to individual cells, at least in the absence of a mechanism that can interpret a temporally varying morphogen level. The structures investigated throughout this paper are those that are stationary relative to cells and the developing embryo as a whole. Relative to the growth/flow boundary they have a velocity that is equal to the growth/flow velocity, *i.e.* $c = \phi$.

In a system described by eqn. (1), a structure that propagates with velocity c_0 in the absence of a flow, will in the presence of a flow typically propagate with a velocity given by:

$$c^{\pm} = \phi \pm c_0. \tag{2}$$

The reaction–diffusion driven propagation is simply decreased when the structure moves against the flow ($c^- < c_0$) and enhanced when it moves with the flow ($c^+ > c_0$).

To obtain structures that move with velocity $c = \phi$ relative to the upstream boundary it is required that $c_0 = 0$. Thus, we are looking for reaction–diffusion schemes that support stationary structures in the absence of a flow. They fall into three generic classes: (1) (nearly) stationary phase waves in media with oscillatory local kinetics, (2) media with steady or oscillatory local kinetics that support Turing structures and (3) media with bistable or excitable local kinetics that support stationary interfaces. When formed in the presence of a flow, we refer to them as flow-distributed structures (FDS).

The paper has two main sections. The idea pursued in Section 2 is that cells within the TGZ in Fig. 1(c) oscillate in synchrony and thereby impose a periodic forcing, $b(t) = b(t + T')$, on the shed daughter cells. We show how the periodic boundary forcing in the three generic scenarios may entrain the spatio-temporal dynamics and give rise to FDS with velocity $c = \phi$. In Section 3, we review the developmental process in vertebrates known as somitogenesis. Experiments have demonstrated that gene expression has wave-like character during somitogenesis in fish, birds and mammals and involves gene expression waves that propagate through the unsegmented tissue with decreasing width and velocity. Eventually, the velocity and wavelength coincides with what is predicted by each of the schemes investigated in Section 2. The spatio-temporal wave behaviour indicates that the local kinetics within cells change smoothly rather than abruptly, as assumed in Section 2. By allowing for a spatial variation of the kinetic parameters, we demonstrate that each of the three schemes account well for the observed wave behaviour.

2 Flow-distributed structures: general aspects

2.1 Phase-wave FDS

We first consider the spatio-temporal phase dynamics of a medium whose local kinetics supports an oscillation with period T_0. To derive the phase function, $\theta(x, t)$, *i.e.* how the oscillation phase depends on space and time, we introduce a new space co-ordinate $x/(\phi T_0)$. This yields a new velocity $\phi/(\phi T_0) = 1/T_0$. Eqn. (1) is then re-scaled by introducing the co-ordinate $z = x/(\phi T_0) - t/T_0$ that moves with the flow. In this reference frame, eqn. (1) transforms into:

$$\frac{\partial u}{\partial t} = f(u) + \varepsilon \frac{\partial^2 u}{\partial z^2} \equiv h(u, \varepsilon), \quad \text{with } \varepsilon = \frac{D}{\phi^2 T_0^2}, \tag{3}$$

The diffusion coefficients are set equal to D for all the species. The kinetics in the co-moving reference frame thus approaches that of the local kinetics as ε approaches zero, $h(u, \varepsilon) \to f(u)$ for $\varepsilon \to 0$. When $\varepsilon = 0$, diffusive interactions are non-existent and the cells oscillate independently of each other. This means that the time periodic solution of the local kinetics, via the above change of variables, translates to a space-periodic solution in eqn. (1). When diffusion cannot be ignored, but ε remains low,[23] it is reasonable to expect that cells oscillate with a period T such that $T(\varepsilon) \to T_0$ for $\varepsilon \to 0$.

Suppose that the boundary forcing, $b(t)$, has a period T' and that daughter cells (or volume elements) have oscillation phase $\theta_0(\tau) = \tau/T'$ when they leave the growth zone (or reactor inlet) at $x = 0$. As a cell moves downstream, its oscillation phase advances linearly. After a time-period a, it is located at $x = a\phi$ and the phase has advanced by a/T. Hence, the phase value at distance x and time $t = a + \tau$ is given by the phase function:

$$\theta(x, t) = \theta_0(t - a) + \frac{a}{T} = \frac{t - a}{T'} + \frac{a}{T} = \frac{x}{\phi}\left(\frac{1}{T} - \frac{1}{T'}\right) + \frac{t}{T'}. \tag{4}$$

Differentiation of eqn. (4) with respect to t reveals that each fixed spatial point oscillates with the same period as the boundary T'. *The boundary forcing thus imposes itself onto the whole system.* The wavelength, λ, and the velocity, c, of the resulting phase waves are readily obtained from eqn. (4)[21] as:

$$c = \frac{-\phi}{R - 1}, \quad \lambda = \frac{\phi T'}{|R - 1|} \tag{5}$$

where $R = T'/T$. Phase dynamics, eqn. (5), thus predicts stationary waves, $c = 0$, for $R = \infty$ (constant boundary forcing), upstream travelling waves, $c < 0$, for $R > 1$, homogeneous oscillations, $c = \infty$, for $R = 1$ and downstream travelling waves, $c > 0$, for $R < 1$. The waves that are relevant here, i.e. those with $c = \phi$, are obtained when $R = 0$. This corresponds to the arrest, $T \to \infty$, of the intrinsic oscillation, for instance by the arrest of the local oscillators, $T_0 \to \infty$. The resulting phase-wave FDS has a wavelength that is given by $\lambda = \phi T'$.

Note that the diffusion term in eqn. (1) cannot be ignored when the local kinetics comes to a halt: $f(u) \to 0$ for $T_0 \to \infty$. As a result, the FDS is a transient when formed for $T_0 = \infty$. Axial dispersion (diffusion) eventually homogenises the system and a non-uniform structure cannot be maintained. The phase-wave scenario is however still relevant for morphogenesis since the transient structure may be sufficiently long lived to induce cell differentiation. This will be discussed in Section 3.2. In addition, there are mechanisms other than setting $T_0 = \infty$ that gives rise to an arrest of the intrinsic oscillation, i.e. $T = \infty$, as shown below.

2.2 Linear analysis

To investigate alternatives to the above phase-wave scheme, we now briefly discuss some useful insight provided by linear stability analysis. It predicts when a space-periodic sinusoidal perturbation, a mode, applied to a homogeneous state is amplified. To show that a linearly amplified mode gives rise to a persistent space-periodic structure requires a non-linear analysis and may involve secondary bifurcations from an already established pattern.[24] Such an analysis is beyond the scope of the present paper and will be pursued elsewhere. Despite these shortcomings, a linear analysis indicates when the relevant FDS may arise from an initially homogeneous state. Their amplification and stability are investigated by numerical simulations.[25]

We focus on systems with two interacting species, an activator u and an inhibitor v. The RDA equations are:

$$\frac{\partial u}{\partial t} = \gamma f(u, v) - \phi \frac{\partial u}{\partial x} + D_u \frac{\partial^2 u}{\partial x^2},$$

$$\frac{\partial v}{\partial t} = \gamma g(u, v) - \phi \frac{\partial v}{\partial x} + D_v \frac{\partial^2 v}{\partial x^2}, \quad (6)$$

where γ determines the time-scale of the kinetic terms $f(u,v)$ and $g(u,v)$. The linear stability of a steady state (u_0, v_0) is determined by the temporal evolution of a small displacement $U = u - u_0$ and $V = v - v_0$ away from it. We look for solutions of the form:

$$U(x, t) = U_0 \exp^{\omega t + ikx}, \quad V(x, t) = V_0 \exp^{\omega t + ikx}, \quad (7)$$

where w determines the temporal evolution of the displacement, $\text{Im}(\omega)$ determines the local frequency and k is the wavenumber. U_0 and V_0 are arbitrary constants.

A simplified description is obtained by expanding eqn. (6) in a Taylor series using eqn. (7) and keeping only the linear terms. Valid solutions to the resulting linear equations must satisfy a dispersion relation. It is given by:

$$2\omega = \text{Tr} - k^2(D_v + D_u) - 2ik\phi \pm \sqrt{\Delta}$$

$$\Delta = \text{Tr}^2 - 4 \det + 2k^2(D_v - D_u)(a_{11} - a_{22}) + k^4(D_v - D_u)^2, \quad (8)$$

where $\text{Tr} = a_{11} + a_{22}$, $\det = a_{11}a_{22} - a_{12}a_{21}$ and a_{ij} are the elements of the Jacobian matrix at the steady state of the local kinetics.

The real part of ω, $\text{Re}(\omega)$, is independent of ϕ since the wave number k is real and ϕ only enters into eqn. (8) through the term $ik\phi$. Furthermore, the imaginary part of ω is given by $\text{Im}(\omega) = k\phi$ when the square root term in eqn. (8) is real (i.e. when $\Delta > 0$). Since the phase velocity of a mode is given by $c = \text{Im}(\omega)/k$, the modes amplified have the desired velocity $c = \phi$ when $\Delta > 0$.

To provide a specific example, we use a model of the FitzHugh–Nagumo (FHN) type.[26,27] It involves the kinetic terms:

$$f(u, v) = \varepsilon(u - u^3 - v), \quad g(u, v) = -v + \alpha u + \beta. \quad (9)$$

The parameters ε, α and β are associated with the time-scale separation between u and v, the coupling-strength of the inhibitor v, and the asymmetry of the phase plane structure, respectively. As shown in Fig. 2(a) the phase plane structure is symmetric for $\beta = 0$ and the local kinetics has a steady state at the origin, $(u_0, v_0) = (0,0)$. It is a focus for $\alpha > 1$ and changes stability through a

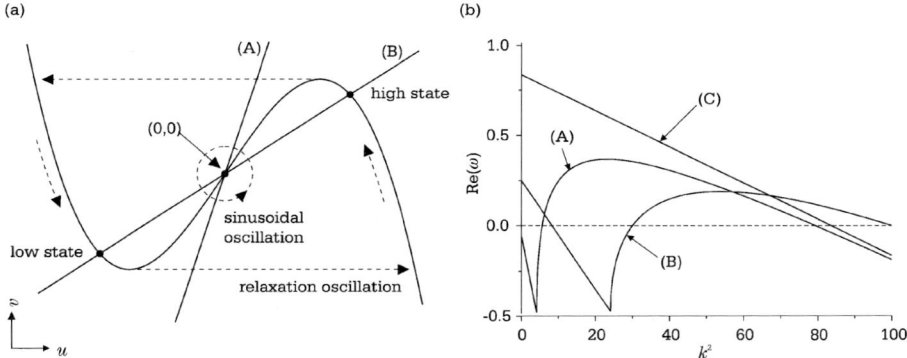

Fig. 2 (a) Phase plane portrait of the FHN kinetics for $\beta = 0$. (A): The steady state at (0,0) is a focus for $\alpha > 1$ and gives rise to relaxation oscillations for $\varepsilon \gg 1$ and sinusoidal oscillations for $\varepsilon > 1$. The steady state is stable for $\varepsilon < 1$. (B): The steady state at (0,0) is a saddle point for $\alpha < 1$. Bistability is ensured when $\alpha < 2/3$. (b) The largest real part of the eigenvalues as a function of the wave number. (A): Local steady state is a stable focus ($\varepsilon = 0.95$, $\alpha = 2$, $D_u = 0.01$, $D_v = 0.1$). (B): Local steady state is an unstable focus ($\varepsilon = 1.5$, $\alpha = 2$, $D_u = 0.01$, $D_v = 0.1$). (C): Local steady state is a saddle point ($\varepsilon = 1$, $\alpha = 0.3$, $D_u = D_v = 0.01$). Other parameter values are: $\gamma = 1$, $\beta = 0$.

super-critical Hopf bifurcation when ε passes through a value of unity. It is stable for $\varepsilon < 1$ and unstable for $\varepsilon > 1$. The steady state is a saddle point for $\alpha < 1$ and bistability is ensured when $\alpha < 2/3$ (for $\beta = 0$). Excitability requires that ε and β be sufficiently large.

The dependence of Re(ω) on k^2 is illustrated in Fig. 2(b) for the cases where the local kinetics at $(u_0,v_0) = (0,0)$ is (A) a stable focus, (B) an unstable focus and (C) a saddle point. These three different cases are discussed below. The modes where Re($\omega[k^2]$) is positive are linearly unstable, meaning that a low amplitude spatial perturbation with wavelength $\lambda = 2\pi/k$ is amplified.

2.3 Turing-type FDS

The steady state solution of the *local* kinetics is a stable focus when Tr < 0 and Tr^2 − 4det < 0. The *homogeneous* steady state undergoes a Turing bifurcation when the inhibitor diffusion coefficient is increased relative to that of the activator and Δ in eqn. (8) becomes positive. The onset of the Turing instability is unaffected by the flow (since ϕ appears only in the imaginary part of ω) and the linearly unstable modes are the same as in the absence of a flow. The critical wavenumber, k_c, is given by:[28]

$$k_c^2 = \frac{a_{11}D_v + a_{22}D_u}{2D_u D_v} \quad (10)$$

and the limits of the range of unstable wave numbers, $k \in [k_-; k_+]$, are:

$$k_\pm^2 = k_c^2 \pm \frac{\sqrt{(a_{11}D_v + a_{22}D_u)^2 - 4D_u D_v \det}}{2D_u D_v}. \quad (11)$$

A perturbation with a wavenumber within this range may give rise to FDS with velocity $c = \phi$ (since $c = \text{Im}(\omega)/k$).

2.3.1 Absolute and convective instability.
Before turning to the case of periodic boundary forcing, we briefly discuss some key general features of open flow systems using an open flow of Turing-unstable medium as example.

An open flow of a linearly unstable reaction–diffusion medium is either *absolutely* or *convectively* (spatially) unstable. If the flow velocity is low, a locally applied perturbation to the homogeneous state causes an inhomogeneity to develop and the edges of this structure spread both upstream and downstream. For equal flow coefficients, they have velocities given by eqn. (2) if they spread at velocity c_0 in the absence of a flow. The system is said to be (linearly) *absolutely unstable* if $c^- < 0$ and the slowest moving edge spreads upstream. In this case, the wavenumber and the velocity are selected intrinsically and they can be obtained from the dispersion relation (see *e.g.* ref. 29). An example of a Turing-type FDS formed under absolutely unstable conditions is shown in Fig. 3(a).

On the other hand, if the flow velocity is high and $c^- > 0$, both edges of the spreading structure are pushed downstream and eventually out of the system. As illustrated for the Turing-type FDS in Fig. 3(b), the initial perturbation is amplified only in the downstream direction. The system

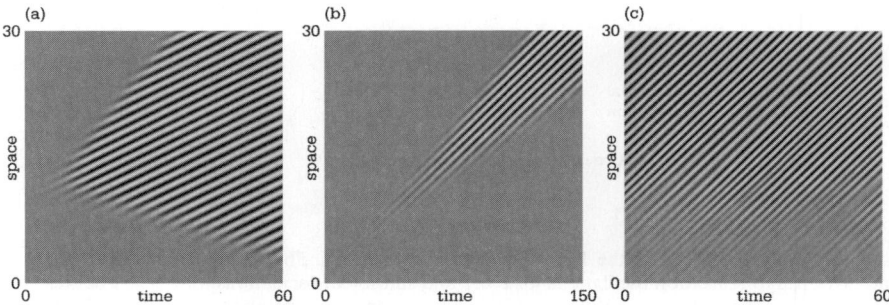

Fig. 3 FDS of the Turing-type without periodic boundary forcing. (a) Absolutely unstable flow conditions for $\phi = 0.1$. (b) Convectively unstable flow conditions for $\phi = 0.5$. (c) Noise-sustained structures (see text). Parameter values used are: $\gamma = 1$, $\varepsilon = 0.95$, $\alpha = 2$, $\beta = 0$, $D_u = 0.01$, $D_v = 0.1$. In (c), the time-step is 0.01 and the average distance between grid points is 0.05.

returns to the homogeneous state when the intrinsically selected spreading structure eventually is washed out. The system is now said to be *convectively unstable*.

A key feature of convectively unstable open flows is that they support *noise-sustained structures*.[12] They arise by the amplification of the small amplitude fluctuations that are present at all times in a real system. The fluctuations at the boundary are of key importance since they have the most time to grow. They can typically be decomposed into a spectrum of temporal frequencies, ω', which in the presence of a flow gives rise to a spectrum of spatial modes, $k' = \phi\omega'$. It is typically the one with the highest growth rate, $\max(\text{Re}[\omega(k')])$, that ultimately is selected. Fig. 3(c) shows a simulation of a noise-sustained Turing-type FDS. In the simulation, the boundary-value of activator is at each time-step set to a random value chosen from a normal distribution with standard deviation of 0.1.

The Turing-type FDS that arises at a low flow velocity (absolutely unstable conditions) or as a noise-sustained structure at high flow velocity (convectively unstable conditions) has the correct velocity, $c = \phi$. A traditional Turing mechanism with an intrinsically selected wavelength may thus be operational in axially growing systems. As we show next, the wavelength of the FDS may also be selected extrinsically. This is important in relation to fundamental biological problems such as size adaptation and regulation of developmental dynamics.

2.3.2 Forced convective FDS.
The amplification of boundary-modes with wavelength $\lambda = \phi T'$ allows for the entrainment of the spatio-temporal dynamics by applying a periodic boundary forcing with period T' (see *e.g.* ref. 20). It is required that the forcing amplitude is sufficiently higher than that of the background noise and that the imposed wavelength is within the appropriate range.

The space–time plots presented in Fig. 4 illustrate the effect of periodic boundary forcing on the spatio-temporal dynamics of a Turing unstable medium under convectively unstable flow conditions. The forcing mimics the Turing structure formed in the absence of a flow (not shown). The three simulations use the forcing periods: (a) $T' = 2.1$, (b) $T' = 0.6$ and (c) $T' = 3$, all at $\phi = 1$. The wavelengths of the unstable modes lie between 0.70 and 2.6 for the chosen parameter values. The imposed periodicity is outside the range amplified by the medium in Fig. 4(b) ($\lambda = 0.6$) and in Fig. 4(c) ($\lambda = 3$).

In Fig. 4(a) the wavelength lies within the range supported by the medium and a new "segment" is added to the FDS each time an oscillation is completed at the boundary. The wavelength of the persistent Turing-type FDS is equal to 2.1, as expected. Other simulations have indicated that persistent structures with $c = \phi$ exist for wavelengths $\lambda = \phi T'$ within the spectrum of unstable modes.

Fig. 4(b) and (c) illustrate what may happen when the imposed wavelength lies outside the range supported by the medium. In Fig. 4(b) the imposed periodicity is too short and it decays to the homogeneous state. In Fig. 4(c) the imposed periodicity is too long. Now the amplitude of each

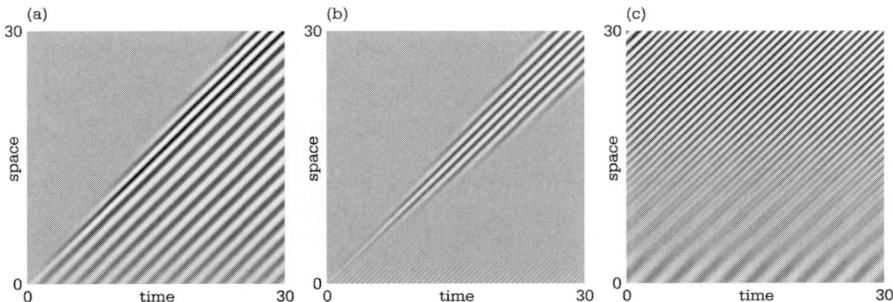

Fig. 4 Forced Turing-type FDS. The boundary forcing is $u(x = 0,t) = 0.2 \times \sin(2\pi t/T')$, $v(x = 0,t) = 0.1 \sin(2\pi t/T')$ and $\phi = 1$. (a) The imposed periodicity, $\lambda = \phi T'$, is within the range of wavelengths supported by the medium ($T' = 2.1$). (b) The imposed structure is too short ($T' = 0.6$) and decays. (c) The imposed structure is too long ($T' = 3$) and the decaying structure splits to give an FDS with a decreased wavelength. Parameter values are: $\varepsilon = 0.9$, $\alpha = 2$, $\beta = 0$, $\gamma = 1$, $D_u = 0.01$, $D_v = 0.2$.

imposed segment initially decays before the segment splits into smaller ones. The wavelength is decreased and lies after the splitting within the range of wavelengths that are amplified.

Turing-type FDS with velocity $c = \phi$ are also supported when the steady state of the local kinetics is an unstable focus. As shown in Fig. 2(b), the homogeneous steady state is in this case unstable to long-wavelength perturbations. The eigenvalues in eqn. (8) are complex conjugate at low values of k (since $\Delta < 0$) and the structures that arise in this range do not have the relevant velocity (since $\text{Im}(\omega) \neq \phi k$). However, as the wavenumber increases, the homogeneous state first becomes stable as $\text{Tr} - k^2(D_v + D_u)$ becomes negative. As indicated in Fig. 2(b), it then becomes unstable (since $\Delta(k^2)$ becomes positive). Unstable modes with $\Delta(k^2) > 0$ have the relevant velocity (since $\text{Im}(\omega) = \phi k$). Simulations have confirmed that these modes give persistent Turing-type FDS.

In summary, we have demonstrated that Turing-unstable media, be it with steady or oscillatory local kinetics, may give FDS with velocity $c = \phi$ either under absolutely unstable conditions or under convectively unstable conditions. In the latter case, periodic boundary forcing may give rise to FDS that have the same properties as those predicted by phase dynamics for $R = 0$, namely velocity $c = \phi$ and wavelength $\lambda = \phi T'$.

2.4 Interface-type FDS

The local kinetics of a bistable system typically supports two stable steady states that are separated in phase space by a saddle point. As shown in Fig. 2(b), a homogeneous system located at the saddle point has unstable modes with wavenumbers between $k = 0$ and some upper critical value k_+. Hence, *any* imposed structure that has a wavelength longer than a critical value, λ_c, is initially amplified. From eqn. (8), the value of λ_c is obtained as:

$$\lambda_c^2 = 8\pi^2 D(\text{Tr} + \sqrt{\text{Tr}^2 - 4 \det})^{-1} \qquad (12)$$

when $D_u = D_v = D$. The unstable modes all have the relevant velocity $c = \text{Im}(\omega)/k = \phi$, since $\Delta(k^2) > 0$ for $k < k_+$. Which mode is ultimately selected depends, of course, on the nonlinear terms in the local kinetics.

Hagberg and Meron[30] did a detailed study of the bistable FHN kinetics in the absence of a flow. For $\beta = 0$ and equal diffusion coefficients, there exists a stable stationary interface in a certain range of ε. This is exactly the case relevant here: a space–periodic interface-structure that is stationary in the absence of a flow is carried downstream at velocity $c = \phi$ in the presence of a flow. When the diffusion coefficients are equal, a space-periodic interface-type FDS is supported when $\varepsilon > 1$ and its wavelength is sufficiently long.[31]

In the presence of a flow, a boundary forcing that periodically switches between the two basins of attraction may give rise to a persistent FDS with wavelength $\lambda = \phi T'$ and $c = \phi$. For this to occur, it is required that a critical nucleus[32] of each state is established at the boundary. An example of an appropriate forcing is one that is symmetric around the unstable steady state. In addition, the system must be nonlinearly convectively unstable[33] in order to avoid a competition between the boundary-mode and the intrinsically selected $k = 0$ mode (*i.e.* one of the stable steady states). An example of a persistent FDS formed with equal diffusion coefficients is shown in Fig. 5(a).

In the absence of a flow, the stationary interface-solution is destroyed for $\beta \neq 0$.[30] The resulting spatio-temporal dynamics is illustrated in the space–time plot shown in Fig. 5(b). The parameters are the same as in Fig. 5(a) with the exception of an increased value of β (from zero to 0.1). It is observed that the spatial periodicity imposed by the boundary is initially amplified. However, an interface with $c_0 \neq 0$ ($c = \phi$) is not supported and the boundary-imposed FDS is only transient.

In the absence of a flow, increasing the diffusion coefficients makes stationary interfaces possible in the case where $\beta \neq 0$. Localised stationary structures with $c_0 = 0$ maintained by rapid inhibitor diffusion have been studied intensively in both bistable and excitable reaction-diffusion media (see for instance ref. 27 and 31 and references therein). In the presence of a flow, a long-range inhibitor is required for persistent FDS with velocity $c = \phi$ when the two steady states have different relative stability. An example of persistent interface FDS formed for $\beta = 0.05$ is shown in Fig. 5(c).

Note, that excitable media support the same structures (*i.e.* $c_0 = 0$) as do bistable media when the inhibitor diffusion coefficient is sufficiently large.[31] Simulations have confirmed that appropri-

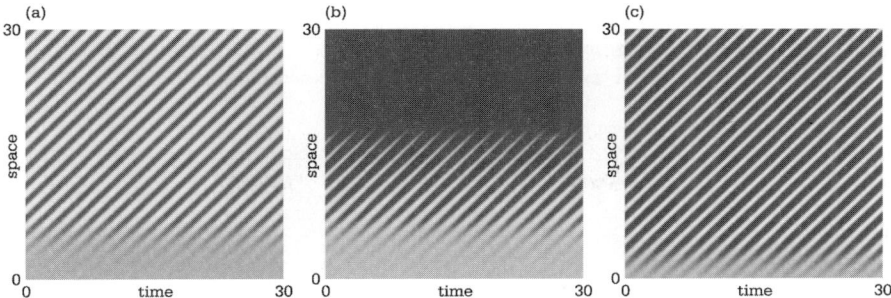

Fig. 5 FDS of the interface-type. (a) The two stable steady states are equally stable ($\beta = 0$). A persistent FDS with wavelength $\lambda = \phi T'$ and $c = \phi$ can be imposed by periodic boundary forcing ($u(0, t) = 0.01 \sin(\pi t)$ and $v(0, t) = 0$) with equal diffusion coefficients. (b) For $\beta = 0.1$, the more stable (dark) phase absorbs the less stable one ($D_u = 0.01$, $D_v = 0.05$). (c) The less stable phase (light) is stabilised when inhibitor diffusion is increased to $D_v = 0.2$. Forcing amplitude in u is 0.1 and $\beta = 0.05$. Other parameter values are: $\varepsilon = 10$, $\alpha = 0.3$, $\phi = 0.5$.

ate periodic boundary forcing and sufficiently rapid inhibitor diffusion can give rise to FDS with velocity $c = \phi$ in both excitable media and in oscillatory media of the relaxation-type.

In summary, FDS with velocities $c = \phi$ and wavelengths $\lambda = \phi T'$ can arise robustly in periodically boundary-forced open flows. When the inhibitor diffusion is sufficiently rapid, they arise with mono-stable local kinetics (Turing unstable or excitable media), with bistable local kinetics or when the local kinetics supports oscillations. In addition, a long-range inhibitor is not strictly required: the transient phase waves that arise when $T_0 \to \infty$ may be long-lived and persistent FDS arise when the local kinetics supports two steady states with the same relative stability.

3 Flow-distributed structures in biology

The most compelling experimental evidence for the operation of an FDS or FDS-like scheme is found during the vertebrate developmental process of somitogenesis. As shown below, the waves of gene expression observed during somitogenesis are not exactly of the type demonstrated in the previous section. The difference is that they have widths and velocities that change gradually as they propagate farther away from the growth zone. In this section, we show that the observed wave behaviour can be accounted for by either of the three generic schemes in Section 2, by allowing for a spatial variation of the kinetic parameters in eqn. (9).

3.1 Somitogenesis

Somitogenesis is the process that establishes the blueprint of the vertebral column by the periodic formation of cell-clusters (the somites) along the body axis of the developing embryo. Fig. 6(a) shows a schematic drawing of a developing chick embryo. The structure at the top is the developing head and the spheres are the somites. The structure flanked by the somites is the neural tube (the precursor to the central nervous system). Structures that are repeated along the body axis, such as the vertebrae, ribs, skeletal muscles *etc*, derive from the differentiated cells of the somites.

The somites on each side of the neural tube derive from rods of approximately cylindrical tissue called the presomitic mesoderm (PSM). The somites form by reorganisation of cells at the anterior (head-most) PSM boundary. As cells are removed from the anterior end of the PSM, new cells are continuously added to its posterior (tail-most) end and the length of the PSM remains relatively constant (10–12 somite-lengths in the chick). This axial growth (downward in Fig. 6) is required for continuous somite formation in many vertebrates[34] and seems to be facilitated, at least in part, by a stem-cell population residing in the tissue prior to the posterior PSM boundary.[35,36] The division of cells within the PSM seems mainly to increase its depth rather than its length or width.[37] For the present purpose, each PSM half is viewed as a rigid one-dimensional structure where non-differentiated cells enter at the posterior end and leave it at the anterior end.

The formation of somites at the anterior PSM boundary is preceded by intricate gene expression patterns believed to serve an important regulatory purpose (see ref. 22 for a discussion and

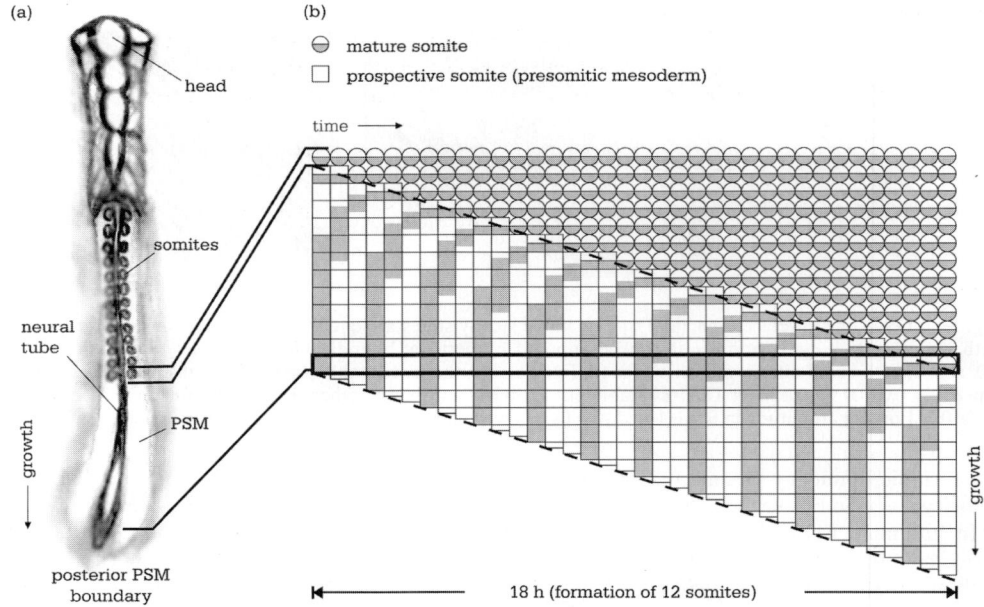

Fig. 6 Somitogenesis in the chick. (a) Schematic drawing of the chick embryo. The spheres are the somites. (b) Spatio-temporal dynamics of gene expression waves during the formation of twelve somite pairs (adapted from ref. 36). Each vertical column represents a snap-shot taken at intervals of 30 min. Gray represents regions where *c-hairy1* mRNA is expressed. Somites are represented by circles and prospective somites within the PSM by squares. The anterior and posterior PSM boundaries are marked by slanted dashed lines. The solid horizontal box marks the position of a group of cells within the PSM. While it remains stationary with respect to the embryo, the movement of the PSM boundaries causes it to be transported across the entire length of the PSM during the 18 h period.

references). The spatio-temporal dynamics of *c-hairy1* mRNA expression in the chick embryo[38] is illustrated schematically in the space–time plot shown in Fig. 6(b). Regions where *c-hairy1* mRNA is expressed are marked in grey.

Fig. 6(b) illustrates the downward movement of the posterior and anterior PSM boundaries (marked by slanted dashed lines). Since cells remain fairly stationary relative to the embryo as a whole, the movement of the PSM boundaries is equivalent to a flow of cells through the PSM. The horizontal box outlines a group of cells (a prospective somite) that enters the PSM at its posterior end (at the left in Fig. 6(b)) and matures into a somite (circle) at the anterior end of the PSM (at the right in Fig. 6(b)). The net result of cells being stationary and boundaries that move is that this group of cells is transported across the entire length of the PSM over an 18 h time-period. Meanwhile, each cell completes about 10 expression cycles before the expression becomes constant near the anterior PSM boundary. At this point, the gene expression waves are stationary relative to cells and the wavelength of the gene expression pattern is one somite length l.

The operation of a cellular segmental clock[39] within the PSM indicates that the gene expression waves are phase waves, at least in the posterior-most part of the PSM. This is further substantiated by the observations[38] that they (1) are unaffected by transverse cuts through the PSM and (2) continue a predetermined program after small fragments of the PSM are separated from the rest of the embryo. In addition, cells enter the PSM with a periodically recurring phase of the gene expression cycle and are synchronised either before or while entering the PSM. It has been reported that the oscillation is present before the cells are added to the PSM.[40,41] The period of "boundary forcing" in the posterior region of the PSM is about 90 min in the chick.

We have demonstrated experimentally that the gene expression waves can be mimicked in an open flow of the oscillating ferroin-catalysed Belousov–Zhabotinsky reaction medium.[21] Blue bands were periodically initiated by a homogeneous oscillation near the reactor inlet and propagated downstream with a decreasing width and velocity, just like the bands in Fig. 6(b). This spatio-temporal phase wave behaviour is the result of a gradually increasing intrinsic oscillation period as volume elements are carried away from the reactor inlet by the flow. Based on this scenario, we proposed in ref. 22 a simple phase model of somitogenesis where the period of the segmental clock gradually increases as cells move away from the posterior boundary. As a result, the value of R (see Section 2.1) decreases smoothly from a value of one at the posterior boundary to a value of zero at the anterior boundary, giving rise to a decreasing width and velocity of the phase waves.

While phase dynamics accounts for a large number of experimental observations,[22] the most compelling evidence for the involvement of an FDS or FDS-like mechanism is the experimentally observed wavelength within the array of mature somites. The length of the PSM is fairly constant and it grows by approximately one somite length, l, for each somite formed. A somite emerges every time a posterior oscillation cycle is completed and the velocity of the growth-induced flow is thus $\phi = l/T'$. With this flow velocity, the scenarios investigated in Section 2 all predict that the wavelength should be $\lambda = \phi T' = l$, exactly as observed. In the subsequent sections, we demonstrate that the three RDA schemes also account for the wave behaviour within the PSM when there is a gradual change in kinetic parameters.

3.2 RDA models of somitogenesis

In the RDA models of somitogenesis, we describe the segmental clock using FHN kinetics in eqn. (9) and the PSM by eqn. (6). No-flux boundary conditions are imposed at $x = 0$. This implies that the stem cells feeding the PSM are synchronised with the daughter cells just after the latter are formed. This is fully equivalent to having a periodic forcing with $T' = T$ at the posterior PSM boundary.

The spatial variation of the oscillation period is obtained by introducing a third dynamic variable, $w(a)$, which, to preserve the kinematic nature of the regulation,[22] depends only on the time, a, a cell has spent in the PSM. The idea is that the properties of the segmental clock also change with cell age. A simple way to model the evolution of w is through a Fisher–Kolmogorov type logistic equation:[42]

$$\frac{dw}{da} = \varepsilon_w w(1 - w^p), \tag{13}$$

where a is the age of a cell. This is equivalent to a scenario where a factor, such as a protein or DNA transcription factor, is present in low levels within the stem-cell population and its starts to accumulate when cells enter the PSM.

The solution of eqn. (10) is for $p = 1$ given by:

$$w(a) = \frac{w_0 \exp(\varepsilon_w a)}{1 - w_0 + w_0 \exp(\varepsilon_w a)}. \tag{14}$$

where $0 < w_0 \ll 1$ is the level of w when a cell enters the PSM. It is a relatively steep saturation curve that connects the initial value $w(0) = w_0$ to the steady state at $w = 1$ for large a. Nonlinear response functions with a shape similar to that of eqn. (11) are frequently found in covalent modification cascades, genetic regulation and signalling pathways (see *e.g.* ref. 43). For the present purposes, a more elaborate model of the regulatory network is not required.

Our investigation is intended to demonstrate that each of the three generic FDS scenarios, in principle, may be operational in somitogenesis. Very little is known about the kinetics of the segmental clock, its spatio-temporal regulation or the role of cell–cell communication. We have not attempted to fit the model parameters to experimental observations and have only considered the simplest couplings. An overview of the different scenarios investigated is given in Table 1.

In the reference frame where the posterior PSM boundary is stationary, eqn. (10) is modelled by the RDA equation:

$$\frac{\partial w}{\partial t} = \varepsilon_w w(1 - w^p) - \phi \frac{\partial w}{\partial x} + D_w \frac{\partial^2 w}{\partial x^2}. \tag{15}$$

A non-uniform spatial profile, $w(x)$, arises along the length of the PSM when w is kept at a constant value w_0 at $x = 0$. While w is intended to be strictly cell intrinsic, a small diffusive term,

Table 1 Summary of FDS scenarios investigated and the couplings used. T_0 is the oscillation period of the local FHN kinetics for $w = 0$ and δ is a constant

Section and scenario	Coupling	Effect on kinetics
3.2 Phase-wave FDS	$\gamma(a) = T_0[1 - w(a)]$	Arrest of oscillation
3.3 Turing-type FDS	$\varepsilon(a) = (\varepsilon_0 - \delta)[1 - w(a)] + \delta$	Oscillatory to mono-stable or
	$\gamma(a) = T_0[1 - \delta w(a)]$	slowing down of oscillation
3.4 Interface-type FDS	$\alpha(a) = \alpha_0[1 - \delta w(a)]$	Oscillatory to bistable

D_w, is needed to avoid numerical instabilities in the simulations. The deviation between the solution of eqn. (12) and eqn. (11) (with $a = x/\phi$) is however inconsequential when $D_w \leqslant 0.01$. Note that the case of periodic boundary forcing is recovered when w changes abruptly from zero at $x = 0$ to one for $x > 0$.

The following scaling is used: unit time is taken to be the oscillation period, T_0, of the FHN kinetics. The value of γ is then set equal to T_0 such that eqn. (6) oscillates with a period $T' = 1$ at $x = 0$. For instance, the oscillation period of the FHN kinetics is $T_0 = 1.79988$ when $w = 0$, $\varepsilon = 10$, $\alpha = 2$, $\gamma = 1$ and by setting $\gamma = T_0$ the oscillation period becomes one. This scaling is implicitly assumed in all simulations in the subsequent sections. Unit length, l, is chosen such that the re-scaled flow velocity is $\phi T'/l = 1$. This corresponds to an axial extension of one somite length for each completed posterior oscillation. With this choice of space- and timescales, the diffusion coefficients are re-scaled by a factor equal to T'/l^2. In all the simulations, we take $\varepsilon_w = 1$, $p = 1$ and $w_0 = 5 \times 10^{-5}$. The FHN kinetics is symmetric, i.e. $\beta = 0$, unless otherwise stated.

3.3 Phase-wave scenario

The first scheme is a RDA version of our earlier phase-mode.[22] It involves an oscillation period that approaches infinity as the cell age increases. As $w(a)$ increases from w_0 at $x = 0$ towards a value of one far from it, the value of R (see Section 2.1) is made to change from one to zero by setting $\gamma(a) = T_0[1 - w(a)]$. The diffusion coefficients are equal.

The simulations in Fig. 7 shows the spatio-temporal phase dynamics at different values of the diffusion coefficients. It is observed, that each boundary oscillation induces a wave that travels in the anterior direction with decreasing width and velocity, in agreement with the experimentally recorded wave behaviour. When the diffusion coefficients are low, the gradual arrest of the oscillation gives rise to a long-lived space-periodic structure, as shown in Fig. 7(a) for $D = 0.001$. As demonstrated in Fig. 7(b), the lifetime of the structure is decreased when the diffusion coefficient is increased, as expected. If diffusion is too high the phase waves are dispersed before the FDS with velocity $c = \phi$ is established. This is illustrated in Fig. 7(c).

While the relevant structure with velocity $c = \phi$ is transient for $D \neq 0$, it may still be sufficiently long-lived to induce cell differentiation. Expression of segmental-clock genes is in the chick

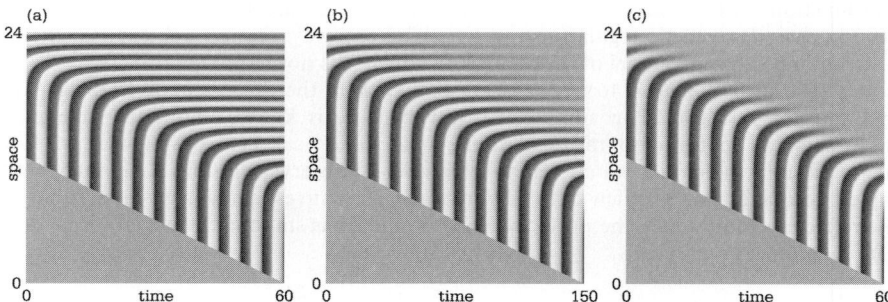

Fig. 7 Spatio-temporal phase dynamics in the presence of diffusion. Note that the space–time plot is shown in the reference frame where the boundary moves downward. (a) A long-lived structure for $D = 0.001$. (b) A transient phase wave for $D = 0.01$. (c) The relevant structure fails to establish for $D = 0.1$. In all simulations: $\gamma(a) = T_0[1 - w(a)]$, $\varepsilon = 10$, $\alpha = 2$.

observed to last for at least 15 h within the mature somite,[38] which may be sufficient time to activate genes that are involved in cell differentiation.

The simulations in Fig. 7 shows that the diffusion coefficient must thus be lower than 0.1 for a stationary transient pattern to arise. In the chick embryo, unit time is equal to about 90 min while unit length is roughly 120 μm. Assuming that the FHN kinetics gives a reasonable approximation to the real gene expression cycle, that is, with dimensional-dependent variables and dimensionless dynamic variables, the diffusion coefficients within the chick embryo must be less than about 10^{-9} cm^2 s^{-1}. This value is quite low (by a factor of 100) compared to diffusion coefficients of macromolecules in water. It is not unrealistically low, however, since the molecules may be embedded in the cell membrane or confined to the interior of the cells. Then it is the Brownian motion of the cells that leads to dispersion of the pattern, not the Fickian diffusion of molecules in water.

3.4 Turing scenario

Based on the investigation in Section 2.2, it is expected that the relevant Turing-type FDS arises in a regulatory scheme where ε gradually decreases as cells age. This is verified by the simulation presented in Fig. 8(a) where ε depends on cell age through $\varepsilon(a) = 9.05[1 - w(a)] + 0.95$. Hence, ε is equal to 10 (oscillatory local kinetics) near the boundary and decreases to a value of 0.95 as cells move downstream. This corresponds to a slow passage through the supercritical Hopf bifurcation at $\varepsilon = 1$. The homogeneous state is made Turing-unstable by increasing the diffusion coefficient of the inhibitor.

The situation in Fig. 8(a) is a bit more complicated than that investigated in Section 2.2 since a Turing-type FDS is supported for the parameter values in the boundary region. However, it does not spontaneously replace the homogenous oscillation. As $\varepsilon(a)$ decreases, the increasing oscillation period establishes the beginning of a space-periodic pattern and the homogenous oscillation is converted into a Turing-type FDS. This occurs before ε passes through the Hopf bifurcation of the local kinetics. Hence, the local steady state needs not to become stable for a space-periodic pattern to be formed. This is illustrated in Fig. 8(b) where the decrease in ε stops short of the Hopf bifurcation.

The slowed-down oscillation used in the phase-wave scenario may also be replaced by a Turing-type FDS in the presence of a long-ranged inhibitor. Hence, it is not required that the local kinetics is arrested, *i.e.* $T_0 \to 0$, as assumed in Section 3.2. Suppose that the period of the intrinsic oscillation is increased through a decrease in γ, but that it does not decrease to zero. In the absence of differential diffusion, the result would be travelling waves whose velocities are greater than the flow velocity. However, the presence of a long-ranged inhibitor may cause arrest of oscillation, $T = \infty$, without the arrest of the local kinetics, $T_0 = \infty$. This scenario is illustrated in the space–time plot in Fig. 8(c).

Both the phase-wave and the Turing schemes reproduce the spatio-temporal behaviour of the gene expression waves quite well. Based on the space–time plots, it is not possible to determine which one is more likely to be operating during somitogenesis. However, a Turing-type FDS will

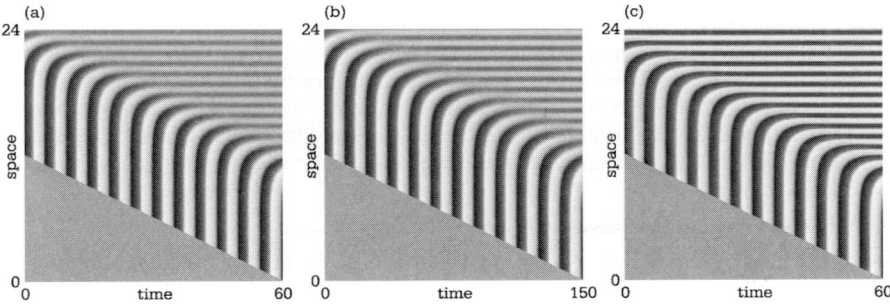

Fig. 8 Spatio-temporal behaviour in Turing-unstable media. (a) The activator kinetics is gradually slowed down: $\varepsilon(a) = 9.05[1 - w(a)] + 0.95$. (b) Passage through the Hopf-bifurcation does not occur: $\varepsilon(a) = 8.5[1 - w(a)] + 1.5$. (c) Arrest of oscillation caused by differential diffusion: $\gamma(a) = T_0[1 - 0.8w(a)]$. In all simulations: $D_u = 0.01$, $D_v = 0.1$. Other parameter values are given in Fig. 7.

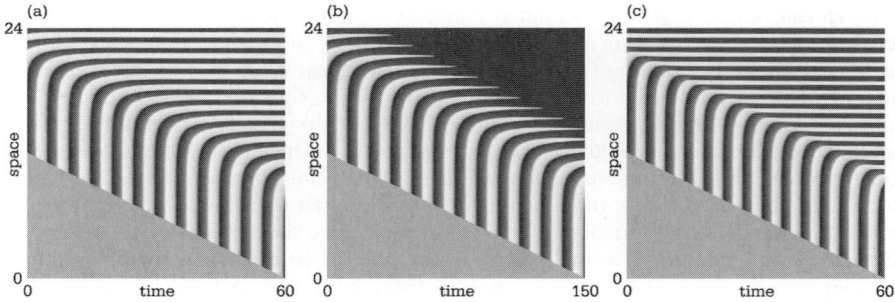

Fig. 9 Spatio-temporal behaviour in bistable media for $\varepsilon = 10$. (a) Persistent structures with equally stable states: $\alpha(a) = 2[1 - w(a)]$, $\beta = 0$, $D_u = D_v = 0.01$ (b) The low activator state absorbs the high activator state: $\alpha(a) = 1.7[1 - w(a)] + 0.3$, $\beta = 0.05$, $D_u = D_v = 0.01$ (c) Persistent structures when the inhibitor diffusion coefficient is increased to $D_v = 0.2$ with the other parameters as in (b).

spread spatially. Recent experiments on *zebrafish* by Jiang *et al.*[44] indicate that stable stripes may coexist with tissue that contains no stripes. The authors interpret the experimental observations as arising from the de-phasing of the posterior synchronous oscillation. This interpretation implies that cells after a while enter the PSM with random phases. One would then expect a noise-sustained Turing-type FDS under convectively unstable conditions, and a space-filling Turing-type FDS under absolutely unstable conditions (see Section 2.2). Since this is not what is observed, a Turing-like mechanism is probably not involved in the regulation of the gene expression in *zebrafish*. The phase-wave scenario on the other hand produces a number of stripes while the posterior oscillation is coherent and no stripes when it has desynchronized.

3.5 Interface scenario

The FHN kinetics in eqn. (9) is bistable when $\alpha < 2/3$ and $\beta = 0$. When α is equal to zero, the inhibitor is independent of the activator and converges rapidly to the stable steady state given by $v = \beta$. For $\beta = 0$, the states with high and low activator levels have the same relative stability and interfaces with velocity $c = \phi$ are supported. An example of persistent FDS that arise for equal diffusion coefficients when $\alpha(a) = 2[1 - w(a)]$ is shown in Fig. 9(a). There is no qualitative difference between the interface-type FDS and those observed in Fig. 7 (phase-wave FDS) and in Fig. 8 (Turing-type FDS).

If β is greater than zero, the state with low activator state is more stable than the high activator state. The space–time plot obtained when β is 0.05 and the value of α changes according to $\alpha(a) = 1.7[1 - w(a)] + 0.3$ is shown in Fig. 9(b). The diffusion coefficients are equal and the state with high activator levels is gradually absorbed by the state with low activator levels. However, increasing the diffusion coefficient, *i.e.* adding long-range inhibition, allows for the formation of persistent interface-type FDS, as in Section 2.3. The space–time plot shown in Fig. 9(c) is obtained from a simulation where the parameters are the same as in Fig. 9(b), with the exception that the ratio of inhibitor and activator diffusion coefficients is 10 : 1.

FDS of the interface-type typically consist of localised and self-sustained elements that do not spread spatially. The homogeneous state is stable and stripes may form only in one part of the PSM. If the coherence of the boundary oscillation is gradually lost (as in the interpretation by Jiang *et al.* mentioned above), a number of stripes would be formed early in somitogenesis. None would however form at a later stage when the boundary synchronisation has deteriorated.

4 Discussion

The spatio-temporal dynamics of open flow systems is often dictated by the temporal dynamics at the origin of the flow, *i.e.* at the inflow boundary. Given appropriate conditions, the boundary dynamics imposes itself onto the system as a whole and forces it into a variety of space-periodic wave-patterns. This is the result of the flow, ϕ, converting a temporal boundary-periodicity, T', into a spatial periodicity, $\phi T'$. The boundary-imposed spatial mode can be modified or maintained by the intrinsic kinetics of the flowing medium. For instance, boundary forcing of flows of oscillatory media can result in upstream and downstream travelling waves whose wavelengths are different from the imposed periodicity. The spatio-temporal dynamics is nevertheless dictated by the period of the boundary forcing (see *e.g.* ref. 18).

In this paper, we have focussed on the special but biologically relevant case where the entrainment of the dynamics leads to space-periodic wave patterns that are stationary relative to the flowing medium. They have velocities $c = \phi$ relative to the upstream boundary and arise in periodically boundary-forced RDA systems when the local kinetics is monostable, bistable or oscillatory. The three schemes investigated involve: (1) phase waves with oscillatory local kinetics and a very long period, (2) Turing structures with monostable or oscillatory local kinetics and (3) interfacial structures with bistable or excitable local kinetics. In all three cases, the boundary-imposed spatial periodicity may be maintained, resulting in space-periodic flow-distributed structures (FDS) with velocities $c = \phi$ and wavelengths $\lambda = \phi T'$.

The entrainment of dynamics in open flows has important implications for axial segmentation in biology. We have shown how axial growth in developing organisms is equivalent to an open flow. It is not important whether a "real" flow is actually present since the open flow concept applies to all system where there is a relative motion between elements and an upstream boundary. The flow may for instance arise from proliferation of stem cells within a TGZ, as is observed in many annelids, arthropods and plants and in some cases of limb development. The progeny cells remain fairly stationary relative to the embryo as a whole and the open flow arises as the distance between a cell and the TGZ increases in time. Alternatively, cells from the surrounding tissue may be recruited to a growth zone at the tip of an elongating structure. The early development of birds and mammals, for example, is known to involve such an organising centre located prior to the axially growing PSM.

We have demonstrated how the operation of an oscillator (a segmental clock) that is synchronised at the growth boundary gives rise to space-periodic variation in a morphogen level along the growth-axis of the developing embryo. The involvement of axial growth and a segmental clock in the formation of space-periodic stripes of gene expression has been demonstrated experimentally during somitogenesis in fish, birds and mammals. The stripes become stationary relative to cells at the anterior end of the PSM, just before a somite is formed.[37] The wavelength of the gene expression wave is observed to be equal to $\lambda = \phi T'$, which provides evidence that it is imposed by the combination of a growth-induced flow and an oscillation that is synchronised at the growth boundary.

While the above described scenarios relate to somitogenesis in vertebrates, the axial segmentation observed in many invertebrates is also consistent with the formation of FDS. It is well established that during the development of the intermediate germ-band insect *Tribolium*, stripes of gene expression appear one after the other at a short distance from a TGZ of proliferating stem cells. However, gene expression waves have not been experimentally recorded for sequential segmentation in annelids and arthropods, but this may be attributed to a lack of research so far directed towards recording spatio-temporal sequences of gene expression.[45] The key issue, namely whether a synchronised oscillation is present within the TGZ has yet to be investigated.

Another important prediction is that an organism can achieve a global control of its developmental dynamics by regulating the rate of axial growth, ϕ, and the period of the segmental clock, T'. For instance, if the shedding of progeny cells from a stem cell population facilitates axial growth, the rate of growth would be proportional to the number of stem cells. By microsurgery, Cooke[46] reduced the number of cells residing at the terminal end of the developing frog embryo. He observed that the number of somites remained the same, but that they were reduced in size. This is consistent with our prediction that the wavelength, $\lambda = \phi T'$, is decreased when the growth rate is reduced by lowering the number of feeding stem cells. In addition, it seems logical that the number of feeding stem cells scales linearly with the overall size of the embryo. In this case, our prediction accounts for size adaptation: a small embryo would naturally produce that same number of segments as a large embryo, but the size of the segments would scale proportionally to the overall size of embryo.

Which one of the scenarios is actually operating may be difficult to determine experimentally since they give rise to the same qualitative wave behaviour. In an evolutionary context, this allows for the transformation from one scenario into another. Consider a biological system that forms stripes due to an age-dependent arrest of the segmental clock. Strong cell–cell signalling (modelled here as a diffusive process) causes the stationary stripes to become transient and may prevent their formation altogether. However, strong coupling spatial couplings (*i.e.* high D_u and D_v) near the growth zone is allowed and could, for instance, ensure that boundary oscillation be properly synchronised among cells. A mutation that substantially reduces the activator signalling (*i.e.* D_u) could then result in stripe formation by a different mechanism, namely the Turing-type FDS. The organism may survive since the wavelength of stationary pattern, $\lambda = \phi T'$, may be the same before and after the mutation. Alternatively, an ancient species could have survived a series of mutations that gradually increases the mobility of the inhibitor (*e.g.* a cell-signalling molecule) while the mobility of the activator (*e.g.* a cell-internal factor) is kept low. In the same vein, mutations from a periodically forced oscillating medium to a bistable medium may result in viable embryos with the same resulting stationary pattern.

The scenarios that involve FDS of the phase wave and the Turing type are closely related to other models of biological morphogenesis. For instance, a number of current models of somitogenesis (see discussion in ref. 22) implicitly or explicitly involve phase dynamics. However, only the recent clock-and-trail model by Kerszberg and Wolpert[47] explicitly considers that axial growth combined with an oscillation at the posterior boundary may entrain the developmental dynamics. The authors suggest that the stem cells feeding the growing structure oscillate in synchrony and that daughter cells preserve a "snapshot" of the oscillation phase. This model corresponds to periodic boundary forcing of a medium whose local kinetics has an infinite period of oscillation (*i.e.* $R = 0$). Our earlier phase model[22] involved the same principle, but with a value of R that decreases gradually from one to zero as cells age. Indeed, the phase-wave scenario without cell–cell communication (diffusion) accounts well for the experimentally observed spatio-temporal behaviour of gene expression waves and can be made consistent with a large number of other experimental observations (see ref. 22 for details).

To preserve the cell-intrinsic regulation of the segmental clock in the chick, we assumed in Section 3 that its kinetics, through $w(a)$, depends only on the time, a, that a cell has spent in the PSM. However, the regulation of the segmental clock could involve cell-external factors and w could, for instance, represent the level of a cell-signalling molecule. The regulation of w in eqn. (11) may readily be adapted to model this situation because it allows a propagation front-solution. Suppose that the system initially is in the $w = 0$ steady state and that the flow is absent. A front connecting the steady states at $w = 0$ and $w = 1$ may develop if a perturbation in w is applied locally. For $p = 1$ it propagates at a velocity $c_0 = 2\sqrt{\varepsilon_w D_w}$.[42] If the local kinetics depends on the level of w, this would lead to sequential segmentation along the body-axis if the front propagates from the anterior towards the posterior end of the embryo. The wavelength would be given by $\lambda = c_0 T'$, where T' is the oscillation period ahead of the front. This scenario corresponds to the classic clock and wave front model by Cooke and Zeeman,[48] which may very well account for sequential segmentation in the absence of growth, as observed, for instance, during the early stages of amphibian somitogenesis.

The involvement of an independent wave front seems however unlikely in species where the growth zone and the segments are separated by an unsegmented tissue of constant length, such as the PSM in many vertebrates. This would require that the front velocity, c_0, be linked to the growth rate, ϕ, since a constant length of the unsegmented tissue can arise only if $c_0 = \phi$. This is also required to give the wavelength $\lambda = \phi T'$ observed during somitogenesis in the chick.

For the same reason, a mechanism where a space-periodic structure is formed by the spreading of an intrinsically selected (*i.e.* Turing-like) structure would require that the velocity of the spreading structure, also denoted by c_0, be coupled to the rate of growth. As pointed out by Meinhardt,[2] such a mechanism may be operational in axially growing systems where segments arise close to the growth boundary and in non-growing systems where segments arise sequentially. In the former case, the relevant Turing-type FDS arises when $c_0 > \phi$ or as noise-sustained structure when $c_0 < \phi$. A segmental clock at the boundary is therefore not necessary. However, its presence would seem advantageous as it allows a more flexible control of the developmental dynamics.

In contrast to phase waves and Turing structures, bistable local kinetics is not commonly found in models of biological morphogenesis. This is probably because a medium with bistable local kinetics does not develop space-periodic structures spontaneously. In reaction–diffusion systems, one of the stable homogeneous steady states is typically selected. The operation of a segmental clock at the TGZ may however induce a space-periodic pattern if it periodically switches between the basins of attraction of the stable steady states. The result is actually one of the most robust ways to produce persistent FDS. Any wavelength above a critical wavelength can be imposed and a long-ranged inhibitor is not required if the two steady states have the same relative stability. In this context, it is worth noting that the view of cell differentiation as a dynamic process involving multiple steady-state attractors is gaining momentum (see *e.g.* ref. 49). It is possible that the purpose of the waves produced by the segmental clock is to determine the fate of cells by switching between the basins of attraction of steady states in a cell regulatory network. A particular cell state would then be induced once the waves have become stationary relative to cells. A switch that determines cell fate and is influenced by segmental clock genes was recently demonstrated in mouse somitogenesis.[50,51]

In summary, we have demonstrated how the combination of axial growth and periodic boundary forcing robustly gives rise to space-periodic structures in developing embryos. All that is required to produce these waves is:

(1) An open flow established by axial growth.
(2) A segmental clock that is synchronised at the growth boundary, and
(3) Means for maintaining the boundary-imposed periodicity.

We have described three generic scenarios for the formation of waves in boundary-forced open flows of active media and discussed their relevance to axial segmentation. They involve local kinetics and spatial interactions that are feasible in chemical systems and represents therefore some of the simplest mechanisms to achieve self-organisation in biology.

References

1 A. M. Turing, *Philos. Trans. R. Soc. London, Ser. B*, 1952, **237**, 37.
2 H. Meinhardt, *Models of Biological Pattern Formation*, Academic Press, New York, 1982.
3 V. Castets, E. Dulos, J. Boissonade and P. DeKepper, *Phys. Rev. Lett.*, 1990, **64**, 2953.
4 Q. Ouyang and H. L. Swinney, *Nature*, 1991, **352**, 612.
5 J. E. Pearson, *Science*, 1993, **261**, 189.
6 K. J. Lee, W. D. McCormick, J. E. Pearson and H. L. Swinney, *Nature*, 1994, **369**, 215.
7 J. Boissonade, *Nature*, 1994, **369**, 188.
8 S. Kondo and R. Asai, *Nature*, 1995, **376**, 765.
9 P. K. Maini, K. J. Painter and H. N. P. Chau, *J. Chem. Soc., Faraday Trans.*, 1997, **93**, 3601.
10 A. Hunding, in *On Growth and Form*, ed. M. A. J. Chaplain, G. D. Singh and J. C. McLachan, Wiley, New York, 1999, pp. 75–88.
11 R. Satnoianu, M. Menzinger and P. K. Maini, *J. Math. Biol.*, 2000, **41**, 493.
12 R. J. Deissler, *J. Stat. Phys.*, 1985, **40**, 371.
13 V. Z. Yakhnin, A. B. Rovinsky and M. Menzinger, *Can. J. Chem. Eng.*, 1995, **74**, 647.
14 S. P. Kuznetsov, E. Mosekilde, G. Dewel and P. Borckmans, *J. Chem. Phys.*, 1997, **106**, 7609.
15 R. Satnoianu, J. H. Merkin and S. K. Scott, *Physica D*, 1998, **124**, 345.
16 P. Andresén, M. Bache, E. Mosekilde, G. Dewel and P. Borckmans, *Phys. Rev. E*, 1999, **60**, 297.
17 M. Kærn and M. Menzinger, *Phys. Rev. E*, 1999, **60**, 3471.
18 M. Kærn and M. Menzinger, *Phys. Rev. E*, 2000, **61**, 3335.
19 R. Satnoianu and M. Menzinger, *Phys. Rev. E*, 2000, **62**, 113.
20 R. Satnoianu, J. H. Merkin and S. K. Scott, *Dyn. Sys. Appl.*, 1999, **14**, 275.
21 M. Kærn, M. Menzinger and A. Hunding, *Biophys. Chem.*, 2000, **87**, 121.
22 M. Kærn, M. Menzinger and A. Hunding, *J. Theor. Biol.*, 2000, **207**, 473.
23 M. Kærn and M. Menzinger, *Phys. Rev. E*, 2000, **62**, 2994.
24 A. Hunding and M. Brøns, *Physica D*, 1990, **44**, 285.
25 J. G. Blom and P. A. Zegeling, *Trans. Math. Software*, 1994, **20**, 197.
26 M. T. M. Koper, *J. Chem. Phys.*, 1994, **102**, 5278.
27 S. P. Dawson, M. V. D'Angelo and J. E. Pearson, *Phys. Lett. A*, 2000, **265**, 346.
28 A. Hunding and R. Engelhardt, *J. Theor. Biol.*, 1995, **173**, 401.
29 J. H. Merkin, R. Satnoianu and S. K. Scott, *Dyn. Sys. Appl.*, 2000, **15**, 209.
30 A. Hagberg and E. Meron, *Nonlinearity*, 1994, **7**, 805.
31 A. Hagberg, E. Meron and T. Passot, *Phys. Rev. E*, 2000, **61**, 6471.
32 A. S. Mikhailov, *Foundations of Synergetics I*, Springer-Verlag, New York, 1990.
33 J. M. Chomaz, *Phys. Rev. Lett.*, 1992, **69**, 1931.
34 P. P. L. Tam, D. Goldman, A. Camus and G. C. Schoenwolf, *Curr. Top. Dev. Biol.*, 2000, **47**, 1.
35 K. J. Dale and O. Pourquié, *Bioessays*, 2000, **22**, 72.
36 C. D. Stern and D. Vasiliauskas, *Curr. Top. Dev. Biol.*, 2000, **47**, 107.
37 C. D. Stern, S. E. Frase, R. J. Keynes and D. R. N. Primmett, *Development*, 1988, **104**, 231.

38 I. Palmeirim, D. Henrique, D. Ish-Horowicz and O. Pourquié, *Cell*, 1997, **91**, 639.
39 Y. Jiang, L. Smithers and J. Lewis, *Curr. Biol.*, 1998, **8**, 868.
40 I. D. Barrantes, A. J. Elia, K. Wünsch, M. Hrabê de Angelis, W. T. Mak, J. Rossant, R. A. Conlon, A. Gossler and J. L. de la Pompa, *Curr. Biol.*, 1999, **9**, 470.
41 O. Pourquié, *Curr. Top. Dev. Biol.*, 2000, **47**, 81.
42 J. D. Murray, *Mathematical Biology*, Springer-Verlag, New York, 2nd edn., 1993.
43 J. E. Ferrell, *Trends Biochem. Sci.*, 1997, **22**, 288.
44 Y. J. Jiang, B. L. Aerne, L. Smithers, C. H. C. D. Ish-Horowicz and J. Lewis, *Nature*, 2000, **408**, 475.
45 W. G. M. Damen, M. Weller and D. Tautz, *Proc. Nat. Acad. Sci. USA*, 2000, **97**, 4515.
46 J. Cooke, *Nature*, 1975, **254**, 196.
47 M. Kerszberg and L. Wolpert, *J. Theor. Biol.*, 2000, **205**, 505.
48 J. Cooke and E. C. Zeeman, *J. Theor. Biol.*, 1976, **58**, 455.
49 S. Huang and D. E. Ingber, *Exp. Cell Res.*, 2000, **261**, 91.
50 Y. Takahashi, K. Koizumi, A. Takagi, S. Kitajima, T. Inoue, H. Koseki and Y. Saga, *Nature Genet.*, 2000, **25**, 390.
51 C. D. Stern, *Nature Genet.*, 2000, **25**, 368.

Excitability in chemical and biochemical pH-autocatalytic systems

J. Zagora, M. Voslař, L. Schreiberová and I. Schreiber*

Department of Chemical Engineering & Center for Nonlinear Dynamics of Chemical and Biological Systems, Prague Institute of Chemical Technology, Technická 5, 166 28 Prague 6, Czech Republic. E-mail: schrig@vscht.cz

Received 19th April 2001
First published as an Advance Article on the web 22nd November 2001

Using two different kinds of pH systems—the papain catalyzed hydrolysis of N-benzoyl-L-arginine ethyl ester in a membrane reactor and the bromate–sulfite–ferrocyanide (BSF) reaction in the CSTR—we study the relation among excitability, oscillations and bistability, and the ability of the system to respond to external periodic perturbations. Excitable properties of dynamical systems are examined in terms of a threshold set which is used to characterise dynamics in the reactor subject to external periodic stimuli. A precise definition and a method of calculating the threshold set are formulated. Two kinds of excitability distinguished by either direct or indirect initiation of the activatory process are found in both pH systems. Periodic pulsed perturbations of the BSF system display a nontrivial dependence of an excitation number on the forcing period. We examined this system also in oscillatory mode by looking at the phase shifts caused by single-pulse perturbations and constructing the phase transition curves (PTCs).

1 Introduction

A distinct feature of excitable systems is their sensitivity to perturbations. Of particular interest in this paper are excitable chemical and biochemical processes. There are many examples of excitable systems in chemistry and biology,[1] ranging from the Belousov–Zhabotinskii chemical reaction[2] to calcium spiking in many types of cells[3] to action potentials in specialized neural cells.[4] A typical excitatory event follows a superthreshold stimulus applied to the system at rest which triggers an autocatalytic increase of some intermediate(s) until an inhibitory decay dominates, leading eventually to the original rest state. The stimulus may be repeated periodically, either by externally controlled conditions (such as drug administration) or by some internal oscillatory process in the neighborhood of the cell transduced inside or an oscillatory subsystem within the cell. Periodic forcing induces repeated firings which may or may not catch up with the pace of stimuli depending on the strength and period of the stimulus. The dynamics may become complex when the cell is unable to respond by an excitation to each external stimulus.[5–8]

In the first section of the paper we examine the threshold dynamics, identify the threshold set, formulate a boundary value problem to find it in two-variable systems and show how to use it in characterising pulsed systems. In the second section, a biochemical papain-catalyzed pH model system[9] is used as an example for determination of the threshold set. In the third section, a response to pulsed external perturbations of an inorganic excitable pH system—the BSF reaction[10] is examined both experimentally and by calculations.

DOI: 10.1039/b103534g

2 Excitability

Following ref. 11 we consider a threshold set as locally separating the phase space into a set of trajectories amplifying initial perturbations and a set of trajectories decaying to the steady state soon after the perturbation. However, this definition is only qualitative and it is the purpose of this section to give a precise meaning to the threshold set in the presence of a unique steady state and formulate an iterative procedure for numerically accurate location of the threshold. This in turn allows us to understand how excitability vanishes and how it is interlinked with nontransient dynamical phenomena, such as bistability and oscillations.

2.1 Threshold sets

Let us assume a dynamical system described by ordinary differential equations:

$$\frac{dx}{dt} = v(x), \quad x \in R^n. \tag{1}$$

Although not explicitly stated, the vector field v is assumed to depend on external parameters. In general, the threshold set $\mathcal{T}$ should be a smooth codimension one semi-open surface which is bounded from one side by a codimension two surface providing an axis about which trajectories of (1) corresponding to excitations wind (for example, a semi-curve in R^2 bounded by a point, a semi-sheet in R^3 bounded by a curve *etc.*). Accordingly, it is convenient to assume that $\mathcal{T}$ is invariant under the flow $\varphi(x, t)$ of eqn. (1) in the negative time direction, *i.e.*, $\varphi(\mathcal{T}, t) \subseteq \mathcal{T}$ for all $t \leq 0$ where $\varphi(\mathcal{T}, t = 0)$ is the "end" of $\mathcal{T}$. Therefore the threshold set must be a (negatively directed) semiorbit in a two-variable system, a smooth one-parameter family of semiorbits in a three-variable system, *etc.*

For simplicity, we will assume that the system has two dynamical variables, $n = 2$, extension to more than two variables is possible. Let x_S be the steady state point. To locate the threshold, we formulate a boundary value problem for a finite segment of the threshold orbit beginning at a point x_L and terminating at another point x_R. The point x_L is specified by applying a perturbation shifting the initial steady state x_S to x_L so that

$$f_L = v(x_L)(x_L - x_S) = 0. \tag{2}$$

Since we assume that the response to this perturbation is amplified, the orbit passing through x_L is locally at a minimum distance P from the steady state which provides the size (or amplitude) of the pulse:

$$P = \|x_L - x_S\|. \tag{3}$$

Thus P is a minimal perturbation amplitude for the given orbit to cause an excitation. The system (1) responds to the perturbation by a motion along the orbit based at x_L, the perturbation becomes amplified, eventually reaching a point x_R such that

$$f_R = v(x_R)(x_R - x_S) = 0 \tag{4}$$

The response amplitude R,

$$R = \|x_R - x_S\|, \tag{5}$$

at that point is, by virtue of eqn. (4), at its maximum. Eqn. (2) and (4) are in fact identical but each of them defines a separate part of the same curve in the state space of (1); the two parts meet at a point where the pertubation and response amplitudes are equal. P depends on x_L according to eqn. (3); conversely, given P, the point x_L is found by solving eqn. (2) and (3). R depends on x_R which in turn depends on x_L via the flow, $x_R = \varphi(x_L, \tau(x_L))$, where τ is a time necessary to reach x_R from x_L.

Let us define a relative amplification as the increase in amplitude from x_L to x_R relative to the amplitude of perturbation:

$$r = \frac{R(x_R(x_L)) - P(x_L)}{P(x_L)}. \tag{6}$$

As x_L varies along the curve defined by eqn. (2) both r and P vary accordingly so that the first derivative of r with respect to P is

$$\frac{dr}{dP} = \frac{\text{grad } r \cdot \boldsymbol{d}^o}{\text{grad } P \cdot \boldsymbol{d}^o}, \qquad (7)$$

where $\text{grad} = \partial/\partial x_L$ and $\boldsymbol{d}^o$ is an arbitrarily normalized vector tangent to the curve defined by eqn. (2) at x_L:

$$\text{grad } f_L \cdot \boldsymbol{d}^o = 0, \quad \|\boldsymbol{d}^o\| = 1. \qquad (8)$$

The derivative in eqn. (7) captures the sensitivity of the response to variations in perturbation amplitude and allows for a characterisation of the threshold because r is expected to grow most significantly with P just on the orbit segment lying on $\mathcal{T}$. We use this property to single out the orbit segment from x_L to x_R so that the sensitivity coefficient dr/dP is at maximum:

$$\frac{dr}{dP} \stackrel{!}{=} \max. \qquad (9)$$

The threshold set $\mathcal{T}$ is then defined as the negative semiorbit starting at x_R—the endpoint of $\mathcal{T}$—and passing through x_L further extending to arbitrary negative times. As an interesting byproduct, the constraint (9) is complemented by a condition for an orbit with smallest possible dr/dP. The associated orbit, in a sense, represents a typical excitatory response and therefore we refer to it as a *characteristic excitation*. Between the threshold and the characteristic excitation is an orbit with a maximal relative amplification r. Since the Euclidean distance depends on the scaling we need to assume that the variables in eqn. (1) are given in their natural scale. However, the results do not change significantly upon rescaling, unless it is very poor, so eqn. (9) may still be conveniently used as an operational definition of the threshold.

In summary, we have a boundary value problem for an orbit segment, satisfying the boundary conditions (2) and (4) and the steady state condition

$$\boldsymbol{v}(x_S) = 0; \qquad (10)$$

the unknowns are x_S, x_L and τ. For $n = 2$, there is one more unknown than the equations therefore one unknown may be taken as a free parameter and this problem is solved by combining shooting and continuation methods.[12,13] The continuation provides a one-parameter family of orbit segments. The sensitivity coefficient (7) is calculated at each point along this curve and searched for maximum and minimum values, indicating the threshold set and the characteristic excitation, respectively. When repeated continuations with sequentially varying external parameters are made, the threshold set and the characteristic excitation can meet and the minimum and maximum value of dr/dP plotted against P merge and disappear. If that happens, the excitability vanishes. Since the threshold trajectory is strongly unstable, a multiple shooting method with nonequidistant time subintervals is used in practical calculations rather than the simple shooting approach outlined above.

2.2 Dynamics in a periodically pulsed system

By applying pulsed perturbations we can examine how the presence of a threshold set influences the dynamics and its changes when the excitability vanishes. We assume that the state point x can be changed instantaneously as a consequence of an external pulsed perturbation. Periodic forcing can be modelled as a periodic series of delta pulses shifting repetitively the value of x, and eqn. (1) now becomes

$$\frac{d\boldsymbol{x}}{dt} = \boldsymbol{v}(\boldsymbol{x}) + \boldsymbol{h}(\boldsymbol{x}) \sum \delta(t - kT), \quad k = 1, \ldots, \qquad (11)$$

where k counts the number of pulses, T is the period of pulse deliveries and $\boldsymbol{h}(\boldsymbol{x})$ is the change in $\boldsymbol{x}$ due to the pulse. In chemical systems, the pulsed addition of a particular species may affect only that species (one component of $\boldsymbol{h}(\boldsymbol{x})$ is nonzero) or more species if instantaneous reactions such as acid–base equilibria are involved.

The excitatory dynamics is well characterised by introducing an excitation (or firing) number v,

$$v = \frac{\text{number of excitatory responses}}{\text{number of pulses}}, \qquad (12)$$

in the limit of a large number of pulses k. Thus v is an average number of excitations per one pulse.

Here we can conveniently use the notion of the threshold set introduced earlier to distinguish between excitatory and nonexcitatory responses. When a pulse is delivered, the jumpwise change of x may penetrate the threshold set. A response is considered as excitatory if the pulse penetrates the threshold set or, equivalently, if the trajectory loops around the endpoint of the threshold set. Periodic orbits have $v = p/q$ where p, q are integers and qT is the period.

3 Papain system—a biochemical pH excitator

The first model we examine is the papain oscillator—a biochemical model involving the enzyme papain. This system describes the enzyme catalyzed hydrolysis of a substrate (N-α-benzoyl-L-arginin ethylester) in a compartment connected to a reservoir.[9,14] The model is expressed in terms of two independent variables s (dimensionless substrate concentration) and h (dimensionless concentration of hydrogen ions) whose temporal dynamics are described by the following dynamical mass balance equations[15]

$$\frac{ds}{d\tau} = d_s(s_0 - s) - r(s, h), \qquad (13)$$

$$\left(1 + \frac{1}{h^2}\right)\frac{dh}{d\tau} = d_h(h_0 - h) - \left(\frac{1}{h_0} - \frac{1}{h}\right) + r(s, h), \qquad (14)$$

where the reaction kinetics term is

$$r(s, h) = Da \frac{s}{(f_1 + f_2)s + 10^{5.736}f_1},$$

with

$$f_1 = 1 + 10^{-2.71}h + 10^{-1.49}h^{-1}, f_2 = 3.213(1 + 10^{-3.09}h).$$

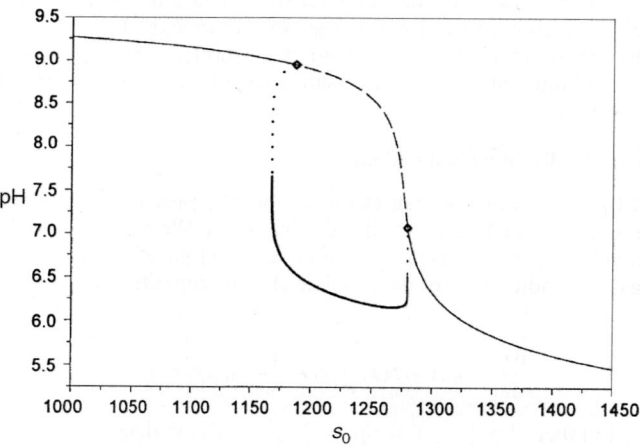

Fig. 1 Dependence of time averaged value of pH for steady states and periodic oscillations in the papain system on the substrate concentration s_0; thin/thick line, steady state/periodic cycle; full/broken line, stable/unstable regimes; diamond, Hopf bifurcation.

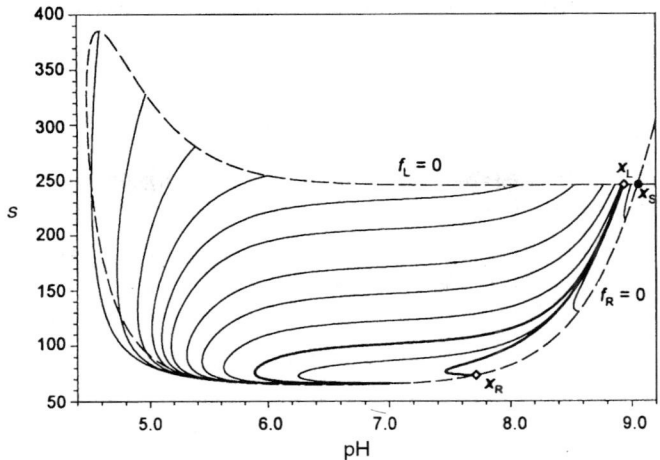

Fig. 2 Phase portrait for activatory excitability at $s_0 = 1150$ showing orbit segments (thin full lines), satisfying boundary conditions (2) and (4) (broken lines); the threshold set and the characteristic excitation are marked with thick lines, the points x_L and x_R identify the threshold set.

Here $d_s = 0.375$ and $d_h = 1.766$ are dimensionless transport coefficients, $Da = 3.5 \times 10^6$ is the Damköhler number and $h_0 = 10^{-2.655}$ and s_0 are dimensionless concentrations in the reservoir; s_0 is used as a variable parameter.

By using the numerical continuation method[12,13] we construct a solution diagram in Fig. 1 showing the steady states and periodic solutions of eqn. (13) and (14) as s_0 is varying. The curve of steady states has a middle unstable part separated from the left and right stable parts by points of oscillatory instability—the Hopf bifurcation. Periodic oscillations bifurcate subcritically from the steady state at the Hopf bifurcation points. The two unstable branches of periodic orbits become stable *via* turning points. Thus there is an overlap of stable periodic oscillations and stable steady states near the Hopf points. The dynamics are excitable in regions adjacent to the region of oscillations.

In the left region, the steady state has low h and high s values and is excitable upon adding h. Addition of h triggers an excitatory event beginning with an activatory (or autocatalytic) process

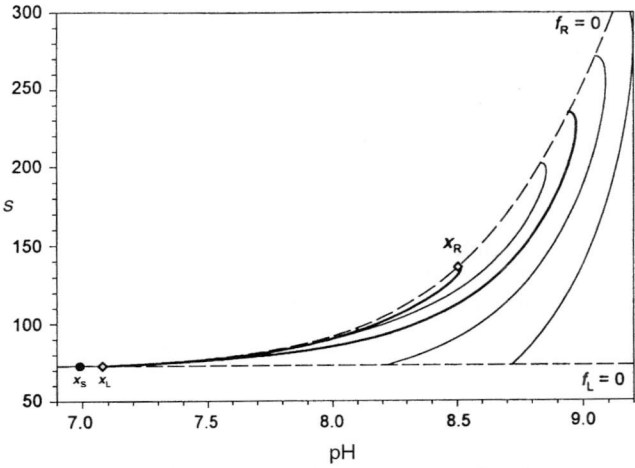

Fig. 3 Phase portrait for inhibitory excitability at $s_0 = 1280$; for further details see Fig. 2.

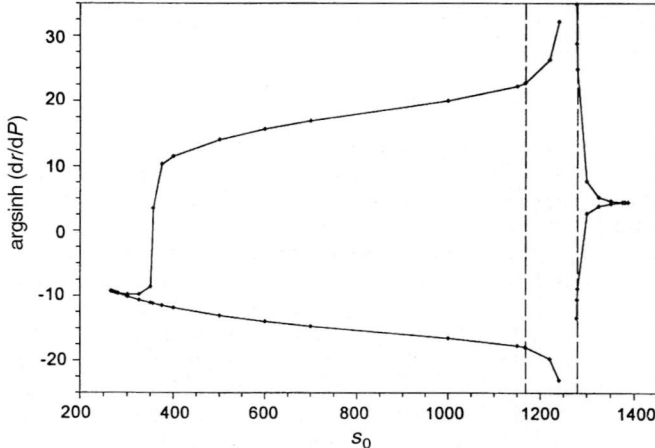

Fig. 4 Maximum and minimum values of dr/dP corresponding to threshold set and characteristic excitation, respectively, plotted against s_0; broken lines mark positions of the two Hopf bifurcation points in Fig. 1.

which rapidly produces h and depletes s until an inhibitory process begins to dominate leading to decrease in h and increase in s. We call this dynamics an *activatory* excitability. The other, somewhat counterintuitive type—*inhibitory* excitability—is associated with a high-h-low-s steady state to the right of the oscillatory region in Fig. 1. The excitation is elicited by removing h (achieved by adding OH$^-$ to the reaction cell) and begins with the inhibitory process, which removes h and simultaneously produces s, and terminates by the autocatalysis. Another classification of excitable systems distinguishes between multiple steady state (type I) and single steady state (type II) phase portraits;[16,17] in the present case we have a unique steady state and therefore the type II excitability.

The threshold set for both activatory and inhibitory excitability near the oscillatory region in Fig. 1 is shown in Fig. 2 and 3, respectively. The orbits in these figures were obtained by continuation of the boundary problem formulated for the threshold set in Section 2.1. The threshold set and characteristic excitation were singled out by monitoring dr/dP. In Fig. 4, the plot of the maximum and minimum values of dr/dP against s_0 determines the region of excitability of either type as well as a point of vanishing excitability where both values merge. The region of activatory excitability is much larger than that of inhibitory excitability which turns out to be a typical case for the papain system. Notice that the threshold set extends well into the region of oscillations, that is, the threshold exists even when the steady state becomes unstable and can no longer be assigned as excitable. This, however, will cause a strong sensitivity of oscillatory phase shifts to perturbations crossing the threshold—excitability of this kind is called type III.[16]

4 BSF system—an inorganic pH excitator

4.1 Experimental set-up

BSF (bromate–sulfite–ferrocyanide) reaction in a flow-through stirred reactor is one of a number of inorganic oscillatory and excitable systems.[18] Edblom et al.[10] observed experimentally periodic oscillations and coexistence of two stable steady states. Other experiments with the BSF system were also done by Rábai et al.[19] None of these results were focused on excitability. Our experimental set-up is a 20 ml CSTR with a jacket for water exchange with a thermostat keeping a constant temperature of 40 °C. The reactants (H_2SO_4, $Fe(CN)_6^{4-}$, BrO_3^- and SO_3^{2-}) were delivered to the reactor *via* a peristaltic pump. The inlet concentrations were: $[H_2SO_4]_0 = 7.5$ mM, $[Fe(CN)_6^{4-}]_0 = 15$ mM, $[BrO_3^-]_0 = 75$ mM; the inflow concentration $[SO_3^{2-}]_0$ and the flow rate k_0 were used as variable parameters. The reactor is provided with a Rushton turbine type of stirrer to ensure best mixing and equipped with a thermocouple for temperature control, and with a pH electrode. Single as well as periodic pulses were generated by injecting a small volume of diluted H_2SO_4 or NaOH by a syringe pump. The pumps were controlled by a computer which was also used for processing of measured signals. Our experimental results are shown along with model calculations below.

4.2 Mechanism

Edblom et al.[10] proposed a mechanism, derived from reaction steps of similar systems $BrO_3^- – SO_3^{2-} – Mn^{2+}$,[20] and $IO_3^- – SO_3^{2-} – Fe(CN)_6^{4-}$,[21] and the Belousov–Zhabotinskii (BZ) reaction,[22] that reproduces observed oscillatory and bistable dynamics. The mechanism assumes reduction of bromate by hydrogen sulfite and ferrocyanide to $HBrO_2$ which is then via several steps, taken from the BZ reaction, converted to free bromine which in turn is reduced by HSO_3^- releasing thereby H^+. Thus the dominant species in the autocatalytic loop are H^+ and Br_2. This model, however, does not account for excitability observed in our experiments.

An alternative model proposed by Rábai et al.[19] considers, according to Williamson and King,[23] successive reduction of bromate by protonated forms of sulfite. These steps are merged to a single-step reduction of bromate to Br^-. Reduction of bromate by ferrocyanide is also taken as a single step. Thus all the reactions from the BZ system considered in the earlier mechanism are left out. Instead of Br_2, the fully protonated form of sulfite, H_2SO_3, plays the role of an autocatalytic species, in addition to H^+. This skeleton mechanism does provide excitability, albeit for unrealistic values of control parameters.

Absence of excitable dynamics rules out the first model. We have decided to re-examine the mechanism due to Rábai et al.[19] in an attempt to improve the fit with our experiments. We included the intermediate BrO_2^- in the successive reduction of bromate as originally proposed in.[23] This intermediate interconnects the sulfite and ferrocyanide reduction pathways which we expect to capture more refined features of the dynamics. Thus our version of the BSF mechanism involves a two-step reduction of bromate by both (protonated) sulfite and ferrocyanide as follows:[17]

$$BrO_3^- + HSO_3^- \xrightarrow{k_1} H^+ + BrO_2^- + SO_4^{2-} \tag{R1}$$

$$BrO_2^- + 2HSO_3^- \xrightarrow{k_2} 2H^+ + Br^- + 2SO_4^{2-} \tag{R2}$$

$$BrO_2^- + 2H_2SO_3 \xrightarrow{k_3} 4H^+ + Br^- + 2SO_4^{2-} \tag{R3}$$

$$BrO_3^- + H_2SO_3 \xrightarrow{k_4} 2H^+ + BrO_2^- + SO_4^{2-} \tag{R4}$$

$$BrO_3^- + 2HFe(CN)_6^{3-} \xrightarrow{k_5} H_2O + BRO_2^- + 2Fe(CN)_6^{3-} \tag{R5}$$

$$BrO_2^- + 4HFe(CN)_6^{3-} \xrightarrow{k_6} 2H_2O + Br^- + 4Fe(CN)_6^{3-} \tag{R6}$$

In addition, the following rapidly equilibrated revesible protonation–deprotonation reactions are considered:

$$HSO_3^- \rightleftharpoons H^+ + SO_3^{2-}, \quad H_2SO_3 \rightleftharpoons H^+ + HSO_3^-$$

$$HFe(CN)_6^{3-} \rightleftharpoons H^+ + Fe(CN)_6^{4-}, \quad HSO_4^- \rightleftharpoons H^+ + SO_4^{2-}$$

All reaction rates are assumed first order with respect to each species. There is a total of 13 dynamical variables describing the system which we reduce by employing conservation constraints for all atoms that take part in chemical changes, electroneutrality condition, instantaneous acid–base equilibria, and pseudo-steady-state assumption for BrO_2^-. Thereby the original 13 mass balance equations are reduced to those describing the dynamics of H^+ and HSO_3^- in a CSTR:

$$\frac{d[H^+]}{dt} = r_H + \Delta r_H + k_0([H^+]_0 - [H^+]) \tag{15}$$

$$\frac{d[HX]}{dt} = r_{HX} + \Delta r_{HX} + k_0([HX]_0 - [HX]), \tag{16}$$

where $HX = HSO_3^-$, r_H and r_{HX} are net rates based on reactions (R1)–(R6) and Δr_H and Δr_{HX} are contributions from the acid–base equilibria, details are elaborated elsewhere.[17] The values of k_4, k_3/k_2, k_6/k_2 are estimated based on data from ref. 10, 19 and 24, the equilibrium constants were all measured[17] (at 40 °C), the values of k_1 and k_5 are based on the values from ref. 9 at 35 °C but were fine-tuned so as to fit our experimentally found periodic oscillations at 40 °C, $k_0 = 1.1 \times 10^{-3}$ s^{-1} and $[SO_3^{2-}]_0 = 70$ mM as shown in Fig. 5. The external parameters are k_0, the flow rate, and $[H^+]_0$, $[HX]_0$, the inflow concentrations of H^+ and HSO_3^- respectively. The latter two are readily calculated from $[H_2SO_4]_0$ and $[SO_3^{2-}]_0$ in the feed. The size (or amplitude) Δc_P of the pulse by a perturbant P is defined as an amount of added P divided by the reactor volume. Due to rapid equilibria, a pulse by adding H^+ or OH^- affects instantaneously both $[H^+]$ and $[HX]$, therefore the function $h(x)$ in eqn. (11) is defined implicitly.[15]

4.3 Bifurcation diagram and excitability

In search of excitable dynamics we generated an experimental bifurcation diagram by sequential variation of the inflow concentration $[SO_3^{2-}]_0$ and the flow rate k_0. We identified points where either oscillations or bistability emerged, see Fig. 6. Simultaneously we used the model (15) and (16) to calculate a bifurcation diagram by using the continuation method.[12,13] Calculated curves of the Hopf bifurcation (oscillatory instability) and curves of saddle-node bifurcation (emergence of multiple steady states) are overlayed with the experimental points in Fig. 6; the observed structure corresponds to a cross-shaped phase diagram. The region of oscillations between the two Hopf curves and the region of bistability below the oscillations fit the experiments rather well. Importantly, two regions of excitability are found adjacent to the oscillatory region in both experiments and calculations.

We identify the region to the left of the oscillatory one with the *inhibitory* excitability (to be compared with the right region in the papain model). The excitation is elicited by adding a small-amplitude pulse of NaOH to the reaction mixture lowering thereby the concentration of the autocatalyst H^+ as displayed by the experimental and calculated time traces in Fig. 7. To the right of the oscillatory region, the *activatory* excitability is found. Here an addition of H^+ initiates directly the autocatalytic process, see Fig. 8. The calculations agree with the measurements quite well. A minor discrepancy is a calculated oscillatory approach to the steady state as opposed to a measured nonoscillatory approach in the case of the inhibitory excitability; also, the experimental

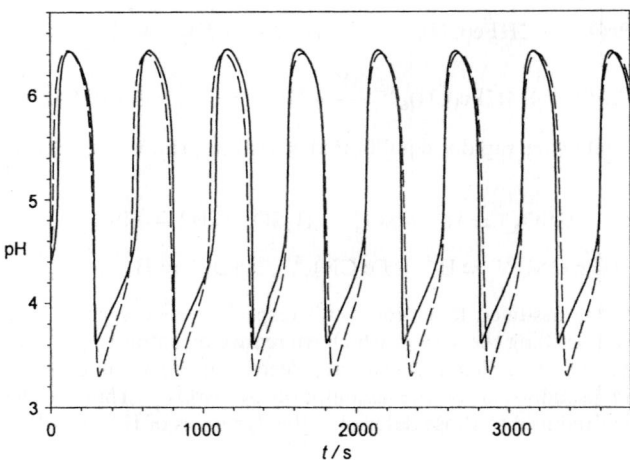

Fig. 5 Regime of periodic oscillations used to optimize parameters of the model; full line, experiments; dashed line, calculation; $k_0 = 1.1 \times 10^{-3}$ s^{-1}, $[SO_3^{2-}]_0 = 70$ mM.

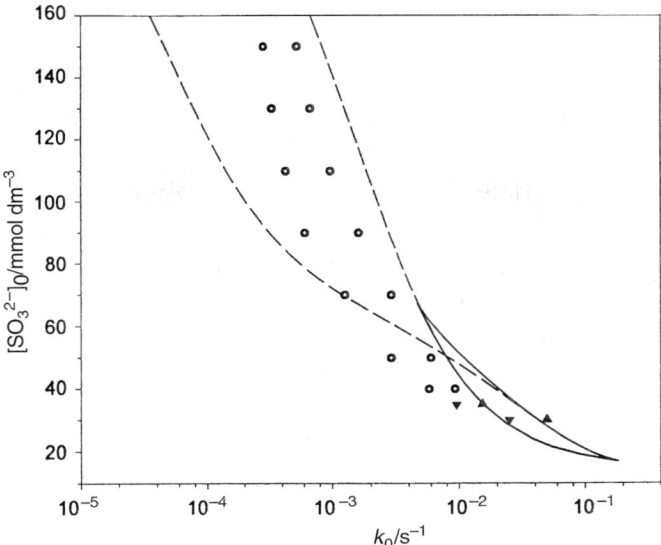

Fig. 6 Bifurcation diagram in the k_0–$[SO_3^{2-}]_0$ plane; points correspond to experiments: circles, transition to oscillations, triangles, transition to bistability; curves correspond to calculations: dashed line, Hopf bifurcation; full line, saddle-node bifurcation.

steady state pH level differs somewhat from the calculated one, see Fig. 7. In the activatory excitability case in Fig. 8, the timescale of the model at larger values of flow rate does not accurately fit the experiments, therefore k_0 was slightly adjusted to correct for the difference.

4.4 Periodic forcing and phase response

Next we examine the response to a periodic pulsed addition of OH^- when the dynamics is chosen to be an inhibitory excitable one. The response is excitatory, and can be characterised by an excitation number as defined in Section 2.2. Since the dynamics is of all-or-none character, excitations in experiments were simply identified as large pH excursions. For the model, a more subtle approach is available in terms of the threshold set. As in the papain model we found the threshold set by methods of Section 2.1 and indicated excitations as loops around its endpoint. Calculations

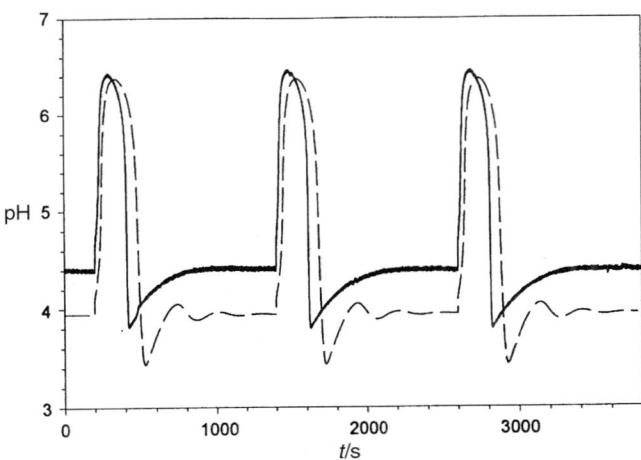

Fig. 7 Several consecutive pulsed additions of NaOH with an amplitude $\Delta_{OH^-} = 0.2$ mM repeatedly exciting a low-pH steady state possessing inhibitory excitability; full line, experiments; dashed line, calculations; $k_0 = 8.0 \times 10^{-4}$ s^{-1}, $[SO_3^{2-}]_0 = 70$ mM.

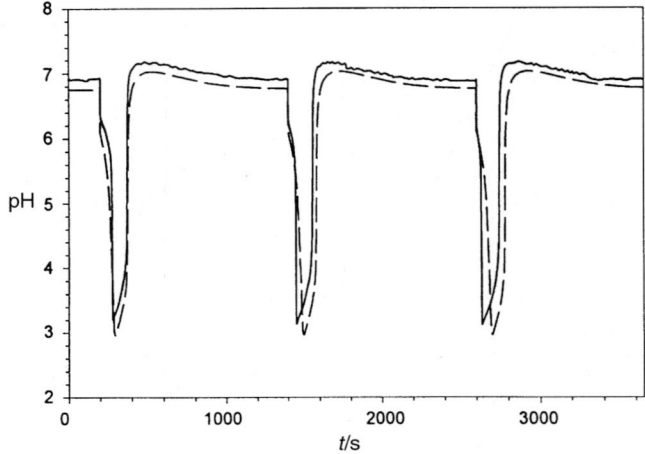

Fig. 8 Several consecutive pulsed additions of H_2SO_4 with an amplitude $\Delta_{H^+} = 7.9$ mM repeatedly exciting a high-pH steady state possessing activatory excitability; full line, experiments with $k_0 = 4.8 \times 10^{-3}$ s^{-1}; dashed line, calculations with $k_0 = 6.0 \times 10^{-3}$ s^{-1}; $[SO_3^{2-}]_0 = 70$ mM.

of excitation number v based on this definition are compared with experimental values in Fig. 9. The staircase-like dependence of v on the forcing period T reflects major resonant p/q regimes arranged as a predominantly nondecreasing function. The calculated local descents in the staircase as T is increased are associated with an emergence of chaotic regimes; that is, the model predicts that the efficiency of generating excitations is lowered by the presence of chaotic dynamics. Experimentally this was not observed; a more dense grid of measurements would be needed to make further conclusions.

The threshold set was shown to exist also in the region of oscillations for the papain model (see Fig. 4) giving rise to type III excitability. In the BSF system, the same behaviour is expected, which can be shown also experimentally by constructing phase resetting curves.[25] A phase Φ of periodic oscillations with period T may be defined as $\Phi = t/T$ (mod 1) provided that we set $\Phi = 0$ when a defined marker event occurs, for instance a sudden decrease in pH at the beginning of the autocatalytic process. By applying a perturbation at a chosen Φ and waiting for the transient to

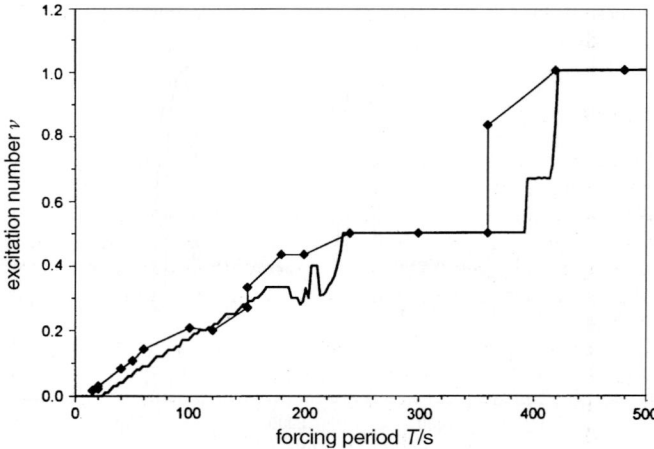

Fig. 9 Dependence of the excitation number v on forcing period T for the inhibitory excitability; diamonds (connected with a thin line), experiments with $\Delta_{OH^-} = 0.1$ mM; thick line, calculations with $\Delta_{OH^-} = 0.2$ mM; $k_0 = 8.0 \times 10^{-4}$ s^{-1}, $[SO_3^{2-}] = 70$ mM.

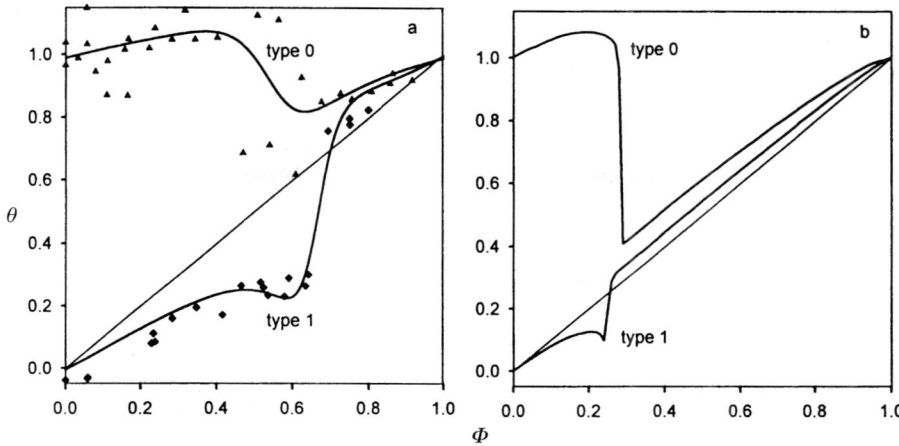

Fig. 10 Phase transition curves of type 1 and type 0 for two different amplitudes $\Delta_{H^+} = 0.36$ mM and $\Delta_{H^+} = 0.9$ mM, respectively; a, experiments with $k_0 = 1.3 \times 10^{-3}$ s^{-1}; b, calculations with $k_0 = 2.3 \times 10^{-3}$ s^{-1}; $[SO_3^{2-}] = 70$ mM.

decay, the periodic oscillations are shifted by a phase difference $\Delta\Phi$ so that the new phase of oscillations $\Theta = \Phi + \Delta\Phi = \mathcal{G}(\Phi)$. The function $\mathcal{G}(\Phi)$ is called a *phase transition curve* (PTC).[25] For small perturbation amplitudes the PTC has an average slope 1 and is said to be of topological type 1; for large enough amplitudes the average slope of the PTC is 0 and the curve is of topological type 0.[25]

If there is a threshold set in the phase space near the periodic cycle then there may exist a critical value of Φ such that the perturbation just hits the threshold. Since the threshold set strongly separates trajectories, perturbations applied at two slightly different phases, which shift the system to opposite sides of the threshold, will lead to very different phases. Hence sharp local gradients of the PTC mark the presence of the threshold set. This feature is indicated in Fig. 10. A regression curve drawn through the experimentally measured points in Fig. 10a makes the gradients less pronounced, yet clearly seen in both type 1 and type 0 PTCs. The calculated curves display very sharp gradients. It is expected, that a nonideal mixing present in the experimental CSTR would slow down the mass transport within the reaction volume and thus cause both the scatter in the experimental data and less steep gradients than calculated. As in Fig. 8, the value of k_0 in calculations was adjusted to get a proper timescale.

5 Discussion and conclusions

A quantitative criterion for the existence of a threshold set in excitable systems has been defined and a boundary value problem for calculating the threshold has been formulated and solved using two models of pH-autocatalytic systems. This enables us to clearly define an excitation number which characterises dynamics of periodically pulsed systems. We focused on the case of a unique excitable steady state (type II excitability). In both studied systems we found that a region of spontaneous periodic oscillations separates adjacent regions with different kinds of excitability— the activatory and inhibitory type.

For the BSF system we formulated an extended version of the mechanism which accounts well for the excitable dynamics and likewise for the oscillations and bistability observed in experiments without using perturbations. When periodic pulses are applied, the system responds by excitations with frequency generally lower than or at most equal to that of the pulses. For the inhibitory excitability, the measured dependence of the excitation number on the forcing period agrees with calculations although adjustment in the amplitude of the pulses had to be made due to a different distance of the threshold from the steady state in experiments and the model, see Fig. 7. The threshold set is shown to exist even within the oscillatory region (type III excitability) and leads to steep gradients in the phase transition curves. This is displayed by the BSF system, both experimentally and by calculations with properly adjusted timescale. Our experiments marked with wide variations of external parameters put a stringent test on the two-variable model which, given the complexity of the BSF mechanism, reproduces the experiments quite well, even though minor adjustments had to be made. In particular, the earlier models do not correctly reproduce excitability—either not at all[10] or for parameter values inconsistent with experiments.[19]

Both systems possess also an excitable steady state coexisting with a pair of unstable steady states (type I excitability). The threshold set for this case consists, in part, of the stable manifold of a the saddle steady state and its calculation is straightforward.

We conclude that both the biochemical and inorganic pH-autocatalytic systems are rather similar in their dynamical behaviour and display a number of interesting dynamical features related to excitability including complex dynamical response to periodic perturbations and phase shifts sensitivity to single pulse perturbations.

Acknowledgements

The authors thank the Grant Agency of the Czech Republic (grant GACR 203/98/1304) and the Ministry of Education of the Czech Republic (project MSM223400007) for financial support.

References

1. *Nonlinear Wave Processes in Excitable Media*, ed. A. V. Holden, M. Markus and H. G. Othmer, Plenum Press, New York, 1990.
2. *Oscillations and Travelling Waves in Chemical Systems*, ed. R. J. Field and M. Burger, Wiley, New York, 1990.
3. A. Goldbeter, *Biochemical Oscillations and Cellular Rhythms*, Cambridge University Press, Cambridge, 1996.
4. A. L. Hodgkin and A. F. Huxley, *J. Physiol.*, 1954, **117**, 500.
5. V. Votrubová, P. Hasal, L. Schreiberová and M. Marek, *J. Phys. Chem. A*, 1998, **102**, 1318.
6. P. Hasal, V. Nevoral, I. Schreiber, H. Ševčíková, D. Šnita and M. Marek, in *Nonlinear Dynamics, Chaotic and Complex Systems*, ed. E. Infeld, R. Zelazny and A. Galkowski, Cambridge University Press, Cambridge, 1996, pp. 72–98.
7. I. Schreiber, P. Hasal and M. Marek, *CHAOS*, 1999, **9**, 43.
8. V. Nevoral, M. Voslař, I. Schreiber, P. Hasal and M. Marek, *FORMA*, 2000, **15**, 291.
9. S. R. Caplan, A. Naparstek and N. J. Zabusky, *Nature*, 1973, **245**, 364.
10. E. C. Edblom, Y. Luo, M. Orbán, K. Kustin and I. R. Epstein, *J. Phys. Chem.*, 1989, **93**, 2722.
11. J. C. Alexander, E. J. Doedel and H. G. Othmer, *SIAM J. Appl. Math.*, 1990, **50**, 1373.
12. M. Kubíček and M. Marek, *Computational Methods in Bifurcation Theory and Dissipative Structures*, Springer, New York, 1983.
13. M. Marek and I. Schreiber, *Chaotic Behaviour of Deterministic Dissipative Systems*, Cambridge University Press, Cambridge, 1995.
14. A. Naparstek, D. Thomas and S. R. Caplan, *Biochim. Biophys. Acta*, 1973, **323**, 643.
15. M. Voslař, Dissertation Thesis, Prague Institute of Chemical Technology, 2001.
16. M. Marek, J. Finkeová and I. Schreiber, in *Irreversible Processes and Selforganization*, ed. W. Ebeling, Teubner Verlag, Leipzig, 1989, pp. 76–79.
17. H. Ševčíková, I. Schreiber and M. Marek, *J. Phys. Chem.*, 1996, **100**, 19153.
18. Y. Luo and I. R. Epstein, *Adv. Chem. Phys.*, 1990, **79**, 269.
19. G. Rábai, A. Kaminaga and I. Hanazaki, *J. Phys. Chem.*, 1996, **100**, 16441.
20. M. Alamgir, M. Orbán and I. R. Epstein, *J. Phys. Chem.*, 1983, **87**, 3725.
21. E. Edblom, M. Orbán and I. R. Epstein, *J. Am. Chem. Soc.*, 1986, **108**, 2826.
22. R. J. Field, E. Körös and R. M. Noyes, *J. Am. Chem. Soc.*, 1972, **94**, 8649.
23. F. S. Williamson and E. L. King, *J. Am. Chem. Soc.*, 1957, **79**, 5397.
24. G. Rábai and I. R. Epstein, *Inorg. Chem.*, 1989, **28**, 732.
25. A. T. Winfree, *The Geometry of Biological Time*, Springer-Verlag, New York, 1980.

General Discussion

Prof. Westerhoff opened the discussion of Prof. Olsen's paper:† The understanding of the functional relevance of many biochemical oscillators is still limited and therefore I welcome your hypothesis in which decreasing the toxicity of diffusive intermediates is the proposed function. However, in the 'Numerical simulations of the PO reaction' section you suggest that the mechanism of damage by the reactive oxygen species is not important. Yet in the 'Discussion' section your explanation of the protective effect is that although the maximum concentration of superoxide anion is up to 6 times higher in the oscillation model, its average concentration is 8 times lower. When damage is proportional to $[O_2^{\cdot -}]^n$ then your conclusion may not be valid for kinetic orders much exceeding 1 ($n \gg 1$). Because these are defense mechanisms *in vivo*, substained oxygen damage tends to occur only when the superoxide anion exceeds a threshold concentration, which corresponds to a larger n. In short: the mechanism and its kinetics should matter.

Dr Hauser responded: There is perhaps a slight misunderstanding here. We do not suggest that the mechanism of damage by the reactive oxygen species is not important. Our statement refers to reaction (14) of Table 3, which accounts for the action of toxic radicals on the enzyme. Our model does neither allow to decide which of the oxidation states of the enzyme is the most susceptible to reaction with reactive oxygen species, nor does it allow to discriminate which of the reactive oxygen species is the most toxic for the enzyme.

We agree with Prof. Westerhoff that the mechanism and the kinetics of enzyme inactivation should indeed matter. In our first attempt to model the inactivation we have assumed simple second order kinetics (first order in $[O_2^-]$), but more experimental work is needed to resolve the kinetics.

Mr Weimer said: My question refers to the statement in the paper that nature might have invented the principle of oscillating concentration changes of toxic substances for protective purposes in the first instance, but then used these oscillations as a means of encoding information, making the oscillating substance a second messenger. I wonder how far this hypothesis about the evolution of second messenger systems holds for calcium, which you mention in the text.

(i) Aren't high concentrations of intracellular calcium toxic for the cell mainly because calcium is a second messenger? The argument about the evolution of calcium oscillations and the calcium second messenger system would then be circular: calcium becomes a second messenger because there are oscillations in its concentration in the first instance. These oscillations have evolved to protect the cell from calcium's toxic effects. But these toxic effects are due to calcium being a second messenger.

(ii) Do you think that the possibility of calcium being toxic, besides its second messenger effects, is a satisfying explanation? Then one has to explain why there is a toxic concentration in the cell at all in the first instance. Why not just keep calcium out of the cell? This is very different from a toxic intermediate of cellular biochemistry, which can't be avoided.

Could the assumption that calcium already has a signaling function when oscillations evolve for protective reasons solve this problem? Only after this, might calcium have become a signal on an oscillatory level. The evolution of calcium's original signaling function by its mere presence would not then be part of the hypothesis given in the paper.

---

† Dr Hauser presented the paper on behalf of Prof. Olsen and took the proceeding questions.

Dr Hauser replied: It is certainly true that calcium is a different case compared to superoxide radicals or other second messengers in so far that it is not an intermediate of biochemical reactions. Also, our speculation about calcium ion is certainly not aimed at bringing up a universal principle behind the evolution of information processing in the cell. On the other hand it calls for speculation that so many, otherwise toxic substances, have been shown to participate in cell signaling. Among these toxic substances is calcium. If calcium concentration in the cytosol was of the same magnitude as the concentration outside the cell it would cause massive precipitation of inorganic and organic phosphates. Therefore, the cell spends a considerable amount of its metabolic energy to pump calcium out of the cytosol. Nevertheless, it is not possible to completely avoid its entrance into the cell, due to the existence of a number of nonspecific cation channels in the cell membrane.

Dr Satnoianu commented: Given the proven existence of oscillations in the PO reaction several questions come to mind. First how generic is this behaviour; is it expected to occur in metabolic processes? Is this a robust dynamical behaviour (*i.e.* easily obtainable) by varying the parameter values? Finally, is this behaviour expected to occur in/for cell activity/dynamics?

Dr Hauser responded: In the laboratory, the behaviour of the peroxidase–oxidase reaction is very robust and reproducible. It can be observed over a wide range of experimental settings. We also expect that such behaviour will occur in metabolic processes. There is already ample evidence for oscillations of NADH and reactive oxygen species in neutrophils (white blood cells that ingest and kill invading bacteria).[1] Given the fact that a peroxidase (myeloperoxidase) is the most abundant enzyme in these cells, it is not unreasonable to believe that the PO reaction is at least partially involved in these oscillations, although direct evidence for this is still lacking. The oscillations are believed to be important for the function of the cell.[1] Also in the neutrophils the behaviour has been found to be robust: oscillations with periods raging from a few seconds to several minutes have been observed under different metabolic conditions.

1 H. R. Petty, *Immunol. Res.*, 2001, **23**, 85.

Prof. Westerhoff commented: With respect to the possible relevance of oscillations in ROS (reactive oxygen species), I should like to mention some recent experimental evidence from my laboratory (Molecular Cell Physiology, BioCentrum, Amsterdam). Using intensive laser illumination of a rhodamine dye to produce ROS in a small area in an isolated heart cell, we observed a wave of ROS production spreading from that area, all over the cell. Peroxidase, superoxide dismutase and catalase should be relevant and the frequency was similar to the ones reported in this paper. In our experiments the isolated heart cells fully recovered from the initial insult and the wave, and may actually benefit from the latter.

Dr Hauser answered: This is a very nice result! I am pleased to hear that the oscillatory frequency is in good agreement with the one we observe in our experiments.

With respect to ROS oscillations, there is increasing evidence that they are relevant in a physiological context. In addition to your example, I would like to point to a study on neutrophils, where similar waves have been observed.[1]

1 H. R. Petty, R. G. Worth and A. L. Kindzelskii, *Phys. Rev. Lett.*, 2000, **84**, 2754.

Prof. Sørensen asked: If the proposed mechanism is an effective method for protecting enzymes from toxic intermediates it should be possible to see these oscillations in living plant cells. Have any observations of this kind been reported?

Dr Hauser replied: Yes, there are some observations. But first, it is important to mention that the levels of peroxidase expression in plants are known to be altered by injuries or by stress conditions, since peroxidases are thought to be part of the plant defence.[1] What has been measured are oscillations of reactive oxygen species—with which peroxidases react—in plant cells. They were observed as responses to stresses or injuries:

(i) Oscillations of the O_2 concentration have been measured in rice leaves.[2] They occur when the plant is infected by some blast fungus. The period of the oscillations is in the range of ~ 2 h.

(ii) Another example has been shown to me by Prof. A. Scheeline (University of Illinois, Urbana-Champain). Here, oscillations of H_2O_2 in horseradish roots have been recorded when the plant is watered with salty water. In this case, too, oscillations with a frequency of minutes occur as a stress response.[3] The oscillations were sustained and they have been monitored continuously. However, these data have never been published since there have been massive doubts if the H_2O_2–solid-state–electrode was indeed only monitoring the concentration of H_2O_2. But, if we are only interested in the oscillations in plant tissue, I think that—in that sense—measurements were convincing, no matter if the oscillations stem exclusively from H_2O_2 or if they originate from multiple sources.

Further support that oscillations might occur in plant cells can be provided by our experiments using horseradish cell extracts (instead of the purified enzyme). In such a setting, we were able to induce damped oscillations upon addition of the substrates of the PO reaction.[4]

Finally, I would like to refer to my previous answer to Dr Satnoianu where I mention oscillations of reactive oxygen species which occur in animal cells (white blood cells), and where I discuss a possible involvement of peroxidases in that reaction.

1 A. Campa, in *Peroxidases in Chemistry and Biology–Vol. II*, ed. J. Everse, K. E. Everse and M. B. Grisham, CRC Press, Boca Raton, 1991, pp. 25–50.
2 Y. Sekizawa, M. Haga, E. Hirabayashi, N. Takeuchi and Y. Takino, *Agric. Biol. Chem.*, 1987, **51**, 763; M. Haga, Y. Sakizawa, M. Ichikawa, H. Hiramatsu, A. Hamamoto, Y. Takino and M. Ameyama, *Agric. Biol. Chem.*, 1986, **50**, 1427.
3 A. Scheeline, personal communication.
4 A. C. Møller, M. J. B. Hauser and L. F. Olsen, *Biophys. Chem.*, 1998, **72**, 63.

Dr Münster said: In heterogeneous catalysis it has been known for a long time that periodic operation of a process may enhance the efficiency of the overall reaction. This observation is related to the suppressed accumulation of catalyst poison which ensures an increased catalytic activity under periodic reaction conditions. Could you please comment on the relation of these findings to the observations you reported in the paper?

Dr Hauser answered: It is indeed interesting that such analogies can be drawn between an enzymic reaction and a technical reaction system. In both cases the oscillatory behaviour protects the enzyme/catalyst from irreversible damage. However, there is a major difference between the two systems under consideration: while the sustained oscillations are an autonomous feature of the enzyme reaction, the periodic operation of technical reactors requires an externally imposed "oscillation", *i.e.* whoever runs the plant imposes periodic operation conditions. For further details on this issue, I should like to refer to the last paragraph of our paper where we discuss this issue in detail.

Prof. Marek commented: The experimental data have been obtained and a model constructed for a well stirred reactor. However, enzymes in living cells are produced and act with a specific spatiotemporal distribution. The time course of local concentrations will then be important. Could you please comment on this?

Dr Hauser replied: Peroxidase from horseradish is a glycoprotein which is mainly located in cell walls.[1] However, a fair amount of the enzyme is solubilized in the cytosol. We believe that our approach of a well stirred "cell" is a viable model to describe the peroxidase-based dynamics occuring in the cytosol.

I agree with Prof. Marek, that the spatial distribution of enzymes in the cell may indeed play a prominent role. I should like to mention that at Magdeburg we are currently immobilising peroxidases on membranes in order to study the spatiotemporal behaviour of the peroxidase–oxidase reaction.

1 A. Campa, in *Peroxidases in Chemistry and Biology–Vol. II*, ed. J. Everse, K. E. Everse and M. B. Grisham, CRC Press, Boca Raton, 1991, pp. 25–50.

Prof. Scott commented: The possibility of exploiting differences between competing dynamical attractors under the same operating conditions has been noted in a different situation elsewhere. Davies and Scott[1] have considered the situation of competing autocatalytic reactions in a CSTR which give rise to two possible products P_1 and P_2. For some conditions, a chaotic attractor can be established. The selectivity S, defined as $[P_1]/([P_1]+[P_2])$ is found to be relatively insensitive to whether the system is run in its autonomous chaotic state or if various periodic states are controlled using the SPF method. There is, however, a co-existing alternative period-1 state and by switching the system to this through appropriate perturbation, a significant change in S can be achieved.

1 M. L. Davies and S. K. Scott, *Chem. Eng. Sci.*, 2001, **56**, 4587.

Prof. Orbán opened the discussion of Dr Hauser's paper: In my opinion, the biochemical relevance of the reported pH oscillator is a bit overemphasised.

The IO_3^-–HSO_3^-–hemin system can be regarded as a 7th member of the family of two-substrate pH oscillators (the first one was developed in 1986).[1] The common substrate in these oscillators is HSO_3^-, the second substrate can be $S_2O_3^{2-}$, $Fe(CN)_6^{4-}$, HCO_3^-, even a biochemical species, hemin. The oscillatory cycle starts with the slow oxidation of HSO_3^- (*e.g.* by H_2O_2, IO_3^-, BrO_3^-) at high pH. This reaction speeds up as H^+ is produced (pH decreases) and it is completed when HSO_3^- is used up. The only role of all second substrates, including hemin, is to participate in a reaction which consumes H^+ when the first reaction terminates and to restore the pH to its original high value. A new cycle can start if HSO_3^- is replenished by flow.

The authors have observed differences in the oscillatory characteristics in the presence and in the absence of air and no explanation was given. The possible reason for the differences is the reaction of HSO_3^- with air oxygen which decreases the concentration of HSO_3^- and shifts the pH to lower values.

The system was tested for oscillations under batch conditions without success. I think a HSO_3^- containing oscillator can not work in batch. The oxidation of HSO_3^- to H_2SO_4 is autocatalytic in $[H^+]$ and once it starts, it proceeds until HSO_3^- is consumed completely. A new portion of HSO_3^- has to be supplied by CSTR for a new cycle to start.

1 E. C. Edblom, M. Orbán and I. R. Epstein, *J. Am. Chem. Soc.*, 1986, **108**, 2826.

Dr Hauser replied: I agree, and we state it in our paper, that the hemin system is a member of the family of pH oscillators that shares the oxidation of HSO_3^- by H_2O_2 as a characteristic feature. What makes the investigation of the hemin–H_2O_2–sulfite system worthwhile—instead of studying "just another" member of that family of pH oscillators—is that hemin is indeed considered as a model compound for enzymes whose reactive centres carry heme groups, like catalases, peroxidase, *etc.*

The important piece of information obtained from the experiments under argon atmosphere is that the reaction system hemin–H_2O_2–sulfite is an oscillator by itself. The differences in the oscillatory characteristics (period and amplitude) in the presence and absence of air can be explained by taking reactions of O_2 and atmospheric CO_2 into account. You just described the possible mechanistic role of oxygen in the hemin system; let me sketch that for CO_2: the reaction system composed of HSO_3^-, H_2O_2 and CO_2 has been shown to be a pH oscillator.[1–3] The role of CO_2 is to consume protons and thus to provide the feed-back step of the reaction mechanism. In experiments under aerobic conditions, not only is the hemin–H_2O_2–sulfite system active. Atmospheric CO_2 will dissolve and participate in the reaction. In addition, oxygen may also be active, as you suggest in your comment. Thus, the additional effects of CO_2 and O_2 are thought to be responsible for the differences between the dynamics observed under aerobic conditions and that obtained under argon atmosphere.

1 Gy. Rábai and I. Hanazaki, *J. Phys. Chem.*, 1996, **100**, 10 615.
2 G. A. Frerichs and R. C. Thompson, *J. Phys. Chem. A*, 1998, **102**, 8142.
3 Gy. Rábai, N. Okazaki and I. Hanazaki, *J. Phys. Chem. A*, 1999, **103**, 7224.

Mr Hantz said: Oscillatory chemical systems often lead to the formation of travelling waves when they are investigated at convection-free circumstances. Have you observed or do you expect such a behaviour?

Dr Hauser answered: So far, we have only studied the temporal behaviour of the hemin system in stirred reactors. However, it should be possible to observe some spatiotemporal behaviour when examining the reaction system in an extended, convection-free medium. The standard approach is to run the reaction in a gel or to immobilise hemin onto a support. In both cases the diffusion coefficient of the "activator" H^+ exceeds that of the "inhibitor" hemin, so that dynamic patterns such as fronts may be formed, while stationary Turing-type patterns are not expected to occur.

Prof. Scott asked: Have the authors developed their computational study based on their proposed mechanism and, in particular, do these reproduce the damped oscillatory batch behaviour?

Dr Hauser responded: Numerical simulations are currently under progress. We aim to assess the relevance of the H^+-consuming reactions (6) and (7), which provide the feed-back steps, for the reaction mechanism. Reactions (1)–(4) are used to model the autocatalytic oxidation of HSO_3^- by H_2O_2. When using reaction (6) as the exclusive feed-back step, we only observed periodic oscillations in the pH value of the medium (Fig. 1(a) overleaf). However, the dynamics become much more complex upon inclusion of reaction (7). Under appropriate values of the rate constants k_7 and k_{-7}, even complex bursting dynamics, such as the one shown in Fig. 1(e) of the paper can be simulated (Fig. 1(b) overleaf). Details on the dynamics shown by the reaction mechanism will be published elsewhere.[1]

So far, our main objective was to identify the roles played by the different H^+-consuming reactions in a CSTR. Simulations of the dynamical behaviour of the hemin system under semi-batch conditions has not yet been performed.

1 M. J. B. Hauser, N. Fricke, U. Storb and S. C. Müller, *Z. Phys. Chem.*, in press.

Prof. Westerhoff commented: Since in the abstract of this contribution you mention hemin as "a well known enzyme model compound", I should like to ask for some more specification of its catalytic role. You seem to be adding it to a final concentration of some 13 µM (or should I not take the dissolved fraction to equal the concentration of NaOH) or lower. This should be in keeping with the H^+ concentration not exceeding 1 µM. However, did you check to see if the reaction rates were proportional to the hemin concentration? Did increases in hemin concentration increase the oscillation frequency proportionally? I also noted that (pH) buffers were absent from your experiments, whereas they will be present in any biological setting, possibly interfering with any oscillations that depend on hemin. What was the effect of added pH buffers?

Dr Hauser responded: We have made measurements using 16–33 µM hemin; within this concentration range the periods of the oscillations did not show significant changes. As for other "reaction characteristics", *e.g.* rate constants, we have not yet performed kinetic experiments to determine them (if that is what you mean by "reaction rates") and their dependence on the hemin concentration.

The objective of the presented study is to demonstrate that the hemin–H_2O_2–sulfite system shows pH oscillations. Therefore, we omitted any additional buffers. However, I expect buffers

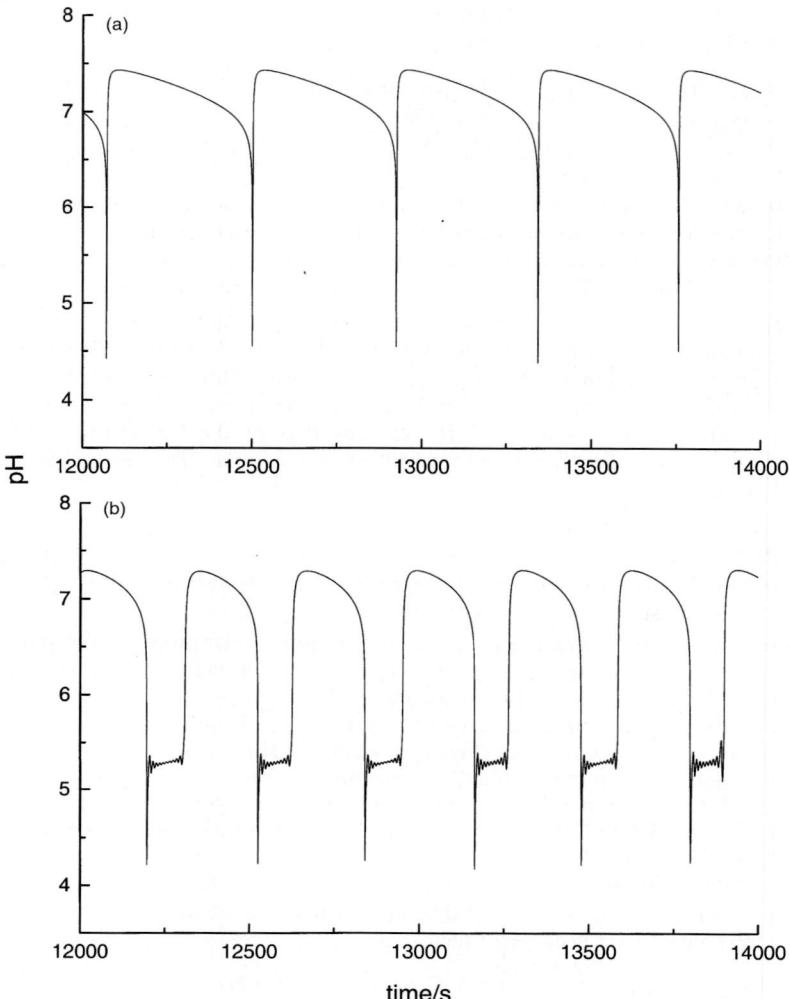

Fig. 1 Simulated time-series of the hemin–H_2O_2–sulfite reaction using the mechanism suggested in the paper. (a) Periodic oscillations obtained at a flow-rate $k_0 = 3.3 \times 10^{-4}$ s^{-1} and (b) bursting oscillations at rate $k_0 = 3.6 \times 10^{-4}$ s^{-1}. The rate constants used were $k_1 = 1.0 \times 10^{10}$ s^{-1}, $k_{-1} = 1.0 \times 10^3$ M^{-1} s^{-1}, $k_2 = 0.2$ M^{-1} s^{-1}, $k_3 = 1.5$ M^{-1} s^{-1}, $k_4 = 6.5 \times 10^6$ M^{-2} s^{-1}, $k_6 = 2.5 \times 10^4$ M^{-1} s^{-1}, $k_{-6} = 1.1 \times 10^{-2}$ s^{-1}, $k_7 = 1.0$ M^{-1} s^{-1}, and $k_{-7} = 1.0 \times 10^7$ M^{-1} s^{-1}.

that buffer at pH ∼6.5 to have a pronounced effect on the hemin system, depending on their concentration. But we have not yet looked into this question. In forthcoming experiments we shall take up your suggestion and investigate the effect of added buffers onto the hemin system.

Prof. Jonnalagadda commented: 1. Fig. 4 of the paper shows, due to bubbling of Ar under an inert atmosphere, that the period length and induction period are increasing. How do you explain this behaviour?

2. If expelling the CO_2 and O_2 are responsible for the longer induction and period length, does saturating the system with CO_2 reduce the induction times and shorten period lengths?

3. Does your model accommodate the above observed results?

Dr Hauser replied: 1. We assume that in the experiments under an inert argon atmosphere we observe the generic dynamics of the "pure" hemin–H_2O_2–sulfite system. The shorter periods observed in this system under aerobic conditions for otherwise identical reaction conditions are attributed to effects of CO_2 and O_2. On the one hand, oxygen may oxidise HSO_3^-, and hence reduce its concentration, as mentioned by Prof. Orbán. On the other hand, CO_2 has been shown to be actively involved in some pH oscillators, which also comprise the oxidation of sulfite by H_2O_2. In fact, the system composed of H_2O_2, sulfite, and CO_2 shows pH oscillations.[1–3] For our hemin system under aerobic conditions, this means that some of the atmospheric CO_2 will be solubilized in the reaction mixture, and than act as an additional source for H^+-consumption (along the lines of the arguments put forward in refs. 1–3). This provides an enforcement of the feed-back steps of the mechanism, that will lead to shorter oscillatory periods in the reaction under air.

2. We did not attempt to saturate our system with CO_2. Doing so would mean that we would change the reaction system under consideration from the hemin–H_2O_2–sulfite system (as discussed here) to the H_2O_2–sulfite–CO_2–system. The latter has already been intensively studied.[1–3]

3. Our reaction model does not accommodate for any effect induced by CO_2 or O_2. Our mechanism attempts to reproduce the experimental observations by a minimum number of chemical reactions, in order to provide a basic understanding of the underlying dynamics of the "pure" hemin–H_2O_2–sulfite system. For the hemin oscillator, CO_2 has been found to be a nonessential species, and it is therefore not included in our mechanism. However, the mechanistic contribution of CO_2 to oscillators of the H_2O_2–sulfite "family" is well established. Therefore an inclusion of the reactions involving CO_2 (as proposed in refs. 1–3) into our reaction mechanism can easily be done; such a modification should account for the effects of CO_2 in our hemin–H_2O_2–sulfite system.

1 Gy. Rábai and I. Hanazaki, *J. Phys. Chem.*, 1996, **100**, 10615.
2 G. A. Frerichs and R. C. Thompson, *J. Phys. Chem. A*, 1998, **102**, 8142.
3 Gy. Rábai, N. Okazaki and I. Hanazaki, *J. Phys. Chem. A*, 1999, **103**, 7224.

Prof. Scott said: There has been much recent interest in exploiting pH oscillator systems in biological systems for pulsatile drug delivery and perhaps also for chemo-mechanical systems. Would the present system suggest that it might be possible to produce an *in vivo* oscillator—perhaps requiring an alternative to H_2O_2 as the reductant?

Dr Hauser answered: *In vivo* biochemical pH oscillators are known, *e.g.* glycolytic oscillations are accompanied by pH oscillations[1,2] and waves,[3] and, recently, travelling pH fronts were also observed in neutrophil cells.[4]

Your question addresses the issue whether the hemin system—or modifications thereof—may be deliberately induced into *in vivo* systems to act as a pH oscillator for some desired, manipulative purpose. My guess is that modified versions of the hemin oscillator might indeed be candidates for this task. In its present form, the concentration of H_2O_2 used exceeds the levels usually tolerated by biological systems. Thus considerably lower concentrations of H_2O_2 are required to suit biological conditions. In addition, cells and tissues quite effectively remove the cytotoxic H_2O_2 by the enzymes catalase and peroxidase. Therefore, it seems desirable to find an alternative reductant to replace H_2O_2 in the biological context. Another important issue is to find out how the hemin oscillator can cope with isotonically buffered media. In any case, a good deal of research still has to be done, before this goal may be achieved.

1 B. Hess, A. Boiteux and J. Krüger, *Adv. Enzyme Regul.*, 1968, **7**, 149.
2 C. G. Hocker, I. R. Epstein, K. Kustin and K. Tornheim, *Biophys. Chem.*, 1994, **51**, 21.
3 T. Mair and S. C. Müller, *J. Biol. Chem.*, 1996, **271**, 627.
4 H. R. Petty, R. G. Worth and A. L. Kindzelskii, *Phys. Rev. Lett.*, **84**, 2754.

Prof. Westerhoff commented: You quite rightly said that in its present formulation the reaction system of this paper cannot occur in biology. One reason you give is the hydrogen peroxide concentration. Indeed you add an awful lot of H_2O_2 (some 10% I guess from the Experimental section of your paper). Could you not reduce the H_2O_2 concentration, yet maintain the rates of reactions (2)–(4) by adding catalysts (*e.g.* metal ions) such as may be present in biological systems?

Dr Hauser replied: The maximum concentration of H_2O_2 in the reactor (*i.e.* under the assumption that no reaction takes place) is 0.033 M (corresponding to a 0.1% w/w aqueous solution of H_2O_2). This is, indeed, an "awful lot" for biological systems. We are at present performing experiments to find out reaction conditions that allows us to obtain pH oscillations at (hopefully much) lower H_2O_2 concentrations. We hope to be able to do it without adding metal ions as catalysts, since metal ions may have other biologically undesired side effects.

Dr Wang said: In your experiments, the pH value oscillates around 7. However, in your numerical simulations, the pH value can barely reach above 7. Since the pH value of water is about 7, do you think that the discrepancy between experiments and numerical studies is because the function of water is underestimated in your model? Also, I think hydrogen ion is the autocatalysis in your system. If that is correct, can you comment about the possible impacts of introducing an additional buffer in your system to reversibly bind hydrogen ion?

Dr Hauser responded: In both, experiments and numerical simulations, the maximum pH of the oscillations attains a value of about 7.5 (Figs. 1 and 5 of the paper); here, experiments and simulations are in good agreement. There is also a good agreement in the line forms of the oscillations. However, the simulations predict peaks in the acidic region that are 1.5–2 pH units below the values measured in the experiments. This discrepancy can be explained by two factors: the transition from slightly alkaline to acidic domains is induced by the autocatalysis in H^+ which therefore is a very fast process. In addition, the system spends only a very brief time at the low pH peak. Given that the pH is measured by a glass electrode which has a response time of $\sim$5 s, this response time will tend to dampen very fast and highly transient dynamics. This is especially the case for the peaks in the acidic domain. Another explanation for the quantitative mismatch lies in the strategy adopted for the numerical modelling: we aim to describe the dynamics with a minimum number of reactions, in order to obtain a fundamental understanding of the reaction system. While we obtain good qualitative agreement between experiments and simulations, we "pay the price" of having quantitative mismatches by adopting our approach.

Concerning your question on the effect of additional buffers on the dynamics of the hemin oscillator, I should like to point out that we have not yet conducted such experiments. However, I expect buffers to have an effect on the dynamics of the hemin oscillator, provided that the rate constants of the protonation and deprotonation steps are in an adequate range. An example is equilibrium (7) of our reaction mechanism, which can act as a buffer. If we omit reaction (7), only periodic oscillations can be obtained in our calculations (similar to Fig. 1a of the paper). If it is considered, the simulations may yield more complex oscillatory behaviour, as shown in Fig. 1(b) of the paper. We have found out that the dynamics are highly sensitive to the value of the rate constants k_7 and k_{-7} of the deprotonation and protonation steps.

Prof. Westerhoff opened the discussion of Prof. Hanke's paper: In the 'Discussion' section of your paper you propose a mechanism for the effect of the electric field on the excitability. This is that the field translates into a current which again translates into a voltage drop of some mV across the plasma membrane of any cell. However, I should expect a drop on one side of the cell and an increase on the other side, relative to the preexisting plasma membrane potential. Therefore the total effect in this mechanism might be nil. I assume that your hypothesis contains some additional features not specified in your paper. What are they?

Prof. Hanke responded: In neuronal tissue, including the chicken retina, the cells have a very complicated geometry including highly specialized structures. According to this, the distribution of membrane proteins and thus functions is strictly inhomogeneous, although details about the topology are not that well known. Very often even the membrane potential is not homogeneous over the cell membrane. Consequently, it can be excluded that the mentioned effect will completely take place (at least, a partial inhomogeneity will remain).

Prof. Müller commented: You mention in the 'Results' section of your paper that gravity effects are likely to have in the spreading depression phenomenon, because there is evidence for gravitational influence on ion channels in model membranes. Could you specify what studies you are referring to?

Prof. Hanke replied: Detailed studies about this question have been carried out and are published. Some citations are given in refs. 1–5.

1 W. Hanke, *Adv. Space Res.*, 1995, **6/7**, 143.
2 N. Klinke, M. Goldermann, H. Rahmann and W. Hanke, *Space Forum*, 1998, **2**, 203.
3 N. Klinke, M. Goldermann and W. Hanke, *Acta Astronautica*, 2000, **47**, 771.
4 M. Goldermann, N. Klinke and W. Hanke, *Der Einfluß der Gravitation auf Künstliche Porenbildner*, in *Bilanzsymposium Forschung unter Weltraumbedingungen*, ed. M. H. Keller and P. R. Sahm, RWTH-Aachen, 2000, pp. 509–515.
5 M. Goldermann and W. Hanke, *J. Microgravity Sci. Technol.*, 2001, in press.

Mr Hantz asked: Did you observe a definite decrease in the excitability of the neuronal tissue when you applied a weak external perturbation on it? If so, can your experiments be considered as an example of stochastic resonance?

Prof. Hanke replied: We did first experiments according to changes of the excitability upon application of weak external forces and indeed found that it decreased in the absence of gravity (μ-gravity in parabolic flights). Changes of excitability induced by other forces or drugs were also found. Details are presently under investigation.

According to the, at present, unknown mechanisms underlying the excitability of neuronal tissue it seems difficult to consider our experiments as evidence of stochastic resonance, however, it cannot be excluded.

Prof. Gáspár commented: You said that there are differences in the effect of gravity on SD waves and BZ waves, and it is related to the different mechanisms underlying these waves. Could you please explain this in more detail.

Prof. Hanke answered: The BZ waves are mainly due to chemical reactions and diffusion- and convection-processes. The retinal SD waves are also due to diffusion of extracellular ions, but only over short distances (from one to the surrounding cells). Mainly, in the retinal SD changes of membrane parameters and a reversible redistribution of different ions and other compounds is involved. Recovery in the SD is also not due to chemical reactions but to membrane pumps and other membrane processes, driven by metabolic energy. Details about the mechanisms of SD can be found for example in ref. 1.

1 V. M. Fernandes de Lima, M. Goldermann and W. Hanke, *The Retinal Spreading Depression*, Shaker Verlag, Aachen, Germany, 1999.

Prof. Gáspár said: I would like to point out that inspite of the just explained differences in mechanisms of SD and BZ waves, there are situations when both waves can be characterised by the same rules. For example, we have found that the dispersion relation data of SD waves by Müller *et al.* fit to the universal dispersion relation we have observed for the BZ-waves. I think the reason for that is the single exponential decay in the recovery period in both excitable systems. Am I right?

Prof. Hanke responded: The data about the dispersion relation of SD waves investigated, at present, in detail according to the fit to theoretical curves to my knowledge are in agreement with this statement (the correct citation about the SD waves is not Müller et al. but Brand et al.[1]) However, more data have been published later (i.e. ref. 2) or are available in our lab, which have to be tested according to the above statement.

1 S. Brand, M. Dahlem, V. M. Fernandes de Lima, S. C. Müller and W. Hanke, *Int. J. Bifurcation Chaos*, 1992, **7**(6), 1359; the data were measured in our lab.
2 S. Brand, V. M. Fernandes de Lima and W. Hanke, *Naunyn-Schmiedberg's Arch. Pharmacol.*, 1998, **357**, 419.

Prof. Noszticzius commented: Experiments were carried out with isolated electrodes "to decide whether the effects measured were due to currents or to field induced polarization effects...". It is well known that no electric field can be established in an electric conductor without electric current and *vice versa*. Biological tissues are good ionic conductors and not insulators, consequently no electric field can be established in them with isolated electrodes. When high voltages up to 5000 V are applied even in this case the electric field will be limited to the insulating air gaps and will not penetrate into the tissue. Thus it is a correct result that "no reducible effects" were observed.

Prof. Hanke responded: The statement is correct, but the experiments being carried out as controls exclude artefacts in the range that the current experiments were done.

According to this, one should also have in mind the complex structure of biological tissue, being composed of equivalent circuits of resistors, capacities and gradients of ions and of batteries (inductivities). Thus the real field distribution on a small scale in space is anyhow difficult to obtain.

Dr Hauser said: In Fig. 8 of the paper the effect of increased gravity on the velocity of rSD waves is shown. While the relative wave velocities increase continuously for horizontally and downwards travelling waves, propagating upwards present a significant jump at ~ 3 g. Do you have any explanation for that "sudden" increase in wave velocity in the upward motion?

Prof. Hanke replied: This increase can be fitted in the simplest approach by a sigmoid curve. We found a comparable result in the dependency of the mean open-state probability of ion channels under the influence of gravity.[1] According to the fact that properties of ion-channels are important parameters in the behaviour of SD waves, this might be a hint for an explanation.

1 M. Golderman and W. Hanke, *J. Microgravity Sci. Technol.*, 2001, in press.

Prof. Epstein commented: In our studies of gravity effects in chemical systems, Nagypál et al.[1] showed that such effects must arise from a change in density, which may result either from a temperature change or from a change in composition. Since temperature changes seem unlikely in this system, can you suggest any possible origin of composition-based density change in spreading depression? Is it possible, for example, that conformational changes in the opening of membrane channels expose charges that more tightly attract neighboring water molecules, thereby increasing the density?

1 I. Nagypál, G. Bazsa and I. R. Epstein, *J. Am. Chem. Soc.*, 1986, **108**, 3635.

Prof. Hanke responded: Concomitant with the SD waves is a complete redistribution of, for example, the ion gradients in the tissue. According to this, at least local changes in density of the intra- and extracellular space solutions can adequately take place. However, other effects of gravity must also be taken into account, for example direct changes of the parameters of ion channels as has been demonstrated in ref. 1.

1 M. Goldermann and W. Hanke, *J. Microgravity Sci. Technol.*, 2001, in press.

Dr Danø said: I would like to ask you two questions. First, what measures have you taken to assure yourself that the effects seen at high g forces are indeed gravitational effects and not merely a result of mechanical deformation of the sample?

Secondly, a more general question. As seen from the discussion here, it is very hard to say anything certain about the mechanism of the gravitational effect. I believe that the mechanisms by which pharmacological manipulations affect the nervous tissue are much better understood. Have you tried to use such tools in your studies?

Prof. Hanke replied: 1. In our video recordings of the eye-cups in the centrifuge up to about 8 g we did not find evidence of significant mechanical deformations (we are using a closed chamber system, which is especially designed to minimize mechanical artefacts). Thus, I think mechanical artefacts can be excluded.

2. At least the direct interaction of gravity with ion-channels[1] might be a mechanism. However, it is certainly true that the pharmacological control of SD waves has been studied to a much greater extent, for example, due to medical relevance (just to give one example from our group see ref. 2).

Nevertheless, to understand the mechanisms of SD waves, it is useful to know its dependency on weak external forces. The effects of electromagnetic fields on SD waves are furthermore of some more general interest due to the question of electrosmog and its interaction with neuronal tissue, here using the retinal SD as a tool to investigate this question.

1   M. Goldermann and W. Hanke, *J. Microgravity Sci. Technol.*, 2001, in press.
2   M. Wiedemann, V. M. Fernandes de Lima and W. Hanke, *Nauny's-Schmiedeberg's Arch. Pharmacol.*, 1996, **353**, 1.

Prof. Westerhoff opened the discussion of Prof. Müller's paper: 1. As much as I like the beautiful work you presented, I should like to take issue with the oversimplified view of glycolysis you gave us, understandably, at the beginning of your oral presentation. Granted, simple models can have the important purpose of illustrating general principles or of showing phenomena that might happen. However, by now we know what might happen in glycolysis and the purpose of models should be to help to verify/falsify our hypothesis concerning glycolysis. Consequently they should be more realistic. Understanding of glycolysis has proceeded such that PFK is not the sole pace maker. Because the experimental biochemists know this, they may not take us seriously enough if we continue to work with oversimplified and falsified models. The same problem reigns in the field of development biology, where principles of self-organisation are no longer taken seriously because they continue to be illustrated with falsified models.

2. In a similar vein we should be more precise when discussing the relevance of the extract system for the actual biology of the living cell. The extract has everything diluted by a factor of perhaps 5, which should reduce the frequency and affect the wavelength. Indeed the extract frequency of 12 min is some 15 times lower than the *in vivo* frequency. And, the wavelength of 2 nm by far exceeds cellular dimensions. I think that your addition of Mg^{2+} simply compensates for this dilution effect. Should you not actually add more of all cofactors and coreagents (*e.g.* NAD, ATP)?

Prof. Müller responded: 1. There are two parts to this question: cellular metabolism is indeed a complex network and the use of simplified models may be regarded with some criticism, if one wants to describe the manifold interactions and dynamics. However, from a viewpoint of self-organization we investigate the basic system properties that result from autocatalytic reactions within a reaction sequence, *i.e.* PFK activity in glycolysis. Experimental as well as theoretical results with other autocatalytic systems from physics, chemistry or biology have shown that they have some basic properties in common, that depend on the systems state. We want to unravel the possible function of these properties for biological information processing. In agreement with the well known principles of self-organization, our results demonstrate that, in fact, the PFK plays a prominent role for control of the dynamics of glycolytic reaction/diffusion waves, as evidenced by the effects of adenine nucleotides and F-2,6-P_2 on the wave dynamics.

The use of simplified models does not mean that these models are oversimplified. Goldbeter and his group have used a simplified model of glycolysis, based on PFK-dynamics, and could reproduce all known oscillatory dynamics of glycolysis. Moreover, by the early seventies they could already predict the occurrence of glycolytic waves in a yeast extract. Twenty years later we confirmed the theoretical results by our experiments with a yeast extract and found an excellent agreement regarding wave velocity and wavelength, providing evidence that the model is neither oversimplified nor wrong. One should take into consideration that the kind of spatio-temporal pattern formation considered does underly regulatory principles which cannot be explained by enzyme-to-enzyme interactions (manifold interactions of many enzymes), but by only a few mechanisms based on autocatalysis and reaction/diffusion coupling. It is just this universal principle which offers new horizons for biological informations processing. In this sense, PFK plays a prominent role for control.

2. The relevance of experiments with yeast extract is based on unravelling the particular mechanisms for glycolytic self-organization. Such experiments cannot be done with living cells, because cellular control mechanisms often compensate for experimental manipulations that are required for mechanistic studies. For example, overexpression of a single glycolytic enzyme in yeast cells is very difficult, because this leads to the concomitant increased expression of other glycolytic enzymes, and hence does not allow one to investigate the mechanistic meaning of a single enzyme in living cells. Changing ATP concentrations would affect PFK activity and hence the dynamics of the patterns.

Prof. Hunding asked: Is it possible to create spatial Turing patterns with the glycolytic oscillator coupled to a gel? Coupling of an oscillator to a gel was the method used when the CIMA reaction was allowed to create stationary spatial patterns. John Pearson worked out the theory. The characteristic Turing wavenumber

$$k^2 = \frac{a_{11}D_2 + a_{22}D_1}{2D_1D_2}$$

is not calculated using an effective diffusion coefficient due to a slowing down of activator effective diffusion by coupling to the gel. Rather it is the actual D's, which may be almost equal, from which Pearson[1] derives k^2 proportional to Tr/D. Since Tr is related to the period of the oscillation, it should be possible to get a rough estimate of the expected Turing wavelength.

1 J. E. Pearson, *Physica A*, 1992, **188**, 178.

Prof. Müller answered: This is a very good idea and there are already experimental efforts under way to look for spatial Turing patterns in glycolysis. In collaboration with P. de Kepper and E. Dulos from Bordeaux, we have constructed an open spatial reactor in order to investigate patterns in a gel-fixed yeast extract. In fact, oscillations can be induced in the gel system, but many problems arise that are related to the transport properties of the membrane separating the gel layer from the substrate supplies. We have been able to solve most of these problems and expect that Turing patterns of the requested kind will be observed in the near future. No prediction has been done, however, on the expected Turing wavelength.

Dr Satnoianu commented: The idea of the authors that the wave in NADH could be used as an information messenger is very provocative and interesting. It reminds me of behaviour known in smaller organisms, *e.g.* bacteria which sometimes secrete various chemoattractants to coordinate their population behaviour. In this context it is of interest to know how the wave characteristics such as wave speed, amplitude or wavelengths depend on the amount of initial stimulus to the cells, for example the amount of sugar used. I believe that there is an optimal amount of initial sugar for example that would give the maximum response for the yeast cells. Beyond this concentration of sugar in water there is probable a more limited stimulus to the cells. This could be extended to understand what type of stimulus is able to give the maximum cellular response and so might be used to increase the efficiency of industrial processes involving yeast production.

Dr Danø commented: when considering intact yeast cells and their biotechnological use, it is of course important to realise that sugars mainly play the role of substrates rather than "signal modifiers". Still, it is well documented that the glucose concentration plays a major role in the regulation of yeast kinetics, especially *via* glucose repression of a large number of genes. Therefore, it is clear that an improvement of the detailed understanding of yeast cell kinetics will be of biotechnological importance.

Prof. Müller said: your comments about the possible functions and industrial applications of glycolytic waves correspond exactly to the goal of our research. In fact, it would be important to test which factors (including glucose) change the wave characteristics. However, such experiments can only be done in an open spatial reactor (just mentioned in the previous discussion), because only then is it possible to adjust the input concentrations to the desired values, *i.e.* to prevent an overflow of glycolysis with glucose.

Prof. Nicolis commented: One way to explore the relevance of the results of this paper in living cells would be to carry out experiments on cell extracts embedded in a heterogeneous medium constituted, for instance, of micelles, Langmuir–Blodgett films, or membranes. This would allow one to assess the robustness of the oscillations and of the spatio-temporal patterns toward the size and the geometry of the microreactor. Non-trivial effects induced by the spatial dimensionality and the coordination number of the support on which nonlinear kinetic processes take place can indeed be expected, as shown recently both experimentally[1] and on a number of representative model systems.[2,3]

1 R. Kopelman, *Science,* 1988, **241**, 1620.
2 R. Ziff, E. Gulari and Y. Barshad, *Phys. Rev. Lett.*, 1986, **56**, 2553.
3 A. Tretyakov, A. Provata and G. Nicolis, *J. Phys. Chem.*, 1995, **99**, 2770; F. Baras, F. Vikas and G. Nicolis, *Phys. Rev. E*, 1999, **60**, 3797.

Prof. Müller replied: In fact, spatial heterogeneity is a common property of biological systems and should have significant impacts on the pattern dynamics. We have already incorporated the yeast extract in hydrogels, where we can induce spatial inhomogeneity. These experiments are in progress.

Prof. Jonnalagadda said: In the mid seventies a lot of work was done by the research groups of Prof. Augustin Betz[1] and Benno Hess[2] on the cell free extracts of Yeast. Yeast extracts can be frozen in liquid N_2 and thawed even after 2 months to conduct experiments. The extract's remain perfect showing signature oscillations upon warming.

As Ca^{2+} ions play a vital role in the living cells when Mg^{2+} is deficient, it will be interesting to look at the response of PFK to the added Ca^{2+}. How about the response of added Na^+ analogues to K^+?

1 S. B. Jonnalagadda, J. U. Becker, E. E. Selkov and E. Betz, *Biosystems*, 1982, **15**, 49.
2 B. Hess and A. Boiteux, *Biological and Biochemical Oscillators*, ed. B. Chance, E. K. Pye, A. K. Ghosh and B. Hess, Academic Press, 1973, pp. 229–241.

Prof. Müller responded: Calcium does not have important regulatory functions on yeast PFK. However, for PFK from animal tissues, it has been shown that calcium together with calmodulin can change the PFK activity. This results probably from phosphorylation of PFK by calmodulin-dependent protein kinases.

So far, we have not tested the effect of sodium on glycolytic oscillations. But the ion-dependent glycolytic enzymes require potassium and not sodium for optimal activity.

Prof. Marek commented: The possibility of transferring the results of observations on nonlinear dynamics of glycolytic oscillations and waves from the yeast extract to observations on yeast cells with critically depend on the presence of transport resistances at the level wall/membrane and within the cell itself. If significant transport resistances exist then the direct transfer of results is difficult. Could you comment?

Prof. Müller replied: Yeast cells display glycolytic oscillations when fed with external glucose. Hence, there is no transport resistance at the level of the cell wall or membrane, which prevents oscillatory behaviour. Whether intracellular transport resistances prevent the formation of traveling waves is not known. However, experiments with other cells of similar size (neutrophil cells, length 10 μm, see ref. 28 in our paper) demonstrate that travelling NADH and proton waves develop as the result of oscillatory glycolysis. Therefore, we are confident to demonstrate similar wave phenomena also in yeast cells.

Dr Hauser opened the discussion of Dr Danø's paper: When elaborating your detailed model, you use a combinatorial approach. How unique is the model you obtain, bearing in mind that you tested hundreds of millions of permutations? If it is not unique, which are the criteria you applied to discriminate between "realistic" and "non-realistic" solutions?

Dr Danø responded: This is a very relevant question. As you point out, we search parameter space for parameter combinations which result in a model that agrees with as many experimental observations as possible. With this procedure, we are faced with the fact that parameter space is so high-dimensional that one cannot possibly scan all of it densely, even if one used the fastest computers available. Therefore, we cannot guarantee that the model (*i.e.* point in parameter space) which we arrive at is the best one possible. In that sense, our model is not unique. On the other hand, it is not just an arbitrary choice among many good candidates. In fact, models with even a limited agreement with experimental findings are very rare in parameter space. The present model represents several months of improvements starting from such a model.

By "realistic model" we mean a model which is based on detailed biochemical kinetics and which has as many properties and mechanistic parameters in quantitative agreement with the experimental system as possible. By construction, the candidate models all have stationary concentrations and size and distribution of the metabolic flux in agreement with experiments. For each model, its "realism" is validated by comparing a range of properties with experiments: presence, location and direction of a supercritical Hopf bifurcation, frequency of oscillations, quenching data for acetaldehyde and glucose, relative amplitudes of 13 metabolites, phases of 9 metabolites, and 24 enzyme kinetic parameters.

Prof. Menzinger commented: The preceding discussion touched upon the issue of "establishing" the detailed biochemical mechanism of glycolytic oscillations. Let us remember the Damocles' sword of kinetic modeling: a given set of experimental kinetic data is generally consistent with a large number of mechanisms. Therefore a given mechanism is never unique and can never be proved but it can be only *falsified* by additional experiments. Researchers dealing with mechanistic modeling, *i.e.* many of us, are in the unenviable position that their brain-children are generally only *valid until further notice*.

A well known example is the $H_2 + I_2 \rightarrow 2HI$ reaction which was for many years "the" textbook example of an elementary bimolecular reaction—until a generation of further experiments by Rosenbaum and Hogness (1934), Benson and Srinivasan (1955) and Sullivan (1959, 1962) unmasked it as a complex chain reaction involving H and I radical intermediates. For a review see Laidler.[1]

1 K. Laidler, *Chemical Kinetics*, McGraw Hill, New York, 2nd edn., 1965, pp. 116ff.

Prof. Sørensen replied: The primary purpose of the optimization calculus was to find a set of parameters which at the Hopf point for the yeast cells is consistent with all available stationary and dynamic experimental measurements as well as with known rate expressions and Michaelis parameters for the glycolytic enzymes. This we have accomplished to an extent that it is no longer clear if the remaining discrepancies are due to omitted reaction steps, inappropriate rate expressions, or experimental errors. To obtain a better model we need more and better experimental data preferably at a different operating point.

Prof. Müller commented: When comparing model and measurements, it would be interesting, of course, to have experimental data on single cell behaviour, because the model in Fig. 6 of the paper has been developed for one cell only, whereas in the experiments one looks at synchronized cell populations. The agents used for synchronization (cyanide) could, after all, have some influence on the single cell behaviour, for example, the transition from aerobic to anaerobic metabolism *via* inhibition of mitochondrial respiration by cyanide.

Prof. Westerhoff responded: I am pleased that you bring up the question whether PFK is crucial for glycolytic dynamics. This issue is close to the heart of our yeast research now. I think that here the problem is perhaps less in finding answers, as it is in defining the question. What do we mean with the question "is PFK important for glycolytic oscillations?" What should be the operational definition of the question? If we take the question to mean that the oscillations stop when PFK is eliminated from the pathway, then PFK is as crucial as hexokinase and glucose transport (or trehalase). If we take it to mean that modulation of the amount of PFK should strongly affect the frequency of the oscillations, then recent experimental work[1] shows that glucose transport may well be more important than PFK. Then there is the issue that the importance of various enzymes for the frequency of the oscillations may differ from their importance for the amplitude and waveform.[2]

1 Reijenga *et al.*, *Biophys. J.*, 2001, **80**, 626.
2 Reijenga *et al.*, *Biophys. J.*, 2002, **81**, in the press.

Prof. Müller said: I agree with you. Indeed the question about "importance" of an enzyme for control is a matter of definition. It is known since more than 30 years, that the input rate of glucose into the yeast extract (*i.e.* the transport) seriously changes amplitude and frequency of glycolytic oscillations. But this only demonstrates that there is a certain "window" for glucose transport that supports and affects the oscillations. But the same studies also demonstrate that there is an autonomous oscillator in glycolysis based on autocatalytic reaction of PFK.

One can entrain the frequency of the autonomous oscillator by the transport characteristics but one can not replace it. Therefore, transport is an important effector but not generator of glycolytic oscillations. For generation of oscillations, an autocatalytic reaction is required.

Prof. Nicolis commented: Significant evidence of the pivotal role of phosphofructokinase and pyruvate kinase in glycolysis is provided by early work of Hess,[1] who deduced from experimental data the chemical potential differences across the different enzymes involved in the pathway (under steady-state conditions). His main result is that the chemical potential difference is practically concentrated in the vicinity of these two regulatory enzymes, whereas in the part of the pathway between these two enzymes there is a "plateau" of chemical potential.

1 B. Hess, in *Energy Transformation in Biological Systems*, Elsevier, Amsterdam, 1975.

Prof. Westerhoff responded: Why should relative distance from equilibrium define the importance of a step since the oscillations PFK, HK, PK are all far from equilibrium and glucose transport, which does control much of the frequency, is much closer to equilibrium.

Dr Schreiber commented: I would like to point out that there is a methodology of categorization of mechanisms and classification of species involved, which is also based on stoichiometric network analysis, and which can help in discriminating between alternative pathways such as involving PFK or not in the glycolysis.

Dr Hauser added: I would suggest use of the stoichiometric analysis technique mentioned by Dr Schreiber to assess the quality of different reduced and extended mechanisms proposed for glycolysis. It would be very interesting to determine how many of the enzymes are "essential" and, in particular, if PFK is one of them.

Prof. Nicolis opened the discussion of Prof. Harrison's and Prof. Hunding's papers: A natural way to model the effects of growth on pattern formation, explored systematically in the 1970s, is to use the system's length L as the principal bifurcation parameter and to identify the new stable (Turing-like) structures that become available for successively higher L values.[1,2] One can also argue that growth induces a ramp in some control parameter, under the effect of which a previously existing pattern is subsequently modified.[3,4] Can you comment on the new insights added by your approaches? Do you predict new effects that cannot be accounted for by this earlier mechanism?

1   A. Babloyantz and J. Hiernaux, *Proc. Natl. Acad. Sci. USA*, 1974, **71**, 1530; A. Babloyantz and J. Hiernaux, *Bull. Math. Biol.*, 1975, **37**, 637.
2   S. Kauffman, R. Shymko and K. Trabert, *Science*, 1978, **199**, 259.
3   M. Herschkowitz-Kaufman and G. Nicolis, *J. Chem. Phys.*, 1972, **56**, 1890.
4   G. Dewel, P. Borckmans, A. DeWit, B. Rudovics, J.-J. Perraud, E. Dulos, J. Boissonade and P. De Kepper, *Physica A*, 1995, **213**, 181.

Prof. Harrison responded: Yes, indeed, the dependence of pattern on a size parameter of the system has been a major feature also of the work in my group since the 1970s, including my earliest publications with Thurston Lacalli (refs. 13 and 30 of the paper) and all my work on the patterns of variable numbers of hairs in the vegetative whorls of *Acetabularia* (refs. 14–16 of the paper). Our work concentrates more on the application of existing theory to biological examples than on additional features of the well-known mechanisms. Nevertheless, section 3.1 of the paper describes something which I believe has not been reported previously regarding pattern selection governed by the radius of a hemispherical shell domain with Dirichlet boundary conditions: at the radius optimal for spherical shell harmonics of $l = 3$, annular and dichotomously branching patterns are selected with about 20/80 probability in a series of computations. This is important in relation to the applicability of reaction–diffusion to a wide range of processes in plant development.

In the other part of our work (results reported in section 3.2), we do indeed report effects that cannot be accounted for by the earlier mechanisms you mention. Instances of tip growth and of repeated branching processes in a system undergoing continuous growth require additions to the chemical mechanisms, such as the Brusselator, to allow the morphogenetic region to modify its own boundaries, as developed in earlier two-dimensional work in my group (refs. 2 and 17 of the paper).

For the kinds of developmental process that are common in plants I have generally found the Brusselator to be one of the most promising reaction–diffusion models to use in its original form or to build on with additional features.

Prof. Maini commented: As you know, the standard reaction–diffusion equation $\partial u/\partial t = D\nabla^2 u + f(u)$ is formulated on a fixed domain. On a growing domain, it takes the form $\partial u/\partial t + \nabla \cdot (\underline{a}u) = D\nabla^2 u + f(u)$ where $\underline{a}$ is the flow velocity field generated by domain growth. How do you account for this in your computations?

Prof. Harrison replied: In the earliest 2D work done in my group on a growing domain (ref. 17 of the paper) we used a "dilution term" for the spreading of existing material over a larger domain, which in effect I think makes some approximation to your "flow velocity field" term. Our assumption, governing choice of parameter values, has been that reaction–diffusion kinetics are very fast compared to the rate of growth of the domain, so that such terms would have very little effect on the reaction–diffusion patterning. We have not used them yet in the 3D work. There is, however, a suggestion in our paper that plants may select dichotomous branching robustly by expressing a reaction–diffusion pattern that has not yet reached steady state. If the pursuit of this concept leads us to choose parameter values for reaction–diffusion and growth more closely matched to each other, we should probably have to use your flow velocity field term.

Prof. Menzinger commented: The early approaches to pattern formation on a growing domain cited by Prof. Nicolis imply an infinitely slow growth and the *adiabatic adjustment* of the pattern to system size. At the finite growth rate considered here, the system is either absolutely stable (passive medium), linearly absolutely or convectively unstable (Turing medium) or nonlinearly absolutely or convectively unstable (bistable medium). Whether the system is absolutely or convectively unstable depends on the relative magnitudes of the growth rate ϕ and the intrinsic spreading rate c_0 of a locally applied perturbation. The $\phi = 0$ cases cited by Prof. Nicolis as well as that mentioned by Prof. Harrison in the preceding comment dealt with *absolute instability* ($\phi < c_0$ in eqn. (2) of our paper) where the spreading pattern is selected from the system's intrinsic modes and its initial conditions. Such absolutely unstable systems are insensitive to forcing at the growth boundary. In contrast to this, the *convectively unstable* media considered in our paper *are extremely sensitive* to periodic or aperiodic boundary forcing. They allow a temporal, local mode at the growth boundary to be converted into a spatial, global mode. Under certain conditions (see Fig. 4a of our paper) this spatial mode is stable and may be selected by the system. In other words, a *localized segmental clock* may act as the system's *organizing center*.

Prof. Maini said: This concerns Prof. Nicolis' comment on growing domains. Although the embryo is of course growing, the presomitic mesoderm, the domain in which somite pattern formation occurs, stays approximately constant in length and simply moves down the head–tail axis of the embryo. This, therefore, is not really a growing domain problem.

Prof. Hunding said: Since Prof. Nicolis addressed his opening question to both papers under discussion, I wish to take this up again. It is fully possible (at least in one dimensional simulations) to make stripes, Turing style, by increasing the domain or changing a gradient. A recent example and source of references is the paper by Crampin *et al.*[1] Here you double the number of stripes by doubling the domain size.

However this kind of pattern formation does not address the recent experiments on gene expression waves. What is seen in the biological experiments is an entirely different mechanism, with waves originating from the growth zone, initially as broad bands which narrow and become quasistationary. This phenomenon was not known, and not addressed, in all the works mentioned above from before 1997. These gene expression waves are explained in our studies as phase waves. In the extension presented here, we show that the size of the segments (the wave length of the emerging spatial pattern) is set by the period of the forcing oscillator in the growth zone times the growth rate. This is found both for a medium of oscillators with increasing oscillating period, or for initial oscillators, which change to bistable or Turing systems. Specifically, we show that the wavelength in the Turing case is *not* given by what is known as the intrinsic, or critical, Turing wavelength. Rather, the wavelength is set by the above relation (forcing oscillator period times growth rate), provided this wavelength is within the excitable Turing wavelength range.

1   E. J. Crampin, E. A. Gaffney and P. K. Maini, *Bull. Math. Biol.*, 1999, **61**(6), 1093.

Prof. Nicolis responded: This clarification comes right to the point. I suppose that one is then entitled to argue that the effects of the flow are similar to those of the ramp, and that the new ingredient that makes the difference is the local oscillation in the growth region.

Dr Kærn said: Our work opens up a number of new approaches to pattern formation in biology. The work referred to by Prof. Nicolis, addresses *isotropically* growing domains where the flow-term, which inevitably arises as the result of domain growth,[1] can be eliminated. Such systems are adequately modeled by reaction–diffusion systems with zero-flux boundary conditions. In *anisotropically* growing systems, on the other hand, the flow-term cannot be eliminated. As a result, for growing systems in general, the chemical basis of morphogenesis is to be sought in reaction–diffusion–advection equations. With a noticeable exception given in ref. 2, the effect of the flow-term has been largely ignored in earlier work. Furthermore, in the systems that we consider, there is a continuous influx of medium (cells) across the domain boundary. For obvious reasons, a no-flux boundary condition does not apply. The system is inevitably subject to a boundary forcing and the state of the cells when they leave the growth zone and enters the growing tissue *must* be considered explicitly. Due to the equivalence of the growth-mode that we consider to an open flow, linearly or nonlinearly convective *versus* absolute unstable flow conditions become an issue. In the convectively unstable regime, the boundary forcing allows for a local, flexible and very powerful control of the global segmental pattern (see the previous comment by Prof. Hunding). Such control is not possible in the models referred to by Prof. Nicolis. Finally, our mechanism works not only in the Turing regime but also in bistable, excitable and even in passive media.

1 E. J. Crampin, E. A. Gaffney and P. K. Maini, *Bull. Math. Biol.*, 1999, **61**, 1093.
2 K. J. Painter, P. K. Maini and H. G. Othmer, *Proc. Natl. Acad. Sci. USA*, 1999, **96**, 5549.

Prof. Menzinger commented: As Fig. 2(a) of Prof. Harrison's paper shows, the early development of *Acetabularia* involves the axial growth of a stalk—presumably from a growth-zone located at the apex of the structure—which goes hand in hand with the periodic formation of vegetative whorls that fall off and give rise to "scars".

This scenario is really the same described in our paper on the segmentation of axially growing structures. It involves axial growth ϕ that originates at the apical growth zone, in addition to a coherent, time-periodic process with period T' in the growth zone. Suppose that no autonomous oscillations occur in the growing stalk, i.e. $T' = \infty$, eqn. (5) predicts a velocity of the resulting gene expression waves $c = \phi$ relative to the growth tip, *i.e.* stationary structures relative to the cells in the stalk, and a wavelength or distance between the whorls or scars given by $\lambda = \phi T'$. Since the growth is inherent and the whorls arise at certain time-intervals it is inescapable that they are induced by this simple growth-and-clock-induced kinematic mechanism. A similar mechanism is operative in plants, such as bamboo.

Prof. Harrison responded: Yes, Fig. 2(a) of our paper indeed indicates a succession of "nodes" and "internodes" along the growing stalk somewhat analogous to the formation of regions given those names along the stalk of a multicellular plant such as bamboo. But perhaps my diagrams have given you a suggestion of regular time-periodicity that is not what usually happens in the growth of *Acetabularia*. In an informal discussion with me, you indicated a rotation around the stages of our Fig. 2(c) at a constant rate. It is in fact usually far from constant. The interval between formation of one whorl and the next is quite variable in the widely-studied *A. acetabulum*. Something more like the regularity you envisage may occur in *A. caliculus*,[1] but I don't think this species has been extensively studied in regard to developmental details.

If *A. acetabulum* is grown in red light and then given a few minutes of exposure to blue light, whorl formation begins (*i.e.* Fig. 2(c) of our paper, stage A, annulus) 12 h afterwards, reproducibly.[2] The identity of the photoreceptor is not known. I am inclined to think that whorl initiation probably involves events outside the growing tip that control the supply of substrate S (Fig. 2(c)) or enzymes that govern the rate constants of the patterning mechanism. These events may be not only outside the growing tip, but sometimes outside the organism, *e.g.* the supply of light.

1 S. Berger and M. J. Kaever, *Dasycladales, An Illustrated Monograph of a Fascinating Algal Order*, Georg Thieme Verlag, Stuttgart and New York, 1992, p. 155, Fig. 3.67(a).
2 R. Schmid, E. M. Idziak and M. Tünnermann, *Planta*, 1987, **171**, 96.

Prof. Merkin said: This is a question to Prof. Hunding about tying the growth model to the flow model. The flow model is an open system with the flow rate ϕ determined and controlled by the experiment. How is the flow rate for the growth model determined? Does it come as part of the solution, with then some extra condition in the model needed to fix it, or is ϕ set by some external 'forcing'?

A second point concerns the possible amplification of any 'noise' in the input. This arises in the flow model from slight oscillations in the pump, for example, and must be present in the growth model. These systems which develop spatio-temporal patterns through a convective instability are strong amplifiers of 'noisy' conditions. Could this have a significant effect on the patterns formed in the growth model? There is a suggestion (worry?) that patterns seen in our experiments on the DIFICI in a BZ system could well be noise-generated structures.

Prof. Hunding responded: Regarding the first part of your question: The growth rate ϕ and the period T' of the posterior forcing oscillator are known from the biological experiments. T' is known to be about 90 min in the chick. It is an experimental observation that one somite (with length a, say) is added anteriorly in the same time span. Since the PSM is constant, the posterior growth zone thus advances with an observed growth rate $\phi = a/T'$.

Our theory predicts a (quasi)stationary spatial pattern with wavelength $\lambda = \phi \times T'$. Using the experimental observables, we obtain $\lambda = (a/T') \times T' = a$ that is exactly the length of one somite. This not a circular argument. If the theory had predicted a wavelength of, say, $\lambda = \pi \times \phi \times T'$, we would have found a predicted somite length different from the one observed.

Dr Kærn commented: Prof. Merkin raised the important question about the robustness of FDS waves to boundary noise. As pointed out, the growing system is in the convectively unstable regime a strong noise-amplifier, since the temporal periodicities in the boundary noise imposes a spectrum of spatial modes onto the flowing medium. The modes that have positive growth-rates are then amplified as the medium is carried downstream. This gives rise to noise-generated structures.

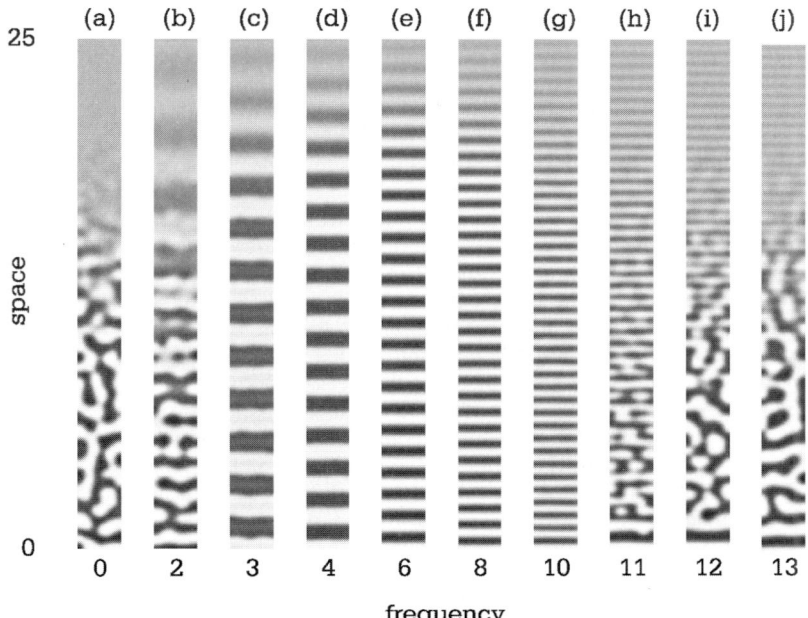

Fig. 2 The effect of boundary- and velocity-additive noise on Turing-type FDS waves. Panel (a) shows the noise-sustained Turing structure that arises in the absence of boundary forcing. The remaining panels show the patterns that arise with the indicated forcing frequencies and amplitude $A = 0.1$. Parameter values are: $\alpha = 1.2$, $\beta = 0$, $\gamma = 0.9$, $D_u = 0.005$, $D_v = 25 D_u$.

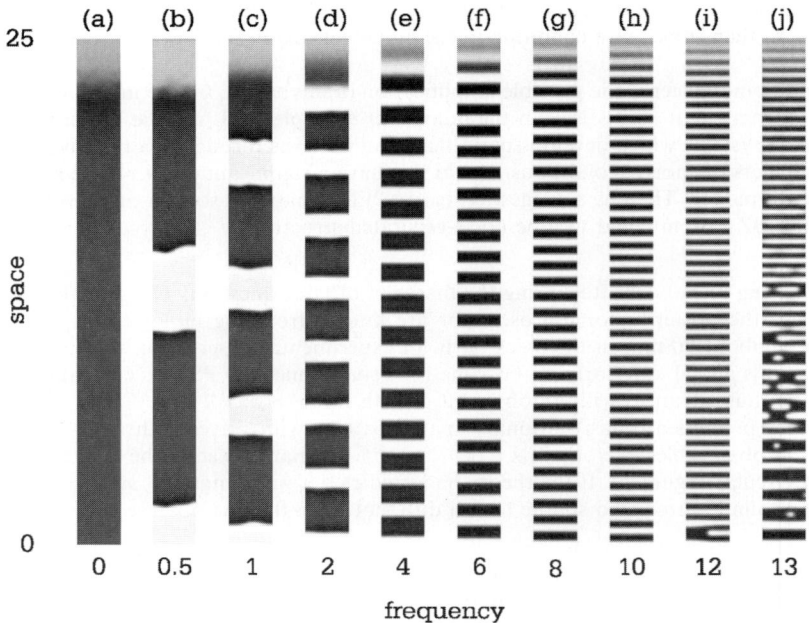

Fig. 3 FDS waves of the interface type in the presence of boundary- and velocity-additive noise. No patterns are formed in (a) where the amplitude of the boundary forcing is zero. The remaining panels shows the effect of periodic boundary forcing at the indicated frequencies and amplitude $A = 0.2$. Parameter values are: $\alpha = 0.5$, $\beta = 0.05$, $\gamma = 2$, $D_u = 0.005$, $D_v = 25 D_u$.

Hence, in a real system, where boundary fluctuations are inevitably present, there is a competition between modes is imposed by the boundary noise and the mode imposed by the boundary forcing. It is very likely that the DIFICI waves observed experimentally are in fact both noise-generated and noise-sustained.

To address the question of robustness of FDS waves to noise, we have investigated numerically the FHN system with velocity-additive noise (variance $\sigma_i^2 = 0.01$) as well as noise in the boundary forcing (variance $\sigma_t^2 = 0.01$) on a rectangular domain with a linearly increasing length $L_x(t) = \phi t$ and constant width L_y. The noise is Gaussian white noise with zero mean. The boundary forcing is sinusoidal with zero mean, amplitude A and frequency ω.

The 10 snapshots in Fig. 2 show the patterns that arise in a Turing-unstable medium under (linearly) convective unstable conditions at different forcing frequencies. Each snapshot was obtained by integrating the system for 25 time units with a flow/growth rate of $\phi = 1$. The width of the growing domain is $L_y = 2$. Fig. 2(a) shows the noise-structure that arises in the absence of periodic boundary forcing ($A = 0$). Note that the intrinsically selected pattern is not transversely uniform. Hence, it does not correspond to the transverse stripes observed experimentally.

With the parameters listed in the caption of Fig. 2, linear stability analysis predicts that low amplitude preturbations with wavenumbers between 1.3 and 13 are amplified. For $\phi = 1$ this corresponds to forcing frequencies between 1.3 and 13. The nine snapshots in Fig. 2(b)–(j) use frequencies within this range. For values of ω between 3 and 10 (wavelengths 0.6 and 2.5) the irregular intrinsically selected Turing pattern is replaced by stable transverse stripes. The imposed periodicity is also amplified at low [Fig. 2(b)] and high frequencies [Fig. 2(h)–(j)], but the presence of noise in the rate equations combined with a slow growth rate of the imposed mode causes the irregular Turing structure to prevail.

The 10 snapshots presented in Fig. 3 were obtained in the same way as those in Fig. 2, but with bistable local kinetics. Bistable (and excitable) systems are absolutely stable and do not support noise-generated structures. The system decays to a homogeneous steady state in the absence of periodic boundary forcing [Fig. 3(a)]. Space-periodic structures arise readily with the appropriate periodic boundary forcing. Fig. 3(b)–(j) shows the transverse stripes that arise with forcing amplitude $A = 0.2$ at various forcing frequencies. In contrast to the Turing case, the bistable system supports stripes even at very low forcing frequencies (long wavelengths). There is no lower limit on the imposed periodicity. There is however a limit on how closely the individual units may be squeezed together and pattern defects are observed when the wavelength is short [Fig. 3(i) and (j)].

The simulations presented in Fig. 2 and 3 demonstrate the robustness of transversely uniform FDS stripes to boundary as well as intrinsic noise. Both are inevitably present in real systems. The range of imposed wavenumbers that can be maintained is narrower than predicted by linear stability analysis. Boundary selected modes that have low growth rates are replaced by an intrinsically selected modes as exemplified by Fig. 2(b), 2(h)–(j) and 3(i)–(j). The amplitude of the boundary forcing relative to the boundary noise is also important. If the noise-amplitude is relatively high, the noise-sustained structure will prevail regardless of the forcing frequency. Conversely, relatively high forcing amplitude makes the FDS waves more robust to fluctuations. Maximal robustness is achieved when the forcing amplitude, *i.e.* the amplitude of the oscillation within the growth zone, is comparable to the amplitude of the space-periodic pattern, as it is observed in the biological experiments.

Prof. Harrison said: The authors mention that there are two ways to model axial growth, one of which is to keep the cells fixed while allowing the growth boundary to move. A few years ago, I gave a Master's student, Jefferey J. Orchard, the problem of studying by computation whether a Brusselator reaction–diffusion model could be persuaded to do something like somite formation. This did not involve an explicit representation of morphological growth, but only the increase in height of X morphogen concentration peaks, successively in a travelling zone.

The biological data we used as a starting point were for the axolotl, *Ambystoma mexicanum*, a Mexican salamander. We collaborated with an experimental biologist, Dr J. B. Armstrong (University of Ottawa), who was studying developmental phenomena in this animal. In its somite formation, there is a travelling zone, called the cohesive zone[1] in which about 6 somites are at graded stages of formation. Our model has not yet been published other than in Jeff Orchard's MSc thesis.[2] It supposes that a somite is fully formed when a morphogen X peak reaches a threshold height (Fig. 4, lightly dotted line) at which it then stops growing. (Morphogen X, upper solid line; morphogen Y, lower and heavier solid line.) The last fully-formed somite acts as a source of a fixed concentration ($bB = 0.61$) of reactant B or an enzyme controlling rate constant b, which diffuses and decays first-order, setting up an exponential gradient of itself (heavily dotted curve). 5 or 6 wavelengths downstream (to the right) parameter values cross the Turing instability boundary into a non-patterning region. By computation, we showed that this model was quite adequate to produce "somites" (X peaks) successively from a zone that jumps to the right when a new somite is fully formed. Fig. 4 shows 4 stages in the development, at times $t = 0$, 45, 48.6 and 192.75. An additional experiment on the model was to simulate a surgical cut by inserting a diffusion barrier. If this was done in the region of a peak about half-formed, such as the second from the left in Fig. 4(c), somitogenesis continued to the right of the cut.

Our model was inspired by earlier work of Thurston Lacalli,[3] who showed that a 1D gradient (in his work, in rate constant c of the autocatalytic step) would persuade the Brusselator to make stripes sequentially along a narrow 2D rectangular zone. All we did was to add a mechanism to make the top of the gradient move, stepwise, as each new somite formed.

I would like to ask Prof. Hunding and Dr Kærn (who has been reading Jeff Orchard's thesis) whether they think this model captures much of the dynamic nature of somite formation, or whether it is negated by more recent experimental evidence.

1   J. B. Armstrong and A. C. Graveson, *Develop. Biol.*, 1988, **126**, 1.
2   J. J. Orchard, *Reaction–diffusion modelling of somite formation: computed dynamics and bifurcation analysis*, MSc Thesis, University of British Columbia, 1996.
3   T. C. Lacalli, D. A. Wilkinson and L. G. Harrison, *Development*, 1988, **104**, 105.

Dr Kærn commented: The reaction–diffusion model by Orchard involves the formation of a space-periodic Turing structure (the somites) behind a moving zone. This moving zone, which is

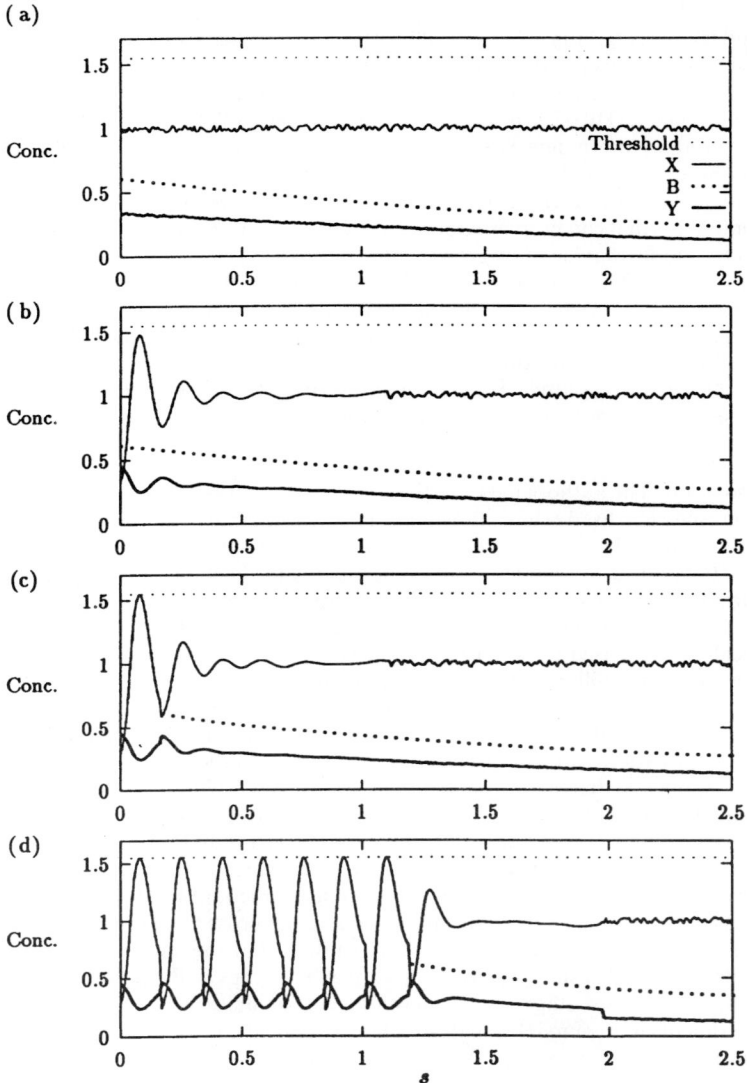

Fig. 4 Four stages in a somite formation computation; see text for details.

equivalent to the presomitic mesoderm (PSM) in the chick, contains a gradient in a parameter value such that the system is in a steady state at the posterior boundary and supports Turing structures at the anterior boundary. This model cannot account for the gene expression waves observed in the chick, the mouse and in the zebra-fish. Furthermore, the observation that the direction of somite formation is reversed when a part of the PSM is inverted is incompatible with the gradient being established and maintained solely by a diffusion-type process (see ref. 1 for a detailed discussion). However, to the best of our knowledge, a segmental clock has not been identified in reptiles and it cannot be ruled out that reptiles rely on a different developmental mechanism.

Orchard's model involves a domain of constant length and a zone that moves through this domain at a certain velocity ϕ' (see Fig. 4). The velocity at which the zone moves is determined by the velocity of the spreading Turing structure. A somewhat related model was suggested by

Meinhardt.[2] In his model, a Turing-type pattern spreads into a region of homogeneous oscillations. In many vertebrates, however, the distance between the posterior growth zone and the space-periodic pattern (the somites) is observed to remain more or less constant. If segmentation in these species rely on the spreading of a Turing pattern, the growth rate must exactly match the velocity at which the Turing structure spreads. Hence, the models cannot robustly reproduce the constant length of the non-segmented tissue. A constant distance between the growth zone and the space-periodic structure would imply that the system is located exactly at the critical point between convectively and absolutely unstable flow conditions. On the other hand, if the gradient is correlated with the time since cells were added to the growing structure, as in the mechanism we propose, it will automatically move with the same velocity as the growth boundary.

1   M. Kærn, A. Hunding and M. Menzinger. *J. Theor. Biol.*, 2000, **207**, 473.
2   H. Meinhardt, *Models of Biological Pattern formation*, Academic Press, New York, 1982.

Prof. Gáspár said: I think we should try to find a better explanation of the difference between Turing-structure and flow distributed pattern—and answer Prof. Nicolis' original question fully. When I think of Turing patterns, I always imagine a 2D structure which starts to appear on the whole area of interest as the initial perturbation grows in time. On the other hand, FDS is a 1D pattern, and it develops step-by-step as the waves carried by the flow stock up along the tube one-by-one.

Prof. Nicolis responded: A Turing structure invades the entire reaction space when the parameters are uniformly distributed in space (and have, of course, values above the critical threshold). The situation may be different in the presence of a ramp in a control parameter as long as part of the reaction space is now found in subcritical conditions. One observes, then, a step by step development of the pattern[1,2] in a way similar to FDS.

1   M. Herschkowitz-Kaufman and G. Nicolis, *J. Chem. Phys.*, 1972, **56**, 1890.
2   J. F. G. Auchmuty and G. Nicolis, *Bull. Math. Biol.*, 1975, **37**, 323.

Prof. Gáspár commented: I would like to remind you of our poster in which we show that in a flow distributed system complex transient behaviour may also occur. They are similar to the resonant patterns observed in other RD systems. However, even if there is a failure in initiating a wave, our results show that the final pattern is being established after all. We think that this might be of biological importance, for example, during axial segmentation presented and discussed in the paper by Hunding *et al.*

Dr Satnoianu said: The flow distributed oscillator mechanism or FDO is probably one of the most interesting ideas of application of the behaviour from chemical reactors to biology. Our paper shows that it can successfully explain the formation of periodic structures in vertebrates, the typical example being the formation of somites. FDO is a particular instance of FDS or flow and diffusion structures mechanism introduced earlier by us.[1] It is a more direct and illuminating idea than the Turing scenario for example.

However, there are also some important differences between the chemistry setup and biology which need to be underlined. As far as the experimental evidence goes, there is no finding of diffusion playing any major role in the transmission of the waves of gene products in the presomitic mesoderm (or PSM). Therefore any modelling in terms of reaction–diffusion–advection (RDA) systems necessarily requires that the flow dominates the diffusion regime. In this limit we recover the notion of FDO. Furthermore the coupling of boundary conditions with flow assures the selection in a robust manner of the wavelength required in the somite formation and this is explained in detail in our paper.

Profs. Merkin and Gáspár commented that they observed complex spatiotemporal behaviour in RDA when using BZ medium and thus possibly implying a limitation of the present mechanism making it more prone to errors in the developmental process. In response to this criticism I note that their observed behaviour is located in the diagram of their work in a regime with low flow rates and therefore where both diffusion and flow play important roles (both have comparable strengths). This is therefore outside of the FDO regime.

1 R. Satnoianu and M. Menzinger, *Phys. Rev. E*, 2000, **62**, 113.

Prof. Menzinger commented: Walking along the St. Lawrence river I recently observed the following hydrodynamic example (Fig. 5) of a flow-distributed oscillation which nicely illustrates the kinematic origin of the process described by eqn. (5) and $T' = \infty$ (*i.e.* constant forcing) of our paper. The flow ϕ encounters a fixed obstacle just below the surface, which sets in motion a damped oscillation with period T in the flow downstream of the obstacle. The resulting waves are standing waves since the phase of the oscillation is tied to the obstacle, and their wavelength is equal to $\lambda = \phi T$, in agreement with eqn. (5).

Prof. Noszticzius opened the discussion of Dr Schreiber's paper: Beside the phase plane picture of an excitable system with one fixed point shown here there is another model of excitability which contains three fixed points.[1-3] My question is how the present generalized concept of excitability can deal with the latter model?

1 W. D. McCormick, Z. Noszticzius and H. L. Swinney, *J. Chem. Phys.*, 1991, **94**, 2159.
2 H. Sevcikova and M. Marek, *Physica D*, 1991, **49**, 114.
3 V. Petrov, S. K. Scott and K. Showalter, *Philos. Trans. R. Soc. London, Ser. A*, 1994, **347**, 631.

Dr Schreiber responded: As described in the references which you cited, the threshold in an excitable planar system with three fixed points has to do with a stable separatrix of the saddle fixed point. The present concept extends to the three-fixed-points case as drawn schematically below (Fig. 6). A transition from a unique fixed point to three occurs *via* a saddle-node bifurcation; a critical point emerges directly on the threshold set and subsequently splits into a saddle and an unstable node. In this way, a branch of the stable separatrix connecting the node with the saddle is inserted into the phase portrait and the threshold set is now composed of three parts: two branches of stable separatrix of the saddle and a piece of trajectory coming out from the node to the point x_R. Clearly, the latter part can be defined in a similar way as the threshold for the unique fixed point system, that is, by maximizing an amplification ratio, but this point remains to be formulated more precisely.

Prof. Gáspár said: I would like to congratulate the authors on their success in modifying the model to show excitability. I am sure you have tried several options and variations before you arrived at the model presented in the paper. Based on your experience, could you, please,

Fig. 5 A section of the St. Lawrence river.

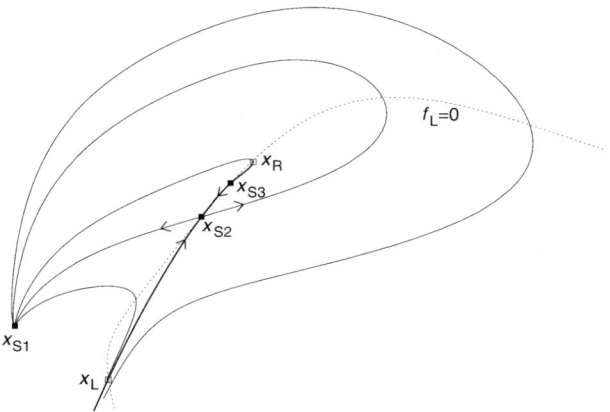

Fig. 6 Phase portrait with three steady states x_{S1} (stable node), x_{S2} (saddle) and x_{S3} (unstable node), and a threshold set connecting the points x_L and x_R.

summarise what should be the essential features of a model—or a chemical system—to show excitability? In other words, how should we design an excitable chemical system? What kind of chemical reactions should we have in a system to show excitability?

Dr Schreiber replied: Thank you. Based on the kind of bifurcation behaviour we observed in the two alternative mechanisms for the BSF reaction, the essential feature seems to be a cross-shaped bifurcation diagram, that is, a saddle-node bifurcation curve having a cusp is interlinked with a pair of Hopf bifurcation curves, that usually cross each other, *via* Bogdanov–Takens bifurcation points. One of the mechanisms showed a region of bistable steady states and adjacent to it was a region of coexisting limit cycle and a steady state. The oscillations were born from a Hopf bifurcation unrelated to the saddle-node bifurcation marking the multiple steady states region. The other mechanism did show a cross-shaped diagram, even though very distorted. By extending and modifying the latter scheme we eventually obtained a good fit with experiments. There are, however, models that do not possess multiple steady states yet they do display excitability; for example the one-pool model of cytosolic calcium oscillations by Dupont and Goldbeter. Here the excitability seems to be linked with two bifurcation features: (a) a subcritical Hopf bifurcation, (b) sudden amplitude expansion also called a canard phenomenon. Since these two features are regular companions of the cross-shaped bifurcation diagram they seem to be the basic requirements for excitability. As to your other question concerning the design of a chemical excitable system and kind of reactions it should involve: excitability is obviously linked to oscillations and in many cases to bistability of steady states. Thus mechanistic features such as autocatalysis and a negative feedback are required, and hence the mechanistic classification of oscillators[1] provides requirements for excitability as well. However, the subcriticality and the canard phenomenon are more subtle conditions and their relation to mechanistic features is so far unclear, although vastly different timescales of various reaction steps doubtless play an important role.

1 M. Eiswirth, A. Freund and J. Ross, *Adv. Chem. Phys.*, 1991, **80**, 127.

Dr Feigin commented: In the paper you have introduced an excitation number as a characteristic of excitability of the system forced by external pulsed perturbations. Is this quantity a qualitative characteristic reflecting qualitative properties of the system, or this is a method to characterise the system reaction on the concrete periodic forcing? It seems to be quite obvious that excitation number introduced by you depends not only on distinctive properties of the forced system but also on period and amplitude of the forcing, and what is more, on the "direction" of the perturbation in the phase space of the system.

Dr Schreiber answered: The excitation number is defined as an average number of excitations per forcing period for a particular initial state of the system. So it is a characteristic of a particular trajectory of the forced system and as such depends on the properties of the unforced system, on the forcing amplitude and period, and in fact, also on the initial state of the system. However, all initial states within a basin of attraction are expected to yield the same excitation number, even though this point becomes subtle, when the attractor is chaotic. Then ergodicity implies the same average for almost every initial state. Another point is that the average may not be a unique number but rather a limit set such as an interval. Nonetheless, the excitation number characterizes an attractor or another invariant set (including trajectories asymptotic to it) associated with a response of the system under specified conditions to a concrete periodic forcing including the "direction" of the perturbation.

Prof. Müller asked: I wonder whether the different kinds of excitability and the corresponding threshold and recovery behaviour have a direct meaning for biosystems, especially with respect to the refractory period, which in biological tissue tends to be rather prolonged and complex. How do these findings fit into the common nullcline representation of excitability that is frequently used and emphasizes the analogous behavior of biological and chemical excitable media?

Dr Schreiber responded: Activatory/inhibitory excitability is a prerequisite to activatory and inhibitory chemical pulse waves. For example, in the Belousov–Zhabotinsky reaction they correspond to oxidation and reduction waves, respectively. The latter are rather uncommon, but this may not be so in biological systems. As modelling suggests, the calcium waves are of activatory type, but models of neural activity or muscle contraction provide readily both types of excitability and related waves. Although general features of dynamical response to repeated perturbations are the same in both kinds of excitability, the inhibitory one is less robust. This reflects itself also in more complex refractory behaviour—including peculiar dynamics of waves such as crossing rather than annihilation. As to the nullcline representation, the association of the threshold with the middle branch of the nullcline for the autocatalyst is a very useful idea providing a unifying mathematical description which is, in addition, simple to calculate. Yet it lacks some desired features, mainly the (backward) time-invariance in the phase space (and an arbitrary extent in backward time). I would call the nullcline method a zeroth order operational definition while the one presented here a first order approximation.

Prof. Menzinger commented: As the dimensionality of dynamical systems increases, so does the dimension of their dynamical manifolds, and with it their topological complexity. One should be on guard to let the intuition derived from two-dimensional dynamic models act as a guide for expectations in real, high-dimensional systems.

As a case in point, consider the hysteresis of the bistable BZ reaction in a CSTR.[1] A response diagram (*e.g.* an electrode response as function of flow rate k_0) is drawn schematically below (Fig. 7).

Ali *et al.*[1] observed that bistability transitions may be induced, in addition to the intuitively well understood perturbation 0–1 of the dynamical (response) variable and the 0–2 perturbation of the control parameter, also somewhat surprisingly by the perturbation 1–1p of the response variable or by the 0–2p perturbation of the control parameter. Obviously, the repelling manifold that constitutes the basin-boundary is crossed in all these transitions and it must have a peculiar topology in this higher-dimensional system. Indeed, the basin boundary constructed from the three-dimensional skeleton model of the BZ system due to Field and Noyes[2] agrees well with the observed behaviour.[1]

My question to Dr Schreiber is whether one should also expect to encounter threshold sets with a similarly complex topology in high-dimensional excitable systems.

1   F. Ali, P. Strizhak and M. Menzinger, *J. Phys. Chem. A*, 1997, **101**, 6048.
2   R. J. Field and R. M. Noyes, *J. Phys. Chem.*, 1974, **60**, 1877.

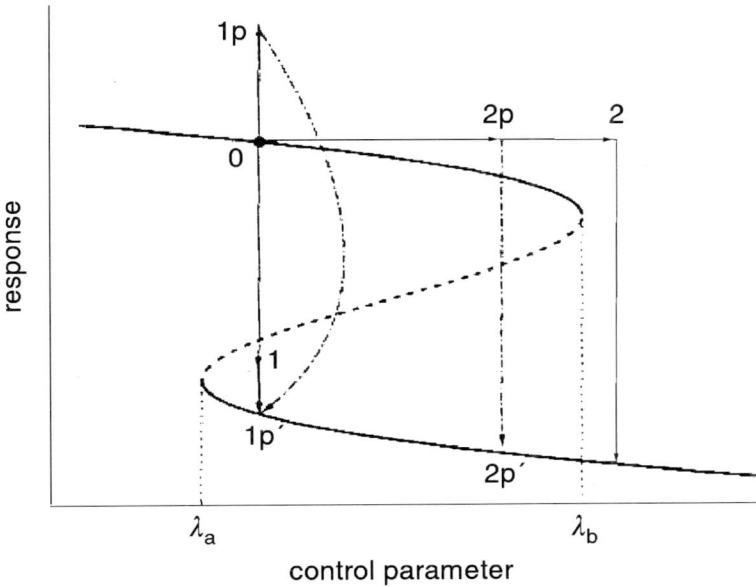

Fig. 7 Response diagram for hysteresis of the bistable BZ reaction in a CSTR.

Dr Schreiber answered: Within the framework introduced here, the threshold set in a high-dimensional space is a backward invariant codimension one surface locally dividing the phase space, and in this regard it works in the same way as a boundary of an attractor basin. Indeed, the threshold is expected to be a curved surface with possibly complex topology. Actually, even in the two-variable papain example the threshold set may have an unexpected topology. For example, a time-backward extension of the threshold trajectory in Fig. 2 of our paper in this volume makes a loop around the steady state x_s (not shown in the figure) so that the activatory excitation may be initiated by either adding H^+ or adding the substrate or—unexpectedly—by removing H^+.

Dr Simon said: The question concerns the mathematical definition of the threshold set. In order to define it one has to assume that the point x_L is determined uniquely by the distance P via eqns. (2) and (3) of the paper. (Here the notations of the paper are used.) Then the relative amplification r can be expressed in terms of P, and the threshold set can be defined as the half-trajectory for which dr/dP is maximal. Can a condition be formulated on the vector field in mathematical terms that ensures that x_L is a function of P?

Dr Schreiber replied: Geometrically, eqns. (2) and (3) represent a curve and a circle, respectively (provided that we deal with two-variable systems). So, in the simplest nontrivial case there are two intersections. One of them is expected to be the point x_R (for which actually eqn. (4) should be used instead of eqn. (2)), the other is the point x_L. Hence P uniquely defines x_L in this case. As seen in Fig. 2 of the paper, the papain model admits two different x_L points (and two different x_R points) because the steady state is a focus but there is no maximum of dr/dP and hence no threshold associated with the other pair (x_L, x_R). If there were, though, it would be a legitimate threshold and the system would possess multiple thresholds. A necessary condition to rule out multiple x_L points would seem to be that the steady state is a node. However, imposing such conditions is unnecessary, rather, one should search for each x_L point and check if there are multiple thresholds.

Spatial bistability and waves in a reaction with acid autocatalysis

J. Boissonade,* E. Dulos, F. Gauffre, M. N. Kuperman† and P. De Kepper*

Centre de Recherche Paul Pascal, C.N.R.S. Bordeaux, Avenue Schweitzer, F-33600 Pessac, France

Received 10th April 2001
First published as an Advance Article on the web 25th October 2001

The phenomenon of spatial bistability has recently been proposed for a comprehensive understanding of a number of chemical patterns observed in open spatial reactors consisting of thin films of gel diffusively fed from one side. We study experimentally and numerically this phenomenon in the tetrathionate–chlorite reaction characterized by an acid superautocatalysis. We focus on the similarities and differences with previous studies on the chlorine dioxide–iodide reaction. In addition, we show that this reaction, which is only bistable in a continuous stirred tank reactor, can exhibit oscillatory and traveling waves when diffusion comes into play. Our computations suggest that the nonstationary behaviour originates from differential diffusive transport.

Introduction

Sustained laboratory reaction–diffusion patterns are commonly produced in gel slabs fed with fresh reactants by diffusion from boundaries to hinder convective transport. Spatial bistability is a rather general phenomenon which occurs in these systems when a reaction which is bistable in homogeneous conditions, exhibits two different stable concentration patterns for the same set of boundary conditions. Although this phenomenon has been previously described in toy models[1] and in real experiments,[2,3] its importance has been recognized only recently and a detailed analysis was worked out in relation to the development of open one-side-fed reactors (OSFR), a handy class of devices for the study of nonequilibrium reaction–diffusion patterns.[4–6]

These reactors are made of a thin film of a chemically resistant gel, of thickness l, which is kept in contact on one face with the contents of a continuous stirred tank reactor (CSTR) and with the other face pressed against an impermeable wall. The coupled dynamical equations for the concentrations in the CSTR and in the gel are respectively:

$$\frac{\partial c_{si}}{\partial t} = f_i(\mathbf{c}_s) + \frac{(c_{i0} - c_{si})}{\tau} + \rho_V \frac{D_i}{l}\left(\frac{\partial c_i}{\partial r}\right)_{r=0} \quad (1)$$

and

$$\frac{\partial c_i}{\partial t} = f_i(\mathbf{c}) + D_i \nabla^2 c_i \quad (2)$$

† Permanent address: Centro Atómico Bariloche and Instituto Balseiro, 8400 San Carlos de Bariloche, Argentina.

DOI: 10.1039/b103240m

where c_{i0}, c_{si} and c_i are the concentrations of species i respectively in the input flow, in the CSTR, and inside the gel, D_i is the corresponding diffusion coefficient, τ the residence time of the reactor, ρ_V the ratio of the volume of the gel to the volume of the CSTR, and r the distance to the CSTR/gel interface. The f_is are the reaction rates. In eqn. (2), the Laplacian operator reduces to $\partial^2/\partial r^2$ in one-dimensional calculations where one considers the sole direction orthogonal to the faces. In eqn. (1), the second term represents the input and output flows of the species. It contains all the expandable control parameters of the system. The third term results from the diffusive flux of the species at the interface between the gel and the CSTR and represents the feedback of the gel contents on the CSTR dynamics. When the volume of the CSTR is large in regard to the volume of the gel ($\rho_v \ll 1$), this last term can be dropped so that the chemical state of the CSTR is independent of the state of the gel and the concentrations in the CSTR become a Dirichlet boundary condition for eqn. (2) at $r = 0$, whereas a no-flux boundary condition is applied at $r = l$.

Let us consider a reaction in a CSTR at the stationary state. At high flow rates, the residence time is small and the extent of reaction is small. The concentrations in the tank are close to those in the input flow. The state belongs to the "flow branch" (branch F). At low flows, the extent of reaction is large, the reaction is almost completed and close to chemical equilibrium. The state belongs to the "thermodynamic branch" (branch T). In standard reactions, these two branches connect smoothly at intermediate flows but when the reaction mechanism includes destabilizing processes such as autocatalysis or substrate inhibition, both branches can be simultaneously stable over a finite range of flow rates (or of other control parameters). Many of these reactions exhibit "clock" behaviour in batch conditions, *i.e.* a more or less long induction period followed by a fast switch to the equilibrium. When such a bistable reaction is kept in the flow state F in the CSTR, one can observe spatial bistability in an OSFR.[4] For a given flow, the amount of fresh reactants that can reach a given point in the gel by diffusing from the CSTR interface depends on the distance r to this boundary. Close to the interface, this amount is high, the extent of reaction remains small and the system remains close to the flow state, forming a boundary layer. Far from the CSTR, the amount of unreacted input species is low, the extent of reaction is high and the reaction is almost complete (T state). The fast switch that would be observed in a batch reactor leads now to the formation of a steep front between the boundary layer (in the F state) and a T state. We call such a chemical state in the gel an FT or "mixed" state. This is the normal situation when l is large. When l is small, the boundary layer can extend to the whole system which is then altogether in a flow state. By extension, we call F such a state of the gel. In the same way that, for intermediate values of the flow rate, the CSTR can present two stable states, the F and FT states of the gel can both be stable for intermediate values of l for the same state of the CSTR. This is a practical realization of a spatial bistability. A third stable state exists at low input flow in which the CSTR and the gel contents are both in a T state. Note that the input flow can be replaced by another control parameter which controls the extent of reaction, generally the concentration of an input species in the feed flow. By extension, in the gel, we continue to call "flow state" the state with a small extent of reaction and "thermodynamic state" the state with a large extent of reaction. The phenomenon is of major importance in OSRFs since domains of different stable states can be observed in the gel for a given F state of the CSTR. The competition between these domains can involve complex front dynamics and eventually lead to the formation of patterns.

Beyond this intuitive approach, a more extensive and quantitative theory of the spatial bistability in OSFRs has been developed in ref. 6 and is summarized in ref. 5. This theory has been supported by experiments and numerical simulations within the framework of the chlorine dioxide–iodide (CDI) reaction.[5] The theoretical and experimental diagrams of stationary states in the plane $([I_2]_0, l)$ are in good quantitative agreement and the different stable concentration profiles have been observed in an annular OSFR. Studies of the competition between both states have also been performed in the experiments. In the present paper, we extend this study to the tetrathionate–chlorite (C–T) reaction which exhibits a strong acid autocatalysis. This reaction has been essentially studied in closed systems for the determination of stoichiometric and kinetic aspects[7,8] as well as a source for propagating cellular fronts.[9-11] We demonstrate the existence of spatial bistability and new dynamical behaviour in this reaction and underline both the similarities and the differences with the cases studied previously.

The reaction kinetics can be approximated by the following overall balance equation

$$7ClO_2^- + 2S_4O_6^{2-} + 6H_2O \rightarrow 7Cl^- + 8SO_4^{2-} + 12H^+ \qquad (3)$$

with a reaction rate $v = k[ClO_2^-][S_4O_6^{2-}][H^+]^2$.[7] The constant k was evaluated to $k = 10^9$ $M^{-3} s^{-1}$ in the pH range 4.7–5.5. The reaction exhibits quadratic autocatalysis in $[H^+]$. When in batch, the reaction is started in basic solution, the induction time can be considered as quasi-infinite with regard to reasonable experimental times. This is an important difference from the CDI reaction which presents a finite induction time in batch.

Experiments

The thin disc-shaped OSFRs are very appealing because they enable observation of quite dramatic quasi-2D reaction–diffusion patterns in planes parallel to the feed surface[12–16] but do not permit resolution of the concentration profiles in the depth of the disc. In many instances, this missing information can be crucial for a clear understanding of the patterning mechanisms in these reactors. In order to complement observations in a disc-shaped OSFR, we have developed one-side-fed flat annular reactors, that is annular OSFRs. Complete descriptions and technical details of such reactors can be found in ref. 5. The annular OSFRs used in this series of experiments are made of 2% agarose gels. They have a fixed outer radius $R = 12.5$ mm and a width l that ranges from 1 to 3 mm. The inner rim of the annulus is the impermeable boundary while the outer rim is in contact with the contents of a CSTR of 25 cm^3, fed through a single inlet port with appropriately premixed solutions of reagents stored in separated reservoirs. The residence time of the reactor $\tau = 600$ s, the temperature $T = 25\,°C$ and the input flow concentrations of sodium chlorite $[NaClO_2]_0 = 1.9 \times 10^{-2}$ M and potassium tetrathionate $[K_2S_4O_6]_0 = 0.5 \times 10^{-2}$ M were kept fixed during all the experiments. Besides the two above-mentioned reagents, variable amounts of perchloric acid and sodium hydroxide were added to the input flow of reagents to control the pH of the premixed feed solution. This is done by pumping different volume ratios of acid and base whilst keeping the overall flow rate constant. For experimental and graphic convenience, we characterize the acid and base flow concentrations by a control parameter α on which they depend linearly: with $[HClO_4]_0 = \alpha \times 0.67 \times 10^{-2}$ M and $[NaOH]_0 = (1-\alpha) \times 3.33 \times 10^{-2}$ M. The pH of the feed mixture decreases as α increases from 0 to 1.

A rapid exploration of the dynamics of the C–T reaction operated in the sole CSTR only exhibits a large domain of bistability between an F branch (with pH values typically above 9) and a T branch (with pH values typically below 3) as a function of $[NaClO_2]_0$ and $[K_2S_4O_6]_0$ as well as the pH of the input feed. To explore this bistable system, the pH of the input feed is the most convenient parameter. All the studies reported here were performed as a function of α, the parameter which effectively controls the pH of the feed.

If one starts with low values of α, the contents of the CSTR exhibit a high pH value (F branch) until $\alpha = 0.83 \pm 0.01$ whereafter the pH drops to ~ 2. If, now, one decreases α, the system remains on this low pH branch (T branch) down to $\alpha = 0.18 \pm 0.01$ below which the pH of the solution suddenly increases to about 12.5. In our experiments we used Bromophenol Blue, at a concentration $[BB]_0 = 2 \times 10^{-4}$ M, as a pH indicator. The high and low pH branches are then characterized respectively by purple–blue and deep yellow solutions.

Let us now consider the annular gel reactor in contact with the contents of this CSTR. When one changes α and occasionally delivers local acid or base perturbations onto the gel, one can observe different stationary or traveling concentration profiles across the width and in the azimuthal direction of the annulus. Let us first consider the case of stationary profiles. Starting with the CSTR in the blue F state, at low values of α, the whole gel annulus is always uniformly blue. Now, when α is gradually increased, nothing particular happens in the gel until the CSTR switches to the yellow T state. This occurs at the same value as in the absence of the gel annulus. When the CSTR contents are on the T branch, so is the whole gel annulus which is then uniformly yellow. This situation remains until the critical value of α for the stability of this branch in the CSTR is reached. Here again, the presence of the gel does not detectably affect this critical value. Beyond the critical value, the CSTR and the gel return to a uniformly blue state. However, when the CSTR is in the blue F state, one can expect that for some values of α and/or width l of the gel, a mixed state as described in the introduction will be observed. Contrary to the situation described in ref. 4 and 5, this mixed state cannot be reached by a gradual change of a control parameter of the CSTR but only by introducing a local acid perturbation on the gel at high enough values of α or by making the CSTR suddenly switch from the low pH branch to the high pH branch by injecting an appropriate amount of base into the CSTR, at high enough values of α. In both cases, while the CSTR is in the high-pH state, the inner part of the gel annulus may remain in a stable clear low-pH state. One can then probe the stability range of this mixed state as a function of α. All critical stability values for different widths l of the annular reactor are reported in Table 1. The accuracy on these limits is better than 2%.

The stability limits of the mixed state lie inside the limits of the uniform F or T states. Since the F state is stable for all values of α lower than 0.83 (Table 1), two different concentration profiles can be observed over the whole range of stability of the mixed state. Fig. 1 exemplifies the visual aspect of the two different stable states observed when the CSTR contents belong to the flow branch.

Table 1 Experimental stability limits

(l/α) /mm	F excitability limit	Mixed state range	F state limit
1	0.51	0.70–0.81	0.83
2	0.48	0.70–0.81	0.83
3	0.46	0.69–0.80	0.83

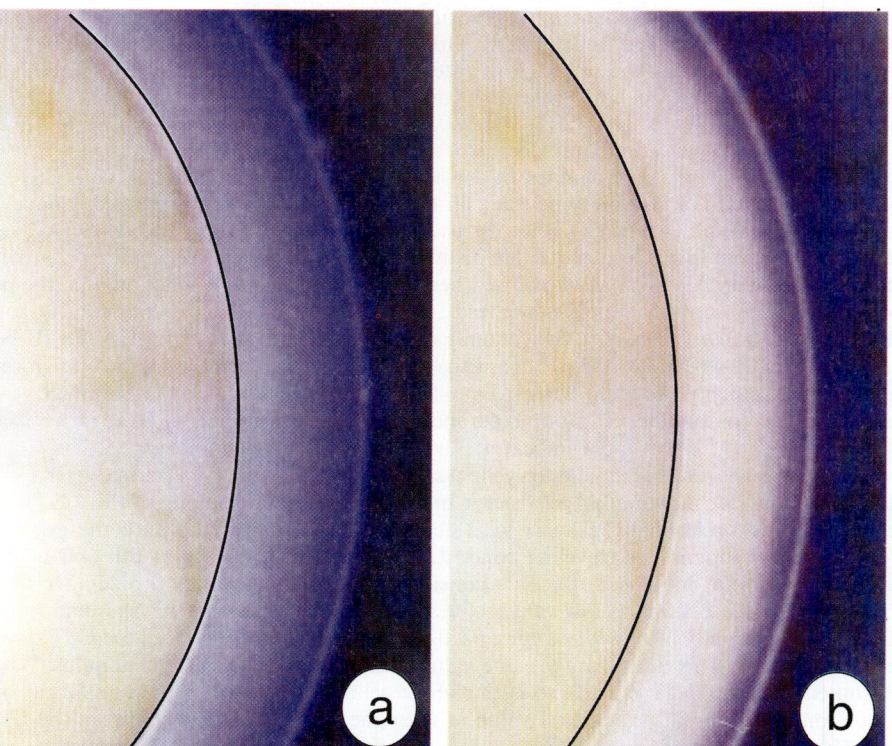

Fig. 1 Stable states of the spatial bistability of the chlorite–tetrathionate reaction operated in an annular OSFR: (a) quasi-uniform basic F state; (b) mixed basic/acid state. The white and black curves respectively delineate the CSTR interface and the impermeable wall. Experimental conditions: $l = 2$ mm, $\alpha = 0.71$.

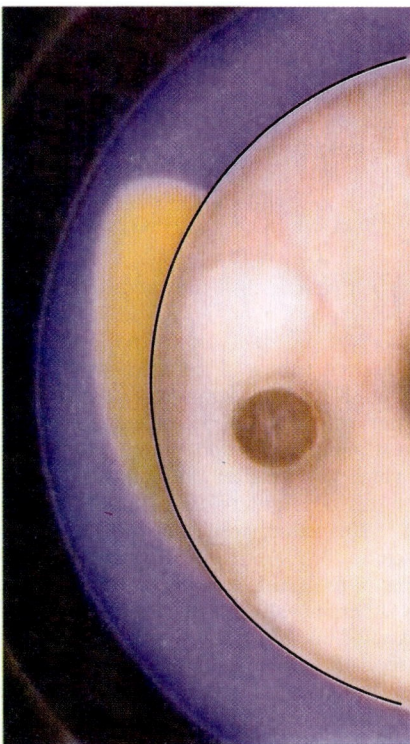

Fig. 2 Excitation wave. Yellow acid front propagating upwards in an asymptotically stable blue basic F state. Experimental conditions: $l = 3$ mm, $\alpha = 0.48$.

At high values of α, the mixed state disappears at a value of α where the gel and the CSTR switch simultaneously to the T state. In the domain of spatial stability, one can study the relative stability of the uniform F state and of the mixed state in the gel annulus. An interface between the two states can be produced by a local transient acid perturbation of the blue F state. Contrary to the situation observed for the CDI reaction, no reversal of the propagation direction of this interface is observed for the C–T reaction. Here, whatever are the values of α within the spatial bistability range, the mixed state always invades the blue F state, even in the immediate vicinity of the stability limit of the mixed state at low values of α. Furthermore, at a value of α below the stability limit of the mixed states, where only the uniform blue F state is asymptotically stable, a local acid perturbation of this blue state triggers a pair of yellow waves that propagate undamped, at a velocity of the order of a few mm min^{-1}, in opposite directions along the annulus. They take the form of a yellow acid state bubble propagating in a blue basic state (Fig. 2). These waves have all the properties of triggered waves in an excitable medium. They are formed of a leading front which connects perpendicularly to the impermeable boundary and curves at the approach of the feed boundary; the front is followed by a more or less elongated recovery tail. The recovery tail is shortest at the lowest values of α and gradually stretches as α is increased and can become at least of the order of the perimeter of the annulus.

The lower critical value of α for which triggered wave propagation is observed is reported in Table 1. This limit slightly decreases with increasing values of l. The propagation velocity of the wave is lowest at this limit and increases when the value of α increases. Note that triggering an acid front in the uniform blue F state in the excitability or the spatial bistability domains made no fundamental difference to the leading part of the front. The only difference lies in what happens behind the front: either the blue uniform state is recovered or the system sets in the mixed state.

Numerical simulations and discussion

To perform our numerical simulations of the reaction kinetics we have used eqn. (3) complemented by the two fast acid–base equilibria

$$H^+ + OH^- \rightleftharpoons H_2O \qquad (4)$$

$$H^+ + SO_4^{2-} \rightleftharpoons HSO_4^- \qquad (5)$$

which play an important role in the dynamics. All the fixed parameters were set at the experimental values.

In a first step, we have computed the bistability limits in the CSTR alone by continuous change of the control parameter α for different values of the constant k. The limit at which the high-pH state F disappears is practically independent of k and equal to $\alpha_f = 0.833$. This value corresponds to the point where the input solution switches from basic to acid values where the pH changes very steeply with α. The value at which the thermodynamic state disappears was found equal to α_t 0.13 for $k = 10^9$ M^{-3} s^{-1}, the value proposed in the literature, and increases when k decreases. We use this property to adjust this constant to fit with the experimental value 0.18 which is reached for $k = 5 \times 10^6$ M^{-3} s^{-1}. This value, significantly lower than that formerly proposed, was used in all our numerical simulations.

One-dimensional computations of the stable concentration profiles in the width of the gel were performed as a function of α at different values of l. The reaction–diffusion eqn. (1) and (2) were integrated by the standard method of lines with the Deufhlard semi-implicit midpoint method for stiff systems.[17] A dynamically reconstructed non-uniform grid was used to keep high spatial accuracy around the fronts. Taking into account the exact diffusion of each species is a complex question. It is known that H$^+$ and, to a lesser extent OH$^-$, diffuse intrinsically much faster than the other species but accounting for the electroneutrality condition drastically reduces these differences. Moreover, the diffusion, strongly controlled by the slowest ions, is a function of the local concentrations.[18] Solving reaction–diffusion equations that include these ionic effects is a formidable task which has only been performed on models.[19] In this paper, we shall first assume that all diffusion coefficients are equal to $D = 1.5 \times 10^{-5}$ cm^2 s^{-1}, a value slightly larger than most of the coefficients in order to account for the presence of the fast ions.

A significant difference with the DCI reaction is that, since in the experimental flow condition (high pH) the induction time is quasi-infinite, the system cannot switch from the flow state to the mixed state by a continuous change of α or of any other input. As already mentioned in the Experimental section, the transition has to be nucleated from outside. Nevertheless, in a numerical simulation, it is easy to prepare a mixed state by starting from a T state and setting temporarily the CSTR in a very high flow state to initiate the F/T front.

In Fig. 3a, we give an example of spatial bistability. We present the two different stable concentration profiles (flow state and mixed state) that correspond to the same set of control parameters. In Fig. 3a, we plot, on a logarithmic scale, the concentration of H$^+$ as a function of r. In Fig. 3b, we report the corresponding concentration profile of OH$^-$. In this case, we have used a linear scale because this species actually plays the role of a substrate, for which the theory predicts a linear profile between the CSTR and the transition front.[5,6] This can be clearly seen in this figure. The small difference in concentrations in the CSTR between the two states will be discussed later.

In Fig. 4a, we report the state diagram in the plane (α, l). Note that an increase in [H$^+$], the autocatalytic species, corresponds to an increase in α. The flow state F corresponds to small values of α. The spatial bistability domain is located between $\alpha_1(l)$, the (lower) limit where the mixed state loses its stability and $\alpha_2(l)$, the upper limit where the F state loses its stability. Let us now discuss successively these two limits.

The limit α_1 exhibits a normal behaviour, analogous to that found in the CDI reaction. When the size of the system increases, it becomes more and more difficult for fresh reactants to reach the impermeable boundary of the gel, which favours the mixed state. Below a critical value of l (here $l \simeq 0.02$ mm), the system is too short to sustain a mixed state when the CSTR is in a flow state F. This limit is beyond the experimental capabilities of the device.

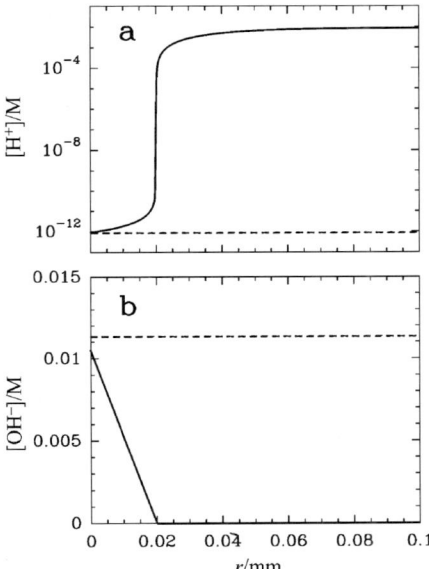

Fig. 3 Stable concentration distribution in the gel for a given set of parameters. Full lines: mixed state. Dotted lines: flow state. $l = 0.1$ mm, $\alpha = 0.55$. (a) $[H^+]$, logarithmic scale; (b) $[OH^-]$, linear scale.

One might expect that α_2 should decrease when l increases, since it becomes more difficult to sustain the flow state. However, this does not happen. When the system is in a high-pH state, the induction time is almost infinite, the reaction never starts and when one increases the size of the system, it never switches into the mixed state. Thus, in contrast with the CDI reaction case where

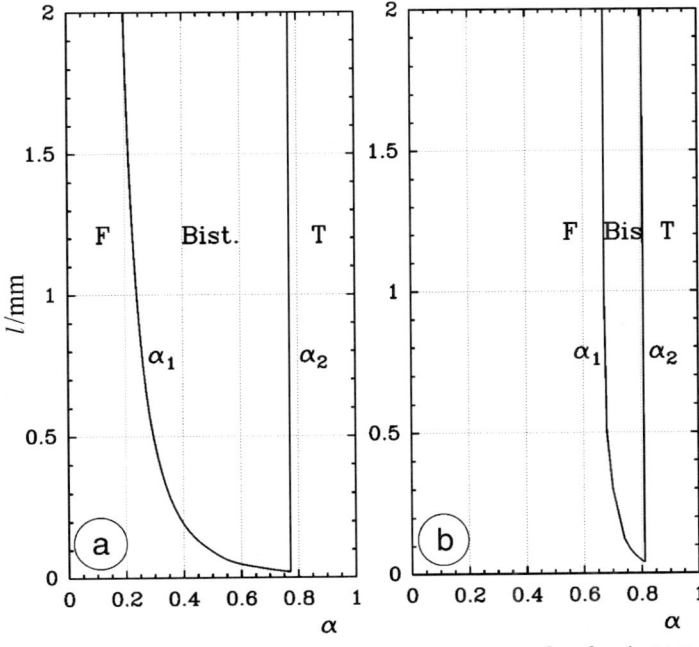

Fig. 4 Nonequilibrium phase diagram in the plane (α, l) (a) $D = 1.5 \times 10^{-5}$ cm^2 s^{-1}; (b) $D_1 = 1 \times 10^{-5}$ cm^2 s^{-1}, $D_2 = 3 \times 10^{-5}$ cm^2 s^{-1}.

α_1 and α_2 merge at large values of l, the domain of spatial bistability does not close up for experimentally accessible values of l. The bistability limit α_2 is the limit for which, starting from a mixed state and increasing α, the CSTR itself cannot be kept in the the flow state. However, this value is slightly different from α_f. Actually, although the ratio of the gel volume to the CSTR volume is small, when the system is in the mixed state the gradients of concentration at the CSTR/gel interface are large so that the last term in eqn. (2) is not completely negligible. There is a backflow of the gel contents into the CSTR which decreases the stability of the CSTR state F. For instance, in Fig. 3b, one notices that the concentration of OH^- in the CSTR is slightly smaller when the gel is in the mixed state. Thus, if the gel is in a mixed state, the transition of the CSTR contents from the flow to the thermodynamic state is advanced to a smaller value. As the theory of spatial bistability[5,6] also predicts that, in a mixed state, the gradients at the boundary and the related corrections are practically independent of the system size, the limit α_2 is almost constant and equal to $\alpha_2 = 0.772$, a value slightly lower than $\alpha_f = 0.833$.

It is clear that this diagram is not in quantitative agreement with the experimental data. Besides the crudeness of the kinetic description, these discrepancies question the validity of our diffusion assumptions. As an alternative, we have fixed all the diffusion coefficients to $D_1 = 10^{-5}$ cm^2 s^{-1} except those of the two fast ions H^+ and OH^- that were fixed to $D_2 = 3 \times 10^{-5}$ cm^2 s^{-1} and rebuilt a diagram (Fig. 4b). Although we have kept the rudimentary kinetic description, one obtains a quasi-quantitative agreement for the limits α_1 and α_2 in the domain of width l explored in our experiments. For technical reasons, values with $l \ll 1$ mm could not be experimentally tested, so that the comparison is limited to the values where these limits are almost constant. The limit α_1 is shifted to 0.81 (as in the experiments), which indicates a smaller feedback effect of the gel contents on the CSTR. The limit for $\alpha_2 \simeq 0.67$ for $l \geq 1$ mm, much lower than with equal diffusion coefficients, is close to the experimental values (0.69–0.70). Moreover, for these values of l, in a narrow range of α (less than 1%) located just before the transition at α_1, the front of the mixed state is no longer stationary but exhibits periodic oscillations of position in time. If the transverse direction (azimuthal) were to be taken into account, any initial shift of phase should give rise to traveling waves. This oscillatory destabilization of the mixed state could account for the experimentally observed excitability properties of the F state beyond the domain of spatial bistability. The domain and amplitude of oscillations decrease with l as does the domain of traveling waves in the experiments. Moreover, preliminary two-dimensional computations with a simpler model have produced excitation waves analogous to those actually observed in the experiments. Since the system only exhibits bistability in homogeneous conditions or when the diffusion coefficients are all equal, this indicates that the excitability and traveling wave originate in the introduction of additional timescales by differential diffusion. These timescales correspond to the different times taken by the different species to diffuse up to a point at a distance from the CSTR/gel boundary, typically the distance to the front. More extended computations and careful data analysis are presently in progress, both on the real kinetics and on toy models.

In this paper, we have provided a new example of spatial bistability which presents significant differences with the formerly studied systems, in particular with the prototypical CDI reaction. First, the domain of spatial bistability extends to values of the gel width l—the gel depth in thin film disc reactors—out of reach of experiments. This is related to the quasi-infinite induction time of the reaction. The second remarkable property concerns the relative stability of the F and FT states. In the CDI reaction, within the spatial bistability domain, one or the other state can propagate into the other according to the control parameter distance from the stability limits, a classical behaviour analogous to the behaviour of potential systems. On the other hand, in the C–T reaction, the mixed state always propagates, either to invade the F state, or under the form of traveling waves or excitation pulses when the mixed state is no longer stable. Most interestingly, we have reported experimental and numerical evidence for the periodic destabilization and oscillations of the front position in a system which cannot exhibit oscillations in the homogeneous case.

Acknowledgements

This work has been partly supported by a France-Argentina grant ECOS-SCyT.

References

1. P. Gray and S. Scott, *Chemical Oscillations and Instabilities*, Clarendon Press, Oxford, 1990.
2. J. Boissonade, in *Dynamics and stochastic processes*, ed. R. Lima, L. Streit and R. Vilela Mendes, Springer, Berlin, 1990, p. 76.
3. Q. Ouyang, V. Castets, J. Boissonade, J. C. Roux, P. De Kepper and H. L. Swinney, *J. Chem. Phys.*, 1991, **95**, 351.
4. P. Blanchedeau and J. Boissonade, *Phys. Rev. Lett.*, 1998, **81**, 5007.
5. P. Blanchedeau, J. Boissonade and P. De Kepper, *Physica D*, 2000, **147**, 283.
6. P. Blanchedeau, PhD Thesis, Bordeaux, 2000.
7. I. Nagypal and I. R. Epstein, *J. Phys. Chem.*, 1986, **90**, 6285.
8. Á. Tóth, D. Horváth and A. Siska, *J. Chem. Soc., Faraday Trans.*, 1997, **93**, 73.
9. Á. Tóth, I. Lagzi and D. Horváth, *J. Phys. Chem.*, 1996, **100**, 14837.
10. D. Horváth and Á. Tóth, *J. Chem. Phys.*, 1998, **108**, 1447.
11. M. Fuentes, M. N. Kuperman and P. De Kepper, *J. Phys. Chem.*, 2001, **105**, 6769.
12. W. Y. Tam, W. Horsthemke, Z. Noszticzius and H. L. Swinney, *J. Chem. Phys.*, 1991, **88**, 3395.
13. R. D. Vigil, Q. Ouyang and H. L. Swinney, *Physica A*, 1992, **188**, 17.
14. B. Rudovics, E. Barillot, P. W. Davies, E. Dulos, J. Boissonade and P. De Kepper, *J. Phys. Chem. A*, 1999, **103**, 1790.
15. K. J. Lee, D. McCormick, Q. Ouyang and H. Swinney, *Science*, 1993, **261**, 192.
16. P. W. Davies, P. Blanchedeau, E. Dulos and P. De Kepper, *J. Phys. Chem. A*, 1998, **102**, 8236.
17. P. Deuflhard, *SIAM Rev.*, 1985, **27**, 505.
18. E. M. Cussler, *Diffusion: Mass Transfer in Fluid Systems*, Cambridge University Press, Cambridge, 1997.
19. D. Snita and M. Marek, *Physica D*, 1994, **75**, 521.

Pattern formation and spatial self-entrainment in bistable chemical systems

G. Dewel,* M. Bachir, S. Métens† and P. Borckmans

Service de Chimie-Physique & CENOLI, CP 231, Campus Plaine, Université libre de Bruxelles, Blvd. du Triomphe, B-1050 Brussels, Belgium. E-mail: gdewel@ulb.ac.be

Received 17th April 2001
First published as an Advance Article on the web 12th November 2001

We describe the formation of spatial structures generated by diffusive instabilities in bistable systems. The coupling between the different spatial modes emanating from the two homogeneous steady states can then give rise to self-parametric instabilities favoring the occurrence of resonant rhombic or quasiperiodic structures such as superlattices or quasicrystalline patterns.

1. Introduction

Diffusive instabilities provide an important pattern forming mechanism for a wide variety of non-equilibrium systems.[1] In this framework, the experimental realization of sustained, well controlled, Turing structures in open gel reactors[2,3] has stimulated a large amount of work, both experimental and theoretical. Most of the theoretical studies have been devoted to systems possessing a single homogeneous steady state (HSS). As a control parameter is varied this state is eventually destabilized by inhomogeneous perturbations with a wavenumber lying in a narrow band around its critical value q_c. Above threshold the ubiquitous competition between hexagonal and striped patterns is then recovered.[4] This has indeed been observed in scores of physically dissimilar systems. A characteristic feature of reaction–diffusion systems lies in the fact that a single kinetic system can generally exhibit multiple instabilities of different natures besides the diffusive type. Some experimentally observed patterns then result from the interaction between such bifurcations.[5] For example, we have shown that the synergy between Turing and Hopf bifurcations can give rise to a new type of wave.[6]

Various experimental and numerical works have now studied diffusive structures in systems that further exhibit the phenomenon of bistability between HSS.[7–13] Turing instabilities can then appear on both the upper and lower homogeneous branches and can interact with the saddle-node bifurcations that determine the limits of the bistability domain. As the hysteresis loop is generically asymmetric the corresponding critical wavenumbers q_c^u and q_c^l are different and generate two critical circles of wave vectors of respective radii q_c^u and q_c^l in spatially extended two-dimensional (2D) systems. In this contribution, we show that the coupling between these two families of spatial modes can give rise to nontrivial resonant effects.

† Present address: Laboratoire de Physique Théorique de la Matière Condensée (Tour 24/14), Université de Paris 7, 2 place Jussieu, 75251 Parix cedex 05, France.

DOI: 10.1039/b103430h

2. Model and linear stability analysis

To be concrete, we illustrate the various behaviours by results obtained from the numerical integration of a variant of the FitzHugh–Nagumo model (FHN) defined by the following equations

$$\partial_t u = u - u^3 + \beta uv - v - \nabla^2 u$$
$$\partial_t v = \varepsilon[\gamma u - v - a] + d\nabla^2 v \quad (1)$$

The results presented below have however a wider application. Here $d = D_u/D_v$ and $\varepsilon = T_u/T_v$ are respectively the ratio of the diffusion coefficients and the characteristic chemical relaxation times of the activator u and inhibitor v concentrations. The parameters a, $\gamma(>0)$ and $\beta(>0)$ control the relative position and thus the number of intersections of the nullclines. In the following, we consider only cases where ε is such that a Hopf bifurcation never comes into play in agreement with the experimental situations we want to describe.

As the control parameter a is varied, bistability arises for $\gamma < \gamma_c$ through two back-to-back saddle-node bifurcations at a_u and a_l, linked by their unstable manifold and creating a hysteresis loop between two HSS inside the cusp region shown in Fig. 1. When $\beta = 0$, the bistability domain is symmetric with respect to the critical point ($a_c = 0$, $\gamma_c = 1$). An increase of the parameter β induces a distortion of this symmetric hysteresis loop in the $[u$ (or v), $a]$ plane.

In the presence of a differential diffusion process ($d > 1$), these HSS can be destabilized by inhomogeneous perturbations. In general, two Turing instabilities are created at (a_T, γ_T). In the $[a, \gamma]$ plane, when γ is decreased below γ_T, they migrate, invade the cusp region and approach the limit points. The location of the Turing thresholds respectively on the upper and lower branch are denoted a_T^u and a_T^l. For a given value of γ, the diffusive instability comes closest to the saddle-node bifurcation whose curvature K_i ($i = u, l$) is the largest and the corresponding wavenumber is the smallest. In our model, we have $K_u > K_l$, and thus $q_c^u < q_c^l$; the Turing bifurcation on the upper branch is also the first to come in coincidence with the limit point a_u at $\gamma = \gamma_u^L$. For $\gamma < \gamma_u^L$, the pattern forming instability only appears on the lower branch and when $\gamma < \gamma_l^L$, no more diffusive instability occurs on the HSS. The results of this linear stability analysis which are common to many bistable systems are summarized in Fig. 1.

In the experiments on the monostable chlorite–iodide–malonic acid (CIMA) reaction, the fast formation of a complex between iodide and the colour indicator introduces the necessary separation of timescales between the activator and the inhibitor that allows the onset of a Turing instability.[14] However this complex formation does not affect bifurcations between steady states such

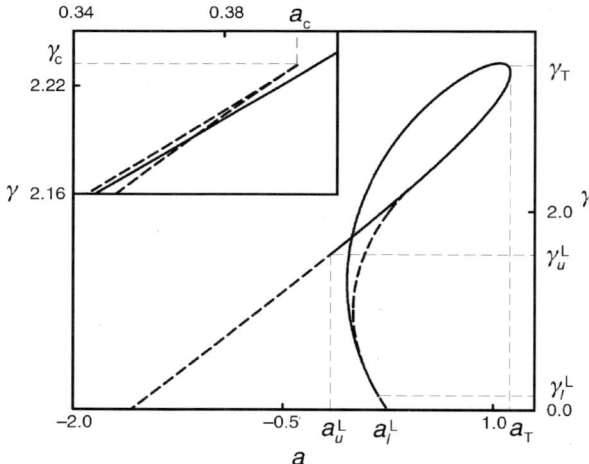

Fig. 1 Schematic phase space $[a, \gamma]$ of the asymmetric FitzHugh–Nagumo model [eqn. (1)] for $\beta = 0.99$, $d = 7.5$ and $\varepsilon = 1.5$. Dashed and solid lines respectively denote the loci of the homogeneous saddle-node bifurcations originating from the cusp at γ_c and of the Turing instabilities given by (γ_u^L γ_T γ_l^L).

as saddle-nodes of HSS or Turing instabilities. The inequality of diffusion coefficients therefore remains a necessary requirement for a diffusive instability to occur in bistable systems at least in the absence of Hopf bifurcations.

3. Pattern selection in the strong asymmetric case

Spatially extended systems wherein the boundaries are too far away to play a role in the nucleation mechanism of the structures exhibit orientational degeneracy. In asymmetric bistable systems,[15] one then faces a pattern selection problem with the presence of two critical wavevector circles of radii q_c^u and q_c^l. All modes with wavevectors lying on these circles are indeed equally amplified.

We therefore approximate the concentration fields by a linear superposition of the homogeneous mode A_0, m_j critical modes associated with the diffusive instability on the lower branch ($\{A_j\}$) and m_k spatial modes pertaining to the instability occurring at a_T^u ($\{B_k\}$):

$$c = e_1 A_0 + e_2 \sum_j A_j \exp[i\boldsymbol{q}_j^l \cdot \boldsymbol{r}] + e_3 \sum_k A_k \exp[i\boldsymbol{q}_k^u \cdot \boldsymbol{r}] + \text{c.c} + \text{h.o.t.} \quad (2)$$

where c is the concentration vector (here built on u and v), $|\boldsymbol{q}_j| = q_c^l$; $|\boldsymbol{q}_k| = q_c^u$, and c.c and h.o.t are complex conjugate and higher order terms, respectively. The coefficients e_m are the eigenvalues of the corresponding Jacobian matrix. The zero mode A_0, that becomes marginal at the hysteresis limits, is included in the set of active modes, as we have discussed previously, to obtain a global description of the bifurcation diagram.[10,16]

Applying standard methods,[17] coupled evolution equations for these complex amplitudes can be derived for a given reaction–diffusion model, in particular the FHN model we treat here. Generically, chemical systems possess quadratic nonlinearities that, in these amplitude equations, generate interactions between triplets of spatial modes forming a triangle $\boldsymbol{q}_j + \boldsymbol{q}_k + \boldsymbol{q}_l = 0$. Equilateral triangles $|\boldsymbol{q}_i| = q_c$ correspond to the ever present hexagonal patterns that have also been observed in bistable systems.

The amplitude equations generally have a complex structure. However they greatly simplify in two specific cases. Symmetric systems ($\beta \approx 0$ and $q_c^u \approx q_c^l$) present a (1:1) resonance. Then the patterned solutions emanating from the two Turing instability points are interconnected in the bifurcation diagram ($\{A_j\} = \{B_k\}$).[10] Similarly, in the strong asymmetry limit, the Turing instability is very close to the limit point on the upper HSS (it may even disappear altogether). The branches of inhomogeneous states issued from this point cannot be stabilized anymore and therefore a single family of spatial modes $\{A\}$ comes into play.[15]

Mixed modes of hexagonal symmetry for which $A_0 \neq 0$ appear near a_T^l. The sum of the phases of the modes forming the equilateral triangles of sides $|\boldsymbol{q}_i| = q_c^l$ satisfy the following equation

$$d_t \Phi = [gA_0 - \alpha]R \sin \Phi \quad (3)$$

where $\Phi = \phi_1 + \phi_2 + \phi_3$ with $A_i = R \exp(i\phi_i)$ and α and g are positive.

It is thus the sign of $\Gamma = (gA_0 - \alpha)$ that determines the nature of these structures. Near a_T^l, A_0 is small and for the values of the parameters we have considered, Γ is then negative. Thereby Φ relaxes to zero giving rise to a honeycomb lattice. Since A_0 is an increasing function of the control parameter a, Γ vanishes somewhere in the bistable region thus allowing for the occurrence of stripes. Further increasing a, Γ becomes positive and a triangular lattice comes into play. The following sequence, HSS1/hexagons/stripes/re-entrant hexagons/HHS2 is thus obtained as the unfolding parameter a is increased through the bistable region. This is the distorted version, because of the presence of the asymmetric term β, of the sequence that was observed in the symmetric case. As shown in Fig. 2 the stability domain of the re-entrant hexagons ($\Phi = \pi$) is widened with respect to that of the hexagonal structures created at a_T^l on the lower branch which can even disappear when β is sufficiently large, leading to the non-standard sequence HSS1/stripes/hexagons/HHS2. These sequences present generic features since they have been observed in physically very dissimilar systems and in particular in experiments with the CIMA reaction in boundary-fed reactors.[18] The latter observations might indicate that for such experimental conditions the system operates near a cusp point.

The chlorine dioxide–iodide reaction exhibits a region of bistability in a CSTR.[19] The HSS in

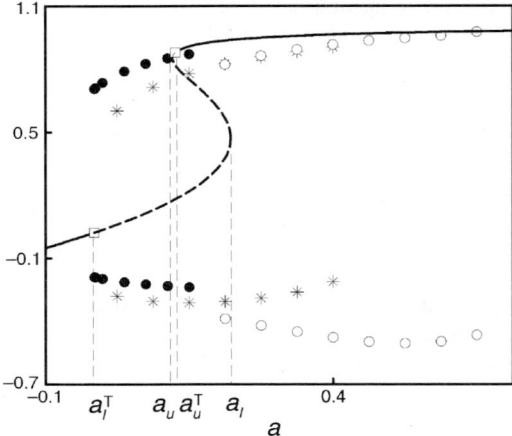

Fig. 2 Bifurcation diagram obtained by numerical integration of the FitzHugh–Nagumo model [eqn. (1)] for $\beta = 0.99$, $\gamma = 1.9$, $\varepsilon = 1.5$ and $d = 7.5$. The extrema of the amplitudes of the various Turing structures are given as a function of the control parameter a. The black circles, the stars and the white circles respectively correspond to hexagons, stripes and re-entrant hexagons.

this zone are however strongly affected by the presence of bifurcations (*e.g.* Hopf) leading to time dependent oscillations of the concentrations. Nevertheless the oscillatory region can again be shrunk by the use of a complexing agent for the I_3^- species thereby exposing a range of parameters where the system exhibits "clean" bistability of HSS. Fig. 3 exhibits a typical Turing bifurcation diagram[15] obtained in this region when allowing for the CFUR approximation.[20] It constitutes an illustration of a strongly asymmetric case as the Turing instability on the upper branch almost straddles the corresponding saddle-node bifurcation. We thus recover the HSS1/hexagons/stripes/re-entrant hexagons/HHS2 sequence. We furthermore uncovered a branch of harmonic hexagonal patterns characterized by the activation of the triplet of $\sqrt{3}\, q_c^l$ modes besides the basic q_c^l hexagonal triad.

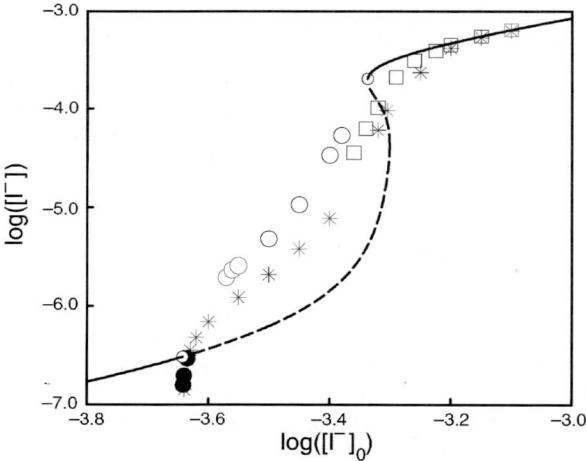

Fig. 3 Bifurcation diagram obtained for a model of the chlorine dioxide–iodide reaction in a region of parameter space where the system is bistable. The following structures are obtained starting from the smaller values of control parameter $[I^-]_0$ where there is a stable HSS of low $[I^-]$ undergoing a Turing bifurcation. Hexagons (black circles), stripes (stars), re-entrant hexagons (white circles) and harmonic hexagons, sometimes named 'black eyes' (squares). One finally recovers a stable HSS of higher $[I^-]$. Various hysteresis are visible as well as the differences in subcriticality or extensions of existence of the various structures due to the asymmetric effects. The other parameters are $[ClO_2]_0 = 1.0 \times 10^{-4}$ M, $[Complexant]_0 = 4.5 \times 10^{-3}$ M, $k_0 = 10^{-2.66}$ s^{-1} and the ratio of diffusion coefficients $d = 5$.

4. Rhombic, superlattice and quasicrystalline resonant patterns

In asymmetric bistable systems, one can also form resonant triads combining critical wavevectors from the two different critical circles and forming isosceles triangles. One can distinguish two types of such structures:

(i) $c = e_1 A_0 + e_2[A_1 \exp(i\boldsymbol{q}_1^l \cdot \boldsymbol{r}) + A_2 \exp(i\boldsymbol{q}_2^l \cdot \boldsymbol{r})] + e_3 B_3 \exp(i\boldsymbol{q}_3^u \cdot \boldsymbol{r}) + \text{c.c}$

$$\text{with } \boldsymbol{q}_1^l + \boldsymbol{q}_2^l + \boldsymbol{q}_3^u = 0 \quad (4)$$

(ii) $c = e_1 A_0 + e_2[A_1 \exp(i\boldsymbol{q}_1^u \cdot \boldsymbol{r}) + A_2 \exp(i\boldsymbol{q}_2^u \cdot \boldsymbol{r})] + e_3 B_3 \exp(i\boldsymbol{q}_3^l \cdot \boldsymbol{r}) + \text{c.c}$

$$\text{with } \boldsymbol{q}_1^u + \boldsymbol{q}_2^u + \boldsymbol{q}_3^l = 0 \quad (5)$$

The corresponding amplitude equations can be written in the following form (case (i))

$$\frac{dA_1}{dt} = \mu_1(A_0)A_1 + (v - g_{ND} A_0)A_2^* B_3^* - g_D|A_1|^2 A_1 - g_{ND}[|A_2|^2 + |B_3|^2]A_1 \quad (6)$$

The equation for A_2 can easily be obtained from eqn. (6) by permutation of indices 1 and 2.

$$\frac{dB_3}{dt} = \mu_2(A_0)B_3 + (v - g_{ND} A_0)A_1^* A_2^* - g_D|B_3|^2 B_3 - g_{ND}[|A_1|^2 + |A_2|^2]B_3 \quad (7)$$

$$\frac{dA_0}{dt} = f(A_0) + (v - g_{ND} A_0)[|A_1|^2 + |A_2|^2 + |B_3|^2] - g_{ND}[A_1 A_2 B_3 + A_1^* A_2^* B_3^*] \quad (8)$$

Where $f(A_0) = 0$ determines the HSS. The linear growth rates $\mu_i(A_0)$, which are nonlinear functions of A_0, and the coupling constants can be expressed in terms of the parameters of the model. Similar equations can be derived for the case (ii) by permuting the amplitudes A and B.

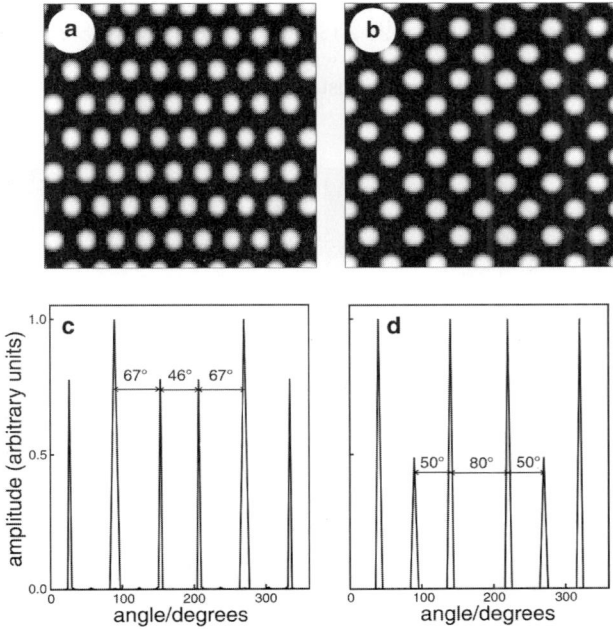

Fig. 4 2D rhombic patterns obtained by numerical integration of the FitzHugh–Nagumo model [eqn. (1)] for $a = -0.1$, $\beta = 0.7$, $\gamma = 0.9$, $\varepsilon = 2.5$ and $d = 120$. Rhombs (a) and (b) respectively correspond to cases (i) and (ii) in the text for which the sum of the phases of the modes $\Phi = 0$. (c) and (d) are the corresponding angular amplitude distribution in spatial frequency space.

In real space such triads generate *rhombic* patterns. In fact they correspond to the stretching or compression of an hexagonal array along one of its symmetry axis. As a consequence the corresponding Fourier transforms present either two large and four small peaks or the opposite (Fig. 4). Such rhombs can be shown to coexist with the regular hexagonal patterns in the vicinity of the Turing instability points.

These resonant patterns should not be confused with the unstable rhombic planforms characterized by two critical wavevectors of length $q_c^i(i = u$ or $l)$ and making an angle such that their sum does not correspond to another active mode. In the latter the phases of the modes are arbitrary. In contrast, as is clear from eqn. (6)–(8), the phases of the modes forming isosceles triangles adjust in such a way as to favor the occurrence of resonant rhombs. Again two conjugated rhomb solutions arise as solutions of the corresponding amplitude equations. Their competition with the standard hexagonal patterns, has been observed in experiments with the CIMA reaction.[21] The origin of nonequilateral triangles has also been ascribed to the existence of non-gradient quadratic terms that include spatial derivatives in the envelope equations.[22,23]

When the critical wavenumbers are sufficiently different, more complex structures can also be generated. They correspond to patterns that have periodicity on two length scales $2\pi/q_c^u$ and $2\pi/q_c^l$ dictated by the instabilities appearing on the HSS. In Fourier space, they are characterized by wave vectors combining resonant triplets (equilateral or isosceles) of the types discussed above. In Fig. 5 an example is shown of a *superlattice* constructed on 9 modes corresponding to wave vectors arranged by decorating an equilateral triangle q_c^l of modes with three isosceles triangles whose two equal sides are q_c^u. The corresponding planform is then

$$c = e_1 A_0 + e_2 \sum_{j=1}^{3} A_j \exp[i\boldsymbol{q}_j^l \cdot \boldsymbol{r}] + e_3 \sum_{k=1}^{6} A_k \exp[i\boldsymbol{q}_k^u \cdot \boldsymbol{r}] + \text{c.c} \qquad (9)$$

with $\boldsymbol{q}_1^l + \boldsymbol{q}_2^l + \boldsymbol{q}_3^l = 0$, $\boldsymbol{q}_1^l + \boldsymbol{q}_2^u + \boldsymbol{q}_3^u = 0$, $\boldsymbol{q}_2^l + \boldsymbol{q}_3^u + \boldsymbol{q}_4^u = 0$ and $\boldsymbol{q}_3^l + \boldsymbol{q}_5^u + \boldsymbol{q}_6^u = 0$.

As preceedingly, coupled amplitude equations can be derived. Here again, the sign of the quadratic term determines the sum of the phases in each resonant triad.

Dynamic superlattice structures have been obtained in nonlinear optical devices[24,25] and in two-frequency Faraday experiments.[26,27] More recently, the stability of stationary superlattices generated by a single Turing instability on a regular lattice has also been studied.[28] The corresponding wavelength is then chosen in such a way that the lattice can accommodate a pattern of given symmetry. On the other hand, in the bistable systems we have considered, the two wave-

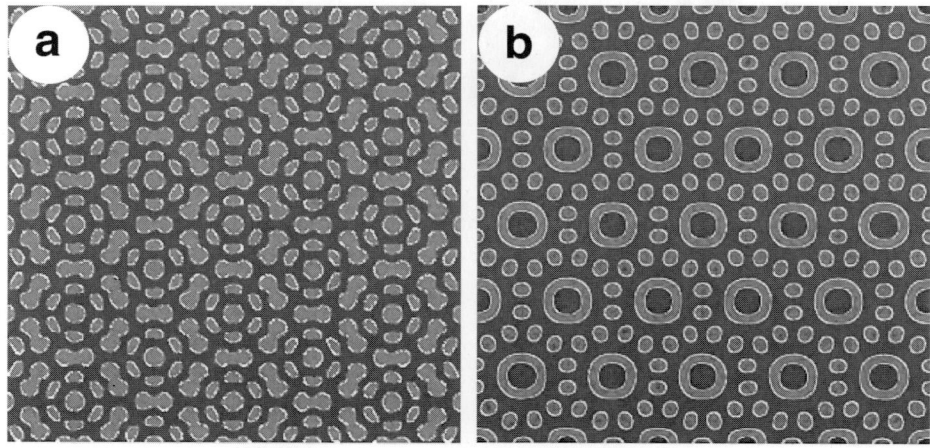

Fig. 5 Superlattice structures obtained by numerical integration of the FitzHugh–Nagumo model [eqn. (1)] for $a = -0.1$, $\beta = 0.7$, $\gamma = 0.9$, $\varepsilon = 2.5$ and $d = 120$ ($q_c^l/q_c^u = 1.3$). They are characterized by 9 pairs of critical modes forming an equilateral triangle decorated by three isosceles triangles on each side of the first as discussed in the text. The summit angle θ of the isosceles triangles is respectively larger than $\pi/3$ (a) or smaller than $\pi/3$ (b).

Fig. 6 Quasicrystalline structures obtained by numerical integration of the FitzHugh–Nagumo model [eqn. (1)]. It is characterized in Fourier space by two stars of wave vectors: one of six vectors of length q_c^u and the other of twelve vectors of length q_c^l symmetrically distributed on the two critical circles.

numbers are *naturally* associated to the two Turing instabilities on the HSS. However our scenario does not reduce to the models introduced preceedingly from a study of *ad hoc*[29,30] reaction–diffusion systems because of the intervention of the zero mode A_0.

Varying parameters in such a way as to modify the summit angle θ of the isosceles triangles one generates a sequence of superlattice patterns. When $\theta < \pi/3$, the Fourier spectrum exhibits two stars: one of 6 vectors placed symmetrically on the critical circle of radius q_c^u, the other of twelve vectors on the circle q_c^l. Their relative orientation is selected to favor resonant interactions between the modes. The reverse situation arises when $\theta > \pi/3$. However when $\theta = \pi/6$, a *quasicrystalline* pattern arises when both families of vectors are symmetrically disposed on the two critical circles (Fig. 6). Such complex structures have also been proposed to explain some aspects of the skin patterns of certain reptiles.[31] In that case they were obtained by the inclusion of mechanochemical contributions that generate linear dispersion curves with two minima at different wavenumbers for a single Turing instability in a model presenting a single HSS.

Finally resonant interactions between the two families of critical modes also occur in the case of structures periodic in one direction (stripes) when the critical wave numbers satisfy the condition: $q_c^l = 2q_c^u$. Only the patterns with the largest wave number q_c^l then survive. This provides a nice example of self-entrainment of spatial modes emanating from different branches of the bistable. It should not be confused with 1:2 resonance that occurs when the first superharmonic of the critical mode becomes active in a system exhibiting a pattern forming instability with a flat dispersion curve.[32] In bistable systems, the coupling with the zero mode prevents the onset of the drift instability and reinforces the stability of the pattern with wave number $q_c^l = 2q_c^u$. A similar self-resonance effect again occurs when $q_c^l = 3q_c^u$; two types of patterns with wavenumbers q_c^u and $3q_c^u$ are then obtained according to the initial conditions.

The mechanisms presented above are still operative in the monostable region close to the critical point.

5. Conclusion

We have shown that asymmetric bistable systems naturally possess the two ingredients necessary for the occurrence of resonant quasiperiodic structures: two different critical wavenumbers and quadratic nonlinearities favoring interactions among triplets of active modes. These patterns thus appear through self-parametric instabilities wherein the spatial modes emanating from one HSS play the role of forcing for the others.

Acknowledgements

This work was supported by Grants from the CGRI-FNRS/CNRS and the "Fondation Universitaire Van Buuren" (Belgium). P.B. and G.D. received support from the FNRS (Belgium) and M.B. from the "Direction de l'Enseignement Supérieur" (Morocco).

References

1. D. Walgraef, *Spatio-Temporal Pattern Formation*, Springer, Berlin, 1997.
2. V. Castets, E. Dulos, J. Boissonade and P. De Kepper, *Phys. Rev. Lett.*, 1990, **64**, 2953.
3. Q. Ouyang and H. L. Swinney, *Nature*, 1991, **352**, 610.
4. P. Borckmans, G. Dewel, A. De Wit and D. Walgraef, in *Chemical Waves and Patterns*, ed. R. Kapral and K. Showalter, Kluwer, Dordrecht, 1995, pp. 323–364; J. Boissonade, E. Dulos and P. De Kepper, in *Chemical Waves and Patterns*, ed. R. Kapral and K. Showalter, Kluwer, Dordrecht, 1995, p. 221.
5. G. Dewel, A. De Wit, S. Métens, J. Verdasca and P. Borckmans, *Phys. Scr. T*, 1996, **67**, 51.
6. J.-J. Perraud, A. De Wit, E. Dulos, P. De Kepper, G. Dewel and P. Borckmans, *Phys. Rev. Lett.*, 1993, **71**, 1272.
7. A. A. Afanas'ev, Yu. A. Logvin, A. M. Samson and B. A. Samson, *Opt. Commun.*, 1995, **115**, 559.
8. W. Breazeal, K. M. Flynn and E. G. Gwinn, *Phys. Rev. E*, 1995, **52**, 1503.
9. T. Ackemann, *Phys. Rev. Lett.*, 1995, **75**, 3450.
10. S. Métens, G. Dewel, P. Borckmans and R. Engelhardt, *Europhys. Lett.*, 1997, **37**, 109.
11. D. Michaelis, U. Peschel and F. Lederer, *Phys. Rev. A*, 1997, **56**, R3366.
12. D. Horvath and A. Toth, *J. Chem. Soc., Faraday Trans.*, 1997, **93**, 4301.
13. M. Hildebrand, A. S. Mikhailov and G. Ertl, *Phys. Rev. E*, 1998, **58**, 5483.
14. I. Lengyel and I. R. Epstein, *Proc. Natl. Acad. Sci. USA*, 1992, **89**, 3977.
15. M. Bachir, PhD Thesis, Université Libre de Bruxelles, 2000.
16. G. Dewel, S. Métens, M'F. Hilali, P. Borckmans and C. B. Price, *Phys. Rev. Lett.*, 1995, **74**, 4647.
17. M. C. Cross and P. C. Hohenberg, *Rev. Mod. Phys.*, 1993, **65**, 854.
18. Q. Ouyang and H. L. Swinney, in *ChemicalWaves and Patterns*, ed. R. Kapral and K. Showalter, Kluwer, Dordrecht, 1995, pp. 269–295.
19. I. Lengyel, J. Li and I. R. Epstein, *J. Phys. Chem.*, 1992, **96**, 7032.
20. J. A. Vastano, J. E. Pearson, W. Horsthemke and H. L. Swinney, *J. Chem. Phys.*, 1988, **88**, 6175.
21. Q. Ouyang, G. H. Gunaratne and H. L. Swinney, *Chaos*, 1993, **3**, 707.
22. E. A. Kuznetsov, A. A. Nepomnyashchy and L. M. Pismen, *Phys. Lett. A*, 1995, **205**, 261.
23. B. Peña and C. Perez-Garcia, *Europhys. Lett.*, 2000, **51**, 300.
24. E. Pampaloni, S. Residori, S. Soria and F. T. Arrecchi, *Phys. Rev. Lett.*, 1997, **78**, 1042.
25. Z. H. Musslimani and L. M. Pismen, *Phys. Rev. E*, 2000, **62**, 389.
26. A. Kudrolli, B. Pier and J. P. Gollub, *Physica D*, 1998, **123**, 99.
27. H. Arbell and J. Fineberg, *Phys. Rev. Lett.*, 1998, **81**, 4384.
28. S. L. Judd and M. Silber, *Physica D*, 2000, **136**, 45.
29. T. Frish and G. Sonnino, *Phys. Rev. E*, 1995, **51**, 1169.
30. R. Lifshitz and D. M. Petrich, *Phys. Rev. Lett.*, 1997, **79**, 1261.
31. L. J. Show and J. D. Murray, *SIAM J. Appl. Math.*, 1990, **50**, 628.
32. M. R. E. Proctor and C. A. Jones, *J. Fluid Mech.*, 1988, **188**, 301.

Turbulent fronts in resonantly forced oscillatory systems

Christopher Hemming and Raymond Kapral

Chemical Physics Theory Group, Department of Chemistry, University of Toronto, Toronto, ON M5S 3H6, Canada

Received 10th April 2001
First published as an Advance Article on the web 7th November 2001

Phase fronts in the forced complex Ginzburg–Landau equation, a model of a resonantly forced oscillatory reaction–diffusion system, are studied in the 3 : 1 resonance regime. The focus is on the turbulent (Benjamin–Feir-unstable) regime of the corresponding unforced system; in the forced system, phase fronts between spatially uniform phase-locked states exhibit complex dynamics. In one dimension, for strong forcing, phase fronts move with constant velocity. As the forcing intensity is lowered there is a bifurcation to oscillatory motion, followed by a bifurcation to a regime in which fronts multiply *via* the nucleation of domains of the third homogeneous phase in the front. In two dimensional systems, rough fronts with turbulent, complex internal structure may arise. For a critical value of the forcing intensity there is a nonequilibrium phase transition in which the turbulent interface grows to occupy the entire system. The phenomena we explore can be probed by experiments on periodically forced light sensitive reaction–diffusion systems.

1 Introduction

Resonant forcing of oscillatory reaction–diffusion systems can lead to a variety of chemical patterns not seen in unforced systems. In a well-stirred system, an external periodic perturbation, sufficiently close to the resonant frequency, causes phase-locking of the oscillatory kinetics. At the $n : m$ resonance stable oscillations may occur only with one of n different phases. These n stable phase states are equivalent under phase shifts of $2\pi/n$.

In a reaction–diffusion system, a spatially-distributed field of these local n-stable oscillators is coupled by diffusion. The system tends to form spatially uniform domains inside which the local oscillations are synchronized to one of the phase-locked states. The boundaries between domains, where the phase of the oscillations shifts, are domain walls or "phase fronts". Such phase fronts have been observed in experimental studies of the ruthenium-catalyzed Belousov–Zhabotinsky reaction in a continously-fed open reactor, illuminated with a periodic light source.[1–4]

In many cases the dynamics of patterns in the system can be understood in terms of the dynamics of the phase fronts. For example, a transition between labyrinthine and non-labyrinthine stationary two-phase patterns in experiments in 2 : 1 forced systems was attributed to a lateral instability in the phase front.[2] Travelling loop structures in 2 : 1 experiments were ascribed to the presence of a nonequilibrium Ising–Bloch front bifurcation.[5] Earlier theoretical work had predicted that such a bifurcation could exist in 2 : 1 phase fronts.[6]

In this paper, we describe phase fronts in the resonantly forced complex Ginzburg–Landau

DOI: 10.1039/b103237m

(FCGL) system

$$\frac{\partial A(\mathbf{r}, t)}{\partial t} = (\mu + i\nu)A - (1 + i\beta)|A|^2 A + \gamma \bar{A}^{n-1} + (1 + i\alpha)\nabla^2 A. \qquad (1)$$

This equation describes the envelope of oscillations of a forced system at the $n:m$ resonance (n, m are coprime integers, $n \leqslant 4$) near the Hopf bifurcation.[7,8] The parameter γ is the amplitude of the forcing. The reagent concentrations $c(\mathbf{r}, t)$ in the original reaction–diffusion system which the complex Ginzburg–Landau equation models are related to the complex amplitude $A(\mathbf{r}, t)$ according to

$$c(\mathbf{r}, t) - c_0 \approx A\mathbf{u}e^{i\omega_f t/n} + \bar{A}\bar{\mathbf{u}}e^{-i\omega_f t/n}, \qquad (2)$$

where $\mathbf{u}$ and $\bar{\mathbf{u}}$ are the critical eigenvectors involved in the Hopf bifurcation and $\omega_f \approx n\omega_0/m$ is the forcing frequency. As well as being the normal form of the bifurcation, eqn. (1) is widely used as a generic model of resonantly forced oscillatory reaction–diffusion systems. In the remainder of this paper we focus on the $n = 3$ case corresponding to the 3:1 resonance.

The ordinary differential equation (ODE) describing the local dynamics of the medium in the absence of diffusion, $dA/dt = (\mu + i\nu)A - (1 + i\beta)|A|^2 A + \gamma \bar{A}^{n-1}$, has only the trivial fixed point $A = 0$ for $\gamma = 0$. As γ increases, there is a bifurcation where n pairs of fixed points appear. From eqn. (2) we see that these fixed point solutions correspond to phase-locked oscillatory solutions in the original forced reacting system that is modelled by the reduced FCGL dynamics. The phase shift symmetry under $t \to t + 2\pi/\omega_f$ in the reacting system system appears in the FCGL equation as the phase shift symmetry $A \to Ae^{2i\pi/n}$; hence, we consider the FCGL equation to describe the dynamics of the original system viewed stroboscopically in Poincaré planes taken at intervals $\Delta t = 2n\pi/\omega_f$. The FCGL equation exhibits spatially uniform phase-locked domains and phase fronts which correspond to those observed in resonantly-forced reaction–diffusion systems.

The unforced CGL equation (eqn. (1) with $\gamma = 0$), exhibits phase turbulence in the Benjamin–Feir-unstable regime, which is the region of parameter space in which $1 + \alpha\beta < 0$. If forcing is added to the CGL equation in the Benjamin–Feir-unstable regime (*i.e.* if γ is taken to be nonzero) the synchronizing effect of the forcing opposes the desynchronizing effect of the turbulent CGL dynamics. Coullet and Emilsson investigated eqn. (1) in this regime.[9–11] In the one-dimensional case, when the forcing intensity is high, one observes phase fronts which either may be stationary or may translate monotonically with a constant velocity v and have a front profile $A(x, t) = A(x - vt)$. This behavior is identical to the Benjamin–Feir-stable case.[12] For $n = 2$, as γ is lowered, one observes a Hopf bifurcation in which an oscillatory motion is superposed on the translating (or stationary) front motion. As γ is decreased further a series of period-doubling bifurcations occurs. Eventually chaotic motion of the front ensues. At still lower γ, a regime in which a turbulent region is nucleated at the front is observed, and ultimately the turbulent region expands to fill the system. The dynamics of phase fronts in the $n = 2$, $d = 1$ system was also investigated by Mizuguchi and Sasa[13] and Battogtokh and Browne[14]. Phase fronts in two dimensions exhibit spontaneous roughening of the front profile.[9]

Phase front roughening in two dimensions has been observed in a coupled map lattice (CML) system which may be regarded as a model for a periodically forced oscillatory system.[15–18] If the uncoupled map is chosen such that it possesses a superstable period-3 cyclic solution $1 \to 2 \to 3 \to 1$, there are three cyclic solutions differing in phase by one discrete time step. In this sense the map is a model for a 3:1 resonantly forced oscillator. In two dimensions, the fronts exhibit a variety of other dynamical phenomena, including a nonequilibrium phase transition to a regime of strong turbulence. Before the transition, interfaces between spatially uniform regions are rough. As a control parameter is varied, the intrinsic width of the interface increases and complex, irregular structure develops. As the transition is approached the width of the complex, disordered interfacial zone increases. The intrinsic interface width diverges at the transition; beyond the transition the spatially uniform phases are unstable relative to the turbulent phase; hence, an initially present interface will grow to fill the entire system with the turbulent phase. In this paper we show that qualitatively similar phenomena are found in the 3:1 FCGL system.

Section 2 is devoted to a study of the one-dimensional, $n = 3$, FCGL equation. In Section 3 we consider phase fronts in two-dimensional systems and in Section 4 we discuss the strong turbulence regime found beyond the nonequilibrium phase transition. The conclusions of the paper are given in Section 5.

2 One-dimensional fronts

We consider first the simplest case: eqn. (1) with μ, ν, α and β such that the unforced ($\gamma = 0$) equation is in the Benjamin–Feir-stable regime. For $\gamma > [2((1 + \beta^2)(\mu^2 + \nu^2))^{1/2} - 2(\mu + \beta\nu)]^{1/2}$, the corresponding ODE $dA/dt = (\mu + i\nu)A - (1 + i\beta)|A|^2 A + \gamma \bar{A}^2$ possesses 3 stable fixed points A_0^i, $i = 1, 2, 3$ (and 3 corresponding unstable ones, as well as the unstable $A = 0$ fixed point) which are also solutions of the spatially distributed system. These solutions are linearly stable to spatially homogeneous perturbations, and in the Benjamin–Feir-stable regime are stable to spatially inhomogeneous perturbations. In one dimension, the next simplest solution is a front joining two of these spatially uniform states. These phase fronts travel at a constant velocity v with constant shape, $A(x, t) = A(x - vt)$.

When the unforced system is Benjamin–Feir-unstable, such uniformly translating front solutions are stable for large values of γ. Fig. 1 shows such a front solution for the parameter values $\mu = 1$, $\nu = 1.55$, $\alpha = -1.3$, $\beta = 1.5$ and $\gamma = 0.63$. Throughout this paper we fix μ, ν, α and β at these values; γ is taken to be the control parameter.†

As γ is lowered, a bifurcation to oscillatory front motion occurs at γ_c. In general, these oscillating fronts have a non-zero average velocity. In the $n = 2$ case, a Hopf bifurcation to oscillatory motion was found.[9,10,13] However, in the $n = 3$ system the bifurcation is not a Hopf bifurcation: the period T of the oscillations diverges as $T \sim |\gamma_c - \gamma|^{-1/2}$ as $\gamma \to \gamma_c^-$ (Fig. 2 (left)). At the bifurcation, the average front velocity, scales as $v - v(\gamma_c) \sim a_\pm |\gamma_c - \gamma|^{1/2}$ (Fig. 2 (right)), where $a_\pm$ depends on the sign of $\gamma_c - \gamma$.

For yet lower γ a bifurcation to a regime of weak turbulence occurs (see Fig. 3). The mechanism for the appearance of weak turbulence involves the nucleation of a finite-sized domain of the third phase at the front. For example letting [12] denote a front separating phases 1 and 2, nucleation of phase 3 at the front leads to a [13···32] front pair, which travels in the opposite direction. These newly-formed fronts subsequently undergo similar nucleation events; in addition fronts may collide and annihilate. Fig. 3 illustrates a number of such front-splitting, reversal, and annihilation events. One can see that the number of fronts increases and that the size of the region in which fronts are found increases. In this weakly turbulent region the front splitting and reversal events exhibit complex dynamics. A front-splitting regime of this kind has not been reported for the

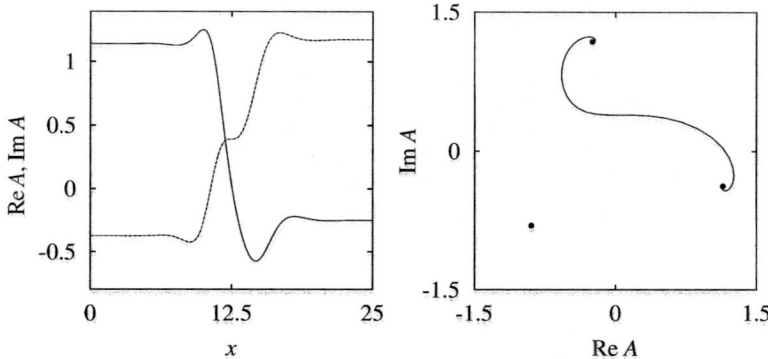

Fig. 1 A travelling front solution in the one-dimensional $n = 3$ FCGL equation for $\gamma = 0.63$. Other parameters are given in the text. Left: Re A (solid line) and Im A (dashed line) against x. Right: The front plotted as a trajectory in the A-plane (solid line); the three filled circles are the three stable spatially uniform states.

---

† Numerical integration of the FCGL equation was performed using explicit forward differencing with a time step size of $\Delta t = 0.01$. The lattice spacing was $\Delta x = 0.25$ in one dimension. In two dimensions, $\Delta x = 0.25$ or 0.5 and a second-order discrete Laplacian was used. Time intervals and lengths are reported in the absolute time and space units in which eqn. (1) is written.

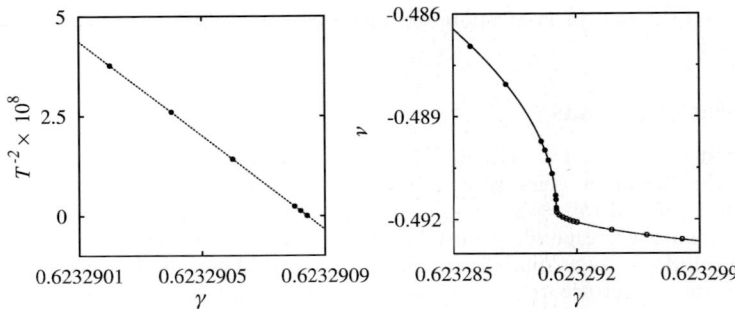

Fig. 2 Left: Plot of T^{-2} against γ, where T is the period of oscillations in the one-dimensional FCGL equation. The filled circles are simulation values, the dotted line is $T^{-2} = 0.05888176(\gamma_c - \gamma)$, where $\gamma_c = 0.6232908414$. Right: Front velocity v against γ. Circles, simulation values; hollow circles, non-oscillatory regime and filled circles, average velocity in the oscillatory regime. The solid line is the curve $v = v_c + 2.196(\gamma_c - \gamma)^{1/2}$; the dashed line is the curve $v = v_c - 0.3192(\gamma - \gamma_c)^{1/2}$, where $v_c = -0.491742$.

$n = 2$ case. The repeated formation of new fronts and spatially homogenous domains observed here bears some resemblance to a mechanism giving rise to spatiotemporal chaos that has been described in a reaction–diffusion system with one stable homogenous state.[19]

Continuing to decrease γ, we encounter windows of regular dynamics within the domain of weakly turbulent dynamics. An interesting localized structure observed was an oscillating hole with zero average velocity within a single phase-locked domain (Fig. 4). This was found at a γ value for which phase fronts are unstable. The character of the spatiotemporal dynamics for

Fig. 3 Space–time plot of the phase ϕ of the complex amplitude $A(x, t) = Re^{i\phi}$ in the one-dimensional FCGL equation for $\gamma = 0.60210562$. Time is on the ordinate and increases downward. The time interval shown is 2500 time units. The system size is 1250 space units, the plot shows a portion with length 750 space units. The initial conditions were $A(x, 0) = \theta(-x)A_0^L + \theta(x)A_0^R$, where $\theta(x)$ is the Heaviside function and A_0^L and A_0^R, (L, R $\in \{1, 2, 3\}$, $L \neq R$) are the fixed points of the ODE represented by the medium and light grey shades, respectively. The third fixed point is represented by the dark grey shade.

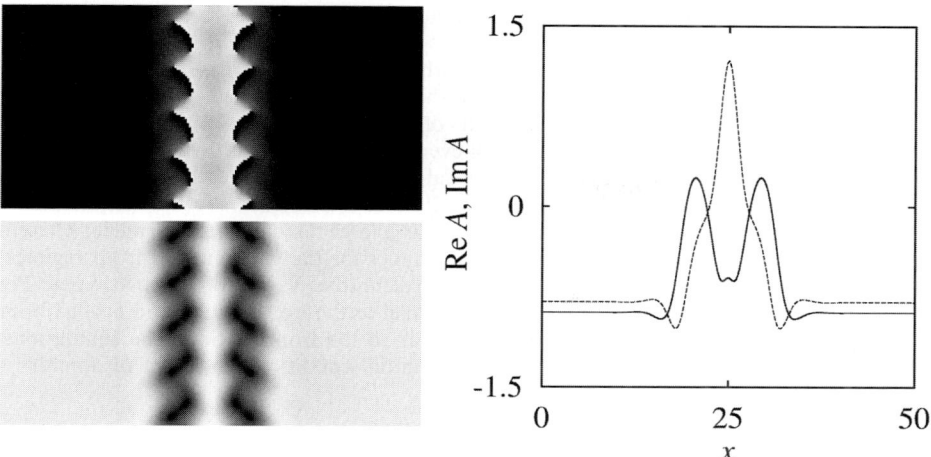

Fig. 4 A zero-velocity hole solution for $\gamma = 0.573$. On the left are space–time plots of the phase (upper panel) and amplitude (lower panel) of A. The plots show an interval of 20 time units. The system size is 1250 space units, the plots show a portion with length 50 units. Right: Graph of Re A (solid line) and Im A (dashed line) against position, recorded at the time at which the space–time plots begin.

$\gamma \approx 0.2$ takes the form of strong turbulence where both the phase and amplitude vary irregularly on short distance and time scales. An example of turbulence of this type is shown in Fig. 5. One can see in this figure that the homogeneous regions within the turbulent phase are rare and short-lived.

Having given this brief overview of the phenomenology of fronts in the one-dimensional system; we consider two-dimensional fronts where new phenomena arise.

3 Two-dimensional fronts

In two spatial dimensions front structure and evolution exhibit a number of distinctive characteristics. For sufficiently high γ planar two-dimensional fronts exist which are similar to those seen in one-dimensional systems. For lower γ, the front profile roughens but the intrinsic width of the front is small and comparable to that of the planar front. The front position oscillates locally but

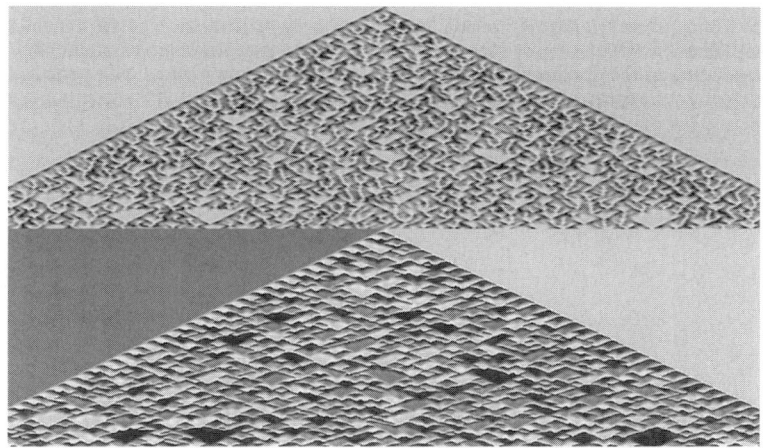

Fig. 5 Space–time plot of the amplitude (upper panel) and phase (lower panel) of the dynamics in the strong turbulence regime for $\gamma = 0.2$. The system size is 1250 space units, the plots show a portion 500 units in length. The time interval is 300 units. The initial conditions were of the form described in the Fig. 3 caption.

these oscillations lose phase-synchronization along the front. Fig. 6 shows typical examples of fronts of this kind for $\gamma = 0.59$.

Decreasing γ further, the average intrinsic width of the fronts increases and complex internal structure develops. In this regime we may regard the front as a turbulent phase separating the two spatially homogenous phases. Fig. 7 shows fronts of this type for $\gamma = 0.46$.

The average width of the interface increases, eventually diverging as $\gamma \to \gamma^*$. For our parameter values $\gamma^* \simeq 0.457$. If $\gamma < \gamma^*$ the turbulent interfacial phase is no longer confined but grows without bound, eventually filling the entire system, regardless of system size. We may regard this "front explosion" as the loss of stability of the homogeneous phases relative to the turbulent phase.

We have statistically characterized the front dynamics for the two confined front regimes mentioned above, using $\gamma = 0.58$ and $\gamma = 0.49$ as representative values of the two cases. Periodic boundary conditions were used in the direction parallel to the front motion and no-flux boundary conditions were imposed on the edges perpendicular to the front motion. The simulations were performed in a frame moving with the front. The initial conditions consisted of domains of the

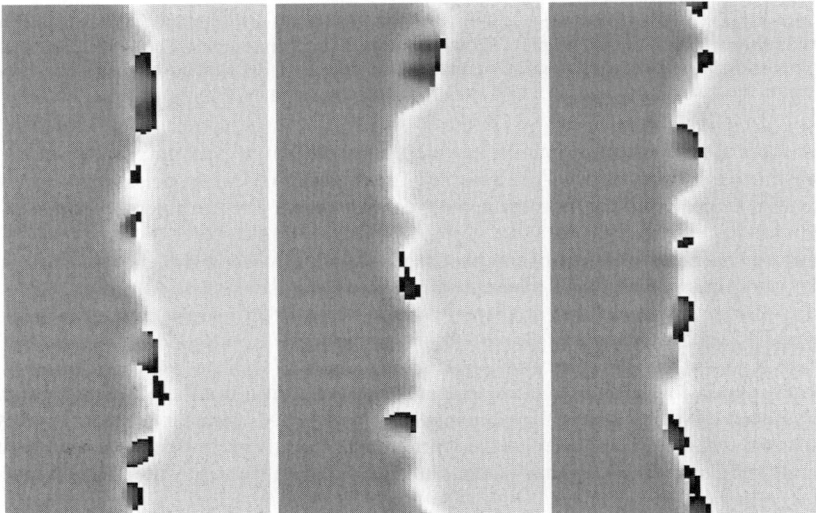

Fig. 6 Typical fronts in the "thin front" regime taken from a single realization of the dynamics at three well separated times. Here $\gamma = 0.59$. The grey-scale indicates the phase ϕ of the complex amplitude $A = Re^{i\phi}$, with $\phi = -\pi$ corresponding to the darkest shade and $\phi = +\pi$ corresponding to the lightest shade. Points around which all shades of grey appear are phase defects. The system size is 200×200 space units, a portion with dimensions 100×200 is shown.

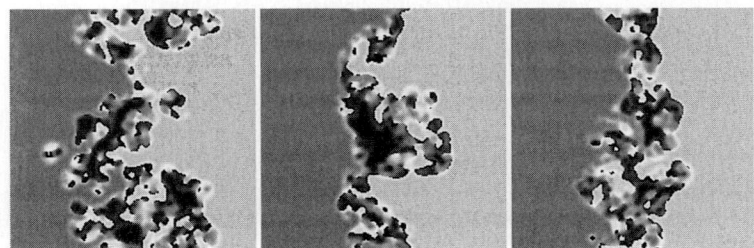

Fig. 7 The turbulent front in a system with $\gamma = 0.46$ at three different times. Grey-scale coding is the same as in Fig. 6. The system size is 800×200 space units, a portion with dimensions 200×200 is shown.

spatially uniform stable states separated by a randomly seeded planar strip. More precisely,

$$A(x, y, 0) = \begin{cases} A_0^L, & -\infty < x < -W/2 \\ R_0 \, e^{i\phi(x, y)}, & -W/2 \leqslant x \leqslant W/2 \\ A_0^R, & W/2 < x < +\infty \end{cases} \quad (3)$$

where W is the width of the randomly seeded strip, $\phi(x, y)$ is a random variable uniformly distributed on $[-\pi, +\pi)$, A_0^L and A_0^L ($L, R \in \{1, 2, 3\}$) are fixed points of the ODE and $R_0 = |A_0^L| = |A_0^R|$.

We define the position of the right profile $h_R(y, t)$ to be the greatest value of x for which $|A(x, y, t) - A_0^R| \geqslant \varepsilon$. An analogous criterion is used to define the left profile: $h_L(y, t)$ is the smallest value for which $|A(x, y, t) - A_0^L| \geqslant \varepsilon$. The value $\varepsilon = 0.02$ was chosen; comparison of the front profiles determined by this method with the images of the phase and amplitude fields confirmed that this value for ε defined the boundaries of the interfacial zone well.

We also define the mean positions of the left and right profiles, $\bar{h}_{L/R}(t) = L^{-1} \int_0^L dy \, h_{L/R}(y, t)$, the deviations from the mean, $\delta h_{L/R}(y, t) = h_{L/R}(y, t) - \bar{h}_{L/R}(t)$, and the widths of the left and right profiles,

$$w_{L/R}(t) = \left\{ \frac{1}{L} \int_0^L dy \, \delta h_{L/R}^2(y, t) \right\}^{1/2}. \quad (4)$$

Diffusively rough interfaces typically obey the following scaling relations:[20] at small t the ensemble average width $\langle w(t) \rangle \sim t^{\hat{\beta}}$, until it saturates at $w_{\text{sat}} \sim L^{\hat{\alpha}}$. One may rescale the $\langle w(t) \rangle$ vs. t curve by these scaling exponents to obtain $\langle w(t) \rangle / L^{\hat{\alpha}}$ vs. $t/L^{\hat{\alpha}/\hat{\beta}}$ which is independent of the system size. Fronts described by the Edwards–Wilkinson (EW) equation[21] have $\hat{\alpha} = 1/2$ and $\hat{\beta} = 1/4$ while Kardar–Parisi–Zhang (KPZ) fronts[22] have $\hat{\alpha} = 1/2$ and $\hat{\beta} = 1/3$.

For $\gamma = 0.58$ ("thin front" case), using both EW and KPZ exponents, neither the left nor the right interfacial profiles yielded data collapse. However, rescaling by the exponents $\hat{\alpha} = 1/2$ and $\hat{\beta} = \hat{\alpha}/3.0 \simeq 0.17$ leads to a reasonable collapse of the $\langle w(t) \rangle$ data (see Fig. 8 (left)), except for the smallest system size, $L = 200$. The oscillations which are clearly visible at early times in the $L = 200$ curve are also present in the data for larger system sizes but are collapsed by the rescaling to a small time interval near the origin and are not visible in the figure. The fact that KPZ exponent scaling fails to provide a data collapse may possibly be ascribed to the correlated nature of the turbulent dynamics driving the front roughening process in the FCGL system. The oscillations present at early times are an indication that the spectrum of the deterministic noise in this system is unlikely to be white.

The above results on the width scaling structure are not generally valid for all fronts observed in this system and the scaling exponents are found to be a function of γ. For example, for $\gamma = 0.49$,

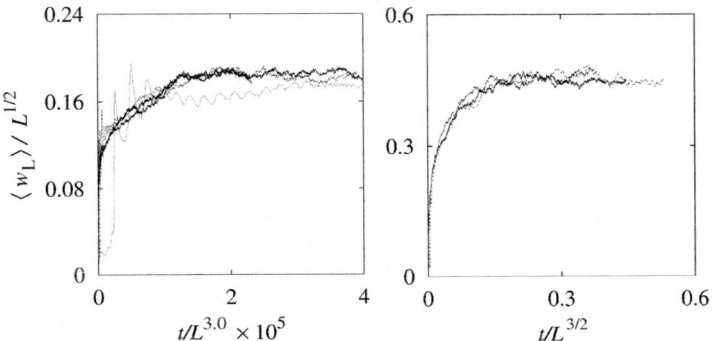

Fig. 8 Left: The scaled left interfacial profile width vs. time, $\langle w_L(t) \rangle / L^{1/2}$ vs. $t/L^{3.0}$, for system sizes $L = 600$ (solid line), $L = 500$ (long-dashed line), $L = 400$ (short-dashed line) and $L = 200$ (dotted line) for $\gamma = 0.58$. Right: Scaled plot of the average width of the left profile, $\langle w_L(t) \rangle / L^{1/2}$ vs. $t/L^{3/2}$ for system sizes $L = 500$ (solid curve), 400 (dashed curve) and 200 (dotted curve) for $\gamma = 0.49$. Each curve is an average over 100 realizations from initial conditions eqn. (3).

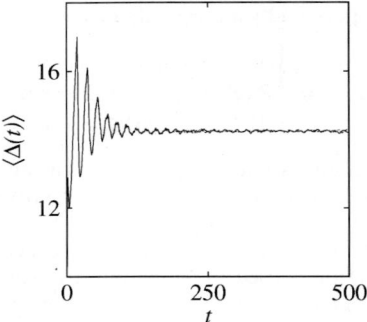

Fig. 9 $\langle \Delta(t) \rangle$ vs. t for $L = 400$ (solid line) and $L = 600$ (dashed line) for $\gamma = 0.58$. The average is taken over 100 realizations from initial conditions eqn. (3).

KPZ scaling is found to produce good data collapse (see Fig. 8 (right)). The interface for this parameter value is thick and the internal turbulent structure is more fully developed. As a result the deterministic chaos likely has shorter spatial and temporal correlations making a KPZ approximation to the interfacial dynamics reasonable.

In addition to the average interfacial profile width discussed above we may examine the structure of the intrinsic width of the front $\Delta(y, t)$ which is defined by $\Delta(y, t) = h_R(y, t) - h_L(y, t)$. A plot of $\langle \Delta(t) \rangle$ against t is shown in Fig. 9 for $\gamma = 0.58$. Here $\langle \cdot \rangle$ is an average over the front length and realizations. One observes oscillatory decay to a constant asymptotic value. Note that the curves obtained are essentially indistinguishable for the two different system sizes considered in the figure. The spatial and temporal autocorrelation functions of the intrinsic width are defined by

$$C_\Delta(x) = \langle \delta\Delta(y, t')\delta\Delta(y + y', t')\rangle/\langle \delta\Delta^2 \rangle, \qquad (5)$$

$$C_\Delta(t) = \langle \delta\Delta(y', t)\delta\Delta(y', t + t')\rangle/\langle \delta\Delta^2 \rangle, \qquad (6)$$

where the $\langle \cdot \rangle$ represents an average over y' and t' after the front has reached its statistically stationary regime and $\delta\Delta = \Delta - \langle \Delta \rangle$. The temporal and spatial autocorrelation functions are plotted in Fig. 10. From this figure we see that both temporal and spatial autocorrelations decay rapidly and both the correlation time and correlation length increase with decreasing γ. From an examination of instantaneous pictures of the front in Fig. 7 one can see the formation of large fluctuations in the intrinsic interfacial width which extend over long distances. The sizes of these extended turbulent domains increase as γ decreases and this is signalled in the increased correlation length. Similarly, the timescale on which these fluctuations decay increases for lower γ values.

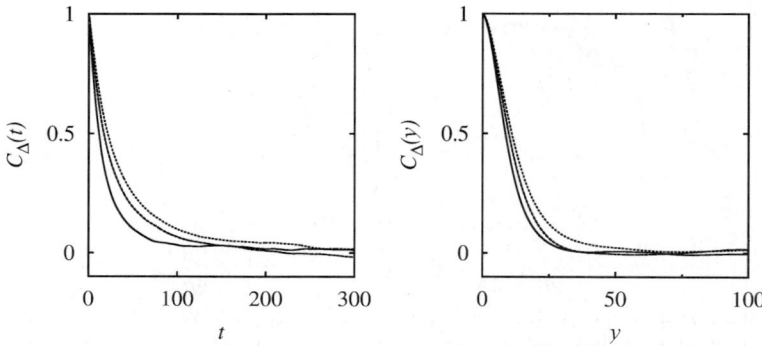

Fig. 10 (left) Temporal autocorrelation function $C_\Delta(t)$ and (right) spatial autocorrelation function $C_\Delta(y)$ for several different γ values: $\gamma = 0.49$ (solid line), $\gamma = 0.48$ (dashed line) and $\gamma = 0.475$ (dotted line). These were measured in simulations in systems of size $L = 200$.

4 The strong turbulence regime in two dimensions

Below $\gamma^* \simeq 0.457$, starting from a planar front separating two stable homogeneous phases, the width Δ increases without bound; thus, the turbulent phase eventually fills the entire system. The time evolution of the front for $\gamma = 0.45$ is presented in Fig. 11 and shows the unbounded growth of the turbulent phase for this γ value. For two-dimensional systems we call this the "strong turbulence" regime. The largest Lyapunov exponent λ in a system containing only the turbulent phase was measured[23] for $\gamma = 0.45$ and found to be $\lambda \simeq 0.31$, signalling the presence of deterministic chaos in this model. Consistent with this, spatial

$$C_A(r) = \langle \bar{A}(r+r')A(r') \rangle / \langle |A|^2 \rangle, \tag{7}$$

and temporal

$$C_A(t) = \langle \bar{A}(t+t')A(t') \rangle / \langle |A|^2 \rangle \tag{8}$$

correlations decay rapidly within the turbulent phase (see Fig. 12).

We have shown in Fig. 11 that the turbulent phase invades the stable homogeneous phases. In this regime there is only one interfacial profile to consider since the turbulent phase is not confined and it is no longer necessary to distinguish left and right interfacial profiles. If the turbulent phase is on the left and the homogeneous phase is on the right, the profile $h(y, t)$ is defined identically to the right profile $h_R(y, t)$ above, i.e., as the greatest x such that $|A(x, y, t) - A_0| > \varepsilon$. It is interesting to investigate the scaling properties of the width of this interfacial profile. The scaling studies were performed starting from initial conditions of the form

$$A(x, y, 0) = \begin{cases} \tilde{A}(x, y), & x \leqslant 0 \\ A_0, & x > 0. \end{cases} \tag{9}$$

where A_0 is one of the stable spatially uniform states, and $\tilde{A}(x, y)$ is a turbulent state prepared by allowing initial conditions of the form

$$A(x, y) = R_0 \, e^{i\phi(x, y)} \tag{10}$$

to evolve under the FCGL dynamics for $t = 200$ time units. In eqn. (10), $R_0 = |A_0|$, and $\phi(x, y)$ is a random variable uniformly distributed on $[-\pi, \pi)$.

Fig. 11 The interface in a system with $\gamma = 0.45$, in the strong turbulence regime, at $t = 500, 1500$ and 3000. The initial condition was a planar interface with a thin randomly-seeded strip. The grey-scale indicates the phase field ϕ of the complex amplitude $A = Re^{i\phi}$ with the darkest shade corresponding to $\phi = -\pi$ and the lightest corresponding to $\phi = +\pi$. The system size is 800×200.

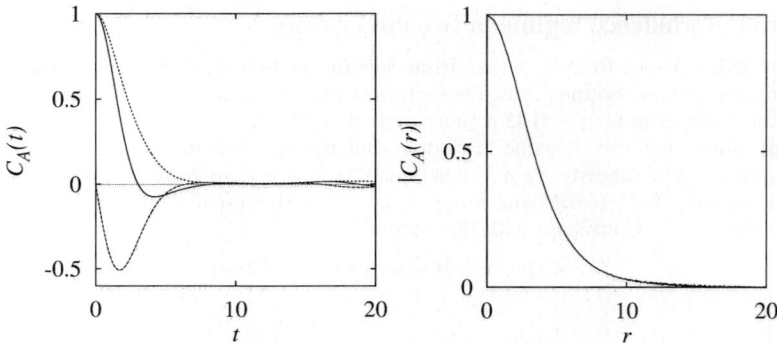

Fig. 12 Left: Temporal autocorrelation function. Real (solid line) and imaginary (dashed line) parts for $\gamma = 0.40$. The magnitude $|C_A(t)|$ is shown as a dotted line. Right: Spatial autocorrelation function for the same γ value. The imaginary part is close to zero and the real part is indistinguishable from $|C_A(r)|$ when plotted.

The results of this calculation are shown in Fig. 13. One can see that the width has a KPZ scaling structure.

5 Discussion and conclusions

We have found phase front dynamics in the 3 : 1 FCGL equation that are qualitatively similar to those reported in a coupled map lattice with period-3 local dynamics. In particular, we observed front roughening, thick fronts with complex internal structure and a "front-explosion", nonequilibrium phase transition.[16,18] In the CML these phenomena were associated with stable chaos. In contrast, we have shown that the FCGL equation exhibits deterministic chaos. When the internal turbulent structure is sufficiently developed, the fronts obey KPZ scaling. When the fronts are thin, however, the dynamics possess temporal and spatial correlations and the fronts no longer have KPZ scaling properties.

The CML dynamics of the thick fronts separating two homogeneous phases and the critical properties of the nonequilibrium phase transition were described well by a phenomenological stochastic model in which a turbulent phase was identified in the interfacial zone.[15] Consequently, the thick front may be decomposed into two interacting fronts (the left and right profiles of the thick front) separating the turbulent phase from the homogeneous phases. In this model Edwards–Wilkinson equations for the left and right profiles were coupled by terms describing the interaction between these profiles. Currently, investigations are in progress to characterize the critical properties of the FCGL phase transition. This should allow one to determine the validity of

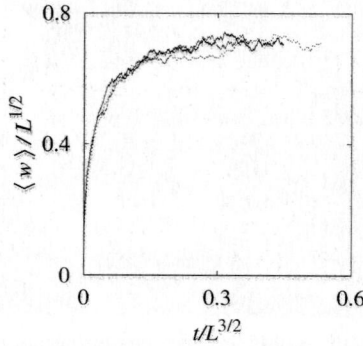

Fig. 13 Plot of $\langle w_h(t)\rangle/L^{1/2}$ vs. $t/L^{3/2}$ for system sizes $L = 500$ (solid curve), 400 (dashed curve) and 200 (dotted curve) for $\gamma = 0.43$. The $L = 400$ and $L = 500$ curves are averages over 100 realizations from initial conditions (9), the $L = 200$ curve is an average over 200 realizations.

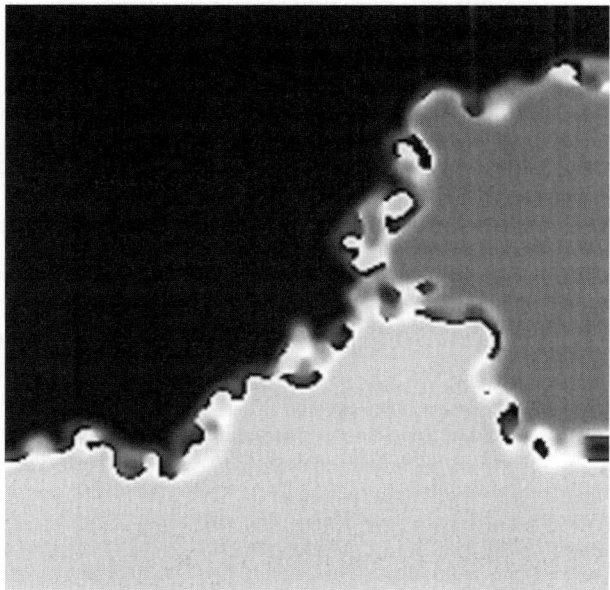

Fig. 14 The phase field ϕ in a simulation with $\gamma = 0.49$ from initial conditions containing a phase defect. The system size is 200 × 200.

phenomenological models of the form described above for the CML, and to shed light on the nature of the front explosion and its description as a nonequilibrium phase transition.

The phenomenology of the 3 : 1 FCGL equation in the Benjamin–Feir-unstable regime is remarkably rich. Some phenomena are similar to those previously found in the 2 : 1 FCGL equation while others have no known counterparts; however, there appears to be no reason why corresponding behaviour should not exist at other resonances.

The results presented here open a number of avenues for further investigation, both theoretical and experimental. Worthy candidates for investigation are the transition between what are apparently weakly turbulent and strongly turbulent regimes in the one-dimensional system, the zero-velocity holes and the origins of the non-KPZ scaling in the two-dimensional "thin front" regime. The nature of the bifurcation from constant-velocity motion to oscillatory motion in one-dimension is also an interesting question.

The FCGL equation is a sufficiently faithful model, as far as qualitative phenomenology is concerned, that there is a good likelihood that the phenomena described here also will exist in experimental 3 : 1 resonantly forced oscillatory systems. An experimental report of a 3-phase, phase-locked pattern in a light-sensitive Belousov–Zhabotinsky reaction shows 3 : 1 phase fronts that resemble the rough fronts in this study.[1]

Finally, we note that the foregoing study has been concerned entirely with phase fronts and has not considered other patterns that may exist in two-dimensional resonantly forced systems. Where three phases meet, one obtains three-armed spiral waves in which rough interfaces separate the arms (Fig. 14). Due to the constant nucleation and annihilation of phase defects in the turbulent interfaces, the situation is different from systems with regular dynamics where the single phase defect present may be identified with the spiral core. Instead, in systems with complex interfacial structure the core is a region with finite spatial extent. Since spiral dynamics plays an important role in reaction–diffusion systems, the investigation of the dynamics of these unusual spiral waves is a promising area for future research.

Acknowledgements

This work was supported in part by a grant from the Natural Sciences and Engineering Research Council of Canada.

References

1. V. Petrov, Q. Ouyang and H. L. Swinney, *Nature*, 1997, **388**, 655.
2. V. Petrov, M. Gustaffson and H. L. Swinney, in *Conference Proceedings of the 4th Experimental Chaos Conference*: August 6–8, 1997, Boca Raton, FL, USA, ed. M. Ding, W. Ditto, L. Pecora, S. Vohra and M. Spano, World Scientific, Singapore, 1998.
3. A. L. Lin, V. Petrov, H. L. Swinney, A. Ardelea and G. F. Carey, in *Pattern Formation in Continous and Coupled Systems*, The IMA Volumes in Mathematics and Its Applications, Vol. 115, ed. M. Golubitsky, D. Luss and S. H. Strogatz, Springer-Verlag, New York, 1999, p. 193.
4. A. L. Lin, A. Hagberg, A. Ardelea, M. Bertram, H. L. Swinney and E. Meron, *Phys. Rev. E*, 2000, **62**, 3790.
5. T. Kawagishi, T. Mizuguchi and M. Sano, *Phys. Rev. Lett.*, 1995, **75**, 3768.
6. P. Coullet, J. Lega, B. Houchmanzadeh and J. Lajzerowicz, *Phys. Rev. Lett.*, 1990, **65**, 1352.
7. J. M. Gambaudo, *J. Diff. Eqns.*, 1985, **57**, 172.
8. C. Elphick, G. Iooss and E. Tirapegui, *Phys. Lett. A*, 1987, **120**, 459.
9. P. Coullet and K. Emilsson, in *Instabilities and Nonequilibrium Structures V*, ed. E. Tirapegui and W. Zeller, Kluwer, Dordrecht, 1996, p. 55.
10. P. Coullet and K. Emilsson, *Physica D*, 1992, **61**, 119.
11. P. Coullet and K. Emilsson, *Physica A*, 1992, **188**, 190.
12. For examples of studies of phase fronts in the Benjamin–Feir-stable regime see: (a) C. Elphick, A. Hagberg, B. A. Malomed and E. Meron, *Phys. Lett. A*, 1997, **230**, 33; (b) C. Elphick, A. Hagberg and E. Meron, *Phys. Rev. Lett.*, 1998, **80**, 5007.
13. T. Mizuguchi and S. Sasa, *Prog. Theor. Phys.*, 1993, **89**, 599.
14. D. Battogtokh and D. A. Browne, *Phys. Lett. A*, 2000, **266**, 359.
15. R. Kapral, R. Livi and A. Politi, *Phys. Rev. Lett.*, 1997, **79**, 2277.
16. R. Kapral, R. Livi, G.-L. Oppo and A. Politi, *Phys. Rev. E*, 1994, **49**, 2009.
17. Y. Cuche, R. Livi and A. Politi, *Physica D*, 1997, **103**, 369.
18. A. Politi, R. Livi, G.-L. Oppo and R. Kapral, *Europhys. Lett.*, 1993, **22**, 571.
19. J. H. Merkin, V. Petrov, S. K. Scott and K. Showalter, *Phys. Rev. Lett.*, 1996, **76**, 546.
20. H. L. Barabási and H. E. Stanley, *Fractal Concepts in Surface Growth*, Press Syndicate of the University of Cambridge, New York, 1995.
21. S. F. Edwards and D. R. Wilkinson, *Proc. R. Soc. London, Ser. A*, 1982, **381**, 17.
22. M. Kardar, G. Parisi and Y.-C. Zhang, *Phys. Rev. Lett.*, 1986, **56**, 889.
23. G. Benettin, L. Galgani and J.-M. Strelcyn, *Phys. Rev. A*, 1976, **14**, 2338.

Experimental and theoretical studies of feedback stabilization of propagating wave segments

Eugene Mihaliuk,[a] Tatsunari Sakurai,[a] Florin Chirila[a] and Kenneth Showalter*[a,b]

[a] *Department of Chemistry, West Virginia University, Morgantown, WV 26506 USA*
[b] *Department of Physical Chemistry, Fritz-Haber-Institut der Max-Planck-Gesellschaft, Faradayweg 4-6, 14195 Berlin, Germany*

Received 17th April 2001
First published as an Advance Article on the web 12th November 2001

Experimental and theoretical studies of the excitability boundary for spiral wave behavior are presented. The boundary is defined by unstable wave segments, which are stabilized by using a negative-feedback control algorithm. A kinematic description of the constant-size, constant-shape wave segments is presented.

I. Introduction

Propagating waves in chemical and biological excitable media arise from the coupling of a positive feedback process, such as autocatalysis, with some form of mass transport, such as molecular diffusion. A wide variety of spatiotemporal patterns, from expanding circular and spiral waves[1–3] to highly disordered structures that resemble turbulence,[4,5] arise from the interplay of these processes. Spiral waves play a special role in spatiotemporal dynamics because they are self-sustaining wave sources that serve to organize their surroundings.

The features of propagating waves are largely determined by the excitability of the medium. Two excitability limits are typically identified when considering the possible wave behavior in an active medium. The lower limit is defined by the excitability below which the propagation of 1D waves (or unbounded 2D planar waves) is no longer possible. A second limit, which occurs at higher excitability, characterizes the behavior of 2D waves with free ends. Below this limit, waves contract at their ends and spiral waves are no longer possible.

In this paper, we present a detailed characterization of the excitability limit for spiral waves and show that it corresponds to a range of excitabilities rather than to a unique value. The evolution of waves with free ends in this range is not only influenced by the excitability of the medium but also by the wave size, such that smaller wave segments contract laterally until they disappear while larger waves expand to form spirals. There is a critical wave size separating these two outcomes that is an unstable waveform inherent to the medium, which will either grow or decay when perturbed. Different critical wave sizes exist for different excitabilities in the boundary regime, and the size of these waves increases with decreasing excitability. The locus of these unstable waves as a function of excitability forms a perturbation threshold, for perturbations in the form of different sized wave segments, separating an attractor characterized by spiral waves from an attractor characterized by the uniform steady state.

We study the locus of critical wave sizes by stabilizing the unstable waves with a feedback control algorithm. The medium excitability is continually adjusted according to the wave area such that it is increased when the wave area decreases and *vice versa*, thereby stabilizing the wave segment at its critical size. We describe experimental and numerical studies implementing this

DOI: 10.1039/b103431f

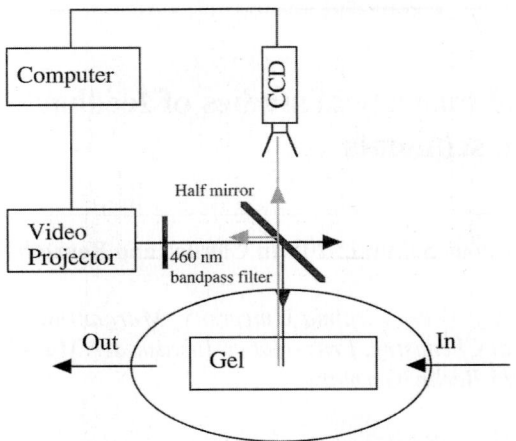

Fig. 1 Schematic diagram of the photochemical feedback experiment.

control method in the Belousov–Zhabotinsky reaction and present a theoretical study based on a kinematic analysis.

II. Experimental study

Experiments were carried out with the photosensitive Belousov–Zhabotinsky (BZ) reaction,[6,7] which was illuminated with 460 nm light to produce a desired excitability. The experiments utilized a 0.3 mm layer of silica gel in which ruthenium(II)-bipyridyl, a light-sensitive catalyst for the BZ reaction, was immobilized. The gel was cast onto a microscope slide and mounted in a reactor that was continuously fed with fresh, catalyst-free BZ solution (reactor residence time 3.0 min) maintained at 9.0 °C. A new gel containing ruthenium(II)-bipyridyl was cast for each experiment. The composition of the catalyst-free BZ solution was 0.28 M $NaBrO_3$, 0.05 M malonic acid, 0.165 M bromomalonic acid and 0.36 M H_2SO_4. The silica gel medium (0.3 × 20 × 30 mm^3) was prepared by acidifying an aqueous solution of 10% (w/w) Na_2SiO_3 and 2.0×10^{-3} M $Ru(bpy)_3^{2+}$ with H_2SO_4.

The experimental set-up is shown in Fig. 1. Images of the chemical wave were captured with a video camera and computer generated images were projected onto the gel medium using a video projector with optics modified to generate an image of approximately 20 mm × 30 mm. The beam was passed through a bandpass filter with a center wavelength of 460 nm, which is close to the

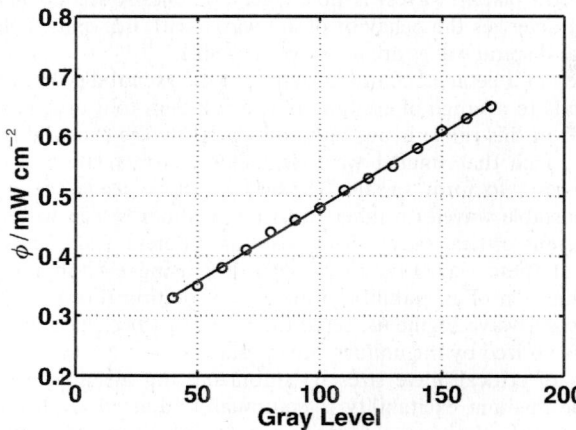

Fig. 2 Relationship between the gray level and the illumination intensity in mW cm^{-2}.

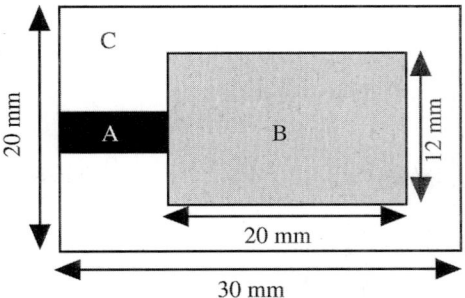

Fig. 3 Image of the illumination pattern projected onto the gel.

excitation wavelength of ruthenium(II)-bipyridyl. The gray scale between 0 and 255 was calibrated at the gel surface to illumination intensity ϕ in mW cm^{-2}, as shown in Fig. 2, and the spatial scale was calibrated at 0.0726 mm (pixel)$^{-1}$.

The reaction zone was generated by a pattern of illumination intensities produced by the video projector, as shown in Fig. 3. The excitability decreases with increasing light intensity, and wave propagation is not possible in the high intensity boundary region C (white) surrounding the reaction zone B (gray). Waves were initiated in the excitable dark region A (black) and allowed to propagate into the reaction zone.

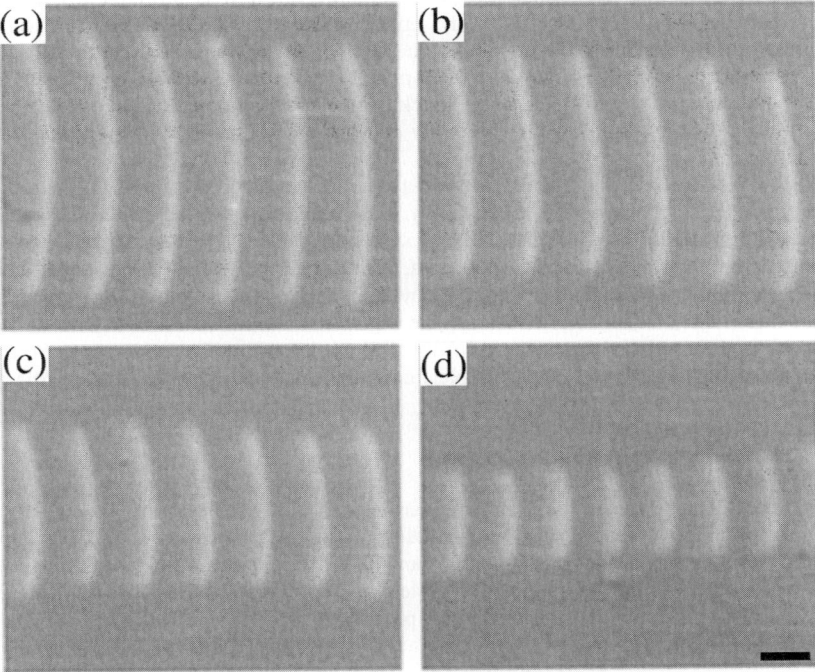

Fig. 4 Examples of typical wave segments stabilized by the feedback algorithm for four different values of the offset parameter b: -0.0744 (a), -0.0248 (b), 0.0248 (c) and 0.0744 mW cm^{-2} (d). The feedback parameter a is 0.375 in all experiments. The interval between the superimposed snapshots is 40.0 s, and the scale bar in (d) is 1.0 mm.

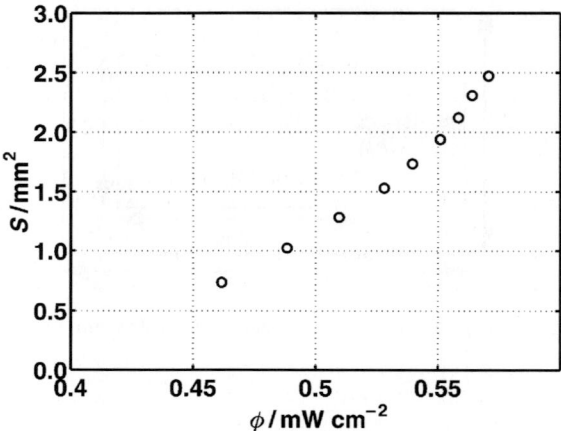

Fig. 5 Relationship between light intensity and the steady-state wave area S.

Wave stabilization is realized by adjusting the incident light ϕ in reaction zone B each iteration according to the negative feedback algorithm

$$\phi = a \times \text{area} + b$$
$$\text{area} = \sum_{x,y} \Theta(p(x, y) - p_{\text{th}})$$
$$p_{\text{th}} = 1.1\overline{p(x, y)} \quad (1)$$

The gray level is measured at each pixel $p(x, y)$, and the threshold p_{th} is subsequently set to a value slightly above the average gray level over the image. Applying the Heaviside function $\Theta(\cdot)$, we count the pixels above the threshold to obtain the area of the wave, corresponding to the region containing the oxidized ruthenium catalyst, Ru(bpy)$_3^{3+}$. This area is used every 2.0 s to calculate the new illumination intensity ϕ, with the feedback coefficient a and offset b. This adaptive threshold scheme allows the wave area to be accurately determined in the presence of unavoidable light intensity fluctuations.

According to the feedback algorithm, the excitability of the medium is reduced or increased as the wave becomes larger or smaller, respectively. Examples of waves stabilized by applying the feedback algorithm are shown in Fig. 4. The dependence of the wave area S on the total illumination intensity is shown in Fig. 5. The wave area, which is proportional to the length of the wave for all but very short wave segments, increases with decreasing excitability. The curve gives the illumination intensity at which an unstable wave segment of a particular area can be stabilized. The stabilized state has the characteristics of a saddle point, since the wave segment will either grow or decay in the presence of unavoidable fluctuations if it is not stabilized by the feedback algorithm.

III. Numerical simulations

The ruthenium catalyst of the reaction, which is embedded in the gel matrix, is excited by 460 nm light. The excited catalyst reacts with bromomalonic acid to produce bromide, an inhibitor of autocatalysis.[8] The excitability of the medium can therefore be controlled by varying the illumination intensity. The essential features of the system are described by a two-variable Oregonator model[9,10] modified to include the photochemical pathway:[11]

$$\frac{\partial u}{\partial t} = \frac{1}{\varepsilon}\left(u - u^2 - (fv + \phi)\frac{u-q}{u+q}\right) + D_u \nabla^2 u$$

$$\frac{\partial v}{\partial t} = (u - v) \quad (2)$$

where the dimensionless variables u and v correspond to $HBrO_2$ and the Ru(III) catalyst concentrations, respectively.[12] The activator diffusion coefficient is D_u, while the corresponding diffusion term for v is absent, since the catalyst is immobilized in the gel. The parameter ϕ corresponds to the light intensity and controls the excitability of the medium. The kinetic parameters ε, q and f are fixed at values such that the medium is excitable but nonoscillatory at $\phi = 0$ and can be made nonexcitable by increasing ϕ.

The feedback is implemented using a control algorithm that is essentially the same as the algorithm used in the experimental study. The size of the wave is determined each time step by counting the number of grid points with $u > u_{th}$, which corresponds to the area of the wave according to a predetermined measurement threshold u_{th}. The area averaged over 50 consecutive iterations is used to calculate the feedback signal with the gain a and offset b:

$$\phi = a \times \text{area} + b \qquad (3)$$

When the wave decreases (increases) in size, the feedback increases (decreases) the excitability of the medium, thus restoring the steady-state area. With the gain chosen appropriately, deviations from the steady state quickly become negligibly small, with only tiny variations in the excitability continually returning the wave to its intrinsic steady-state area corresponding to a given excitability. For control to be effective, a should be sufficiently large with respect to the inverse of the slope of the system response. Conversely, an oscillatory response occurs when a is made too large.[13]

As in the experiments, the initial conditions consist of a small wave segment of a particular length. Applying the control algorithm, the propagating wave segment eventually assumes a constant steady-state size and shape, as shown in Fig. 6(a). The saddle character of the stabilized wave

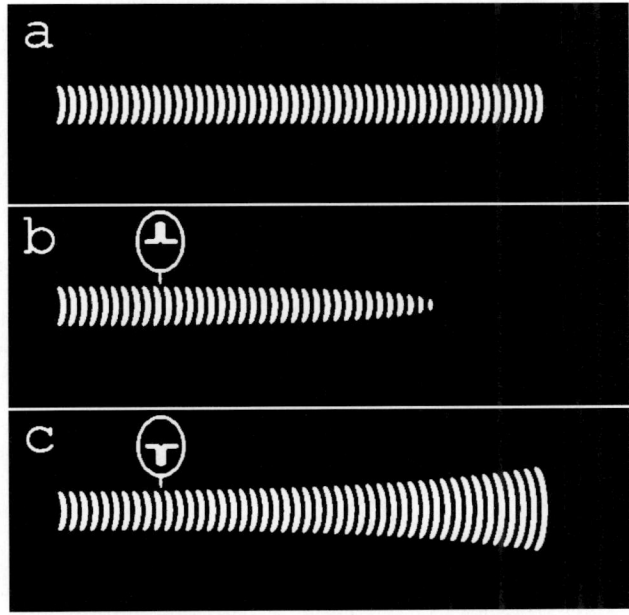

Fig. 6 Simulation of wave segment stabilization using eqn. (2). Panel (a) shows the evolution of the wave segment under feedback control with the parameters $a = 2.5 \times 10^{-4}$, $b = -0.14$, giving $\phi \approx 0.0909$. In panels (b) and (c), the control was discontinued and a small perturbation of duration $\Delta t = 0.1$ and amplitude $\Delta\phi = \pm 1.0 \times 10^{-3}$ was applied, after which the light intensity was fixed at the value applied prior to the perturbation. Due to the perturbation, the wave size becomes slightly smaller in (b) and slightly larger in (c), which is further amplified by the dynamical instability as described in the text. Eqn. (2) was integrated using an Euler method with a time step of 5.0×10^{-4} and a grid size of 0.02 on an array of 1000 × 500 grid points with zero-flux boundary conditions. The values of the kinetic parameters are $\varepsilon = 0.01$, $f = 2.5$ and $q = 0.002$, and the diffusion coefficient is $D_u = 0.1$. The image in each panel represents an overlay of snapshots taken every 0.25 time units.

is shown in panels (b) and (c), where, in the absence of control, the wave either contracts and eventually disappears or grows until it reaches the boundaries of the medium. Small perturbations were applied to accelerate the departure from the unstable steady state; however, an infinitesimal perturbation would ultimately cause the unstable wave to grow or decay.

Wave segments of different stationary sizes, as shown in Fig. 7, are obtained by changing the offset term b, which determines the resulting excitability of the medium. The steady-state wave area S increases with decreasing excitability, as shown in Fig. 8, with the area diverging at a critical value of the light intensity ϕ_∞. It should be noted that the medium remains sufficiently excitable to support the propagation of 1D waves (or unbounded 2D planar waves) at excitabilities below this critical point. The critical light intensity for propagation of 1D waves, ϕ_{1D}, occurs at a significantly lower excitability. At light intensities above ϕ_{1D}, the medium is unable to support wave propagation.

The steady-state size S as a function of excitability, shown in Fig. 8 between ϕ_S and ϕ_∞, is an intrinsic property of the medium, reflecting the kinetics of the chemical reaction and the species diffusivities. The locus of S is not influenced by the parameters of the control loop. At light intensities lower than ϕ_S, the medium is excitable and waves with free ends evolve into spiral waves, provided they are larger than a critical size. We were able to determine that wave segments smaller than a critical size contract and disappear, while larger wave segments expand to fill the

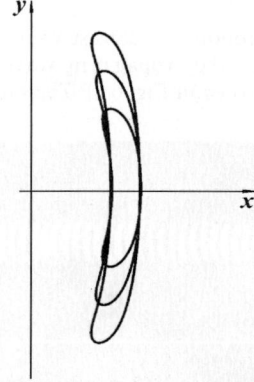

Fig. 7 Stationary contours of stabilized waves for different values of light intensity ϕ calculated with eqn. (2). The contours correspond to $u = u_{th}$, with $u_{th} = 0.25$, the same as in the area calculation step of the control algorithm. The gain $a = 2.5 \times 10^{-4}$ and the offset $b = -0.1, -0.2, -0.3$ in the small, medium and large wave segments. Model parameters and numerical integration are the same as in Fig. 6.

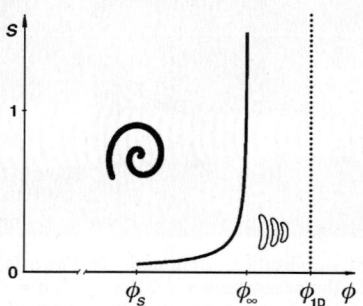

Fig. 8 Dependence of critical wave size S on light intensity ϕ between $\phi_S = 0.081$ and $\phi_\infty = 0.0915$ calculated with eqn. (2). For each value of the medium excitability (determined by the light intensity ϕ) there is a unique stationary wave size S. At ϕ_∞ the wave size diverges to infinity. The dark system with $\phi = 0$, corresponding to the nonoscillatory but excitable experimental system, lies just above the Hopf bifurcation at $\phi_H = -5.0 \times 10^{-4}$. The 1D wave propagation limit $\phi_{1D} = 0.0977$ marks the point beyond which no propagation is possible even for unbounded waves. Model parameters and numerical integration are the same as in Fig. 6.

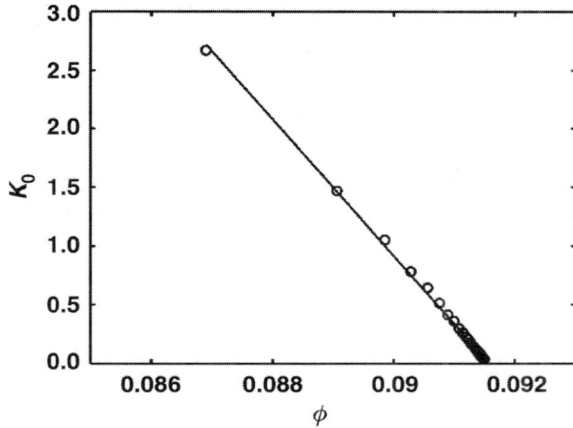

Fig. 9 The curvature K_0 at the midpoint of the wave contour as a function of light intensity ϕ. The curvature K_0 varies linearly with excitability of the medium. Model parameters and numerical integration are the same as in Fig. 6. The contours are defined as in Fig. 7.

medium with spiral waves; however, we have not characterized the critical nucleation size for spiral waves throughout the excitable regime in the present study.

Finally, we note that the curvature at the midpoint of the leading front of steady-state wave segments is a linear function of excitability, a finding that plays an important role in our subsequent theoretical analysis. This is quantified by measuring the curvature K_0 at the midpoint of the contours of different sized wave segments, such as those presented in Fig. 7, as a function of the medium excitability, as shown in Fig. 9.

IV. Kinematic model

We consider a wave segment propagating with constant size and shape through a medium with a particular excitability. Propagation of the excitation is mediated by diffusion processes, and it is therefore natural to define a contour of the excited region such that the direction normal to the contour designates the gradient of the autocatalyst and thus the direction of propagation. We reduce the problem of the wave motion to that of the motion of this contour.

Since the front is symmetrical, we consider only the upper half of the leading front, as shown in Fig. 10. The midpoint at $y = 0$ defines the motion of the wave segment as a whole, with velocity v_0 in the positive x-direction, and we therefore define this point as the 'origin' of the curve s representing the half-front. We shall use the terms 'front' and 'wave' to refer to this curve, understanding that a complete leading front can be obtained by adjoining a reflection of s with respect to the x-axis. The normal to the front at any given point makes an angle α with the x-direction, and α becomes $\pi/2$ at the extremum of the curve. We define the length of the wave as the length of the curve L at this point.

Consider the wave front at two moments t and $t + dt$, as shown in Fig. 11. To maintain shape stationarity, every point A of the curve $s(t)$ must become some point B of its rigidly translated replica $s(t + dt)$, while moving in the normal direction with the velocity $v_\perp$. This condition can be written as

$$v_\perp = v_0 \cos(\alpha) \tag{4}$$

Thus, the normal velocity changes from the maximum value v_0 at the midpoint of the wave, $l = 0$, to 0 at its endpoint, $l = L$. The different values of the normal velocity at different points along the front are due to the effect of the local curvature K on the spreading of the diffusive flow. The stationarity condition, expressed by the above equation, provides a direct connection between the functional dependence $v_\perp = f(K)$ and the stationary wave shape $\alpha(l)$.

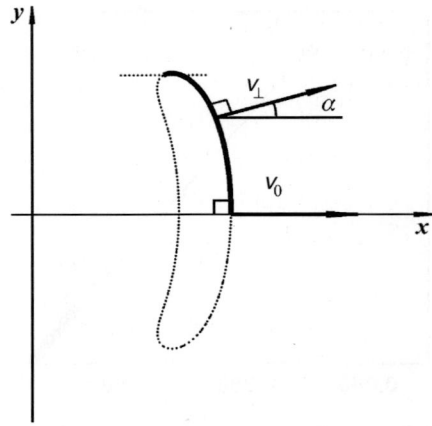

Fig. 10 The kinematic model considers one-half of the leading front of a wave segment propagating with velocity v_0 in the x-direction. The angle between the normal direction to the front and the x-direction is α, and $v_\perp$ is the normal velocity at this point.

Away from the endpoint, the curvature is sufficiently small that the normal velocity changes with the local curvature of the front according to a linear eikonal relationship:

$$v_\perp = v_\infty - DK, \qquad (5)$$

where v_∞ is the velocity of the planar front, K is the curvature, and D is a coefficient associated with the diffusivity of the autocatalyst. The velocity v_0 of the wave at the midpoint is similarly given by

$$v_0 = v_\infty - DK_0, \qquad (6)$$

where K_0 is the curvature at the midpoint. We also recall that curvature is defined by the derivative of α with respect to l. Combining eqn. (4) with eqn. (5) and using the definition $K = \alpha'$ gives

$$v_0 \cos(\alpha) = v_\infty - D\alpha' \qquad (7)$$

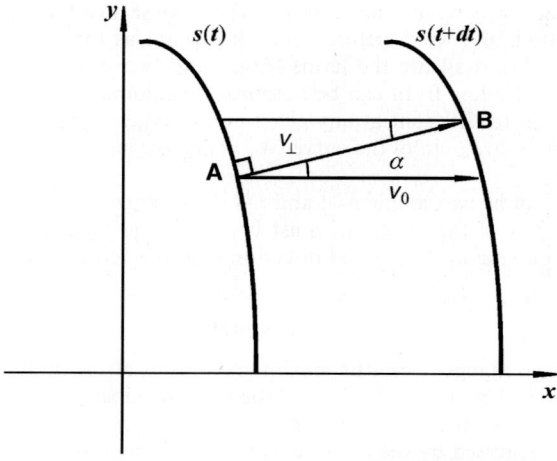

Fig. 11 Every point A of the wave front at time t becomes some point B of the wave front at time $t + dt$.

Contour lines obtained in numerical simulations confirm the validity of the above assumption (except very close to the endpoint), giving a linear relationship between the curvature α' and $\cos(\alpha) \propto v_\perp$. Expressing v_∞ in eqn. (7) in terms of K_0 and v_0 according to eqn. (6) we obtain

$$v_0 \cos(\alpha) = v_0 + DK_0 - D\alpha' \tag{8}$$

This is a differential equation with respect to α and can be easily solved without further simplifications (being a table integral):

$$\frac{v_0}{D} \int dl = \int \frac{d\alpha}{(1 + DK_0/v_0) - \cos(\alpha)} \tag{9}$$

The larger part of the front corresponds to small values of α, and we can therefore approximate $\cos(\cdot)$ with the first two terms of its Taylor series, $\cos(\alpha) \approx 1 - \alpha^2/2$ for $\alpha \ll \pi/2$, to obtain

$$\frac{v_0}{D} \int dl \approx \int \frac{d\alpha}{DK_0/v_0 + \alpha^2/2} \tag{10}$$

with the solution, considering that $\alpha = 0$ at $l = 0$,

$$\alpha = \lambda K_0 \tan(l/\lambda) \tag{11}$$

where,

$$\lambda = \sqrt{\frac{2D}{v_0 K_0}} \tag{12}$$

A comparison of the solution (11) with the partial differential equation (PDE) simulations and with the solution of the full equation (9) justifies the approximation, since there is very good agreement between both of the solutions and the simulations. Since the approximate solution gives a faithful description of the wave shape, we will use it in place of the complete solution to simplify the development.

At the endpoint $\alpha = \pi/2$, which, when substituted into eqn. (11), gives an expression for the length of the wave L:

$$L = \lambda \arctan\left(\frac{\pi}{2K_0 \lambda}\right) \tag{13}$$

Typically, $v_0 \gg DK_0$, and to a first approximation

$$L \approx \frac{\pi \lambda}{2} \tag{14}$$

since $\arctan(x) \approx \pi/2$ for $x \gg 1$. With the definition of λ from eqn. (12), a linear relationship between the square of the wave length and its radius of curvature at the midpoint is obtained:

$$L^2 \propto \frac{\pi^2 D}{2v_0} \frac{1}{K_0} \tag{15}$$

The above equations contain the parameter D/v_0 as well as the curvature at the midpoint of the wave K_0. Although a curve defined by eqn. (11) satisfies the stationarity condition and the eikonal eqn. (5) for any value of K_0, we know there exists only one solution for the wave front given a particular set of parameters for the medium. An additional constraint is therefore required to relate K_0 to the properties of the medium and select a unique solution from this parametric family. Measurements of curvature variation as a function of excitability in the PDE simulations, shown in Fig. 9, demonstrate a linear dependence of K_0 on ϕ:

$$K_0 = (\phi_\infty - \phi)/c, \tag{16}$$

Substituting this empirical relation into eqn. (15) gives

$$L^2 \propto \frac{\pi^2 D}{2v_0} \frac{c}{(\phi_\infty - \phi)} \tag{17}$$

A comparison of the PDE simulations with the prediction of eqn. (17) is shown in Fig. 12.

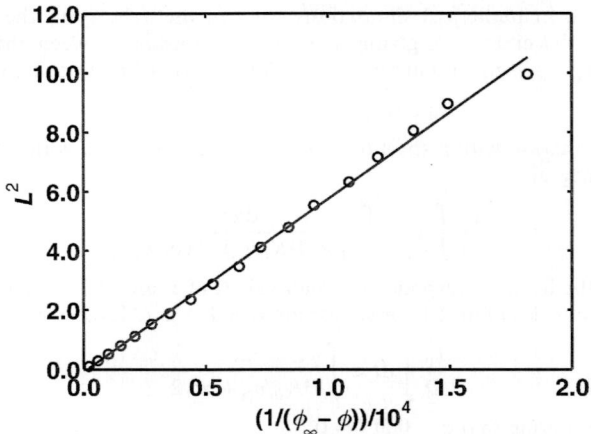

Fig. 12 The square of the length, L^2, as a function of the inverse difference between ϕ and $\phi_\infty = 0.0915$ calculated with eqn. (2). The dependence is linear for all but the shortest wave segments.

V. Discussion

We have presented a characterization of the excitability boundary, common to all active media, defining the lower excitability limit for spiral wave behavior. The boundary gives the perturbation magnitude as a function of excitability that yields either spiral patterns or the homogenous stationary state. We have studied this boundary by examining unstable wave segments that would either grow or decay were they not stabilized by the negative feedback algorithm. The unstable waves are saddle-like in character and form a one-dimensional separatrix as a function of excitability, delineating perturbations that take the system to different spatiotemporal attractors. The perturbation boundary is analogous to that in a subcritical Hopf bifurcation (for a two-variable ordinary differential equation system), where the unstable limit cycle defines the threshold for perturbations causing the system to change from steady-state to oscillatory behavior. In the spatially distributed system, the perturbation is a wave segment of a particular size and shape, which either decays to give the uniform stationary state or grows to form spiral waves that represent an oscillatory state. The distributed system requires a particular spatial perturbation; other perturbations, such as point perturbations, do not take the system to the spiral state.

From the perspective of control theory, the stabilization of unstable wave segments is at once a very simple control problem and one that is complex and fascinating. The simplicity is obvious: a negative feedback loop adjusts the excitability to offset any growth or decay of the wave segment. The complexity is more subtle: a spatiotemporal perturbation in the form of a wave segment is high dimensional, *i.e.*, a wave can take on any shape and size. However, a wave segment under control assumes a natural waveform, which is uniquely determined by the properties of the medium.

Extensive theoretical studies have been carried out on the dynamics of spiral waves,[14-20] and the lower excitability limit for spiral wave behavior has been described. The wavelength[21] of a spiral wave increases with decreasing excitability until it becomes infinite at the excitability boundary. At this critical point, the unbounded spiral wave has completely opened up to become an unbounded planar wave with a free end. As the excitability is further decreased, the planar wave contracts at its free end. The excitability limit is defined as the point where the wave becomes planar and is stationary in the moving-coordinate system. This is exactly the same excitability limit identified by the divergence of the steady-state wave segments shown in Fig. 8. The light intensity where the steady-state wave segment becomes unbounded defines a critical excitability for two-dimensional media: media with excitabilities lower than this do not support asymptotic wave propagation of waves with free ends.

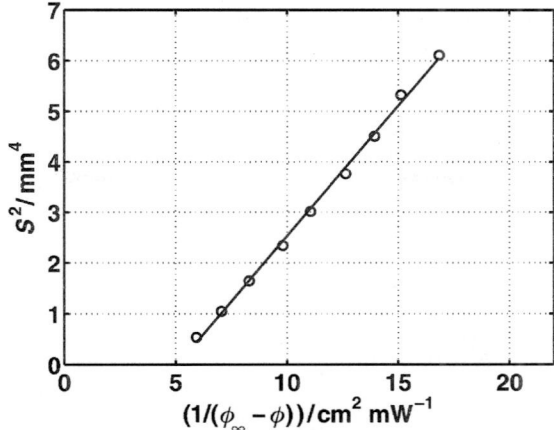

Fig. 13 Experimental data from Fig. 5, with the square of the size, S^2, plotted as a function of the inverse difference between ϕ and $\phi_\infty = 0.63$.

The kinematic description for wave segment length as a function of excitability is in excellent agreement with the numerical simulations, as shown in Fig. 12. We also note that the kinematic description is in good agreement with the experimental measurements. Fig. 13 shows the experimental data from Fig. 5, where S^2 is plotted as a function of the inverse difference between ϕ and ϕ_∞. We see very good agreement with the prediction of eqn. (17) from the kinematic description, where we note that S and L are proportional except for the smallest wave segments.

As discussed in the Introduction, there are two distinct excitability limits for wave propagation in excitable media: the boundaries for 1D and for 2D wave propagation. Fig. 8 gives a perspective of where these limits lie for the photosensitive BZ system described by eqn. (2), where excitability decreases with increasing light intensity. The regime of excitable media extends from the Hopf bifurcation, at $\phi \approx 0$, to the boundary regime for spiral wave behavior. Beyond the excitability limit, 2D waves with free ends do not propagate asymptotically, but, rather, contract and ultimately disappear. However, 2D waves without free ends, such as unbounded planar waves or circular waves larger than a critical size, are supported at these excitabilities.[22]

We have defined the excitability boundary for spiral wave behavior by the locus of steady-state wave segments between ϕ_S and ϕ_∞ in Fig. 8. The value of ϕ_∞ clearly corresponds to the divergence of this locus; however, the value of ϕ_S is somewhat arbitrarily defined according to our method of area measurement in the control algorithm. At this point it becomes necessary to redefine the size of the wave, since the amplitude of the excitatory variable decreases as the length of the wave approaches its width. We also note that the kinematic model does not accurately describe very short wave segments, for which higher-order terms in the eikonal equation become significant. The locus of steady-state wave segments extends into the excitable regime, which represents the critical nucleation size for spiral waves. We note that molecular dynamics studies of waves in this regime have been carried out.[23,24] We will report on experimental and theoretical studies of the critical nucleation size for spiral waves in a forthcoming paper.

Acknowledgements

We thank the National Science Foundation (Grant No. CHE-9974336) and the Office of Naval Research for supporting this research. K.S. thanks Alexander Mikhailov for his hospitality at the Fritz-Haber-Institut der Max-Planck-Gesellschaft and for many useful discussions that benefitted this study.

References

1 A. T. Winfree, *When Time Breaks Down*, Princeton University Press, Princeton, 1987.
2 *Chemical Waves and Patterns*, ed. R. Kapral and K. Showalter, Kluwer, Dordrecht, 1995.

3   I. R. Epstein and K. Showalter, *J. Phys. Chem.*, 1996, **100**, 13132.
4   M. C. Cross and P. C. Hohenberg, *Rev. Mod. Phys.*, 1993, **65**, 851.
5   M. Bär and M. Eiswirth, *Phys. Rev. E*, 1993, **48**, R1635.
6   A. N. Zaikin and A. M. Zhabotinsky, *Nature*, 1970, **225**, 535.
7   L. Kuhnert, *Nature*, 1986, **319**, 393.
8   S. Kádár, T. Amemiya and K. Showalter, *J. Phys. Chem.*, 1997, **101**, 8200.
9   R. J. Field and R. M. Noyes, *J. Chem. Phys.*, 1973, **60**, 1877.
10 J. J. Tyson and P. C. Fife, *J. Chem. Phys.*, 1980, **73**, 2224.
11 H.-J. Krug, L. Pohlmann and L. Kuhnert, *J. Phys. Chem.*, 1990, **94**, 4862.
12 I. Sendiña-Nadal, E. Mihaliuk, J. Wang, V. Pérez-Muñuzuri and K. Showalter, *Phys. Rev. Lett.*, 2001, **86**, 1646.
13 An implementation of control without averaging is sufficient to achieve wave stabilization; however, the discrete nature of the simulation grid causes a wave of constant size to produce slightly varying area counts as the position of the wave contour changes with respect to the simulation grid. Averaging over successive iterations greatly reduces this effect. It should be noted that averaging introduces a small delay in the control loop, which can be shown to have a negligible effect by comparisons to simulations without averaging.
14 A. S. Mikhailov, *Foundations of Synergetics. I. Distributed Active Systems*, Springer-Verlag, Berlin, 1990.
15 V. A. Davydov, A. S. Mikhailov and V. S. Zykov, *Research Reports in Physics, Nonlinear Waves in Active Media*, Springer-Verlag, Berlin, 1989.
16 V. A. Davydov, V. S. Zykov and A. S. Mikhailov, *Sov. Phys. Usp.*, 1991, **34**(8), 665.
17 A. S. Mikhailov and V. S. Zykov, *Physica D*, 1991, **52**, 379.
18 E. Meron, *Phys. Rep.*, 1992, **218**, 1.
19 A. Karma, *Phys. Rev. Lett.*, 1991, **66**, 2274.
20 V. Hakim and A. Karma, *Phys. Rev. E*, 1999, **60**, 5073.
21 Here we refer to the traditional meaning of wavelength, rather than the length of a wave segment.
22 Á. Tóth, V. Gáspár and K. Showalter, *J. Phys. Chem.*, 1994, **98**, 522.
23 L. Schimansky-Geier, M. Mieth, H. Rose and H. Malchow, *Phys. Lett. A*, 1995, **207**, 140.
24 H. Hempel, L. Schimansky-Geier and J. Garcia-Ojalvo, *Phys. Rev. Lett.*, 1999, **82**, 3713.

Control and coupling of spiral waves in excitable media

Michael Seipel, Friedemann W. Schneider and Arno F. Münster*

Institute of Physical Chemistry, University of Würzburg, Am Hubland, D-97074 Würzburg, Germany. E-mail: phch030@phys-chemie.uni-wuerzburg.de

Received 10th April 2001
First published as an Advance Article on the web 13th November 2001

We report the complex dynamics of spiral waves observed in the ferroin-catalyzed BZ reaction. The reaction is run in an open unstirred reactor (CFUR) with the catalyst immobilized on a polysulfone membrane. The catalyst-loaded membrane is placed between two well stirred compartments which are fed with solutions of sulfuric acid/malonic acid/bromide and sulfuric acid/bromate, respectively. An electrical field perpendicular to the membrane can be applied via Pt-ring electrodes or, alternatively, via transparent electrodes made of ITO-coated glass. In the field-free case relatively simple target and spiral patterns are observed in the membrane. If an alternating electrical field is applied the spiral core drifts through the membrane. The actual trajectory of the spiral tip depends on the amplitude and frequency of the applied electrical field. If the perturbation parameters are chosen properly the wave fronts break up and new spiral cores emerge under the influence of the alternating field. Complex spatio-temporal patterns may be induced which are reminiscent of "spiral-chaos". After switching off the perturbation the system returns to its previous, "simple" behaviour. Our experimental observations are confirmed by model calculations based on the Barkley model of spiral waves. The technique of using modulated excitability to control the dynamics of spiral waves is further extended to the coupling of two spirals in two CFURs. We present numerical simulations based on two identical excitable reaction–diffusion (RD) systems which are mutually coupled. The coupling is based on the observation of an arbitrarily chosen point inside each of the RD systems: If a chemical wave passes the point of optical observation in system 1 an electric field is applied to system 2 and vice versa. Thus, a local observation made in one system is transformed in a global perturbation of the second CFUR. We report the observation of CFUR states where the two spiral waves are spatially and temporally coupled to each other.

Introduction

Chemical waves are generic spatio-temporal structures which emerge in excitable media. These structures are involved in a multitude of biological and chemical as well as technical processes.[1–5] An important problem concerns the deliberate control of the dynamical behaviour of chemical waves. In particular, the control of spiral waves has attracted attention since these structures may act as pacemakers emitting waves into the excitable system. Here, periodic perturbations of the underlying nonlinear reaction may be used to control the motion of the wave. In the light-sensitive variant of the Belousov–Zhabotinsky (BZ) reaction a periodic modulation of the illumination intensity has been used by several authors to control the motion of the spiral tip.[6–9] In this

DOI: 10.1039/b103239a

reaction the production rate of the inhibitory species Br⁻ is increased by light and the excitability of the reaction–diffusion system is thus reduced. Further examples of periodically forced spiral waves have been reported in catalytic surface reactions, *e.g.* ref. 10. Moreover, some theoretical studies on forced spiral waves have been performed.[11,12] These investigations show various types of nonlinear responses such as resonance-induced drift, entrainment and irregular responses of the spiral tip motion.

The majority of chemical and biological systems where pattern formation has been observed involve ions, *i.e.* electrically charged reactants or intermediates are important. In systems involving ionic species in the pattern formation process an externally applied electrical field may be used as a control parameter. Experimental investigations of electrical field effects on chemical waves in one- and two-dimensional excitable media have been performed with several chemical reactions, *e.g.* the BZ reaction,[13–15] the iodate–arsenous acid reaction[16,17] and the mercury(II) chloride–potassium iodide system.[18] In all of these studies the electric field was applied along the plane of wave propagation. The migration of ions together with molecular diffusion of chemical species has to be considered since both mechanisms of transport contribute to the system's behaviour.

In this contribution we show spiral waves which emerge in an open unstirred membrane reactor based on the ferroin-catalyzed BZ reaction. In contrast to previous investigations an electrical field is applied *perpendicular* to the plane of the membrane *via* ring or transparent electrodes. Here the electric field controls the rate of exchange of chemical species between the excitable membrane and the surrounding solutions.

Experimental

A membrane made of polysulfone (pore size 0.45 μm) carrying the immobilized catalyst ferroin[23] was fixed between two stirred flow cells of 25 ml volume each. Catalyst-free reactant solutions were slowly pumped through the cells. The solutions A and B were fed separately into the corresponding stirred flow cell. The concentration set 1 consisted of Solution A: 0.5 M $NaBrO_3$, 5.0 M H_2SO_4 and Solution B: 1.5 M malonic acid, 0.3 M NaBr, 5.0 M H_2SO_4, while set 2 contained Solution A: 0.4 M $NaBrO_3$, 3.6 M H_2SO_4 and Solution B: 1.24 M malonic acid, 0.2 M NaBr, 3.6 M H_2SO_4.

All chemicals used were of analytical grade and were used without further purification. The flow velocity of reactant solutions through both cells was adjusted to 3.0 ml h⁻¹ using a multichannel peristaltic pump. Stirring was accomplished by Teflon-coated magnetic stirrers at 750 rpm in each cell. However, the waves emerging inside the membrane were not noticeably affected by variations of the stirring rate between 500 and 1000 rpm. Observation of the membrane from both sides confirmed that only a single pattern emerges inside the catalyst loaded membrane. Using a computer controlled potentiostat (EG&G Instruments model 363) an electric field was applied across the membrane *via* Pt-ring electrodes or, alternatively, *via* transparent electrodes made of indium-tin-oxide (ITO) coated glass. The electrodes were fixed at a distance of 30 mm from each other, *i.e.* symmetrically at 15 mm distance from the membrane. The dynamics of the spiral waves emerging inside the membrane was monitored by an 8-bit black-and-white CCD camera and the motion of the spiral tip was tracked from the digitized video images. The set-up allows us to modulate the imposed electrical field according to the equation

$$U = U_0 \sin(2\pi\nu t) = U_0 \sin(2\pi t/T). \tag{1}$$

Here, U denotes the voltage applied, U_0 is the perturbation amplitude, ν is the perturbation frequency and T the period of the perturbation.

All experiments were performed at a constant temperature of 298 ± 0.3 K.

Reactor design

In Fig. 1 a sketch of the experimental set-up is depicted. Two main features of the open unstirred reactor used are of particular importance: the use of a thin membrane made of polysulfone which contains the immobilized catalyst and the use of transparent (or ring) electrodes to apply an electric field perpendicular to the reactive layer. The membrane is in contact with reactant solutions from both sides and the reaction inside the membrane is thus kept far from chemical equilibrium. On the other hand, the electric field allows us to deliberately control the exchange of ions (such as bromate and bromide) between the membrane and the feeding solutions. In addition to

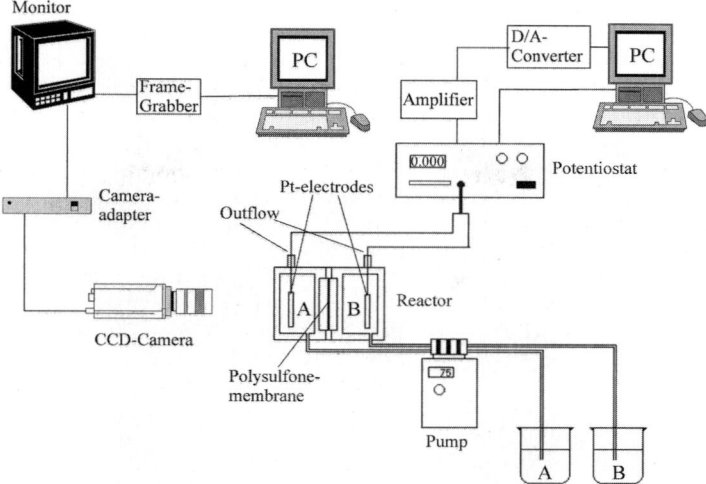

Fig. 1 Sketch of the experimental set-up. A catalyst-loaded porous membrane is fixed between two stirred flow cells holding the reactant solutions.

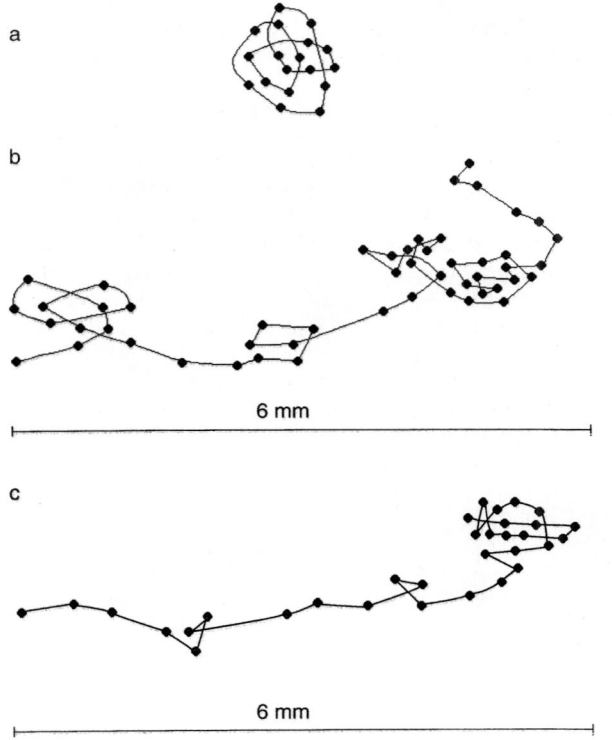

Fig. 2 Trajectories of the spiral tip. (a) Unperturbed meandering motion; (b) and (c) examples of resonance drift at $U_0 = 2.0$ V and $T = 30$ s. The position was measured in intervals of 30 s. The distance from the starting point (right margin) to the end point was 5.5 mm in (b) and 4.9 mm in (c).

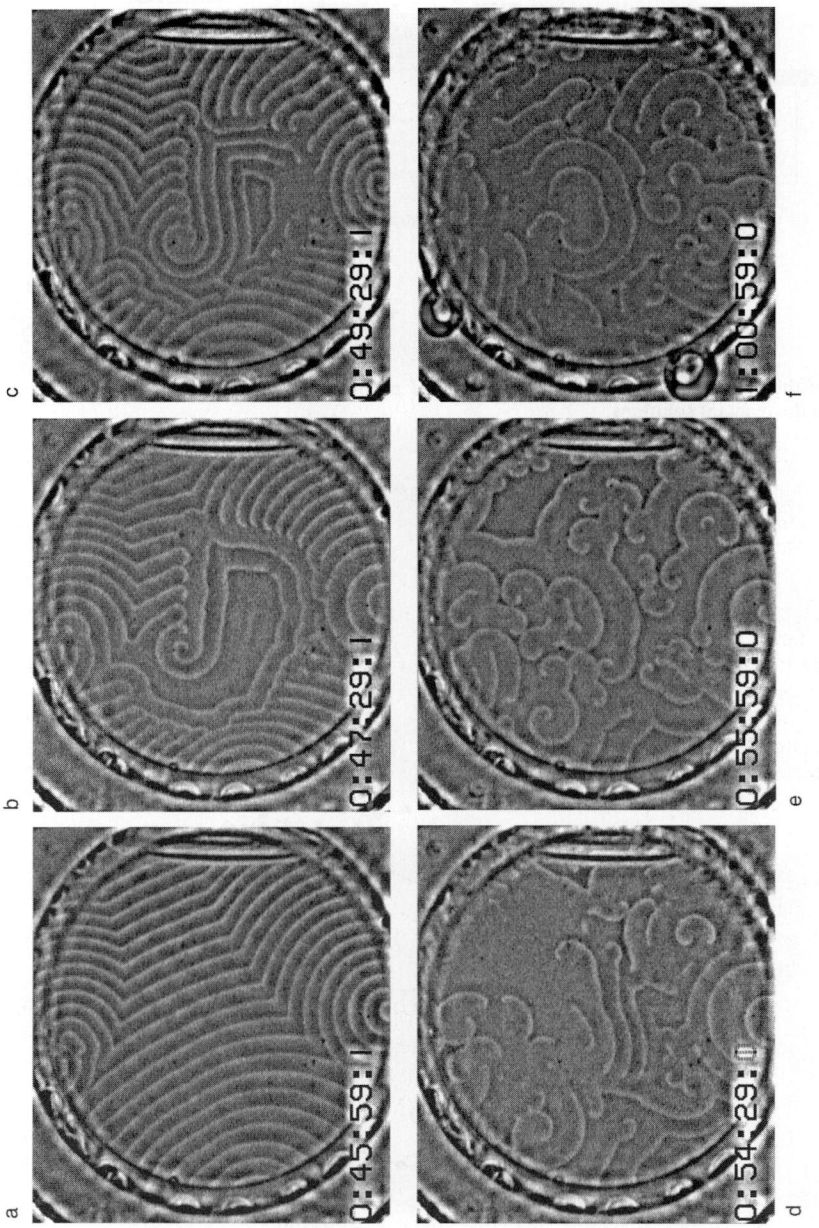

Fig. 3 Snapshots illustrating the transition to complex spatio-temporal behaviour induced by alternating current. After the alternating electrical field ($U_0 = 3.0$ V, $T = 240$ s) has been switched on at $t = 45$ min new spiral centers emerge.

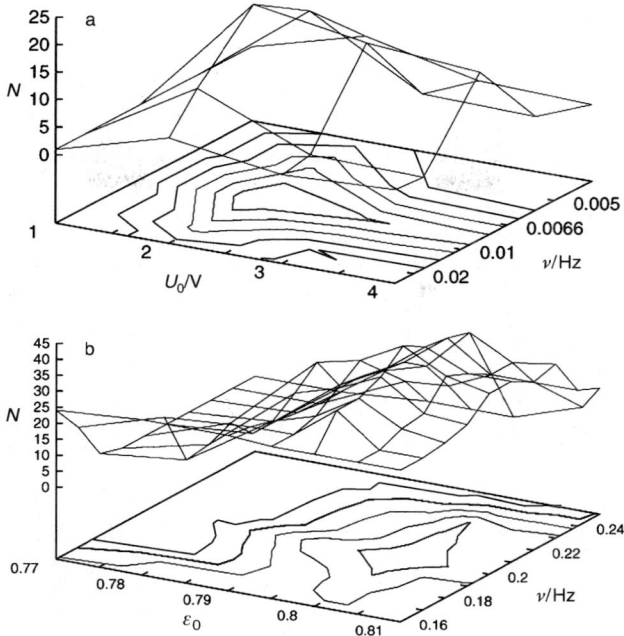

Fig. 4 Average number of spiral cores observed at different alternating current amplitudes U_0 and frequencies v. (a) Experimental results, (b) numerical simulation.

diffusion and migration of ions the electroosmotic flux evoked by the applied electric field contributes to the exchange of mass between the membrane and its ambient. Therefore not only the exchange of ions but also that of neutral molecules is affected by the electric field. As a result, the excitability of the system can be effectively controlled by applying an electric field oriented perpendicular to the plane of the wave propagation. In contrast to previous studies based on the light sensitive ruthenium-catalyzed BZ reaction a control of the excitability *via* an electric field is applicable to chemical waves in a variety of nonlinear reactions which need not be light sensitive.

Results

In Fig. 2 the motion of a spiral tip at resonance-drift conditions is demonstrated. For the feed concentrations of set 1, stirring rate and temperature given in the Experimental section the tip of the *un*perturbed wave meanders along the epicyclic closed trajectory displayed in Fig. 2a. In the

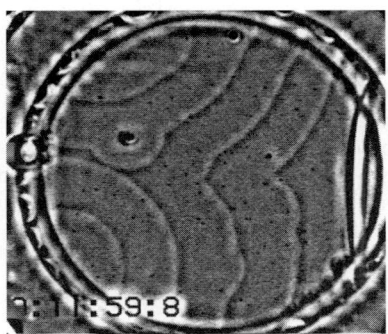

Fig. 5 Destabilizing effect of an alternating electric field of $U_0 = 1.0$ V and $T = 15$ s. The wave front displays fluctuating wrinkles.

experiments shown in Fig. 2b and c the amplitude of the imposed alternating field was $U_0 = 2.0$ V and its period was $T = 30.0$ s corresponding to a perturbation frequency $v = 0.03$ Hz. After the alternating electric field is switched on the spiral tip leaves its original path and approaches an orbit with a drastically enlarged radius. The Fig. 2b and c give two typical examples of this field-induced resonance drift; the points at the *right* margin of the trajectories represent the starting points where the alternating electrical field was switched on. The time interval between two successive points was 30 s.

The effects of an imposed alternating current are not limited to the field-induced resonance drift described above. In chemical oscillators run in a well-stirred flow reactor complex behaviour—such as entrainment of different order, quasiperiodicity and chaos—induced by an externally applied periodic perturbation has been widely studied.[5,19] The experiments described in this contribution revealed complex behaviour in the periodically forced, spatially extended system described above. Here, complex spatio-temporal patterns are evoked by the imposed alternating electric field. Without an externally applied electrical field experiments using the concentration set 2 as given in the Experimental section showed steadily rotating spiral waves. The sequence of images in Fig. 3a–f gives a typical example for the development of chemical waves exposed to an alternating electrical field. The perturbation parameters were $U_0 = 3.0$ V and $T = 240$ s. Here the initially observed smooth wave fronts break up and new spiral cores emerge: the perturbation leads to a complex spatio-temporal pattern which is reminiscent of spiral-chaos.[20] In the process of spiral core generation, islands of low excitability emerge which cause the initially smooth wave fronts to break. These islands spontaneously appear in different regions of the reaction–diffusion system. They may be attributed to an increased supply of the inhibitor bromide. The typical lifetime of islands is approximately half a perturbation period (compare Fig. 3b, c and d); this reflects the actual polarity at the electrodes. At the boundary between the islands of low excitability and the surrounding excitable medium wave fronts are cut apart resulting in a large local curvature. The excitability increases again during the following half-period of the alternating current. Bromide is removed from the membrane at the now reversed polarity of the electrodes. New spiral cores are formed along the island boundaries according to the well-established relation between curvature and normal velocity of the wave front.[21] The simple behaviour of steadily rotating spirals was recovered after the alternating field was switched off. This observation confirms that the observed transition to complex behaviour may indeed be attributed to the imposed periodic perturbation and not to any aging effects of the membrane or the solutions.

Fig. 4a shows the average number of spiral cores observed in the reaction–diffusion system at different perturbation parameters U_0 and v. The number of spiral cores observed displays a broad maximum at $U_0 = 2.0$ V and $v = 0.0083$ Hz ($T = 120$ s). Interestingly, this perturbation period approximately matches the period of damped oscillations observed in batch experiments performed with the same concentration set of educt solutions in a well stirred reactor.

For small perturbation amplitudes and high frequencies of the imposed alternating field no breaking of wave fronts has been observed in our experiments. However, the imposed perturbation destabilized the planar wave fronts resulting in flucuating wrinkles along the front. In Fig. 5 an example of such a wrinkled wave front observed at $U_0 = 1.0$ V and $v = 0.07$ Hz ($T = 15$ s) is shown.

Numerical simulations

We performed numerical simulations based on a simple general model of an excitable chemical system, the so-called Barkley model.[22] This model describes an excitable two-variable system with cubic and linear nullclines where the cubic nullcline is approximated by piecewise linear segments. Thus, the model can very efficiently be implemented on a computer. Two coupled differential equations describe the nonlinear kinetics of the model:

$$\frac{\partial x}{\partial t} = \nabla^2 x + \frac{1}{\varepsilon} x(1-x)(x - x_{\text{th}}(z)) \qquad (2)$$

$$\frac{\partial z}{\partial t} = D_z \nabla^2 z + x - z; \quad x_{\text{th}} = \frac{z+b}{a} \qquad (3)$$

Here the kinetic parameter ε determines the threshold of excitability. In our simulations of complex spiral patterns we sinusoidally modulated the value of ε:

$$\varepsilon = \varepsilon_0(1 + \alpha \sin(2\pi\nu t)) \quad (4)$$

The threshold of excitability was thus perturbed in analogy to the field-induced modulation of the excitability in the experiments. The simple model does not take migration or electroosmosis into account; the effects caused by the electric field in the experiments are summarized in the modulation of ε. Despite its simplicity the model qualitatively reproduces complex behaviour similar to the experimental observations. A comparison of Fig. 6 and 3 confirms the similarity between the numerical and experimental results. In Fig. 6 we depict the spatial distribution of the autocatalytic species x during one single perturbation cycle. Fig. 4b shows the average number of spiral cores found in the simulations as a function of the amplitude α and frequency ν of the imposed perturbation. The simulation is in qualitative agreement with the corresponding experimental observations shown in Fig. 4a.

Coupling of two independent spirals in two distinct reaction–diffusion systems may be achieved in a similar way by modulating the value of ε in one system according to the local state of the model in the other. Here the perturbation is represented by eqn. (5) which now replaces eqn. (4) used in the simulations of complex spiral patterns.

$$\varepsilon_i = \varepsilon_{0,i}(1 - \alpha x_j), \quad i, j = 1, 2 \quad (5)$$

This method of coupling requires a point of reference to be chosen in both systems; the value of x at this selected locus is used to determine ε in the other system and *vice versa*. Note that the amplitude of x in the model always falls into an interval [0,1]; thus, the excitability in system i is decreased as a wave with high value of x crosses the point of observation in system j. For the

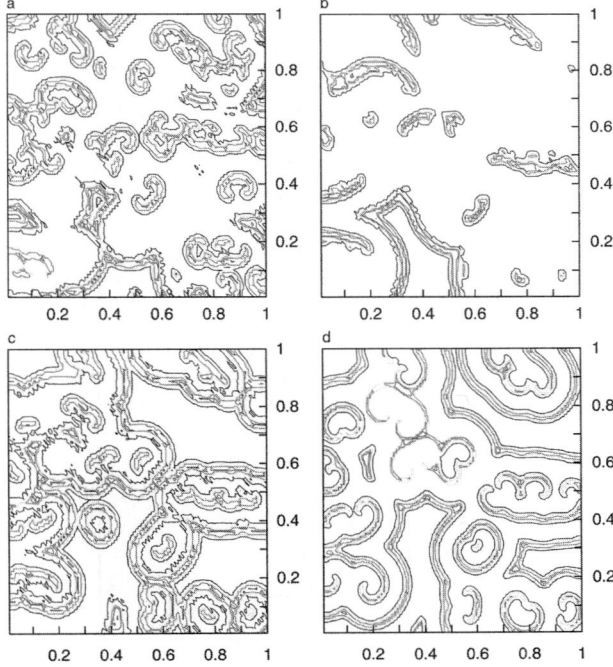

Fig. 6 Spatial distribution of the variable x in the Barkley model of an excitable system with periodic perturbation of the parameter ε. The dimensionless parameters[22] were $a = 0.75$, $b = 0.01$, $\varepsilon_0 = 50$, length = 140. The species z was assumed to be immobile and D_x was set to unity. Perturbation parameters were $\alpha = 0.8$ and $\nu = 0.2$. The initial condition of the simulation was a single spiral wave with its core located in the center of the system.

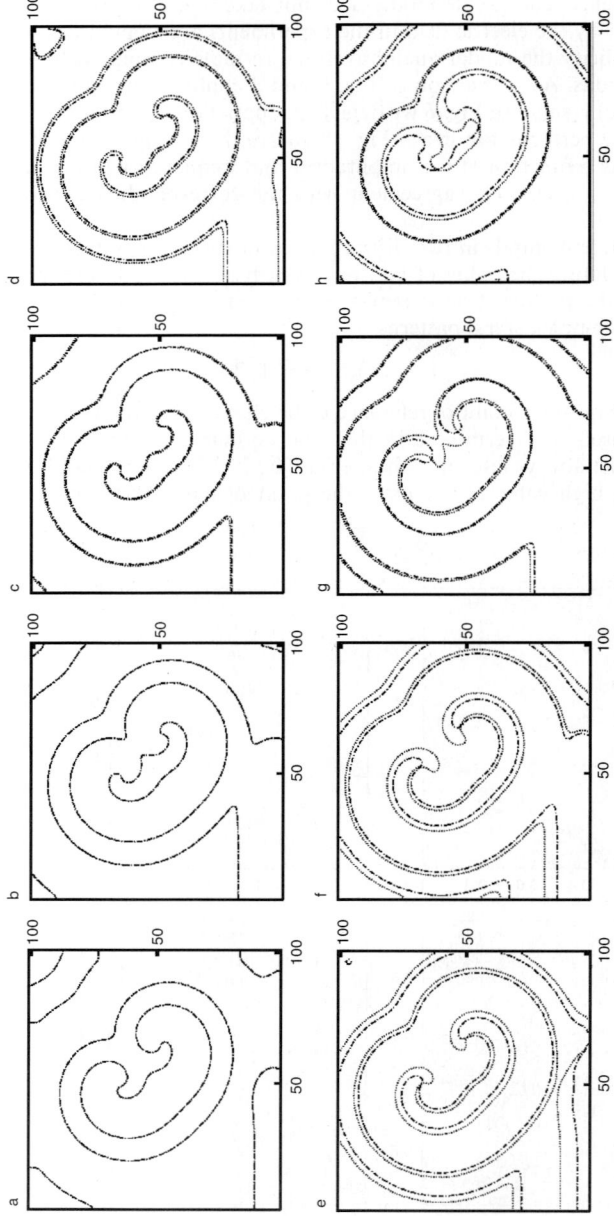

Fig. 7 Evolution of coupled spiral waves in the Barkley model starting from almost identical initial conditions. An overlay of waves in the two coupled RD systems is shown. Here the value of ε in one system is modulated according to the value of x at locations (20,20) and (80,80), respectively. Snap shots were taken in intervals of 0.4 time units.

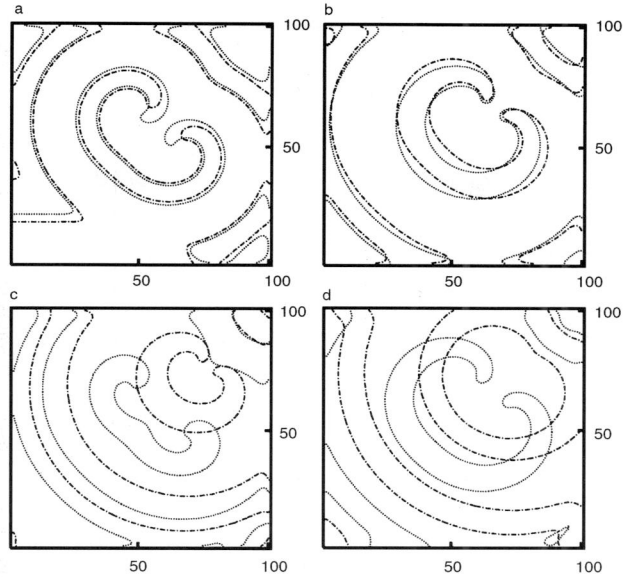

Fig. 8 Long-term evolution of the waves in Fig. 7. The images were taken at $t = 2$, $t = 10$, $t = 20$ and $t = 30$.

simulations below the following parameter values were used: $a = 0.75$, $b = 0.01$, $\varepsilon_0 = 100$ and $\alpha = 0.5$; the length of the system was set to $L = 100$.

Fig. 7 depicts a sequence of snapshots taken at intervals of 0.4 time units. The contour lines indicate the distribution of x in both systems; to facilitate comparison an overlay of waves in systems 1 and 2 is shown in the same frame. Similar initial conditions were chosen for both systems. The initial pattern is almost symmetric with respect to a diagonal line; however, true symmetry of the initial pattern was avoided in order to allow instable states to evolve. The points

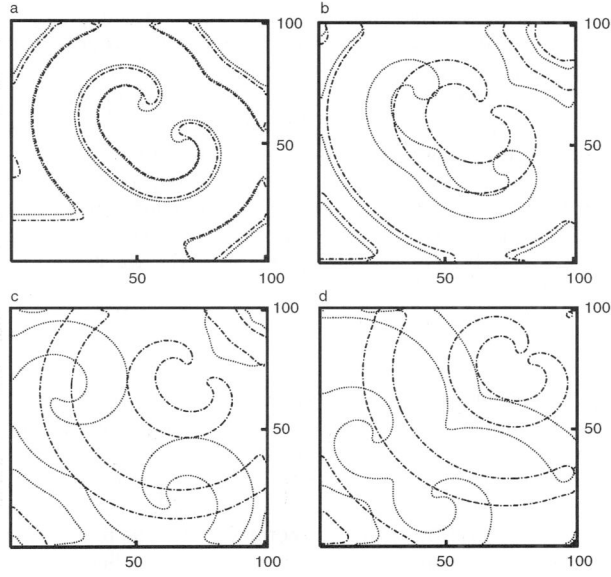

Fig. 9 Evolution of waves with observation points at (20,20) in system 1 and (80,20) in system 2. The images were taken at $t = 2$, $t = 10$, $t = 20$ and $t = 30$.

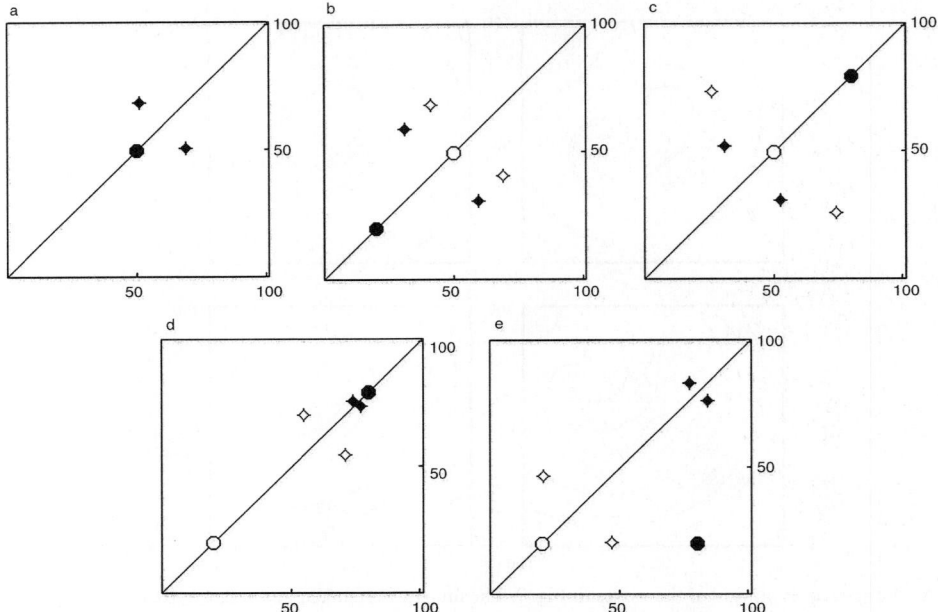

Fig. 10 The location of the observation point determines the direction of the spiral core drift. Open diamonds indicate the spiral tip at $t = 30$ in system 1, filled diamonds in system 2. Correspondingly, the points of observation are marked by an open circle in system 1 and by a filled circle in system 2.

of observation in Fig. 7 were fixed on the line of (approximate) symmetry with coordinates (20,20) and (80,80), respectively. During one period of spiral rotation the waves in both systems are driven apart from each other by the coupling. In Fig. 8 we show the evolution of the same waves up to 30 time units. The spiral cores in both systems perform a meandering motion towards the upper right corner of the frame.

It is obvious that the choice of reference points in the coupled systems will influence the resulting dynamical behaviour. If the observation point in system 1 is adjusted to coordinates (20, 20)—as in the case described above—and to (80,20) in system 2 the spiral cores drift towards opposite corners of the system. Fig. 9 depicts the temporal evolution of waves under those conditions. In Fig. 10, finally, the actual position of the spiral tips after 30 time units is marked for different locations of the points of observation. If the position of the observation point is chosen to be identical in both systems (an example is shown in Fig. 10a) the two spiral waves display the same behaviour, as expected. If, however, the positions of the reference points do not coincide the two spirals drift apart from each other. In their asymptotic state the spirals meander around the tip positions shown in Fig. 10. Hence, the final position of the waves can be controlled by the choice of observation points.

Discussion

The control of chemical waves in an excitable reaction–diffusion system requires an experimental method which allows one to adjust the excitability of the system. In light sensitive excitable reactions such as the $Ru(bipy)_3$-catalyzed BZ reaction the illumination intensity can be used as an adjustable parameter. Alternatively, an electric field may be utilized to adjust the exchange of chemical compounds between the excitable system and its ambient. An alternating electrical field induces resonance drift of the spiral tip as well as complex patterns made up from a multitude of interacting spirals. The technique proposed in this paper can be applied to other nonlinear chemical reactions which are not light sensitive. Thus, a wide range of nonlinear chemical reactions can be investigated with respect to controlling chemical waves and patterns.

In numerical simulations two spiral waves can be coupled by modulating the excitability in one system according to the actual state of the other system. Experimentally this type of coupling can be realized by modulating the electric field intensity applied to reactor 1 according to the video signal intensity read out from a chosen pixel of the video camera observing reactor 2 (and *vice versa*). Future experiments along these lines will contribute to a better understanding of interacting chemical waves.

References

1. *Chemical Waves and Patterns*, ed. R. Kapral and K. Showalter, Kluwer, Dordrecht, 1995.
2. *Self-Organization in Activator-Inhibitor-Systems: Semiconductors, Gas-Discharge and Chemical Active Media*, ed. H. Engel, F.-J. Niedernostheide, H.-G. Purwins and E. Schöll, Wissenschaft & Technik Verlag, Berlin, 1996.
3. *Chaos*, 1994, **4**(3), 439 ff.
4. *Transport and Structure—Their Competitive Roles in Biophysics and Chemistry*, ed. S. C. Müller, J. Parisi and W. Zimmermann, Lecture Notes in Physics, vol. 532, Springer, Berlin, 1999.
5. F. W. Schneider and A. F. Münster, *Nichtlineare Dynamik in der Chemie*, Spektrum Akademischer Verlag, Heidelberg, 1996.
6. O. Steinbock, V. Zykov and S. C. Müller, *Nature*, 1993, **366**, 322.
7. V. Zykov, O. Steinbock and S. C. Müller, *Chaos*, 1994, **4**, 509.
8. S. Grill, V. Zykov and S. C. Müller, *J. Phys. Chem.*, 1996, **100**, 19082.
9. M. Braune, A. Schrader and H. Engel, *Chem. Phys. Lett.*, 1994, **222**, 358.
10. S. Nettesheim, A. von Oerzten, H. Rorthermund and G. Ertl, *J. Chem. Phys.*, 1993, **98**, 9977.
11. R. M. Mantel and D. Barkley, *Phys. Rev. E*, 1996, **54**, 4791.
12. A. Schrader, M. Braune and H. Engel, *Phys. Rev. E*, 1995, **52**, 98.
13. B. Schmidt and S. C. Müller, *Phys. Rev. E*, 1997, **55**, 4390.
14. H. Sevcikova, M. Marek and S. C. Müller, *Science*, 1992, **257**, 951.
15. S. C. Müller and V. Zykov, in ref. 4, pp. 74–79.
16. L. Forstova, H. Sevcikova, M. Marek and J. H. Merkin, *Chem. Eng. Sci.*, 1999, **55**, 233.
17. L. Forstova, H. Sevcikova, M. Marek and J. H. Merkin, *J. Phys. Chem. A*, 2000, **104**, 9136.
18. I. Das and A. Bhattacharjee, *J. Phys. Chem.*, 1990, **94**, 8968.
19. M. Marek and I. Schreiber, *Chaotic Behaviour of Deterministic Dissipative Systems*, Academia, Prague, 1991.
20. M. Bär and M. Eiswirth, *Phys. Rev. E*, 1993, **48**, 1635.
21. A. S. Mikhailov, *Foundations of Synergetics I*, Springer, Berlin, 1990, p. 40.
22. D. Barkley, M. Kness and L. S. Tuckermann, *Phys. Rev. A*, 1990, **42**, 2489.
23. A. Lazar, Z. Noszticzius, H.-D. Försterling and Z. Nagy-Ungvarai, *Physica D*, 1994, **84**, 112.

General Discussion

Dr Tóth opened the discussion of Dr Boissonade's paper: The chlorite–tetrathionate reaction is a simple, highly reproducible system. Reaction fronts, convective instabilities, diffusion-driven cellular fronts and now, using the inherent diffusion coefficients, excitability and oscillating patterns are exhibited. In the light of these, I would expect to see further interesting spatiotemporal patterns. Would you agree with that?

Dr Boissonade responded: Up to now, studies of spatial patterns have been performed on two classes of reaction–diffusion reactions: (i) those in which the patterns originate in the oscillating or excitable properties of the reaction in the homogeneous system, such as the target patterns or traveling waves in the BZ reaction; and (ii) those in which the patterns originate in long range inhibition, such as the Turing patterns in the CIMA reaction.

It is clear that the study of systems with long range activation open up a new experimental field in the search for spatiotemporal patterns. Since, in aqueous solutions, the proton is the species which exhibits the largest self-diffusion coefficient, reactions activated by pH are the best candidates. The chlorite–tetrathionate reaction (C–T) is a convenient prototype of such a reaction since this system presents a superautocatalysis in H^+ and since the kinetics are rather well described by a unique simple rate law. Moreover, this reaction offers the ability to explore a large spectrum of situations from long range activation to long range inhibition in open spatial reactors.

Prof. Scott commented: The oscillatory fronts you have shown seem typically to be associated with travelling fronts in the experimental system. It is clear that the oscillators are not caused by the fronts. For instance, can the oscillators be observed in computations?

Dr Boissonade replied: To produce oscillations, one must have at least two feedback processes with different time scales. In classical oscillating reactions (*i.e.* which oscillate in homogeneous conditions), these time scales are provided by the complexity of the kinetic mechanism that exhibit competition between different intermediate reaction steps. In the C–T reaction, the kinetics are well approximated by a unique rate equation (thus a unique time scale) and can be reduced to a one-variable system. This excludes any oscillating or excitable behaviour. The reaction can only be bistable in an homogeneous open reactor (CSTR). This is supported by our numerical calculations which also take into account the main fast equilibria which are present in the system.

Here, the time scales difference actually originates in the difference of the diffusion coefficients between the activatory species and the other reactants. The chemical transformation takes place essentially inside the front. The input flux of fresh reactants from the CSTR/gel boundary to this reaction zone and the reverse output flux of products is controlled by the rate of diffusive transport between the boundary and the front. The exchange rate for each species—quantified by its diffusion coefficients —defines the inverse of a time scale. A consequence of these different time scales is that, in a narrow domain of control parameter values located close to the limits of spatial bistability, the front can become oscillatory. This is observed both in our experiments and in our numerical simulations when the ratio of the diffusion coefficient is large enough (of the order of 3). Furthermore, numerical excitation waves in a 2D ring, analogous to those of our experiments, could also be obtained by Fuentes and Kuperman (private communication), using a simplified model. All details will be provided in a forthcoming publication.

Prof. Westerhoff asked: Since the proton and hydroxyl diffusion coefficients seem to matter for the results, have you added pH buffers to the gel or system (in small amounts) so as to affect the effective diffusion coefficients?

If so, did this just slow down the phenomena, or also alter them, vis-a-vis differential effects on H^+ and OH^- diffusion?

Dr Boissonade answered: The control of the effective diffusion by a complexing agent is an interesting challenge. On can even expect to switch from a long range activation to a long range inhibition if the complexation is large enough. One can thus expect to switch from oscillations to stationary patterns. Nevertheless this should also involve other parameters like the electroneutrality of the medium. The frequency of the oscillations does not depend only on the ratio of the diffusion coefficient but also on the distance from the spatial bistability limit. In the computations, the period strongly increases (diverges?) when one approaches this limit. Decreasing the differences between effective diffusion coefficients would more likely decrease the extent of the oscillatory domain.

Prof. Jonnalagadda said: If I understand your system correctly, the system is excitable by increasing the length of the gel and characteristics of the gel. Have you provision in the model to accommodate the variations in the characteristics of the gel?

Dr Boissonade responded: In our experiments, we use a soft agarose gel, 98% water content. The gel can be considered as an inert medium which only avoids convective phenomena. The diffusion coefficients are quite likely very close to those in pure water. The length plays only a geometrical role. Parameters which would include gel properties in the model are irrelevant in this context. The excitability does not correspond to an increase of the length (actually the thickness). In the parameter space it corresponds to a finite domain beyond the high pH of spatial bistability.

Prof. Gáspár commented: A few years ago Lee and Swinney[1] reported on the formation of patterns in another bistable system, the Edblan–Orbán–Epstein reaction. We have failed to repeat these experiments. In the context of your paper, one reason for our failure could be the different lengths of the gel, I think. Your reactor is different than that of Lee and Swinney. What phenomenon would you expect to occur in your reactor with the system of Lee and Swinney?

1 K. J. Lee and H. Swinney, *Phys. Rev. E*, 1995, **51**, 1899.

Dr Boissonade replied: In the experiments you are referring to,[1] the system is able to exhibit two stationary stable uniform states in a gel disc OSFR. This is the signature of spatial bistability and the thickness of the gel becomes an essential control parameter. However, in the quoted paper, the system is modelled in the CFUR approximation which is a different type of phenomenological description.

Besides the geometrical aspects, *i.e.* the thickness of the gel on which we have focused, there are a number of possible reasons for the experiments performed with the EOE reaction not being reproducible in different laboratories. The Texas group uses a polyacrylamide gel which can hydrolyse and is probably not inert in the reaction. This hydrolysis leads to the formation of carboxylic groups that would interfere with the acid catalysis of the reaction by complexing protons. Thus the pattern development should be very sensitive to the exact composition and network of the polymer. Membranes are also often used to support the gel mechanically, as in the Texas experiments. Their nature, their thickness and their porosity have to be taken into account since they control the boundary condition at the CSTR/gel interface. Note that in our experiments on the C–T reaction, there is no membrane between the gel and the CSTR.

1 K. J. Lee and H. Swinney, *Phys. Rev. E*, 1995, **51**, 1899.

Prof. Scott commented. In view of the importance of the gel depth parameter, could you speculate on the response of an elliptic gel region surrounding a circular boundary.

For some conditions one might expect FT states at the thicker 'ends' and only F states along the thinner 'sides'. Would this allow a steady spatial structure or will the FT states act as wave sources?

Dr Boissonade responded: Preliminary experiments show that gradients of thickness can either lead to the development of fixed wave sources at the F/FT states interface or to the stabilisation of this interface. Nevertheless, such a stationary spatial structure would result from an external spatial forcing and would not result from a genuine self-organisation process. More challenging is the possible stabilisation of the front, or, on the contrary, the destabilisation of a stationary state by performing the chemical reaction in a chemically responsive gel. Some gels are known to be able to swell by a large amount (up to one thousand times their volume) as a function of pH. Temporal or spatio-temporal structures could result from these chemo-mechanical effects that couple the geometry of the system with the reaction. Theoretical and experimental developments, taking spatial bistability and pH mechanical sensitivity as a starting point, are presently in progress in our group.

Dr Trevelyan said: The reactor configuration could be realigned so that the impermeable wall is non-equidistant from the reactor wall. This may achieve regions which can support different states. It may be interesting to observe the transitions around the gel from one state to the next.

Prof. Showalter said: I would like to ask Prof. Marek whether experimental or theoretical studies have been carried out to examine the potential developed across a propagating front? This would be particularly interesting in fronts where the autocatalyst is hydrogen ion, since it has such a high diffusivity.

Dr Šnita and Prof. Marek replied: In our earlier theoretical studies, *cf.* refs. 1–6, we have described how the moving autocatalytic fronts are coupled with the moving pattern of nonzero charge density (electric double layer) in the case of different diffusion coefficients of the reacting species.

It would probably be possible to study the evolution of spatial patterns of the local charge density using the molecular potential sensitive probes (dyes). As far as it is known to us, no such studies for chemical propagating fronts have been reported.

1 D. Šnita and M. Marek, *Physica D*, 1994, **75**, 531.
2 D. Šnita, H. Ševčíková, M. Marek and J. H. Merkin, *J. Phys. Chem.*, 1996, **100**, 18 740.
3 D. Šnita, H. Ševčíková, M. Marek and J. H. Merkin, *Proc. R. Soc. London Ser. A*, 1997, **453**, 2325.
4 D. Šnita, H. Ševčíková, J. Lindner, M. Marek and J. H. Merkin. *J. Chem. Soc., Faraday Trans.*, 1998, **94**(2), 213.
5 D. Šnita, Lindner, M. Marek and J. H. Merkin, *Math. Comput. Modell.*, 1998, **27**(2), 1.
6 J. H. Merkin, H. Ševčíková, D. Šnita and M. Marek, *IMA J. Appl. Math.* 1998, **60**, 1.

Dr Mayama opened the discussion of Prof. Dewel's paper: What is your suggestion in your model to generate superlattice structure in experiments using a chemical diffusion system? I am asking you the direct relationship between your model and the experiment.

Prof. Dewel answered: The mechanism I have presented does not rely on the learning model we have used in the paper to illustrate the formation of Turing superlattices. The two necessary ingredients for the occurrence of such quasiperiodic structure are: (1) the existence of two disparate critical wavelengths and (2) the presence of quadratic nonlinearities favoring the interaction between three different spatial modes. The generality of this scenario encompasses also hydrodynamic systems and nonlinear optical devices.

Dr Simon commented: This question concerns the relationship of the saddle-node bifurcation curve and the Turing bifurcation curve. Relationships between the saddle-node bifurcation curve and another general bifurcation curve (*e.g.* the Hopf bifurcation curve) are revealed in a forthcoming paper by Simon et al.,[1] entitled 'Relationships between the discriminant curve and other bifurcation diagrams'. In that paper a general 2D system of ordinary differential equations depending on two parameters is investigated. The saddle-node bifurcation curve (SN-curve) is introduced together with another bifurcation curve that is the loci of some bifurcation of the steady states, *e.g.* Hopf bifurcation (we will refer to it as H-curve). The following general relationships of the two bifurcation curves are established and proved rigorously.

1. If the SN-curve and the H-curve have a common point (belongings to the same steady state), then they have a common tangent at that point. This is a Takens–Bogdanov bifurcation point if H is the loci of Hopf bifurcation points.

2. The cusp points of the H-curve are on the SN-curve.

3. Let us assume that the H-curve has a loop that has no common point with the SN-curve. Then there is a cusp point of the SN-curve in the loop.

4. If the SN-curve and the H-curve depend on a third parameter u, and at a certain value of u they have a common point (with common tangent), then this common point does not disappear as u is varied just moves along the bifurcation curves.

(The exact formulation of these statements can be found in the above paper.) The first and third properties can be observed in Fig. 1 of the paper under discussion. Can these relationships be observed in other systems as well? That would suggest that the above relationships could be established also between the SN and Turing bifurcation curves generally.

1 P. L. Simon, E. Hild and H. Farkas, *J. Math. Chem.*, 2001, **29**, 245.

Prof. Dewel replied: The bifurcation diagram of Fig. 1 presents generic features. The same behavior has been observed in physically dissimilar systems:

(i) A single step autocatalytic reaction that exhibits Turing patterns,[1] (ii) a model of surface reaction that can give rise to nanostructures,[2] and (iii) nonlinear optical devices that present both the phenomenon of optical bistability and diffraction patterns.[3] In all these examples, the first and the third properties cited above are satisfied. I am therefore convinced that general relationships between the saddle-node and the Turing bifurcations could be proved rigorously using a general two variable system of partial differential equations.

1 D. Horvath and A. Toth, *J. Chem. Soc., Faraday Trans.*, 1997, **93**, 4301.
2 M. Hildebrand, A. S. Mikhailov and Ertl, *Phys. Rev. E*, 1998, **58**, 5483.
3 F. T. Arecchi, S. Boccaletti and P. L. Ramazza, *Phys. Rep.*, 1999, **318**, 1.

Prof. Sørensen asked: How robust are the superlattices shown in Fig. 5 of the paper and how can they be recognized in an experiment situation where control of initial conditions are less than in numerical calculations?

Prof. Dewel responded: Stable superlattices generally coexist with regular hexagons (multistability) in the vicinity of the instability points. They are stabilized by resonant triads; therefore, when these quadatic contributions are dominant, superlattices can be the first structures to appear subcritically; they can then be reached by finite perturbations of the homogeneous steady state.

In spatially extended systems, domains of superlattice structures can appear immersed in a sea of more regular patterns; the fronts between these domains being pinned as a result of nonadiabatic effects. Such localized structures have for instance been observed in experiments on the two-frequency Faraday instability.[1]

1 A. Kudrolli, B. Pier and J. P. Gollub, *Physica D*, 1998, **123**, 99.

Prof. Epstein commented: The coexistence of the superlattice patterns with simpler patterns is likely to make the former difficult to obtain experimentally in reaction–diffusion systems. Perhaps the most promising approach would be to use the photosensitive BZ reaction and, once a parameter region for the existence of superlattices has been identified, use light to imprint the system with a pattern close to the desired one. We have used this technique[1] in studies of spatial forcing of hexagonal and striped Turing patterns.

1   M. Dolnik, I. Berenstein, A. M. Zhabotinsky and I. R. Epstein, *Phys. Rev. Lett.*, 2001, **87**, 238 301.

Prof. Dewel replied: The periodic spatial forcing of a photosensitive chemical reaction that presents a diffusive instability indeed provides a promising approach to produce Turing superlattices. In that case, it is the competition between the external and intrinsic length scales that can give rise to stable resonant structures. A similar procedure has been used to obtain the first example of incommensurate quasiperiodic pattern in a nonequilibrium system: an electrohydrodynamic instability in a nematic subjected to a spatially periodic electric field.[1]

1   M. Lowe, J. P. Gollub and T. Lubensky, *Phys. Rev. Lett.*, 1983, **51**, 786.

Dr Boissanade said: In your analysis, you also predict situations which could be simpler to observe than quasiperiodic patterns, such as the coexistence of different types of hexagonal patterns for the same control parameters. Did you try to apply your approach to a genuine chemical model?

Prof. Dewel answered: We have also considered the case where a single Turing instability appears on a branch of a bistable system that exhibits a strongly asymmetric hysteresis loop. As explained in the paper, the coupling with the homogeneous mode then generates the following characteristic sequence of structures HSS1/hexagons/stripes/re-entrant hexagons/HSS2. We have illustrated this behavior on a simple reaction–diffusion model of the chlorine dioxide–iodide reaction. A complexing agent has also been introduced in the model to clear cut a region of bistability where Turing patterns are allowed to develop. The results of the numerical integration of this system are summarized in Fig. 3 of the paper.

Dr Satnoianu commented: 1. It would be interesting to see what is the separate influence of the quadratic and cubic nonlinearity terms in the formation of the superlattice patterns presented. This is known to be an important factor in the selection of Turing patterns with spots being preferably selected against the stripes in general in the Turing case.

2. Is it possible to generate such superlattice patterns with wavelengths corresponding to one oscillatory mode (but spatially non-uniform so not Hopf) and the other one a stationary mode?

Prof. Dewel responded: The stability of the superlattices discussed in the paper proceeds from the competition between nonresonant cubic interactions which are nondistributive and therefore favor stripe patterns and quadratic terms that induce resonant triads between different spatial modes. It is important to note that superlattices can also appear in systems lacking quadratic nonlinearities. Square superlattices can for instance be stabilized by resonant tetrads. A nice example is provided by the experiment on vertically oscillated Rayleigh–Benard convection.[1] It is also possible to construct dynamic quasicrystalline patterns by superposing Turing and waves modes. This problem has been studied in details in connection with transverse patterns in non-linear optical devices.[2]

1   J. L. Rogers, M. F. Schatz, O. Brausch and W. Pesch, *Phys. Rev. Lett.*, 2000, **85**, 4281.
2   Z. H. Musslimani and I. M. Pismen, *Phys. Rev. E*, 2000, **62**, 389.

Dr Boissonade commented: These superlattice structures are defined in very narrow domain of parameters values and generally coexist in the phase diagram with simpler structures. It is probably impossible, from a practical point of view, to observe such superlattices in a chemical system, since we do not have the same facilities as those offered by the Faraday experiment in hydrodynamics.
  The coupling between the Hopf and Turing instability in chemical systems has been extensively studied in the nineties both theoretically[1,2] and experimentally.[3,4]

1   A. Rovinsky and M. Menzinger, *Phys. Rev. A*, 1992, **46**, 6315.
2   A. De Wit, D. Lima, G. Dewel and P. Borckmans, *Phys. Rev. E*, 1996, **54**, 261.
3   J. J. Perraud, A. De Wit, E. Dulos, P. De Kepper, G. Dewel and P. Borckmans, *Phys. Rev. Lett.*, 1993, **71**, 1272.
4   P. De Kepper, J. J. Perraud, B. Rudovics and E. Dulos, *Int. J. Bifur. Chaos*, 1994, **4**, 1215.

Prof. Scott asked: Could the authors comment of the effect of concentration gradients perpendicular to the plane of the pattern that might arise in 2D/3D systems.

Prof. Dewel replied: In boundary-fed systems, gradients in the concentrations of the main chemical species generally develop along the width of the reactor. These gradients confine the structures in a more or less thick stratum of the system. The simplest principle that springs to the mind is that a pattern will develop in a region of space where the local value of the bifurcation parameter allows it to be stable in the corresponding uniform system. This leads immediately to the prediction of spatial coexistence of structures with different symmetries; one observes then the unfolding in space of the uniform bifurcation diagram. This simple locally uniform approach has proved useful in the qualitative interpretation of many experiments.

Dr Boissonade also responded: The gradients are controlled by the thickness of the gel. Dulos *et al.* have built a reactor which exhibits a continuous change of thickness in one direction of space.[1] This allows for the unfolding of a pattern sequence in the direction where the confinement is progressively relaxed.

In most experiments on Turing patterns, gradients are chosen to confine the structure in a stratum the width of which is limited to one wavelength. Although these "monolayers" are quasi 2D systems, the selection of patterns can be significantly modified by the presence of the gradient. Dufiet and Boissonade have used a toy model of the Fitzhugh–Nagumo type to study the selection of patterns close to onset in such a monolayer.[2]

1   E. Dulos, P. Davies, B. Rudovics and P. De Kepper, *Physica D*, 1996, **98**, 53.
2   V. Dufiet and J. Boissonade, *Phys. Rev. E*, 1996, **53**, 4883.

Prof. Harrison commented: The question of quadratic *versus* cubic terms in chemical kinetics (linearized about homogeneous steady state), as governing spotting *versus* striping tendencies in pattern formation, was raised in discussion of Prof. Dewel's paper, and absence of quadratic terms from chemistry was indicated as somewhat unlikely. I wish to make three points about this:

(1) In my early work on reaction–diffusion models, I devised a hypothetical mechanism which I called "hyperchirality" which, on linearization about the homogeneous state, completely lacks quadratic terms in the nonlinearities. It was based on a possible (even if rather improbable) Flatland chirality in alternative geometries of attachment of a protein tetramer to a membrane. The model has strongly striping behaviour.[1,2]

(2) A phenomenon involving both chemistry and physics (of lateral interactions across an array of synapses during their formation) is the formation of ocular dominance stripes of connections of right and left optic nerves to one side of the brain in the primates in which R and L connections are numerically equal, to either side of the visual cortex. Here again there is an aspect of chiral symmetry in suggested mechanisms. My student Michael Lyons showed[2] that a generalized mechanism of this symmetry gave, algebraically, exactly the same stability analysis as reaction–diffusion kinetics lacking quadratic terms.

(3) The CIMA reaction has been observed to give stripes or spots for different initial values of the chlorite/malonate ratio. For the published mechanism of Lengyel and Epstein[3] change of this ratio in fact changes the ratio of cubic to quadratic terms in the right direction to account for this. This can be seen from eqns. (4) to (8) of Lengyel and Epstein[3] by the standard procedure of linearizing about the homogeneous steady state X_0, Y_0 and looking at the new forms of the nonlinear terms in the variables $U = X - X_0$ and $V = Y - Y_0$, where $X = [\text{I}^-]$ and $Y = [\text{ClO}_2]$. I obtain terms in UV, U^2 and U^2V, with the ratios of cubic to quadratic terms:

coefficient of U^2V/coefficient of $UV \propto [\text{ClO}_2]_0/[\text{MA}]_0$;

coefficient of U^2V/coefficient of $U^2 \propto [\text{ClO}_2]_0/[\text{MA}]_0$,

where MA is malonic acid. Since $[\text{ClO}_2]_0$ is probably proportional to $[\text{I}^-]_0$, these dependencies are in accord with the observation of Ouyang and Swinney[4] that: "stripes ... were observed at high $[\text{I}^{-1}]_0$ or low $[\text{MA}]_0$". Either of those conditions will give a high ratio of cubic term to quadratic terms. Lyons and Harrison[2] showed by computation that the transition from stripes to spots is not sudden but goes *via* intermediate patterns as the ratio of cubic to quadratic terms decreases.

1   L. G. Harrison and T. C. Lacalli, *Proc. R. Soc. London, Ser. B*, 1978, **202**, 361.
2   M. J. Lyons and L. G. Harrison, *Dev. Dyn.*, 1992, **195**, 201.
3   I. Lengyel and I. R. Epstein, *Science*, 1991, **251**, 650.
4   Q. Ouyang and H. L. Swinney, *Nature*, 1991, **352**, 610.

Prof. Merkin opened the discussion of Mr Hemming's paper: In 1D simulations of a model system we found parameter ranges where there was spatiotemporal chaos. This resulted from the spatially uniform system having a strongly stable (unreacted) steady state and a weakly unstable state, with the existence of travelling waves in the spatial system which could propagate into the unreacted state. The spatio-temporal dynamics gave regions of the unreacted state which were 'eaten away' by the travelling waves. The effect of the saddle-point in the kinetics was to cause 'information to be lost' as the system was returned to the unreacted state. This led to an irregular response somewhat similar in appearance to the patterns you show in Fig. 5. Is there a similar mechanism present in your system?

Mr Hemming replied: The similarity between the 'weak turbulence' we describe and the phenomena seen in your system (ref. 19 of the paper) is that there are repeated domain nucleation events. In terms of the phase space corresponding to the 'reaction term', *i.e.* the ode (ordinary differential equation) arising by dropping the diffusion term, the mechanism is not similar. In your system there is a single stable state and a weakly unstable state, and domains of the stable state nucleate in domains of the unstable state. In our system domains of a third stable state nucleate at the front between two stable states when the oscillations of the front cause it to come sufficiently near the basin of attraction for the third stable state. In ref. 19 you describe your system as fulfilling a Shilnikov condition in which the front reinjects the stable state into the weakly unstable focus. This arises because the front travels in the direction such that the unstable state propagates into the stable state. It is not likely that there is a comparable mechanism present in our system.

Prof. Showalter asked: What are the main differences between the 2 : 1 and 3 : 1 resonances?

Mr Hemming responded: A full investigation of the 2 : 1 system remains to be carried out. In our investigations in the two-dimensional case using the same parameter values as used for the 3 : 1 case (apart from γ), we saw a transition from regular fronts to rough fronts. As the forcing intensity was lowered the fronts became increasingly rough and there was considerable fragmentation of the front. We did not observe a thick, turbulent interface like that in the 3 : 1 system at the forcing intensities we studied. The greater tendency to front fragmentation in the 2 : 1 system could be associated with the interface having zero average velocity. Fronts in the 2 : 1 FCGL equation either have zero average velocity or exist as a chiral pair with opposite velocities due to the non-equilibrium Ising–Bloch bifurcation which breaks the chiral symmetry (ref. 6 of the paper). In one-dimension the bifurcation structure of the 2 : 1 and 3 : 1 cases are quite different. In the 2 : 1 system the onset of oscillations is through a Hopf bifurcation and there is a period doubling cascade which leads to chaotic front motion (see refs. 9–11 of the paper). It seems unlikely that something analogous to the 'weak turbulence' we find in the 3 : 1 system could exist in the 2 : 1 system because it involves nucleation of domains of the third state at the front.

Dr Satnoianu commented: The mechanism of turbulence presented is very interesting. In my previous work (see ref. 1). I encountered the case of amplitude turbulence in a cubic Ginzburg–Landau context. In that case the dynamics led to the appearance first of a phase turbulence (weak turbulence) which later developed into a full amplitude regime (strong turbulence). It would therefore be of interest to study whether a similar scenario does apply for the current problem with forcing.

1 J. H. Merkin, R. A. Satnoianu and S. K. Scott, *J. Chem. Soc., Faraday Trans.*, 1998, **94**, 1211.

Mr Hemming responded: The phenomena you describe underlie the behavior seen in our forced CGL model. In particular, the fact that, for the CGL parameters chosen, the unforced CGL equation exhibits phase and amplitude turbulence is the source of turbulence seen in our interfacial dynamics. The full investigation of such front behavior when the underlying CGL dynamics is either phase or amplitude turbulent is an interesting topic that merits further investigation.

At the Benjamin–Feir instability, one may reduce the FCGL equation to a phase equation which is the damped Kuramoto–Sivashinsky equation (see refs. 10 and 11 of the paper), and this may be a useful equation if one wishes to understand the role of phase turbulence alone.

Prof. Merkin said: In your paper you mention that the exponent of the forcing term $n \leqslant 4$. Is this just a requirement from how the CGL equation arises or is there a deeper mathematical reason for this restriction?

Mr Hemming answered: In the reduction of the Hopf bifurcation to the CGL equation the leading nonlinear term is of order 3. Forcing at the $n : m$ resonance gives rise to a term of order $n - 1$. The strong resonances $n = 1,2,3,4$ are those for which this term is of the same or lower order as the cubic term. A discussion of strong *versus* weak resonances may be found, for example, in the PhD thesis of Bernd Krauskopf, Rijksuniversiteit Groningen, 1995, and references therein.

Dr Satnoianu said: I observe that the form of the Ginzburg–Landau equation with the conjugate term due to forcing could be thought of as being similar to a quintic Ginzburg–Landau equation without forcing. It would therefore be of interest, and I suggest this to the authors, to study whether the latter quintic equation can display the same dynamical behaviour (*i.e.* turbulence) as the cubic Ginzburg problem with forcing.

Mr Hemming answered: The quintic CGL equation models a sub-critical Hopf bifurcation in a spatially distributed oscillatory system without forcing. Our cubic FCGL equation is a model for a resonantly forced system. In the system modelled by the quintic CGL equation there is no external source of an additional frequency to give rise to a resonance with the unforced oscillation frequency. Additionally, like the cubic Stuart–Landau equation, in the quintic Stuart–Landau equation the evolution of the R variable is independent of ϕ, and one observes only limit-cycles (and approaches to them) in the phase plane. Without the complex conjugate terms to introduce ϕ-dependence into the R dynamics there is no means of obtaining non-trivial fixed points in the phase portrait.

Mr Hantz said: You have mentioned in your lecture that dynamical phase transition occurs in your model. In what sense did you use the term "dynamical phase transition"?

Mr Hemming responded: As a parameter value (the forcing amplitude) is changed, the system makes a transition between the "confined interface" and "exploding interface" regimes at a critical value of this parameter. Properties of the system such as the mean interface width, the correlation length and time of the interface width, and the mean interface velocity, display power-law scaling behavior approaching this value. Consider a system in the "exploding interface" domain composed of the turbulent phase. As one changes the parameter to move the system into the "confined interface" domain, patches of the homogeneous phase form, separated by the turbulent phase. The separation between homogeneous patches corresponds to the interfacial width. The power-law scaling of this process has the hallmarks of a continuous nonequilibrium phase transition.

Prof. Noszticzius opened the discussion of Prof. Showalter's paper: How was the bromomalonic acid produced?

As the rate of enolisation of bromomalonic acid (R13 in Table 1 of the MBM mechanism from our Discussion paper) is about five times faster than that of malonic acid (R29 in the MBM scheme of our paper) further bromination of bromomalonic acid cannot be avoided when malonic acid is brominated by bromine. Such a bromination in an aqueous medium yields a mixture of malonic, bromomalonic and dibromomalonic acids. This is not a problem when a reaction mixture is prepared for chemical wave experiments because a mixture of these organic substrates works well, but for a better reproducibility of the experimental results it can be useful to give a brief description of the preparation.

Prof. Showalter responded: All stock solutions were prepared using doubly distilled water and were stored under refrigeration. Sodium bromide solution was added to a 500 ml stirred reaction mixture containing $NaBrO_3$, H_2SO_4, and malonic acid in an ice bath at a rate of 1.0 ml min^{-1} using a controlled syringe pump. The resulting reaction mixture was stored in an ice bath during the experiment.

Prof. Noszticzius asked: How uniform was the wave velocity in the reactor area B? I am asking this question because in a nonhomogeneous medium we could explain certain wave shapes with the geometrical theory of waves.[1]

1 A. Lázár, H. D. Försterling, H. Farkas, P. Simon, A. Volford and Z. Noszticzius, *Chaos*, 1997, **7**, 731.

Prof. Showalter replied: The medium excitability, which was primarily determined by the light intensity, was carefully checked for uniformity. No boundary effects were exhibited, as the boundaries were avoided during the wave segment propagation.

Prof. Müller commented: The results of this detailed study are an important step forward in characterizing the two excitability limits for the propagation of 2 dimensional waves with free ends, that have been first identified experimentally in an investigation of aging Ce-catalyzed BZ medium.[1]

So far, the experiments in the light-sensitive BZ reaction focus on stability properties of wave segments around the "second" limit at higher excitability. As supported by simulations (Fig. 8 of the paper) the critical wave size decreases with the excitability, until the lower limit ϕ_{1D} is reached and wave propagation is then completely suppressed.

Has this lower limit also been characterized by experiments under similar conditions in the ruthenium-catalyzed BZ system?

1 Zs. Nagy-Ungvárai, J. Ungvarai and S. C. Müller, *Chaos*, 1993, **3**, 15.

Prof. Showalter answered: All waves with free ends contract tangentially and eventually disappear in the excitability range between ϕ_∞ and ϕ_{1D}. Hence, spiral waves are not supported by the medium in this range. Waves without free ends may be supported in this range, depending on the curvature; however, all waves collapse at excitabilities below ϕ_{1D}. We are currently experimentally and theoretically investigating the excitability boundary at ϕ_{1D}.

Prof. Scott said: With reference to Figs. 12 and 13 of your paper, could you explain the equivalence of wave segment length L and the area S in these experiments.

Prof. Showalter replied: Wave segment length L and area S are directional proportional except for the smallest wave segments.

Prof. Sørensen commented: I like very much the well controlled experiments shown in Figs. 3 and 5 which are qualitatively explained by the two-variable photochemical Oregonator. With biological experiments in mind I wish to get your opinion on the possibility of using measurements of the spatial dynamics to get mechanistic information on the BZ in the gel. These type of measurements could be essential for revealing the mechanism of reactions inside a living cell. As demonstrated in the case of glycolysis it is sometimes problematic to take experimental results for homogeneous extracts and apply them directly to the processes inside a cell. The spatial BZ system seems to be an almost ideal test bed for the development of such methods. Here the results could be compared directly with a well known mechanism.

Prof. Showalter responded: I completely agree with your suggestion that it should be possible to obtain mechanistic information from controlled spatiotemporal systems using perturbation methods. Such methods have been developed for homogeneous chemical systems, in which information on the mechanism is obtained from the response of the system to perturbations.[1] There is no reason that similar methods could not be applied to spatiotemporal systems, which might offer an important tool for investigating mechanistic aspects of biological systems.

1 E. Mihaliuk, H. Skødt, F. Hynne, P. G. Sørensen and K. Showalter, *J. Phys. Chem.*, 1999, **103**, 8246.

Prof. Gáspár commented: In Fig. 8 of the paper, we see that at light intensities less than ϕ_S the medium remains excitable, and waves with free ends evolve into spiral waves. I would like to call your attention to a paper—coauthored by On-Uma, Zykov and Müller from Magdeburg and myself—that has recently appeared in *Phys. Chem. Chem. Phys.*[1] In this paper we report on studying the dynamics of these spiral waves as a function of light intensity less than ϕ_S. We measured the size of the spiral core, the wave velocity, and the rotation period as a function of ϕ. It has been found that these dependencies can be also well described by the kinematic theory of chemical waves. I am delighted to see that so many interesting experiments have been done by the photosensitive BZ system discovered by Beck, Bazsa and myself in 1983 in Debrecen. I would also like to mention that we have found a second method to generate wave segments. These observations have been published in *J. Chem Soc., Faraday Trans.* a few years ago.[2] In the experiments, we applied a bathoferroin loaded membrane following the method of Lázár et al.,[3] and wavetrains were initiated by using a silver electrode. We have found that if the period of wave initiation is very close to the length of the refractory period, wave segments are formed instead of nice circular waves. By applying a slightly smaller initiation period, we observed the formation of one armed spiral waves. However, these phenomena are not understood even today.

1 O.-U. Kheowan, V. Gáspár, V. S. Zykov and S. C. Müller, *Phys. Chem. Chem. Phys.*, 2001, **3**, 4747.
2 J. Dajka, T. Károly, I. Nagy, V. Gáspár and Z. Noszticzius, *J. Chem. Soc., Faraday Trans.*, 1996, **92**, 2897.
3 A. Lázár, H.-D. Försterling, A. Volford and Z. Noszticzius, *J. Chem. Soc., Faraday Trans.*, 1996, **92**, 2903.

Dr Hauser commented: The required light-intensity is recalculated and updated every 2 s. Did you investigate whether the location of the boundary of spiral waves is sensitive to the duration of the illumination 'steps'?

Prof. Showalter answered: The boundary for spiral wave behavior is comprised of the locus of unstable wave segment size as a function of light intensity. For each value of light intensity, which determines the medium excitability, there is an unstable wave segment of a particular size. The time interval in the algorithm may affect the ability to achieve control; however, once control is successful, the stabilized wave segment size is a function only of the medium excitability.

Prof. Müller commented. The illumination of layers of photosensitive BZ reaction may easily lead—due to the light absorption across the layer—to three dimensional aspects of the evolving patterns. From our experience one has to reduce the layer thickness to 0.3 mm or less to avoid three-dimensionality. This fits nicely to the thickness you have used in your experiments. Did you make specific efforts to ensure that two dimensionality is, in fact, ensured?

Prof. Showalter replied: Three-dimensional wave behavior in thin layers of BZ gel medium is certainly possible. In fact, we used 0.5 mm layers of gel to study highly uniform and long lived scroll waves, which were initiated by a discontinuous change in illumination intensity.[1,2] We have found well defined two-dimensional wave behavior in our experiments using 0.3 mm layers of gel with continuous illumination.

1 T. Amemiya, S. Kádár, P. Kettunen and K. Showalter, *Phys. Rev. Lett.*, 1996, **77**, 3244.
2 T. Amemiya, P. Kettunen, S. Kádár, T. Yamaguchi and K. Showalter, *Chaos*, 1998, **8**, 872.

Prof. Scott commented: The terminology "subexcitable media" has been used recently, particularly in regard to important work of relevance to wave propagation in neuronal tissue analogues. Are we in a position to provide a cleaver definition of this term?

Prof. Showalter answered: As shown in Fig. 8 of our paper, the locus of unstable wave segment size as a function of light intensity defines the boundary for spiral wave behavior. It is important to note that excitability decreases as light intensity increases and *vice versa*. The locus diverges at ϕ_∞, corresponding to a wave segment of infinite length. This excitability also corresponds to an unbounded spiral wave with an infinite period. The period of a spiral wave increases as the excitability of the medium is decreased, and when the excitability is given by ϕ_∞, the spiral has opened up to become an unbounded planar wave with a free end. For excitabilities between the boundary for spiral waves, ϕ_∞, and the 1D excitability limit, ϕ_{1D}, the medium does not support waves with free ends, such as spiral waves. Thus, at excitabilities between ϕ_∞ and ϕ_{1D}, the medium is unconditionally subexcitable, *i.e.*, waves with free ends contract tangentially and eventually disappear. Below the 1D wave excitability limit, all waves collapse and disappear. At excitabilities higher than that defined by ϕ_∞, the fate of a wave with a free end depends on the wave size, as defined by the locus in Fig. 8: smaller waves collapse, while larger waves grow to fill the medium with spiral waves.

Dr Schreiber commented: In the present context subexcitability seems to refer to a situation when an initial wave segment develops into a specific wave pattern depending on its size or shape, and on external parameters. When generalized, this may be viewed as an excitable system with multiple thresholds. For example, in a simple system of two coupled excitable one-variable subsystems we have found (unpublished work) two different thresholds. A medium-size perturbation crosses the first threshold, elicits an excitation in the first cell which then propagates to the second cell, while a larger perturbation causes an excitation in the first cell only. When the coupling strength is decreased, the two threshold sets partly merge so that a superthreshold perturbation of the first cell gives rise to an excitation that never propagates to the second cell. Thus this system may be called conditionally subexcitable for a stronger coupling and unconditionally subexcitable when the coupling is weak.

Dr Hauser opened the discussion of Dr Münster's paper: Fig. 2 of the paper shows the resonant trajectories of spirals. The points were sampled at equidistant times. Shouldn't the lengths (or arc lengths) of this resonant trajectories between two points be equally long? What is the reason for the big variation in those lengths? Is it a sampling problem, which could be overcome by a higher sampling frequency?

Dr Münster responded: Indeed the overall sampling time and sampling rate in both experiments shown was the same. However, the number of loops and hence the length is different in both cases.

Moreover, the velocity of the spiral tip is not constant as can be seen from the distances between the individual points. The different lengths of the resonant trajectories reflect the limitations of quantitative reproducibility of the complex spiral tip motion. Higher sampling rates were employed in other series of experiments but they do not change the observed behaviour.

Mr Sielewiesiuk asked: As regards the experimental setup: can you apply the electric field to a selected area of the sample? Can this area be of desired shape and size? How accurately can you control the size of this area? What is the smallest area to which you can apply the electric field?

Dr Münster replied: The electric field can be applied to selected areas by the use of shaped electrodes located close to the membrane surface. Some control of shape and size of the perturbed area can be achieved this way. The accuracy, however, will certainly be inferior compared to *e.g.* the use of illumination through a mask—but our technique can be used to perturb systems which are not light sensitive! Regarding the last part of your question: using a planar electrode on one side of the membrane and a needle-like tip-electrode on the other will affect quite small areas.

Prof. Müller commented: In the experimental section of your paper you mention Pt-ring electrodes and transparent electrodes used for applying an electric field. How homogeneous is the spatial distribution of the field across the membrane in these geometrical arrangements? Are there advantages for using the one or the other?

Dr Münster answered: The transparent electrodes are made of ITO (indium tin oxide)-coated glass and are planar. In a coplanar arrangement they generate a fairly homogeneous electrical field. The ring electrodes lead to an inhomogeneous field, particularly if they are mounted close to the membrane. However, the advantage of the Pt-ring electrodes over the ITO-electrodes is their stability to the BZ reaction mixture: platinum is inert with respect to sulfuric acid and bromate whereas the ITO-layer is slowly decomposed by the reaction mixture. Therefore the ITO-electrodes must be replaced after 8 to 10 h; this is an expensive and time-consuming procedure.

Prof. Müller asked: In Fig. 3, did you use the ring or the transparent electrode?

The spontaneous appearance of islands of low excitability with more or less pronounced boundaries looks rather mysterious to me. Could you give a more detailed reasoning why such islands should form? Is there any correspondence to that in the simulations?

Dr Münster responded: All the experiments shown in the contribution were obtained with planar ITO-coated electrodes. The qualitative appearance of the patterns, however, was very similar if ring electrodes were used.

Regarding the second part of your question: in the simulations the mechanism of island formation can be seen more clearly than in the experiments: as the excitability decreases some of the initially smooth, curved waves break apart leaving edges which are located along a more or less bent line. During the following half cycle of the perturbation new spiral cores form at these edges. The new cores are located in a close neighborhood to each other and start to emit waves. At some distance from the newly formed cores the waves annihilate and a refractory region is created. This region cannot be penetrated by other waves until the excitability increases in the next perturbation cycle. Basically, that is how the observed "islands" form.

Prof. Showalter commented: I wish to point out a study that is complementary to your coupled pattern experiments. We carried out a study of coupled spatiotemporal patterns in which bromous acid served as the messenger species through a Nafion membrane.[1] These studies demonstrated that spatiotemporal synchronization occurs with increasing coupling.

1 D. Winston, M. Arora, J. Maselko, V. Gáspár and K. Showalter, *Nature*, 1991, **351**, 132.

Dr Münster answered: These experiments are complementary to our work since the coupling there is realized by a diffusive exchange of matter between the subsystems, if I remember the work correctly. The coupling therefore is proportional to the difference of concentrations in the two coupled systems and its magnitude decreases as the patterns become synchronized. In our method of coupling the coupling does not involve any difference of both patterns and hence the result of the coupling should be a repulsion of the waves as observed.

Prof. Merkin said: You model the electric field effects with a time-varying excitability (eqn. (14)). This is an interesting idea, however, the effect of the electric field will involve the electromigration of the ionic species, which will give additional, convective-like, terms in the model. How good an approximation is it to model these effects by varying ε, and does this approach just really model your experimental results qualitatively?

Dr Münster responded: The idea behind the simulations was to investigate the effect of periodic perturbations of the excitability since the effect of the field in the experiments is apparently the same. A detailed simulation of the electrical field effect should not only be based on a more detailed chemical model of the BZ reaction, it should also include diffusive and convective terms to describe the exchange between the membrane and the stirred cells. Here, not only migration of ions should be taken into account but also electroosmosis should be considered since the membrane is charged. Such simulations should be carried out in three space dimensions in order to capture the spatial gradients inside and perpendicular to the membrane. We are currently working on simulations of that kind.

Prof. Müller said: Please specify the role and the different impact of migration *versus* osmosis in your system.

Dr Münster replied: Migration in the applied electric field acts on ionic species such as bromide and bromate. These components are particularly crucial for the propagation of waves inside the membrane and they are supplied into different chambers. Depending on the actual polarity at the electrodes, the supply of bromate is increased by the field while bromide is removed from the membrane or *vice versa*. In addition to migration, electroosmosis occurs since the membrane carries the immobilized catalyst and hence is charged. The electroosmotic flux (EOF) acts not only on charged but also on electrically neutral species. For example, the autocatalytic species $HBrO_2$ is removed from the membrane by EOF; this leads to a decreased excitability regardless of the polarity at the electrodes.

Prof. Müller commented: The statement in the 'Numerical Simulations' section that there is a similarity between the experimental patterns of Fig. 3 and the simulations in Fig. 6 is a qualitative one which relies on our visual impression. Are there better, more quantitative criteria to underline this statement?

Dr Münster responded: In other work on complex spatiotemporal patterns we used the Karhunen–Loeve decomposition technique for an analysis of chemical patterns. In the case of target or spiral waves this method is computationally expensive—the decomposition must be performed for a sequence of spatially two-dimensional images—and, moreover, it is not very useful for these kinds of patterns. In a similar way, the spatial Fourier transforms do not give insights which cannot be seen from the original images as well. In Fig. 4 we plotted the average number of spiral cores observed at a given perturbation amplitude and frequency. The experiments as well as the simulations show a maximum number of cores if the amplitude and frequency is varied. This observation relates the experimental to the numerical findings in a more quantitative way than the visual comparison of Figs. 3 and 6.

Concluding Remarks

Nonlinear chemical kinetics: past, present and future

Irving R. Epstein

Department of Chemistry, MS 015, Brandeis University, Waltham, MA 02454-9110, USA

Received 17th October 2001
First published as an Advance Article on the web 8th January 2002

Introduction

This has been a memorable Discussion, both for the quality of the science presented and for the tragic events that took place in New York and Washington while we were meeting.

In order to add some context to the papers presented here, I should like to begin with a brief historical overview, focusing on some of the major meetings that have shaped the study of nonlinear chemical kinetics. This approach seems particularly appropriate in view of the role (see below) that this Discussion's predecessor, Faraday Symposium 9, played in the development of the field.

Some history

Perhaps the first truly influential meeting on this topic was held in Prague in 1968.[1] The Prague meeting helped to establish glycolysis, a system about which we have heard much at this Discussion, as the prototypical biological oscillator. At least as important was the opportunity for Western scientists to learn of an obscure reaction, discovered serendipitously in the Soviet Union, that had the ability to generate both temporal oscillation and spatial pattern formation. The Belousov–Zhabotinsky reaction soon became the workhorse and the prototype for the study of nonlinear chemical kinetics, a role which it continues to hold.

After the Prague meeting, interest in these fascinating reactions grew rapidly, and in 1974, Faraday Symposium 9 was organised by Peter Gray, who is here with us today, on the topic of "Physical Chemistry of Oscillatory Phenomena".[2] That Symposium generated a palpable sense of excitement, particularly among some of the younger researchers who attended. The organizers also displayed considerable perspicacity in their choice of papers. A strong case can be made that the structure of that meeting, with sessions on inorganic oscillators; thermokinetic oscillations; membranes, heterogeneous and biological systems; and theory of excitable media, laid out the structure of the field for the next several decades. In Table 1, I compare the numbers of papers in each area for Faraday Symposium 9 and Faraday Discussion 120. The similarity is remarkable; the shift away from inorganic toward biological and heterogeneous systems accurately reflects the direction of recent activity.

Two other sets of conferences deserve mention here. A series of meetings[3–5] organized in Bordeaux in the late 1970's and early 1980's helped to solidify the international nature of the nonlinear chemical kinetics community and served as a vehicle for introducing the continuous flow stirred tank reactor as a key tool in experimental studies. The Gordon Conferences on Chemical Oscillations, the first of which was held in Plymouth, New Hampshire, USA, in 1982, have provided a regular venue for the growing number of practitioners in the field to come together, at first every third, and now every second, year to share their latest results and ideas.

DOI: 10.1039/b109477g

Table 1 Comparison of numbers of papers given at Faraday Symposium 9 (1974) and Faraday Discussion 120 (2001)

	1974	2001
Introductory and inorganic oscillators	6	3
Thermokinetic oscillations	4	5 (3 combustion + 2 atmospheric)
Membranes, heterogeneous and biological systems	5	11 (8 biological + 3 heterogeneous)
Theory of excitable media	4	5
Total	19	24

Where we stand

Having sketched some of the past of nonlinear chemical kinetics, I turn now to the present, both the work that has been presented at this Discussion and the current state of the art. It seems clear that nonlinear chemical dynamics has grown remarkably in both breadth and depth. One might argue that in 1974 the subject stood at the margins of chemistry and had little overlap with the "core" of that discipline. I shall return at the end to the question of whether nonlinear kinetics has yet reached the core, but at this Discussion we have discussed work that touches on nearly all of the most exciting and fastest-growing areas of chemistry and related sciences. Among these are: surface science, electrochemistry, combustion, polymer science, single molecule chemistry, and biology, including neurobiology, development and metabolism. Despite this branching out of our field (or perhaps the recognition by other fields that nonlinear chemical kinetics has something to offer them), we have also continued to cultivate "our roots". We have also heard at this Discussion about some of the topics from which nonlinear chemical kinetics sprang: the Belousov–Zhabotinsky reaction, glycolysis, traveling waves, excitability, multiple stationary states and mechanistic studies. In all cases, the work presented here represents significant advances in the depth of understanding and the sophistication of the techniques employed over what was possible a few decades, or even a few years, ago.

It is useful for those of us who have been working in this area for some time, as well as for those only recently embarked on the study of nonlinear chemical kinetics, to step back and reflect on how far we have come. The development of flow reactors, both stirred and unstirred, the application of powerful methods of spectroscopy and microscopy, the extension of our studies to membranes and catalytic surfaces, are only some of the striking experimental advances. On the theoretical side, we have benefited from the many orders of magnitude enhancement in computational capability that has occurred over the past quarter century, but we have also seen the development of new and powerful theoretical methods involving amplitude equations, bifurcation theory (analytical and numerical) and stochastic approaches. Where at one time developing a mechanism or a model that gave oscillations of any sort was considered a triumph, today we have multiple alternatives to choose from, and sophisticated experiments and algorithms are often invoked to distinguish among them.

The diversity and range of systems and phenomena that the field comprises has grown exponentially. The number of oscillating reactions, to cite just one example, has increased from two or three serendipitously discovered systems to literally dozens, most of them systematically designed. In addition to temporal oscillation and the occasional study of spiral waves or multistability, we now investigate Turing patterns, chaos, standing waves, clusters, scroll waves and many more exotic phenomena. In 1984, the organizer of the final Bordeaux meeting noted that since 1967 "over 26 meetings have been held, and their yearly number is increasing".[5] Pacault's concern was well-founded. Today, there may well be 26 meetings each year on nonlinear chemical kinetics and related subjects, and it is difficult for any single meeting to have the impact of the Faraday Symposium of 27 years ago.

One feature that distinguishes this area from many others in science is a special culture, spawned, I suspect, by the fact that until recently nearly all of its practitioners were "immigrants" from other fields, who had not been trained in nonlinear chemical kinetics. To an unusual degree, the field is interdisciplinary. A typical meeting will be attended not only by chemists, but also by physicists, mathematicians, engineers, biologists, and, occasionally, geologists, economists, astronomers or physicians. By comparison with many other fields, the atmosphere in nonlinear chemical kinetics tends to be much more one of collaboration than of competition. The number of papers that come out of multiple laboratories is striking. The degree of cross-fertilization between theory and experiment is also atypical. Theories beget experiments and *vice versa*, and people tend to have mastery of, or at least respect for, both theory and experiment to a much greater extent than is common in most other fields. Finally, a very special aspect of this field is the striking visual phenomena that characterize it. The power of an oscillating reaction or a spiral wave in a petri dish to capture the attention and the imagination of prospective students of nonlinear chemical kinetics cannot be overstated.

The future

One of the most salient lessons that we learn from tracking the progress from Faraday Symposium 9 to Faraday Discussion 120 is how perilous it is to attempt to predict the future development of a vibrant field of science. Nonetheless, while avoiding the details, which I would undoubtedly get wrong, it seems to me that the work presented here enables one to make some general statements about the likely directions in which nonlinear chemical kinetics is heading.

The future of this subject is likely to include more and more of what might be termed "chemistry plus", that is, chemical reactions coupled to other phenomena that are typically considered to belong to the domains of physics or biology. This trend is already evident in some of the work presented here. The sorts of interactions I have in mind include hydrodynamics, surface tension and fluid flows; growth, both of living systems and of materials; mechanical forces; external fields, particularly electromagnetic and gravitational; and heterogeneous media—biological cells, membranes, micelles, emulsions, surfaces...

Our attention is increasingly likely to be focused on "big systems". These might include systems in which many subsystems are coupled together, such as a dish full of yeast or a microemulsion containing 10^{17} nanodroplets of water; any living organism; the atmosphere; combustion problems; materials science; and the oceans. Fortunately, we may anticipate further experimental advances, and we are already becoming adept at generating and solving very large models, some containing hundreds of equations, and at developing methods to reduce those models to smaller sets of equations that offer some hope for intuitive understanding to complement the numerical results obtained from their more detailed precursors.

The future of the field will inevitably bring it into closer touch with "real world" problems. Issues that have effects outside of the laboratory not only offer increased prospects for research funding, but they are often scientifically fascinating and almost invariably challenging. Some areas in which ideas and techniques from nonlinear chemical kinetics and related fields are already beginning to be applied include the development of gels for drug delivery, meteorological analysis, the effects of weak electromagnetic fields on living tissue, and the prediction of currency fluctuations.

As was pointed out in the Introductory Lecture, further development of nonlinear chemical kinetics will necessitate a multi-scale approach to problems and the building of bridges between the micro- and macroscopic views of phenomena. Successful attack on this aspect of our field will ultimately require the development of new experimental and theoretical techniques. As one example, I cite recent work on the BZ system in a microemulsion consisting of water, oil (octane) and a surfactant, sodium bis(2-ethylhexyl) sulfosuccinate (AOT).[6] As Fig. 1 shows, this BZ–AOT system exhibits a remarkable range of spatio-temporal phenomena. The medium consists of nanometer-sized droplets of water surrounded by a monolayer of AOT, floating in a sea of octane. Since the key species in the BZ reaction are polar and are therefore confined almost exclusively to the water, the reaction takes place in the droplets. On average, each droplet contains fewer than a dozen molecules of the ferrous phenanthroline catalyst. The chemistry is occurring in a sub-femtoliter (actually

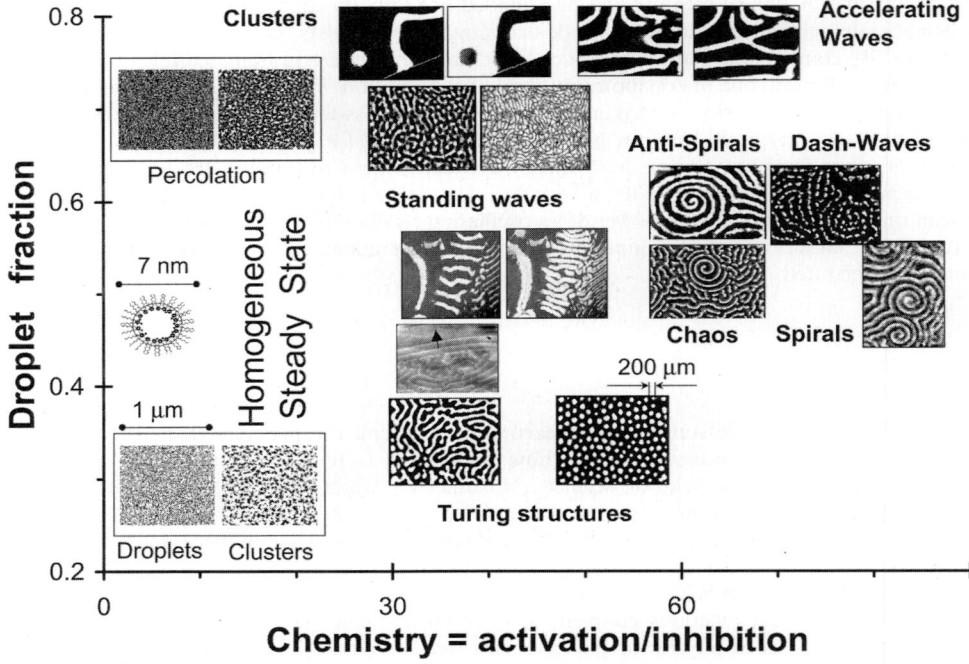

Fig. 1 A schematic phase diagram summarising complex spatio-temporal behavior in the BZ–AOT system.

micro-femtoliter) volume! This situation poses at least two fascinating challenges, one theoretical, the other experimental. The theoretical problem is to develop approaches that allow us to deal with large numbers of interacting units in which statistical fluctuations are very large without having to invoke a mean field approximation. A typical experiment in a petri dish involves 10^{17} water droplets. The structures shown in Fig. 1 have length scales of a few tenths of a millimeter, five orders of magnitude larger than the size of a droplet. It is likely that interesting structures arise at much smaller length scales than we have been able to observe thus far, but the tools remain to be developed.

I mention the BZ–AOT system merely as one example of the sorts of phenomena that are likely to arise in the years to come and the exciting problems that they will bring with them. The field is still growing and will continue to do so in directions that are impossible to predict today. One cautionary note, however, is that, to be completely honest, nonlinear chemical kinetics has still not gained a place in the central core of chemical science. For all its growth and respectability, its inclusion in textbooks, it is still considered a peripheral topic by many, perhaps most, chemistry departments. In the end, we will have become a truly central field when every major chemistry department feels that, just as it needs a spectroscopist, a crystallographer, and a synthetic organic chemist, it also needs a practitioner of nonlinear chemical kinetics. I think that day is drawing nearer, but only time will tell.

Acknowledgements and verse

My participation in this meeting was made possible by a grant from the Chemistry Division of the National Science Foundation. I should like to thank Steve Scott, Ian Smith and the staff of the Faraday Division for making possible such a successful Discussion under the most trying of external circumstances. I offer the following sonnet as a summary of our discussions.

On Nonlinear Chemical Kinetics
This meeting we have just attended,
Deftly crafted by the great Scott.
Now we're sorry that it's ended.
We depart enlightened but not
Even close to saturated. We've
Viewed data, models, theories Steve
Chose for our illumination
In five minutes plus explication—
Surface, microgravity, BZ,
Even glycolysis in yeast.
We've had a scientific feast
Of fine nonlinear chemistry.
New insights we've received aplenty
At this, the Faraday 120.

References

1. *Biological and Biochemical Oscillators*, ed. B. Chance, A. K. Ghosh, E. K. Pye and B. Hess, Academic Press, New York, 1973.
2. *J. Chem. Soc., Faraday Trans.*, 1975, **13**.
3. *Far from Equilibrium: Instabilities and Structures*, ed. A. Pacault and C. Vidal, Springer-Verlag, Berlin, 1979.
4. *Nonlinear Phenomena in Chemical Dynamics*, ed. C. Vidal and A. Pacault, Springer-Verlag, Berlin, 1981.
5. *Non-Equilibrium Dynamics in Chemical Systems*, ed. C. Vidal and A. Pacault, Springer-Verlag, Berlin, 1984.
6. V. Vanag and I. R. Epstein, *Phys. Rev. Lett.*, 2001, **87**, 228301.

List of Posters

Formation of localized wave initiation sites in aggregating dictyostelium discoideum **T. Godula, J. Lindner** and **H. Ševčíková**, *Prague Institute of Chemical Technology, Czech Republic*

Determination of the activity of antioxidants using the Briggs–Rauscher oscillating reaction **R. Cervellati** and **S. Costa**, *Università di Bologna, Italy*, **K. Hoener** and **C. Neddens**, *TU Braunschweig, Germany*, **S. D. Furrow**, *Penn State University, USA*

Gravity dependence of waves in gel-type Belousov–Zhabotinsky systems and in the retinal spreading depression **W. Hanke** and **M. Wiedemann**, *Universität Hohenheim, Germany*, **V. M. Fernandes de Lima**, *University of Sao Paulo, Brazil*

Macroscopic and microscopic patterns formed in reaction–diffusion systems of inorganic compounds **P. Hantz**, *Eötvös Loránd University, Hungary*

Temperature-controlled cellular fronts **É. Jakab, D. Horváth** and **Á. Tóth**, *University of Szeged, Hungary*

Nonlinear dynamics of safranine-O reaction with acidic bromate and role of bromide **S. B. Jonnalagadda, M. Shezi** and **N. R. Gollapalli**, *University of Durban-Westville, South Africa*

Brilliant cresyl blue–acid chlorite nonlinear reaction **S. B. Jonnalagadda**, *University of Durban-Westville, South Africa*

Complex voltammetric response in the course of hydrogen oxidation on Pt **H. Varela** and **K. Krischer**, *Fritz-Haber-Institut der Max-Planck-Gesellschaft, Germany*

Pattern formation during periodate reduction on Au(111) film electrode in $NaClO_4$ and $NaClO_4$–camphor electrolytes **Y.-J. Li, J. Oslonovitch, N. Mazouz, F. Plenge** and **K. Krischer**, *Fritz-Haber-Institut der Max-Planck-Gesellschaft, Germany*

Effect of pulsed illumination on the Belousov–Zhabotinsky reaction catalyzed with tris(bipyridine)ruthenium(II) in continuous stirred tank reactor **T. Matsumura-Inoue** and **T. Nakamura**, *Nara University of Education, Japan*, **Y. Mori**, *Ochanomizu University, Japan*

Pattern formation in globally coupled electrochemical systems with a S-shaped current potential curve **F. Plenge, N. Mazouz, J.-Y. Li** and **K. Krischer**, *Fritz-Haber-Institut der Max-Planck-Gesellschaft, Germany*

Bifurcation diagrams for the bromate–sulphite–ferrocyanide reaction **L. Schreiberová, J. Zagora, M. Voslař** and **I. Schreiber**, *Prague Institute of Chemical Technology, Czech Republic*

Complex response to perturbations in the CSTR with pH-oscillating reaction **J. Zagora, P. Kašpar, L. Schreiberová** and **I. Schreiber**, *Prague Institute of Chemical Technology, Czech Republic*

Mechanistic studies on the bromate–1,4-cyclohexanedione–ferroin oscillatory system **I. Szalai, K. Kurin-Csörgei** and **M. Orbán**, *Eötvös Loránd University, Hungary*

Convective instability in the chlorite–tetrathionate reaction **Á. Tóth** and **D. Horváth**, *University of Szeged, Hungary*, **P. D. Ronney**, *University of Southern California, USA*

Electrochemical turbulence during the hydrogen oxidation on Pt in the presence of Cl^- and Cu^{2+} **H. Varela, A. Bonnefont** and **K. Krischer**, *Fritz-Haber-Institut der Max-Planck-Gesellschaft, Germany*

Simulating the effect of polyethylene glycol on the BZ reaction with the MBM model **M. Wittmann, K. Pelle** and **Z. Noszticzius**, *Budapest University of Technology and Economics, Hungary*, **R. Lombardo** and **M. L. Turco Liveri**, *Università di Palermo, Italy*

Studies concerning the use of a pH oscillator to drive a self-oscillating polymer gel **A. J. Ryan, R. A. L. Jones, A. Cadby, A. Pryke** and **C. J. Crook**, *University of Sheffield, UK*

Bifurcation of spatio-temporal stripe patterns formed by the Ag/Sb co-electrodeposition **Y. Nagamine** and **M. Hara**, *Institute of Physical and Chemical Research, Japan*

Excitable media in a chaotic flow **Z. Neufeld**, *University of Cambridge, UK*

A chaotic traveling pulse in discrete dissipative systems **Y. Nishiura** and **T. Yanagita**, *Hokkaido University, Japan*, **D. Ueyama**, *Hiroshima University, Japan*

On the direct processing of a chemical signal **J. Sielewiesiuk** and **J. Górecki**, *Polish Academy of Sciences, Poland*

Dynamics of a vertical falling film in the presence of a first order exothermic reaction **P. M. J. Trevelyan, S. Kalliadasis, J. H. Merkin** and **S. K. Scott**, *University of Leeds, UK*

Density fingering of chlorite–tetrathionate fronts **J. Yang** and **S. Kalliadasis**, *University of Leeds, UK*, **A. De Wit** and **A. D'Onofril**, *Université Libre de Bruxelles, Belgium*

Quenching of flame propagation with heat loss **P. L. Simon, S. Kalliadasis, J. H. Merkin** and **S. K. Scott**, *University of Leeds, UK*

Complex oscillations in the fed-batch reactor: a work in progress **M. L. Davies** and **S. K. Scott**, *University of Leeds, UK*, **I. Schreiber**, *Prague Institute of Chemical Technology, Czech Republic*,

Dynamic evolution of flow distributed oscillations: the movie **J. R. Bamforth** and **S. K. Scott**, *University of Leeds, UK*, **R. Tóth** and **V. Gáspár**, *University of Debrecen, Hungary*,

Cellular rhythms and network morphology of cell shape in physarum **T. Nakagaki**, *Hokkaido University, Japan*

Wave onset in central gray matter. Its intrinsic optical signal and phase transitions in extracellular polymers **V. M. Fernandes de Lima**, *University of Sao Paulo, Brazil*, **J. E. Kogler, J. Bennaton** and **W. Hanke**, *Universität Hohenheim, Germany*

Macroscopic behaviour of a population of coupled oscillators **S. De Monte** and **F. d'Ovidio**, *Danish Technical University, Denmark*, **S. Danø** and **P. G. Sørensen**, *University of Copenhagen, Denmark*

Kinetics of temperature driven phase separation in low molecular binary mixtures **D. Vollmer** and **A. Wagner**, *University of Edinburgh, UK*, **J. Vollmer**, *Universität Mainz, Germany*

Quantitative modelling of glycolytic oscillations **F. Hynne, S. Danø** and **P. G. Sørensen**, *University of Copenhagen, Denmark*

List of Participants

Dr G. M. Allcock, *Royal Society of Chemistry, UK*
Mr J. R. Bamforth, *University of Leeds, UK*
Mrs A. C. Bennett, *Royal Society of Chemistry, UK*
Dr J. F. Boissonade, *Centre de Recherche Paul Pascal, France*
Dr M. Britton, *University of Cambridge, UK*
Prof. Dr R. Cervellati, *Università di Bologna, Italy*
Ms A. Chernova, *Mathematical Institute, UK*
Dr C. J. Crook, *University of Sheffield, UK*
Mr F. d'Ovidio, *Danish Technical University, Denmark*
Dr S. Danø, *University of Copenhagen, Denmark*
Mr M. L. Davies, *University of Leeds, UK*
Dr B. P. J. De Lacy Costello, *University of the West of England, UK*
Ms S. De Monte, *Danish Technical University, Denmark*
Prof. D. G. Dewel, *Université Libre de Bruxelles, Belgium*
Prof. I. R. Epstein, *Brandeis University, USA*
Dr A. M. Feigin, *Russian Academy of Sciences, Russia*
Prof. V. M. Fernandes de Lima, *University of Sao Paulo, Brazil*
Prof. V. Gáspár, *University of Debrecen, Hungary*
Prof. P. Gray, *University of Cambridge, UK*
Prof. J. Griffiths, *University of Leeds, UK*
Miss C. L. Hall, *Royal Society of Chemistry, UK*
Prof. W. Hanke, *Universität Hohenheim, Germany*
Mr J. S. Hansen, *Roskilde University, Denmark*
Mr P. Hantz, *Eötvös Loránd University, Hungary*
Prof. M. Hara, *Institute of Physical and Chemical Research, Japan*
Prof. L. G. Harrison, *University of British Columbia, Canada*
Dr M. J. B. Hauser, *Otto-von-Guericke Universität, Germany*
Mr C. Hemming, *University of Toronto, Canada*
Prof. A. Hunding, *University of Copenhagen, Denmark*
Prof. F. Hynne, *University of Copenhagen, Denmark*
Ms É. Jakab, *University of Szeged, Hungary*
Prof. S. Jonnalagadda, *University of Durban-Westville, South Africa*
Dr M. Kærn, *University of Toronto, Canada*
Dr K. Krischer, *Fritz-Haber-Institut der Max-Planck-Gesellschaft, Germany*
Dr K. Kurin-Csörgei, *Eötvös Loránd University, Hungary*
Dr A. Lang, *Reckitt Benckiser plc., UK*
Mr Y.-J. Li, *Fritz-Haber-Institut der Max-Planck-Gesellschaft, Germany*
Mr R. Lombardo, *Università di Palermo, Italy*
Prof. P. K. Maini, *Mathematical Institute, UK*
Prof. M. Marek, *Prague Institute of Chemical Techology, Czech Republic*
Dr H. Mayama, *Kyoto University, Japan*
Dr A. G. McDonald, *Trinity College Dublin, Ireland*
Prof. M. Menzinger, *University of Toronto, Canada*
Prof. J. H. Merkin, *University of Leeds, UK*
Dr Y. Mori, *Ochanomizu University, Japan*
Prof. S. C. Müller, *Otto-von-Guericke Universität, Germany*
Dr A. Münster, *Universität Würzburg, Germany*
Dr Y. Nagamine, *Institute of Physical and Chemical Research, Japan*
Prof. T. Nakagaki, *Hokkaido University, Japan*
Dr Z. Neufeld, *University of Cambridge, UK*
Prof. G Nicolis, *Université Libre de Bruxelles, Belgium*
Prof. Y. Nishiura, *Hokkaido University, Japan*
Prof. Z. Noszticzius, *Budapest University of Technology and Economics, Hungary*
Prof. M. Orbán, *Eötvös Loránd University, Hungary*
Mr F. Plenge, *Fritz-Haber-Institut der Max-Planck-Gesellschaft, Germany*
Dr N. M. Ratcliffe, *University of the West of England, UK*

Dr M. Rustici, *Università degli Studi di Sassari, Italy*
Dr R. Salazar, *Eindhoven University of Technology, Netherlands*
Dr R. A. Satnoianu, *Mathematical Institute, UK*
Dr I. Schreiber, *Prague Institute of Chemical Technology, Czech Republic*
Prof. S. K. Scott, *University of Leeds, UK*
Dr H. Ševčíková, *Prague Institute of Chemical Technology, Czech Republic*
Prof. K. C. Showalter, *West Virginia University, USA*
Mr J. Sielewiesiuk, *Polish Academy of Sciences, Poland*
Dr P. L. Simon, *University of Leeds, UK*
Mr H. Skødt, *University of Copenhagen, Demark*
Dr M. M. Slin'ko, *Institute of Chemical Physics, Russia*
Prof. I. W. Smith, *University of Birmingham, UK*
Dr D. A. Šnita, *Prague Institute of Chemical Technology, Czech Republic*
Prof. P. G. Sørensen, *University of Copenhagen, Denmark*
Dr I. Szalai, *Eötvös Loránd University, Hungary*
Dr A. F. Taylor, *University of Leeds, UK*
Dr A. S. Tomlin, *University of Leeds, UK*
Dr Á. Tóth, *University of Szeged, Hungary*
Mrs R. Tóth, *University of Debrecen, Hungary*
Dr P. M. J. Trevelyan, *University of Leeds, UK*
Dr M. L. Turco Liveri, *Università di Palermo, Italy*
Mr H. B. Varela, *Fritz-Haber-Institut der Max-Planck-Gesellschaft, Germany*
Dr D. Vollmer, *University of Edinburgh, UK*
Dr J. Wang, *University of Lethbridge, Canada*
Mr M. S. Weimer, *Universität Hohenheim, Germany*
Prof. H. V. Westerhoff, *Vrije Universiteit, Netherlands*
Miss L. E. Whitehouse, *University of Leeds, UK*
Dr S. B. Wilkes, *Royal Society of Chemistry, UK*
Dr M. Wittmann, *Budapest University of Technology and Economics, Hungary*
Mr J. Yang, *University of Leeds, UK*

Index of Contributors*

Bachir, M., 363
Bakos, R., **229**
Boissonade J. F., 202, 205, **353**, 407, 408, 409, 411, 412
Borckmans, P., **363**
Chirila, F., **383**
d'Ovidio, F., **261**
Danø, S., **261**, 335, 337, 338
De Kepper, P., **353**
De Monte, S., **261**
Dewel, D. G., 89, 102, 209, **363**, 409, 410, 411, 412
Dulos, E., **353**
Epstein, I. R., **11**, 86, 89, 95, 334, 410, **421**
Fairlie, R., **147**
Feigin, A. M., 100, **105**, 197, 198, 199, 200, 201, 203, 211, 349
Fernandes de Lima, V. M., 103, **237**
Field, R. J., **21**
Försterling, H.-D., **21**
Gáspár, V., 89, 90, 92, 94, 101, 198, 209, 333, 347, 348, 408, 416
Gauffre, F., **353**
Grauel, P., **165**
Griffiths, J. F., **147**, 205, 206, 207
Hanke, W., 87, 96, 100, 207, **237**, 332, 333, 334, 335
Hantz, P., 329, 333, 414
Harrison, L. G., 85, **277**, 340, 342, 345, 412
Hauser, M. J. B., 94, **215**, **229**, 325, 326, 327, 328, 329, 331, 332, 334, 338, 340, 416, 417
Hegedűs, L., **21**
Hemming, C., **371**, 413, 414
Holloway, D. M., **277**
Hunding, A., **295**, 336, 341, 343
Hynne, F., **261**
Jaeger, N. I., **179**
Jonnalagadda, S., 330, 337, 408
Kærn, M., **295**, 341, 343, 345
Kapral, R., **371**
Kosek, J., **53**
Krischer, K., **165**, 208, 209, 210, 213
Kummer, U., **215**
Kuperman, M. N., **353**
Kurin-Csörgei, K., **11**
Larsen, A. Z., **215**
Li, Y.-J., 200
Lindner, J., **53**
Lombardo, R., **39**, 99
Loskutov, E. M., **105**
Lowe, R., **125**
Maini, P. K., 340, 341
Mair, T., **249**
Mangone, M., **39**
Marek, M., **53**, 209, 213, 327, 338, 409
Mayama, H., **67**, 102, 103, 104, 409
Menzinger, M., 89, 100, 102, **295**, 338, 341, 342, 348, 350
Merkin, J. H., 206, 342, 413, 414, 419
Métens, S., **363**
Mihaliuk, E., **383**

Molkov, Y. I., **105**
Mukhin, D. N., **105**
Müller, S. C., 93, 101, 104, 209, **229**, **249**, 333, 335, 336, 337, 338, 339, 350, 415, 417, 418, 419
Münster, A. F., 98, 207, 211, 327, **395**, 417, 418, 419
Nagy-Ungvarai, Z., **229**
Nicolis, G., **1**, 85, 86, 93, 95, 197, 199, 201, 207, 211, 337, 339, 340, 341, 347
Nosztíczius, Z., **21**, 87, 88, 90, 92, 93, 94, 95, 99, 101, 208, 334, 348, 415
Olsen, L. F., **215**
Orbán, M., **11**, 87, 88, 89, 90, 93, 99, 328
Pačes, M., **53**
Peskov, N. V., **179**
Pilling, M. J., **125**
Rustici, M., **39**
Sbriziolo, C., **39**
Sakurai, T., **383**
Satnoianu, R. A., 96, 102, 203, **295**, 326, 336, 347, 411, 413, 414
Schreiber, I., 88, 99, 204, 210, 212, **313**, 339, 348, 349, 350, 351, 417
Schreiberová, L., **313**
Scott, S. K., 88, 89, 90, 200, 201, 203, 204, 205, 206, 207, 328, 329, 331, 407, 408, 412, 415, 417
Seipel, M., **395**
Schneider, F. W., **395**
Showalter, K. C., 88, 206, **383**, 409, 413, 415, 416, 417, 418
Sielewiesiuk, J., 418
Simon, P. L., 351, 409
Sirimungkala, A., **21**
Slin'ko, M. M., **179**, 198, 211, 212, 213
Šnita, D. A., **53**, 99, 100, 101, 409
Sørensen, P. G., 90, 92, 98, 102, 201, 210, **261**, 326, 339, 410, 416
Strich, A., **229**
Tomlin, A. S., **125**, 199, 200, 201, 202, 203, 204, 205, 207, 407
Tóth, Á., 207, 407
Trevelyan, P. M. J., 409
Turco Liveri, M. L., **39**, 95, 96, 97, 98, 99
Ukharskii, A. A., **179**
Varela, H. B., **165**
Voslař, M., **313**
Wang, J., 94, 332
Warnke, C., **249**
Wehner, S., **277**
Weimer, M. S., 325
Westerhoff, H. V., 85, 92, 95, 103, 197, **261**, 325, 326, 329, 331, 332, 335, 339, 408
Whitehouse, L. E., **125**
Wiedemann, M., **237**
Wittmann, M., **21**, 97
Yan, S., **21**
Yoshikawa, K., **67**
Zagora, J., **313**
Zambrano, V., **39**
Zhabotinsky, A. M., **11**

* The page numbers in **bold** type indicate papers submitted for discussions.

General Discussions of the Faraday Society/Faraday Discussions of the Chemical Society

Date	Subject	Volume
1907	Osmotic Pressure	Trans. 3
1907	Hydrates in Solution	3
1910	The Constitution of Water	6
1911	High Temperature Work	7
1912	Magnetic Properties of Alloys	8
1913	Colloids and their Viscosity	9
1913	The Corrosion of Iron and Steel	9
1913	The Passivity of Metals	9
1914	Optical Rotary Power	10
1914	The Hardening of Metals	10
1915	The Transformation of Pure Iron	11
1916	Methods and Appliances for the Attainment of High Temperatures in a Laboratory	12
1916	Refractory Materials	12
1917	Training and Work of the Chemical Engineer	13
1917	Osmotic Pressure	13
1917	Pyrometers and Pyrometry	13
1918	The Setting of Cements and Plasters	14
1918	Electric Furnaces	14
1918	Co-ordination of Scientific Publication	14
1918	The Occlusion of Gases by Metals	14
1919	The Present Position of the Theory of Ionization	15
1919	The Examination of Materials by X-Rays	15
1920	The Microscope: Its Design, Construction and Applications	16
1920	Basic Slags: Their Production and Utilization in Agriculture	16
1920	Physics and Chemistry of Colloids	16
1920	Electrodeposition and Electroplating	16
1921	Capillarity	17
1921	The Failure of Metals under Internal and Prolonged Stress	17
1921	Physico-Chemical Problems Relating to the Soil	17
1921	Catalysis with special reference to Newer Theories of Chemical Action	17
1922	Some Properties of Powders with special reference to Grading by Elutriation	18
1922	The Generation and Utilization of Cold	18
1923	Alloys Resistant to Corrosion	19
1923	The Physical Chemistry of the Photographic Process	19
1923	The Electronic Theory of Valency	19
1923	Electrode Reactions and Equilibria	19
1923	Atmospheric Corrosion. First Report	19
1924	Investigation on Oppau Ammonium Sulphate-Nitrate	20
1924	Fluxes and Slags in Metal Melting and Working	20
1924	Physical and Physico-Chemical Problems relating to Textile Fibres	20
1924	The Physical Chemistry of Igneous Rock Formation	20
1924	Base Exchange in Soils	20
1925	The Physical Chemistry of Steel-Making Processes	21
1925	Photochemical Reactions of Liquids and Gases	21
1926	Explosive Reactions in Gaseous Media	22
1926	Physical Phenomena at Interfaces, with special reference to Molecular Orientation	22
1927	Atmospheric Corrosion, Second Report	23
1927	The Theory of Strong Electrolytes	23
1927	Cohesion and Related Problems	24
1928	Homogeneous Catalysis	24
1929	Crystal Structure and Chemical Constitution	25
1929	Atmospheric Corrosion of Metals, Third Report	25
1929	Molecular Spectra and Molecular Structure	26
1930	Colloid Science Applied to Biology	26
1931	Photochemical Processes	27
1932	The Adsorption of Gases by Solids	28
1932	The Colloid Aspect of Textile Materials	29

Date	Subject	Volume
1933	Liquid Crystals and Anisotropic Melts	29
1933	Free Radicals	30
1934	Dipole Moments	30
1934	Colloidal Electrolytes	31
1935	The Structure of Metallic Coatings, Films and Surfaces	31
1935	The Phenomena of Polymerization and Condensation	32
1936	Disperse Systems in Gases: Dust, Smoke and Fog	32
1936	Structure and Molecular Forces in (*a*) Pure Liquids, and (*b*) Solutions	33
1937	The Properties and Function of Membranes, Natural and Artificial	33
1937	Reaction Kinetics	34
1938	Chemical Reactions Involving Solids	34
1938	Luminescence	35
1939	Hydrocarbon Chemistry	35
1939	The Electrical Double Layer (owing to the outbreak of the war the meeting was abandoned, but the papers were printed in the *Transactions*)	35
1940	The Hydrogen Bond	36
1941	The Oil-Water Interface	37
1941	The Mechanism and Chemical Kinetics of Organic Reactions in Liquid Systems	37
1942	The Structure and Reactions of Rubber	38
1943	Modes of Drug Action	39
1944	Molecular Weight and Molecular Weight Distribution in High Polymers (Joint Meeting with the Plastics Group, Society of Chemical Industry)	40
1945	The Application of Infra-red Spectra to Chemical Problems	41
1945	Oxidation	42
1946	Dielectrics	42 A
1946	Swelling and Shrinking	42 B
1947	Electrode Processes	Disc. 1
1947	The Labile Molecule	2
1947	Surface Chemistry (Jointly with the Sociéité de Chimie Physique at Bordeaux Published by Butterworths Scientific Publications Ltd	
1947	Colloidal Electrolytes and Solutions	Trans. 43
1948	The Interaction of Water and Porous Materials	Disc. 3
1948	The Physical Chemistry of Process Metallurgy	4
1949	Crystal Growth	5*
1949	Lipo-proteins	6
1949	Chromatographic Analysis	7
1950	Heterogeneous Catalysis	8
1950	Physico-chemical Properties and Behaviour of Nuclear Acids	Trans. 46
1950	Spectroscopy and Molecular Structure and Optical Methods of Investigating Cell Structure	Disc. 9
1950	Electrical Double Layer	Trans. 47
1951	Hydrocarbons	Disc. 10
1951	The Size and Shape Factor in Colloidal Systems	11
1952	Radiation Chemistry	12
1952	The Physical Chemistry of Proteins	13
1952	The Reactivity of Free Radicals	14
1953	The Equilibrium Properties of Solutions on Non-electrolytes	15
1953	The Physical Chemistry of Dyeing and Tanning	16
1954	The Study of Fast Reactions	17
1954	Coagulation and Flocculation	18
1955	Microwave and Radio-frequency Spectroscopy	19
1955	Physical Chemistry of Enzymes	20
1956	Membrane Phenomena	21
1956	Physical Chemistry of Processes at High Pressures	22
1957	Molecular Mechanism of Rate Processes in Solids	23
1957	Interactions in Ionic Solutions	24
1958	Configurations and Interactions of Macromolecules and Liquid Crystals	25
1958	Ions of the Transition Elements	26
1959	Energy Transfer with special reference to Biological Systems	27
1959	Crystal Imperfections and the Chemical Reactivity of Solids	28
1960	Oxidation-Reduction Reactions in Ionizing Solvents	29
1960	The Physical Chemistry of Aerosols	30
1961	Radiation Effects in Inorganic Solids	31
1961	The Structure and Properties of Ionic Melts	32
1962	Inelastic Collisions of Atoms and Simple Molecules	33
1962	High Resolution Nuclear Magnetic Resonance	34
1963	The Structure of Electronically Excited Species in the Gas Phase	35
1963	Fundamental Processes in Radiation Chemistry	36
1964	Chemical Reactions in the Atmosphere	37
1964	Dislocations in Solids	38
1965	The Kinetics of Proton Transfer Processes	39

Date	Subject	Volume
1965	Intermolecular Forces	40
1966	The Role of the Absorbed State in Heterogeneous Catalysis	41
1966	Colloid Stability in Aqueous and Non-aqueous Media	42
1967	The Structure and Properties of Liquids	43
1967	Molecular Dynamics of the Chemical Reactions of Gases	44
1968	Electrode Reactions of Organic Compounds	45
1968	Homogeneous Catalysis with Special Reference to Hydrogenation and Oxidation	46
1969	Bonding in Metallo-organic Compounds	47
1969	Motions in Molecular Crystals	48
1970	Polymer Solutions	49
1970	The Vitreous State	50
1971	Electrical Conduction in Organic Solids	51
1971	Surface Chemistry of Oxides	52
1972	Reactions of Small Molecules in Excited States	53
1972	The Photoelectron Spectroscopy of Molecules	54
1973	Molecular Beam Scattering	55
1973	Intermediates in Electrochemical Reactions	56
1974	Gels and Gelling Processes	57
1974	Photo-effects in Adsorbed Species	58
1975	Physical Adsorption in Condensed Phases	59
1975	Electron Spectroscopy of Solids and Surfaces	60
1976	Precipitation	61
1977	Potential Energy Surfaces	62
1977	Radiation Effects in Liquids and Solids	63
1977	Ion–Ion and Ion–Solvent Interactions	64
1978	Colloid Stability	65
1978	Structures and Motion in Molecular Liquids	66
1979	Kinetics of State Selected Species	67
1979	Organization of Macromolecules in the Condensed Phase	68
1980	Phase Transitions in Molecular Solids	69
1980	Photoelectrochemistry	70
1981	High Resolution Spectroscopy	71
1981	Selectivity in Heterogeneous Catalysis	72
1982	Van der Waals Molecules	73
1982	Electron and Proton Transfer	74
1983	Intramolecular Kinetics	75
1983	Concentrated Colloidal Dispersions	76
1984	Interfacial Kinetics in Solution	77
1984	Radicals in Condensed Phases	78
1985	Polymer Liquid Crystals	79
1985	Physical Interactions and Energy Exchange at the Gas–Solid Interface	80
1986	Lipid Vesicles and Membranes	81
1986	Dynamics of Molecular Photofragmentation	82
1987	Brownian Motion	83
1987	Dynamics of Elementary Gas-phase Reactions	84
1988	Solvation	85
1988	Spectroscopy at Low Temperatures	86
1989	Catalysis by Well Characterised Materials	87
1989	Charge Transfer in Polymeric Systems	88
1990	Structure of Surfaces and Interfaces as studied using Synchrotron Radiation	89
1990	Colloidal Dispersions	90
1991	Structure and Dynamics of Reactive Transition States	91
1991	The Chemistry and Physics of Small Metallic Particles	92
1992	Structure and Activity of Enzymes	93
1992	The Liquid/Solid Interface at High Resolution	94
1993	Crystal Growth	95
1993	Dynamics at the Gas/Solid Interface	96
1994	Structure and Dynamics of Van der Waals Complexes	97
1994	Polymers at Surfaces and Interfaces	98
1994	Vibrational Optical Activity: From Fundamentals to Biological Applications	99
1995	Atmospheric Chemistry: Measurements, Mechanisms and Models	100
1995	Gels	101
1995	Unimolecular Reaction Dynamics	102
1996	Hydration Processes in Biological and Macromolecular Systems	103
1996	Complex Fluids at Interfaces	104
1996	Catalysis and Surface Science at High Resolution	105
1997	Solid State Chemistry: New Opportunities from Computer Simulations	106
1997	Interactions of Acoustic Waves with Thin Films and Interfaces	107*
1997	Dynamics of Electronically Excited States in Gaseous, Clusters and Condensed Media	108*
1998	Chemistry and Physics of Molecules and Grains in Space	109*
1998	Chemical Reaction Theory	110*

Date	Subject	Volume
1998	Molecular Interactions of Biomembranes	111*
1999	Physical Chemistry in the Mesoscopic Regime	112*
1999	Stereochemistry and Control in Molecular Reaction Dynamics	113*
1999	The Surface Science of Metal Oxides	114*
2000	Molecular Photoionisation	115*
2000	Bioelectrochemistry	116*
2000	Excited States at Surfaces	117*
2001	Cluster Dynamics	118*
2001	Combustion Chemistry: Elementary Reactions to Macroscopic Processes	119*

* *Available for purchase, for current information on prices* etc. *please contact the Sales and Promotion Department, The Royal Society of Chemistry, Thomas Graham House, Science Park, Milton Road, Cambridge, UK CB4 0WF.*